LPF	low-pass filter
LSSB	lower single sideband
MAP	maximum a posteriori (criterion)
MPSK	*M*-ary phase shift keying
MQAM	*M*-ary quadrature amplitude modulation
MSK	minimum-shift keying
NBFM	narrowband frequency modulation
NRZ	nonreturn-to-zero
OFDM	orthogonal frequency division multiplexing
OOK	on-off keying
OQPSK	offset quadrature phase shift keying
OSI	open system interconnection protocol model
PAM	pulse amplitude modulation
PCM	pulse code modulation
PCS	personal communication service
PD	phase detection
PDF	probability density function
PEP	peak envelope power
PLL	phase-locked loop
PM	phase modulation
PPM	pulse position modulation
PSD	power spectral density
PSK	phase shift keying
PTM	pulse time modulation
PWM	pulse width modulation
QAM	quadrature amplitude modulation
QPSK	quadrature phase-shift keying
rms	root-mean-square
RF	radio frequency
RT	remote telephone terminal
RZ	return-to-zero
SAW	surface acoustics wave
SDLC	synchronous data link control protocol
SDTV	standard definition digital television
S/N or SNR	signal-to-noise ratio
SS	spread spectrum (system)
SSB	single sideband
TDM	time-division multiplexing
TDMA	time-division multiplex access
TELCO	telephone company
THD	total harmonic distortion
TTL	transistor-transistor logic
TV	television
TVRO	TV receive only terminal
TWT	traveling-wave tube
UHF	ultra high frequency
USSB	upper single sideband
VCO	voltage-controlled oscillator
VF	voice frequency
VHF	very high frequency
VSB	vestigial sideband
WBFM	wideband frequency modulation

Sixth Edition

DIGITAL AND ANALOG COMMUNICATION SYSTEMS

LEON W. COUCH II

*Professor Emeritus of Electrical
and Computer Engineering
University of Florida, Gainesville*

PRENTICE HALL, Upper Saddle River, New Jersey 07458

Library of Congress Cataloging-in-Publication Data

Couch, Leon W.
    Digital and analog communication systems / Leon W. Couch II. —
6th ed.
        p.   cm.
    Includes bibliographical references and index.
    ISBN 0-13-081223-4
    1. Telecommunication systems.   2. Digital communications.
I. Title.
TK5101.C69   2000
621.382--dc21
                              00-029131
                              CIP

Vice president and editorial director: MARCIA HORTON
Publisher: TOM ROBBINS
Associate editor: ALICE DWORKIN
Editorial assistant: JESSICA POWER
Senior marketing manager: DANNY HOYT
Marketing manager: ERIC WEISSER
Managing editor: DAVID A. GEORGE
Executive managing editor: VINCE O'BRIEN
Cover designer: JAYNE CONTE
Production/editorial supervision: PREPARÉ, Inc.
Manufacturing buyer: DAWN MURRIN
Manufacturing manager: TRUDY PISCIOTTI
Assistant vice president of production and manufacturing: DAVID RICCARDI

© 2001, 1997 by Prentice-Hall, Inc.
Upper Saddle River, NJ 07458

The author and publisher of this book have used their best efforts in preparing this book. These efforts include the development, research, and testing of the theories and programs to determine their effectiveness. The author and publisher make no warranty of any kind, expressed or implied, with regard to these programs or the documentation contained in this book. The author and publisher shall not be liable in any event for incidental or consequential damages in connection with, or arising out of, the furnishing, performance, or use of these programs.

MATLAB® is a registered trademark of The MathWorks, Inc.
For MATLAB product information, please contact:
The MathWorks, Inc.
3 Apple Hill Drive
Natick, MA, 01760-2098 USA
Tel: 508-647-7000
Fax: 508-647-7101
E-mail: info@mathworks.com
Web: www.mathworks.com

Printed in the United States of America

10 9 8 7 6 5 4 3

ISBN 0-13-081223-4

Reprinted with corrections July, 2002.

Pearson Education LTD.
Pearson Education Australia PTY, Limited
Pearson Education Singapore, Pte. Ltd
Pearson Education North Asia Ltd
Pearson Education Canada, Ltd.
Pearson Educación de Mexico, S.A. de C.V.
Pearson Education -- Japan
Pearson Education Malaysia, Pte. Ltd
Pearson Education, Upper Saddle River, New Jersey

To my wife,
Margaret Wheland Couch,
and
to our children,
Leon III, Jonathan, and Rebecca

CONTENTS

PREFACE

Continuing the tradition of the first to fifth editions of this book, this new edition provides the latest up-to-date treatment of digital communication systems. It includes a number of new study-aid examples and homework problems, many of which require solutions via a personal computer. It is written as a textbook for junior or senior engineering students and is also appropriate for an introductory graduate course or as a modern technical reference for practicing electrical engineers.

To learn about communication systems, it is essential to first understand *how communication systems work*. Based on the principles of communications (power, frequency spectra, and Fourier analysis) that are covered in the first five chapters of this book, this understanding is motivated by the use of extensive examples, study-aid problems, and the inclusion of adopted standards. Especially interesting is the material on wire and wireless communication systems. Also of importance is the effect of noise on these systems, since, without noise (described by probability and random processes), one could communicate to the limits of the universe with negligible transmitted power. In summary, this book covers the essentials needed for the understanding of wire and wireless communication systems and includes adopted standards. These essentials are

- How communication systems work: Chapters 1 through 5.
- The effect of noise: Chapters 6 and 7.
- Wire and Wireless Communication Systems: Chapter 8.

This book is ideal for either a one-semester or a two-semester course. For a one-semester course, the basics of how communication systems work may be taught by using the first five chapters (with selected readings from Chapter 8). For a two-semester course, the whole book is used.

This book covers *practical aspects* of communication systems developed from a sound *theoretical basis*.

THE THEORETICAL BASIS

- Digital and analog signals
- Magnitude and phase spectra
- Fourier analysis
- Orthogonal function theory
- Power spectral density
- Linear systems
- Nonlinear systems
- Intersymbol interference
- Complex envelopes
- Modulation theory
- Probability and random processes
- Matched filters
- Calculation of SNR
- Calculation of BER
- Optimum systems
- Block and convolutional codes

THE PRACTICAL APPLICATIONS

- PAM, PCM, DPCM, DM, PWM, and PPM baseband signaling
- OOK, BPSK, QPSK, MPSK, MSK, OFDM, and QAM bandpass digital signaling
- AM, DSB-SC, SSB, VSB, PM, and FM bandpass analog signaling
- Time-division multiplexing and the standards used
- Digital line codes and spectra
- Circuits used in communication systems
- Bit, frame, and carrier synchronizers
- Software radios
- Frequency-division multiplexing and the standards used
- Telecommunication systems
- Telephone systems
- Digital subscriber lines
- Satellite communication systems

- Effective input-noise temperature and noise figure
- Link budget analysis
- SNR at the output of analog communication systems
- BER for digital communication systems
- Fiber-optic systems
- Spread-spectrum systems
- AMPS, GSM, iDEN, TDMA, and CDMA cellular telephone and PCS systems
- Digital and analog television systems
- Technical standards for AM, FM, TV, DTV, and CATV
- Protocols for computer communications
- Technical standards for computer communications
- MATLAB M files
- Mathematical tables
- Study-aid examples
- Over 550 homework problems with selected answers
- Over 60 computer-solution homework problems
- Extensive references
- Emphasis on the design of communication systems

Many of the equations and homework problems are marked with a personal computer symbol, 🖳 , which indicates that the given equation or problem has a MATLAB and MATH-CAD solution on an available floppy disk or via the Internet at www.couch.ece.ufl.edu or www.prenhall.com/couch.

This book is an outgrowth of my teaching at the University of Florida and is tempered by my experiences as an amateur radio operator (K4GWQ). I believe that the reader will not understand the technical material unless he or she works some homework problems. Consequently, over 550 problems have been included. Some of them are easy, so that the beginning student will not become frustrated, and some are difficult enough to challenge the more advanced students. All the problems are designed to provoke thought about, and understanding of, communication systems.

I appreciate the help of the many persons who contributed to this book and the very helpful comments that have been provided by the reviewers—in particular, Marvin Siegel of the Department of Electrical Engineering at the University of Michigan and J. B. O'Neal of North Carolina State University. I also appreciate the help of my colleagues at the University of Florida. I thank my wife, Dr. Margaret Couch, who typed the original and revised manuscripts.

Leon W. Couch, II
Gainesville, Florida
couch@ece.ufl.edu

LIST OF SYMBOLS

There are not enough symbols in the English and Greek alphabets to allow the use of each letter only once. Consequently, some symbols may be employed to denote more than one entity, but their use should be clear from the context. Furthermore, the symbols are chosen to be generally the same as those used in the associated mathematical discipline. For example, in the context of complex variables, x denotes the real part of a complex number (i.e., $c = x + jy$), whereas in the context of statistics, x might denote a random variable.

Symbols

a_n	a constant
a_n	quadrature Fourier series coefficient
A_c	level of modulated signal of carrier frequency f_c
A_e	effective area of an antenna
b_n	quadrature Fourier series coefficient
B	baseband bandwidth

B_p	bandpass filter bandwidth
B_T	transmission (bandpass) bandwidth
c	a complex number ($c = x + jy$)
c	a constant
c_n	complex Fourier series coefficient
C	channel capacity
C	capacitance
$°C$	degrees Celsius
dB	decibel
D	dimensions/s, symbols/s ($D = N/T_0$), or baud rate
D_f	frequency modulation gain constant
D_n	polar Fourier series coefficient
D_p	phase modulation gain constant
e	error
e	the natural number 2.7183
E	modulation efficiency
E	energy
$\mathscr{E}(f)$	energy spectral density (ESD)
E_b/N_0	ratio of energy per bit to noise power spectral density
f	frequency (Hz)
$f(x)$	probability density function (PDF)
f_c	carrier frequency
f_i	instantaneous frequency
f_0	a (frequency) constant; the fundamental frequency of a periodic wavefonn
f_s	sampling frequency
F	noise figure
$F(a)$	cumulative distribution function (CDF)
$g(t)$	complex envelope
$\tilde{g}(t)$	corrupted complex envelope
G	power gain
$G(f)$	power transfer function
h	Planck's constant, 6.2×10^{-34} joule-s
$h(t)$	impulse response of a linear network
$h(x)$	mapping function of x into $h(x)$
H	entropy
$H(f)$	transfer function of a linear network
i	an integer
I_j	information in the jth message
j	the imaginary number $\sqrt{-1}$
j	an integer
k	Boltzmann's constant, 1.38×10^{-23} joule/K
k	an integer
$k(t)$	complex impulse response of a bandpass network
K	number of bits in a binary word that represents a digital message
K	degrees Kelvin ($°C + 273$)

l	an integer
ℓ	number of bits per dimension or bits per symbol
L	inductance
L	number of levels permitted
m	an integer
m	mean value
$m(t)$	message (modulation) waveform
$\widetilde{m}(t)$	corrupted (noisy received) message
M	an integer
M	number of messages permitted
n	an integer
n	number of bits in message
$n(t)$	noise waveform
N	an integer
N	number of dimensions used to represent a digital message
N	noise power
N_0	level of the power spectral density of white noise
$p(t)$	an absolutely time-limited pulse waveform
$p(t)$	instantaneous power
$p(m)$	probability density function of frequency modulation
P	average power
P_e	probability of bit error
$P(C)$	probability of correct decision
$P(E)$	probability of message error
$\mathcal{P}(f)$	power spectral density (PSD)
$Q(z)$	integral of Gaussian function
$Q(x_k)$	quantized value of the kth sample value, x_k
$r(t)$	received signal plus noise
R	data rate (bits/s)
R	resistance
$R(t)$	real envelope
$R(\tau)$	autocorrelation function
$s(t)$	signal
$\widetilde{s}(t)$	corrupted signal
S/N	ratio of signal power to noise power
t	time
T	a time interval
T	absolute temperature (Kelvin)
T_b	bit period
T_e	effective input-noise temperature
T_0	duration of a transmitted symbol or message
T_0	period of a periodic waveform
T_0	standard room temperature (290 K)
T_s	sampling period
u_{11}	covariance

$v(t)$	a voltage waveform
$v(t)$	a bandpass waveform or a bandpass random process
$w(t)$	a waveform
$W(f)$	spectrum (Fourier transform) of $w(t)$
x	an input
x	a random variable
x	real part of a complex function or a complex constant
$x(t)$	a random process
y	an output
y	an output random variable
y	imaginary part of a complex function or a complex constant
$y(t)$	a random process
α	a constant
β	a constant
β_f	frequency modulation index
β_p	phase modulation index
δ	step size of delta modulation
δ_{ij}	Kronecker delta function
$\delta(t)$	impulse (Dirac delta function)
ΔF	peak frequency deviation (Hz)
$\Delta\theta$	peak phase deviation
ϵ	a constant
ϵ	error
η	spectral efficiency [(bits/sec)/Hz]
$\theta(t)$	phase waveform
λ	dummy variable of integration
λ	wavelength
$\Lambda(r)$	likelihood ratio
π	3.14159
ρ	correlation coefficient
σ	standard deviation
τ	independent variable of autocorrelation function
τ	pulse width
$\varphi_j(t)$	orthogonal function
ϕ_n	polar Fourier series coefficient
ω_c	radian carrier frequency, $2\pi f_c$
$\equiv$	mathematical equivalence
$\triangleq$	mathematical definition of a symbol

DEFINED FUNCTIONS

$J_n(\cdot)$	Bessel function of the first kind, nth order
$\ln(\cdot)$	natural logarithm
$\log(\cdot)$	base 10 logarithm

$\log_2(\cdot)$ base 2 logarithm
$Q(z)$ integral of a Gaussian probability density function
$\mathrm{Sa}(z)$ $(\sin z)/z$
$u(\cdot)$ unit step function
$\Lambda(\cdot)$ triangle function
$\Pi(\cdot)$ rectangle function

OPERATOR NOTATION

$\mathrm{Im}\{\cdot\}$ imaginary part of
$\mathrm{Re}\{\cdot\}$ real part of
$\overline{[\cdot]}$ ensemble average
$\langle[\cdot]\rangle$ time average
$[\cdot] * [\cdot]$ convolution
$[\cdot]^*$ conjugate
$\underline{/[\cdot]}$ angle operator or angle itself, see Eq. (2–108)
$|[\cdot]|$ absolute value
$[\hat{\cdot}]$ Hilbert transform
$\mathscr{F}[\cdot]$ Fourier transform
$\mathscr{L}[\cdot]$ Laplace transform
$[\cdot] \cdot [\cdot]$ dot product

CHAPTER *1*

INTRODUCTION

The subject of communication systems is immense. It is not possible to include all topics and keep one book of reasonable length. In this book, the topics are carefully selected to accentuate basic communication principles. Moreover, the reader is motivated to appreciate these principles by the use of many practical applications. Often, practical applications are covered before the principles are fully developed. This provides "instant gratification" and motivates the reader to learn the basic principles well. The goal is to experience the joy of understanding how communication systems work and to develop an ability to design new communication systems.

What is a communication system? Moreover, what is electrical and computer engineering (ECE)? ECE is concerned with solving problems of two types: (1) production or transmission of electrical energy and (2) transmission or processing of information. *Communication systems are designed to transmit information.*

It is important to realize that communication systems and electric energy systems have markedly different sets of constraints. In electric energy systems, the waveforms are usually *known*, and one is concerned with designing the system for *minimum energy loss.*

In communication systems, the waveform present at the receiver (user) is *unknown* until after it is received—otherwise, no information would be transmitted, and there would be no

1

need for the communication system. More information is communicated to the receiver when the user is "more surprised" by the message that was transmitted. That is, the transmission of information implies the communication of messages that are not known ahead of time (a priori).

Noise limits our ability to communicate. If there were no noise, we could communicate messages electronically to the outer limits of the universe by using an infinitely small amount of power. This has been intuitively obvious since the early days of radio. However, the theory that describes noise and the effect of noise on the transmission of information was not developed until the 1940s, by such persons as D. O. North [1943], S. O. Rice [1944], C. E. Shannon [1948], and N. Wiener [1949].

Communication systems are designed to transmit information bearing waveforms to the receiver. There are many possibilities for selecting waveforms to represent the information. For example, how does one select a waveform to represent the letter A in a typed message? Waveform selection depends on many factors. Some of these are bandwidth (frequency span) and center frequency of the waveform, waveform power or energy, the effect of noise on corrupting the information carried by the waveform, and the cost of generating the waveform at the transmitter and detecting the information at the receiver.

The book is divided into eight chapters and four appendices. Chapter 1 introduces some key concepts, such as the definition of information, and provides a method for evaluating the information capacity of a communication system. Chapter 2 covers the basic techniques for obtaining the spectrum bandwidth and power of waveforms. Baseband waveforms (which have frequencies near $f = 0$) are studied in Chapter 3, and bandpass waveforms (frequencies in some band not near $f = 0$) are examined in Chapters 4 and 5. The effect of noise on waveform selection is covered in Chapters 6 and 7. Case studies of wire and wireless communications, including personal communication systems (PCS) are emphasized in Chapter 8. The appendices include mathematical tables, a short course on probability and random variables, standards for computer communications, and an introduction to MATLAB. Standards for communications systems are included, as appropriate, in each chapter. The personal computer is used as a tool to plot waveforms, compute spectra of waveforms, and analyze and design communications systems.

In summary, communication systems are designed to transmit information. Communication system designers have four main concerns:

1. Selection of the information-bearing waveform
2. Bandwidth and power of the waveform
3. Effect of system noise on the received information
4. Cost of the system.

1–1 HISTORICAL PERSPECTIVE

A time chart showing the historical development of communications is given in Table 1–1. The reader is encouraged to spend some time studying this table to obtain an appreciation for the chronology of communications. Note that although the telephone was developed late in the 19th century, the first transatlantic telephone cable was not completed until 1954. Previous to that date, transatlantic calls were handled via shortwave radio. Similarly,

TABLE 1–1 IMPORTANT DATES IN COMMUNICATIONS

Year	Event
Before 3000 B.C.	Egyptians develop a picture language called *hieroglyphics*.
A.D. 800	Arabs adopt our present number system from India.
1440	Johannes Gutenberg invents movable metal type.
1752	Benjamin Franklin's kite shows that lightning is electricity.
1827	Georg Simon Ohm formulates his law ($I = E/R$).
1834	Carl F. Gauss and Ernst H. Weber build the electromagnetic telegraph.
1838	William F. Cooke and Sir Charles Wheatstone build the telegraph.
1844	Samuel F. B. Morse demonstrates the Baltimore, MD, and Washington, DC, telegraph line.
1850	Gustav Robert Kirchhoff first publishes his circuit laws.
1858	The first transatlantic cable is laid and fails after 26 days.
1864	James C. Maxwell predicts electromagnetic radiation.
1871	The Society of Telegraph Engineers is organized in London.
1876	Alexander Graham Bell develops and patents the telephone.
1883	Thomas A. Edison discovers a flow of electrons in a vacuum, called the "Edison effect," the foundation of the electron tube.
1884	The American Institute of Electrical Engineers (AIEE) is formed.
1887	Heinrich Hertz verifies Maxwell's theory.
1889	The Institute of Electrical Engineers (IEE) forms from the Society of Telegraph Engineers in London.
1894	Oliver Lodge demonstrates wireless communication over a distance of 150 yards.
1900	Guglielmo Marconi transmits the first transatlantic wireless signal.
1905	Reginald Fessenden transmits speech and music by radio.
1906	Lee deForest invents the vacuum-tube triode amplifier.
1907	The Society of Wireless Telegraph Engineers is formed in the United States.
1909	The Wireless Institute is established in the United States.
1912	The Institute of Radio Engineers (IRE) is formed in the United States from the Society of Wireless Telegraph Engineers and the Wireless Institute.
1915	Bell System completes a U.S. transcontinental telephone line.
1918	Edwin H. Armstrong invents the superheterodyne receiver circuit.
1920	KDKA, Pittsburgh, PA, begins the first scheduled radio broadcasts.
1920	J. R. Carson applies sampling to communications.
1923	Vladimir K. Zworykin devises the "iconoscope" television pickup tube.
1926	J. L. Baird, (England) and C. F. Jenkins (United States) demonstrate television.
1927	The Federal Radio Commission is created in the United States.
1927	Harold Black develops the negative-feedback amplifier at Bell Laboratories.
1928	Philo T. Farnsworth demonstrates the first all-electronic television system.
1931	Teletypewriter service is initiated.
1933	Edwin H. Armstrong invents FM.
1934	The Federal Communication Commission (FCC) is created from the Federal Radio Commission in the United States.
1935	Robert A. Watson-Watt develops the first practical radar.
1936	The British Broadcasting Corporation (BBC) begins the first television broadcasts.
1937	Alex Reeves conceives pulse code modulation (PCM).

TABLE 1–1   *(cont.)*

Year	Event
1941	John V. Atanasoff invents the digital computer at Iowa State College.
1941	The FCC authorizes television broadcasting in the United States.
1945	The ENIAC electronic digital computer is developed at the University of Pennsylvania by John W. Mauchly.
1947	Walter H. Brattain, John Bardeen, and William Shockley devise the transistor at Bell Laboratories.
1947	Steve O. Rice develops a statistical representation for noise at Bell Laboratories.
1948	Claude E. Shannon publishes his work on information theory.
1950	Time-division multiplexing is applied to telephony.
1950s	Microwave telephone and communication links are developed.
1953	NTSC color television is introduced in the United States.
1953	The first transatlantic telephone cable (36 voice channels) is laid.
1957	The first Earth satellite, *Sputnik I*, is launched by USSR.
1958	A. L. Schawlow and C. H. Townes publish the principles of the laser.
1958	Jack Kilby of Texas Instruments builds the first germanium integrated circuit (IC).
1958	Robert Noyce of Fairchild produces the first silicon IC.
1961	Stereo FM broadcasts begin in the United States.
1962	The first active satellite, *Telstar I*, relays television signals between the United States and Europe.
1963	Bell System introduces the touch-tone phone.
1963	The Institute of Electrical and Electronic Engineers (IEEE) is formed by merger of the IRE and AIEE.
1963–66	Error-correction codes and adaptive equalization for high-speed error-free digital communications are developed.
1964	The electronic telephone switching system (No. 1 ESS) is placed into service.
1965	The first commercial communications satellite, *Early Bird*, is placed into service.
1968	Cable television systems are developed.
1971	Intel Corporation develops the first single-chip microprocessor, the 4004.
1972	Motorola demonstrates the cellular telephone to the FCC.
1976	Personal computers are developed.
1979	64-kb random access memory ushers in the era of very large-scale integrated (VLSI) circuits.
1980	Bell System FT3 fiber-optic communication is developed.
1980	Compact disk is developed by Philips and Sony.
1981	IBM PC is introduced.
1982	AT&T agrees to divest its 22 Bell System telephone companies.
1984	Macintosh computer is introduced by Apple.
1985	FAX machines become popular.
1989	Global positioning system (GPS) using satellites is developed.
1995	The Internet and the World Wide Web become popular.
2000–present	Era of digital signal processing with microprocessors, digital oscilloscopes, digitally tuned receivers, megaflop personal computers, spread spectrum systems, digital satellite systems, digital television (DTV), and personal communications systems (PCS).

although the British began television broadcasting in 1936, transatlantic television relay was not possible until 1962, when the *Telstar I* satellite was placed into orbit. Digital transmission systems—embodied by telegraph systems—were developed in the 1850s before analog systems—the telephone—in the 20th century. Now, digital transmission is again becoming the preferred technique.

1–2 DIGITAL AND ANALOG SOURCES AND SYSTEMS

DEFINITION.   A *digital information source* produces a finite set of possible messages.

A typewriter is a good example of a digital source. There is a finite number of characters (messages) that can be emitted by this source.

DEFINITION.   An *analog information source* produces messages that are defined on a continuum.

A microphone is a good example of an analog source. The output voltage describes the information in the sound, and it is distributed over a continuous range of values.

DEFINITION.   A *digital communication system* transfers information from a digital source to the intended receiver (also called the sink).

DEFINITION.   An *analog communication system* transfers information from an analog source to the sink.

Strictly speaking, a *digital waveform* is defined as a function of time that can have only a discrete set of amplitude values. If the digital waveform is a binary waveform, only two values are allowed. An *analog waveform* is a function of time that has a continuous range of values.

An electronic *digital* communication system usually has voltage and current waveforms that have digital values; however, it *may* have analog waveforms. For example, the information from a binary source may be transmitted to the receiver by using a sine wave of 1,000 Hz to represent a binary 1 and a sine wave of 500 Hz to represent a binary 0. Here the digital source information is transmitted to the receiver by the use of analog waveforms, but the system is still called a digital communication system. From this viewpoint, we see that a *digital* communication engineer needs to know how to analyze analog circuits as well as digital circuits.

Digital communication has a number of advantages:

- Relatively inexpensive digital circuits may be used.
- Privacy is preserved by using data encryption.
- Greater dynamic range (the difference between the largest and smallest values) is possible.
- Data from voice, video, and data sources may be merged and transmitted over a common digital transmission system.
- In long-distance systems, noise does not accumulate from repeater to repeater.
- Errors in detected data may be small, even when there is a large amount of noise on the received signal.
- Errors may often be corrected by the use of coding.

Digital communication also has disadvantages:

- Generally, more bandwidth is required than that for analog systems.
- Synchronization is required.

The advantages of digital communication systems usually outweigh their disadvantages. Consequently, digital systems are becoming more and more popular.

1–3 DETERMINISTIC AND RANDOM WAVEFORMS

In communication systems, we are concerned with two broad classes of waveforms: deterministic and random (or stochastic).

> **DEFINITION.**   A *deterministic waveform* can be modeled as a completely specified function of time.

For example, if

$$w(t) = A \cos(\omega_0 t + \varphi_0) \tag{1-1}$$

describes a waveform, where A, ω_0, and φ_0 are known constants, this waveform is said to be deterministic because, for any value of t, the value $w(t)$ can be evaluated. If any of the constants are unknown, then the value of $w(t)$ cannot be calculated, and consequently, $w(t)$ is not deterministic.

> **DEFINITION.**   A *random waveform* (or stochastic waveform) cannot be completely specified as a function of time and must be modeled probabilistically.[†]

Here we are faced immediately with a dilemma when analyzing communication systems. We know that the waveforms that represent the source cannot be deterministic. For example, in a digital communication system, we might send information corresponding to any one of the letters of the English alphabet. Each letter might be represented by a deterministic waveform, but when we examine the waveform that is emitted from the source, we find that it is a random waveform because we do not know exactly which characters will be transmitted. Consequently, we really need to design the communication system by using a random signal waveform. Noise would also be described by a random waveform. This requires the use of probability and statistical concepts (covered in Chapters 6 and 7) that make the design and analysis procedure more complicated. However, if we represent the signal waveform by a "typical" deterministic waveform, we can obtain most, but not all, of the results we are seeking. That is the approach taken in the first five chapters of this book.

1–4 ORGANIZATION OF THE BOOK

Chapters 1 to 5 use a deterministic approach in analyzing communication systems. This approach allows the reader to grasp some important concepts without the complications of statistical analysis. It also allows the reader who is not familiar with statistics to obtain a basic

---

[†]A more complete definition of a random waveform, also called a *random process*, is given in Chapter 6.

understanding of communication systems. However, the important topic of performance of communication systems in the presence of noise cannot be analyzed without the use of statistics. These topics are covered in Chapters 6 and 7 and Appendix B.[†] Chapter 8 gives practical case studies of wire and wireless communication systems.

This textbook is designed to be reader friendly. To aid the student in problem solving, several study-aid problems with abbreviated solutions are included at the end of each chapter. In addition, the personal computer (PC) is used to solve problems as appropriate.

The book is also useful as a reference source for mathematics (Appendix A), statistics (Appendix B and Chapter 6), computer communication systems (Appendix C), and MATLAB (Appendix D) and as a reference listing communication systems standards that have been adopted (Chapters 3, 4, 5, and 8 and Appendix C).

Communications is an exciting area in which to work. The reader is urged to browse through Chapter 8, looking at case-study topics that are of special interest, such as personal communication systems (PCSs).

1–5 USE OF A PERSONAL COMPUTER AND MATLAB

This textbook is designed so that a PC may be used as a tool to plot waveforms; compute spectra (using the fast Fourier transform); evaluate integrals; and, in general, help the reader to understand, analyze, and design communication systems. MATLAB was chosen as the program language since it is very efficient at these tasks and a student version is available at a reasonable cost. For a brief summary of MATLAB programming concepts, see Appendix D–2 ("Programming in MATLAB").

To solve a problem using a MATLAB, the MATLAB program is first run on the PC. MATLAB is an interpretive program. That is, results are computed after each line of code is entered. One has the option of keying in MATLAB statements one line at a time for immediate execution; or, alternately, a script file containing MATLAB code statements may be called up and run by MATLAB. The script or text file is also called an M file because the filename has the form xxxx.M. For programs with more than a couple of lines of code, the M-file method is usually used. The computed results may be shown in tabulated or graphical form. The M-files may be created by the MATLAB text editor or by another text editor, such as Notepad (running under Windows on the PC).

M-files are provided for solving selected equations and study-aid problems. The selected equation are marked with a PC (🖥) symbol. The M-files can be downloaded from the World Wide Web at the Internet site

http://www.couch.ece.ufl.edu

or

http://www.prenhall.com/couch

or, by using anonymous file transfer protocol (FTP) from

ftp.ece.ufl.edu

[†] Appendix B covers the topic of probability and random variables and is a complete chapter in itself. This allows the reader who has not had a course on this topic to learn the material before Chapters 6 and 7 are studied.

under the subdirectory /pub/COUCH/6ed/MATLAB. Alternatively, the M-files may be obtained from your instructor. MATHCAD template files can also be downloaded from the previouly listed sites. (Additional MATLAB and MATHCAD files for the homework problems marked with the PC (🖥) symbol are made available to the instructor on a floppy disk that is included with the *Solutions Manual*.) For instructions on running the M-files, see Appendix D–1 ("Quick Start for Running M-files"). As an example, run the file *e1_006.m*, which computes the results for Equation (1–6) of this chapter. *Table 2_3.m,* is the M-file shown in Table 2–3 and produces the MATLAB plots shown in Fig. 2–21.

1–6 BLOCK DIAGRAM OF A COMMUNICATION SYSTEM

Communication systems may be described by the block diagram shown in Fig. 1–1. Regardless of the particular application, all communications systems involve three main subsystems: the *transmitter*, the *channel*, and the *receiver*. Throughout this book, we use the symbols as indicated in this diagram so that the reader will not be confused about where the signals are located in the overall system. The message from the source is represented by the information input waveform $m(t)$. The message delivered by the receiver is denoted by $\tilde{m}(t)$. The [~] indicates that the message received may not be the same as that transmitted. That is, the message at the sink, $\tilde{m}(t)$, may be corrupted by noise in the channel, or there may be other impairments in the system, such as undesired filtering or undesired nonlinearities. The message information may be in analog or digital form, depending on the particular system, and it may represent audio, video, or some other type of information. In multiplexed systems, there may be multiple input and output message sources and sinks. The spectra (or frequencies) of $m(t)$ and $\tilde{m}(t)$ are concentrated about $f = 0$; consequently, they are said to be *baseband* signals.

The signal-processing block at the transmitter conditions the source for more efficient transmission. For example, in an analog system, the signal processor may be an analog low-pass filter that is used to restrict the bandwidth of $m(t)$. In a hybrid system, the signal processor may be an analog-to-digital converter (ADC), which produces a "digital word" that represents samples of the analog input signal (as described in Chapter 3 in the section on pulse code modulation). In this case, the ADC in the signal processor is providing *source coding* of the input signal. In addition, the signal processor may add parity bits to the digital word to provide *channel coding* so that error detection and correction can be used by the signal processor in the receiver to reduce or eliminate bit errors that are caused by noise

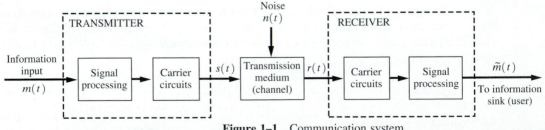

Figure 1–1   Communication system.

in the channel. The signal at the output of the transmitter signal processor is a baseband signal, because it has frequencies concentrated near $f = 0$.

The transmitter carrier circuit converts the processed baseband signal into a frequency band that is appropriate for the transmission medium of the channel. For example, if the channel consists of a fiber-optic cable, the carrier circuits convert the baseband input (i.e., frequencies near $f = 0$) to light frequencies, and the transmitted signal, $s(t)$, is light. If the channel propagates baseband signals, no carrier circuits are needed, and $s(t)$ can be the output of the processing circuit at the transmitter. Carrier circuits are needed when the transmission channel is located in a band of frequencies around $f_c \gg 0$. (The subscript denotes "carrier" frequency.) In this case, $s(t)$ is said to be a *bandpass,* because it is designed to have frequencies located in a band about f_c. For example, an amplitude-modulated (AM) broadcasting station with an assigned frequency of 850 kHz has a carrier frequency of $f_c = 850$ kHz. The mapping of the baseband input information waveform $m(t)$ into the bandpass signal $s(t)$ is called *modulation.* [$m(t)$ is the audio signal in AM broadcasting.] In Chapter 4, it will be shown that any bandpass signal has the form

$$s(t) = R(t) \cos[\omega_c t + \theta(t)] \tag{1–2}$$

where $\omega_c = 2\pi f_c$. If $R(t) = 1$ and $\theta(t) = 0$, $s(t)$ would be a pure sinusoid of frequency $f = f_c$ with *zero* bandwidth. In the modulation process provided by the carrier circuits, the baseband input waveform $m(t)$ causes $R(t)$ or $\theta(t)$ or both to change as a function of $m(t)$. These fluctuations in $R(t)$ and $\theta(t)$ cause $s(t)$ to have a nonzero bandwidth that depends on the characteristics of $m(t)$ and on the mapping functions used to generate $R(t)$ and $\theta(t)$. In Chapter 5, practical examples of both digital and analog bandpass signaling are presented.

Channels may be classified into two categories: wire and wireless. Some examples of *wire* channels are twisted-pair telephone lines, coaxial cables, waveguides, and fiber-optic cables. Some typical *wireless* channels are air, vacuum, and seawater. Note that the general principles of digital and analog modulation apply to all types of channels, although channel characteristics may impose constraints that favor a particular type of signaling. In general, the channel medium attenuates the signal so that the noise of the channel or the noise introduced by an imperfect receiver causes the delivered information $\tilde{m}$ to be deteriorated from that of the source. The channel noise may arise from natural electrical disturbances (e.g., lightning) or from artificial sources, such as high-voltage transmission lines, ignition systems of cars, or switching circuits of a nearby digital computer. The channel may contain active amplifying devices, such as repeaters in telephone systems or satellite transponders in space communication systems. These devices are necessary to help keep the signal above the noise level. In addition, the channel may provide undesirable *multiple paths* between its input and output that have different time delays and attenuation characteristics. Even worse, these characteristics may vary with time, which makes the signal fade at the channel output. You have probably observed this type of fading when listening to distant shortwave stations.

The receiver takes the corrupted signal at the channel output and converts it to a baseband signal that can be handled by the receiver baseband processor. The baseband processor "cleans up" this signal and delivers an estimate of the source information $\tilde{m}(t)$ to the communication system output.

The goal is to design communication systems that transmit information to the receiver with as little deterioration as possible while satisfying design constraints, of allowable transmitted energy, allowable signal bandwidth, and cost. In digital systems, the measure of deterioration is usually taken to be the *probability of bit error* (P_e)—also called the *bit error rate* (BER)—of the delivered data $\tilde{m}$. In analog systems, the performance measure is usually taken to be the signal-to-noise ratio at the receiver output.

1–7 FREQUENCY ALLOCATIONS

Wireless communication systems often use the atmosphere for the transmission channel. Here, interference and propagation conditions are strongly dependent on the transmission frequency. Theoretically, any type of modulation (e.g., amplitude modulation, frequency modulation, single sideband, phase-shift keying, frequency-shift keying, etc.) could be used at any transmission frequency. However, to provide some semblance of order and to minimize interference, government regulations specify the modulation type, bandwidth, power, and type of information that a user can transmit over designated frequency bands.

Frequency assignments and technical standards are set internationally by the International Telecommunications Union (ITU). The ITU is a specialized agency of the United Nations, and the ITU administrative headquarters is located in Geneva, Switzerland, with a staff of about 700 persons (see http:/www.itu.ch). This staff is responsible for administering the agreements that have been ratified by about 200 member nations of the ITU. The ITU is structured into three sectors. The Radiocommunication Sector (ITU-R) provides frequency assignments and is concerned with the efficient use of the radio frequency spectrum. The Telecommunications Standardization Section (ITU-T) examines technical, operating, and tariff questions. It recommends worldwide standards for the public telecommunications network (PTN) and related radio systems. The Telecommunication Development Sector (ITU-D) provides technical assistance, especially for developing countries. This assistance encourages a full array of telecommunication services to be economically provided and integrated into the worldwide telecommunication system. Before 1992, the ITU was organized into two main sectors: the International Telegraph and Telephone Consultative Committee (CCITT) and the International Radio Consultative Committee (CCIR).

Each member nation of the ITU retains sovereignty over the spectral usage and standards adopted in its territory. However, each nation is expected to abide by the overall frequency plan and standards that are adopted by the ITU. Usually, each nation establishes an agency that is responsible for the administration of the radio frequency assignments within its borders. In the United States, the Federal Communications Commission (FCC) regulates and licenses radio systems for the general public and state and local government (see http://www.fcc.gov). In addition, the National Telecommunication and Information Administration (NTIA) is responsible for U.S. government and U.S. military frequency assignments. The international frequency assignments are divided into subbands by the FCC to accommodate 70 categories of services and 9 million transmitters. Table 1–2 gives a general listing of frequency bands, their common designations, typical propagation conditions, and typical services assigned to these bands.

TABLE 1–2 FREQUENCY BANDS

Frequency Band[a]	Designation	Propagation Characteristics	Typical Uses
3–30 kHz	Very low frequency (VLF)	Ground wave; low attenuation day and night; high atmospheric noise level	Long-range navigation; submarine communication
30–300 kHz	Low frequency (LF)	Similar to VLF, slightly less reliable; absorption in daytime	Long-range navigation and marine communication radio beacons
300–3000 kHz	Medium frequency (MF)	Ground wave and night sky wave; attenuation low at night and high in day; atmospheric noise	Maritime radio, direction finding, and AM broadcasting
3–30 Mhz	High frequency (HF)	Ionospheric reflection varies with time of day, season, and frequency; low atmospheric noise at 30 Mhz	Amateur radio; international broadcasting, military communication, long- distance aircraft and ship communication, telephone, telegraph, facsimile
30–300 MHz	Very high frequency (VHF)	Nearly line-of-sight (LOS) propagation, with scattering because of temperature inversions, cosmic noise	VHF television, FM two-way radio, AM aircraft communication, aircraft navigational aids
0.3–3 GHz	Ultrahigh frequency (UHF)	LOS propagation, cosmic noise	UHF television, cellular telephone, navigational aids, radar, GPS, microwave links, personal communication systems
	Letter designation		
1.0–2.0	L		
2.0–4.0	S		
3–30 GHz	Superhigh frequency (SHF)	LOS propagation; rainfall attenuation above 10 GHz, atmospheric attenuation because of oxygen and water vapor, high water vapor absorption at 22.2 GHz	Satellite communication, radar microwave links
	Letter designation		
2.0–4.0	S		
4.0–8.0	C		
8.0–12.0	X		
12.0–18.0	Ku		
18.0–27.0	K		
27.0–40.0	Ka		
26.5–40.0	R		
30–300 GHz	Extremely high frequency (EHF)	Same; high water-vapor absorption at 183 GHz and oxygen absorption at 60 and 119 GHz	Radar, satellite, experimental

[a] kHz $= 10^3$ Hz; MHz $= 10^6$ Hz; GHz $= 10^9$ Hz.

TABLE 1–2   FREQUENCY BANDS *(cont.)*

Frequency Band[a]	Designation	Propagation Characteristics	Typical Uses
	Letter designation		
27.0–40.0	Ka		
26.5–40.0	R		
33.0–50.0	Q		
40.0–75.0	V		
75.0–110.0	W		
110–300	mm (millimeter)		
10^3–10^7 GHz	Infrared, visible light, and ultraviolet	LOS propagation	Optical communications

1–8 PROPAGATION OF ELECTROMAGNETIC WAVES

The propagation characteristics of electromagnetic waves used in wireless channels are highly dependent on the frequency. This situation is shown in Table 1–2, where users are assigned frequencies that have the appropriate propagation characteristics for the coverage needed. The propagation characteristics are the result of changes in the radio-wave velocity as a function of altitude and boundary conditions. The wave velocity is dependent on air temperature, air density, and levels of air ionization.

Ionization (i.e., free electrons) of the rarified air at high altitudes has a dominant effect on wave propagation in the medium-frequency (MF) and high-frequency (HF) bands. The ionization is caused by ultraviolet radiation from the sun, as well as cosmic rays. Consequently, the amount of ionization is a function of the time of day, season of the year, and activity of the sun (sunspots). This results in several layers of varying ionization density located at various heights surrounding the Earth.

The dominant ionized regions are D, E, F_1, and F_2 layers. The D layer is located closest to the Earth's surface at an altitude of about 45 or 55 miles. For $f > 300$ kHz, the D layer acts as a radio-frequency (RF) sponge to absorb (or attenuate) these radio waves. The attenuation is inversely proportional to frequency and becomes small for frequencies above 4 MHz. For $f < 300$ kHz, the D layer provides refraction (bending) of RF waves. The D layer is most pronounced during the daylight hours, with maximum ionization when the sun is overhead, and almost disappears at night. The E layer has a height of 65 to 75 miles, has maximum ionization around noon (local time), and practically disappears after sunset. It provides reflection of HF frequencies during the daylight hours. The F layer ranges in altitude between 90 and 250 miles. It ionizes rapidly at sunrise, reaches its peak ionization in early afternoon, and decays slowly after sunset. The F region splits into two layers, F_1 and F_2, during the day and combines into one layer at night. The F region is the most predominant medium in providing reflection of HF waves. As shown in Fig. 1–2, the electromagnetic spectrum may be divided into three broad bands that have one of three dominant propagation characteristics: ground wave, sky wave, and line of sight (LOS).

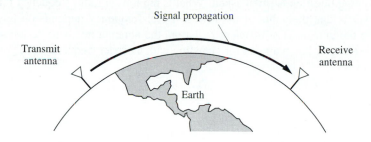

(a) Ground–Wave Propagation (Below 2 MHz)

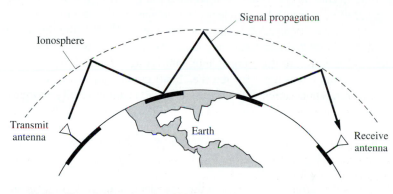

(b) Sky–Wave Propagation (2 to 30 MHz)

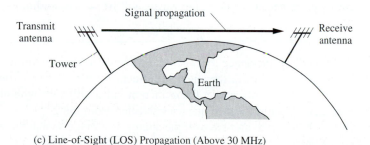

(c) Line-of-Sight (LOS) Propagation (Above 30 MHz)

Figure 1–2   Propagation of radio frequencies.

Ground-wave propagation is illustrated in Fig. 1–2a. It is the dominant mode of propagation for frequencies below 2 MHz. Here, the electromagnetic wave tends to follow the contour of the Earth. That is, diffraction of the wave causes it to propagate along the surface of the Earth. This is the propagation mode used in AM broadcasting, where the local coverage follows the Earth's contour and the signal propagates over the visual horizon. The following question is often asked: What is the lowest radio frequency that can be used? The answer is that the value of the lowest useful frequency depends on how long you want to make the antenna. For efficient radiation, the antenna needs to be longer than one-tenth of a wavelength. For example, for signaling with a carrier frequency of $f_c = 10$ kHz, the wavelength is

$$\lambda = \frac{c}{f_c}$$

$$\lambda = \frac{(3 \times 10^8 \text{ m/s})}{10^4} = 3 \times 10^4 \text{ m} \tag{1-3}$$

where c is the speed of light. (The formula $\lambda = c/f_c$ is distance = velocity $\times$ time, where the time needed to traverse one wavelength is $t = 1/f_c$.) Thus, an antenna needs to be at least 3,000 m in length for efficient electromagnetic radiation at 10 kHz.

Sky-wave propagation is illustrated in Fig. 1–2b. It is the dominant mode of propagation in the 2- to 30-MHz frequency range. Here, long-distance coverage is obtained by reflecting the wave at the ionosphere, and at the Earth's boundaries. Actually, in the ionosphere the waves are refracted (i.e., bent) gradually in an inverted U shape, because the index of refraction varies with altitude as the ionization density changes. The refraction index of the ionosphere is given by [Griffiths, 1987; Jordan and Balmain, 1968]

$$n = \sqrt{1 - \frac{81N}{f^2}} \tag{1-4}$$

where n is the refractive index, N is the free-electron density (number of electrons per cubic meter), and f is the frequency of the wave (in hertz). Typical N values range between 10^{10} and 10^{12}, depending on the time of day, the season, and the number of sunspots. In an ionized region $n < 1$ because $N > 0$, and outside the ionized region $n \approx 1$ because $N \approx 0$. In the ionized region, because $n < 1$, the waves will be bent according to Snell's law; viz,

$$n \sin \varphi_r = \sin \varphi_i \tag{1-5}$$

where φ_i is the angle of incidence (between the wave direction and vertical), measured just below the ionosphere, and φ_r is the angle of refraction for the wave (from vertical), measured in the ionosphere. Furthermore, the refraction index will vary with altitude within the ionosphere because N varies. For frequencies selected from the 2- to 30-MHz band, the refraction index will vary with altitude over the appropriate range so that the wave will be bent back to Earth. Consequently, the ionosphere acts as a reflector. The transmitting station will have coverage areas as indicated in Fig. 1–2b by heavy black lines along the Earth's surface. The coverage near the transmit antenna is due to the ground-wave mode, and the other coverage areas are due to sky wave. Notice that there are areas of no coverage along the Earth's surface between the transmit and receive antennas. The angle of reflection and the loss of signal at an ionospheric reflection point depend on the frequency, the time of day, the season of the year, and the sunspot activity [Jordan, 1985, Chap. 33].

During the daytime (at the ionospheric reflection points), the electron density will be high, so that $n < 1$. Consequently, sky waves from distant stations on the other side of the world will be heard on the shortwave bands. However, the D layer is also present during the day. This absorbs frequencies below 4 MHz.

This is the case for AM broadcast stations, where distant stations cannot be heard during the day, but at night the layer disappears, and distant AM stations can be heard via sky-wave propagation. In the United States, the FCC has designated some frequencies within the AM band as *clear channels* (as shown in Table 5–1). On these channels, only one or two high-power 50-kw stations are assigned to operate at night, along with a few low-power stations. Since these channels are relatively free of interfering stations, night sky-wave signals of the dominant 50-kw station can often be heard at distances up to 1,500 miles from the station. For example, some clear-channel 50-kw stations are WSM, Nashville, on 650 kHz; WCCO, Minneapolis, on 830 kHz; and WHO, Des Moines, on 1040 kHz.

Sky-wave propagation is caused primarily by reflection from the F layer (90 to 250 miles in altitude). Because of this layer, international broadcast stations in the HF band can be heard from the other side of the world almost anytime during the day or night.

LOS propagation (illustrated in Fig. 1–2c) is the dominant mode for frequencies above 30 MHz. Here, the electromagnetic wave propagates in a straight line. In this case, $f^2 \gg 81N$, so that $n \approx 1$, and there is very little refraction by the ionosphere. In fact, the signal will propagate *through* the ionosphere. This property is used for satellite communications.

The LOS mode has the disadvantage that, for communication between two terrestrial (Earth) stations, the signal path has to be above the horizon. Otherwise, the Earth will block the LOS path. Thus, antennas need to be placed on tall towers so that the receiver antenna can "see" the transmitting antenna. A formula for the distance to the horizon, d, as a function of antenna height can be easily obtained by the use of Fig. 1–3. From this figure,

$$d^2 + r^2 = (r + h)^2$$

or

$$d^2 = 2rh + h^2$$

where r is the radius of the Earth and h is the height of the antenna above the Earth's surface. In this application, h^2 is negligible with respect to $2rh$. The radius of the Earth is 3,960

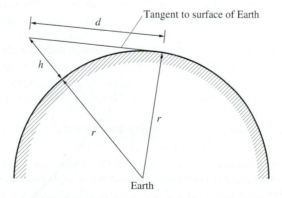

Figure 1–3   Calculation of distance to horizon.

statute miles. However, at LOS radio frequencies the effective Earth radius[†] is $\frac{4}{3}$ (3,960) miles. Thus, the distance to the radio horizon is

$$d = \sqrt{2h} \text{ miles} \qquad (1\text{--}6)$$

where conversion factors have been used so that h is the antenna height measured in feet and d is in statute miles. For example, television stations have assigned frequencies above 30 MHz in the VHF or UHF range (see Table 1–2), and the fringe-area coverage of high-power stations is limited by the LOS radio horizon. For a television station with a 1,000-ft tower, d is 44.7 miles. For a fringe-area viewer who has an antenna height of 30 ft, d is 7.75 miles. Thus, for these transmitting and receiving heights, the television station would have fringe-area coverage out to a radius of 44.7 + 7.75 = 52.5 miles around the transmitting tower.

In addition to the LOS propagation mode, it is possible to have *ionospheric scatter propagation*. This mode occurs over the frequency range of 30 to 60 MHz, when the radio frequency signal is scattered because of irregularities in the refractive index of the lower ionosphere (about 50 miles above the Earth's surface). Because of the scattering, communications can be carried out over path lengths of 1,000 miles, even though that is beyond the LOS distance. Similarly, *tropospheric scattering* (within 10 miles above the Earth's surface) can propagate radio frequency signals that are in the 40-MHz to 4-GHz range over paths of several hundred miles.

For more technical details about radio-wave propagation, the reader is referred to textbooks that include chapters on ground-wave and sky-wave propagation [Griffiths, 1987; Jordan and Balmain, 1968] and to a radio engineering handbook [Jordan, 1985]. A very readable description of this topic is also found in the ARRL handbook [ARRL, 1997], and personal computer programs, (e.g., MINIMUF, and PropMan) that predict sky-wave propagation conditions [Rose, 1982, 1984; Rockwell, 1995] are available.

1–9 INFORMATION MEASURE

As we have seen, the purpose of communication systems is to transmit information from a source to a receiver. However, what exactly is information, and how do we measure it? We know qualitatively that it is related to the surprise that is experienced when we receive the message. For example, the message "The ocean has been destroyed by a nuclear explosion" contains more information than the message "It is raining today."

> **DEFINITION.**   The *information* sent from a digital source when the jth message is transmitted is given by
>
> $$I_j = \log_2\left(\frac{1}{P_j}\right) \text{ bits} \qquad (1\text{--}7a)$$
>
> where P_j is the probability of transmitting the jth message.[††]

[†] The refractive index of the atmosphere decreases slightly with height, which causes some bending of radio rays. This effect may be included in LOS calculations by using an effective Earth radius that is four-thirds of the actual radius.

[††] The definition of probability is given in Appendix B.

From this definition, we see that messages that are less likely to occur (smaller value for P_j) provide more information (larger value of I_j). We also observe that the information measure depends on only the likelihood of sending the message and does not depend on possible interpretation of the content as to whether or not it makes sense.

The base of the logarithm determines the units used for the information measure. Thus, for units of "bits," the base 2 logarithm is used. If the natural logarithm is used, the units are "nats" and for base 10 logarithms, the unit is the "hartley," named after R. V. Hartley, who first suggested using the logarithm measure in 1928 [Hartley, 1948].

In this section, the term *bit* denotes a unit of information as defined by Eq. (1–7a). In later sections, particularly in Chapter 3, *bit* is also used to denote a unit of binary data. These two different meanings for the word *bit* should not be confused. Some authors use *binit* to denote units of data and use *bit* exclusively to denote units of information. However, most engineers use the same word (*bit*) to denote both kinds of units, with the particular meaning understood from the context in which the word is used. This book follows that industry custom.

For ease of evaluating I_j on a calculator, Eq. (1–7a) can be written in terms of the base 10 logarithm or the natural logarithm:

$$I_j = -\frac{1}{\log_{10} 2} \log_{10} P_j = -\frac{1}{\ln 2} \ln P_j \qquad (1\text{–}7b)$$

In general, the information content will vary from message to message because the P_j's will not be equal. Consequently, we need an average information measure for the source, considering all the possible messages we can send.

DEFINITION.   The *average information* measure of a digital source is

$$H = \sum_{j=1}^{m} P_j I_j = \sum_{j=1}^{m} P_j \log_2 \left(\frac{1}{P_j}\right) \text{ bits} \qquad (1\text{–}8)$$

where m is the number of possible different source messages and P_j is the probability of sending the jth message (m is finite because a digital source is assumed). The average information is called *entropy*.

Example 1–1 EVALUATION OF INFORMATION AND ENTROPY

Find the information content of a message that consists of a digital word 12 digits long in which each digit may take on one of four possible levels. The probability of sending any of the four levels is assumed to be equal, and the level in any digit does not depend on the values taken on by previous digits.

In a string of 12 symbols (digits), where each symbol consists of one of four levels, there are $4 \cdot 4 \cdots 4 = 4^{12}$ different combinations (words) that can be obtained. Because each level is equally likely, all the different words are equally likely. Thus,

$$P_j = \frac{1}{4^{12}} = \left(\frac{1}{4}\right)^{12}$$

or

$$I_j = \log_2 \left(\frac{1}{\left(\frac{1}{4}\right)^{12}}\right) = 12 \log_2(4) = 24 \text{ bits}$$

In this example, we see that the information content in every one of the possible messages equals 24 bits. Thus, the average information H is 24 bits.

Suppose that only two levels (binary) had been allowed for each digit and that all the words were equally likely. Then the information would be $I_j = 12$ bits for the binary words, and the average information would be $H = 12$ bits. Here, all the 12-bit words gave 12 bits of information, because the words were equally likely. If they had not been equally likely, some of the 12-bit words would contain more than 12 bits of information and some would contain less, and the average information would have been less than 12 bits. For example, if half of the 12-bit words (2,048 of the possible 4,096) have probability of occurence of $P_j = 10^{-5}$ for each of these words (with a corresponding $I_j = 16.61$ bits) and the other half have $P_j = 4.78 \times 10^{-4}$ (for a corresponding $I_j = 11.03$ bits), then the average information is $H = 11.14$ bits.

The rate of information is also important.

DEFINITION. The *source rate* is given by

$$R = \frac{H}{T} \text{ bits/s} \tag{1–9}$$

where H is evaluated by using Eq. (1–8) and T is the time required to send a message.

The definitions previously given apply to digital sources. Results for analog sources can be approximated by digital sources with as much accuracy as we desire.

1–10 CHANNEL CAPACITY AND IDEAL COMMUNICATION SYSTEMS

Many criteria can be used to measure the effectiveness of a communication system to see if it is ideal or perfect. For digital systems, the optimum system might be defined as the system that minimizes the probability of bit error at the system output subject to constraints on transmitted energy and channel bandwidth. Thus, bit error and signal bandwidth are of prime importance and are covered in subsequent chapters. This raises the following question: Is it possible to invent a system with no bit error at the output even when we have noise introduced into the channel? This question was answered by Claude Shannon in 1948–1949 [Wyner and Shamai, 1998; Shannon, 1948, 1949]. The answer is yes, under certain assumptions. Shannon showed that (for the case of signal plus white Gaussian noise) a channel capacity C (bits/s) could be calculated such that if the rate of information R (bits/s) was less than C, the probability of error would approach zero. The equation for C is

$$C = B \log_2\left(1 + \frac{S}{N}\right) \tag{1–10}$$

where B is the channel bandwidth in hertz (Hz) and S/N is the signal-to-noise power ratio (watts/watts, not dB) at the input to the digital receiver. Shannon does not tell us how to build this system, but he proves that it is theoretically possible to have such a system. Thus, Shannon gives us a theoretical performance bound that we can strive to achieve with practical communication systems. Systems that approach this bound usually incorporate error-correction coding.

In analog systems, the optimum system might be defined as the one that achieves the largest signal-to-noise ratio at the receiver output, subject to design constraints such as channel bandwidth and transmitted power. Here, the evaluation of the output signal-to-noise ratio is of prime importance. We might ask the question, Is it possible to design a system with infinite signal-to-noise ratio at the output when noise is introduced by the channel? The answer is no. The performance of practical analog systems with respect to that of Shannon's ideal system is illustrated in Chapter 7. (See Fig. 7–27.)

Other fundamental limits for digital signaling were discovered by Nyquist in 1924 and Hartley in 1928. Nyquist showed that if a pulse represents one bit of data, noninterfering pulses could be sent over a channel no faster than $2B$ pulses/s, where B is the channel bandwidth in hertz. This is now known as the dimensionality theorem and is discussed in Chapter 2. Hartley generalized Nyquist's result for the case of multilevel pulse signaling, as discussed in Chapters 3 and 5.

The following section describes the improvement that can be obtained in digital systems when coding is used and how these coded systems compare with Shannon's ideal system.

1–11 CODING

If the data at the output of a digital communication system have errors that are too frequent for the desired use, the errors can often be reduced by the use of either of two main techniques:

- Automatic repeat request (ARQ)
- Forward error correction (FEC)

In an ARQ system, when a receiver circuit detects parity errors in a block of data, it requests that the data block be retransmitted. In an FEC system, the transmitted data are encoded so that the receiver can correct, as well as detect, errors. These procedures are also classified as *channel coding* because they are used to correct errors caused by channel noise. This is different from *source coding*, described in Chapter 3, where the purpose of the coding is to extract the essential information from the source and encode it into digital form so that it can be efficiently stored or transmitted using digital techniques.

The choice between using the ARQ or the FEC technique depends on the particular application. ARQ is often used in computer communication systems because it is relatively inexpensive to implement and there is usually a duplex (two-way) channel so that the receiving end can transmit back an acknowledgment (ACK) for correctly received data or a request for retransmission (NAC) when the data are received in error. (See Appendix C, Section C–4, for examples of ARQ signaling.) FEC techniques are used to correct errors on simplex (one-way) channels, where returning of an ACK/NAC indicator (required for the ARQ technique) is not feasible. FEC is preferred on systems with large transmission delays, because if the ARQ technique were used, the effective data rate would be small; the transmitter would have long idle periods while waiting for the ACK/NAC indicator, which is retarded by the long transmission delay. Since ARQ systems are covered in Appendix C, we concentrate on FEC techniques in the remainder of this section.

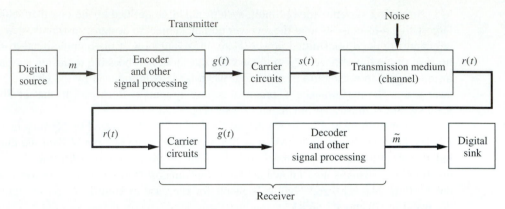

Figure 1–4   General digital communication system.

Communication systems with FEC are illustrated in Fig. 1–4, where encoding and decoding blocks have been designated. Coding involves adding extra (redundant) bits to the data stream so that the decoder can reduce or correct errors at the output of the receiver. However, these extra bits have the disadvantage of increasing the data rate (bits/s) and, consequently, increasing the bandwidth of the encoded signal.

Codes may be classified into two broad categories:

- *Block codes*. A block code is a mapping of k input binary symbols into n output binary symbols. Consequently, the block coder is a *memoryless* device. Because $n > k$, the code can be selected to provide redundancy, such as *parity bits*, which are used by the decoder to provide some error detection and error correction. The codes are denoted by (n, k), where the code rate $R^\dagger$ is defined by $R = k/n$. Practical values of R range from $\frac{1}{4}$ to $\frac{7}{8}$, and k ranges from 3 to several hundred [Clark and Cain, 1981].

- *Convolutional codes*. A convolutional code is produced by a coder that has *memory*. The convolutional coder accepts k binary symbols at its input and produces n binary symbols at its output, where the n output symbols are affected by $v + k$ input symbols. Memory is incorporated because $v > 0$. The code rate is defined by $R = k/n$. Typical values for k and n range from 1 to 8, and the values for v range from 2 to 60. The range of R is between $\frac{1}{4}$ and $\frac{7}{8}$ [Clark and Cain, 1981]. A small value for the code rate R indicates a high degree of redundancy, which should provide more effective error control at the expense of increasing the bandwidth of the encoded signal.

Block Codes

Before discussing block codes, several definitions are needed. The *Hamming weight* of a code word is the number of binary 1 bits. For example, the code word 110101 has a Hamming weight of 4. The *Hamming distance* between two code words, denoted by d, is the number of positions by which they differ. For example, the code words 110101 and 111001

---

†Do not confuse the code rate (with units of bits/bits) with the data rate or information rate (which has units of bits/s).

have a distance of $d = 2$. A received code word can be checked for errors. Some of the errors can be detected and corrected if $d \geq s + t + 1$, where s is the number of errors that can be detected and t is the number of errors that can be corrected ($s \geq t$). Thus, a pattern of t or fewer errors can be both detected and corrected if $d \geq 2t + 1$.

A general code word can be expressed in the form

$$i_1 i_2 i_3 \cdots i_k p_1 p_2 p_3 \cdots p_r$$

where k is the number of information bits, r is the number of parity check bits, and n is the total word length in the (n, k) block code, where $n = k + r$. This arrangement of the information bits at the beginning of the code word followed by the parity bits is most common. Such a block code is said to be *systematic*. Other arrangements with the parity bits interleaved between the information bits are possible and are usually considered to be equivalent codes.

Hamming has given a procedure for designing block codes that have single error-correction capability [Hamming, 1950]. A Hamming code is a block code having a Hamming distance of 3. Because $d \geq 2t + 1, t = 1$, and a single error can be detected and corrected. However, only certain (n, k) codes are allowable. These allowable Hamming codes are

$$(n, k) = (2^m - 1, 2^m - 1 - m) \tag{1–11}$$

where m is an integer and $m \geq 3$. Thus, some of the allowable codes are (7, 4), (15, 11), (31, 26), (63, 57), and (127, 120). The code rate R approaches 1 as m becomes large.

In addition to Hamming codes, there are many other types of block codes. One popular class consists of the cyclic codes. *Cyclic codes* are block codes, such that another code word can be obtained by taking any one code word, shifting the bits to the right, and placing the dropped-off bits on the left. These types of codes have the advantage of being easily encoded from the message source by the use of inexpensive linear shift registers with feedback. This structure also allows these codes to be easily decoded. Examples of cyclic and related codes are Bose–Chaudhuri–Hocquenhem (BCH), Reed-Solomon, Hamming, maximal–length, Reed–Müller, and Golay codes. Some properties of block codes are given in Table 1–3 [Bhargava, 1983].

TABLE 1–3 PROPERTIES OF BLOCK CODES

Property	Code[a]			
	BCH	Reed-Solomon	Hamming	Maximal Length
Block length	$n = 2^m - 1$ $m = 3, 4, 5, \ldots$	$n = m(2^m - 1)$ bits	$n = 2^m - 1$	$n = 2^m - 1$
Number of parity bits		$r = m2t$ bits	$r = m$	
Minimum distance	$d \geq 2t + 1$	$d = m(2t + 1)$ bits	$d = 3$	$d = 2^m - 1$
Number of information bits	$k \geq n - mt$			$k = m$

[a] m is any positive integer unless otherwise indicated; n is the block length; k is the number of information bits.

Convolutional Codes

A convolutional encoder is illustrated in Fig. 1–5. Here k bits (one input frame) are shifted in each time, and, concurrently, n bits (one output frame) are shifted out, where $n > k$. Thus, every k-bit input frame produces an n-bit output frame. Redundancy is provided in the output, because $n > k$. Also, there is memory in the coder, because the output frame depends on the previous K input frames, where $K > 1$. The *code rate* is $R = k/n$, which is $\frac{3}{4}$ in this illustration. The *constraint length*, K, is the number of input frames that are held in the kK-bit shift register.[†] Depending on the particular convolutional code that is to be generated, data from the kK stages of the shift register are added (modulo 2) and used to set the bits in the n-stage output register.

For example, consider the convolutional coder shown in Fig. 1–6. Here, $k = 1$, $n = 2$, $K = 3$, and a commutator with two inputs performs the function of a two-stage output shift register. The convolutional code is generated by inputting a bit of data and then giving the commutator a complete revolution. The process is repeated for successive input bits to produce the convolutionally encoded output. In this example, each $k = 1$ input bit produces $n = 2$ output bits, so the code rate is $R = k/n = \frac{1}{2}$. The code tree of Fig. 1–7 gives the encoded sequences for the convolutional encoder example of Fig. 1–6. To use the code tree, one moves up if the input is a binary 0 and down if the input is a binary 1. The corre-

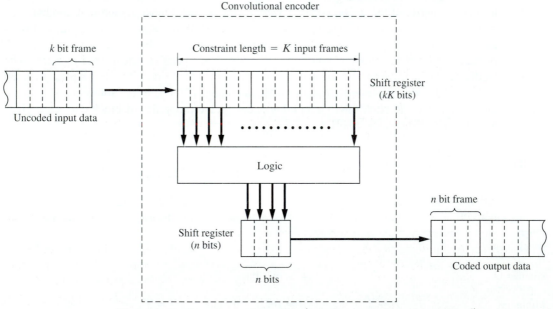

Figure 1–5 Convolutional encoding $\left(k = 3,\, n = 4,\, K = 5,\, \text{and } R = \frac{3}{4}\right)$.

[†] Several different definitions of constraint length are used in the literature [Blahut, 1983; Clark and Cain, 1981; Proakis, 1995].

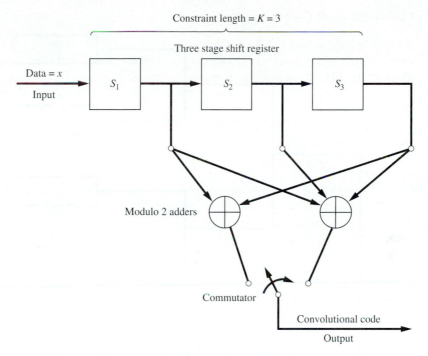

Constraint length = $K = 3$

Three stage shift register

Data = x

Input

S_1 S_2 S_3

Modulo 2 adders

Commutator

Convolutional code

Output

Figure 1–6   Convolutional encoder for a rate $\frac{1}{2}$, constraint length 3 code.

sponding encoded bits are shown in parentheses. For example, if the input sequence $x_{11} = 1010$ is fed into the input (with the most recent input bit on the right), the corresponding encoded output sequence is $y_{11} = 11010001$, as shown by path A in Fig. 1–7.

A convolutionally encoded signal is decoded by "matching" the encoded received data to the corresponding bit pattern in the code tree. In sequential decoding (a suboptimal technique), the path is found like that of a driver who occasionally makes a wrong turn at a fork in a road but discovers the mistake, goes back, and tries another path. For example, if $y_{11} = 11010001$ was received, path A would be the closest match, and the decoded data would be $x_{11} = 1010$. If noise was present in the channel, some of the received encoded bits might be in error, and then the paths would not match exactly. In this case, the match is found by choosing a path that will minimize the Hamming distance between the selected path sequence and the received encoded sequence.

An optimum decoding algorithm, called *Viterbi decoding*, uses a similar procedure. It examines the possible paths and selects the best ones, based on some conditional probabilities [Forney, 1973]. The Viterbi procedure can use either soft or hard decisions. A *soft-decision* algorithm first decides the result on the basis of the test statistic[†] being above or below a decision threshold and then gives a "confidence" number that specifies how close the test statistic was to the threshold value. In *hard decisions*, only the decision output is

---

[†] The test statistic is a value that is computed at the receiver, based on the receiver input during some specified time interval. [See Eq. (7–4).]

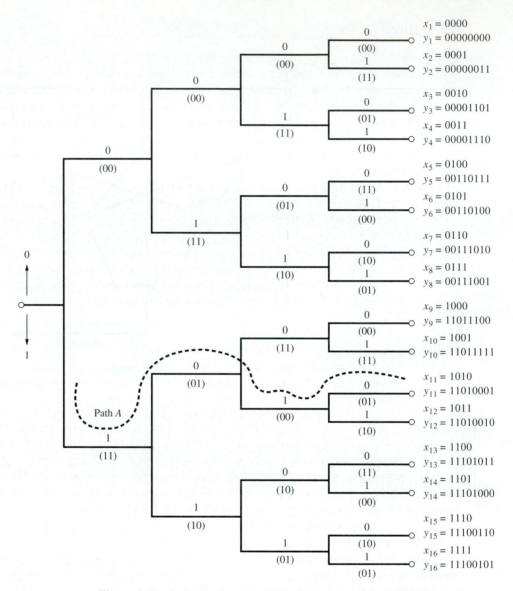

Figure 1–7 Code tree for convolutional encoder of Figure 1–6.

known, and it is not known if the decision was almost "too close to call" (because the test value was almost equal to the threshold value). The soft-decision technique can translate in-to a 2-dB improvement (decrease) in the required receiver input E_b/N_0 [Clark and Cain, 1981]. E_b is the received signal energy over a 1-bit time interval, and $N_0/2$ is the power spectral density (PSD) of the channel noise at the receiver input. Both E_b and N_0 will be defined in detail in later chapters. [For example, see Eq. (7–24b) or Eq. (8–44).]

Code Interleaving

In the previous discussion, it was assumed that if no coding was used, the channel noise would cause random bit errors at the receiver output that are more or less isolated (that is, not adjacent). When coding was added, redundancy in the code allowed the receiver decoder to correct the errors so that the decoded output was almost error free. However, in some applications, large, wide pulses of channel noise occur. If the usual coding techniques are used in these situations, bursts of errors will occur at the decoder output because the noise bursts are wider than the "redundancy time" of the code. This situation can be ameliorated by the use of code interleaving.

   At the transmitting end, the coded data are interleaved by shuffling (i.e., like shuffling a deck of cards) the coded bits over a time span of several block lengths (for block codes) or several constraint lengths (for convolutional codes). The required span length is several times the duration of the noise burst. At the receiver, before decoding, the data with error bursts are deinterleaved to produce coded data with isolated errors. The isolated errors are then corrected by passing the coded data through the decoder. This produces almost error-free output, even when noise bursts occur at the receiver input. There are two classes of interleavers—block interleavers and convolutional interleavers [Sklar, 1988].

Code Performance

The improvement in the performance of a digital communication system that can be achieved by the use of coding is illustrated in Fig. 1–8. It is assumed that a digital signal plus channel noise is present at the receiver input. The performance of a system that uses binary-phase-shift-keyed (BPSK) signaling is shown both for the case when coding is used and for the case when there is no coding. For the no-code case, the optimum (matched filter) detector circuit is used in the receiver, as derived in Chapter 7 and described by Eq. (7–38). For the coded case, a (23, 12) Golay code is used. P_e is the *probability of bit error*—also called the *bit error rate* (BER)—that is measured at the receiver output. E_b/N_0 is the energy-per-bit/noise-density ratio at the receiver input (as described in the preceding section). For $E_b/N_0 = 7$ dB, Fig. 1–8 shows that the BER is 10^{-3} for the uncoded case and that the BER can be reduced to 10^{-5} if coding is used.

   The *coding gain* is defined as the reduction in E_b/N_0 (in decibels) that is achieved when coding is used, when compared with the E_b/N_0 required for the uncoded case at some specific level of P_e. For example, as can be seen in the figure, a coding gain of 1.33 dB is realized for a BER of 10^{-3}. The coding gain increases if the BER is smaller, so that a coding gain of 2.15 dB is achieved when $P_e = 10^{-5}$. This improvement is significant in space communication applications, where every decibel of improvement is valuable. The figure also shows noted that there is a *coding threshold* in the sense that the coded system actually provides *poorer* performance than the uncoded system when E_b/N_0 is less than the threshold value. In this example, the coding threshold is about 3.5 dB. A coding threshold is found in all coded systems.

   For optimum coding, Shannon's channel capacity theorem, Eq. (1–10), gives the E_b/N_0 required. That is, if the source rate is below the channel capacity, the optimum code

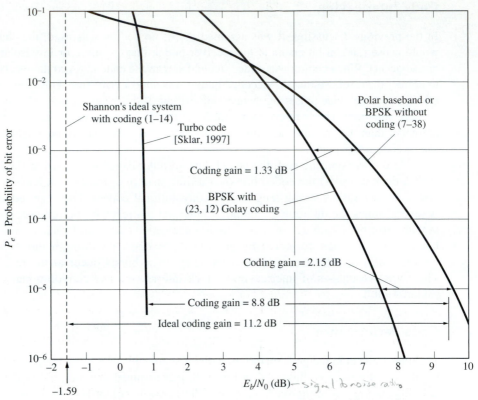

Figure 1–8   Performance of digital systems—with and without coding.

will allow the source information to be decoded at the receiver with $P_e \to 0$ (i.e., $10^{-\infty}$), even though there is some noise in the channel. We will now find the E_b/N_0 required so that $P_e \to 0$ with the optimum (unknown) code. Assume that the optimum encoded signal is not restricted in bandwidth. Then, from Eq. (1–10),

$$C = \lim_{B \to \infty} \left\{ B \log_2 \left(1 + \frac{S}{N} \right) \right\} = \lim_{B \to \infty} \left\{ B \log_2 \left(1 + \frac{E_b/T_b}{N_0 B} \right) \right\}$$

$$= \lim_{x \to 0} \left\{ \frac{\log_2[1 + (E_b/N_0 T_b)x]}{x} \right\}$$

where T_b is the time that it takes to send one bit and N is the noise power that occurs within the bandwidth of the signal. The power spectral density (PSD) is $\mathscr{P}_n(f) = N_0/2$, and, as shown in Chapter 2, the noise power is

$$N = \int_{-B}^{B} \mathscr{P}_n(f)df = \int_{-B}^{B} \left(\frac{N_0}{2} \right) df = N_0 B \qquad (1\text{--}12)$$

where B is the signal bandwidth. L'Hospital's rule is used to evaluate this limit:

$$C = \lim_{x \to 0} \left\{ \frac{1}{1 + (E_b/N_0 T_b)x} \left(\frac{E_b}{N_0 T_b} \right) \log_2 e \right\} = \frac{E_b}{N_0 T_b \ \ln \ 2} \qquad (1\text{–}13)$$

If we signal at a rate approaching the channel capacity, then $P_e \to 0$, and we have the maximum information rate allowed for $P_e \to 0$ (i.e., the optimum system). Thus, $1/T_b = C$, or, using Eq. (1–13),

$$\frac{1}{T_b} = \frac{E_b}{N_0 T_b \ \ln \ 2}$$

or

$$E_b/N_0 = \ln 2 = -1.59 \text{ dB} \qquad (1\text{–}14)$$

This minimum value for E_b/N_0 is $-$ 1.59 dB and is called *Shannon's limit*. That is, if optimum coding/decoding is used at the transmitter and receiver, error-free data will be recovered at the receiver output, provided that the E_b/N_0 at the receiver input is larger than −1.59 dB. This "brick wall" limit is shown by the dashed line in Fig. 1–8, where P_e jumps from 0 ($10^{\ -\infty}$) to $\frac{1}{2}(0.5 \times 10^0)$ as E_b/N_0 becomes smaller than −1.59 dB, assuming that the ideal (unknown) code is used. Any practical system will perform worse than this ideal system described by Shannon's limit. Thus, the goal of digital system designers is to find practical codes that approach the performance of Shannon's ideal (unknown) code.

When the performance of the optimum encoded signal is compared with that of BPSK without coding (10^{-5} BER), it is seen that the optimum (unknown) coded signal has a coding gain of $9.61 - (-1.59) = 11.2$ dB. Using Fig. 1–8, compare this value with the coding gain of 8.8 dB that is achieved when a turbo code is used. Table 1–4 shows the gains that can be obtained for some other codes.

Since their introduction in 1993, *turbo codes* have become very popular because they can perform near Shannon's limit, yet they also can have reasonable decoding complexity [Sklar, 1997]. Turbo codes are generated by using the parallel concatenation of two simple convolutional codes, with one coder preceded by an interleaver [Benedetto and Montorsi, 1996]. The interleaver ensures that error-prone words received for one of the codes corresponds to error-resistant words received for the other code.

All of the codes described earlier achieve their coding gains at the expense of *bandwidth expansion*. That is, when redundant bits are added to provide coding gain, the overall data rate and, consequently, the bandwidth of the signal are increased by a multiplicative factor that is the reciprocal of the code rate; the bandwidth expansion of the coded system relative to the uncoded system is $1/R = n/k$. Thus, if the uncoded signal takes up all of the available bandwidth, coding cannot be added to reduce receiver errors, because the coded signal would take up too much bandwidth. However, this problem can be ameliorated by using trellis-coded modulation (TCM).

TABLE 1–4   CODING GAINS WITH POLAR BASEBAND, BPSK OR QPSK

Coding Technique Used	Coding Gain (dB) at 10^{-5} BER	Coding Gain (dB) at 10^{-8} BER	Data Rate Capability
Ideal coding	11.2	13.6	
Turbo code [Sklar, 1997]	8.8		
Concatenated[a] Reed–Solomon and convolution (Viterbi decoding)	6.5–7.5	8.5–9.5	Moderate
Convolution with sequential decoding (soft decisions)	6.0–7.0	8.0–9.0	Moderate
Block codes (soft decision)	5.0–6.0	6.5–7.5	Moderate
Concatenated[a] Reed–Solomon and short block	4.5–5.5	6.5–7.5	Very high
Convolutional with Viterbi decoding	4.0–5.5	5.0–6.5	High
Convolutional with sequential decoding (hard decisions)	4.0–5.0	6.0–7.0	High
Block codes (hard decisions)	3.0–4.0	4.5–5.5	High
Block codes with threshold decoding	2.0–4.0	3.5–5.5	High
Convolutional with threshold decoding	1.5–3.0	2.5–4.0	Very high

[a]Two different encoders are used in series at the transmitter (see Fig. 1–4), and the corresponding decoders are used at the receiver.

Source: Bhargava [1983] and [Sklar, 1997].

Trellis-Coded Modulation

Gottfried Ungerboeck has invented a technique called *trellis-coded modulation* (TCM) that combines multilevel modulation with coding to achieve coding gain without bandwidth expansion [Benedetto, Mondin, and Montorsi, 1994; Biglieri, Divsalar, McLane, and Simon, 1991; Ungerboeck, 1982, 1987]. The trick is to add the redundant coding bits by increasing the number of levels (amplitude values) allowed in the digital signal without changing the pulse width. (The bandwidth will remain the same if the pulse width is not changed, since the bandwidth is proportional to the reciprocal of the pulse width.) This technique is called multilevel signaling and is first introduced in Section 3–4. For example, the pulses shown in Fig. 3–14a represent $L = 4$ multilevel signaling, where each level carries two bits of information as shown in Table 3–3. Now add one redundant coding bit to the two information bits to provide eight amplitude levels for the pulses, but maintaining the same pulse width so that the waveform would have the same bandwidth. Then the redundant bit, due to the coding, could be accommodated without any increase in bandwidth. This concept can be generalized to complex-valued multilevel signaling, as shown at the end of Sec. 5–10. In summary, TCM integrates waveform modulation design, with coding design, while maintaining the bandwidth of the uncoded waveform.

When a convolutional code of constraint length $K = 3$ is implemented, this TCM technique produces a coding gain of 3 dB relative to an uncoded signal that has the same bandwidth and information rate. Almost 6 dB of coding gain can be realized if coders of constraint length 9 are used. The larger constraint length codes are not too difficult to generate, but the corresponding decoder for a code of large constraint length is very complicated. However, very-high-speed integrated circuits (VHSIC) make this such a decoder feasible.

The 9,600-bit/s CCITT V.32 (Table C–7), 14,400-bit/s CCITT V.33bis (Table C–8), and 28,800-bit/s CCITT V.34 (Table C–5) computer modems use TCM. The CCITT V.32 modem has a coding gain of 4 dB and is described by Example 4 of Wei's paper [Wei, 1984; CCITT Study Group XVII, 1984].

For further study about coding, the reader is referred to several excellent books on the topic [Blahut, 1983; Clark and Cain, 1981; Gallagher, 1968; Lin and Costello, 1983; McEliece, 1977; Peterson and Weldon, 1972; Sweeney, 1991; Viterbi and Omura, 1979].

1–12 PREVIEW

From the previous discussions, we see the need for some basic tools to understand and design communication systems. Some prime tools that are required are mathematical models to represent signals, noise, and linear systems. Chapter 2 provides these tools. It is divided into the broad categories of properties of signal and noise, Fourier transforms and spectra, orthogonal representations, bandlimited representations, and descriptions of linear systems. Measures of bandwidth are also defined.

1–13 STUDY-AID EXAMPLES

SA1–1 Evalution of Line of Site (LOS)   The antenna for a television (TV) station is located at the top of a 1,500-foot transmission tower. Compute the LOS coverage for the TV station if the receiving antenna (in the fringe area) is 20 feet above ground.

Solution: Using Eq. (1–6), we find that the distance from the TV transmission tower to the radio horizon is

$$d_1 = \sqrt{2h} = \sqrt{2(1,500)} = 54.8 \text{ miles}$$

The distance from the receiving antenna to the radio horizon is

$$d_2 = \sqrt{2(20)} = 6.3 \text{ miles}$$

Then, the total radius for the LOS coverage contour (which is a circle around the transmission tower) is

$$d = d_1 + d_2 = 61.1 \text{ miles}$$

SA1–2 Information Data Rate A telephone touch-tone keypad has the digits 0 to 9, plus the * and # keys. Assume that the probability of sending * or # is 0.005 and the probability of sending 0 to 9 is 0.099 each. If the keys are pressed at a rate of 2 keys/s, compute the data rate for this source.

Solution: Using Eq. (1–8), we obtain

$$H = \Sigma P_j \log_2 \left(\frac{1}{P_j} \right)$$

$$= \frac{1}{\log_{10}(2)} \left[10(0.099) \log_{10} \left(\frac{1}{0.099} \right) + 2(0.005) \log_{10} \left(\frac{1}{0.005} \right) \right]$$

or

$$H = 3.38 \text{ bits/key}$$

Using Eq. (1–9), where $T = 1/(2 \text{ keys/s}) = 0.5$ s/key, yields

$$R = \frac{H}{T} = \frac{3.38}{0.5} = 6.76 \text{ bits/s}$$

SA1–3 Maximum Telephone Line Data Rate A computer user plans to buy a higher-speed modem for sending data over his or her analog telephone line. The telephone line has a signal-to-noise ratio (SNR) of 25 dB and passes audio frequencies over the range from 300 to 3,200 Hz. Calculate the maximum data rate that could be sent over the telephone line when there are no errors at the receiving end.

Solution: In terms of a power ratio, the SNR is $S/N = 10^{(25/10)} = 316.2$ (see dB in Chapter 2), and the bandwidth is $B = 3,200 - 300 = 2,900$ Hz. Using Eq. (1–10), we get

$$R = B \log_2 \left(1 + \frac{S}{N} \right) = 2,900 \ [\log_{10}(1 + 316.2)]/\log_{10}(2),$$

or

$$R = 24,097 \text{ bits/s}$$

Consequently, a 28.8-kbit/s modem signal would not work on this telephone line; however, a 14.4-kbit/s modem signal should transmit data without error.

PROBLEMS

1–1 A high-power FM station of frequency 96.9 MHz has an antenna height of 1200 ft. If the signal is to be received 60 miles from the station, how high does a prospective listener need to mount his or her antenna in this fringe area?

1–2 Using geometry, prove that Eq. (1–6) is correct.

1–3 A terrestrial microwave system is being designed. The transmitting and receiving antennas are to be placed at the top of equal-height towers, with one tower at the transmitting site and one at the receiving site. The distance between the transmitting and receiving sites is 25 miles. Calculate the minimum tower height required for an LOS transmission path.

1–4 A cellular telephone cell site has an antenna located at the top of a 60-ft tower. A typical cellular telephone user has his or her antenna located 4 ft above the ground. What is the LOS radius of coverage for this cell site to a distant user?

1–5 A digital source emits −1.0- and 0.0-V levels with a probability of 0.2 each and +3.0- and + 4.0-V levels with a probability of 0.3 each. Evaluate the average information of the source.

1–6 Prove that base 10 logarithms may be converted to base 2 logarithms by using the identity $\log_2(x) = [1/\log_{10}(2)] \log_{10}(x)$.

1–7 If all the messages emitted by a source are equally likely (i.e., $P_j = P$), show that Eq. (1–8) reduces to $H = \log_2(1/P)$.

1–8 For a binary source:
(a) Show that the entropy H is a maximum when the probability of sending a binary 1 is equal to the probability of sending a binary 0.
(b) Find the value of maximum entropy.

1–9 A single-digit, seven-segment liquid crystal display (LCD) emits a 0 with a probability of 0.25; a 1 and a 2 with a probability of 0.15 each; 3, 4, 5, 6, 7, and 8 with a probability of 0.07 each; and a 9 with a probability of 0.03. Find the average information for this source.

1–10 (a) A binary source sends a binary 1 with a probability of 0.3. Evaluate the average information for the source.
(b) For a binary source, find the probability for sending a binary 1 and a binary 0, such that the average source information will be maximized.

1–11 A numerical keypad has the digits 0, 1, 2, 3, 4, 5, 6, 7, 8, and 9. Assume that the probability of sending any one digit is the same as that for sending any of the other digits. Calculate how often the buttons must be pressed in order to send out information at the rate of 2 bits/s.

1–12 Refer to Example 1–1 and assume that words, each 12 digits in length, are sent over a system and that each digit can take on one of two possible values. Half of the possible words have a probability of being transmitted that is $\left(\frac{1}{2}\right)^{13}$ for each word. The other half have probabilities equal to $3\left(\frac{1}{2}\right)^{13}$. Find the entropy for this source.

1–13 Evaluate the channel capacity for a teleprinter channel that has a 300-Hz bandwidth and an SNR of 30 dB.

1–14 Assume that a computer terminal has 110 characters (on its keyboard) and that each character is sent by using binary words.
(a) What are the number of bits needed to represent each character?
(b) How fast can the characters be sent (characters/s) over a telephone line channel having a bandwidth of 3.2 kHz and an SNR of 20 dB?
(c) What is the information content of each character if each is equally likely to be sent?

1–15 An analog telephone line has an SNR of 45 dB and passes audio frequencies over the range of 300 to 3,200 Hz. A modem is to be designed to transmit and receive data simultaneously (i.e., full duplex) over this line without errors.
(a) If the frequency range 300 to 1,200 Hz is used for the transmitted signal, what is the maximum transmitted data rate?
(b) If the frequency range 1,500 to 3,200 Hz is used for the signal being simultaneously received, what is the maximum received data rate?
(c) If the whole frequency range of 300 to 3,200 Hz is used simultaneously for transmitting and receiving (by the use of a hybrid circuit as described in Chapter 8, Fig. 8–4), what are the maximum transmitting and receiving data rates?

1–16 Using the definitions for terms associated with convolutional coding, draw a block diagram for a convolutional coder that has rate $R = \frac{2}{3}$ and constraint length $K = 3$.

1-17 For the convolutional encoder shown in Fig. P1–17, compute the output coded data when the input data is $\mathbf{x} = [10111]$. (The first input bit is the leftmost element of the $\mathbf{x}$ row vector.)

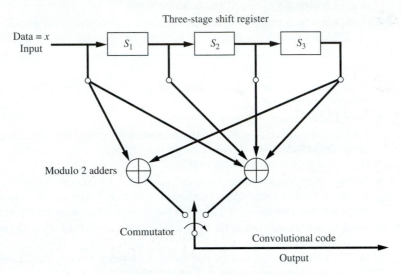

Figure P1–17

CHAPTER 2

Chapter Objectives

- Basic signal properties (dc, rms, dBm, and power)

- Fourier transform and spectra

- Linear systems and linear distortion

- Bandlimted signals and sampling

- Discrete Fourier transform

- Bandwidth of signals

SIGNALS AND SPECTRA

2–1 PROPERTIES OF SIGNALS AND NOISE

In communication systems, the received waveform is usually categorized into the desired part containing the information and the extraneous or undesired part. The desired part is called the *signal*, and the undesired part is called *noise*.

This chapter develops mathematical tools that are used to describe signals and noise from a deterministic waveform point of view. (The random waveform approach is given in Chapter 6.) The waveforms will be represented by direct mathematical expressions or by the use of orthogonal series representations such as the Fourier series. Properties of these waveforms, such as their dc value, root-mean-square (rms) value, normalized power, magnitude spectrum, phase spectrum, power spectral density, and bandwidth, will also be established. In addition, effects of linear filtering will be studied.

The waveform of interest may be the voltage as a function of time, $v(t)$, or the current as a function of time, $i(t)$. Often, the same mathematical techniques can be used when one is working with either type of waveform. Thus, for generality, waveforms will be denoted simply as $w(t)$ when the analysis applies to either case.

Physically Realizable Waveforms

Practical waveforms that are *physically realizable* (i.e., measurable in a laboratory) satisfy several conditions:[†]

1. The waveform has significant nonzero values over a composite time interval that is finite.
2. The spectrum of the waveform has significant values over a composite frequency interval that is finite.
3. The waveform is a continuous function of time.
4. The waveform has a finite peak value.
5. The waveform has only real values. That is, at any time, it cannot have a complex value $a + jb$, where b is nonzero.

The first condition is necessary because systems (and their waveforms) appear to exist for a finite amount of time. Physical signals also produce only a finite amount of energy. The second condition is necessary because any transmission medium—such as wires, coaxial cable, waveguides, or fiber-optic cable—has a restricted bandwidth. The third condition is a consequence of the second, and will become clear from spectral analysis as developed in Sec. 2–2. The fourth condition is necessary because physical devices are destroyed if voltage or current of infinite value is present within the device. The fifth condition follows from the fact that only real waveforms can be observed in the real world, although *properties* of waveforms, such as spectra, may be complex. Later, in Chapter 4, it will be shown that complex waveforms can be very useful in representing real bandpass signals mathematically.

Mathematical models that violate some or all of the conditions listed previously are often used, and for one main reason—to simplify the mathematical analysis. However, if we are careful with the mathematical model, the correct result can be obtained when the answer is properly interpreted. For example, consider the digital waveform shown in Fig. 2–1. The mathematical model waveform has discontinuities at the switching times. This situation violates the third condition—that the physical waveform be continuous. The physical waveform is of finite duration (decays to zero before $t = \pm\infty$), but the duration of the mathematical waveform extends to infinity.

In other words, this mathematical model assumes that the physical waveform has existed in its steady-state condition for all time. Spectral analysis of the model will approximate the correct results, except for the extremely high-frequency components. The average power that is calculated from the model will give the correct value for the average power of the physical signal that is measured over an appropriate time interval. The total energy of the mathematical model's signal will be infinity because it extends to infinite time, whereas that of the physical signal will be finite. Consequently, the model will not give the correct value for the total energy of the physical signal without the use of some additional information. However, the model can be used to evaluate the energy of the physical signal over some finite time interval. This mathematical model is said to be a *power signal* because it has the property of finite power (and infinite energy), whereas the physical waveform is said to be an *energy signal*

[†] For an interesting discussion relative to the first and second conditions, see the paper by Slepian [1976].

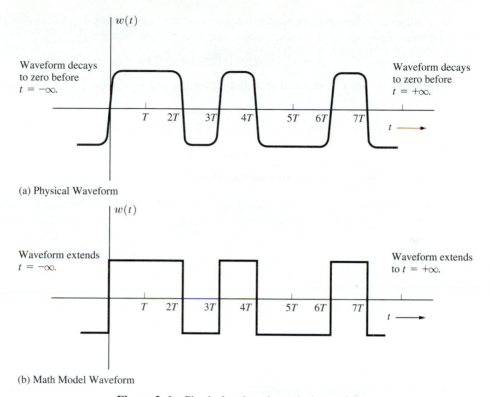

(a) Physical Waveform

(b) Math Model Waveform

Figure 2–1   Physical and mathematical waveforms.

because it has finite energy. (Mathematical definitions for power and energy signals will be given in a subsequent section.) All physical signals are energy signals, although we generally use power signal mathematical models to simplify the analysis.

In summary, waveforms may often be classified as signals or noise, digital or analog, deterministic or nondeterministic, physically realizable or nonphysically realizable, and belonging to the power or energy type. Additional classifications, such as periodic and nonperiodic, will be given in the next section.

Time Average Operator

Some useful waveform characteristics are the direct "current" (dc) value, average power, and root-mean-square (rms) value. Before these concepts are reviewed, the time average operator needs to be defined.

DEFINITION.   The *time average operator*† is given by

$$\langle [\,\cdot\,] \rangle = \lim_{T \to \infty} \frac{1}{T} \int_{-T/2}^{T/2} [\,\cdot\,]\, dt \tag{2–1}$$

† In Appendix B, the ensemble average operator is defined.

It is seen that this operator is a *linear* operator, since, from Eq. (2–1), the average of the sum of two quantities is the same as the sum of their averages:[†]

$$\langle a_1 w_1(t) + a_2 w_2(t) \rangle = a_1 \langle w_1(t) \rangle + a_2 \langle w_2(t) \rangle \qquad (2\text{–}2)$$

Equation (2–1) can be reduced to a simpler form given by Eq. (2–4) if the operator is operating on a *periodic* waveform.

DEFINITION.   A waveform $w(t)$ is *periodic* with period T_0 if

$$w(t) = w(t + T_0) \qquad \text{for all } t \qquad (2\text{–}3)$$

where T_0 is the smallest positive number that satisfies this relationship.[‡]

For example, a sinusoidal waveform of frequency $f_0 = 1/T_0$ hertz is periodic, since it satisfies Eq. (2–3). From this definition, it is clear that a periodic waveform will have significant values over an infinite time interval $(-\infty, \infty)$. Consequently, physical waveforms cannot be truly periodic, but they can have periodic values over a finite time interval. That is, Eq. (2–3) can be satisfied for t over some finite interval, but not for all values of t.

THEOREM.   *If the waveform involved is periodic, the time average operator can be reduced to*

$$\langle [\cdot] \rangle = \frac{1}{T_0} \int_{-T_0/2+a}^{T_0/2+a} [\cdot] \, dt \qquad (2\text{–}4)$$

where T_0 is the period of the waveform and a is an arbitrary real constant, which may be taken to be zero.

Equation (2–4) readily follows from Eq. (2–1) because, referring to Eq. (2–1), integrals over successive time intervals T_0 seconds wide have identical areas, since the waveshape is periodic with period T_0. As these integrals are summed, the total area and T are proportionally larger, resulting in a value for the time average that is the same as just integrating over one period and dividing by the width of that interval, T_0.

In summary, Eq. (2–1) may be used to evaluate the time average of any type of waveform, whether or not it is periodic. Equation (2–4) is valid only for periodic waveforms.

Dc Value

DEFINITION.   The *dc* (direct "current") value of a waveform $w(t)$ is given by its time average, $\langle w(t) \rangle$.

Thus,

$$W_{\text{dc}} = \lim_{T \to \infty} \frac{1}{T} \int_{-T/2}^{T/2} w(t) \, dt \qquad (2\text{–}5)$$

[†] See Eq. (2–130) for the definition of linearity.

[‡] Nonperiodic waveforms are called *aperiodic* waveforms by some authors.

For any physical waveform, we are actually interested in evaluating the dc value only over a finite interval of interest, say, from t_1 to t_2, so that the dc value is

$$\frac{1}{t_2 - t_1} \int_{t_1}^{t_2} w(t) \, dt$$

However, if we use a mathematical model with a steady-state waveform of infinite extent, we will obtain the correct result by using our definition, Eq. (2–5), which involves the limit as $T \to \infty$. This will be demonstrated subsequently by Example 2–1. Moreover, as will be shown in Chapter 6, for the case of ergodic stochastic waveforms, the time averag operator $\langle [\cdot] \rangle$ may be replaced by the ensemble average operator $\overline{[\cdot]}$.

Power

In communication systems, if the received (average) signal power is sufficiently large compared to the (average) noise power, information may be recovered. This concept was demonstrated by the Shannon channel capacity formula, Eq. (1–10). Consequently, average power is an important concept that needs to be clearly understood. From physics, it is known that power is defined as work per unit time, voltage is work per unit charge, and current is charge per unit time. This is the basis for the definition of power in terms of electrical quantities.

> **DEFINITION.**   Let $v(t)$ denote the voltage across a set of circuit terminals, and let $i(t)$ denote the current into the terminal, as shown in Fig. 2–2. The *instantaneous power* (incremental work divided by incremental time) associated with the circuit is given by
>
> $$p(t) = v(t)i(t) \qquad (2\text{–}6)$$
>
> where the instantaneous power flows into the circuit when $p(t)$ is positive and flows out of the circuit when $p(t)$ is negative. The *average power* is
>
> $$P = \langle p(t) \rangle = \langle v(t)i(t) \rangle \qquad (2\text{–}7)$$

Example 2–1 EVALUATION OF POWER

Let the circuit of Fig. 2–2 contain a 120-V, 60-Hz fluorescent lamp wired in a high-power–factor configuration. Assume that the voltage and current are both sinusoids and in phase (unity power factor), as shown in Fig. 2–3.[†] The dc value of this (periodic) voltage waveform is

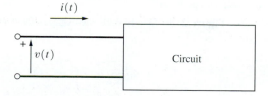

Figure 2–2   Polarity convention used for voltage and current.

[†] Two-lamp fluorescent circuits can be realized with a high-power–factor ballast that gives an overall power factor greater than 90% [Fink and Beaty, 1978].

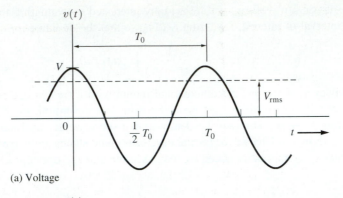

(a) Voltage

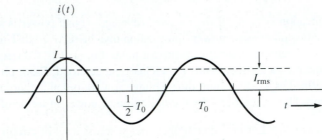

(b) Current

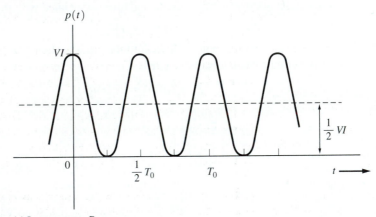

(c) Instantaneous Power

Figure 2–3   Steady-state waveshapes for Example 2–1.

$$V_{dc} = \langle v(t) \rangle = \langle V \cos \omega_0 t \rangle$$

$$= \frac{1}{T_0} \int_{-T_0/2}^{T_0/2} V \cos \omega_0 t \, dt = 0 \qquad (2\text{–}8)$$

where $\omega_0 = 2\pi/T_0$ and $f_0 = 1/T_0 = 60$ Hz. Similarly, $I_{dc} = 0$. The instantaneous power is

$$p(t) = (V \cos \omega_0 t)(I \cos \omega_0 t) = \tfrac{1}{2} VI(1 + \cos 2\omega_0 t) \qquad (2\text{–}9)$$

The average power is

$$P = \langle \tfrac{1}{2}VI(1 + \cos 2\omega_0 t)\rangle$$

$$= \frac{VI}{2T_0} \int_{-T_0/2}^{T_0/2} (1 + \cos 2\omega_0 t)\, dt$$

$$= \frac{VI}{2} \tag{2–10}$$

As can be seen from Eq. (2–9) and Fig. 2–3c, the power (i.e., light emitted) occurs in pulses at a rate of $2f_0 = 120$ pulses per second. (In fact, this lamp can be used as a stroboscope to "freeze" mechanically rotating objects.) The peak power is VI, and the average power is $\tfrac{1}{2}VI$, where V is the peak voltage and I is the peak current. Furthermore, for this case of sinusoidal voltage and sinusoidal current, we see that the average power could be obtained by multiplying $V/\sqrt{2}$ with $I/\sqrt{2}$.

Rms Value and Normalized Power

DEFINITION.   The *root-mean-square (rms)* value of $w(t)$ is

$$W_{\text{rms}} = \sqrt{\langle w^2(t)\rangle} \tag{2–11}$$

THEOREM.   *If a load is resistive (i.e., with unity power factor), the average power is*

$$P = \frac{\langle v^2(t)\rangle}{R} = \langle i^2(t)\rangle R = \frac{V_{\text{rms}}^2}{R}$$

$$= I_{\text{rms}}^2 R = V_{\text{rms}} I_{\text{rms}} \tag{2–12}$$

where R is the value of the resistive load.

Equation (2–12) follows from Eq. (2–7) by the use of Ohm's law, which is $v(t) = i(t)R$, and Eq. (2–11).

Continuing Example 2–1, $V_{\text{rms}} = 120$ V. It is seen from Eq. (2–11), when sinusoidal waveshapes are used, that $V_{\text{rms}} = V/\sqrt{2}$ and $I_{\text{rms}} = I/\sqrt{2}$. Thus, using Eq. (2–12), we see that the average power is $\tfrac{1}{2}VI$, which is the same value that was obtained in the previous discussion.

The concept of *normalized* power is often used by communications engineers. In this concept, R is assumed to be 1 Ω, although it may be another value in the actual circuit. Another way of expressing this concept is to say that the power is given on a per-ohm basis. In the signal-to-noise power ratio calculations, R will automatically cancel out, so that normalized values of power may be used to obtain the correct ratio. If the actual value for the power is needed, say, at the end of a long set of calculations, it can always be obtained by "denormalization" of the normalized value. From Eq. (2–12), it is also realized that *the square root of the normalized power is the rms value.*

DEFINITION.   The *average normalized power* is

$$P = \langle w^2(t) \rangle = \lim_{T \to \infty} \frac{1}{T} \int_{-T/2}^{T/2} w^2(t)\, dt \tag{2–13}$$

where $w(t)$ represents a real voltage or current waveform.

Energy and Power Waveforms[†]

DEFINITION. $w(t)$ is a *power waveform* if and only if the normalized average power P is finite and nonzero (i.e., $0 < P < \infty$).

DEFINITION.   The *total normalized energy* is

$$E = \lim_{T \to \infty} \int_{-T/2}^{T/2} w^2(t)\, dt \tag{2–14}$$

DEFINITION. $w(t)$ is an *energy waveform* if and only if the total normalized energy is finite and nonzero (i.e., $0 < E < \infty$).

From these definitions, it is seen that if a waveform is classified as either one of these types, it cannot be of the other type. That is, if $w(t)$ has finite energy, the power averaged over infinite time is zero, and if the power (averaged over infinite time) is finite, the energy is infinite. Moreover, mathematical functions can be found that have both infinite energy and infinite power and, consequently, cannot be classified into either of these two categories. One example is $w(t) = e^{-t}$. Physically realizable waveforms are of the energy type, but we will often model them by infinite-duration waveforms of the power type. Laboratory instruments that measure average quantities—such as the dc value, rms value, and average power—average over a finite time interval. That is, T of Eq. (2–1) remains finite instead of approaching some large number. Thus, nonzero average quantities for finite energy (physical) signals can be obtained. For example, when the dc value is measured with a conventional volt-ohm-milliamp meter containing a meter movement, the time-averaging interval is established by the mass of the meter movement that provides damping. Hence, the average quantities calculated from a power-type mathematical model (averaged over infinite time) will give the results that are measured in the laboratory (averaged over finite time).

Decibel

The *decibel* is a base 10 logarithmic measure of power ratios. For example, the ratio of the power level at the output of a circuit compared with that at the input is often specified by the decibel gain instead of the actual ratio.

DEFINITION.   The *decibel gain* of a circuit is[‡]

$$\text{dB} = 10 \log \left(\frac{\text{average power out}}{\text{average power in}} \right) = 10 \log \left(\frac{P_{\text{out}}}{P_{\text{in}}} \right) \tag{2–15}$$

[†] This concept is also called *energy signals* and *power signals* by some authors, but it applies to noise as well as signal waveforms.

[‡] Logarithms to the base 10 will be denoted by $\log(\cdot)$, and logarithms to the base e will be denoted by $\ln(\cdot)$. Note that both dB and the ratio $P_{\text{out}}/P_{\text{in}}$ are dimensionless quantities.

This definition gives a number that indicates the *relative* value of the *power out* with respect to the *power in* and does not indicate the actual magnitude of the power levels involved. If resistive loads are involved, Eq. (2–12) may be used to reduce Eq. (2–15) to

$$ dB = 20 \log \left(\frac{V_{rms\ out}}{V_{rms\ in}} \right) + 10 \log \left(\frac{R_{in}}{R_{load}} \right) \tag{2–16} $$

or

$$ dB = 20 \log \left(\frac{I_{rms\ out}}{I_{rms\ in}} \right) + 10 \log \left(\frac{R_{load}}{R_{in}} \right) \tag{2–17} $$

Note that the same value for decibels is obtained regardless of whether power, voltage, or current [Eq. (2–15), Eq. (2–16), or Eq. (2–17)] was used to obtain that value. That is, decibels are defined in terms of the logarithm of a power ratio, but may be evaluated from voltage or current ratios.

If normalized powers are used,

$$ dB = 20 \log \left(\frac{V_{rms\ out}}{V_{rms\ in}} \right) = 20 \log \left(\frac{I_{rms\ out}}{I_{rms\ in}} \right) \tag{2–18} $$

This equation does not give the true value for decibels, unless $R_{in} = R_{load}$; however, it is common engineering practice to use Eq. (2–18) even if $R_{in} \neq R_{load}$ and even if the number that is obtained is not strictly correct. Engineers understand that if the correct value is needed, it may be calculated from this pseudovalue if R_{in} and R_{load} are known.

If the dB value is known, the power ratio or the voltage ratio can be easily obtained by inversion of the appropriate equations just given. For example, if the power ratio is desired, Eq. (2–15) can be inverted to obtain

$$ \frac{P_{out}}{P_{in}} = 10^{dB/10} \tag{2–19} $$

The decibel measure can also be used to express a measure of the ratio of signal power to noise power, as measured at some point in a circuit.

DEFINITION.   The *decibel signal-to-noise ratio* is[†]

$$ (S/N)_{dB} = 10 \log \left(\frac{P_{signal}}{P_{noise}} \right) = 10 \log \left(\frac{\langle s^2(t) \rangle}{\langle n^2(t) \rangle} \right) \tag{2–20} $$

Because the signal power is $\langle s^2(t) \rangle / R = V_{rms\ signal}^2 / R$ and the noise power is $\langle n^2(t) \rangle / R = V_{rms\ noise}^2 / R$, this definition is equivalent to

$$ (S/N)_{dB} = 20 \log \left(\frac{V_{rms\ signal}}{V_{rms\ noise}} \right) \tag{2–21} $$

[†] This definition involves the ratio of the *average signal* power to the *average noise* power. An alternative definition that is also useful for some applications involves the ratio of the *peak* signal power to the average noise power. See Sec. 6–8.

The decibel measure may also be used to indicate absolute levels of power with respect to some reference level.

DEFINITION.   The *decibel power level* with respect to 1 mW is

$$\text{dBm} = 10 \log \left(\frac{\text{actual power level (watts)}}{10^{-3}} \right)$$

$$= 30 + 10 \log[\text{actual power level (watts)}] \qquad (2\text{--}22)$$

where the "m" in the dBm denotes a milliwatt reference. Laboratory RF signal generators are usually calibrated in terms of dBm.

Other decibel measures of absolute power levels are also used. When a 1-W reference level is used, the decibel level is denoted dBW; when a 1-kW reference level is used, the decibel level is denoted dBk. For example, a power of 5 W could be specified as +36.99 dBm, 6.99 dBW, or −23.0 dBk. The telephone industry uses a decibel measure with a "reference noise" level of 1 picowatt (10^{-12} W) [Jordan, 1985]. This decibel measure is denoted dBrn. A level of 0 dBrn corresponds to −90 dBm. The cable television (CATV) industry uses a 1-millivolt rms level across a 75-Ω load as a reference. This decibel measure is denoted dBmV, and it is defined as

$$\text{dBmV} = 20 \log \left(\frac{V_{\text{rms}}}{10^{-3}} \right) \qquad (2\text{--}23)$$

where 0 dBmV corresponds to −48.75 dBm.

It should be emphasized that the general expression for evaluating power is given by Eq. (2–7). That is, Eq. (2–7) can be used to evaluate the average power for any type of waveshape and load condition, whereas Eq. (2–12) is useful only for resistive loads. In other books, especially in the power area, equations are given that are valid only for sinusoidal waveshapes.

Phasors

Sinusoidal test signals occur so often in electrical engineering problems that a shorthand notation called *phasor notation* is often used.

DEFINITION.   A complex number c is said to be a *phasor* if it is used to represent a *sinusoidal* waveform. That is,

$$w(t) = |c| \cos[\omega_0 t + \underline{/c}\,] = \text{Re}\{ce^{j\omega_0 t}\} \qquad (2\text{--}24)$$

where the phasor $c = |c|e^{j\underline{/c}}$ and Re$\{\cdot\}$ denotes the real part of the complex quantity $\{\cdot\}$.

We will refer to $ce^{j\omega_0 t}$ as the *rotating* phasor, as distinguished from the phasor c. When c is given to represent a waveform, it is *understood* that the actual waveform that appears in the circuit is a sinusoid as specified by Eq. (2–24). Because the phasor is a complex number, it can be written in either Cartesian form or polar form; that is,

$$c \triangleq x + jy = |c|e^{j\varphi} \qquad (2\text{--}25)$$

where x and y are real numbers along the Cartesian coordinates and $|c|$ and $\underline{/c} = \varphi$ $= \tan^{-1}(y/x)$ are the length and angle (real numbers) in the polar coordinate system. Shorthand notation for the polar form on the right-hand side of Eq. (2–25) is $|c|\ \underline{/\varphi}$.

For example, $25 \sin(2\pi 500t + 45°)$ could be denoted by the phasor $25 \;\underline{/-45°}$, since, from Appendix A, $\sin x = \cos(x - 90°)$ and, consequently, $25 \sin(\omega_0 t + 45°) = 25 \cos(\omega_0 t - 45°) = \text{Re}\{(25e^{-j\pi/4}) e^{j\omega_0 t}\}$, where $\omega_0 = 2\pi f_0$ and $f_0 = 500$ Hz. Similarly, $10 \cos(\omega_0 t + 35°)$ could be denoted by $10 \;\underline{/35°}$.

Some other authors may use a slightly different definition for the phasor. For example, $w(t)$ may be expressed in terms of the imaginary part of a complex quantity instead of the real part as defined in Eq. (2–24). In addition, the phasor may denote the rms value of $w(t)$, instead of the peak value [Kaufman and Seidman, 1979]. In this case, $10 \sin(\omega_0 t + 45°)$ should be denoted by the phasor $7.07 \;\underline{/45°}$. Throughout this book, the definition as given by Eq. (2–24) will be used. Phasors can represent only *sinusoidal* waveshapes.

2–2 FOURIER TRANSFORM AND SPECTRA

Definition

How does one find the frequencies that are present in a waveform? Moreover, what is the definition of frequency? For waveforms of the sinusoidal type, we know that we can find the frequency by evaluating $f_0 = 1/T_0$, where T_0 is the period of the sinusoid. That is, frequency is the rate of occurrence of the sinusoidal waveshape. All other nonsinusoidal waveshapes have more than one frequency.[†]

In most practical applications, the waveform is not periodic, so there is no T_0 to use to calculate the frequency. Consequently, we still need to answer this question: Is there a general method for finding the frequencies of a waveform that will work for any type of waveshape? The answer is yes. It is the Fourier transform (FT). It finds the sinusoidal-type components in $w(t)$

DEFINITION.   The *Fourier transform* (FT) of a waveform $w(t)$ is

$$W(f) = \mathcal{F}[w(t)] = \int_{-\infty}^{\infty} [w(t)]e^{-j2\pi ft} \, dt \qquad (2\text{–}26)$$

where $\mathcal{F}[\cdot]$ denotes the Fourier transform of $[\cdot]$ and f is the frequency parameter with units of Hz (i.e., $1/s$).[‡] This *defines* the term *frequency* which is the parameter f in the Fourier transform.

$W(f)$ is also called a *two-sided spectrum* of $w(t)$, because both positive and negative frequency components are obtained from Eq. (2–26). It should be clear that the spectrum of a voltage (or current) waveform is obtained by a mathematical calculation and that it does not appear physically in an actual circuit. f is just a parameter (called *frequency*) that determines which point of the spectral function is to be evaluated.

[†] The constant-voltage or constant-current *dc* waveform has one frequency, $f = 0$. It is a special case of a cosine wave (i.e. sinusoidal-type waveform) where $T_0 \to \infty$ and $f_0 \to 0$. A periodic square wave has an infinite number of odd-harmonic frequencies, as shown by Example 2–12.

[‡] Some authors define the FT in terms of the frequency parameter $\omega = 2\pi f$ where ω is in radians per second. That definition is identical to (2–26) when ω is replaced by $2\pi f$. I prefer Eq. (2–26) because spectrum analyzers are usually calibrated in units of hertz, not radians per second.

The FT is used to find the frequencies in $w(t)$. One chooses some value of f, say, $f = f_0$, and evaluates $|W(f_0)|$. If $|W(f_0)|$ is not zero, then the frequency f_0 is present in $w(t)$. In general, the FT integral is evaluated again and again for all possible values of f over the range $-\infty < f < \infty$ to find all of the frequencies in $w(t)$.

It is easy to demonstrate why the FT finds the frequencies in $w(t)$. For example, suppose that $w(t) = 1$. For this dc waveform, the integrand of Eq. (2–26) is $e^{-j2\pi ft} = \cos 2\pi ft - j \sin 2\pi ft$, and consequently, the FT integral is zero (provided that $f \neq 0$), because the area under a sinusoidal wave over multiple periods is zero. However, if one chooses $f = 0$, then the FT integral is not zero. Thus, as expected, Eq. (2–26) identifies $f = 0$ as the frequency of $w(t) = 1$. In another example, let $w(t) = 2 \sin 2\pi f_0 t$. For this case, the integrand of Eq. (2–26) is $\sin 2\pi(f_0 - f)t + \sin 2\pi(f_0 + f)t - j \cos 2\pi(f_0 - f)t + \cos 2\pi(f_0 + f)t$ and the integral is nonzero when $f = f_0$ or when $f = -f_0$.[†] In this way, the FT finds the frequencies in $w(t)$.

Direct evaluation of the FT integral can be difficult, so a list of alternative evaluation techniques is very helpful. The FT integral, Eq. (2–26), can be evaluated by the use of

1. Direct integration.[‡] *See Example 2–2.*
2. Tables of Fourier transforms or Laplace transforms. *See Table 2–2 and Example 2–9.*
3. FT theorems. *See Table 2–1 and Example 2–3.*
4. Superposition to break the problem into two or more simple problems. *See study-aid example SA–5.*
5. Differentiation or integration of $w(t)$. *See Example 2–6.*
6. Numerical integration of the FT integral on the PC via MATLAB or MathCAD integration functions. *See Example 2–5 and file E2_055N.M.*
7. Fast Fourier transform (FFT) on the PC via MATLAB or MathCAD FFT functions. *See Fig. 2–21, file FIG2_21.M, and Fig. 2–22, file FIG2_22.M.*

These evaluation techniques are developed throughout the remainder of this chapter.

From Eq. (2–26), since $e^{-j2\pi ft}$ is complex, $W(f)$ is a complex function of frequency. $W(f)$ may be decomposed into two real functions $X(f)$ and $Y(f)$ such that

$$W(f) = X(f) + jY(f) \qquad (2\text{--}27)$$

which is identical to writing a complex number in terms of pairs of real numbers that can be plotted in a two-dimensional Cartesian coordinate system. For this reason, Eq. (2–27) is sometimes called the *quadrature* form, or *Cartesian* form. Similarly, Eq. (2–26) can be written equivalently in terms of a polar coordinate system, where the pair of real functions denotes the magnitude and phase:

$$W(f) = |W(f)|e^{j\theta(f)} \qquad (2\text{--}28)$$

[†] The frequency of the sine wave is f_0, but the FT finds both f_0 and its mirror image $-f_0$, as explained in the discussion following Example 2–4.

[‡] Contour integration, covered in a mathematics course in complex variables, is also a helpful integration technique.

That is,

$$|W(f)| = \sqrt{X^2(f) + Y^2(f)} \quad \text{and} \quad \theta(f) = \tan^{-1}\left(\frac{Y(f)}{X(f)}\right) \tag{2–29}$$

This is called the *magnitude–phase* form, or *polar* form. To determine whether certain frequency components are present, one would examine the magnitude spectrum $|W(f)|$, which engineers sometimes loosely call the *spectrum*.

The time waveform may be calculated from the spectrum by using the *inverse Fourier transform*

$$w(t) = \int_{-\infty}^{\infty} W(f)e^{j2\pi ft}\, df \tag{2–30}$$

The functions $w(t)$ and $W(f)$ are said to constitute a *Fourier transform pair*, where $w(t)$ is the time domain description and $W(f)$ is the frequency domain description. In this book, the time domain function is usually denoted by a lowercase letter and the frequency domain function by an uppercase letter. Shorthand notation for the pairing between the two domains will be denoted by a double arrow: $w(t) \leftrightarrow W(f)$.

The waveform $w(t)$ is Fourier transformable (i.e., sufficient conditions) if it satisfies both *Dirichlet conditions*:

- Over any time interval of finite width, the function $w(t)$ is single valued with a finite number of maxima and minima, and the number of discontinuities (if any) is finite.
- $w(t)$ is absolutely integrable. That is,

$$\int_{-\infty}^{\infty} |w(t)|\, dt < \infty \tag{2–31}$$

Although these conditions are sufficient, they are not necessary. In fact, some of the examples given subsequently do not satisfy the Dirichlet conditions, and yet the Fourier transform can be found.

A weaker sufficient condition for the existence of the Fourier transform is

$$E = \int_{-\infty}^{\infty} |w(t)|^2\, dt < \infty \tag{2–32}$$

where E is the normalized energy [Goldberg, 1961]. This is the finite-energy condition that is satisfied by all physically realizable waveforms. Thus, all physical waveforms encountered in engineering practice are Fourier transformable.

It should also be noted that mathematicians sometimes use other definitions for the Fourier transform rather than Eq. (2–26). However, in these cases, the equation for the corresponding inverse transforms, equivalent to Eq. (2–30), would also be different, so that when the transform and its inverse are used together, the original $w(t)$ would be recovered. This is a consequence of the Fourier integral theorem, which is

$$w(t) = \int_{-\infty}^{\infty}\int_{-\infty}^{\infty} w(\lambda)e^{j2\pi f(t-\lambda)}d\lambda\, df \tag{2–33}$$

Equation (2–33) may be decomposed into Eqs. (2–26) and (2–30), as well as other definitions for Fourier transform pairs. The Fourier integral theorem is strictly true only for well-behaved functions (i.e., physical waveforms). For example, if $w(t)$ is an ideal square wave, then at a discontinuous point of $w(\lambda)$, denoted by λ_0, $w(t)$ will have a value that is the average of the two values that are obtained for $w(\lambda)$ on each side of the discontinuous point λ_0.

Example 2–2 SPECTRUM OF AN EXPONENTIAL PULSE

Let $w(t)$ be a decaying exponential pulse that is switched on at $t = 0$. That is,

$$w(t) = \begin{cases} e^{-t}, & t > 0 \\ 0, & t < 0 \end{cases}$$

Directly integrating the FT integral, we get

$$W(f) = \int_0^\infty e^{-t} e^{-j2\pi ft}\, dt = \frac{e^{-(1+j2\pi f)t}}{1 + j2\pi f}\Big|_0^\infty$$

or

$$W(f) = \frac{1}{1 + j2\pi f}$$

In other words, the FT pair is

$$\begin{cases} e^{-t}, & t > 0 \\ 0, & t < 0 \end{cases} \leftrightarrow \frac{1}{1 + j2\pi f} \qquad (2\text{--}34)$$

The spectrum can also be expressed in terms of the quadrature functions by rationalizing the denominator of Eq. (2–34); thus,

$$X(f) = \frac{1}{1 + (2\pi f)^2} \quad \text{and} \quad Y(f) = \frac{-2\pi f}{1 + (2\pi f)^2}$$

The magnitude–phase form is

$$|W(f)| = \sqrt{\frac{1}{1 + (2\pi f)^2}} \quad \text{and} \quad \theta(f) = -\tan^{-1}(2\pi f)$$

More examples will be given after some helpful theorems are developed.

Properties of Fourier Transforms

Many useful and interesting theorems follow from the definition of the spectrum as given by Eq. (2–26). One of particular interest is a consequence of working with real waveforms. In any physical circuit that can be built, the voltage (or current) waveforms are real functions (as opposed to complex functions) of time.

THEOREM. *Spectral symmetry of real signals. If $w(t)$ is real, then*

$$W(-f) = W^*(f) \qquad (2\text{--}35)$$

(The superscript asterisk denotes the conjugate operation.)

Proof. From Eq. (2–26), we get

$$W(-f) = \int_{-\infty}^{\infty} w(t)e^{j2\pi ft}\, dt \tag{2–36}$$

and taking the conjugate of Eq. (2–26) yields

$$W^*(f) = \int_{-\infty}^{\infty} w^*(t)e^{j2\pi ft}\, dt \tag{2–37}$$

But since $w(t)$ is real, $w^*(t) = w(t)$, and Eq. (2–35) follows because the right sides of Eqs. (2–36) and (2–37) are equal. It can also be shown that if $w(t)$ is real and happens to be an even function of t, $W(f)$ is *real*. Similarly, if $w(t)$ is real and is an odd function of t, $W(f)$ is *imaginary*.

Another useful corollary of Eq. (2–35) is that, for $w(t)$ real, the magnitude spectrum is even about the origin (i.e., $f = 0$), or

$$|W(-f)| = |W(f)| \tag{2–38}$$

and the phase spectrum is odd about the origin:

$$\theta(-f) = -\theta(f) \tag{2–39}$$

This can easily be demonstrated by writing the spectrum in polar form:

$$W(f) = |W(f)|e^{j\,\theta(f)}$$

Then

$$W(-f) = |W(-f)|e^{j\,\theta(-f)}$$

and

$$W^*(f) = |W(f)|e^{-j\,\theta(f)}$$

Using Eq. (2–35), we see that Eqs. (2–38) and (2–39) are true.

In summary, from the previous discussion, the following are some properties of the Fourier transform:

- f, called frequency and having units of hertz, is just a parameter of the FT that specifies what frequency we are interested in looking for in the waveform $w(t)$.
- The FT looks for the frequency f in the $w(t)$ over *all* time, that is, over

$$-\infty < t < \infty$$

- $W(f)$ can be complex, even though $w(t)$ is real.
- If $w(t)$ is real, then $W(-f) = W^*(f)$.

Parseval's Theorem and Energy Spectral Density

Parseval's Theorem.

$$\int_{-\infty}^{\infty} w_1(t)w_2^*(t)\,dt = \int_{-\infty}^{\infty} W_1(f)W_2^*(f)\,df \qquad (2\text{--}40)$$

If $w_1(t) = w_2(t) = w(t)$, then the theorem reduces to

Rayleigh's energy theorem, which is

$$E = \int_{-\infty}^{\infty} |w(t)|^2\,dt = \int_{-\infty}^{\infty} |W(f)|^2\,df \qquad (2\text{--}41)$$

Proof. Working with the left side of Eq. (2–40) and using Eq. (2–30) to replace $w_1(t)$ yields.

$$\int_{-\infty}^{\infty} w_1(t)w_2^*(t)\,dt = \int_{-\infty}^{\infty} \left[\int_{-\infty}^{\infty} W_1(f)e^{j2\pi ft}\,df\right]w_2^*(t)\,dt$$

$$= \int_{-\infty}^{\infty}\int_{-\infty}^{\infty} W_1(f)w_2^*(t)e^{j2\pi ft}\,df\,dt$$

Interchanging the order of integration on f and t[†] gives

$$\int_{-\infty}^{\infty} w_1(t)w_2^*(t)\,dt = \int_{-\infty}^{\infty} W_1(f)\left[\int_{-\infty}^{\infty} w_2(t)e^{-j2\pi ft}\,dt\right]^*\,df$$

Using Eq. (2–26) produces Eq. (2–40). Parseval's theorem gives an alternative method for evaluating the energy by using the frequency domain description instead of the time domain definition. This leads to the concept of the energy spectral density function.

DEFINITION. The *energy spectral density* (*ESD*) is defined for energy waveforms by

$$\mathcal{E}(f) = |W(f)|^2 \qquad (2\text{--}42)$$

where $w(t) \leftrightarrow W(f)$. $\mathcal{E}(f)$ has units of joules per hertz.

Using Eq. (2–41), we see that the total normalized energy is given by the area under the ESD function:

$$E = \int_{-\infty}^{\infty} \mathcal{E}(f)\,df \qquad (2\text{--}43)$$

For power waveforms, a similar function called the *power spectral density* (*PSD*) can be defined. This is developed in Sec. 2–3 and in Chapter 6.

There are many other Fourier transform theorems in addition to Parseval's theorem. Some are summarized in Table 2–1. These theorems can be proved by substituting the corresponding time function into the definition of the Fourier transform and reducing the result

---

[†] Fubini's theorem states that the order of integration may be exchanged if all of the integrals are absolutely convergent. That is, these integrals are finite valued when the integrands are replaced by their absolute values. We assume that this condition is satisfied.

to that given in the rightmost column of the table. For example, the scale-change theorem is proved by substituting $w(at)$ into Eq. (2–26). We get

$$\mathcal{F}[w(at)] = \int_{-\infty}^{\infty} w(at)e^{-j2\pi ft}\, dt$$

Letting $t_1 = at$, and assuming that $a > 0$, we obtain

$$\mathcal{F}[w(at)] = \int_{-\infty}^{\infty} \frac{1}{a}\, w(t_1)e^{-j2\pi(f/a)t_1}\, dt_1 = \frac{1}{a} W\left(\frac{f}{a}\right)$$

For $a < 0$, this equation becomes

$$\mathcal{F}[w(at)] = \int_{-\infty}^{\infty} \frac{-1}{a}\, w(t_1)e^{-j2\pi(f/a)t_1}\, dt_1 = \frac{1}{|a|} W\left(\frac{f}{a}\right)$$

Thus, for either $a > 0$ or $a < 0$, we get

$$w(at) \leftrightarrow \frac{1}{|a|} W\left(\frac{f}{a}\right)$$

The other theorems in Table 2–1 are proved in a similar straightforward manner, except for the integral theorem, which is more difficult to derive because the transform result involves a Dirac delta function $\delta(f)$. This theorem may be proved by the use of the convolution theorem, as illustrated by Prob. 2–36. The bandpass signal theorem will be discussed in more detail in Chapter 4. It is the basis for the digital and analog modulation techniques covered in Chapters 4 and 5. In Sec. 2–8, the relationship between the Fourier transform and the discrete Fourier transform (DFT) will be studied.

As we will see in the examples that follow, these theorems can greatly simplify the calculations required to solve Fourier transform problems. The reader should study Table 2–1 and be prepared to use it when needed. After the Fourier transform is evaluated, one should check that the easy-to-verify properties of Fourier transforms are satisfied; otherwise, there is a mistake. For example, if $w(t)$ is real,

- $W(-f) = W^*(f)$, or $|W(f)|$ is even and $\theta(f)$ is odd.
- $W(f)$ is real when $w(t)$ is even.
- $W(f)$ is imaginary when $w(t)$ is odd.

Example 2–3 SPECTRUM OF A DAMPED SINUSOID

Let the damped sinusoid be given by

$$w(t) = \begin{cases} e^{-t/T}\sin \omega_0 t, & t > 0,\ T > 0 \\ 0, & t < 0 \end{cases}$$

The spectrum of this waveform is obtained by evaluating the FT. This is easily accomplished by

TABLE 2–1   SOME FOURIER TRANSFORM THEOREMS[a]

Operation	Function	Fourier Transform
Linearity	$a_1 w_1(t) + a_2 w_2(t)$	$a_1 W_1(f) + a_2 W_2(f)$
Time delay	$w(t - T_d)$	$W(f) \, e^{-j\omega T_d}$
Scale change	$w(at)$	$\dfrac{1}{\lvert a \rvert} W\left(\dfrac{f}{a}\right)$
Conjugation	$w^*(t)$	$W^*(-f)$
Duality	$W(t)$	$w(-f)$
Real signal frequency translation [$w(t)$ is real]	$w(t) \cos(w_c t + \theta)$	$\frac{1}{2}[e^{j\theta} W(f - f_c) + e^{-j\theta} W(f + f_c)]$
Complex signal frequency translation	$w(t) \, e^{j\omega_c t}$	$W(f - f_c)$
Bandpass signal	$\text{Re}\{g(t) \, e^{j\omega_c t}\}$	$\frac{1}{2}[G(f - f_c) + G^*(-f - f_c)]$
Differentiation	$\dfrac{d^n w(t)}{dt^n}$	$(j2\pi f)^n W(f)$
Integration	$\displaystyle\int_{-\infty}^{t} w(\lambda) d\lambda$	$(j2\pi f)^{-1} W(f) + \frac{1}{2} W(0) \, \delta(f)$
Convolution	$w_1(t) * w_2(t) = \displaystyle\int_{-\infty}^{\infty} w_1(\lambda)$ $\cdot w_2(t - \lambda) \, d\lambda$	$W_1(f) W_2(f)$
Multiplication[b]	$w_1(t) w_2(t)$	$W_1(f) * W_2(f) = \displaystyle\int_{-\infty}^{\infty} W_1(\lambda) \, W_2(f - \lambda) \, d\lambda$
Multiplication by t^n	$t^n w(t)$	$(-j2\pi)^{-n} \dfrac{d^n W(f)}{df^n}$

[a] $\omega_c = 2\pi f_c$.

[b] $*$ denotes convolution as described in detail by Eq. (2–62).

using the result of the previous example plus some of the Fourier theorems. From Eq. (2–34), and using the scale-change theorem of Table 2–1, where $a = 1/T$, we find that

$$\begin{cases} e^{-t/T}, & t > 0 \\ 0, & t < 0 \end{cases} \leftrightarrow \frac{T}{1 + j(2\pi fT)}$$

Using the real signal frequency translation theorem with $\theta = -\pi/2$, we get

$$W(f) = \frac{1}{2} \left\{ e^{-j\pi/2} \frac{T}{1 + j2\pi T(f - f_0)} + e^{j\pi/2} \frac{T}{1 + j2\pi T(f + f_0)} \right\}$$

$$= \frac{T}{2j} \left\{ \frac{1}{1 + j2\pi T(f - f_0)} - \frac{1}{1 + j2\pi T(f + f_0)} \right\} \tag{2–44}$$

where $e^{\pm j\pi/2} = \cos(\pi/2) \pm j \sin(\pi/2) = \pm j$. This spectrum is complex (i.e., neither real nor imaginary), because $w(t)$ does not have even or odd symmetry about $t = 0$.

As expected, Eq. (2–44) shows that the peak of the magnitude spectrum for the damped sinusoid occurs at $f = \pm f_0$. Compare this with the peak of the magnitude spectrum for the exponential decay (Example 2–2) that occurs at $f = 0$. That is, the $\sin \omega_0 t$ factor caused the spectral peak to move from $f = 0$ to $f = \pm f_0$.

Dirac Delta Function and Unit Step Function

DEFINITION.   The *Dirac delta function* $\delta(x)$ is defined by

$$\int_{-\infty}^{\infty} w(x) \, \delta(x) \, dx = w(0) \tag{2–45}$$

where $w(x)$ is any function that is continuous at $x = 0$.

In this definition, the variable x could be time or frequency, depending on the application. An alternative definition for $\delta(x)$ is

$$\int_{-\infty}^{\infty} \delta(x) \, dx = 1 \tag{2–46a}$$

and

$$\delta(x) = \begin{cases} \infty, & x = 0 \\ 0 & x \neq 0 \end{cases} \tag{2–46b}$$

where both Eqs. (2–46a) and (2–46b) need to be satisfied. The Dirac delta function is not a true function, so it is said to be a singular function. However, $\delta(x)$ can be defined as a function in a more general sense and treated as such in a branch of mathematics called *generalized functions* and the *theory of distributions*.

From Eq. (2–45), the *sifting* property of the δ function is

$$\int_{-\infty}^{\infty} w(x) \, \delta(x - x_0) \, dx = w(x_0) \tag{2–47}$$

That is, the δ function sifts out the value $w(x_0)$ from the integral.

In some problems, it is also useful to use the equivalent integral for the δ function, which is

$$\delta(x) = \int_{-\infty}^{\infty} e^{\pm j2\pi xy} \, dy \tag{2–48}$$

where either the $+$ or the $-$ sign may be used as needed. This assumes that $\delta(x)$ is an even function: $\delta(-x) = \delta(x)$. Equation (2–48) may be verified by taking the Fourier transform of a delta function

$$\int_{-\infty}^{\infty} \delta(t) e^{-j2\pi ft} \, dt = e^0 = 1$$

and then taking the inverse Fourier transform of both sides of this equation; Eq. (2–48) follows. (For additional properties of the delta function, see Appendix A.)

Another function that is closely related to the Dirac delta function is the unit step function.

DEFINITION.   The *unit step function* $u(t)$ is

$$u(t) = \begin{cases} 1, & t > 0 \\ 0, & t < 0 \end{cases} \tag{2–49}$$

Because $\delta(\lambda)$ is zero, except at $\lambda = 0$, the Dirac delta function is related to the unit step function by

$$\int_{-\infty}^{t} \delta(\lambda)\, d\lambda = u(t) \tag{2–50}$$

and consequently,

$$\frac{du(t)}{dt} = \delta(t) \tag{2–51}$$

Example 2–4 SPECTRUM OF A SINUSOID

Find the spectrum of a sinusoidal voltage waveform that has a frequency f_0 and a peak value of A volts. That is,

$$v(t) = A \sin \omega_0 t \quad \text{where} \quad \omega_0 = 2\pi f_0$$

From Eq. (2–26), the spectrum is

$$V(f) = \int_{-\infty}^{\infty} A\left(\frac{e^{j\omega_0 t} - e^{-j\omega_0 t}}{2j}\right) e^{-j\omega t}\, dt$$

$$= \frac{A}{2j} \int_{-\infty}^{\infty} e^{-j2\pi(f-f_0)t}\, dt - \frac{A}{2j} \int_{-\infty}^{\infty} e^{-j2\pi(f-f_0)t}\, dt$$

By Eq. (2–48), these integrals are equivalent to Dirac delta functions. That is,

$$V(f) = j\frac{A}{2}\left[\delta(f + f_0) - \delta(f - f_0)\right]$$

Note that this spectrum is imaginary, as expected, because $v(t)$ is real and odd. In addition, a meaningful expression was obtained for the Fourier transform, although $v(t)$ was of the infinite energy type and not absolutely integrable. That is, this $v(t)$ does not satisfy the sufficient (but not necessary) Dirichlet conditions as given by Eqs. (2–31) and (2–32).

The magnitude spectrum is

$$|V(f)| = \frac{A}{2}\delta(f - f_0) + \frac{A}{2}\delta(f + f_0)$$

where A is a positive number. Since only two frequencies ($f = \pm f_0$) are present, $\theta(f)$ is strictly defined only at these two frequencies. That is, $\theta(f_0) = \tan^{-1}(-1/0) = -90°$ and $\theta(-f_0) = \tan^{-1}(1/0) = +90°$. However, because $|V(f)| = 0$ for all frequencies except $f = \pm f_0$ and $V(f) = |V(f)|\, e^{j\theta(f)}$, $\theta(f)$ can be taken to be any convenient set of values for $f \neq \pm f_0$. Thus, the phase spectrum is taken to be

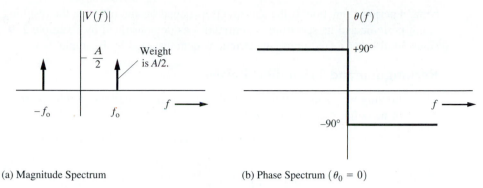

(a) Magnitude Spectrum   (b) Phase Spectrum ($\theta_0 = 0$)

Figure 2–4   Spectrum of a sine wave.

$$\theta(f) = \begin{cases} -\pi/2, & f > 0 \\ +\pi/2, & f < 0 \end{cases} \text{ radians} = \begin{cases} -90°, & f > 0 \\ 90°, & f < 0 \end{cases}$$

Plots of these spectra are shown in Fig. 2–4. As shown in Fig. 2–4a, the *weights* of the delta functions are plotted, since it is impossible to plot the delta functions themselves because they have infinite values. It is seen that the magnitude spectrum is even and the phase spectrum is odd, as expected from Eqs. (2–38) and (2–39).

Now let us generalize the sinusoidal waveform to one with an arbitrary phase angle θ_0. Then

$$w(t) = A \sin(\omega_0 t + \theta_0) = A \sin[\omega_0(t + \theta_0/\omega_0)]$$

and, by using the time delay theorem, the spectrum becomes

$$W(f) = j \frac{A}{2} e^{j\theta_0(f/f_0)} [\delta(f + f_0) - \delta(f - f_0)]$$

The resulting magnitude spectrum is the same as that obtained before for the $\theta_0 = 0$ case. The new phase spectrum is the sum of the old phase spectrum plus the linear function $(\theta_0/f_0)f$. However, since the overall spectrum is zero except at $f = \pm f_0$, the value for the phase spectrum can be arbitrarily assigned at all frequencies except $f = \pm f_0$. At $f = f_0$ the phase is $(\theta_0 - \pi/2)$ radians, and at $f = -f_0$ the phase is $-(\theta_0 - \pi/2)$ radians.

From a *mathematical* viewpoint, Fig. 2–4 demonstrates that two frequencies are present in the sine wave, one at $f = +f_0$ and another at $f = -f_0$. This can also be seen from an expansion of the time waveform; that is,

$$v(t) = A \sin \omega_0 t = \frac{A}{j2} e^{j\omega_0 t} - \frac{A}{j2} e^{-j\omega_0 t}$$

which implies that the sine wave consists of two rotating phasors, one rotating with a frequency $f = +f_0$ and another rotating with $f = -f_0$. From the *engineering* point of view it is said that one frequency is present, namely, $f = f_0$, because for any physical (i.e., real) waveform, Eq. (2–35) shows that for any positive frequency present, there is also a mathematical negative frequency present. The phasor associated with $v(t)$ is $c = 0 - jA = A \angle -90°$. Another interesting observation is that the magnitude spectrum consists of *lines* (i.e., Dirac delta functions). As shown by Eq. (2–109), the lines are a consequence of $v(t)$

being a periodic function. If the sinusoid is switched on and off, then the resulting waveform is not periodic and its spectrum is continuous as demonstrated by Example 2–9. A damped sinusoid also has a continuous spectrum, as demonstrated by Example 2–3.

Rectangular and Triangular Pulses

The following waveshapes frequently occur in communications problems, so special symbols will be defined to shorten the notation.

DEFINITION. Let $\Pi(\cdot)$ denote a single *rectangular pulse*. Then

$$\Pi\left(\frac{t}{T}\right) \triangleq \begin{cases} 1, & |t| \le \dfrac{T}{2} \\[2mm] 0, & |t| > \dfrac{T}{2} \end{cases} \tag{2-52}$$

DEFINITION. $\mathrm{Sa}(\cdot)$ denotes the function[†]

$$\mathrm{Sa}(x) = \frac{\sin x}{x} \tag{2-53}$$

DEFINITION. Let $\Lambda(\cdot)$ denote the triangular function. Then

$$\Lambda\left(\frac{t}{T}\right) \triangleq \begin{cases} 1 - \dfrac{|t|}{T}, & |t| \le T \\[2mm] 0, & |t| > T \end{cases} \tag{2-54}$$

These waveshapes are shown in Fig. 2–5. A tabulation of $\mathrm{Sa}(x)$ is given in Sec. A–9 (Appendix A).

Example 2–5 SPECTRUM OF A RECTANGULAR PULSE

The spectrum is obtained by taking the Fourier transform of $w(t) = \Pi(t/T)$.

$$W(f) = \int_{-T/2}^{T/2} 1 e^{-j\omega t}\, dt = \frac{e^{-j\omega T/2} - e^{j\omega T/2}}{-j\omega}$$

$$= T\frac{\sin(\omega T/2)}{\omega T/2} = T\,\mathrm{Sa}(\pi T f)$$

Thus,

$$\Pi\left(\frac{t}{T}\right) \leftrightarrow T\,\mathrm{Sa}(\pi T f) \tag{2-55}$$

A numerical evaluation of this FT integral is given by file E2_055N.M. The preceding Fourier transform pair is shown in Fig. 2–6a. Note the inverse relationship between the pulse width T

[†] This is related to the sinc function by $\mathrm{Sa}(x) = \mathrm{sinc}(x/\pi)$ because $\mathrm{sinc}(\lambda) \triangleq (\sin \pi\lambda)/\pi\lambda$. The notation $\mathrm{Sa}(x)$ and $\mathrm{sinc}(x)$ represent the same concept, but can be confused because of scaling. In this book, $(\sin x)/x$ will often be used because it avoids confusion and does not take much more text space.

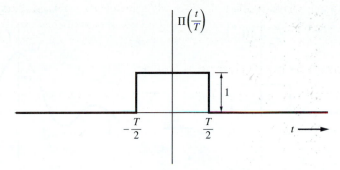

(a) Rectangular Pulse

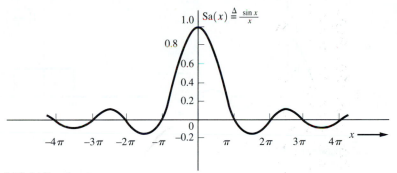

(b) Sa(x) Function

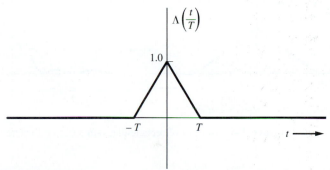

(c) Triangular Function

Figure 2–5   Waveshapes and corresponding symbolic notation.

and the spectral zero crossing $1/T$. Also, by use of the duality theorem (listed in Table 2–1), the spectrum of a $(\sin x)/x$ pulse is a rectangle. That is, realizing that $\Pi(x)$ is an even function, and applying the duality theorem to Eq. (2–55), we get

$$T\,\mathrm{Sa}(\pi T t) \leftrightarrow \Pi\left(-\frac{f}{T}\right) = \Pi\left(\frac{f}{T}\right)$$

Time Domain   Frequency Domain

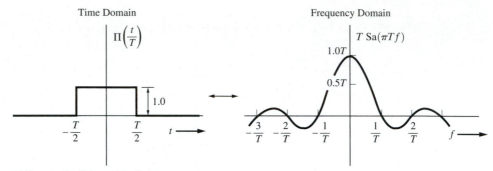

(a) Rectangular Pulse and Its Spectrum

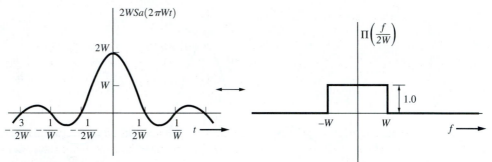

(b) $Sa(x)$ Pulse and Its Spectrum

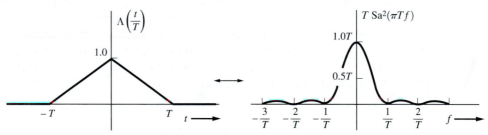

(c) Triangular Pulse and Its Spectrum

Figure 2–6   Spectra of rectangular, $(\sin x)/x$, and triangular pulses.

Replacing the parameter T by $2W$, we obtain the Fourier transform pair.

$$2W\ Sa(2\pi Wt) \leftrightarrow \Pi\left(\frac{f}{2W}\right) \tag{2-56}$$

where W is the absolute bandwidth in hertz. This Fourier transform pair is also shown in Fig. 2–6b.

The spectra shown in Fig. 2–6 are real because the time domain pulses are real and even. If the pulses are offset in time to destroy the even symmetry, the spectra will be complex. For example, let

$$v(t) = \begin{cases} 1, & 0 < t < T \\ 0, & t \text{ elsewhere} \end{cases} = \Pi\left(\frac{t - T/2}{T}\right)$$

Then, using the time delay theorem and Eq. (2–55), we get the spectrum

$$V(f) = Te^{-j\pi fT}\, \text{Sa}(\pi Tf) \qquad (2\text{–}57)$$

In terms of quadrature notation, Eq. (2–57) becomes

$$V(f) = \underbrace{[T\, \text{Sa}(\pi fT)\cos(\pi fT)]}_{X(f)} + \underbrace{j[-T\, \text{Sa}(\pi fT)\sin(\pi fT)]}_{Y(f)} \qquad (2\text{–}58)$$

Examining Eq. (2–57), we find that the magnitude spectrum is

$$|V(f)| = T\left|\frac{\sin \pi fT}{\pi fT}\right| \qquad (2\text{–}59)$$

and the phase spectrum is

$$\theta(f) = \underline{/e^{-j\pi fT}} + \underline{/\text{Sa}(\pi fT)} = -\pi fT + \begin{cases} 0, & \dfrac{n}{T} < |f| < \dfrac{n+1}{T}, \quad n \text{ even} \\[2ex] \pi, & \dfrac{n}{T} < |f| < \dfrac{n+1}{T}, \quad n \text{ odd} \end{cases} \qquad (2\text{–}60)$$

The rectangular pulse is one of the most important and popular pulse shapes, because it is convenient to represent binary one and binary zero data by using rectangular pulses. For example, TTL logic circuits use +5-volt rectangular pulses to represent binary ones and zero volts to represent binary zeros. This is shown in Fig. 3–15a, where $A = 5$. Other examples of rectangular pulse encoding are also shown in Fig. 3–15.

The null bandwidth for rectangular pulse digital signaling is obtained from Eq. (2–55) or Eq. (2–59). That is, *if the pulse width of a digital signal is T seconds, then the bandwidth (i.e., the width of the frequency band where the spectral magnitude is not small) is approximately 1/T Hz.* A complete discussion of signal bandwidth is involved, and this is postponed until Sec. 2–9.

Example 2–6 SPECTRUM OF A TRIANGULAR PULSE

The spectrum of a triangular pulse can be obtained by direct evaluation of the FT integral using equations for the piecewise linear lines of the triangle shown in Fig. 2–6c. Another approach, which allows us to obtain the FT quickly, is to first compute the FT for the second *derivative* of the triangle. We will take the latter approach to demonstrate this technique. Let

$$w(t) = \Lambda(t/T)$$

Then

$$\frac{dw(t)}{dt} = \frac{1}{T}\, u(t+T) - \frac{2}{T}\, u(t) + \frac{1}{T}\, u(t-T)$$

and

$$\frac{d^2 w(t)}{dt^2} = \frac{1}{T}\, \delta(t+T) - \frac{2}{T}\, \delta(t) + \frac{1}{T}\, \delta(t-T)$$

Using Table 2–2, we find that the FT pair for the second derivative is

$$\frac{d^2w(t)}{dt^2} \leftrightarrow \frac{1}{T} e^{j\omega T} - \frac{2}{T} + \frac{1}{T} e^{-j\omega T}$$

which can be rewritten as

$$\frac{d^2w(t)}{dt^2} \leftrightarrow \frac{1}{T} (e^{j\omega T/2} - e^{-j\omega T/2})^2 = \frac{-4}{T} (\sin \pi f T)^2$$

Referring to Table 2–1 and applying the integral theorem twice, we get the FT pair for the original waveform:

$$w(t) \leftrightarrow \frac{-4}{T} \frac{(\sin \pi f T)^2}{(j2\pi f)^2}$$

Thus,

$$w(t) = \Lambda\left(\frac{t}{T}\right) \leftrightarrow T \, \text{Sa}^2(\pi f T) \tag{2–61}$$

This is illustrated in Fig. 2–6c.

Convolution

The convolution operation, as listed in Table 2–1, is very useful. Section 2–6 shows how the convolution operation is used to evaluate the waveform at the output of a linear system.

> **DEFINITION.** The *convolution* of a waveform $w_1(t)$ with a waveform $w_2(t)$ to produce a third waveform $w_3(t)$ is

$$w_3(t) = w_1(t) * w_2(t) \triangleq \int_{-\infty}^{\infty} w_1(\lambda) \, w_2(t - \lambda) \, d\lambda \tag{2–62a}$$

where $w_1(t) * w_2(t)$ is a shorthand notation for this integration operation and $*$ is read "convolved with."

When the integral is examined, we realize that t is a parameter and λ is the variable of integration.

If discontinuous waveshapes are to be convolved, it is usually easier to evaluate the equivalent integral

$$w_3(t) = \int_{-\infty}^{\infty} w_1(\lambda) \, w_2(-(\lambda - t)) \, d\lambda \tag{2–62b}$$

Thus, the integrand for Eq. (2–62b) is obtained by

1. Time reversal of w_2 to obtain $w_2(-\lambda)$,
2. Time shifting of w_2 by t seconds to obtain $w_2(-(\lambda - t))$, and
3. Multiplying this result by w_1 to form the integrand $w_1(\lambda) \, w_2(-(\lambda - t))$.

These three operations are illustrated in the examples that follow.

Example 2–7 CONVOLUTION OF A RECTANGLE WITH AN EXPONENTIAL

Let

$$w_1(t) = \Pi\left(\frac{t - \frac{1}{2}T}{T}\right) \qquad \text{and} \qquad w_2(t) = e^{-t/T} u(t)$$

as shown in Fig. 2–7. Implementing step 3 with the help of the figure, the convolution of $w_1(t)$ with $w_2(t)$ is 0 if $t < 0$ because the product $w_1(\lambda)\, w_2(-(\lambda - t))$ is zero for all values of λ. If $0 < t < T$, Eq. (2-62b) becomes

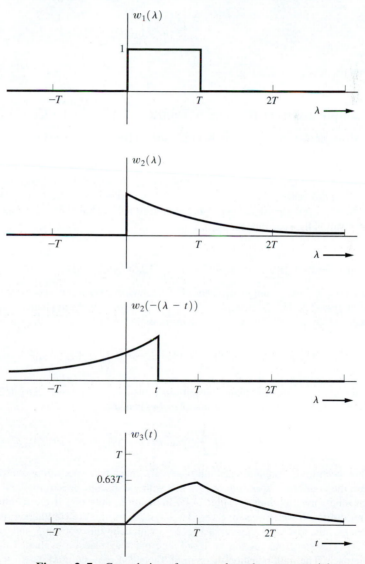

Figure 2–7   Convolution of a rectangle and an exponential.

$$w_3(t) = \int_0^t 1e^{+(\lambda - t)/T} \, d\lambda = T(1 - e^{-t/T})$$

If $t > T$, then Eq. (2–62b) becomes

$$w_3(t) = \int_0^T 1e^{+(\lambda - t)/T} \, d\lambda = T(e - 1) \, e^{-t/T}$$

Thus,

$$w_3(t) = \begin{cases} 0, & t < 0 \\ T(1 - e^{-t/T}), & 0 < t < T \\ T(e - 1)e^{-t/T}, & t > T \end{cases}$$

This result is plotted in Fig. 2–7.

Example 2–8 SPECTRUM OF A TRIANGULAR PULSE BY CONVOLUTION

In Example 2–6, the spectrum of a triangular pulse was evaluated by using the integral theorem. The same result can be obtained by using the convolution theorem of Table 2–1. If the rectangular pulse of Fig. 2–6a is convolved with itself and then scaled (i.e., multiplied) by the constant $1/T$, the resulting time waveform is the triangular pulse of Fig. 2–6c. Applying the convolution theorem, we obtain the spectrum for the triangular pulse by multiplying the spectrum of the rectangular pulse (of Fig. 2–6a) with itself and scaling with a constant $1/T$. As expected, the result is the spectrum shown in Fig. 2–6c.

Example 2–9 SPECTRUM OF A SWITCHED SINUSOID

In Example 2–4, a continuous sinusoid was found to have a line spectrum with the lines located at $f = \pm f_0$. In this example, we will see how the spectrum changes when the sinusoid is switched on and off. The switched sinusoid is shown in Fig. 2–8a and can be represented by

$$w(t) = \Pi\left(\frac{t}{T}\right) A \sin \omega_0 t = \Pi\left(\frac{t}{T}\right) A \cos\left(\omega_0 t - \frac{\pi}{2}\right)$$

Using the FT of the rectangular pulse from Table 2–2 and the real-signal translation theorem of Table 2–1, we see that the spectrum of this switched sinusoid is

$$W(f) = j\frac{A}{2} T \left[\text{Sa}(\pi T(f + f_0)) - \text{Sa}(\pi T(f - f_0))\right] \tag{2–63}$$

This spectrum is continuous and imaginary. The magnitude spectrum is shown in Fig. 2–8. Compare the continuous spectrum of Fig. 2–8 with the discrete spectrum obtained for the continuous sine wave, as shown in Fig. 2–4a. In addition, note that if the duration of the switched sinusoid is allowed to become very large (i.e., $T \to \infty$), the continuous spectrum of Fig. 2–8 becomes the discrete spectrum of Fig. 2–4a with delta functions at f_0 and $-f_0$.

The spectrum for the switched sinusoid may also be evaluated using a convolution approach. The multiplication theorem of Table 2–1 may be used where $w_1(t) = \Pi(t/T)$ and

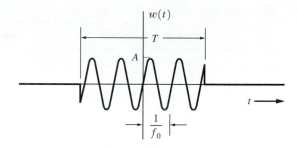

(a) Time Domain

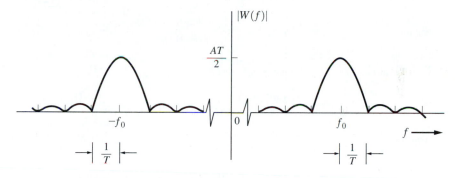

(b) Frequency Domain (Magnitude Spectrum)

Figure 2–8   Waveform and spectrum of a switched sinusoid.

$w_2(t) = A \sin \omega_0 t$. Here the spectrum of the switched sinusoid is obtained by working a convolution problem in the frequency domain:

$$W(f) = W_1(f) * W_2(f) = \int_{-\infty}^{\infty} W_1(\lambda)W_2(f - \lambda)\, d\lambda$$

This convolution integral is easy to evaluate because the spectrum of $w_2(t)$ consists of two delta functions. The details of this approach are left as a homework exercise for the reader.

A summary of Fourier transform pairs is given in Table 2–2. Numerical techniques using the discrete Fourier transform are discussed in Sec. 2–8.

2–3 POWER SPECTRAL DENSITY AND AUTOCORRELATION FUNCTION

Power Spectral Density

The normalized power of a waveform will now be related to its frequency domain description by the use of a function called the *power spectral density* (*PSD*). The PSD is very useful in describing how the power content of signals and noise is affected by filters and other devices in communication systems. In Eq. (2–42), the energy spectral density (ESD) was defined in terms of the magnitude-squared version of the Fourier transform of the waveform.

TABLE 2–2   SOME FOURIER TRANSFORM PAIRS

Function	Time Waveform $w(t)$	Spectrum $W(f)$
Rectangular	$\Pi\left(\dfrac{t}{T}\right)$	$T[\mathrm{Sa}(\pi f T)]$
Triangular	$\Lambda\left(\dfrac{t}{T}\right)$	$T[\mathrm{Sa}(\pi f T)]^2$
Unit step	$u(t) \triangleq \begin{cases} +1, & t>0 \\ 0, & t<0 \end{cases}$	$\frac{1}{2}\delta(f) + \dfrac{1}{j2\pi f}$
Signum	$\mathrm{sgn}(t) \triangleq \begin{cases} +1, & t>0 \\ -1, & t<0 \end{cases}$	$\dfrac{1}{j\pi f}$
Constant	1	$\delta(f)$
Impulse at $t = t_0$	$\delta(t - t_0)$	$e^{-j2\pi f t_0}$
Sinc	$\mathrm{Sa}(2\pi W t)$	$\dfrac{1}{2W}\Pi\left(\dfrac{f}{2W}\right)$
Phasor	$e^{j(\omega_0 t + \varphi)}$	$e^{j\varphi}\,\delta(f - f_0)$
Sinusoid	$\cos(\omega_c t + \varphi)$	$\frac{1}{2}e^{j\varphi}\,\delta(f - f_c) + \frac{1}{2}e^{-j\varphi}\,\delta(f + f_c)$
Gaussian	$e^{-\pi(t/t_0)^2}$	$t_0 e^{-\pi(f t_0)^2}$
Exponential, one-sided	$\begin{cases} e^{-t/T}, & t>0 \\ 0, & t<0 \end{cases}$	$\dfrac{T}{1 + j2\pi f T}$
Exponential, two-sided	$e^{-\lvert t\rvert/T}$	$\dfrac{2T}{1 + (2\pi f T)^2}$
Impulse train	$\displaystyle\sum_{k=-\infty}^{k=\infty} \delta(t - kT)$	$\displaystyle f_0 \sum_{n=-\infty}^{n=\infty} \delta(f - nf_0),$ where $f_0 = 1/T$

The PSD will be defined in a similar way. The PSD is more useful than the ESD, since power-type models are generally used in solving communication problems.

First, we define the truncated version of the waveform by

$$w_T(t) = \begin{cases} w(t), & -T/2 < t < T/2 \\ 0, & t \text{ elsewhere} \end{cases} = w(t)\Pi\left(\frac{t}{T}\right) \tag{2–64}$$

Using Eq. (2–13), we obtain the average normalized power:

$$P = \lim_{T\to\infty} \frac{1}{T} \int_{-T/2}^{T/2} w^2(t)\, dt = \lim_{T\to\infty} \frac{1}{T} \int_{-\infty}^{\infty} w_T^2(t)\, dt$$

By the use of Parseval's theorem, Eq. (2–41), the average normalized power becomes

$$P = \lim_{T\to\infty} \frac{1}{T} \int_{-\infty}^{\infty} \lvert W_T(f)\rvert^2\, df = \int_{-\infty}^{\infty} \left(\lim_{T\to\infty} \frac{\lvert W_T(f)\rvert^2}{T} \right) df \tag{2–65}$$

where $W_T(f) = \mathcal{F}[w_T(t)]$. The integrand of the right-hand integral has units of watts/hertz (or, equivalently, volts^2/hertz or amperes^2/hertz, as appropriate) and can be defined as the PSD.

DEFINITION.   The *power spectral density* (PSD) for a deterministic power wave-form is[†]

$$\mathcal{P}_w(f) \triangleq \lim_{T \to \infty} \left(\frac{|W_T(f)|^2}{T} \right) \tag{2–66}$$

where $w_T(t) \leftrightarrow W_T(f)$ and $\mathcal{P}_w(f)$ has units of watts per hertz.

Note that the PSD is always a real nonnegative function of frequency. In addition, the PSD is not sensitive to the phase spectrum of $w(t)$, because that is lost due to the absolute value operation used in Eq. (2–66). From Eq. (2–65), the normalized average power is[†]

$$P = \langle w^2(t) \rangle = \int_{-\infty}^{\infty} \mathcal{P}_w(f) \, df \tag{2–67}$$

That is, the area under the PSD function is the normalized average power.

Autocorrelation Function

A related function called the *autocorrelation*, $R(\tau)$, can also be defined.[‡]

DEFINITION.   The *autocorrelation* of a real (physical) waveform is[§]

$$R_w(\tau) \triangleq \langle w(t)w(t+\tau) \rangle = \lim_{T \to \infty} \frac{1}{T} \int_{-T/2}^{T/2} w(t)w(t+\tau) \, dt \tag{2–68}$$

Furthermore, it can be shown that the PSD and the autocorrelation function are Fourier transform pairs; that is,

$$R_w(\tau) \leftrightarrow \mathcal{P}_w(f) \tag{2–69}$$

where $\mathcal{P}_w(f) = \mathcal{F}[R_w(\tau)]$. This is called the *Wiener–Khintchine theorem*. The theorem, along with properties for $R(\tau)$ and $\mathcal{P}(f)$, are developed in Chapter 6.

In summary, the PSD can be evaluated by either of the following two methods:

1. *Direct method*, by using the definition, Eq. (2-66).[◊]

2. *Indirect method*, by first evaluating the autocorrelation function and then taking the Fourier transform: $\mathcal{P}_w(f) = \mathcal{F}[R_w(\tau)]$.

[†] Equations (2–66) and (2–67) give normalized (with respect to 1 Ω) PSD and average power, respectively. For unnormalized (i.e., actual) values, $\mathcal{P}_w(f)$ is replaced by the appropriate expression as follows: If $w(t)$ is a voltage waveform that appears across a resistive load of R ohms, the unnormalized PSD is $\mathcal{P}_w(f)/R$ W/Hz, where $\mathcal{P}_w(f)$ has units of volts^2/Hz. Similarly, if $w(t)$ is a current waveform passing through a resistive load of R ohms, the unnormalized PSD is $\mathcal{P}_w(f)R$ W/Hz, where $\mathcal{P}_w(f)$ has units of amperes^2/Hz.

[‡] Here a time average is used in the definition of the autocorrelation function. In Chapter 6, an ensemble (statistical) average is used in the definition of $R(\tau)$, and as shown, there, these two definitions are equivalent if $w(t)$ is ergodic.

[§] The autocorrelation of a complex waveform is $R_w(\tau) \triangleq \langle w^*(t)w(t+\tau) \rangle$.

[◊] The direct method is usually more difficult than the indirect method.

Furthermore, the total average normalized power for the waveform $w(t)$ can be evaluated by using any of the *four* techniques embedded in the following equation:

$$P = \langle w^2(t) \rangle = W^2_{\text{rms}} = \int_{-\infty}^{\infty} \mathcal{P}_w(f) \, df = R_w(0) \tag{2-70}$$

Example 2–10 PSD OF A SINUSOID

Let

$$w(t) = A \sin \omega_0 t$$

The PSD will be evaluated using the indirect method. The autocorrelation is

$$R_w(\tau) = \langle w(t)w(t + \tau) \rangle$$

$$= \lim_{T \to \infty} \frac{1}{T} \int_{-T/2}^{T/2} A^2 \sin \omega_0 t \sin \omega_0 (t + \tau) \, dt$$

Using a trigonometric identity, from Appendix A we obtain

$$R_w(\tau) = \frac{A^2}{2} \cos \omega_0 \tau \lim_{T \to \infty} \frac{1}{T} \int_{-T/2}^{T/2} dt - \frac{A^2}{2} \lim_{T \to \infty} \frac{1}{T} \int_{-T/2}^{T/2} \cos(2\omega_0 t + \omega_0 \tau) \, dt$$

which reduces to

$$R_w(\tau) = \frac{A^2}{2} \cos \omega_0 \tau \tag{2-71}$$

The PSD is then

$$\mathcal{P}_w(f) = \mathcal{F}\left[\frac{A^2}{2} \cos \omega_0 \tau\right] = \frac{A^2}{4} [\delta(f - f_0) + \delta(f + f_0)] \tag{2-72}$$

as shown in Fig. 2–9. The PSD may be compared to the "voltage" spectrum for a sinusoid found in Example 2–4 and shown in Fig. 2–4.

The average normalized power may be obtained by using Eq. (2–67):

$$P = \int_{-\infty}^{\infty} \frac{A^2}{4} [\delta(f - f_0) + \delta(f + f_0)] \, df = \frac{A^2}{2} \tag{2-73}$$

This value, $A^2/2$, checks with the known result for the normalized power of a sinusoid:

$$P = \langle w^2(t) \rangle = W^2_{\text{rms}} = (A/\sqrt{2})^2 = A^2/2 \tag{2-74}$$

It is also realized that $A \sin \omega_0 t$ and $A \cos \omega_0 t$ have exactly the same PSD (and autocorrelation function) because the phase has no effect on the PSD. This can be verified by evaluating the PSD for $A \cos \omega_0 t$, using the same procedure that was used earlier to evaluate the PSD for $A \sin \omega_0 t$.

Thus far, we have studied properties of signals and noise, such as the spectrum, average power, and rms value, but how do we represent the signal or noise waveform itself? The direct approach is to write a closed-form mathematical equation for the waveform itself.

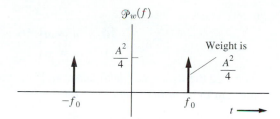

Figure 2–9　Power spectrum of a sinusoid.

Other equivalent ways of modeling the waveform are often found to be useful as well. One, which the reader has already studied in calculus, is to represent the waveform by the use of a Taylor series (i.e., power series) expansion about a point a; that is,

$$w(t) = \sum_{n=0}^{\infty} \frac{w^{(n)}(a)}{n!} (t - a)^n \qquad (2\text{--}75)$$

where

$$w^{(n)}(a) = \left. \frac{d^n w(t)}{dt^n} \right|_{t=a} \qquad (2\text{--}76)$$

From Eq. (2–76), if the derivatives at $t = a$ are known, Eq. (2–75) can be used to reconstruct the waveform. Another type of series representation that is especially useful in communication problems is the orthogonal series expansion. This is discussed in the next section.

2–4 ORTHOGONAL SERIES REPRESENTATION OF SIGNALS AND NOISE

An orthogonal series representation of signals and noise has many significant applications in communication problems, such as Fourier series, sampling function series, and the representation of digital signals. Because these specific cases are so important, they will be studied in some detail in sections that follow.

Orthogonal Functions

Before the orthogonal series is studied, we define orthogonal functions.

DEFINITION.　Functions $\varphi_n(t)$ and $\varphi_m(t)$ are said to be *orthogonal* with respect to each other over the interval $a < t < b$ if they satisfy the condition

$$\int_a^b \varphi_n(t) \varphi_m^*(t) \, dt = 0, \qquad \text{where } n \neq m \qquad (2\text{--}77)$$

Furthermore, if the functions in the set $\{\varphi_n(t)\}$ are orthogonal, then they also satisfy the relation

$$\int_a^b \varphi_n(t) \varphi_m^*(t) \, dt = \begin{cases} 0, & n \neq m \\ K_n, & n = m \end{cases} = K_n \delta_{nm} \qquad (2\text{--}78)$$

where

$$\delta_{nm} \triangleq \begin{cases} 0, & n \neq m \\ 1, & n = m \end{cases} \qquad\qquad (2\text{--}79)$$

Here, δ_{nm} is called the *Kronecker delta function*. If the constants K_n are all equal to 1, the $\varphi_n(t)$ are said to be *orthonormal functions*.[†]

Equation (2–77) is used to test pairs of functions to see if they are orthogonal. Any such pair of functions is orthogonal over the interval (a, b) if the integral of the product of the functions is zero. The zero result implies that these functions are "independent" or in "disagreement." If the result is not zero, they are not orthogonal, and consequently, the two functions have some "dependence" on, or "alikeness" to each other.

Example 2–11 ORTHOGONAL COMPLEX EXPONENTIAL FUNCTIONS

Show that the set of complex exponential functions $\{e^{jn\omega_0 t}\}$ are orthogonal over the interval $a < t < b$, where $b = a + T_0$, $T_0 = 1/f_0$, $\omega_0 = 2\pi f_0$, and n is an integer.

Solution.   Substituting $\varphi_n(t) = e^{jn\omega_0 t}$ and $\varphi_m(t) = e^{jn\omega_0 t}$ into Eq. (2–77), we get

$$\int_a^b \varphi_n(t)\varphi_m^*(t)\,dt = \int_a^{a+T_0} e^{jn\omega_0 t}\,e^{-jm\omega_0 t}\,dt = \int_a^{a+T_0} e^{j(n-m)\omega_0 t}\,dt \qquad (2\text{--}80)$$

For $m \neq n$, Eq. (2–80) becomes

$$\int_a^{a+T_0} e^{j(n-m)\omega_0 t}\,dt = \frac{e^{j(n-m)\omega_0 a}[e^{j(n-m)2\pi} - 1]}{j(n-m)\omega_0} = 0 \qquad (2\text{--}81)$$

since $e^{j(n-m)2\pi} = \cos[2\pi(n-m)] + j\sin[2\pi(n-m)] = 1$. Thus, Eq. (2–77) is satisfied, and consequently, the complex exponential functions are orthogonal to each other over the interval $a < t < a + T_0$, where a is any real constant. These exponential functions are *not orthogonal* over some other intervals, such as $a < t < a + 0.5T_0$.

Also, for the case $n = m$, Eq. (2–80) becomes

$$\int_a^{a+T_0} \varphi_n(t)\,\varphi_n^*(t)\,dt = \int_a^{a+T_0} 1\,dt = T_0 \qquad (2\text{--}82)$$

Using Eq. (2–82) in Eq. (2–78), we find that $K_n = T_0$ for all (integer) values of n. Because $K_n \neq 1$, these $\varphi_n(t)$ are not orthonormal (but they are orthogonal). An orthonormal set of exponential functions is obtained by scaling the old set, where the functions in the new set are

$$\varphi_n(t) = \frac{1}{\sqrt{T_0}}\,e^{jn\omega_0 t}.$$

[†] To normalize a set of functions, we take each old $\varphi_n(t)$ and divide it by $\sqrt{K_n}$ to form the normalized $\varphi_n(t)$.

Orthogonal Series

Assume that $w(t)$ represents some practical waveform (signal, noise, or signal–noise combination) that we wish to represent over the interval $a < t < b$. Then we can obtain an equivalent orthogonal series representation by using the following theorem.

THEOREM.   *$w(t)$ can be represented over the interval (a, b) by the series*

$$w(t) = \sum_n a_n \varphi_n(t) \tag{2–83}$$

where the orthogonal coefficients are given by

$$a_n = \frac{1}{K_n} \int_a^b w(t) \varphi_n^*(t)\, dt \tag{2–84}$$

and the range of n is over the integer values that correspond to the subscripts that were used to denote the orthogonal functions in the complete orthogonal set.

For Eq. (2–83) to be a valid representation for any physical signal (i.e., one with finite energy), the orthogonal set has to be complete. This implies that the set $\{\varphi_n(t)\}$ can be used to represent any function with an arbitrarily small error [Wylie, 1960]. In practice, it is usually difficult to prove that a given set of functions is complete. It can be shown that the complex exponential set and the harmonic sinusoidal sets that are used for the Fourier series in Sec. 2–5 are complete [Courant and Hilbert, 1953]. Many other useful sets are also complete, such as Bessel functions, Legendre polynomials, and the $(\sin x)/x$-type sets [described by Eq. (2–161)].

Proof of Theorem.   Assume that the set $\{\varphi_n(t)\}$ is sufficient to represent the waveform. Then in order for Eq. (2–83) to be correct, we only need to show that we can evaluate the a_n. Accordingly, we operate on both sides of that equation with the integral operator

$$\int_a^b [\cdot]\, \varphi_m^*(t)\, dt \tag{2–85}$$

obtaining

$$\int_a^b [w(t)]\varphi_m^*(t)\, dt = \int_a^b \left[\sum_n a_n \varphi_n(t) \right] \varphi_m^*(t)\, dt$$

$$= \sum_n a_n \int_a^b \varphi_n(t) \varphi_m^*(t)\, dt = \sum_n a_n K_n \delta_{nm}$$

$$= a_m K_m \tag{2–86}$$

Thus, Eq. (2–84) follows.

The orthogonal series is very useful in representing a signal, noise, or signal–noise combination. The orthogonal functions $\varphi_j(t)$ are deterministic. Furthermore, if the waveform

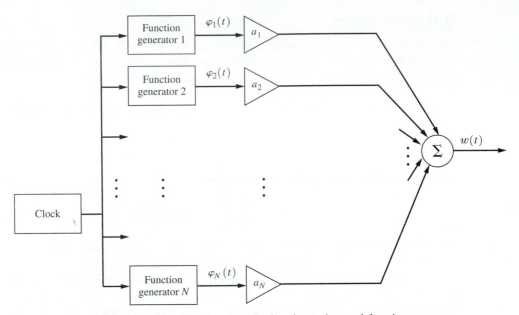

Figure 2–10   Waveform synthesis using orthogonal functions.

$w(t)$ is deterministic, the constants $\{a_j\}$ are also deterministic and may be evaluated using Eq. (2–84). In Chapter 6, we will see that if $w(t)$ is stochastic (e.g., in a noise problem), the $\{a_j\}$ are a set of random variables that give the desired random process $w(t)$.

It is also possible to use Eq. (2–83) to generate $w(t)$ from the $\varphi_j(t)$ functions and the coefficients a_j. In this case, $w(t)$ is approximated by using a reasonable number of the $\varphi_j(t)$ functions. As shown in Fig. 2–10, for the case of real values for a_j and real functions for $\varphi_j(t)$, $w(t)$ can be synthesized by adding up weighted versions of $\varphi_j(t)$, where the weighting factors are given by $\{a_j\}$. The summing-and-gain weighting operation may be conveniently realized by using an operational amplifier with multiple inputs.

2–5 FOURIER SERIES

The Fourier series is a particular type of orthogonal series that is very useful in solving engineering problems, especially communication problems. The orthogonal functions that are used are either sinusoids or, equivalently, complex exponential functions.[†]

Complex Fourier Series

The complex Fourier series uses the orthogonal exponential functions

$$\varphi_n(t) = e^{jn\omega_0 t} \qquad\qquad (2\text{–}87)$$

[†] Mathematicians generally call any orthogonal series a Fourier series.

where n ranges over all possible integer values, negative, positive, and zero; $\omega_0 = 2\pi/T_0$, where $T_0 = (b - a)$ is the length of the interval over which the series, Eq. (2–83), is valid; and, from Example 2–11, $K_n = T_0$. The Fourier series theorem follows from Eq. (2–83).

THEOREM. *A physical waveform (i.e., finite energy) may be represented over the interval $a < t < a + T_0$ by the complex exponential Fourier series*

$$w(t) = \sum_{n=-\infty}^{n=\infty} c_n e^{jn\omega_0 t} \tag{2–88}$$

where the complex (phasor) Fourier coefficients are

$$c_n = \frac{1}{T_0} \int_{a}^{a+T_0} w(t)e^{-jn\omega_0 t} \, dt \tag{2–89}$$

and where $\omega_0 = 2\pi f_0 = 2\pi/T_0$.

If the waveform $w(t)$ is periodic with period T_0, this Fourier series representation is valid *over all time* (i.e., over the interval $-\infty < t < +\infty$), because $w(t)$ and $\varphi_n(t)$ are periodic with the same fundamental period T_0. For this case of periodic waveforms, the choice of a value for the parameter a is arbitrary and is usually taken to be $a = 0$ or $a = -T_0/2$ for mathematical convenience. The frequency $f_0 = 1/T_0$ is said to be the *fundamental* frequency and the frequency nf_0 is said to be the *n*th harmonic frequency, when $n > 1$. The Fourier coefficient c_0 is equivalent to the dc value of the waveform $w(t)$, because, when $n = 0$, Eq. (2–89) is identical to Eq. (2–4).

c_n is, in general, a complex number. Furthermore, it is a phasor, since it is the coefficient of a function of the type $e^{j\omega t}$. Consequently, Eq. (2–88) is said to be a complex or phasor Fourier series.

Some properties of the complex Fourier series are as follows:

1. If $w(t)$ is real,

$$c_n = c_{-n}^* \tag{2–90}$$

2. If $w(t)$ is real and even [i.e., $w(t) = w(-t)$],

$$\text{Im}[c_n] = 0 \tag{2–91}$$

3. If $w(t)$ is real and odd [i.e., $w(t) = -w(-t)$],

$$\text{Re}[c_n] = 0$$

4. Parseval's theorem is

$$\frac{1}{T_0} \int_{a}^{a+T_0} |w(t)|^2 \, dt = \sum_{n=-\infty}^{n=\infty} |c_n|^2 \tag{2–92}$$

(See Eq. (2–125) for the proof.)

5. The complex Fourier series coefficients of a real waveform are related to the quadrature Fourier series coefficients by

$$c_n = \begin{cases} \frac{1}{2} a_n - j\frac{1}{2} b_n, & n > 0 \\ a_0, & n = 0 \\ \frac{1}{2} a_{-n} + j\frac{1}{2} b_{-n}, & n < 0 \end{cases} \tag{2-93}$$

[See Eqs. (2–96), (2–97) and (2–98).]

6. The complex Fourier series coefficients of a real waveform are related to the polar Fourier series coefficients by

$$c_n = \begin{cases} \frac{1}{2} D \; \underline{/\varphi_n}, & n > 0 \\ D_0, & n = 0 \\ \frac{1}{2} D_{-n} \; \underline{/-\varphi_{-n}}, & n < 0 \end{cases} \tag{2-94}$$

[See Eqs. (2–106) and (2–107).]

Note that these properties for the complex Fourier series coefficients are similar to those of the Fourier transform as given in Sec. 2–2.

Quadrature Fourier Series

The *quadrature* form of the Fourier series representing any physical waveform $w(t)$ over the interval $a < t < a + T_0$ is

$$w(t) = \sum_{n=0}^{n=\infty} a_n \cos n\omega_0 t + \sum_{n=1}^{n=\infty} b_n \sin n\omega_0 t \tag{2-95}$$

where the orthogonal functions are $\cos n\omega_0 t$ and $\sin n\omega_0 t$. Using Eq. (2–77), we find that these Fourier coefficients are given by

$$a_n = \begin{cases} \dfrac{1}{T_0} \displaystyle\int_a^{a+T_0} w(t) \, dt, & n = 0 \\ \dfrac{2}{T_0} \displaystyle\int_a^{a+T_0} w(t) \cos n\omega_0 t \, dt, & n \geq 1 \end{cases} \tag{2-96}$$

and

$$b_n = \frac{2}{T_0} \int_a^{a+T_0} w(t) \sin n\omega_0 t \, dt, \qquad n > 0 \tag{2-97}$$

Once again, because these sinusoidal orthogonal functions are periodic, this series is periodic with the fundamental period T_0, and if $w(t)$ is periodic with period T_0, the series will represent $w(t)$ over the whole real line (i.e., $-\infty < t < \infty$).

The complex Fourier series, [Eq. (2–88)], and the quadrature Fourier series, [Eq. (2–95)] are equivalent representations. This can be demonstrated by expressing the complex number c_n in terms of its conjugate parts, x_n and y_n. That is, using Eq. (2–89), we get

$$c_n = x_n + jy_n$$

$$= \left[\frac{1}{T_0} \int_a^{a+T_0} w(t) \cos n\omega_0 t \, dt \right] + j \left[\frac{-1}{T_0} \int_a^{a+T_0} w(t) \sin n\omega_0 t \, dt \right] \tag{2-98}$$

for all integer values of n. Thus,

$$x_n = \frac{1}{T_0} \int_a^{a+T_0} w(t) \cos n\omega_0 t \; dt \tag{2–99}$$

and

$$y_n = \frac{-1}{T_0} \int_a^{a+T_0} w(t) \sin n\omega_0 t \; dt \tag{2–100}$$

Using Eq. (2–96) and (2–97), we obtain the identities

$$a_n = \begin{cases} c_0, & n = 0 \\ 2x_n, & n \geq 1 \end{cases} = \begin{cases} c_0, & n = 0 \\ 2 \operatorname{Re}\{c_n\}, & n \geq 1 \end{cases} \tag{2–101}$$

and

$$b_n = -2y_n = -2 \operatorname{Im}\{c_n\}, \; n \geq 1 \tag{2–102}$$

where $\operatorname{Re}\{\cdot\}$ denotes the real part of $\{\cdot\}$ and $\operatorname{Im}\{\cdot\}$ denotes the imaginary part of $\{\cdot\}$.

Polar Fourier Series

The quadrature Fourier series, Eq. (2–95), may be rearranged and written in a polar (amplitude-phase) form. The *polar* form is

$$w(t) = D_0 + \sum_{n=1}^{n=\infty} D_n \cos(n\omega_0 t + \varphi_n) \tag{2–103}$$

where $w(t)$ is real and

$$a_n = \begin{cases} D_0, & n = 0 \\ D_n \cos \varphi_n, & n \geq 1 \end{cases} \tag{2–104}$$

$$b_n = -D_n \sin \varphi_n, \qquad n \geq 1 \tag{2–105}$$

The latter two equations may be inverted, and we obtain

$$D_n = \begin{cases} a_0, & n = 0 \\ \sqrt{a_n^2 + b_n^2}, & n \geq 1 \end{cases} = \begin{cases} c_0, & n = 0 \\ 2|c_n|, & n \geq 1 \end{cases} \tag{2–106}$$

and

$$\varphi_n = -\tan^{-1}\left(\frac{b_n}{a_n}\right) = \underline{/c_n}, \quad n \geq 1 \tag{2–107}$$

where the angle operator is defined by

$$\underline{/[\cdot]} = \tan^{-1}\left(\frac{\operatorname{Im}[\cdot]}{\operatorname{Re}[\cdot]}\right) \tag{2–108}$$

It should be clear from the context whether $\underline{/\quad}$ denotes the angle operator or the angle itself. For example, $\underline{/90°}$ denotes an angle of 90°, but $\underline{/[1 + j2]}$ denotes the angle operator and is equal to 63.4° when evaluated.

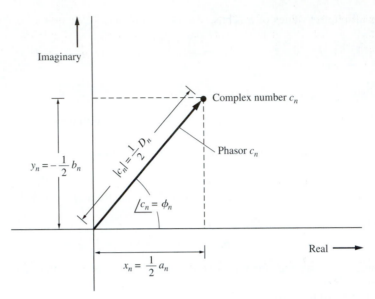

Figure 2–11   Fourier series coefficients, $n \geq 1$.

The equivalence between the Fourier series coefficients is demonstrated geometrically in Fig. 2–11. It is seen that, in general, when a physical (real) waveform $w(t)$ is represented by a Fourier series, c_n is a complex number with a real part x_n and an imaginary part y_n (which are both real numbers), and consequently, a_n, b_n, D_n, and φ_n are real numbers. In addition, D_n is a nonnegative number for $n \geq 1$. Furthermore, all of these coefficients describe the amount of frequency component contained in the signal at the frequency of nf_0 Hz.

In practice, the Fourier series (FS) is often *truncated* to a finite number of terms. For example, 5 or 10 harmonics might be used to approximate the FS series for a square wave. Thus, an important question arises: For the finite series, are the optimum values for the series coefficients the same as those for the corresponding terms in the infinite series, or should the coefficients for the finite series be adjusted to some new values to give the best finite-series approximation? The answer is that the optimum values for the coefficients of the truncated FS are the same as those for the corresponding terms in the nontruncated FS.[†]

As we have seen, the complex, quadrature, and polar forms of the Fourier series are all equivalent, but the question is, *Which is the best form to use?* The answer is that it depends on the particular problem being solved! If the problem is being solved analytically, the complex coefficients are *usually* easier to evaluate. On the other hand, if measurements of a waveform are being made in a laboratory, the polar form is usually more convenient, since measuring instruments, such as voltmeters, scopes, vector voltmeters, and wave analyzers, generally give magnitude and phase readings. Using the laboratory results, engineers often draw *one-sided* spectral plots in which lines are drawn corresponding to each D_n value at $f = nf_0$, where $n \geq 0$ (i.e., positive frequencies only). Of course, this one-sided spec-

[†] For a proof of this statement, see [Couch, 1995].

tral plot may be converted to the two-sided spectrum given by the c_n plot by using Eq. (2–94). It is understood that the two-sided spectrum is defined to be the Fourier transform of $w(t)$. This is demonstrated by Fig. 2–4, where Eq. (2–109), from the next theorem, may be used.

Line Spectra for Periodic Waveforms

For *periodic* waveforms, the Fourier series representations are valid over all time (i.e., $-\infty < t < \infty$). Consequently, the (two-sided) spectrum, which depends on the waveshape from $t = -\infty$ to $t = \infty$, may be evaluated in terms of Fourier coefficients.

> **THEOREM.**   *If a waveform is periodic with period T_0, the spectrum of the waveform $w(t)$ is*

$$W(f) = \sum_{n=-\infty}^{n=\infty} c_n\, \delta(f - nf_0) \tag{2–109}$$

where $f_0 = 1/T_0$ and c_n are the phasor Fourier coefficients of the waveform as given by Eq. (2–89).

> **Proof.**

$$w(t) = \sum_{n=-\infty}^{n=\infty} c_n\, e^{jn\omega_0 t} \qquad *\quad \text{*} \ 4d\ on\ test$$

Taking the Fourier transform of both sides, we obtain

$$W(f) = \int_{-\infty}^{\infty} \left(\sum_{n=-\infty}^{n=\infty} c_n\, e^{jn\omega_0 t} \right) e^{-j\omega t}\, dt$$

$$= \sum_{n=-\infty}^{n=\infty} c_n \int_{-\infty}^{\infty} e^{-j2\pi(f-nf_0)t}\, dt = \sum_{n=-\infty}^{n=\infty} c_n\, \delta(f - nf_0)$$

where the integral representation for a delta function, Eq. (2–48), was used.

This theorem indicates that a periodic function *always* has a line (delta function) spectrum, with the lines being at $f = nf_0$ and having weights given by the c_n values. An illustration of that property was given by Example 2–4, where $c_1 = -jA/2$ and $c_{-1} = jA/2$ and the other c_n's were zero. It is also obvious that there is no dc component, since there is no line at $f = 0$ (i.e., $c_0 = 0$). Conversely, if a function does not contain any periodic component, the spectrum will be continuous (no lines), except for a line at $f = 0$ when the function has a dc component.

It is also possible to evaluate the Fourier coefficients by sampling the Fourier transform of a pulse corresponding to $w(t)$ over a period. This is shown by the following theorem:

> **THEOREM.**   *If $w(t)$ is a periodic function with period T_0 and is represented by*

$$w(t) = \sum_{n=-\infty}^{n=\infty} h(t - nT_0) = \sum_{n=-\infty}^{n=\infty} c_n\, e^{jn\omega_0 t} \tag{2–110}$$

where

$$h(t) = \begin{cases} w(t), & |t| < \dfrac{T_0}{2} \\ 0, & t \ \text{elsewhere} \end{cases} \tag{2-111}$$

then the Fourier coefficients are given by

$$c_n = f_0 H(n f_0) \tag{2-112}$$

where $H(f) = \mathcal{F}[h(t)]$ *and* $f_0 = 1/T_0$.

Proof.

$$w(t) = \sum_{n=-\infty}^{\infty} h(t - nT_0) = \sum_{n=-\infty}^{\infty} h(t) * \delta(t - nT_0) \tag{2-113}$$

where $*$ denotes the convolution operation. Thus,

$$w(t) = h(t) * \sum_{n=-\infty}^{\infty} \delta(t - nT_0) \tag{2-114}$$

But the impulse train may itself be represented by its Fourier series[†]; that is,

$$\sum_{n=-\infty}^{\infty} \delta(t - nT_0) = \sum_{n=-\infty}^{\infty} c_n e^{jn\omega_0 t} \tag{2-115}$$

where all the Fourier coefficients are just $c_n = f_0$. Substituting Eq. (2–115) into Eq. (2–114), we obtain

$$w(t) = h(t) * \sum_{n=-\infty}^{\infty} f_0 e^{jn\omega_0 t} \tag{2-116}$$

Taking the Fourier transform of both sides of Eq. (2–116), we have

$$W(f) = H(f) \sum_{n=-\infty}^{\infty} f_0 \, \delta(f - n f_0)$$

$$= \sum_{n=-\infty}^{\infty} [f_0 H(n f_0)] \, \delta(f - n f_0) \tag{2-117}$$

Comparing Eq. (2–117) with Eq. (2–109), we see that Eq. (2–112) follows.

This theorem is useful for evaluating the Fourier coefficients c_n when the Fourier transform of the fundamental pulse shape $h(t)$ for the periodic waveform is known or can be obtained easily from a Fourier transform table (e.g., Table 2–2).

[†] This is called the *Poisson sum formula*.

Example 2–12 FOURIER COEFFICIENTS FOR A PERIODIC RECTANGULAR WAVE

Find the Fourier coefficients for the periodic rectangular wave shown in Fig. 2–12a, where T is the pulse width and $T_0 = 2T$ is the period. The complex Fourier coefficients, from Eq. (2–89), are

$$c_n = \frac{1}{T_0} \int_0^{T_0/2} A e^{-jn\omega_0 t}\, dt = j\,\frac{A}{2\pi n}\left(e^{-jn\pi} - 1\right) \tag{2–118}$$

which reduce, (using l'Hospital's rule for evaluating the indeterminant form for $n = 0$), to

$$c_n = \begin{cases} \dfrac{A}{2}, & n = 0 \\[2ex] -j\dfrac{A}{n\pi}, & n = \text{odd} \\[2ex] 0, & n \text{ otherwise} \end{cases} \tag{2–119}$$

Equation (2–119) may be verified by using Eq. (2–112) and the result of Example 2–5, where $T_0 = 2T$. That is,

$$c_n = \frac{A}{T_0} V(nf_0) = \frac{A}{2}\, e^{-jn\pi/2}\, \frac{\sin(n\pi/2)}{n\pi/2} \tag{2–120}$$

which is identical to Eq. (2–119). It is realized that the dc value of the waveform is $c_0 = A/2$, which checks with the result by using Eq. (2–4).

The spectrum of the square wave is easily obtained from Eq. (2–109). The magnitude spectrum is illustrated by the solid lines in Fig. 2–12b. Since delta functions have infinite values, they cannot be plotted; but the weights of the delta functions can be plotted as shown by the dashed line in Fig. 2–12b.

Now compare the spectrum for this periodic rectangular wave (solid lines in Fig. 2–12b) with the spectrum for the rectangular pulse shown in Fig. 2–6a. Note that the *spectrum for the periodic wave contains spectral lines*, whereas *the spectrum for the nonperiodic pulse is continuous*. Note that the envelope of the spectrum for both cases is the same $|(\sin x)/x|$ shape, where $x = \pi T f$. Consequently, the null bandwidth (for the envelope) is $1/T$ for both cases, where T is the pulse width. This is a basic property of digital signaling with rectangular pulse shapes. *The null bandwidth is the reciprocal of the pulse width.*

The other types of Fourier coefficients may also be obtained. Using Eqs. (2–101) and (2–102), we obtain the quadrature Fourier coefficients:

$$a_n = \begin{cases} \dfrac{A}{2}, & n = 0 \\[2ex] 0, & n > 0 \end{cases} \tag{2–121a}$$

$$b_n = \begin{cases} \dfrac{2A}{\pi n}, & n = \text{odd} \\[2ex] 0, & n = \text{even} \end{cases} \tag{2–121b}$$

Here all the $a_n = 0$, except for $n = 0$, because if the dc value were suppressed, the waveform would be an odd function about the origin. Using Eqs. (2–106) and (2–107), we find that the polar Fourier coefficients are

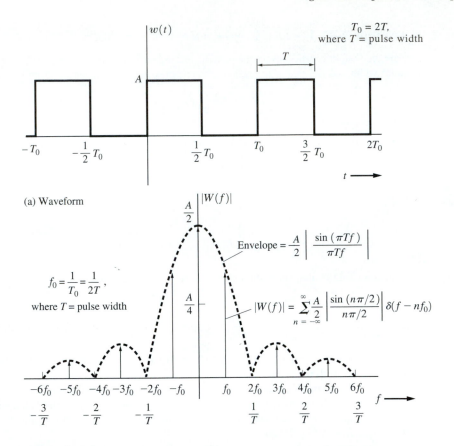

(a) Waveform

(b) Magnitude Spectrum

Figure 2–12   Periodic rectangular wave used in Example 2–12.

$$D_n = \begin{cases} \dfrac{A}{2}, & n = 0 \\[2mm] \dfrac{2A}{n\pi}, & n = 1,\, 3,\, 5,\, \dots \\[2mm] 0, & n \ \text{otherwise} \end{cases} \qquad (2\text{–}122)$$

and

$$\varphi_n = -90° \quad \text{for } n \geq 1 \qquad (2\text{–}123)$$

In communication problems the normalized average power is often needed, and for the case of periodic waveforms, it can be evaluated using Fourier series coefficients.

THEOREM. *For a periodic waveform $w(t)$, the normalized power is given by*

$$P_w = \langle w^2(t) \rangle = \sum_{n=-\infty}^{n=\infty} |c_n|^2 \tag{2-124}$$

where the $\{c_n\}$ are the complex Fourier coefficients for the waveform.

Proof. For periodic $w(t)$, the Fourier series representation is valid over all time and may be substituted into Eq. (2–12) to evaluate the normalized (i.e., $R = 1$) power:

$$P_w = \left\langle \left(\sum_n c_n e^{jn\omega_0 t} \right)^2 \right\rangle = \left\langle \sum_n \sum_m c_n c_m^* e^{jn\omega_0 t} e^{-jm\omega_0 t} \right\rangle$$

$$= \sum_n \sum_m c_n c_m^* \langle e^{j(n-m)\omega_0 t} \rangle = \sum_n \sum_m c_n c_m^* \delta_{nm} = \sum_n c_n c_n^*$$

or

$$P_w = \sum_n |c_n|^2 \tag{2-125}$$

Equation (2–124) is a special case of Parseval's theorem, Eq. (2–40), as applied to power signals.

Power Spectral Density for Periodic Waveforms

THEOREM. *For a periodic waveform, the power spectral density (PSD) is given by*

$$\mathcal{P}(f) = \sum_{n=-\infty}^{n=\infty} |c_n|^2\, \delta(f - nf_0) \tag{2-126}$$

where $T_0 = 1/f_0$ is the period of the waveform and the $\{c_n\}$ are the corresponding Fourier coefficients for the waveform.

Proof. Let $w(t) = \sum_{-\infty}^{\infty} c_n e^{jn\omega_0 t}$. Then the autocorrelation function of $w(t)$ is

$$R(\tau) = \langle w^*(t) w(t + \tau) \rangle$$

$$= \left\langle \sum_{n=-\infty}^{\infty} c_n^* e^{-jn\omega_0 t} \sum_{m=-\infty}^{\infty} c_m e^{jm\omega_0(t+\tau)} \right\rangle$$

or

$$R(\tau) = \sum_{n=-\infty}^{\infty} \sum_{m=-\infty}^{\infty} c_n^* c_m e^{jm\omega_0 \tau} \langle e^{j\omega_0(m-n)t} \rangle$$

But $\langle e^{j\omega_0(n-m)t} \rangle = \delta_{nm}$, so this reduces to

$$R(\tau) = \sum_{n=-\infty}^{\infty} |c_n|^2 e^{jn\omega_0 \tau} \tag{2-127}$$

The PSD is then

$$\mathcal{P}(f) = \mathcal{F}[R(\tau)] = \mathcal{F}\left[\sum_{-\infty}^{\infty} |c_n|^2 e^{jn\omega_0\tau}\right]$$

$$= \sum_{-\infty}^{\infty} |c_n|^2\, \mathcal{F}[e^{jn\omega_0\tau}] = \sum_{-\infty}^{\infty} |c_n|^2\, \delta(f - nf_0) \tag{2–128}$$

Equation (2–126) not only gives a way to evaluate the PSD for periodic waveforms, but also can be used to evaluate the bandwidths of the waveforms. For example, the frequency interval in which 90% of the waveform power was concentrated could be found.

Example 2–13 PSD for a Square Wave

The PSD for the periodic square wave shown in Fig. 2–12a will be found. Because the waveform is periodic, Eq. (2–126) can be used to evaluate the PSD. Consequently this problem becomes one of evaluating the FS coefficients. Furthermore, the FS coefficients for a square wave are given by Eq. (2–120). Thus,

$$\mathcal{P}(f) = \sum_{-\infty}^{\infty} \left(\frac{A}{2}\right)^2 \left(\frac{\sin(n\pi/2)}{n\pi/2}\right)^2 \delta(f - nf_0) \tag{2–129}$$

This PSD is shown by the solid lines in Fig. 2–13 where the delta functions (which have infinite amplitudes) are represented with vertical lines that have length equal to the weight (i.e., area) of the corresponding delta function.

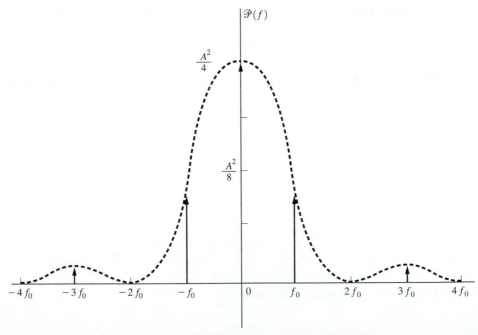

Figure 2–13   PSD for a square wave used in Example 2–13.

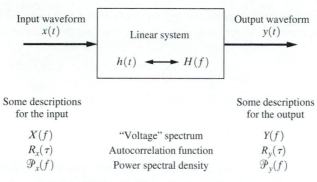

Figure 2–14   Linear system.

2–6 REVIEW OF LINEAR SYSTEMS

Linear Time-Invariant Systems

An electronic filter or system is *linear* when *superposition* holds—that is, when

$$y(t) = \mathcal{L}[a_1 x_1(t) + a_2 x_2(t)] = a_1 \mathcal{L}[x_1(t)] + a_2 \mathcal{L}[x_2(t)] \qquad (2\text{–}130)$$

where $y(t)$ is the output and $x(t) = a_1 x_1(t) + a_2 x_2(t)$ is the input, as shown in Fig. 2–14. $\mathcal{L}[\cdot]$ denotes the linear (differential equation) system operator acting on $[\cdot]$. The system is said to be *time invariant* if, for any delayed input $x(t - t_0)$, the output is delayed by just the same amount $y(t - t_0)$. That is, the *shape* of the response is the same no matter when the input is applied to the system.

A detailed discussion of the theory and practice of linear systems is beyond the scope of this book. That would require a book in itself [Irwin, 1995]. However, some basic ideas that are especially relevant to communication problems will be reviewed here.

Impulse Response

The linear time-invariant system without delay blocks is described by a linear ordinary differential equation with constant coefficients and may be characterized by its impulse response $h(t)$. The *impulse response* is the solution to the differential equation when the forcing function is a Dirac delta function. That is, $y(t) = h(t)$ when $x(t) = \delta(t)$. In physical networks, the impulse response has to be *causal*. That is, $h(t) = 0$ for $t < 0$.[†]

This impulse response may be used to obtain the system output when the input is *not* an impulse. In that case, a general waveform at the input may be approximated by[‡]

$$x(t) = \sum_{n=0}^{\infty} x(n\,\Delta t)[\delta(t - n\,\Delta t)]\,\Delta t \qquad (2\text{–}131)$$

[†] The Paley–Wiener criterion gives the frequency domain equivalence for the causality condition of the time domain. It is that $H(f)$ must satisfy the condition

$$\int_{-\infty}^{\infty} \frac{|\ln\,|H(f)|\,|}{1 + f^2}\,df < \infty$$

[‡] Δt corresponds to dx of Eq. (2–47).

which indicates that samples of the input are taken at Δt-second intervals. Then, using the time-invariant and superposition properties, we find that the output is approximately

$$y(t) = \sum_{n=0}^{\infty} x(n \, \Delta t)[h(t - n \, \Delta t)] \, \Delta t \tag{2–132}$$

This expression becomes the exact result as Δt becomes zero. Letting $n \, \Delta t = \lambda$, we obtain

$$y(t) = \int_{-\infty}^{\infty} x(\lambda) h(t - \lambda) \, d\lambda \equiv x(t) * h(t) \tag{2–133}$$

An integral of this type is called the *convolution* operation, as first described by Eq. (2–62) in Sec. 2–2. That is, the output waveform for a time-invariant network can be obtained by convolving the input waveform with the impulse response for the system. Consequently, the impulse response can be used to characterize the response of the system in the time domain, as illustrated in Fig. 2–14.

Transfer Function

The spectrum of the output signal is obtained by taking the Fourier transform of both sides of Eq. (2–133). Using the convolution theorem of Table 2–1, we get

$$Y(f) = X(f)H(f) \tag{2–134}$$

or

$$H(f) = \frac{Y(f)}{X(f)} \tag{2–135}$$

where $H(f) = \mathcal{F}[h(t)]$ is said to be the *transfer function* or *frequency response* of the network. That is, the impulse response and frequency response are a Fourier transform pair:

$$h(t) \leftrightarrow H(f)$$

Of course, the transfer function $H(f)$ is, in general, a complex quantity and can be written in polar form as

$$H(f) = |H(f)| \, e^{j \underline{/H(f)}} \tag{2–136}$$

where $|H(f)|$ is the *amplitude* (or *magnitude*) response and

$$\theta(f) = \underline{/H(f)} = \tan^{-1} \left[\frac{\text{Im}\{H(f)\}}{\text{Re}\{H(f)\}} \right] \tag{2–137}$$

is the phase response of the network. Furthermore, since $h(t)$ is a real function of time (for real networks), it follows from Eqs. (2–38) and (2–39) that $|H(f)|$ is an even function of frequency and $\underline{/H(f)}$ is an odd function of frequency.

The transfer function of a linear time-invariant network can be measured by using a sinusoidal testing signal that is swept over the frequency band of interest. For example, if

$$x(t) = A \cos \omega_0 t$$

then the output of the network will be

$$y(t) = A|H(f_0)| \cos \left[\omega_0 t + \underline{/H(f_0)} \right] \qquad (2\text{–}138)$$

where the amplitude and phase may be evaluated on an oscilloscope or by the use of a vector voltmeter.

If the input to the network is a periodic signal with a spectrum given by

$$X(f) = \sum_{n=-\infty}^{n=\infty} c_n \, \delta(f - nf_0) \qquad (2\text{–}139)$$

where, from Eq. (2–109), $\{c_n\}$ are the complex Fourier coefficients of the input signal, the spectrum of the periodic output signal, from Eq. (2–134), is

$$Y(f) = \sum_{n=-\infty}^{n=\infty} c_n H(nf_0) \, \delta(f - nf_0) \qquad (2\text{–}140)$$

We can also obtain the relationship between the power spectral density (PSD) at the input, $\mathcal{P}_x(f)$, and that at the output, $\mathcal{P}_y(f)$, of a linear time-invariant network.[†] From Eq. (2–66), we know that

$$\mathcal{P}_y(f) = \lim_{T \to \infty} \frac{1}{T} |Y_T(f)|^2 \qquad (2\text{–}141)$$

Using Eq. (2–134) in a formal sense, we obtain

$$\mathcal{P}_y(f) = |H(f)|^2 \lim_{T \to \infty} \frac{1}{T} |X_T(f)|^2$$

or

$$\mathcal{P}_y(f) = |H(f)|^2 \, \mathcal{P}_x(f) \qquad (2\text{–}142)$$

Consequently, the *power transfer function* of the network is

$$G_h(f) = \frac{\mathcal{P}_y(f)}{\mathcal{P}_x(f)} = |H(f)|^2 \qquad (2\text{–}143)$$

A rigorous proof of this theorem is given in Chapter 6.

Example 2–14 RC LOW-PASS FILTER

An *RC* low-pass filter is shown in Fig. 2–15, where $x(t)$ and $y(t)$ denote the input and output voltage waveforms, respectively. Using Kirchhoff's law for the sum of voltages around a loop, we get

$$x(t) = Ri(t) + y(t)$$

where $i(t) = C \, dy(t)/dt$, or

$$RC \frac{dy}{dt} + y(t) = x(t) \qquad (2\text{–}144)$$

[†] The relationship between the input and output autocorrelation functions $R_x(\tau)$ and $R_y(\tau)$ can also be obtained as shown by (6–82).

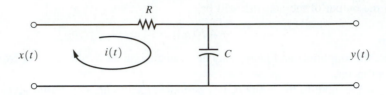

(a) RC Low-Pass Filter

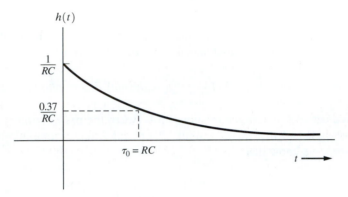

(b) Impulse Response

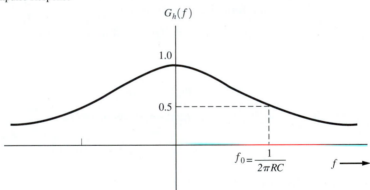

(c) Power Transfer Function

Figure 2–15   Characteristics of an *RC* low-pass filter.

From Table 2–1, we find that the Fourier transform of this differential equation is

$$RC(j2\pi f)Y(f) + Y(f) = X(f)$$

Thus, the transfer function for this network is

$$H(f) = \frac{Y(f)}{X(f)} = \frac{1}{1 + j(2\pi RC)f} \qquad (2\text{–}145)$$

Using the Fourier transform of Table 2–2, we obtain the impulse response

$$h(t) = \begin{cases} \dfrac{1}{\tau_0}\, e^{-t/\tau_0}, & t \geq 0 \\ 0, & t < 0 \end{cases} \qquad (2\text{–}146)$$

where $\tau_0 = RC$ is the *time constant*. When we combine Eqs. (2–143) and (2–145), the power transfer function is

$$G_h(f) = |H(f)|^2 = \frac{1}{1 + (f/f_0)^2} \qquad (2\text{–}147)$$

where $f_0 = 1/(2\pi RC)$.

The impulse response and the power transfer function are shown in Fig. 2–15. Note that the value of the power gain at $f = f_0$ (called the 3-dB *frequency*) is $G_h(f_0) = \frac{1}{2}$. That is, the frequency component in the output waveform at $f = f_0$ is attenuated by 3 dB compared with that at $f = 0$. Consequently, $f = f_0$ is said to be the 3-dB bandwidth of this filter. The topic of bandwidth is discussed in more detail in Sec. 2–9.

Distortionless Transmission

In communication systems, a *distortionless channel* is often desired. This implies that the channel output is just proportional to a delayed version of the input

$$y(t) = Ax(t - T_d) \qquad (2\text{–}148)$$

where A is the gain (which may be less than unity) and T_d is the delay.

The corresponding requirement in the frequency domain specification is obtained by taking the Fourier transform of both sides of Eq. (2–148).

$$Y(f) = AX(f)e^{-j2\pi f T_d}$$

Thus, for distortionless transmission, we require that the transfer function of the channel be given by

$$H(f) = \frac{Y(f)}{X(f)} = Ae^{-j2\pi f T_d} \qquad (2\text{–}149)$$

which implies that, to have no distortion at the output of a linear time-invariant system, two requirements must be satisfied:

1. The amplitude response is flat. That is,

$$|H(f)| = \text{constant} = A \qquad (2\text{–}150a)$$

2. The phase response is a *linear* function of frequency. That is,

$$\theta(f) = \underline{/H(f)} = -2\pi f T_d \qquad (2\text{–}150b)$$

When the first condition is satisfied, there is no *amplitude distortion*. When the second condition is satisfied, there is no *phase distortion*. For distortionless transmission, both conditions must be satisfied.

The second requirement is often specified in an equivalent way by using the time delay. We define the *time delay* of the system as

$$T_d(f) = -\frac{1}{2\pi f}\,\theta(f) = -\frac{1}{2\pi f}\,\underline{/H(f)]} \qquad (2\text{-}151)$$

By Eq. (2–149), it is required that

$$T_d(f) = \text{constant} \qquad (2\text{-}152)$$

for distortionless transmission. If $T_d(f)$ is not constant, there is phase distortion, because the phase response, $\theta(f)$, is not a linear function of frequency.

Example 2–15 DISTORTION CAUSED BY A FILTER

Let us examine the distortion effect caused by the *RC* low-pass filter studied in Example 2–14. From Eq. (2–145), the amplitude response is

$$|H(f)| = \frac{1}{\sqrt{1 + (f/f_0)^2}} \qquad (2\text{-}153)$$

and the phase response is

$$\theta(f) = \underline{/H(f)} = -\tan^{-1}(f/f_0) \qquad (2\text{-}154)$$

The corresponding time delay function is

$$T_d(f) = \frac{1}{2\pi f}\,\tan^{-1}(f/f_0) \qquad (2\text{-}155)$$

These results are plotted in Fig. 2–16, as indicated by the solid lines. This filter will produce some distortion since Eqs. (2–150a) and (2–150b) are not satisfied. The dashed lines give the equivalent results for the distortionless filter. Several observations can be made. First, if the signals involved have spectral components at frequencies below $0.5f_0$, the filter will provide almost distortionless transmission, because the error in the magnitude response (with respect to the distortionless case) is less than 0.5 dB and the error in the phase is less than 2.1° (8%). For $f < f_0$, the magnitude error is less than 3 dB and the phase error is less than 12.3° (27%). In engineering practice, this type of error is often considered to be tolerable. Waveforms with spectral components below $0.50f_0$ would be delayed by approximately $1/(2\pi f_0)$ s, as shown in Fig. 2–16c. That is, if the cutoff frequency of the filter was $f_0 = 1$ kHz, the delay would be 0.2 ms. For wideband signals, the higher-frequency components would be delayed less than the lower-frequency components.

Distortion of Audio, Video, and Data Signals

A linear time-invariant system will produce amplitude distortion if the amplitude response is not flat, and it will produce phase distortion (i.e., differential time delays) if the phase response is not a linear function of frequency.

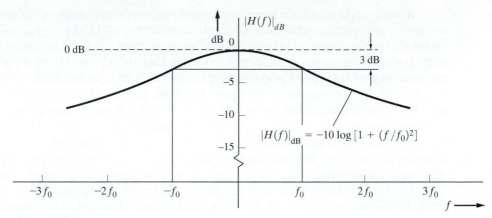

(a) Magnitude Response

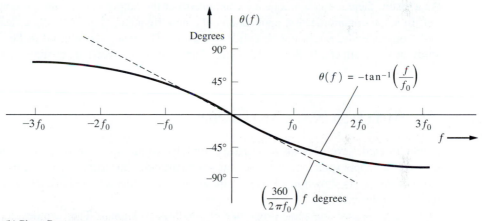

(b) Phase Response

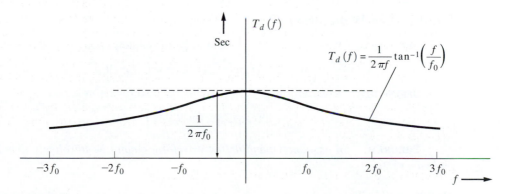

(c) Time Delay

Figure 2–16   Distortion caused by an *RC* low-pass filter.

In audio applications, the human ear is relatively sensitive to amplitude distortion, but insensitive to phase distortion. This is because a phase error of 15° for an audio filter at 15 kHz would produce a variation (error) in time delay of about 3 μ sec. Comparing this error to the duration of a spoken syllable, which is in the range of 0.01 to 0.1 sec, the time delay error due to poor filter phase response is negligible. However, an amplitude error of 3 dB would certainly be detectable by the human ear. Thus, in linear distortion specifications of high-fidelity audio amplifiers, one is interested primarily in nonflat magnitude frequency response characteristics and is not too concerned about the phase response characteristics.

In analog video applications, the opposite is true: The phase response becomes the dominant consideration. This is because the human eye is more sensitive to time delay errors, which result in smearing of an object's edges, rather than errors in amplitude (intensity).

For data signals, a linear filter can cause a data pulse in one time slot to smear into adjacent time slots causing intersymbol interference (ISI). Filter design for minimum ISI is discussed in Sec. 3–6.

If the system is nonlinear or time varying, other types of distortion will be produced. As a result, there are new frequency components at the output that are not present at the input. In some communication applications, the new frequencies are actually a desired result and, consequently, might not be called distortion. The reader is referred to Sec. 4-3 for a study of these effects. In Sec. 4–5, the time delay of bandpass filters is studied, and formulas for group delay and phase delay are developed.

2–7 BANDLIMITED SIGNALS AND NOISE

A bandlimited waveform has nonzero spectra only within a certain frequency band. In this case we can apply some powerful theorems—in particular, the sampling theorem—to process the waveform. As shown in Chapter 3, these ideas are *especially applicable to digital communication* problems.

First we examine some properties of bandlimited signals. Then we develop the sampling and dimensionality theorems.

Bandlimited Waveforms

DEFINITION.   A waveform $w(t)$ is said to be (absolutely) *bandlimited* to B hertz if

$$W(f) = \mathscr{F}[w(t)] = 0, \quad \text{for } |f| \geq B \tag{2–156}$$

DEFINITION.   A waveform $w(t)$ is (absolutely) *time limited* if

$$w(t) = 0, \quad \text{for } |t| > T \tag{2–157}$$

THEOREM.   *An absolutely bandlimited waveform cannot be absolutely time limited, and vice versa.*

This theorem is illustrated by the spectrum for the rectangular pulse waveform of Example 2–5. A relatively simple proof of the theorem can be obtained by contradiction [Wozencraft and Jacobs, 1965].

The theorem raises an engineering paradox. We know that a bandlimited waveform cannot be time limited. However, we believe that a physical waveform *is* time limited because the device that generates the waveform was built at some finite past time and the device will decay at some future time (thus producing a time-limited waveform). This paradox is resolved by realizing that we are modeling a physical process with a mathematical model and perhaps the assumptions in the model are not satisfied—although we believe them to be satisfied. That is, there is some uncertainty as to what the actual time waveform and corresponding spectrum look like—especially at extreme values of time and frequency—owing to the inaccuracies of our measuring devices. This relationship between the physical process and our mathematical model is discussed in an interesting paper by David Slepian [1976]. Although the signal may not be absolutely bandlimited, it may be bandlimited for all practical purposes in the sense that the amplitude spectrum has a negligible level above a certain frequency.

Another interesting theorem states that if $w(t)$ is absolutely *bandlimited*, it is an *analytic* function. An analytic function is a function that possesses finite-valued derivatives when they are evaluated for any finite value of t. This theorem may be proved by using a Taylor series expansion [Wozencraft and Jacobs, 1965].

Sampling Theorem

The sampling theorem is one of the most useful theorems, since it applies to digital communication systems. The sampling theorem is another application of an orthogonal series expansion.

> **SAMPLING THEOREM.**   *Any physical waveform may be represented over the interval $-\infty < t < \infty$ by*

$$w(t) = \sum_{n=-\infty}^{n=\infty} a_n \frac{\sin\{\pi f_s[t - (n/f_s)]\}}{\pi f_s[t - (n/f_s)]} \tag{2–158}$$

where

$$a_n = f_s \int_{-\infty}^{\infty} w(t) \frac{\sin\{\pi f_s[t - (n/f_s)]\}}{\pi f_s[t - (n/f_s)]} \, dt \tag{2–159}$$

and f_s is a parameter that is assigned some convenient value greater than zero. Furthermore, if $w(t)$ is bandlimited to B hertz and $f_s \geq 2B$, then Eq. (2–158) becomes the sampling function representation, where

$$a_n = w(n/f_s) \tag{2–160}$$

That is, for $f_s \geq 2B$, the orthogonal series coefficients are simply the values of the waveform that are obtained when the waveform is sampled every $1/f_s$ seconds.

The series given by Eqs. (2–158) and (2–160) is sometimes called the *cardinal series*. It has been known by mathematicians since at least 1915 [Whittaker, 1915] and to engineers since the pioneering work of Shannon, who connected the series with information theory [Shannon, 1949]. An excellent tutorial paper on this topic has been published in the *Proceedings of the IEEE* [Jerri, 1977].

Proof.   We need to show that

$$\varphi_n(t) = \frac{\sin\{\pi f_s[t - (n/f_s)]\}}{\pi f_s[t - (n/f_s)]} \tag{2-161}$$

form a set of the orthogonal functions. From Eq. (2–77), we must demonstrate that Eq. (2–161) satisfies

$$\int_{-\infty}^{\infty} \varphi_n(t)\varphi_m^*(t)\, dt = K_n\, \delta_{nm} \tag{2-162}$$

Using Parseval's theorem, Eq. (2–40), we see that the left side becomes

$$\int_{-\infty}^{\infty} \varphi_n(t)\varphi_m^*(t)\, dt = \int_{-\infty}^{\infty} \Phi_n(f)\Phi_m^*(f)\, df \tag{2-163}$$

where

$$\Phi_n(f) = \mathcal{F}[\varphi_n(t)] = \frac{1}{f_s}\, \Pi\left(\frac{f}{f_s}\right) e^{-j2\pi(nf/f_s)} \tag{2-164}$$

Hence, we have

$$\int_{-\infty}^{\infty} \varphi_n(t)\varphi_m^*(t)\, dt = \frac{1}{(f_s)^2} \int_{-\infty}^{\infty} \left[\Pi\left(\frac{f}{f_s}\right)\right]^2 e^{-j2\pi(n-m)f/f_s}\, df$$

$$= \frac{1}{(f_s)^2} \int_{-f_s/2}^{f_s/2} e^{-j2\pi(n-m)(f/f_s)}\, df = \frac{1}{f_s}\, \delta_{nm} \tag{2-165}$$

Thus, the $\varphi_n(t)$, as given by (2–161), are orthogonal functions with $K_n = 1/f_s$. Using Eq. (2–84), we see that Eq. (2–159) follows. Furthermore, we will show that Eq. (2–160) follows for the case of $w(t)$ being absolutely bandlimited to B hertz with $f_s \geq 2B$. Using Eq. (2–84) and Parseval's theorem, Eq. (2–40), we get

$$a_n = f_s \int_{-\infty}^{\infty} w(t)\varphi_n^*(t)\, dt$$

$$= f_s \int_{-\infty}^{\infty} W(f)\Phi_n^*(f)\, df \tag{2-166}$$

Substituting (2–164) yields

$$a_n = \int_{-f_s/2}^{f_s/2} W(f)e^{-j2\pi f(n/f_s)}\, df \tag{2-167}$$

But because $W(f)$ is zero for $|f| > B$, where $B \leq f_s/2$, the limits on the integral may be extended to $(-\infty, \infty)$ without changing the value of the integral. This integral with infinite limits is just the inverse Fourier transform of $W(f)$ evaluated at $t = n/f_s$. Consequently, $a_n = w(n/f_s)$, which is (2–160).

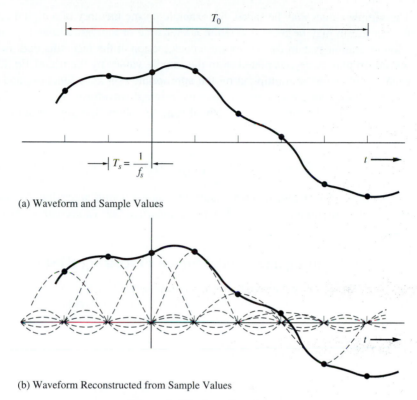

(a) Waveform and Sample Values

(b) Waveform Reconstructed from Sample Values

Figure 2-17   Sampling theorem.

From Eq. (2–167), it is obvious that the minimum sampling rate allowed to reconstruct a bandlimited waveform without error is given by

$$(f_s)_{\min} = 2B \tag{2-168}$$

This is called the *Nyquist frequency*.

Now we examine the problem of reproducing a bandlimited waveform by using N sample values. Suppose that we are only interested in reproducing the waveform over a T_0-s interval as shown in Fig. 2–17a. Then we can truncate the sampling function series of Eq. (2–158) so that we include only N of the $\varphi_n(t)$ functions that have their peaks within the T_0 interval of interest. That is, the waveform can be approximately reconstructed by using N samples. The equation is

$$w(t) \approx \sum_{n=n_1}^{n=n_1+N} a_n \varphi_n(t) \tag{2-169}$$

where the $\{\varphi_n(t)\}$ are described by Eq. (2–161). Figure 2–17b shows the reconstructed waveform (solid line), which is obtained by the weighted sum of time-delayed $(\sin x)/x$ waveforms (dashed lines), where the weights are the sample values $a_n = w(n/f_s)$ denoted by the dots. The waveform is bandlimited to B hertz with the sampling frequency $f_s \geq 2B$.

The sample values may be saved, for example, in the memory of a digital computer, so that the waveform may be reconstructed at a later time or the values may be transmitted over a communication system for waveform reconstruction at the receiving end. In either case, the waveform may be reconstructed from the sample values by the use of Eq. (2–169). That is, each sample value is multiplied by the appropriate $(\sin x)/x$ function, and these weighted $(\sin x)/x$ functions are summed to give the original waveform. This procedure is illustrated in Fig. 2–17b. The minimum number of sample values that are needed to reconstruct the waveform is

$$N = \frac{T_0}{1/f_s} = f_s T_0 \geq 2BT_0 \qquad (2\text{--}170)$$

and there are N orthogonal functions in the reconstruction algorithm. We can say that N is the number of dimensions needed to reconstruct the T_0-second approximation of the waveform.

Impulse Sampling and Digital Signal Processing (DSP)

Another useful orthogonal series is *the impulse-sampled series*, which is obtained when the $(\sin x)/x$ orthogonal functions of the sampling theorem are replaced by an orthogonal set of delta (impulse) functions. The impulse-sampled series is also identical to the impulse-sampled waveform $w_s(t)$: Both can be obtained by multiplying the unsampled waveform by a unit-weight impulse train, yielding

$$w_s(t) = w(t) \sum_{n=-\infty}^{\infty} \delta(t - nT_s)$$

$$= \sum_{n=-\infty}^{\infty} w(nT_s)\, \delta(t - nT_s) \qquad (2\text{--}171)$$

where $T_s = 1/f_s$, as illustrated in Fig. 2–18.[†] In the figure, the weight (area) of each impulse, $w(nT_s)$, is indicated by the height of the impulse.

The spectrum for the impulse-sampled waveform $w_s(t)$ can be evaluated by substituting the Fourier series of the (periodic) impulse train into Eq. (2–171), giving

$$w_s(t) = w(t) \sum_{n=-\infty}^{\infty} \frac{1}{T_s}\, e^{jn\omega_s t} \qquad (2\text{--}172)$$

Taking the Fourier transform of both sides of this equation, we get

$$W_s(f) = \frac{1}{T_s} W(f) * \mathcal{F}\left[\sum_{n=-\infty}^{\infty} e^{jn\omega_s t} \right] = \frac{1}{T_s} W(f) * \sum_{n=-\infty}^{\infty} \mathcal{F}[e^{jn\omega_s t}]$$

$$= \frac{1}{T_s} W(f) * \sum_{n=-\infty}^{\infty} \delta(f - nf_s)$$

[†] For illustrative purposes, we assume that $W(f)$ is real.

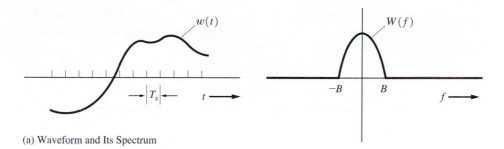

(a) Waveform and Its Spectrum

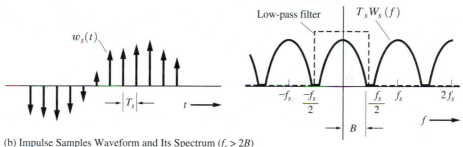

(b) Impulse Samples Waveform and Its Spectrum ($f_s > 2B$)

Figure 2–18   Impulse sampling.

or

$$W_s(f) = \frac{1}{T_s} \sum_{n=-\infty}^{\infty} W(f - nf_s) \qquad (2\text{--}173)$$

As is exemplified in Fig. 2–18b, *the spectrum of the impulse sampled signal is the spectrum of the unsampled signal that is repeated every f_s Hz, where f_s is the sampling frequency (samples/sec).*[†] This quite significant result is one of the basic principles of *digital signal processing* (DSP).

Note that this technique of impulse sampling may be used to translate the spectrum of a signal to another frequency band that is centered on some harmonic of the sampling frequency. A more general circuit that can translate the spectrum to any desired frequency band is called a *mixer*. Mixers are discussed in Sec. 4–11.

If $f_s \geq 2B$, as illustrated in Fig. 2–18, the replicated spectra do not overlap, and the original spectrum can be regenerated by chopping $W_s(f)$ off above $f_s/2$. Thus, $w(t)$ can be reproduced from $w_s(t)$ simply by passing $w_s(t)$ through an ideal low-pass filter that has a cutoff frequency of $f_c = f_s/2$, where $f_s \geq 2B$.

[†] In Chapter 3, this result is generalized to that of instantaneous sampling with a pulse train consisting of pulses with finite width and arbitrary shape (instead of impulses). This is called pulse amplitude modulation (PAM) of the instantaneous sample type.

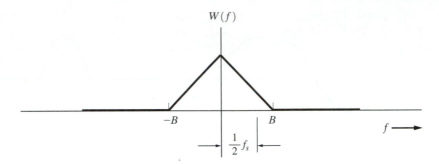

(a) Spectrum of Unsampled Waveform

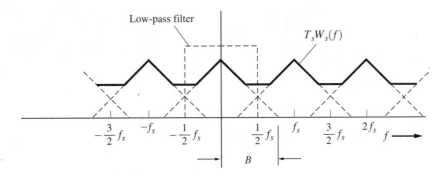

(b) Spectrum of Impulse Sampled Waveform ($f_s < 2B$)

Figure 2–19   Undersampling and aliasing.

If $f_s < 2B$ (i.e., the waveform is undersampled), the spectrum of $w_s(t)$ will consist of overlapped, replicated spectra of $w(t)$, as illustrated in Fig. 2–19.[†] The spectral overlap or tail inversion, is called *aliasing* or *spectral folding*.[‡] In this case, the low-pass filtered version of $w_s(t)$ will not be exactly $w(t)$. The recovered $w(t)$ will be distorted because of the aliasing. This distortion can be eliminated by *prefiltering* the original $w(t)$ before sampling, so that the prefiltered $w(t)$ has no spectral components above $|f| = f_s/2$. The prefiltering still produces distortion on the recovered waveform because the prefilter chops off the spectrum of the original $w(t)$ above $f = f_s/2$. However, from Fig. 2–19, it can be shown that if a prefilter is used, the recovered waveform obtained from the low-pass version of the sample signal will have one-half of the error energy compared to the error energy that would be obtained without using the presampling filter.

[†] For illustrative purposes, we assume that $W(f)$ is real.

[‡] $f_s/2$ is the folding frequency, where f_s is the sampling frequency. For no aliasing, $f_s > 2B$ is required, where $2B$ is the Nyquist frequency.

A physical waveform $w(t)$ has finite energy. From Eqs. (2–42) and (2–43), it follows that the magnitude spectrum of the waveform, $|W(f)|$, has to be negligible for $|f| > B$, where B is an appropriately chosen positive number. Consequently, from a practical standpoint, the physical waveform is essentially bandlimited to B Hz, where B is chosen to be large enough so that the error energy is below some specified amount.

Dimensionality Theorem

The sampling theorem may be restated in a more general way called the *dimensionality theorem* (which is illustrated in Fig. 2–17).

> **THEOREM.**   *When BT_0 is large, a real waveform may be completely specified by*
>
> $$N = 2BT_0 \tag{2–174}$$
>
> *independent pieces of information that will describe the waveform over a T_0 interval. N is said to be the number of dimensions required to specify the waveform, and B is the absolute bandwidth of the waveform.* [Shannon, 1949; Wozencraft and Jacobs, 1965; Wyner and Shamai (Shitz), 1998].

The dimensionality theorem of Eq. (2–174) says simply that the information which can be conveyed by a bandlimited waveform or a bandlimited communication system is proportional to the product of the bandwidth of that system and the time allowed for transmission of the information. The *dimensionality theorem has profound implications in the design and performance of all types of communication systems.* For example, in radar systems, it is well known that the time–bandwidth product of the received signal needs to be large for superior performance.

There are *two distinct ways in which the dimensionality theorem can be applied. First,* if any bandlimited waveform is given and we want to store some numbers in a table (or a computer memory bank) that could be used to reconstruct the waveform over a T_0-s interval, *at least N numbers must be stored*, and furthermore, the average sampling rate must be at least the Nyquist rate. That is,[†]

$$f_s \geq 2B \tag{2–175}$$

Thus, in this first type of application, the dimensionality theorem is used to calculate the number of storage locations (amount of memory) required to represent a waveform.

The *second* type of application is the inverse type of problem. Here the dimensionality theorem is used to estimate the bandwidth of waveforms. This application is discussed in detail in Chapter 3, Sec. 3–4, where the dimensionality theorem is used to give a lower bound for the bandwidth of digital signals.

[†] If the spectrum of the waveform being sampled has a line at $f = \pm B$, there is some ambiguity as to whether the line is included within bandwidth B. The line is included by letting $f_s > 2B$ (i.e., dropping the equality sign).

2–8 DISCRETE FOURIER TRANSFORM

With the convenience of personal computers and the availability of digital signal-processing integrated circuits, the spectrum of a waveform can be easily approximated by using the discrete Fourier transform (DFT). Here we show how the DFT can be used to compute samples of the continuous Fourier transform (CFT), Eq. (2–26), and values for the complex Fourier series coefficients of Eq. (2–94).

DEFINITION. The *discrete Fourier transform (DFT)* is defined by

$$X(n) = \sum_{k=0}^{k=N-1} x(k)e^{-j(2\pi/N)nk} \tag{2-176}$$

where $n = 0, 1, 2, ..., N - 1$, and the *inverse discrete Fourier transform (IDFT)* is defined by

$$x(k) = \frac{1}{N}\sum_{n=0}^{n=N-1} X(n)e^{j(2\pi/N)nk} \tag{2-177}$$

where $k = 0, 1, 2, ..., N - 1$.

Time and frequency do not appear explicitly, because Eqs. (2–176) and (2–177) are just definitions implemented on a digital computer to compute N values for the DFT and IDFT, respectively. One should be aware that other authors may use different (equally valid) definitions. For example, a $1/\sqrt{N}$ factor could be used on the right side of Eq. (2–176) if the $1/N$ factor of Eq. (2–177) is replaced by $1/\sqrt{N}$. This produces a different scale factor when the DFT is related to the CFT. Also, the signs of the exponents in Eq. (2–176) and (2–177) could be exchanged. This would reverse the spectral samples along the frequency axis. The fast Fourier transform (FFT) is a fast algorithm for evaluating the DFT [Ziemer, Tranter, and Fannin, 1998].

MATLAB uses the DFT and IDFT definitions that are given by Eqs. (2–176) and (2–177), except that the elements of the vector are indexed 1 through N instead of 0 through $N - 1$. Thus, the MATLAB FFT algorithms are related to Eqs. (2–176) and (2–177) by[†]

$$\mathbf{X} = fft(\mathbf{x}) \tag{2-178}$$

and

$$\mathbf{x} = ifft(\mathbf{X}) \tag{2-179}$$

where $\mathbf{x}$ is an N-element vector corresponding to samples of the waveform and $\mathbf{X}$ is the N-element DFT vector. N is chosen to be a power of 2 (i.e., $N = 2^m$, where m is a positive integer). If other FFT software is used, the user should be aware of the specific definitions that are implemented so that the results can be interpreted properly.

Two important applications of the DFT will be studied. The first application uses the DFT to approximate the spectrum $W(f)$; that is, the DFT is employed to approximate the continuous Fourier transform of $w(t)$. The approximation is given by Eq. (2–184) and illustrated

---

[†] The MATHCAD algorithms are related to Eqs. (2–176) and (2–177) by $\mathbf{X} = \sqrt{N}\,icfft\ (\mathbf{x})$ and $\mathbf{x} = (1/\sqrt{N})\,cfft\ (\mathbf{X})$.

by Example 2–16. The second application uses the DFT to evaluate the complex Fourier series coefficients c_n. The result is given by Eq. (2–187) and illustrated by Example 2–17.

Using the DFT to Compute the Continuous Fourier Transform

The relationship between the DFT, as defined by Eq. (2–180), and the CFT will now be examined. It involves three concepts: windowing, sampling, and periodic sample generation. These are illustrated in Fig. 2–20, where the left side is time domain and the right side is the corresponding frequency domain. Suppose that the CFT of a waveform $w(t)$ is to be evaluated by use of the DFT. The time waveform is first windowed (truncated) over the interval $(0, T)$ so that only a finite number of samples, N, are needed. The windowed waveform, denoted by the subscript w, is

$$w_w(t) = \begin{cases} w(t), & 0 \leq t \leq T \\ 0, & t \text{ elsewhere} \end{cases} = w(t) \, \Pi \left(\frac{t - (T/2)}{T} \right) \qquad (2\text{--}180)$$

The Fourier transform of the windowed waveform is

$$W_w(f) = \int_{-\infty}^{\infty} w_w(t) e^{-j2\pi ft} \, dt = \int_0^T w(t) e^{-j2\pi ft} \, dt \qquad (2\text{--}181)$$

Now we approximate the CFT by using a finite series to represent the integral, where $t = k \, \Delta t$, $f = n/T$, $dt = \Delta t$, and $\Delta t = T/N$. Then

$$W_w(f)|_{f=n/T} \approx \sum_{k=0}^{N-1} w(k \, \Delta t) e^{-j(2\pi/N)\,nk} \, \Delta t \qquad (2\text{--}182)$$

Comparing this result with Eq. (2–176), we obtain the relation between the CFT and DFT; that is,

$$W_w(f)|_{f=n/T} \approx \Delta t \, X(n) \qquad (2\text{--}183)$$

where $f = n/T$ and $\Delta t = T/N$. The sample values used in the DFT computation are $x(k) = w(k \, \Delta t)$, as shown in the left part of Fig. 2–20c. Also, because $e^{-j(2\pi/N)nk}$ of Eq. (2–176) is periodic in n—in other words, the same values will be repeated for $n = N, N + 1,$ ... as were obtained for $n = 0, 1, ...$—it follows that $X(n)$ is periodic (although only the first N values are returned by DFT computer programs, since the others are just repetitions). Another way of seeing that the DFT (and the IDFT) are periodic is to recognize that because samples are used, the discrete transform is an example of impulse sampling, and consequently, the spectrum must be periodic about the sampling frequency, $f_s = 1/\Delta t = N/T$ (as illustrated in Fig. 2–18 and again in Figs. 2–20c and d). Furthermore, if the spectrum is desired for negative frequencies—the computer returns $X(n)$ for the positive n values of 0, 1, ..., $N - 1$—Eq. (2–183) must be modified to give spectral values over the entire fundamental range of $-f_s/2 < f < f_s/2$. Thus, for positive frequencies we use

$$W_w(f)|_{f=n/T} \approx \Delta t \, X(n), \quad 0 \leq n < \frac{N}{2} \qquad (2\text{--}184a)$$

and for negative frequencies we use

$$W_w(f)|_{f=(n-N)/T} \approx \Delta t \, X(n), \quad \frac{N}{2} < n < N \qquad (2\text{--}184b)$$

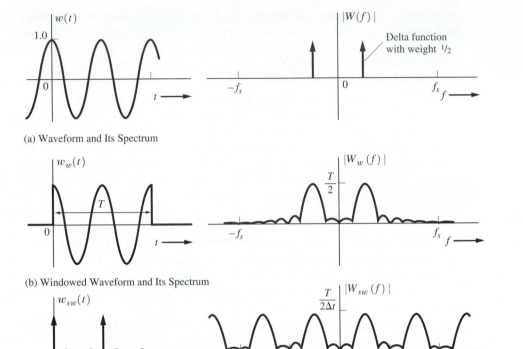

(a) Waveform and Its Spectrum

(b) Windowed Waveform and Its Spectrum

(c) Sampled Windowed Waveform and Its Spectrum ($f_s = 1/\Delta T$)

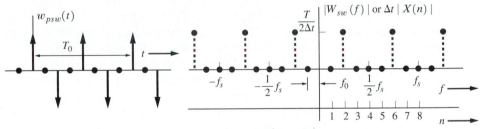

(d) Periodic Sampled Windowed Waveform and Its Spectrum ($f_0 = 1/T$)

Figure 2–20   Comparison of CFT and DFT spectra.

In a similar way, the relationship between the *inverse* CFT and the inverse DFT in described by Eq. (2–207).

Figure 2–20 illustrates the fact that if one is not careful, the DFT may give significant errors when it is used to approximate the CFT. The errors are due to a number of factors that may be categorized into three basic effects: *leakage*, *aliasing*, and the *picket-fence effect*.

The first effect is caused by windowing in the time domain. In the frequency domain, this corresponds to convolving the spectrum of the unwindowed waveform with the spectrum (Fourier transform) of the window function. This spreads the spectrum of the frequency components of $w(t)$, as illustrated in Fig. 2–20b, and causes each frequency component to "leak" into adjacent frequencies. The leakage may produce errors when the DFT is compared with the CFT. The effect can be reduced by increasing the window width T or, equivalently, increasing N for a given Δt. Window shapes other than the rectangle can also be used to reduce the side-lobes in the spectrum of the window function [Harris, 1978; Ziemer, Tranter, and Fannin, 1998]. Large periodic components in $w(t)$ cause more leakage, and if these components are known to be present, they might be eliminated before evaluating the DFT in order to reduce leakage.

From our previous study of sampling, it is known that the spectrum of a sampled waveform consists of replicating the spectrum of the unsampled waveform about harmonics of the sampling frequency. If $f_s < 2B$, where $f_s = 1/\Delta t$ and B is the highest significant frequency component in the unsampled waveform, aliasing errors will occur. The second effect, aliasing error, can be decreased by using a higher sampling frequency or a presampling low-pass filter. Note that the highest frequency component that can be evaluated with an N-point DFT is $f = f_s/2 = N/(2T)$.

The third type of error, the picket-fence effect, occurs because the N-point DFT cannot resolve the spectral components any closer than the spacing $\Delta f = 1/T$. Δf can be decreased by increasing T. If the data length is limited to T_0 seconds, where $T_0 \leq T$, T may be extended by adding zero-value sampling points. This is called *zero-padding* and will reduce Δf to produce better spectral resolution.

The computer cannot compute values that are infinite; therefore, the DFT will approximate Dirac delta functions by finite-amplitude pulses. However, the weights of the delta functions can be computed accurately by using the DFT to evaluate the Fourier series coefficients. This is demonstrated in Example 2–17.

In summary, several fundamental concepts apply when one uses the DFT to evaluate the CFT. First, the waveform is windowed over a $(0, T)$ interval so that a *finite* number of samples is obtained. Second, the DFT and the IDFT are periodic with periods $f_s = 1/\Delta t$ and T, respectively. The parameters Δt, T, and N are selected with the following considerations in mind:

- Δt is selected to satisfy the Nyquist sampling condition, $f_s = 1/\Delta t > 2B$, where B is the highest frequency in the waveform. Δt is the time between samples and is called the *time resolution* as well. Also, $t = k\Delta t$.
- T is selected to give the desired *frequency resolution*, where the frequency resolution is $\Delta f = 1/T$. Also, $f = n/T$.
- N is the number of data points and is determined by $N = T/\Delta t$.

N depends on the values used for Δt and T. The computation time increases as N is increased.[†] The N-point DFT gives the spectra of N frequencies over the frequency interval $(0, f_s)$ where $f_s = 1/\Delta t = N/T$. Half of this frequency interval represents positive frequencies and half represents negative frequencies, as described by Eq. (2–184). This is illustrated in the following example.

[†] The *fast Fourier transform* (FFT) algorithms are fast ways of computing the DFT. The number of complex multiplications required for the DFT is N^2, whereas the FFT (with N selected to be a power of 2) requires only $(N/2) \log_2 N$ complex multiplications. Thus, the FFT provides an improvement factor of $2N/(\log_2 N)$ compared with the DFT, which gives an improvement of 113.8 for an $N = 512$ point FFT.

Example 2–16 DFT FOR A RECTANGULAR PULSE

Table 2–3 lists a MATLAB file that calculates the spectrum for a rectangular pulse that uses the DFT algorithm. The computed results are shown in Fig. 2–21. Note that Eqs. (2–178) and (2–183) are used to relate the DFT results to the spectrum (i.e., the CFT). The parameters M,

TABLE 2–3   MATLAB LISTING FOR CALCULATING THE SPECTRUM OF A RECTANGULAR PULSE, USING THE DFT

```
% File: FIG2_21.M
% Calculate the FFT for a truncated step
% Let tend be the end of the step.
M = 7;
N = 2^M;
n = 0:1:N-1;
tend = 1;
T = 10;
dt = T/N;
t = n*dt;
% Creating time waveform
w = zeros(length(t),1);
for (i = 1:1:length(w))
  if (t(i) <= tend)
    w(i) = 1;
  end;
end;
% Calculating FFT
W = dt*fft(w);
f = n/T;
% Calculating position of 4th NULL
pos = index(f,4/tend);
plot_pr(2);
plot(t,w);
axis([0 T 0 1.5]);
xlabel('t (sec) -->');
ylabel('w(t)');
title('Time Waveform');
pause;
subplot(311);
plot(f(1:pos),abs(W(1:pos)));
xlabel('f (Hz) -->');
ylabel('|W(f)|');
title('MAGNITUDE SPECTRUM out to 4th Null');
subplot(312);
plot(f(1:pos),180/pi*angle(W(1:pos)));
xlabel('f (Hz) -->');
ylabel('theta(f) (degrees)');
title('PHASE SPECTRUM out to 4th Null');
grid;
subplot(313);
plot(f,abs(W));
xlabel('f (Hz) -->');
ylabel('|W(f)|');
title('MAGNITUDE SPECTRUM over whole FFT frequency range');
```

tend, and T are selected so that the computed magnitude and phase-spectral results match the true spectrum of the rectangular pulse as given by Eqs. (2–59) and (2–60) of Example 2–5. (T in Example 2–5 is equivalent to tend in Table 2–3.)

The rectangular pulse is not absolutely bandlimited. However, for a pulse width of tend = 1, the magnitude spectrum becomes relatively small at 5/tend = 5Hz = B. Thus, we need to sample the waveform at a rate of 2 B = 10 Hz or greater. For T = 10 and N = 128, Δt = 0.08 or f_s = $1/\Delta t$ = 12.8 Hz. Therefore, the values of T and N have been selected to satisfy the Nyquist rate of $f_s > 2 B$. The frequency resolution is $\Delta f = 1/T = 0.1$ Hz. Consequently, a good spectral representation is obtained by using the DFT.

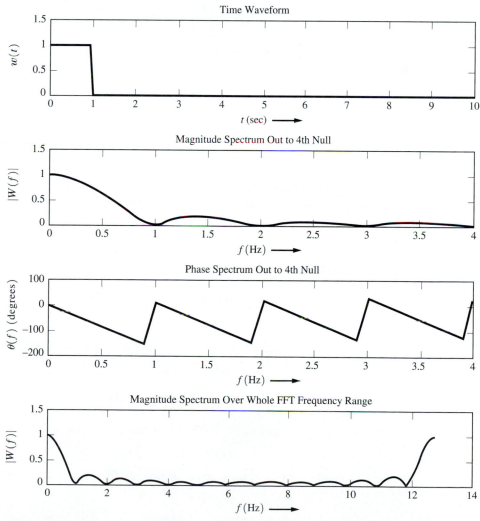

 Figure 2–21   Spectrum for a rectangular pulse, using the MATLAB DFT.

In the bottom plot of Fig. 2–21, the magnitude spectrum is shown over the whole range of the FFT vector—that is, for $0 < f < f_s$, where $f_s = 12.8$ Hz. Because Eq. (2–184b) was not used in the MATLAB program, the plot for $0 < f < 6.8$ ($f_s/2 = 6.8$ Hz) corresponds to the magnitude spectrum of the CFT over positive frequencies, and the plot for $6.4 < f < 12.8$ corresponds to the negative frequencies of the CFT. The reader should try other values of M, tend, and T to see how leakage, aliasing, and picket-fence errors become large if the parameters are not carefully selected. Also, note that significant errors occur if tend $= T$. Why?

Using the DFT to Compute the Fourier Series

The DFT may be also used to evaluate the coefficients for the complex Fourier series. From Eq. (2–89),

$$c_n = \frac{1}{T} \int_0^T w(t) e^{-j2\pi n f_0 t} \, dt$$

We approximate this integral by using a finite series, where $t = k \, \Delta t$, $f_0 = 1/T$, $dt = \Delta t$, and $\Delta t = T/N$. Then

$$c_n \approx \frac{1}{T} \sum_{k=0}^{N-1} w(k \, \Delta t) e^{-j(2\pi/N)nk} \, \Delta t \tag{2–185}$$

Using Eq. (2–176), we find that the Fourier series coefficient is related to the DFT by

$$c_n \approx \frac{1}{N} X(n) \tag{2–186}$$

The DFT returns $X(n)$ values for $n = 0, 1, ..., N - 1$. Consequently, Eq. (2–186) must be modified to give c_n values for negative n. For positive n, we use

$$c_n = \frac{1}{N} X(n), \quad 0 \le n < \frac{N}{2} \tag{2–187a}$$

and for negative n we use

$$c_n = \frac{1}{N} X(N + n), \quad -\frac{N}{2} < n < 0 \tag{2–187b}$$

Example 2–17 USE THE DFT TO COMPUTE THE SPECTRUM OF A SINUSOID

Let

$$w(t) = 3 \sin(\omega_0 t + 20°) \tag{2–188}$$

where $\omega_0 = 2\pi f_0$ and $f_0 = 10$ Hz.

Because $w(t)$ is periodic, Eq. (2–109) is used to obtain the spectrum

$$W(f) = \Sigma \, c_n \, \delta(f - nf_0)$$

where $\{c_n\}$ are the complex Fourier series coefficients for $w(t)$. Furthermore, because $\sin(x) = (e^{jx} - e^{-jx})/(2j)$,

$$3 \sin(\omega_0 t + 20°) = \left(\frac{3}{2j} e^{j20}\right) e^{j\omega_0 t} + \left(\frac{-3}{2j} e^{-j20}\right) e^{-j\omega_0 t}$$

Consequently, the FS coefficients are known to be

$$c_1 = \left(\frac{3}{2j}\ e^{j20}\right) = 1.5\,\underline{/-70°} \tag{2–189a}$$

$$c_{-1} = \left(\frac{-3}{2j}\ e^{-j20}\right) = 1.5\,\underline{/+70°} \tag{2–189b}$$

and the other c_n's are zero. Now see if this known correct answer can be computed by using the DFT. Referring to Table 2–4 and Fig. 2–22, we observe that MATLAB computes the FFT and plots the spectrum. The computed result checks with the known analytical result.

Note that δ functions cannot be plotted, because $\delta(0) = \infty$. Consequently, the weights of the δ functions are plotted instead, to indicate the magnitude spectra. Also, at frequencies where $|W(f)| = 0$, *any* value may be used for $\theta(f)$ because $W(f) = |W(f)|\,\underline{/\theta(f)} = 0$.

These results can also be compared with those of Example 2–4, where the spectrum for a sinusoid was obtained by the direct evaluation of the Fourier transform integral.

2–9 BANDWIDTH OF SIGNALS

The spectral width of signals and noise in communication systems is a very important concept, for two main reasons. First, more and more users are being assigned to increasingly crowded RF bands, so that the spectral width required for each one needs to be considered carefully. Second, the spectral width is important from the equipment design viewpoint, since the circuits need to have enough bandwidth to accommodate the signal but reject the noise. The question is, What is *bandwidth*? As we will see, there are numerous definitions of the term.

As long as we use the same definition when working with several signals and noise, we can compare their spectral widths by using the particular bandwidth definition that was selected. If we change definitions, "conversion factors" will be needed to compare the spectral widths that were obtained by using different definitions. Unfortunately, the conversion factors usually depend on the type of spectral *shape* involved [e.g., $(\sin x)/x$ type of spectrum or rectangular spectrum].

In engineering definitions, the bandwidth is taken to be the width of a *positive* frequency band. (We are describing the bandwidth of real signals or the bandwidth of a physical filter that has a real impulse response; consequently, the magnitude spectra of these waveforms are even about the origin $f = 0$.) In other words, the bandwidth would be $f_2 - f_1$, where $f_2 > f_1 \geq 0$ and f_2 and f_1 are determined by the particular definition that is used. For baseband waveforms or networks, f_1 is usually taken to be zero, since the spectrum extends down to dc ($f = 0$). For bandpass signals, $f_1 > 0$ and the band $f_1 < f < f_2$ encompasses the carrier frequency f_c of the signal. Recall that as we increase the "signaling speed" of a signal (i.e., decrease T) the spectrum gets wider. (See Fig. 2–6.) Consequently, for engineering definitions of bandwidth, we require the bandwidth to vary as $1/T$.

We will give six engineering definitions and one legal definition of bandwidth that are often used:

TABLE 2–4   MATLAB LISTING FOR CALCULATING THE SPECTRUM
OF A SINUSOID, USING THE DFT

```
% File: TABLE2_4.M
% Using the FFT, calculate the spectrum for a sinusoid.
M = 4;
N = 2^M;
fo = 10;
wo = 2*pi*fo;
n = 0:1:N-1;
T = 1/fo;
dt = T/N;
t = n*dt;
% Creating time waveform
w = 3*sin(wo*t + (pi/180*20));
% Compute the FFT data points.
W = fft(w);
W = W(:);
% SINCE THE WAVEFORM IS PERIODIC,
%      USE COMPLEX FOURIER SERIES TO GET SPECTRUM.
% ==> Compute the FS coefficients from the FFT data, using Eq. (2-186).
% Then use Eq. (2-109) to get the spectrum.
n1 = -N/2:1:N/2;
fn1 = n1/T;
fs = 1/dt;
% Generating complex fourier series coefficients
cn = 1/N * W;
% Generating Phase
Theta = (180/pi)*angle(cn + 0.001);
% Converting samples 0,1,2,3,...,N-1 to a positive and negative
% Note that Eq. (2-187) is a built-in command: fftshift
cn = fftshift(cn)';
Theta = fftshift(Theta)';
cn = [cn cn(1)];
Theta = [Theta Theta(1)];
cn = cn(:);
Theta = Theta(:);
% Plotting results
plot_pr(4);
plot(t,w);
xlabel('t (sec) -->');
ylabel('w(t)');
title('Time Waveform');
pause;
plot(n,abs(W),'o');
for (i = 1:1:length(n))
  line([n(i) n(i)], [0 abs(W(i))]);
end;
xlabel('n');
ylabel('|W(n)|');
title('FFT Data Points');
axis([0 16 0 25])
pause;
subplot(211)
```

TABLE 2–4   MATLAB LISTING FOR CALCULATING THE SPECTRUM
OF A SINUSOID, USING THE DFT *(cont.)*

```
plot(fn1,abs(cn),'o');
for (i = 1:1:length(n1))
  line([fn1(i) fn1(i)], [0 abs(cn(i))]);
end;
xlabel('f  (Hz) -->');
ylabel('|c(n)|');
title('MAGNITUDE SPECTRUM, |W(f)|');
axis([-80 80 0 2])
subplot(212)
plot(fn1,zeros(length(fn1),1),'w',fn1,Theta,'o');
for (i = 1:1:length(n1))
  line([fn1(i) fn1(i)], [0 Theta(i)]);
end;
xlabel('f  (Hz) -->');
ylabel('theta(f) (degrees)');
title('PHASE SPECTRUM, Theta(f)');
```

1. *Absolute bandwidth* is $f_2 - f_1$, where the spectrum is zero outside the interval $f_1 < f < f_2$ along the positive frequency axis.

2. *3-dB bandwidth* (or half-power bandwidth) is $f_2 - f_1$, where for frequencies inside the band $f_1 < f < f_2$, the magnitude spectra, say, $|H(f)|$, fall no lower than $1/\sqrt{2}$ times the maximum value of $|H(f)|$, and the maximum value occurs at a frequency inside the band.

3. *Equivalent noise bandwidth* is the width of a fictitious rectangular spectrum such that the power in that rectangular band is equal to the power associated with the actual spectrum over positive frequencies. From Eq. (2–142), the PSD is proportional to the square of the magnitude of the spectrum. Let f_0 be the frequency at which the magnitude spectrum has a maximum; then the power in the equivalent rectangular band is proportional to

$$\text{equivalent power} = B_{eq}|H(f_0)|^2 \qquad (2\text{–}190)$$

where B_{eq} is the equivalent bandwidth that is to be determined. The actual power for positive frequencies is proportional to

$$\text{actual power} = \int_0^\infty |H(f)|^2 \, df \qquad (2\text{–}191)$$

Setting Eq. (2–190) equal to Eq. (2–191), the formula for the *equivalent noise bandwidth* is

$$B_{eq} = \frac{1}{|H(f_0)|^2} \int_0^\infty |H(f)|^2 \, df \qquad (2\text{–}192)$$

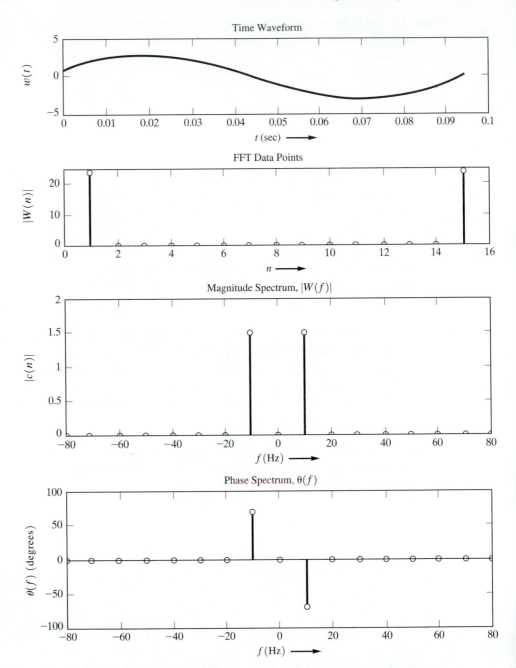

Figure 2–22   Spectrum of a sinusoid obtained by using the MATLAB DFT.

4. *Null-to-null bandwidth* (or zero-crossing bandwidth) is $f_2 - f_1$, where f_2 is the first null in the envelope of the magnitude spectrum above f_0 and, for bandpass systems, f_1 is the first null in the envelope below f_0, where f_0 is the frequency where the magnitude spectrum is a maximum.[†] For baseband systems, f_1 is usually zero.

5. *Bounded spectrum bandwidth* is $f_2 - f_1$ such that outside the band $f_1 < f < f_2$, the PSD, which is proportional to $|H(f)|^2$, must be down by at least a certain amount, say 50 dB, below the maximum value of the power spectral density.

6. *Power bandwidth* is $f_2 - f_1$, where $f_1 < f < f_2$ defines the frequency band in which 99% of the total power resides. This is similar to the FCC definition of *occupied bandwidth*, which states that the power above the upper band edge f_2 is $\frac{1}{2}\%$ and the power below the lower band edge is $\frac{1}{2}\%$, leaving 99% of the total power within the occupied band (*FCC Rules and Regulations*, Sec. 2.202).

7. *FCC bandwidth* is an authorized bandwidth parameter assigned by the FCC to specify the spectrum allowed in communication systems. When the FCC bandwidth parameter is substituted into the FCC formula, the minimum attenuation is given for the power level allowed in a 4-kHz band at the band edge with respect to the total average signal power. Sec. 21.106 of the *FCC Rules and Regulations*: asserts, "For operating frequencies below 15 GHz, in any 4 kHz band, the center frequency of which is removed from the assigned frequency by more than 50 percent up to and including 250 percent of the authorized bandwidth, as specified by the following equation, but in no event less than 50 dB":

$$A = 35 + 0.8(P - 50) + 10 \log_{10}(B) \tag{2–193}$$

(Attenuation greater than 80 dB is not required.) In this equation,

$A =$ attenuation (in decibels) below the mean output power level,

$P =$ percent removed from the carrier frequency,

and

$B =$ authorized bandwidth in megahertz.

The FCC definition (as well as many other legal rules and regulations) is somewhat obscure, but it will be interpreted in Example 2–18. It actually defines a spectral mask. That is, the spectrum of the signal must be less than or equal to the values given by this spectral mask at all frequencies. The FCC bandwidth parameter B is not consistent with the other definitions that are listed, in the sense that it is not proportional to $1/T$, the "signaling speed" of the corresponding signal [Amoroso, 1980]. Thus, the FCC bandwidth parameter B is a *legal* definition instead of an engineering definition. The *rms bandwidth*, which is very useful in analytical problems, is defined in Chapter 6.

[†] In cases where there is no definite null in the magnitude spectrum, this definition is not applicable.

Example 2–18 BANDWIDTHS FOR A BPSK SIGNAL

A binary-phase-shift-keyed (BPSK) signal will be used to illustrate how the bandwidth is evaluated for the different definitions just given.

The BPSK signal is described by

$$s(t) = m(t) \cos \omega_c t \qquad (2\text{–}194)$$

where $\omega_c = 2\pi f_c$, f_c being the carrier frequency (hertz), and $m(t)$ is a serial binary (± 1 values) modulating waveform originating from a digital information source such as a digital computer, as illustrated in Fig. 2–23a. Let us evaluate the spectrum of $s(t)$ for the worst case (the widest bandwidth).

The worst-case (widest-bandwidth) spectrum occurs when the digital modulating waveform has transitions that occur most often. In this case $m(t)$ would be a square wave, as shown in Fig. 2–23a. Here a binary 1 is represented by $+1$ V and a binary 0 by -1 V, and the signaling rate is $R = 1/T_b$ bits/s. The power spectrum of the square-wave modulation can be evaluated by using Fourier series analysis. Equations (2–126) and (2–120) give

$$\mathcal{P}_m(f) = \sum_{n=-\infty}^{\infty} |c_n|^2 \, \delta(f - nf_0) = \sum_{\substack{n=-\infty \\ n \neq 0}}^{\infty} \left[\frac{\sin(n\pi/2)}{n\pi/2} \right]^2 \delta\left(f - n\frac{R}{2} \right) \qquad (2\text{–}195)$$

where $f_0 = 1/(2T_b) = R/2$. The PSD of $s(t)$ can be expressed in terms of the PSD of $m(t)$ by evaluating the autocorrelation of $s(t)$—that is,

$$R_s(\tau) = \langle s(t)s(t + \tau) \rangle$$

$$= \langle m(t)m(t + \tau) \cos \omega_c t \cos \omega_c(t + \tau) \rangle$$

$$= \tfrac{1}{2}\langle m(t)m(t + \tau) \rangle \cos \omega_c \tau + \tfrac{1}{2} \langle m(t)m(t + \tau) \cos (2\omega_c t + \omega_c \tau) \rangle$$

or

$$R_s(\tau) = \tfrac{1}{2}R_m(\tau) \cos \omega_c \tau + \tfrac{1}{2} \lim_{T \to \infty} \frac{1}{T} \int_{-T/2}^{T/2} m(t)m(t + \tau) \cos (2\omega_c t + \omega_c \tau) \, dt \qquad (2\text{–}196)$$

The integral is negligible because $m(t)m(t + \tau)$ is constant over small time intervals, but cos $(2\omega_c t + \omega_c \tau)$ has many cycles of oscillation, since $f_c \gg R$.[†] Any small area that is accumulated by the integral becomes negligible when divided by T, $T \to \infty$. Thus,

$$R_s(\tau) = \tfrac{1}{2}R_m(\tau) \cos \omega_c \tau \qquad (2\text{–}197)$$

The PSD is obtained by taking the Fourier transform of both sides of Eq. (2–197). Using the real-signal frequency transform theorem (Table 2–1), we get

$$\mathcal{P}_s(f) = \tfrac{1}{4}[\mathcal{P}_m(f - f_c) + \mathcal{P}_m(f + f_c)] \qquad (2\text{–}198)$$

Substituting Eq. (2–195) into Eq. (2–198), we obtain the PSD for the BPSK signal:

$$\mathcal{P}_s(f) = \tfrac{1}{4} \sum_{\substack{n=-\infty \\ n \neq 0}}^{\infty} \left[\frac{\sin(n\pi/2)}{n\pi/2} \right]^2$$

$$\times \{ \delta[f - f_c - n(R/2)] + \delta[f + f_c - n(R/2)] \} \qquad (2\text{–}199)$$

This spectrum is shown in Fig. 2–23b.

[†] This is a consequence of the Riemann-Lebesque lemma from integral calculus [Olmsted, 1961].

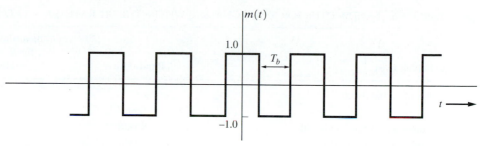

(a) Digital Modulating Waveform

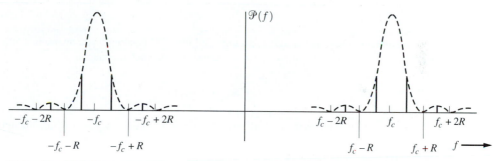

(b) Resulting BPSK Spectrum

Figure 2–23   Spectrum of a BPSK signal.

The spectral shape that results from utilizing this worst-case deterministic modulation is essentially the same as that obtained when random data are used; however, for the random case, the spectrum is continuous. The result for the random data case, as worked out in Chapter 3 where $\mathcal{P}_m(f)$ is given by Eq. (3–41), is

$$\mathcal{P}(f) = \tfrac{1}{4} T_b \left[\frac{\sin \pi T_b (f - f_c)}{\pi T_b (f - f_c)} \right]^2 + \tfrac{1}{4} T_b \left[\frac{\sin \pi T_b (f + f_c)}{\pi T_b (f + f_c)} \right]^2 \qquad (2\text{–}200)$$

when the data rate is $R = 1/T_b$ bits/sec. This is shown by the dashed curve in Fig. 2–23b.

The preceding derivation demonstrates that we can often use (deterministic) square-wave test signals to help us analyze a digital communication system, instead of using a more complicated random-data model.

The bandwidth for the BPSK signal will now be evaluated for each of the bandwidth definitions given previously. To accomplish this, the shape of the PSD for the positive frequencies is needed. From Eq. (2–200), it is

$$\mathcal{P}(f) = \left[\frac{\sin \pi T_b(f - f_c)}{\pi T_b(f - f_c)} \right]^2 \qquad (2\text{–}201)$$

Substituting Eq. (2–201) into the definitions, we obtain the resulting BPSK bandwidths as shown in Table 2–5, except for the FCC bandwidth.

TABLE 2–5   BANDWIDTHS FOR BPSK SIGNALING WHERE THE BIT RATE IS $R = 1/T_b$ BITS/S.

Definition Used	Bandwidth	Bandwidths (kHz) for $R = 9,600$ bits/s
1. Absolute bandwidth	∞	∞
2. 3-dB bandwidth	$0.88R$	8.45
3. Equivalent noise bandwidth	$1.00R$	9.60
4. Null-to-null bandwidth	$2.00R$	19.20
5. Bounded spectrum bandwidth (50 dB)	$201.04R$	1,930.0
6. Power bandwidth	$20.56R$	197.4

The relationship between the spectrum and the FCC bandwidth parameter is a little more tricky to evaluate. To do that, we need to evaluate the decibel attenuation

$$A(f) = -10 \log_{10}\left[\frac{P_{4\text{kHz}}(f)}{P_{\text{total}}}\right] \tag{2–202}$$

where $P_{4\text{kHz}}(f)$ is the power in a 4-kHz band centered at frequency f and P_{total} is the total signal power. The power in a 4-kHz band (assuming that the PSD is approximately constant across the 4-kHz bandwidth) is

$$P_{4\text{kHz}}(f) = 4000\, \mathcal{P}(f) \tag{2–203}$$

and, using the definition of equivalent bandwidth, we find that the total power is

$$P_{\text{total}} = B_{\text{eq}}\mathcal{P}(f_c) \tag{2–204}$$

where the spectrum has a maximum value at $f = f_c$. With the use of these two equations, Eq. (2–202) becomes

$$A(f) = -10 \log_{10}\left[\frac{4000\ \mathcal{P}(f)}{B_{\text{eq}}\mathcal{P}(f_c)}\right] \tag{2–205}$$

where $A(f)$ is the decibel attenuation of power measured in a 4-kHz band at frequency f compared with the total average power level of the signal. For the case of BPSK signaling, using Eq. (2–201), we find that the decibel attenuation is

$$A(f) = -10 \log_{10}\left\{\frac{4000}{R}\left[\frac{\sin \pi T_b(f - f_c)}{\pi T_b(f - f_c)}\right]^2\right\} \tag{2–206}$$

where $R = 1/T_b$ is the data rate. If we attempt to find the value of R such that $A(f)$ will fall below the specified FCC spectral envelope shown in Fig. 2–24 ($B = 30$ MHz), we will find that R is so small that there will be numerous zeros in the $(\sin x)/x$ function of Eq. (2–206) within the desired frequency range, -50 MHz $< (f - f_c) < 50$ MHz. This is difficult to plot, so by replacing $\sin \pi T(f - f_c)$ with its maximum value (unity) we plot the envelope of $A(f)$ instead. The resulting BPSK envelope curve for the decibel attenuation is shown in Fig. 2–24, where $R = 0.0171$ Mbits/s.

It is obvious that the data rate allowed for a BPSK signal to meet the FCC $B = 30$-MHz specification is ridiculously low, because the FCC bandwidth parameter specifies an almost absolutely bandlimited spectrum. To signal with a reasonable data rate, the pulse shape used in the transmitted signal must be modified from a rectangular shape (that gives the BPSK signal) to a rounded pulse such that the bandwidth of the transmitted signal will be almost absolutely bandlimited. Recalling our study of the sampling theorem, we realize that $(\sin x)/x$ pulse shapes are prime candidates, since they have an absolutely bandlimited spectrum. However, the $(\sin x)/x$ pulses are not absolutely timelimited, so that this exact pulse shape cannot be used. Frank Amoroso and others have studied the problem, and a quasi-bandlimited version of the $(\sin x)/x$ pulse shape has been proposed [Amoroso, 1980]. The decibel attenuation curve for this type of signaling, shown in Fig. 2–24, is seen to fit very nicely under the allowed FCC spectral envelope curve for the case of $R = 25$ Mbits/s. The allowable data rate of 25 Mbits/s is a fantastic improvement over the $R = 0.0171$ Mbits/s that was allowed for BPSK. It is also interesting to note that analog pulse shapes [$(\sin x)/x$ type] are required instead of a digital (rectangular) pulse shape, which is another way of saying that *it is vital for digital communication engineers to be able to analyze and design analog systems as well as digital systems.*

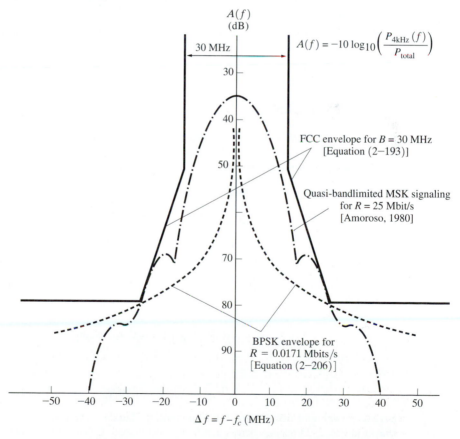

Figure 2–24   FCC-allowed envelope for $B = 30$ MHz.

2–10 SUMMARY

- Signals and noise may be deterministic (their waveform is known), where the signal is the desired part and noise the undesired part; or they may be stochastic (their waveform is unknown, but statistics about the waveform are known). Properties of signals and noise are their spectra, dc value, rms value, and associated power.

- Fourier transforms and spectra were studied in detail. The FT is used to obtain the spectrum of a waveform. The spectrum gives the frequencies of the sinusoidal components of the waveform. The PSD and the autocorrelation function were defined and examined.

- Signals and noise may be represented by orthogonal series expansions. The Fourier series and sampling function series were found to be especially useful.

- Linear systems were reviewed, and the condition for distortionless transmission was found.

- The properties of bandlimited signals and noise were developed. This resulted in the sampling theorem and the dimensionality theorem. The DFT was studied with MATLAB examples.

- The concept of bandwidth was discussed, and seven popular definitions were given.

- The null bandwidth of a rectangular pulse of width T is 1/T. This is a basic concept used in digital communication systems.

2–11 STUDY-AID EXAMPLES

SA2–1 DC and RMS Values for Exponential Signals   Assume that $v(t)$ is a periodic voltage waveform as shown in Fig. 2–25. Over the time interval $0 < t < 1$, $v(t)$ is described by e^t. Find the dc value and the rms value of this voltage waveform.

Solution.   For periodic waveforms, the dc value is

$$V_{dc} = \langle v(t) \rangle = \frac{1}{T_0} \int_0^{T_0} v(t)\, dt = \int_0^1 e^t dt = e^1 - e^0$$

or

$$V_{dc} = e - 1 = 1.72 \text{ V}$$

Likewise,

$$V_{rms}^2 = \langle v^2(t) \rangle = \int_0^1 (e^t)^2\, dt = \frac{1}{2}(e^2 - e^0) = 3.19$$

Thus,

$$V_{rms} = \sqrt{3.19} = 1.79 \text{ V}$$

SA2–2 Power and dBm Value for Exponential Signals   The periodic voltage waveform shown in Fig. 2–25 appears across a 600-Ω resistive load. Calculate the average power dissipated in the load and the corresponding dBm value.

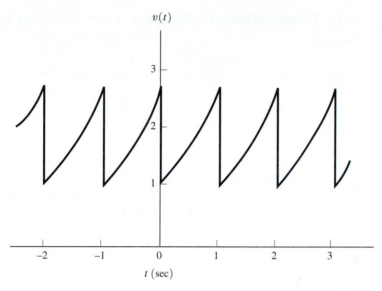

$v(t)$

Figure 2–25

Solution.

$$P = V^2_{\text{rms}}/R = (1.79)^2/600 = 5.32 \text{ mW}$$

and

$$10 \log \left(\frac{P}{10^{-3}}\right) = 10 \log \left(\frac{5.32 \times 10^{-3}}{10^{-3}}\right) = 7.26 \text{ dBm}$$

Note: The peak instantaneous power is

$$\max[p(t)] = \max[v(t)\, i(t)] = \max[v^2(t)/R]$$

$$= \frac{(e)^2}{600} = 12.32 \text{ mW}$$

SA2–3 Evaluation of Spectra by Superposition   Find the spectrum for the waveform

$$w(t) = \Pi\left(\frac{t-5}{10}\right) + 8 \sin(6\pi t)$$

Solution.   The spectrum of $w(t)$ is the superposition of the spectrum of the rectangular pulse and the spectrum of the sinusoid. Using Tables 2–1 and 2–2, we have

$$\mathcal{F}\left[\Pi\left(\frac{t-5}{10}\right)\right] = 10\,\frac{\sin(10\pi f)}{10\pi f}\,e^{-j2\pi f 5}$$

and using the result of Example 2–4, we get

$$\mathcal{F}[8 \sin(6\pi t)] = j\frac{8}{2}\,[\delta(f+3) - \delta(f-3)]$$

Thus,

$$W(f) = 10\,\frac{\sin(10\pi f)}{10\pi f}\,e^{-j10\pi f} + j4[\delta(f+3) - \delta(f-3)]$$

SA2–4 Evaluation of Spectra by Integration   Find the spectrum for $w(t) = 5 - 5e^{-2t}u(t)$.

Solution.

$$W(f) = \int_{-\infty}^{\infty} w(t)e^{-j\omega t}\, dt$$

$$= \int_{-\infty}^{\infty} 5e^{-j2\pi ft}\, dt - 5\int_{0}^{\infty} e^{-2t}e^{-j\omega t}\, dt$$

$$= 5\delta(f) - 5\frac{e^{-2(1+j\pi f)t}}{-2(1+j\pi f)}\Bigg|_{0}^{\infty}$$

or

$$W(f) = 5\delta(f) - \frac{5}{2 + j2\pi f}$$

SA2–5 Evaluation of FT by Superposition   Assume that $w(t)$ is the waveform shown in Fig. 2–26. Find the Fourier transform of $w(t)$.

Solution.   Referring to Fig. 2–26, we can express $w(t)$ as the superposition (i.e., sum) of two rectangular pulses:

$$w(t) = \Pi\left(\frac{t-2}{4}\right) + 2\Pi\left(\frac{t-2}{2}\right)$$

Using Table 2–2 and Table 2–1, we find that the FT is

$$W(f) = 4Sa(\pi f4)e^{-j\omega 2} + 2(2)Sa(\pi f2)e^{-j\omega 2}$$

or

$$W(f) = 4[Sa(4\pi f) + Sa(2\pi f)]e^{-j4\pi f}$$

SA2–6 Orthogonal Functions   Show that $\varphi_1(t) = \Pi(t)$ and $\varphi_2(t) = \sin 2\pi t$ are orthogonal functions over the interval $-0.5 < t < 0.5$.

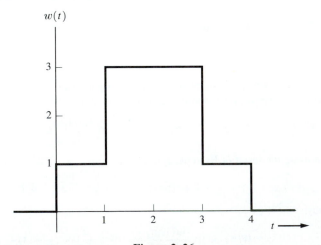

Figure 2–26

Solution.

$$\int_a^b \varphi_1(t)\, \varphi_2(t)\, dt = \int_{-0.5}^{0.5} 1 \sin 2\pi t\, dt = -\left.\frac{\cos 2\pi t}{2\pi}\right|_{-0.5}^{0.5}$$

$$= \frac{-1}{2\pi} [\cos \pi - \cos(-\pi)] = 0$$

Because this integral is zero, Eq. (2–77) is satisfied. Consequently, $\Pi(t)$ and $\sin 2\pi t$ are orthogonal over $-0.5 < t < 0.5$. [*Note:* $\Pi(t)$ and $\sin 2\pi t$ are not orthogonal over the interval $0 < t < 1$, because $\Pi(t)$ is zero for $t > 0.5$. That is, the integral is $1/\pi$ (which is not zero).]

SA2–7 Use FS to Evaluate PSD   Find the Fourier series and the PSD for the waveform shown in Fig. 2–25. Over the time interval $0 < t < 1$, $v(t)$ is described by e^t.

Solution.   Using Eqs. (2–88) and (2–89), where $T_0 = 1$ and $\omega_0 = 2\pi/T_0 = 2\pi$, we get

$$c_n = \int_0^1 e^t e^{-jn2\pi t}\, dt = \left.\frac{e^{(1-j2\pi n)t}}{1 - j2\pi n}\right|_0^1$$

$$= \frac{e - 1}{1 - j2\pi n} = 1.72 \frac{1}{1 - j6.28n}$$

Thus,

$$v(t) = 1.72 \sum_{n=-\infty}^{\infty} \frac{1}{1 - j6.28n}\, e^{j2\pi nt}$$

Because $v(t)$ is periodic, the PSD consists of delta functions as given by Eq. (2–126), where $f_0 = 1/T_0 = 1$. That is,

$$\mathcal{P}(f) = \sum_{-\infty}^{\infty} |c_n|^2\, \delta(f - nf_0)$$

or

$$\mathcal{P}(f) = \sum_{-\infty}^{\infty} \frac{2.95}{1 + (39.48)n^2}\, \delta(f - n)$$

SA2–8 Property of FS   Let $w(t)$ be a periodic function with period T_0, and let $w(t)$ be *rotationally symmetric*. That is, $w(t) = -w(t \pm T_0/2)$. Prove that $c_n = 0$ for even harmonics.

Solution.   Using Eq. (2–89) and $w(t) = -w(t - T_0/2)$, we obtain

$$c_n = \frac{1}{T_0} \int_0^{T_0/2} w(t)\, e^{-jn\omega_0 t}\, dt - \frac{1}{T_0} \int_{T_0/2}^{T} w(t - T_0/2)\, e^{-jn\omega_0 t}\, dt$$

Now we make a change of variables. Let $t_1 = t$ in the first integral and $t_1 = t - T_0/2$ in the second integral. Then

$$c_n = \frac{1}{T_0} \int_0^{T_0/2} w(t_1)\, e^{-jn\omega_0 t}\, dt_1 - \frac{1}{T_0} \int_0^{T_0/2} w(t_1) e^{-jn\omega_0(t_1 + T_0/2)}\, dt_1$$

$$= \frac{1}{T_0} \int_0^{T_0/2} w(t_1)e^{jn\omega_0 t_1}\left(1 - e^{-jn\pi}\right)dt_1$$

But $(1 - e^{jn\pi}) = 0$ for $n = \ldots -2, 0, 2. \ldots$ Thus, $c_n = 0$ if n is even and $w(t) = -w(t - T_0/2)$. Similarly, it can be shown that $c_n = 0$ if n is even and $w(t) = -w(t + T_0/2)$.

SA2–9 Evaluation of the Inverse FT by Use of the Inverse FFT   Equation (2–184) describes how elements of the fast Fourier transform (FFT) vector can be used to approximate the continuous Fourier transform (CFT):

a. In a similar way, derive a formula that shows how the elements of the *inverse discrete Fourier transform* (IDFT) vector can be used to approximate the *inverse continuous Fourier transform* (ICFT).

b. Given the RC low-pass filter transfer function

$$H(f) = \frac{1}{1 + j(f/f_0)}$$

where $f_0 = 1$ Hz, use the inverse fast Fourier transform (IFFT) of MATLAB to calculate the impulse response $h(t)$.

Solution.   (a) From Eq. (2–30), the ICFT is

$$w(t) = \int_{-\infty}^{\infty} W(f)e^{j2\pi ft}\, df$$

Referring to the discussion leading up to Eq. (2–184), the ICFT is approximated by

$$w(k\Delta t) \approx \Sigma W(n\Delta f)e^{j2\pi n\Delta f k\Delta t}\,\Delta f$$

But $\Delta t = T/N$, $\Delta f = 1/T$, and $f_s = 1/\Delta t$, so

$$w(k\Delta t) \approx N\left[\frac{1}{N}\Sigma W(n\Delta f)e^{j(2\pi/N)nk}\right]\Delta f$$

Using the definition of the IDFT as given by Eq. (2–177), we find that the ICFT is related to the IDFT by

$$w(k\Delta t) \approx f_s x(k) \tag{2–207}$$

where $x(k)$ is the kth element of the N-element IDFT vector. As indicated in the discussion leading up to Eq. (2–184), the elements of the **X** vector are chosen so that the first $N/2$ elements are samples of the positive frequency components of $W(f)$, where $f = n\Delta f$, and the second $N/2$ elements are samples of the negative frequency components of $W(f)$.

(b) Run file *SA 2_9.M* for a plot of $h(t)$ as computed by using the IFFT and Eq. (2–207). Compare this IFFT-computed $h(t)$ with the analytical $h(t)$ that is shown in Fig. 2–15b, where $\tau_0 = RC = 1/(2\pi f_0)$.

PROBLEMS

2–1 For a sinusoidal waveform with a peak value of A and a frequency of f_0, use the time average operator to show that the rms value for this waveform is $A/\sqrt{2}$.

2–2 A function generator produces the periodic voltage waveform shown in Fig. P2–2.

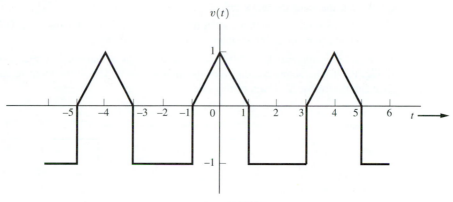

Figure P2–2

(a) Find the value for the dc voltage.
(b) Find the value for the rms voltage.
(c) If this voltage waveform is applied across a 1,000-Ω load, what is the power dissipated in the load?

2–3 The voltage across a load is given by $v(t) = A_0 \cos \omega_0 t$, and the current through the load is a square wave,

$$ i(t) = I_0 \sum_{n=-\infty}^{\infty} \left[\Pi\left(\frac{t - nT_0}{T_0/2}\right) - \Pi\left(\frac{t - nT_0 - (T_0/2)}{T_0/2}\right) \right] $$

where $\omega_0 = 2\pi/T_0$, $T_0 = 1$ sec, $A_0 = 10$ V, and $I_0 = 5$ mA.
(a) Find the expression for the instantaneous power and sketch this result as a function of time.
(b) Find the value of the average power.

2–4 The voltage across a 50-Ω resistive load is the positive portion of a cosine wave. That is,

$$ v(t) = \begin{cases} 10 \cos \omega_0 t, & |t - nT_0| < T_0/4 \\ 0, & t \text{ elsewhere} \end{cases} $$

where n is any integer.
(a) Sketch the voltage and current waveforms.
(b) Evaluate the dc values for the voltage and current.
(c) Find the rms values for the voltage and current.
(d) Find the total average power dissipated in the load.

2–5 For Prob. 2–4, find the energy dissipated in the load during a 1-hr interval if $T_0 = 1$ sec.

2–6 Determine whether each of the following signals is an energy signal or a power signal, and evaluate the normalized energy or power, as appropriate:
(a) $w(t) = \Pi(t/T_0)$.

(b) $w(t) = \Pi(t/T_0) \cos \omega_0 t$.
(c) $w(t) = \cos^2 \omega_0 t$.

2–7 An average-reading power meter is connected to the output circuit of a transmitter. The transmitter output is fed into a 75-Ω resistive load, and the wattmeter reads 67 W.
(a) What is the power in dBm units?
(b) What is the power in dBk units?
(c) What is the value in dBmV units?

2–8 Assume that a waveform with a known rms value V_{rms} is applied across a 50-Ω load. Derive a formula that can be used to compute the dBm value from V_{rms}.

2–9 An amplifier is connected to a 50-Ω load and driven by a sinusoidal current source, as shown in Fig. P2–9. The output resistance of the amplifier is 10 Ω and the input resistance is 2 kΩ. Evaluate the true decibel gain of this circuit.

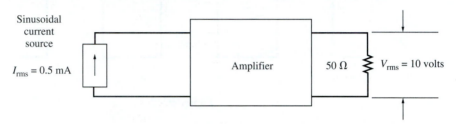

Figure P2–9

2–10 The voltage (rms) across the 300-Ω antenna input terminals of an FM receiver is 3.5 μV.
(a) Find the input power (watts).
(b) Evaluate the input power as measured in decibels below 1 mW (dBm).
(c) What would be the input voltage (in microvolts) for the same input power if the input resistance were 75 Ω instead of 300 Ω?

2–11 What is the value for the phasor that corresponds to the voltage waveform $v(t) = 12 \sin(\omega_0 t - 25°)$, where $\omega_0 = 2000\pi$?

2–12 A signal is $w(t) = 3 \sin(100\pi t - 30°) + 4 \cos(100\pi t)$. Find the corresponding phasor.

2–13 Evaluate the Fourier transform of

$$w(t) = \begin{cases} e^{-at}, & t \geq 1 \\ 0, & t < 0 \end{cases}$$

2–14 Find the spectrum for the waveform $w(t) = e^{-\pi(t/T)^2}$ in terms of the parameter T. What can we say about the width of $w(t)$ and $W(f)$ as T increases?

2–15 Using the convolution property, find the spectrum for

$$w(t) = \sin 2\pi f_1 t \cos 2\pi f_2 t$$

2–16 Find the spectrum (Fourier transform) of the triangle waveform

$$s(t) = \begin{cases} At, & 0 < t < T_0 \\ 0, & t \ \text{elsewhere} \end{cases}$$

in terms of A and T_0.

2–17 Find the spectrum for the waveform shown in Fig. P2–17.

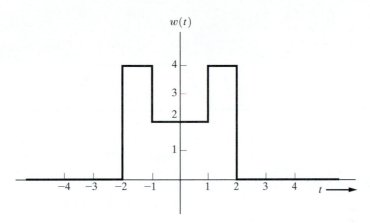

Figure P2–17

 2–18 If $w(t)$ has the Fourier transform

$$W(f) = \frac{j2\pi f}{1 + j2\pi f}$$

find $X(f)$ for the following waveforms:

(a) $x(t) = w(2t + 2)$.

(b) $x(t) = e^{-jt}w(t - 1)$.

(c) $x(t) = 2\dfrac{dw(t)}{dt}$.

(d) $x(t) = w(1 - t)$.

2–19 From Eq. (2–30), find $w(t)$ for $W(f) = A\Pi(f/2B)$, and verify your answer by using the duality property.

2–20 Find the quadrature spectral functions $X(f)$ and $Y(f)$ for the damped sinusoidal waveform

$$w(t) = u(t)e^{-at} \sin \omega_0 t$$

where $u(t)$ is a unit step function, $a > 0$, and $W(f) = X(f) + jY(f)$.

2–21 Derive the spectrum of $w(t) = e^{-|t|/T}$.

 2–22 Find the Fourier transforms for the following waveforms. Plot the waveforms and their magnitude spectra. [*Hint:* Use Eq. (2–184).]

(a) $\Pi\left(\dfrac{t - 3}{4}\right)$.

(b) 2.

(c) $\Lambda\left(\dfrac{t - 5}{5}\right)$.

2–23 By using Eq. (2–184), find the approximate Fourier transform for the following waveform:

$$x(t) = \begin{cases} \sin(2\pi t/512) + \sin(70\pi t/512), & 5 < t < 75 \\ 0, & t \text{ elsewhere.} \end{cases}$$

2–24 Evaluate the spectrum for the trapezoidal pulse shown in Fig. P2–24.

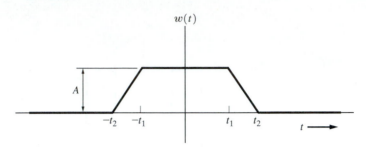

$w(t)$

A

$-t_2$ $-t_1$ t_1 t_2

$t \longrightarrow$

Figure P2–24

2–25 Show that

$$\mathcal{F}^{-1}\{\mathcal{F}[w(t)]\} = w(t)$$

[*Hint:* Use Eq. (2–33).]

2–26 Using the definition of the inverse Fourier transform, show that the value of $w(t)$ at $t = 0$ is equal to the area under $W(f)$. That is, show that

$$w(0) = \int_{-\infty}^{\infty} W(f)\, df$$

2–27 Prove that
  (a) If $w(t)$ is real and an even function of t, $W(f)$ is real.
  (b) If $w(t)$ is real and an odd function of t, $W(f)$ is imaginary.

2–28 Suppose that the spectrum of a waveform as a function of frequency in hertz is

$$W(f) = \frac{1}{2}\, \delta(f - 4) + \frac{1}{2}\, \delta(f + 4) + \frac{j\pi f}{2 + j2\pi f}\, e^{j\pi f}$$

Find the corresponding spectrum as a function of radian frequency, $W(\omega)$.

2–29 The unit impulse can also be defined as

$$\delta(t) = \lim_{a \to \infty} \left[Ka \left(\frac{\sin at}{at} \right) \right]$$

Find the value of K needed, and show that this definition is consistent with those given in the text. Give another example of an ordinary function such that, in the limit of some parameter, the function becomes a Dirac delta function.

2–30 Use $v(t) = ae^{-at}$, $a > 0$, to approximate $\delta(t)$ as $a \to \infty$.
  (a) Plot $v(t)$ for $a = 0.1$, 1, and 10.
  (b) Plot $V(f)$ for $a = 0.1$, 1, and 10.

2–31 Show that

$$\text{sgn}(t) \leftrightarrow \frac{1}{j\pi f}$$

[*Hint:* Use Eq. (2–30) and $\int_0^\infty (\sin x)/x\, dx = \pi/2$ from Appendix A.]

2–32 Show that

$$u(t) \leftrightarrow \tfrac{1}{2}\delta(f) + \frac{1}{j2\pi f}$$

[*Hint:* Use the linearity (superposition) theorem and the result of Prob. 2–31.]

2–33 Show that the sifting property of δ functions, Eq. (2–47), may be generalized to evaluate integrals that involve derivatives of the delta function. That is, show that

$$\int_{-\infty}^{\infty} w(x) \, \delta^{(n)}(x - x_0)dx = (-1)^n w^{(n)}(x_0)$$

where the superscript (n) denotes the nth derivative. (*Hint:* Use integration by parts.)

2–34 Let $x(t) = \Pi\left(\dfrac{t - 0.05}{0.1}\right)$. Plot the spectrum of $x(t)$ using MATLAB with the help of Eqs. (2–59) and (2–60). Check your results by using the FFT and Eq. (2–184).

2–35 If $w(t) = w_1(t)w_2(t)$, show that

$$W(f) = \int_{-\infty}^{\infty} W_1(\lambda)W_2(f - \lambda) \, d\lambda$$

where $W(f) = \mathscr{F}[w(t)]$.

2–36 Show that
(a) $\int_{-\infty}^{t} w(\lambda)d\lambda = w(t) * u(t)$.
(b) $\int_{-\infty}^{t} w(\lambda)d\lambda \leftrightarrow (j2\pi f)^{-1} W(f) + \frac{1}{2}W(0) \, \delta(f)$.
(c) $w(t) * \delta(t - a) = w(t - a)$.

2–37 Show that

$$\frac{dw(t)}{dt} \leftrightarrow (j2\pi f)W(f)$$

[*Hint:* Use Eq. (2–26) and integrate by parts. Assume that $w(t)$ is absolutely integrable.]

2–38 As discussed in Example 2–8, show that

$$\frac{1}{T} \, \Pi\left(\frac{t}{T}\right) * \Pi\left(\frac{t}{T}\right) = \Lambda\left(\frac{t}{T}\right)$$

2–39 Given the waveform $w(t) = A\Pi(t/T) \sin \omega_0 t$, find the spectrum of $w(t)$ by using the multiplication theorem as discussed in Example 2–9.

2–40 Evaluate the following integrals.

(a) $\displaystyle\int_{-\infty}^{\infty} \frac{\sin 4\lambda}{4\lambda} \, \delta(t - \lambda) \, d\lambda$.

(b) $\displaystyle\int_{-\infty}^{\infty} (\lambda^3 - 1) \, \delta(2 - \lambda) \, d\lambda$.

2–41 Prove that

$$M(f) * \delta(f - f_0) = M(f - f_0)$$

2–42 Evaluate $y(t) = w_1(t) * w_2(t)$, where

$$w_1(t) = \begin{cases} 1, & |t| < T_0 \\ 0, & t \text{ elsewhere} \end{cases}$$

and

$$w_2(t) = \begin{cases} [1 - 2|t|], & |t| < \frac{1}{2}T_0 \\ 0, & t \text{ elsewhere} \end{cases}$$

2–43 Given $w(t) = 5 + 12 \cos \omega_0 t$, where $f_0 = 10$ Hz, find
   (a) $R_w(\tau)$.
   (b) $\mathcal{P}_w(f)$.

2–44 Given the waveform

$$w(t) = A_1 \cos(\omega_1 t + \theta_1) + A_2 \cos(\omega_2 t + \theta_2)$$

where A_1, A_2, ω_1, ω_2, θ_1, and θ_2 are constants,
   (a) Find the autocorrelation for $w(t)$ as a function of the constants.
   (b) Find the PSD function for $w(t)$.
   (c) Sketch the PSD for the case of $\omega_1 \neq \omega_2$.
   (d) Sketch the PSD for the case of $\omega_1 = \omega_2$ and $\theta_1 = \theta_2 + 90°$.
   (e) Sketch the PSD for the case of $\omega_1 = \omega_2$ and $\theta_1 = \theta_2$.

2–45 Given the periodic voltage waveform shown in Fig. P2–45,
   (a) Find the dc value for this waveform.
   (b) Find the rms value for this waveform.
   (c) Find the complex exponential Fourier series.
   (d) Find the voltage spectrum for this waveform.

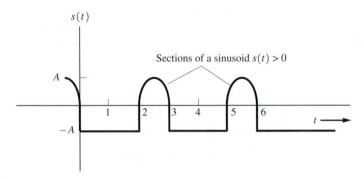

Figure P2–45

2–46 Determine if $s_1(t)$ and $s_2(t)$ are orthogonal over the interval $(-\frac{5}{2}T_2 < t < \frac{5}{2}T_2)$, where $s_1(t) = A_1 \cos(\omega_1 t + \varphi_1)$, $s_2(t) = A_2 \cos(\omega_2 t + \varphi_2)$, and $\omega_2 = 2\pi/T_2$ for the following cases.
   (a) $\omega_1 = \omega_2$ and $\varphi_1 = \varphi_2$.
   (b) $\omega_1 = \omega_2$ and $\varphi_1 = \varphi_2 + \pi/2$.
   (c) $\omega_1 = \omega_2$ and $\varphi_1 = \varphi_2 + \pi$.
   (d) $\omega_1 = 2\omega_2$ and $\varphi_1 = \varphi_2$.
   (e) $\omega_1 = \frac{4}{5}\omega_2$ and $\varphi_1 = \varphi_2$.
   (f) $\omega_1 = \pi\omega_2$ and $\varphi_1 = \varphi_2$.

2–47 Let $s(t) = A_1 \cos(\omega_1 t + \varphi_1) + A_2 \cos(\omega_2 t + \varphi_2)$. Determine the rms value of $s(t)$ in terms of A_1 and A_2 for the following cases.
   (a) $\omega_1 = \omega_2$ and $\varphi_1 = \varphi_2$.
   (b) $\omega_1 = \omega_2$ and $\varphi_1 = \varphi_2 + \pi/2$.
   (c) $\omega_1 = \omega_2$ and $\varphi_1 = \varphi_2 + \pi$.
   (d) $\omega_1 = 2\omega_2$ and φ_1 and φ_2.
   (e) $\omega_1 = 2\omega_2$ and $\varphi_1 = \varphi_2 + \pi$.

2–48 Show that

$$\sum_{k=-\infty}^{\infty} \delta(t - kT_0) \leftrightarrow f_0 \sum_{k=-\infty}^{\infty} \delta(f - nf_0)$$

where $f_0 = 1/T_0$. [*Hint:* Expand $\sum_{k=-\infty}^{\infty} \delta(t - kT_0)$ into a Fourier series and then take the Fourier transform.]

2–49 Three functions are shown in Fig. P2–49.

(a) Show that these functions are orthogonal over the interval $(-4, 4)$.

(b) Find the corresponding orthonormal set of functions.

(c) Expand the waveform

$$w(t) = \begin{cases} 1, & 0 \le t \le 4 \\ 0, & t \text{ elsewhere} \end{cases}$$

into an orthonormal series by using the orthonormal set found in part (b).

(d) Evaluate the mean square error for the orthogonal series obtained in part (c) by evaluating

$$\varepsilon = \int_{-4}^{4} \left[w(t) - \sum_{j=1}^{3} a_j \varphi_j(t) \right]^2 dt$$

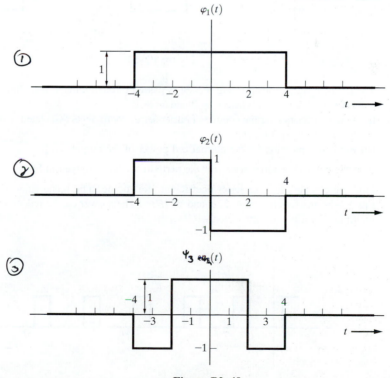

Figure P2–49

(e) Repeat parts (c) and (d) for the waveform

$$w(t) = \begin{cases} \cos\left(\frac{1}{4}\pi t\right), & -4 \leq t \leq 4 \\ 0, & t \text{ elsewhere} \end{cases}$$

Are the three orthonormal functions a complete orthonormal set?

2–50 Show that the quadrature Fourier series basis functions $\cos(n\omega_0 t)$ and $\sin(n\omega_0 t)$, as given in Eq. (2–95), are orthogonal over the interval $a < t < a + T_0$, where $\omega_0 = 2\pi/T_0$.

2–51 Find expressions for the complex Fourier series coefficients that represent the waveform shown in Fig. P2–51.

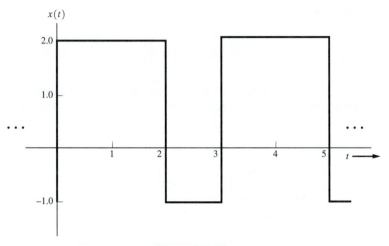

Figure P2–51

2–52 The periodic signal shown in Fig. P2–51 is passed through a linear filter having the impulse response $h(t) = e^{-\alpha t}u(t)$, where $t > 0$ and $\alpha > 0$.
   (a) Find expressions for the complex Fourier series coefficients associated with the output waveform $y(t) = x(t) * h(t)$.
   (b) Find an expression for the normalized power of the output, $y(t)$.

2–53 Find the complex Fourier series for the periodic waveform given in Figure P2–2.

2–54 Find the complex Fourier series coefficients for the periodic rectangular waveform shown in Fig. P2–54 as a function of A, T, b, and τ_0. [*Hint:* The answer can be reduced to a $(\sin x)/x$ form multiplied by a phase-shift factor, $e^{j\theta_n(\tau_0)}$.]

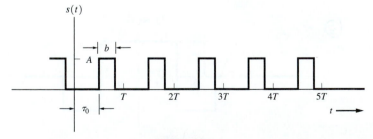

Figure P2–54

2–55 For the waveform shown in Fig. P2–55, find
  (a) The complex Fourier series.
  (b) The quadrature Fourier series.

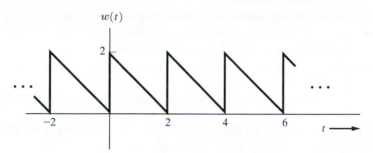

Figure P2-55

2–56 Given a periodic waveform $s(t) = \sum_{n=-\infty}^{\infty} p(t - nT_0)$, where

$$p(t) = \begin{cases} At, & 0 < t < T \\ 0, & t \text{ elsewhere} \end{cases}$$

and $T \leq T_0$,
  (a) Find the c_n Fourier series coefficients.
  (b) Find the $\{x_n, y_n\}$ Fourier series coefficients.
  (c) Find the $\{D_n, \varphi_n\}$ Fourier series coefficients.

2–57 Prove that the polar form of the Fourier series, Eq. (2–103), can be obtained by rearranging the terms in the complex Fourier series, Eq. (2–88).

2–58 Prove that Eq. (2–93) is correct.

2–59 Let two complex numbers c_1 and c_2 be represented by $c_1 = x_1 + jy_1$ and $c_2 = x_2 + jy_2$, where x_1, x_2, y_1 and y_2 are real numbers. Show that $\text{Re}\{\cdot\}$ is a linear operator by demonstrating that

$$\text{Re}\{c_1 + c_2\} = \text{Re}\{c_1\} + \text{Re}\{c_2\}$$

2–60 Assume that $y(t) = s_1(t) + 2s_2(t)$, where $s_1(t)$ is given by Fig. P2–45 and $s_2(t)$ is given by Fig. P2–54. Let $T = 3$, $b = 1.5$, and $\tau_0 = 0$. Find the complex Fourier coefficients $\{c_n\}$ for $y(t)$.

2–61 Evaluate the PSD for the waveform shown in Fig. P2–2.

2–62 Assume that $v(t)$ is a triangular waveform, as shown in Fig. P2–62.

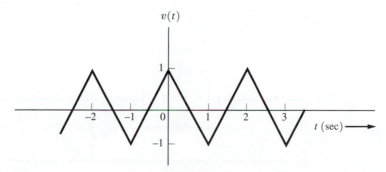

Figure P2–62

(a) Find the complex Fourier series for $v(t)$.
(b) Calculate the normalized average power.
(c) Calculate and plot the voltage spectrum.
(d) Calculate and plot the PSD.

2-63 Let any complex number be represented by $c = x + jy$, where x and y are real numbers. If c^* denotes the complex conjugate of c, show that

(a) $\text{Re}\{c\} = \frac{1}{2}c + \frac{1}{2}c^*$.

(b) $\text{Im}\{c\} = \dfrac{1}{2j}c - \dfrac{1}{2j}c^*$.

Note that when $c = e^{jz}$, part (a) gives the definition of cos z and part (b) gives the definition of sin z.

2-64 Calculate and plot the PSD for the half-wave rectified sinusoid described in Prob. 2–4.

2-65 The basic definitions for sine and cosine waves are

$$\sin z_1 \triangleq \frac{e^{jz_1} - e^{-jz_1}}{2j} \quad \text{and} \quad \cos z_2 \triangleq \frac{e^{jz_2} + e^{-jz_2}}{2}.$$

Show that
(a) $\cos z_1 \cos z_2 = \frac{1}{2}\cos(z_1 - z_2) + \frac{1}{2}\cos(z_1 + z_2)$.
(b) $\sin z_1 \sin z_2 = \frac{1}{2}\cos(z_1 - z_2) - \frac{1}{2}\cos(z_1 + z_2)$.
(c) $\sin z_1 \cos z_2 = \frac{1}{2}\sin(z_1 - z_2) + \frac{1}{2}\sin(z_1 + z_2)$.

2-66 Let two complex numbers be respectively represented by $c_1 = x_1 + jy_1$ and $c_2 = x_2 + jy_2$, where x_1, x_2, y_1, and y_2 are real numbers. Show that

$$\text{Re}\{c_1\}\,\text{Re}\{c_2\} = \frac{1}{2}\,\text{Re}\{c_1 c_2^*\} + \frac{1}{2}\,\text{Re}\{c_1 c_2\}$$

Note that this is a generalization of the cos z_1 cos z_2 identity of Prob. 2–65, where, for the cos z_1 cos z_2 identity, $c_1 = e^{jz_1}$ and $c_2 = e^{jz_2}$.

2-67 Prove that the Fourier transform is a linear operator. That is, show that

$$\mathcal{F}[ax(t) + by(t)] = a\mathcal{F}[x(t)] + b\mathcal{F}[y(t)]$$

2-68 Plot the amplitude and phase response for the transfer function

$$H(f) = \frac{j10f}{5 + jf}$$

2-69 Given the filter shown in Fig. P2–69, where $w_1(t)$ and $w_2(t)$ are voltage waveforms,
(a) Find the transfer function.
(b) Plot the magnitude and phase response.
(c) Find the power transfer function.
(d) Plot the power transfer function.

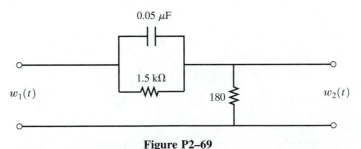

Figure P2–69

2–70 A signal with a PSD of

$$\mathcal{P}_x(f) = \frac{2}{(1/4\pi)^2 + f^2}$$

is applied to the network shown in Fig. P2–70.
(a) Find the PSD for $y(t)$.
(b) Find the average normalized power for $y(t)$.

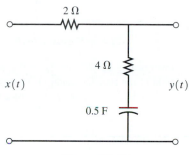

Figure P2–70

2–71 A signal $x(t)$ has a PSD

$$\mathcal{P}_x(f) = \frac{K}{[1 + (2\pi f/B)^2]^2}$$

where $K > 0$ and $B > 0$.
(a) Find the 3-dB bandwidth in terms of B.
(b) Find the equivalent noise bandwidth in terms of B.

2–72 The signal $x(t) = e^{-400\,\pi t}u(t)$ is applied to a brick-wall low-pass filter whose transfer function is

$$H(f) = \begin{cases} 1, & |f| \le B \\ 0, & |f| > B \end{cases}$$

Find the value of B such that the filter passes one-half the energy of $x(t)$.

2–73 Show that the average normalized power of a waveform can be found by evaluating the auto-correlation $R_w(\tau)$ at $\tau = 0$. That is, $P = R_w(0)$.
[*Hint:* See Eqs. (2–69) and (2–70).]

2–74 The signal $x(t) = 0.5 + 1.5 \cos[(\frac{2}{3})\pi t] + 0.5 \sin[(\frac{2}{3})\pi t]$ V is passed through an RC low-pass filter (see Fig. 2–15a) where $R = 1\ \Omega$ and $C = 1$ F.
(a) What is the input PSD, $\mathcal{P}_x(f)$?
(b) What is the output PSD, $\mathcal{P}_y(f)$?
(c) What is the average normalized output power, P_y?

2–75 The input to the RC low-pass filter shown in Fig. 2–15 is

$$x(t) = 0.5 + 1.5 \cos \omega_x t + 0.5 \sin \omega_x t$$

Assume that the cutoff frequency is $f_0 = 1.5f_x$.
(a) Find the input PSD, $\mathcal{P}_x(f)$.
(b) Find the output PSD, $\mathcal{P}_y(f)$
(c) Find the normalized average power of the output, $y(t)$.

2–76 Using MATLAB, plot the frequency magnitude response and phase response for the low-pass filter shown in Fig. P2–76, where $R_1 = 7.5$ kΩ, $R_2 = 15$ kΩ, and $C = 0.1$ μF.

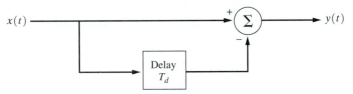

Figure P2–76

2–77 A comb filter is shown in Fig. P2–77. Let $T_d = 0.1$.
   (a) Plot the magnitude of the transfer function for this filter.
   (b) If the input is $x(t) = \Pi(t/T)$, where $T = 1$, plot the spectrum of the output $|Y(f)|$.

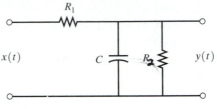

Figure P2–77

2–78 A signal $x(t) = \Pi(t - 0.5)$ passes through a filter that has the transfer function $H(f) = \Pi(f/B)$. Plot the output waveform when
   (a) $B = 0.6$ Hz.
   (b) $B = 1$ Hz.
   (c) $B = 50$ Hz.

2–79 Examine the distortion effect of an RC low-pass filter. Assume that a unity-amplitude periodic square wave with a 50% duty cycle is present at the filter input and that the filter has a 3-dB bandwidth of 1,500 Hz. Using a computer or a programmable calculator, find and plot the output waveshape if the square wave has a frequency of
   (a) 300 Hz.
   (b) 500 Hz.
   (c) 1,000 Hz.
   (*Hint:* Represent the square wave with a Fourier series.)

2–80 Given that the PSD of a signal is flat [i.e., $\mathcal{P}_s(f) = 1$], design an RC low-pass filter that will attenuate this signal by 20 dB at 15 kHz. That is, find the value for the RC of Fig. 2–15 such that the design specifications are satisfied.

2–81 The bandwidth of $g(t) = e^{-0.1t}$ is approximately 0.5 Hz; thus, the signal can be sampled with a sampling frequency of $f_s = 1$ Hz without significant aliasing. Take samples a_n over the time interval (0, 14). Use the sampling theorem, Eq. (2–158), to reconstruct the signal. Plot and compare the reconstructed signal with the original signal. Do they match? What happens when the sampling frequency is reduced?

2–82 A waveform, $20 + 20 \sin(500t + 30°)$, is to be sampled periodically and reproduced from these sample values.
   (a) Find the maximum allowable time interval between sample values.
   (b) How many sample values need to be stored in order to reproduce 1 sec of this waveform?

2–83 Using a computer program, calculate the DFT of a rectangular pulse, $\Pi(t)$. Take five samples of the pulse and pad it with 59 zeros so that a 64-point FFT algorithm can be used. Sketch the resulting magnitude spectrum. Compare this result with the actual spectrum for the pulse. Try other combinations of the number of pulse samples and zero-pads to see how the resulting FFT changes.

2–84 Using the DFT, compute and plot the spectrum of $\Lambda(t)$. Check your results against those given in Fig. 2–6c.

2–85 Using the DFT, compute and plot $|W(f)|$ for the pulse shown in Fig. P2–24, where $A = 1$, $t_1 = 1s$, and $t_2 = 2$ s.

2–86 Let a certain waveform be given by

$$w(t) = 4\sin(2\pi f_1 t + 30°) + 2\cos(2\pi f_2 t - 10°),$$

where $f_1 = 10$ Hz and $f_2 = 25$ Hz,
(a) Using the DFT, compute and plot $|W(f)|$ and $\theta(f)$
(b) Let $\mathcal{P}_w(f)$ denote the PSD of $w(t)$. Using the DFT, compute and plot $\mathcal{P}_w(f)$.
(c) Check your computed results obtained in parts (a) and (b) against known correct results that you have evaluated by analytical methods.

2–87 Using the DFT, compute and plot $|S(f)|$ for the periodic signal shown in Fig. P2–45, where $A = 5$.

2–88 The transfer function of a raised cosine-rolloff filter is

$$H(f) = \begin{cases} 0.5[1 + \cos(0.5\pi f/f_0)], & |f| \le 2f_0 \\ 0, & f \text{ elsewhere} \end{cases}$$

Let $f_0 = 1$ Hz. Using the IFFT, compute the impulse response $h(t)$ for this filter. Compare your computed results with those shown in Fig. 3–26b for the case of $r = 1$.

2–89 Given the low-pass filter shown in Fig. 2–15,
(a) Find the equivalent bandwidth in terms of R and C.
(b) Find the first zero-crossing (null) bandwidth of the filter.
(c) Find the absolute bandwidth of the filter.

2–90 Assume that the PSD of a signal is given by

$$\mathcal{P}_s(f) = \left[\frac{\sin(\pi f/B_n)}{\pi f/B_n} \right]^2$$

where B_n is the null bandwidth. Find the expression for the equivalent bandwidth in terms of the null bandwidth.

2–91 Table 2–5 gives the bandwidths of a BPSK signal in terms of six different definitions. Using these definitions and Eq. (2–201), show that the results given in the table are correct.

2–92 Given a triangular pulse signal

$$s(t) = \Lambda(t/T_0)$$

(a) Find the absolute bandwidth of the signal.
(b) Find the 3-dB bandwidth in terms of T_0.
(c) Find the equivalent bandwidth in terms of T_0.
(d) Find the zero-crossing bandwidth in terms of T_0.

CHAPTER 3

- Analog-to-digital signaling (Pulse code modulation and delta modulation)

- Binary and multilevel digitals signals

- Spectra and bandwidths of digital signals

- Prevention of intersymbol interference

- Time division multiplexing

- Packet transmission

BASEBAND PULSE AND DIGITAL SIGNALING

3–1 INTRODUCTION

This chapter shows how to encode analog waveforms (from analog sources) into baseband digital signals. As we will see, the digital approximation to the analog signal can be made very precise if we wish. In addition, we will learn how to process the digital baseband signals so that their bandwidth is minimized.

Digital signaling is popular because of the low cost of digital circuits and the flexibility of the digital approach. This flexibility arises because digital data from digital sources may be merged with digitized data derived from analog sources to provide a general-purpose communication system.

The signals involved in the analog-to-digital conversion problem are *baseband* signals. We also realize that *bandpass* digital communication signals are produced by using baseband digital signals to modulate a carrier, as described in Chapter 1.

The following are the four main goals of this chapter:

- To study how analog waveforms can be converted to digital waveforms. The most popular technique is called *pulse code modulation* (PCM).
- To learn how to compute the spectrum for digital signals.
- To examine how the filtering of pulse signals affects our ability to recover the digital information at the receiver. This filtering can produce what is called *intersymbol interference* (ISI) in the recovered data signal.
- To study how we can *multiplex* (combine) data from several digital bit streams into one high-speed digital stream for transmission over a digital system. One such technique, called *time-division multiplexing* (TDM), will be studied in this chapter.[†]

Another very important problem in digital communication systems is the effect of noise, which may cause the digital receiver to produce some bit errors at the output. This problem will be studied in Chapter 7 since it involves the use of statistical concepts that are emphasized in the second part of this book.

3–2 PULSE AMPLITUDE MODULATION

Pulse amplitude modulation (PAM) is an engineering term that is used to describe the conversion of the analog signal to a pulse-type signal in which the amplitude of the pulse denotes the analog information. PAM is studied first because the analog-to-PAM conversion process is the first step in converting an analog waveform to a PCM (digital) signal. In a few applications the PAM signal is used directly and conversion to PCM is not required.

The sampling theorem, studied in Chapter 2, provides a way to reproduce an analog waveform by using sample values of that waveform and $(\sin x)/x$ orthogonal functions. The purpose of PAM signaling is to provide *another* waveform that looks like pulses, yet contains the information that was present in the analog waveform. Because we are using pulses, we would expect the bandwidth of the PAM waveform to be wider than that of the analog waveform. However, the pulses are more practical to use in digital systems. We will see, that the pulse rate, f_s, for PAM is the same as that required by the sampling theorem, namely, $f_s \geq 2B$, where B is the highest frequency in the analog waveform and $2B$ is called the *Nyquist rate*.

There are two classes of PAM signals: PAM that uses *natural sampling* (gating) and PAM that uses *instantaneous sampling* to produce a flat-top pulse. These signals are illustrated in Figs. 3–1 and 3–5, respectively. The flat-top type is more useful for conversion to PCM; however, the naturally sampled type is easier to generate and is used in other applications.

Natural Sampling (Gating)

DEFINITION.   If $w(t)$ is an analog waveform bandlimited to B hertz, the *PAM* signal that uses *natural sampling* (gating) is

$$w_s(t) = w(t)\,s(t) \tag{3–1}$$

[†] Other techniques, including frequency division multiplexing, and code division multiplexing, are covered in Chapters 5 and 8.

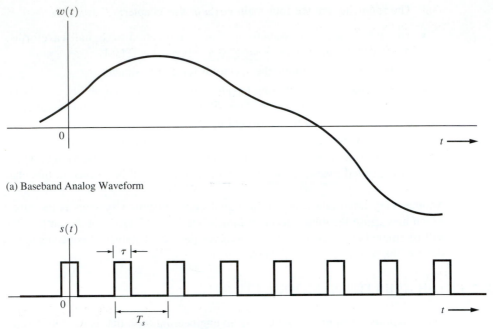

(a) Baseband Analog Waveform

(b) Switching Waveform with Duty Cycle $d = \tau/T_s = 1/3$

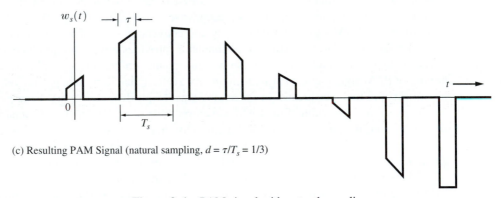

(c) Resulting PAM Signal (natural sampling, $d = \tau/T_s = 1/3$)

Figure 3–1   PAM signal with natural sampling.

where

$$s(t) = \sum_{k=-\infty}^{\infty} \Pi\left(\frac{t - kT_s}{\tau}\right) \tag{3-2}$$

is a rectangular wave switching waveform and $f_s = 1/T_s \geq 2B$.

THEOREM.   *The spectrum for a naturally sampled PAM signal is*

$$W_s(f) = \mathscr{F}[w_s(t)] = d \sum_{n=-\infty}^{\infty} \frac{\sin \pi n d}{\pi n d} W(f - nf_s) \tag{3-3}$$

where $f_s = 1/T_s$, $\omega_s = 2\pi f_s$, the duty cycle of $s(t)$ is $d = \tau/T_s$, and $W(f) = \mathcal{F}[w(t)]$ is the spectrum of the original unsampled waveform.

Proof. Taking the Fourier transform of Eq. (3–1), we get

$$W_s(f) = W(f) * S(f) \tag{3–4}$$

$s(t)$ may be represented by the Fourier series

$$s(t) = \sum_{n=-\infty}^{\infty} c_n e^{jn\omega_s t} \tag{3–5a}$$

where

$$c_n = d \frac{\sin n\pi d}{n\pi d} \tag{3–5b}$$

Since $s(t)$ is periodic Eq. (2–109) may be used to get the spectrum:

$$S(f) = \mathcal{F}[s(t)] = \sum_{n=-\infty}^{\infty} c_n \delta(f - nf_s) \tag{3–6}$$

Then Eq. (3–4) becomes

$$W_s(f) = W(f) * \left(\sum_{n=-\infty}^{\infty} c_n \delta(f - nf_s) \right) = \sum_{n=-\infty}^{\infty} c_n W(f) * \delta(f - nf_s)$$

or

$$W_s(f) = \sum_{n=-\infty}^{\infty} c_n W(f - nf_s) \tag{3–7}$$

This equation becomes Eq. (3–3) upon substituting Eq. (3–5b).

The PAM waveform with natural sampling is relatively easy to generate, since it only requires the use of an analog switch that is readily available in CMOS hardware (e.g., the 4016-quad bilateral switch). This hardware is shown in Fig. 3–2, where the associated waveforms $w(t)$, $s(t)$, and $w_s(t)$ are as illustrated in Fig. 3–1.

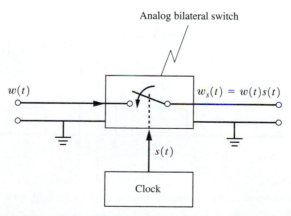

Figure 3–2   Generation of PAM with natural sampling (gating).

The spectrum of the PAM signal with natural sampling is given by Eq. (3–3) as a function of the spectrum of the analog input waveform. This relationship is illustrated in Fig. 3–3 for the case of an input waveform that has a rectangular spectrum, where the duty cycle of the switching waveform is $d = \tau/T_s = \frac{1}{3}$ and the sampling rate is $f_s = 4B$. As expected, the spectrum of the input analog waveform is repeated at harmonics of the sampling frequency. This is similar to the spectrum for impulse sampling that was studied in Sec. 2–7—for example, compare Fig. 2–18 with Fig. 3–3. For this example, where $d = \frac{1}{3}$, the PAM spectrum is zero for $\pm 3f_s$, $\pm 6f_s$, and so on, because the spectrum in these harmonic bands is nulled out by the $(\sin x)/x$ function. From the figure, one sees that the bandwidth of the PAM signal is much larger than the bandwidth of the original analog signal. In fact, for the example illustrated in Fig. 3–3b, the null bandwidth for the envelope of the PAM signal is $3f_s = 12B$; that is, the null bandwidth of this PAM signal is 12 times the bandwidth of the analog signal.

At the receiver, the original analog waveform, $w(t)$, can be recovered from the PAM signal, $w_s(t)$, by passing the PAM signal through a low-pass filter where the cutoff frequency is $B < f_{cutoff} < f_s - B$. This is seen by comparing Fig. 3–3b with Fig. 3–3a. Because the spectrum out of the low-pass filter would have the same shape as that of the original analog signal shown in Fig. 3–3a, the waveshape out of the low-pass filter would be identical

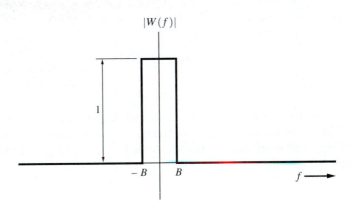

(a) Magnitude Spectrum of Input Analog Waveform

$$|W_s(f)| = \sum_{n=-\infty}^{\infty} \left\{ \left(d \left| \frac{\sin(\pi nd)}{\pi nd} \right| \right) |W(f - nf_s)| \right\}$$

$$\left(d \left| \frac{\sin(\pi nf)}{\pi nf} \right| \right)$$

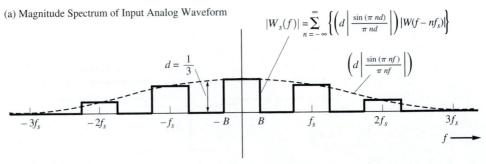

(b) Magnitude Spectrum of PAM (natural sampling) with $d = 1/3$ and $f_s = 4$ B

Figure 3–3   Spectrum of a PAM waveform with natural sampling.

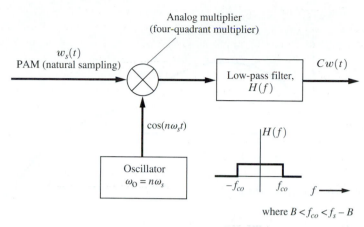

Figure 3–4   Demodulation of a PAM signal (naturally sampled).

to that of the original analog signal, except for a gain factor of d (which could be compensated for by using an amplifier). From Fig. 3–3b, the spectrum out of the low-pass filter will have the same shape as the spectrum of the original analog signal only when $f_s \geq 2B$, because otherwise spectral components in harmonic bands (of f_s) would overlap. This is another illustration of the Nyquist sampling rate requirement. If the analog signal is *undersampled* ($f_s < 2B$), the effect of spectral overlapping is called *aliasing*. This results in a recovered analog signal that is distorted compared to the original waveform. In practice, physical signals are usually considered to be time limited, so that (as we found in Chapter 2) they cannot be absolutely bandlimited. Consequently, there will be some aliasing in a PAM signal. Usually, we prefilter the analog signal before it is introduced to the PAM circuit so we do not have to worry about this problem; however, the effect of aliasing noise has been studied [Spilker, 1977].

It can also be shown (see Prob. 3-4) that the analog waveform may be recovered from the PAM signal by using *product* detection, as shown in Fig. 3–4. Here the PAM signal is multiplied with a sinusoidal signal of frequency $\omega_o = n\omega_s$. This shifts the frequency band of the PAM signal that was centered about nf_s to baseband (i.e., $f = 0$) at the multiplier output. We will study the product detector in Chapter 4. For $n = 0$, this is identical to low-pass filtering, just discussed. Of course, you might ask, Why do you need to go to the trouble of using a product detector when a simple low-pass filter will work? The answer is, because of noise on the PAM signal. Noise due to power supply hum or noise due to mechanical circuit vibration might fall in the band corresponding to $n = 0$, and other bands might be relatively noise free. In this case, a product detector might be used to get around the noise problem.

Instantaneous Sampling (Flat-Top PAM)

Analog waveforms may also be converted to pulse signaling by the use of *flat-top* signaling with instantaneous sampling, as shown in Fig. 3–5. This is another generalization of the impulse train sampling technique that was studied in Sec. 2–7.

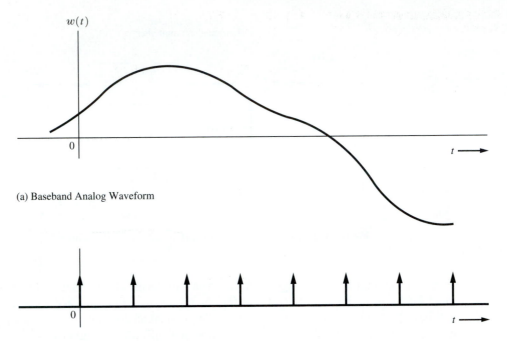

$w(t)$

0

$t \longrightarrow$

(a) Baseband Analog Waveform

0

$t \longrightarrow$

(b) Impulse Train Sampling Waveform

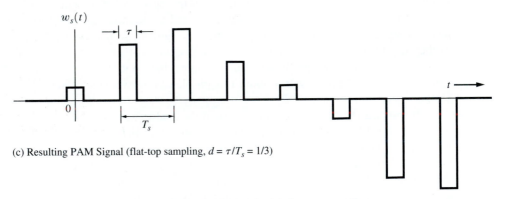

$w_s(t)$

τ

0

$t \longrightarrow$

T_s

(c) Resulting PAM Signal (flat-top sampling, $d = \tau/T_s = 1/3$)

Figure 3–5   PAM signal with flat-top sampling.

DEFINITION.   If $w(t)$ is an analog waveform bandlimited to B hertz, the *instantaneous sampled* PAM signal is given by

$$w_s(t) = \sum_{k=-\infty}^{\infty} w(kT_s) h(t - kT_s) \qquad (3\text{–}8)$$

where $h(t)$ denotes the sampling-pulse shape and, for flat-top sampling, the pulse shape is

$$h(t) = \Pi\left(\frac{t}{\tau}\right) = \begin{cases} 1, & |t| < \tau/2 \\ 0, & |t| > \tau/2 \end{cases} \tag{3–9}$$

where $\tau \le T_s = 1/f_s$ and $f_s \ge 2B$.

THEOREM. *The spectrum for a flat-top PAM signal is*

$$W_s(f) = \frac{1}{T_s}H(f) \sum_{k=-\infty}^{\infty} W(f - kf_s) \tag{3–10}$$

where

$$H(f) = \mathcal{F}[h(t)] = \tau\left(\frac{\sin \pi\tau f}{\pi\tau f}\right) \tag{3–11}$$

This type of PAM signal is said to consist of instantaneous samples, since $w(t)$ is sampled at $t = kT_s$ and the sample values $w(kT_s)$ determine the amplitude of the flat-top rectangular pulses, as demonstrated in Fig. 3–5c. The flat-top PAM signal could be generated by using a sample-and-hold type of electronic circuit.

Another pulse shape, rather than the rectangular shape, could be used in Eq. (3–8), but in this case the resulting PAM waveform would not be flat topped. Note that if the $h(t)$ is of the $(\sin x)/x$ type with overlapping pulses, then Eq. (3–8) becomes identical to the sampling theorem of Eq. (2–158), and this sampled signal becomes identical to the original unsampled analog waveform, $w(t)$.

Proof. The spectrum for flat-top PAM can be obtained by taking the Fourier transform of Eq. (3–8). First we rewrite that equation, using a more convenient form involving the convolution operation:

$$w_s(t) = \sum_k w(kT_s)h(t) * \delta(t - kT_s)$$

$$= h(t) * \sum_k w(kT_s)\delta(t - kT_s)$$

Hence,

$$w_s(t) = h(t) * \left[w(t) \sum_k \delta(t - kT_s)\right]$$

The spectrum is

$$W_s(f) = H(f)\left[W(f) * \sum_k e^{-j2\pi f kT_s}\right] \tag{3–12}$$

But the sum of the exponential functions is equivalent to a Fourier series expansion (in the frequency domain), where the periodic function is an impulse train. That is,

$$\frac{1}{T_s} \sum_k \delta(f - kf_s) = \frac{1}{T_s} \sum_{n=-\infty}^{\infty} c_n e^{j(2\pi nT_s)f}, \tag{3–13a}$$

where

$$c_n = \frac{1}{f_s} \int_{-f_s/2}^{f_s/2} \left[\sum_k \delta(f - kf_s) \right] e^{-j2\pi nT_s f} \, df = \frac{1}{f_s} \tag{3-13b}$$

Using Eq. (3–13a), we find that Eq. (3–12) becomes

$$W_s(f) = H(f) \left[W(f) * \frac{1}{T_s} \sum_k \delta(f - kf_s) \right]$$

$$= \frac{1}{T_s} H(f) \left[\sum_k W(f) * \delta(f - kf_s) \right]$$

which reduces to Eq. (3–10).

The spectrum of the flat-top PAM signal is illustrated in Fig. 3–6 for the case of an analog input waveform that has a rectangular spectrum. The analog signal may be recovered from the flat-top PAM signal by the use of a low-pass filter. However, there is some high-frequency loss in the recovered analog waveform due to the filtering effect, $H(f)$, caused by the flat-top pulse shape. This loss, if significant, can be reduced by decreasing τ

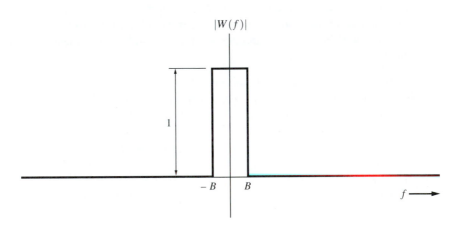

(a) Magnitude Spectrum of Input Analog Waveform

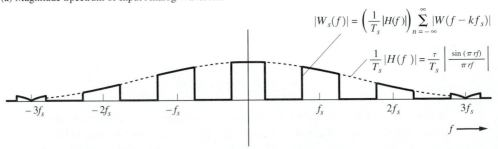

(b) Magnitude Spectrum of PAM (flat-top sampling), $\tau/T_s = 1/3$ and $f_s = 4B$

Figure 3–6   Spectrum of a PAM waveform with flat-top sampling.

or by using some additional gain at the high frequencies in the low-pass filter transfer function. In that case, the low-pass filter would be called an *equalization* filter and have a transfer function of $1/H(f)$. The pulse width τ is also called the *aperture* since τ/T_s determines the gain of the recovered analog signal, which is small if τ is small relative to T_s. It is also possible to use product detection similar to that shown in Fig. 3–4, except that now some prefilter might be needed before the multiplier (to make the spectrum flat in a band centered on $f = nf_s$), to compensate for the spectral loss due to the aperture effect. Once again, f_s needs to be selected so that $f_s \geq 2B$ in order to prevent aliasing.

The transmission of either naturally or instantaneously sampled PAM over a channel requires a very wide frequency response because of the narrow pulse width, which imposes stringent requirements on the magnitude and phase response of the channel. The bandwidth required is much larger than that of the original analog signal, and the noise performance of the PAM system can never be better than that achieved by transmitting the analog signal directly. Consequently, PAM is not very good for long-distance transmission. It does provide a means for converting an analog signal to a PCM signal (as discussed in the next section). PAM also provides a means for breaking a signal into time slots so that multiple PAM signals carrying information from different sources can be interleaved to transmit all of the information over a single channel. This is called time-division multiplexing and will be studied in Sec. 3–9.

3–3 PULSE CODE MODULATION

> **DEFINITION.** *Pulse code modulation* (PCM) is essentially analog-to-digital conversion of a special type where the information contained in the instantaneous samples of an analog signal is represented by digital words in a *serial bit stream*.

If we assume that each of the digital words has n binary digits, there are $M = 2^n$ unique code words that are possible, each code word corresponding to a certain amplitude level. However, each sample value from the analog signal can be any one of an infinite number of levels, so that the digital word that represents the amplitude closest to the actual sampled value is used. This is called *quantizing*. That is, instead of using the exact sample value of the analog waveform $w(kT_s)$, the sample is replaced by the closest allowed value, where there are M allowed values, each corresponding to one of the code words. Other popular types of analog-to-digital conversion, such as delta modulation (DM) and differential pulse code modulation (DPCM), are discussed in later sections.

PCM is very popular because of the many advantages it offers, including the following:

- Relatively inexpensive digital circuitry may be used extensively in the system.
- PCM signals derived from all types of analog sources (audio, video, etc.) may be merged with data signals (e.g., from digital computers) and transmitted over a common high-speed digital communication system. This merging is called time-division multiplexing and is discussed in detail in a later section.

- In long-distance digital telephone systems requiring repeaters, a *clean* PCM waveform can be regenerated at the output of each repeater, where the input consists of a noisy PCM waveform. However, the noise at the input may cause bit errors in the regenerated PCM output signal.
- The noise performance of a digital system can be superior to that of an analog system. In addition, the probability of error for the system output can be reduced even further by the use of appropriate coding techniques as discussed in Chapter 1.

These advantages usually outweigh the main disadvantage of PCM: a much wider bandwidth than that of the corresponding analog signal.

Sampling, Quantizing, and Encoding

The PCM signal is generated by carrying out three basic operations: sampling, quantizing, and encoding (Fig. 3–7). The sampling operation generates a flat-top PAM signal.

The quantizing operation is illustrated in Fig. 3–8 for the case of $M = 8$ levels. This quantizer is said to be *uniform* because all of the steps are of equal size. Since we are approximating the analog sample values by using a finite number of levels ($M = 8$ in this illustration), *error* is introduced into the recovered output analog signal because of the quantizing effect. The error waveform is illustrated in Fig. 3–8c. The quantizing error consists of the difference between the analog signal at the sampler input and the output of the quantizer. Note that the peak value of the error (± 1) is one-half of the quantizer step size (2). If we sample at the Nyquist rate ($2B$) or faster, and there is negligible channel noise, there will still be noise, called *quantizing noise*, on the recovered analog waveform due to this error. The quantizing noise can also be thought of as a round-off error. In Sec. 7–7 statistics of this quantizing noise are evaluated and a formula for the PCM system signal-to-noise ratio is developed. The quantizer output is a *quantized* PAM signal.

The PCM signal is obtained from the quantized PAM signal by encoding each quantized sample value into a digital word. It is up to the system designer to specify the exact code word that will represent a particular quantized level. For a Gray code taken from Table 3–1, the resulting PCM signal is shown in Fig. 3–8d, where the PCM word for each quantized sample is strobed out of the encoder by the next clock pulse. The Gray code was chosen because it has only one bit change for each step change in the quantized level. Consequently, single errors in the received PCM code word will cause minimal errors in the recovered analog level, provided that the sign bit is not in error.

Here we have described PCM systems that represent the quantized analog sample values by *binary* code words. In general, of course, it is possible to represent the quantized analog samples by digital words using a base other than base 2 or, equivalently, to convert the binary signal to a multilevel signal (as discussed in Sec. 3–4). Multilevel signals have the advantage of possessing a much smaller bandwidth than binary signals, but the disadvantage of requiring multilevel circuitry instead of binary circuitry.

Practical PCM Circuits

Three popular techniques are used to implement the analog-to-digital converter (ADC) encoding operation: the *counting or ramp, serial or successive approximation*, and *parallel or flash* encoders.

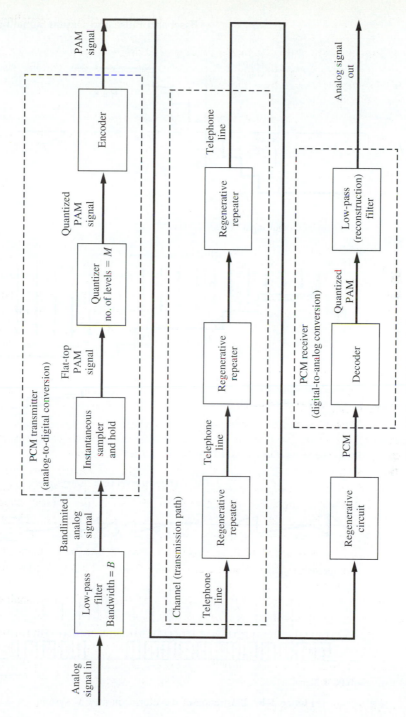

Figure 3–7 PCM trasmission system.

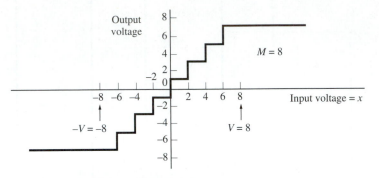

(a) Quantizer Output-Input Characteristics

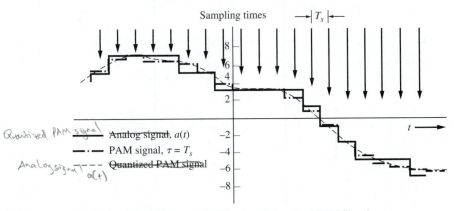

(b) Analog Signal, Flat-top PAM Signal, and Quantized PAM Signal

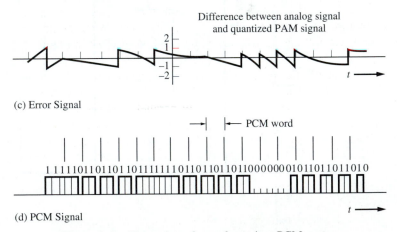

(c) Error Signal

(d) PCM Signal

Figure 3–8   Illustration of waveforms in a PCM system.

In the counting encoder, at the same time that the sample is taken, a ramp generator is energized and a binary counter is started. The output of the ramp generator is continuously compared to the sample value; when the value of the ramp becomes equal to the sample

TABLE 3–1   THREE-BIT GRAY CODE FOR M = 8 LEVELS

Quantized Sample Voltage	Gray Code Word (PCM Output)	
+7	110	
+5	111	
+3	101	
+1	100	
		Mirror image except for sign bit
−1	000	
−3	001	
−5	011	
−7	010	

value, the binary value of the counter is read. This count is taken to be the PCM word. The binary counter and the ramp generator are then reset to zero and are ready to be reenergized at the next sampling time. This technique requires only a few components, but the speed of this type of ADC is usually limited by the speed of the counter. The Maxim ICL7126 CMOS ADC integrated circuit uses a counting encoder.

The serial encoder compares the value of the sample with trial quantized values. Successive trials depend on whether the past comparator outputs are positive or negative. The trial values are chosen first in large steps and then in small steps so that the process will converge rapidly. The trial voltages are generated by a series of voltage dividers that are configured by (on–off) switches controlled by digital logic. After the process converges, the value of the switch settings is read out as the PCM word. This technique requires more precision components (for the voltage dividers) than the ramp technique. The speed of the feedback ADC technique is determined by the speed of the switches. The Analog Devices AD570 and the National Semiconductor ADC0804 8-bit ADC use serial encoding.

The parallel encoder uses a set of parallel comparators with reference levels that are the permitted quantized values. The sample value is fed into all of the parallel comparators simultaneously. The high or low level of the comparator outputs determines the binary PCM word with the aid of some digital logic. This is a fast ADC technique, but requires more hardware than the other two methods. The Harris CA3318 8-bit ADC integrated circuit is an example of the technique. A list of ADCs with their performance characteristics has been published [Walden, 1999].

All of the integrated circuits listed as examples have parallel digital outputs that correspond to the digital word, which represents the analog sample value. For the generation of PCM, the parallel output (digital word) needs to be converted to serial form for transmission over a two-wire channel. This conversion is accomplished by using a parallel-to-serial converter integrated circuit, which is also usually known as a *serial input–output* (SIO) chip. The SIO chip includes a shift register that is set to contain the parallel data (usually 8 or 16 input lines). Then the data are shifted out of the last stage of the shift register bit by bit onto a single output line to produce the serial format. Furthermore, the SIO chips are usually full duplex; that is, they have two sets of shift registers, one that functions for data flowing in each direction. One shift register converts parallel input data to serial output data for transmission over the channel, and simultaneously, the other shift register converts received serial data from another input to parallel data that are available at another output. Three types of SIO chips are available: the *universal asynchronous receiver/transmitter* (UART), the *universal synchronous receiver/transmitter* (USRT), and the *universal synchronous/asynchronous receiver transmitter* (USART). The UART transmits and receives asynchronous serial data, the USRT transmits and receives synchronous serial data, and the USART combines both a UART and a USRT on one chip. (See Sec. 3–9 for a discussion of asynchronous and synchronous serial data lines.)

At the receiving end, the PCM signal is decoded back into an analog signal by using a digital-to-analog converter (DAC) chip. If the DAC chip has a parallel data input, the received serial PCM data are first converted to a parallel form, using an SIO chip as described earlier. The parallel data are then converted to an approximation of the analog sample value by the DAC chip. This conversion is usually accomplished by using the parallel digital word to set the configuration of electronic switches on a resistive current (or voltage) divider network so that the analog output is produced. This chip is called a *multiplying* DAC because the "analog" output voltage is directly proportional to the divider reference voltage multiplied by the value of the digital word. The National Semiconductor DAC0808 8-bit DAC chip is an example of this technique. The DAC chip outputs samples of the quantized analog signal that approximate the analog sample values. Consequently, as the DAC chip is clocked, it generates a quantized PAM signal that is then smoothed by a low-pass reconstruction filter to produce the analog output as illustrated in Fig. 3–7.

Semiconductor companies manufacture several hundred types of ADC and DAC circuits. Data sheets for these ICs can be found on the Web pages of these manufacturers. Many of the ICs are designed for specific applications. For example, the AD 1861 from Analog Devices is a PCM audio DAC that generates analog audio from the PCM data on compact disks. The Motorola MC145503 and the Texas Instruments TCM320AC54 are PCM codecs for encoding and decoding PCM data signals for telephone analog audio applications (as is shown later in Example 3–1).

For more details on ADC, DAC, and PCM circuits, see the *Electrical Engineering Handbook* [Dorf, 1993] and the *Electronics Handbook* [Whitaker, 1996].

Bandwidth of PCM Signals

A good question to ask is, What is the spectrum of a PCM (serial) data waveform? For the case of PAM signaling, the spectrum of the PAM signal could be obtained as a function of the spectrum of the input analog signal because the PAM signal is a linear function of the analog signal. This is not the case for PCM. As shown in Figs. 3–7 and 3–8, the PCM is a *nonlinear* function of the input analog signal. Consequently, the spectrum of the PCM signal is not directly related to the spectrum of the input analog signal (as will be shown in Secs. 3–4 and 3–5).

The bandwidth of (serial) binary PCM waveforms depends on the bit rate and the waveform pulse shape used to represent the data. From Fig. 3–8, the bit rate is

$$R = nf_s \qquad\qquad (3\text{–}14)$$

where n is the number of bits in the PCM word ($M = 2^n$) and f_s is the sampling rate. For no aliasing, we require that $f_s \geq 2B$, where B is the bandwidth of the analog signal (that is to be converted to the PCM signal). In Sec. 3–4, we see that the dimensionality theorem shows that the bandwidth of the binary encoded PCM waveform is bounded by

$$B_{\text{PCM}} \geq \tfrac{1}{2}R = \tfrac{1}{2}nf_s \qquad (3\text{–}15a)$$

the minimum bandwidth of $\tfrac{1}{2}R = \tfrac{1}{2}nf_s$ is obtained only when $(\sin x)/x$ type pulse shape is used to generate the PCM waveform. However, usually a more rectangular type of pulse shape is used, and consequently, the bandwidth of the binary–encoded PCM waveform will be larger than this minimum. Details of line codes and pulse shape selection will be studied later in Sections 3–5 and 3–6. The exact bandwidth that is attained will depend on what selection is used. For example, referring to Fig. 3–15, suppose that one selects the typical case of a rectangular pulse shape and uses a unipolar NRZ, a polar NRZ, or a bipolar RZ PCM waveform as shown in Figs. 3–15b, 3–15c, and 3–15e. These are typical of waveforms generated by popular PCM integrated circuits. Then, as shown in Figs. 3–16a, 3–16b, and 3–16d, the null bandwidth will be the reciprocal of the pulse width, which is $1/T_b = R$ for these cases of binary signaling. Thus, for rectangular pulses, the first null bandwidth is

$$B_{\text{PCM}} = R = nf_s \quad \text{(first null bandwidth)} \qquad (3\text{–}15b)$$

Table 3–2 presents a tabulation of this result for the case of the minimum sampling rate, $f_s = 2B$. Note that the dimensionality theorem of Eq. (3–15a) demonstrates that the bandwidth of the PCM signal has a lower bound given by

$$B_{\text{PCM}} \geq nB \qquad (3\text{–}15c)$$

where $f_s \geq 2B$ and B is the bandwidth of the corresponding analog signal. Thus, for reasonable values of n, *the bandwidth of the serial PCM signal will be significantly larger than the bandwidth of the corresponding analog signal* that it represents. For the example shown in Fig. 3–8 where $n = 3$, the PCM signal bandwidth will be at least three times wider than that of the corresponding analog signal. Furthermore, if the bandwidth of the PCM signal is reduced by improper filtering or by passing the PCM signal through a system that has a poor frequency response, the filtered pulses will be elongated (stretched in width), so that pulses corresponding to any one bit will smear into adjacent bit slots. If this condition becomes too serious, it will cause errors in the detected bits. This pulse-smearing effect is called *inter-symbol interference* (ISI). The filtering specifications for a signal without ISI are discussed in Sec. 3–6.

Effects of Noise

The analog signal that is recovered at the PCM system output is corrupted by noise. Two main effects produce this noise or distortion:

- Quantizing noise that is caused by the M-step quantizer at the PCM transmitter.
- Bit errors in the recovered PCM signal. The bit errors are caused by *channel noise*, as well as improper channel filtering, which causes ISI.

TABLE 3–2   PERFORMANCE OF A PCM SYSTEM WITH UNIFORM QUANTIZING AND NO
CHANNEL NOISE

Number of Quantizer Levels Used, M	Length of the PCM Word, n (bits)	Bandwidth of PCM Signal (First Null Bandwidth)[a]	Recovered Analog Signal-Power-to-Quantizing-Noise Power Ratios (dB)	
			$(S/N)_{\text{pk out}}$	$(S/N)_{\text{out}}$
2	1	$2B$	10.8	6.0
4	2	$4B$	16.8	12.0
8	3	$6B$	22.8	18.1
16	4	$8B$	28.9	24.1
32	5	$10B$	34.9	30.1
64	6	$12B$	40.9	36.1
128	7	$14B$	46.9	42.1
256	8	$16B$	52.9	48.2
512	9	$18B$	59.0	54.2
1,024	10	$20B$	65.0	60.2
2,048	11	$22B$	71.0	66.2
4,096	12	$24B$	77.0	72.2
8,192	13	$26B$	83.0	78.3
16,384	14	$28B$	89.1	84.3
32,768	15	$30B$	95.1	90.3
65,536	16	$32B$	101.1	96.3

[a] B is the absolute bandwidth of the input analog signal.

In addition, as shown in Sections 2–7 and 3–2, the input analog signal needs to be sufficiently bandlimited (with a low-pass antialiasing filter) and sampled fast enough so that the aliasing noise on the recovered analog signal is negligible.

As shown in Chapter 7, under certain assumptions, the ratio of the recovered analog *peak* signal power to the total *average* noise power is given by[†]

$$\left(\frac{S}{N}\right)_{\text{pk out}} = \frac{3M^2}{1 + 4(M^2 - 1)P_e} \tag{3–16a}$$

and the ratio of the *average* signal power to the average noise power is

$$\left(\frac{S}{N}\right)_{\text{out}} = \frac{M^2}{1 + 4(M^2 - 1)P_e} \tag{3–16b}$$

where M is the number of quantized levels used in the PCM system and P_e is the probability of bit error in the recovered binary PCM signal at the receiver DAC before it is converted

[†] This derivation is postponed until Chapter 7 because a knowledge of statistics is needed to carry it out.

back into an analog signal. In Chapter 7, P_e is evaluated for many different types of digital transmission systems. In Chapter 1, it was shown how channel coding could be used to correct some of the bit errors and, consequently, reduce P_e. Therefore, in many practical systems, P_e is negligible. If we assume that there are no bit errors resulting from channel noise (i.e., $P_e = 0$) and no ISI, then, from Eq. (3–16a), the peak SNR resulting from only quantizing errors is

$$\left(\frac{S}{N}\right)_{\text{pk out}} = 3M^2 \qquad (3\text{–}17a)$$

and from Eq. (3–16b), the average SNR due only to quantizing errors is

$$\left(\frac{S}{N}\right)_{\text{out}} = M^2 \qquad (3\text{–}17b)$$

Numerical values for these SNRs are given in Table 3–2.

To realize these SNRs, one critical assumption is that the peak-to-peak level of the analog waveform at the input to the PCM encoder is set to the design level of the quantizer. For example, referring to Fig. 3–8a, this corresponds to the input traversing the range $-V$ to $+V$ volts, where $V = 8$ volts is the design level of the quantizer. Equations (3–16) and (3–17) were derived for waveforms with equally likely values, such as a triangle waveshape, that have a peak-to-peak value of $2V$ and an rms value of $V/\sqrt{3}$, where V is the design peak level of the quantizer.

From a practical viewpoint, the quantizing noise at the output of the PCM decoder can be categorized into four types, depending on the operating conditions. The four types are overload noise, random noise, granular noise, and hunting noise. As discussed earlier, the level of the analog waveform at the input of the PCM encoder needs to be set so that its peak level does not exceed the design peak of V volts. If the peak input does exceed V, then the recovered analog waveform at the output of the PCM system will have flat-tops near the peak values. This produces *overload noise*. The flat-tops are easily seen on an oscilloscope, and the recovered analog waveform sounds distorted since the flat-topping produces unwanted harmonic components.

The second type of noise, *random noise*, is produced by the random quantization errors in the PCM system under normal operating conditions when the input level is properly set. This type of condition is assumed in Eq. (3–17). Random noise has a "white" hissing sound. If the input level is not sufficiently large, the SNR will deteriorate from that given by Eq. (3–17) to one that will be described later by Eq. (3–28a) and Eq. (3–28b); however, the quantizing noise will still be more or less random.

If the input level is reduced further to a relatively small value with respect to the design level, the error values are not equally likely from sample to sample, and the noise has a harsh sound resembling gravel being poured into a barrel. This is called *granular noise*, and it can be randomized (the noise power can be decreased) by increasing the number of quantization levels and, consequently, the PCM bit rate. Alternatively, granular noise can be reduced by using a nonuniform quantizer, such as the μ-law or A-law quantizers that are described in the next section.

The fourth type of quantizing noise that may occur at the output of a PCM system is *hunting noise*. It can occur when the input analog waveform is nearly constant, including when there is no signal (i.e., the zero level). For these conditions, the sample values at the quantizer output (see Fig. 3–8) can oscillate between two adjacent quantization levels, causing an undesired sinusoidal-type tone of frequency $\frac{1}{2}f_s$ at the output of the PCM system. Hunting noise can be reduced by filtering out the tone or by designing the quantizer so that there is no vertical step at the "constant" value of the inputs—such as at zero volts input for the case of no signal. For that case, the hunting noise is also called *idle channel noise*. This kind of noise can be reduced by using a horizontal step at the origin of the quantizer output-input characteristic, instead of a vertical step as shown in Fig. 3–8a.

Recalling that $M = 2^n$, we may express Eqs. (3–17a) and (3–17b) in decibels as

$$\left[\begin{array}{c}\text{from minus full scale}\\ \text{to plus full scale}\end{array}\right] \longrightarrow \left(\frac{S}{N}\right)_{\text{dB}} = 6.02n + \alpha \qquad \begin{array}{c}\text{for uniform distributed}\\ \text{signal}\end{array} \qquad (3\text{–}18)$$

where n is the number of bits in the PCM word, $\alpha = 4.77$ for the peak SNR, and $\alpha = 0$ for the average SNR. This equation—called the *6-dB rule*—points out the significant performance characteristic for PCM: *An additional 6-dB improvement in SNR is obtained for each bit added to the PCM word.* This relationship is illustrated in Table 3–2. Equation (3–18) is valid for a wide variety of assumptions (such as various types of input waveshapes and quantization characteristics), although the value of α will depend on these assumptions [Jayant and Noll, 1984]. It is assumed that there are no bit errors and that the input signal level is large enough to range over a significant number of quantizing levels.

[handwritten margin notes:]
Gaussian
5 bode
if 8 bit A/D
convert
if uniform 48 A/D convert
if Gaussian 40 A/D convert

Example 3–1 DESIGN OF A PCM SIGNAL FOR TELEPHONE SYSTEMS

Assume that an analog audio voice-frequency (VF) telephone signal occupies a band from 300 to 3,400 Hz. The signal is to be converted to a PCM signal for transmission over a digital telephone system. The minimum sampling frequency is $2 \times 3.4 = 6.8$ ksample/sec. In order to allow the use of a low-cost low-pass antialiasing filter with a reasonable transition band, the VF signal is oversampled with a sampling freqency of 8 ksamples/sec. This is the standard adopted by the Unites States telephone industry. Assume that each sample value is represented by 8 bits; then the bit rate of the binary PCM signal is

$$R = (f_s \text{ samples/s})(n \text{ bits/sample})$$

$$= (8\text{k samples/s})(8 \text{ bits/sample}) = 64 \text{ kbits/s} \qquad (3\text{–}19)$$

This 64-kbit/s signal is called a *DS-0* signal (digital signal, type zero).

Referring to the dimensionality theorem, Eq. (3–15a), the minimum absolute bandwidth of this binary PCM signal is

$$(B)_{\min} = \tfrac{1}{2}R = 32 \text{ kHz} \qquad (3\text{–}20)$$

This bandwidth is realized when a $(\sin x)/x$ pulse shape is used to generate the PCM waveform. If rectangular pulse shaping is used, the absolute bandwidth is infinity, and from Eq. (3–15b), the first null bandwith is

$$B_{\text{PCM}} = R = 64 \text{ kHz} \qquad (3\text{–}21)$$

That is, we require a bandwidth of 64 kHz to transmit this digital voice PCM signal, where the bandwidth of the original analog voice signal was, at most, 4 kHz. (In Sec. 3–5, we will see that the bandwidth of this binary PCM signal may be reduced somewhat by filtering without introducing ISI.) Using Eq. (3–17a), we observe that the peak signal-to-quantizing noise power ratio is

$$\left(\frac{S}{N}\right)_{pk\ out} = 3(2^8)^2 = 52.9 \text{ dB} \tag{3–22}$$

where $M = 2^8$. Note that the inclusion of a parity bit does not affect the quantizing noise. However, coding (parity bits) may be used to decrease the number of decoded errors caused by channel noise or ISI. In this example, these effects were assumed to be negligible because P_e was assumed to be zero.

The performance of a PCM system for the most optimistic case (i.e., $P_e = 0$) can easily be obtained as a function of M, the number of quantizing steps used. The results are shown in Table 3–2. In obtaining the SNR, no parity bits were used in the PCM word.

One might use this table to look at the design requirements in a proposed PCM system. For example, high-fidelity enthusiasts are turning to digital audio recording techniques, in which PCM signals, instead of the analog audio signal, are recorded to produce superb sound reproduction. For a dynamic range of 90 dB, at least 15-bit PCM words would be required. Furthermore, if the analog signal had a bandwidth of 20 kHz, the first null bandwidth for a rectangular–bitshape PCM would be 2×20 kHz $\times 15 = 600$ kHz. With an allowance for some oversampling and a wider bandwidth to minimize ISI, the bandwidth needed would be around 1.5 MHz. Consequently, video-type tape recorders are needed to record and reproduce high-quality digital audio signals. Although this type of recording technique might seem ridiculous at first, it is realized that expensive high-quality analog recording devices are hard pressed to reproduce a dynamic range of 70 dB. Thus, digital audio is one way to achieve improved performance, a fact that is being proven in the marketplace with the popularity of the digital compact disk (CD). The CD uses a 16-bit PCM word and a sampling rate of 44.1 kHz on each stereo channel [Miyaoka, 1984; Peek, 1985]. Reed–Solomon coding with interleaving is used to correct burst errors that occur as a result of scratches and fingerprints on the compact disk.

Nonuniform Quantizing: μ-Law and A-Law Companding

Voice analog signals are more likely to have amplitude values near zero than at the extreme peak values allowed. For example, in digitizing voice signals, if the peak value allowed is 1 V, weak passages may have voltage levels on the order of 0.1 V (20 dB down). For signals, such as these, with nonuniform amplitude distribution, the granular quantizing noise will be a serious problem if the step size is not reduced for amplitude values near zero and increased for extremely large values. This technique is called *nonuniform quantizing*, since a variable step size is used. An example of a nonuniform quantizing characteristic is shown in Fig. 3–9a.

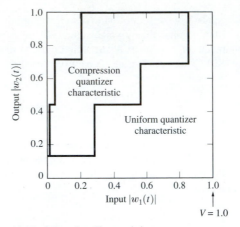

(a) $M = 8$ Quantizer Characteristic

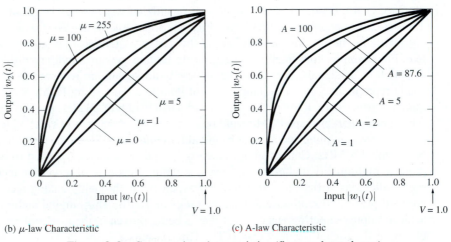

(b) μ-law Characteristic   (c) A-law Characteristic

Figure 3–9   Compression characteristics (first quadrant shown).

The effect of nonuniform quantizing can be obtained by first passing the analog signal through a compression (nonlinear) amplifier and then into a PCM circuit that uses a uniform quantizer. In the United States, a *μ-law* type of *compression* characteristic is used that is defined [Smith, 1957] by

$$|w_2(t)| = \frac{\ln(1 + \mu|w_1(t)|)}{\ln(1 + \mu)} \qquad (3\text{–}23)$$

where the allowed peak values of $w_1(t)$ are ± 1 (i.e., $|w_1(t)| \leq 1$), μ is a positive constant that is a parameter, and ln denotes the natural logarithm. This compression characteristic is shown in Fig. 3–9b for several values of μ, and it is noted that $\mu = 0$ corresponds to linear amplification (uniform quantization overall). In the United States, Canada, and Japan, the telephone companies use a $\mu = 255$ compression characteristic in their PCM systems [Dammann, McDaniel, and Maddox, 1972].

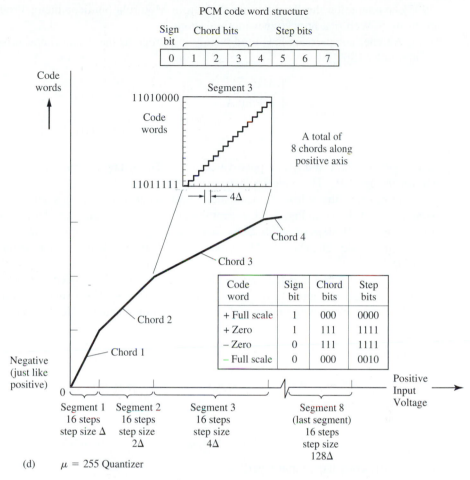

Code word	Sign bit	Chord bits	Step bits
+ Full scale	1	000	0000
+ Zero	1	111	1111
– Zero	0	111	1111
– Full scale	0	000	0010

(d) μ = 255 Quantizer

Figure 3–9   Continued

In practice, the smooth nonlinear characteristics of Fig 3–9b are approximated by piecewise linear chords as shown in Fig. 3–9d for the μ = 255 characteristic [Dammann, McDaniel, and Maddox, 1972]. Each chord is approximated by a uniform quantizer with 16 steps and an input step size that is set by the particular segment number. That is, 16 steps (including a half-width step on each side of zero) of width Δ are used for Segment 1, 16 steps of width 2Δ are used for Segment 2, 16 steps of width 4Δ for Segment 3, etc. The value of Δ is chosen so that the full scale value (last step of Segment 8) matches the peak value of the input analog signal. This segmenting technique is now accepted worldwide for the μ = 255 characteristic. As shown in Fig. 3–9d, the 8-bit PCM code word consists of a sign bit that denotes a positive or negative input voltage, three chord bits that denote the segment number, and four step bits that denote the particular step within the segment. (For more details, see the data sheet for the Motorola MC145500 series of PCM codecs [Motorola,

1995], available on the World Wide Web from Motorola Semiconductor Products at the company's Web site, http://www.motorola.com..)

Another compression law, used mainly in Europe, is the *A-law* characteristic, defined [Cattermole, 1969] by

$$|w_2(t)| = \begin{cases} \dfrac{A|w_1(t)|}{1 + \ln A}, & 0 \le |w_1(t)| \le \dfrac{1}{A} \\[2ex] \dfrac{1 + \ln(A|w_1(t)|)}{1 + \ln A}, & \dfrac{1}{A} \le |w_1(t)| \le 1 \end{cases} \tag{3-24}$$

where $|w_1(t)| \le 1$ and A is a positive constant. The A-law compression characteristic is shown in Fig. 3–9c. The typical value for A is 87.6.

In practice, the A-law characteristic is implemented by using a segmenting technique similar to that shown in Fig. 3–9d, except that, for Segment 1, there are 16 steps of width Δ; for segment 2, 16 steps of width Δ; for Segment 3, 16 steps of width 2Δ; for Segment 4, 16 steps of width 4Δ; etc. (See the Motorola MC 145500 data sheet for more details [Motorola, 1995].)

When compression is used at the transmitter, *expansion* (i.e., decompression) must be used at the receiver output to restore signal levels to their correct relative values. The *expandor* characteristic is the inverse of the compression characteristic, and the combination of a compressor and an expandor is called a *compandor*.

Once again, it can be shown that the output SNR follows the 6-dB law [Couch, 1993]

$$\left(\frac{S}{N}\right)_{dB} = 6.02n + \alpha \tag{3-25}$$

where

$$\alpha = 4.77 - 20 \log(V/x_{rms}) \qquad \text{(uniform quantizing)} \tag{3-26a}$$

or for sufficiently large input levels,

$$\alpha \approx 4.77 - 20 \log[\ln(1 + \mu)] \qquad (\mu\text{-law companding)} \tag{3-26b}$$

or [Jayant and Noll, 1984]

$$\alpha \approx 4.77 - 20 \log[1 + \ln A] \qquad (A\text{-law companding)} \tag{3-26c}$$

and n is the number of bits used in the PCM word. Also, V is the peak design level of the quantizer, and x_{rms} is the rms value of the input analog signal. Notice that the output SNR is a function of the input level for the case of uniform quantizing (no companding), but is relatively insensitive to the input level for μ-law and A-law companding, as shown in Fig. 3–10. The ratio V/x_{rms} is called the *loading factor*. The input level is often set for a loading factor of 4, which is 12 dB, to ensure that the overload quantizing noise will be negligible. In practice, this gives $\alpha = -7.3$ for the case of uniform encoding, as compared to $\alpha = 0$ that was obtained for the ideal conditions associated with Eq. (3–17b). All of these results give a 6-dB increase in the signal-to-quantizing noise ratio for each bit added to the PCM code word.

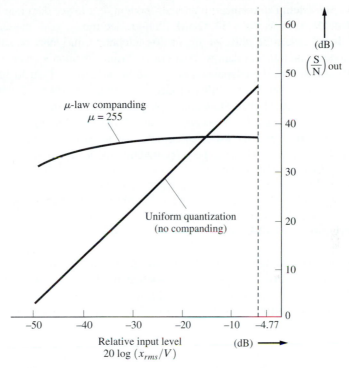

Figure 3–10   Output SNR of 8-bit PCM systems with and without companding.

V.90 56-kb/s PCM Computer Modem

The V.90 personal computer (PC) modem transmits data at 56 kb/s from a PC via an analog signal on a dial-up telephone line. This analog signal is "prequantized" to the step levels of the μ-law PCM quantizer, as shown along the horizontal axis of Fig. 3–9d. The PC modem clock is synchronized to the PCM 8-ksample/sec clock of the telephone company (TELCO), and, effectively, 7 bits of the 8-bit PCM word are used.[†] Thus, a data rate of 56 kb/s is obtained where the digital data at the TELCO are the PCM words as shown in Fig. 3–9d.

This 56-kb/s modem technique will work only if the TELCO has all digital links (i.e., does not convert the data back to analog) within the telephone plant and if the recipient is connected to the TELCO via a digital link (such as a T1 line discussed in Sec. 3–9), since a training signal is sent by the recipient through the TELCO PCM line card. This training signal is used to adapt the V.90 modem to the line card and the line characteristics. Internet Service Providers (ISPs) are usually connected to the TELCO via digital links. Thus, a PC PCM modem can be used to connect to an ISP at 56 kb/s, but the PCM modem cannot connect to another PCM modem that is attached to the TELCO via an analog line. However,

---

[†] For μ-law quantization near the zero-volt input level, the quantization steps are so close together that any noise on the telephone line often prevents recovery of the correct PCM code word. Consequently, some of these steps (i.e., code words) are not used by the V.90 modem.

the V.90 modem can connect to another modem at a lower data rate by switching to a non-PCM mode, such as the V.34, QAM, 28.8-kb/sec mode, as discussed in Appendix C–5.

In all cases, the noise on the analog telephone line must be sufficiently small so that the modem can transmit data without errors. From Shannon's channel capacity formula, Eq. (1–10), the S/N of the telephone line should be at least 51.1 dB to support a data rate of 56 kb/s, assuming that TELCO filtering circuits limit the useful bandwidth of the telephone line to 3,300 Hz. (See Modem Standards in Appendix C–5.) If the S/N is not this large, the modem will fall back to a lower speed, such as 33.3 kb/s, 28.8 kb/s, or 24 kb/s, where the S/N is sufficient to support the lower speed data transmission without errors. Measurements have shown that at least 24 kb/s can be supported by practically all of the lines in the United States [Forney, Brown, et al. 1996].

3–4 DIGITAL SIGNALING

In this section, we will answer the following questions: How do we mathematically represent the waveform for a digital signal, such as the PCM signal of Figure 3–8d, and how do we estimate the bandwidth of the waveform?

The voltage (or current) waveforms for digital signals can be expressed as an orthogonal series with a finite number of terms N. That is, the waveform can be written as

$$w(t) = \sum_{k=1}^{N} w_k \varphi_k(t), \qquad\qquad 0 < t < T_0 \qquad\qquad (3\text{--}27)$$

where w_k represents the digital data and $\varphi_k(t)$, $k = 1, 2, \ldots, N$, are N orthogonal functions that give the waveform its waveshape. (This will be illustrated by examples that follow in the sections titled "Binary Waveforms" and "Multilevel Waveforms.") N is the number of dimensions required to describe the waveform. The term *dimension* arises from the geometric interpretation as described in the next section on vector representation. The waveform $w(t)$, as given by Eq. (3–27) to represent a PCM word or any message of the M message digital source, is assigned a unique set of digital data $\{w_k\}$, $k = 1, 2, \ldots, N$, to represent that message. For example, for a binary source consisting of an ASCII computer keyboard, the letter X is assigned the code word 0001101. (See Appendix C, Table C–2.) In this case, $w_1 = 0$, $w_2 = 0$, $w_3 = 0$, $w_4 = 1$, $w_5 = 1$, $w_6 = 0$, $w_7 = 1$, and $N = 7$. This message (i.e., the letter X) is sent out over a time interval of T_0 seconds, and the voltage (or current waveform) representing the message, as described by Eq. (3–27), would have a time span of T_0 seconds. The data rate could be computed using the following definitions:

DEFINITION.   The *baud (symbol rate)* is[†]

$$D = N/T_0 \text{ symbols/s} \qquad\qquad (3\text{--}28)$$

where N is the number of dimensions used in T_0 s.

[†] In the technical literature, the term *baud rate* instead of baud is sometimes used, even though baud rate is a misnomer, since the term *baud*, by definition, is the symbol rate (symbols/second).

DEFINITION.   The *bit rate* is

$$R = n/T_0 \text{ bits/s} \qquad (3\text{–}29)$$

where n is the number of data bits sent in T_0 s.

For the case when the w_ks have binary values, $n = N$, and $w(t)$ is said to be a *binary signal*. When the w_k's are assigned more than two possible values (i.e., when they are not binary), $w(t)$ is said to be a *multilevel signal*. These two types of signaling are discussed in separate sections later.

A critical question is, If the waveform of Eq. (3–27) is transmitted over a channel and appears at the input to the receiver, how can a receiver be built to detect the data? Because $w(t)$ is an orthogonal series, the formal way to detect the data is for the receiver to evaluate the orthogonal series coefficient. That is, using Eq. (2–84), we get

$$w_k = \frac{1}{K_k} \int_0^{T_0} w(t)\varphi_k^*(t)\,dt, \; k = 1, 2, \ldots N \qquad (3\text{–}30)$$

where $w(t)$ is the waveform at the receiver input and $\varphi_k(t)$ is the known orthogonal function that was used to generate the waveform. It can also be shown that Eq. (3–30) is the optimum way to detect data when the received signal is corrupted by white additive noise; this procedure is called *matched filter* detection using correlation processing and is described in Sec. 6–8. Detection of data is illustrated in Examples 3–3 and 3–4.

Vector Representation

The orthogonal function space representation of Eq. (3–27) corresponds to the orthogonal vector space represented by

$$\mathbf{w} = \sum_{j=1}^{N} w_j\varphi_j \qquad (3\text{–}31\text{a})$$

where the boldface type denotes a vector representing the waveform of Eq. (3–27), $\mathbf{w}$ is an N-dimensional vector in Euclidean vector space, and $\{\varphi_j\}$ is an orthogonal set of N-directional vectors that become a unit vector set if the K_j's of Eq. (2–78) are all unity. A shorthand notation for the vector $\mathbf{w}$ of Eq. (3–52a) is given by a row vector:

$$\mathbf{w} = (w_1, w_2, w_3, \ldots, w_N) \qquad (3\text{–}31\text{b})$$

Example 3–2   VECTOR REPRESENTATION OF A BINARY SIGNAL

Examine the representation for the waveform of a 3-bit (binary) signal shown in Fig. 3–11a. This signal could be directly represented by

$$s(t) = \overset{N=3}{\underset{j=1}{\sum}} d_j p[t - (j - \tfrac{1}{2})T] = \overset{N=3}{\underset{j=1}{\sum}} d_j p_j(t)$$

where $p(t)$ is shown in Fig. 3–11b and $p_j(t) \triangleq p[t - (j - \tfrac{1}{2})T]$.

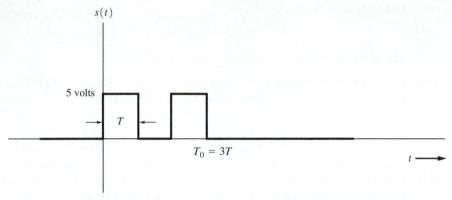

(a) A Three-Bit Signal Waveform

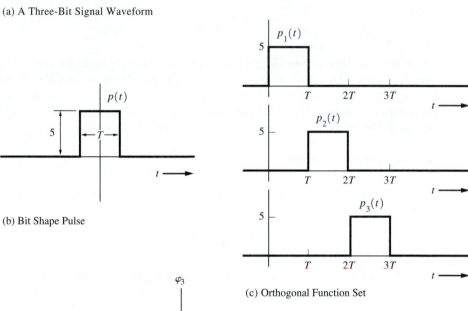

(b) Bit Shape Pulse

(c) Orthogonal Function Set

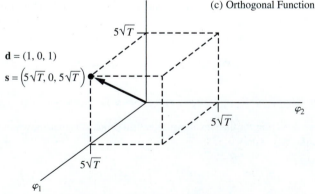

(d) Vector Representation of the 3-Bit Signal

Figure 3–11   Representation for a 3-bit binary digital signal.

The set $\{p_j(t)\}$ is a set of orthogonal functions that are not normalized. The vector

$$\mathbf{d} = (d_1, d_2, d_3) = (1,0,1)$$

is the binary word with 1 representing a binary 1 and 0 representing a binary 0. The function $p(t)$ is the pulse shape for each bit.

Using the orthogonal function approach, we can represent the waveform by

$$s(t) = \sum_{j=1}^{N=3} s_j \varphi_j(t)$$

Let $\{\varphi_j(t)\}$ be the corresponding set of ortho*normal* functions. Then, using Eq. (2–78) yields

$$\varphi_j(t) = \frac{p_j(t)}{\sqrt{K_j}} = \frac{p_j(t)}{\sqrt{\displaystyle\int_0^{T_0} p_j^2(t)\ dt}} = \frac{p_j(t)}{\sqrt{25T}}$$

or

$$\varphi_j(t) = \begin{cases} \dfrac{1}{\sqrt{T}}, & (j-1)T < t < jT \\ 0, & t \text{ otherwise} \end{cases}$$

where $j = 1, 2,$ or 3. Using Eq. (2–84), where $a = 0$ and $b = 3T$, we find that the orthonormal series coefficients for the digital signal shown in Fig. 3–11a are

$$\mathbf{s} = (s_1, s_2, s_3) = (5\sqrt{T},\ 0,\ 5\sqrt{T})$$

The vector representation for $s(t)$ is shown in Fig. 3–11d. Note that for this $N = 3$-dimensional case with binary signaling, only $2^3 = 8$ different messages could be represented. Each message corresponds to a vector that terminates on a vertex of a cube.

Bandwidth Estimation

The lower bound for the bandwidth of the waveform representing the digital signal, Eq. (3–27), can be obtained from the dimensionality theorem. Thus, from Eqs. (2–174) and (3–28), the bandwidth of the waveform $w(t)$ is

$$B \geq \frac{N}{2T_0} = \frac{1}{2}D \text{ Hz} \tag{3–32}$$

If the $\varphi_k(t)$ are of the $\sin(x)/x$ type, the lower bound absolute bandwidth of $N/(2T_0) = D/2$ will be achieved; otherwise (i.e., for other pulse shapes), the bandwidth will be larger than this lower bound. Equation (3–32) is useful for predicting the bandwidth of digital signals, especially when the exact bandwidth of the signal is difficult (or impossible) to calculate. This point is illustrated in Examples 3–3 and 3–4.

Binary Signaling

A waveform representing a binary signal can be described by the N-dimensional orthogonal series of Eq. (3–27), where the orthogonal series coefficients, w_k, take on binary values. More details about binary signaling, including their waveforms, data rate, and waveform bandwidth, are illustrated by the following example.

Example 3–3 BINARY SIGNALING

Let us examine some properties of *binary signaling* from a digital source that can produce $M = 256$ distinct messages. Each message could be represented by $n = 8$-bit binary words because $M = 2^n = 2^8 = 256$. Assume that it takes $T_0 = 8$ ms to transmit one message and that a particular message corresponding to the code word 01001110 is to be transmitted. Then,

$$w_1 = 0, \; w_2 = 1, \; w_3 = 0, \; w_4 = 0, \; w_5 = 1, \; w_6 = 1, \; w_7 = 1, \text{ and } w_8 = 0$$

CASE 1. RECTANGULAR PULSE ORTHOGONAL FUNCTIONS   Assume that the orthogonal functions $\varphi_k(t)$ are given by unity-amplitude rectangular pulses that are $T_b = T_0/n = 8/8 = 1$ ms wide, where T_b is the time that it takes to send 1 bit of data. Then, with the use of Eq. (3–27) and MATLAB, the resulting waveform transmitted is given by Fig. 3–12a.

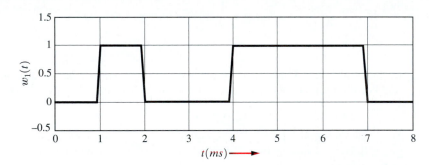

(a) Rectangular Pulse Shape, $T_b = 1$ *ms*

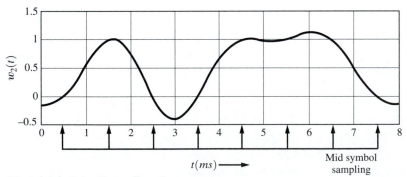

(b) $\sin(x)/x$ Pulse Shape, $T_b = 1$ *ms*

Figure 3–12   Binary signaling (computed).

The data can be detected at the receiver by evaluating the orthogonal-series coefficients as given by Eq. (3–30). For the case of rectangular pulses, this is equivalent to sampling the waveform anywhere within each bit interval.[†] Referring to Fig. 3–12a, we see that sampling within each $T_b = 1$-ms interval results in detecting the correct 8-bit data word 01001110.

The bit rate and the baud (symbol rate) of the binary signal are $R = n/T_0 = 1$ kbit/s, and $D = N/T_0 = 1$ kbaud, because $N = n = 8$ and $T_0 = 8$ ms. That is, the bit rate and the baud are equal for binary signaling.

What is the bandwidth for the waveform in Fig. 3–12a? Using Eq. (3–32), we find that the lower bound for the bandwidth is $\frac{1}{2}D = 500$ Hz. In Sec. 3–5, it is shown that the actual null bandwidth of this binary signal with a rectangular pulse shape is $B = 1/T_s = D = 1,000$ Hz. This is larger than the lower bound for the bandwidth, so the question arises, What is the wave-shape that gives the lower bound bandwidth of 500 Hz? The answer is one with $\sin (x)/x$ pulses, as described in Case 2.

CASE 2. SIN $(X)/X$ PULSE ORTHOGONAL FUNCTIONS   From an intuitive viewpoint, we know that the sharp corners of the rectangular pulse need to be rounded to reduce the bandwidth of the waveform. Furthermore, recalling our study with the sampling theorem, Eq. (2–158), we realize that the $\sin (x)/x$-type pulse shape has the minimum bandwidth. Consequently, we choose

$$\varphi_k(t) = \frac{\sin \left\{ \dfrac{\pi}{T_s} (t - kT_s) \right\}}{\dfrac{\pi}{T_s} (t - kT_s)} \tag{3–33}$$

where $T_s = T_b$ for the case of binary signaling. The resulting waveform that is transmitted is shown in Fig. 3–12b.

Once again, the data can be detected at the receiver by evaluating the orthogonal-series coefficients. Because $\sin (x)/x$ orthogonal functions are used, Eq. (2–160) shows that the data can be recovered simply by sampling[†] the received waveform at the midpoint of each symbol interval. Referring to Fig. 3–12 and sampling at the midpoint of each $T_s = 1$-ms interval, the correct 8-bit word 01001110 is detected.

For Case 2, the bit rate and the baud are still $R = 1$ kbit/s and $D = 1$ kbaud. The absolute bandwidth of Eq. (3–33) can be evaluated with the help of Fig. 2–6b, where $2W = 1/T_s$. That is, $B = 1/(2T_s) = 500$ Hz. Thus, the lower-bound bandwidth (as predicted by the dimensionality theorem) is achieved.

[†] Sampling detection is optimal only if the received waveform is noise free. See the discussion following Eq. (3–30).

Note that when the rectangular pulse shape is used, as shown in Fig. 3–12a, the digital source information is transmitted via a binary *digital waveform*. That is, the digital signal is a digital waveform. However, when the $\sin(x)/x$ pulse shape is used, as shown in Fig. 3–12b, the digital source information is transmitted via an *analog* waveform (i.e., an infinite number of voltage values ranging between -0.4 and 1.2 V are used).

Multilevel Signaling

In the case of binary signaling discussed in Example 3–3, the lower-bound bandwidth of $B = N/(2T_0)$ was achieved. That is, for Case 2, $N = 8$ pulses were required and gave a bandwidth of 500 Hz, for a message duration of $T_0 = 8$ ms. However, this bandwidth could be made smaller if N could be reduced. Indeed, N (and, consequently, the bandwidth) can be reduced by letting the w_k's of Eq. (3–27) take on $L > 2$ possible values (instead of just the two possible values that were used for binary signaling). When the w_k's have $L > 2$ possible values, the resulting waveform obtained from Eq. (3–27) is called a *multilevel signal*.

Example 3–4 $L = 4$ MULTILEVEL SIGNAL

Here, the $M = 256$-message source of Example 3–3 will be encoded into an $L = 4$ multilevel signal, and the message will be sent, once again, in $T_0 = 8$ ms. Multilevel data can be obtained by encoding the ℓ-bit binary data of the message source into L-level data by using a *digital-to-analog* converter (DAC),[†] as shown in Fig. 3–13. For example, one possible encoding scheme for an $\ell = 2$-bit DAC is shown in Table 3–3. $\ell = 2$, bits are read in at a time to produce an output that is one of $L = 4$ possible levels, where $L = 2^\ell$.

TABLE 3–3   A 2-BIT DIGITAL-TO-ANALOG
CONVERTER

Binary Input (ℓ = 2 bits)	Output Level (V)
11	+3
10	+1
00	−1
01	−3

Thus, for the binary code word 01001110, the sequence of four-level outputs would be $-3, -1, +3, +1$. Consequently, the w_k's of Eq. (3–27) would be $w_1 = -3$, $w_2 = -1$, $w_3 = +3$, and $w_4 = +1$, where $N = 4$ dimensions are used. The corresponding $L = 4$-level waveform is shown in Fig. 3–14. Fig. 3–14a gives the multilevel waveform when rectangular pulses are used for $\varphi_k(t)$, and Fig. 3–14b gives the multilevel waveform when $\sin(x)/x$ pulses are used. For either case, the receiver could recover the four-level data corresponding to the w_k values by sampling the received waveform at the middle of the $T_s = 2$-ms symbol intervals (i.e., $T = 1, 3, 5,$ and 7 ms). ‡

[†] The term *analog* in *digital-to-analog converter* is a misnomer because the output is an L-level digital signal. However, these devices are called digital-to-analog converters in data manuals.

‡ Sampling detection is optimal only if the received waveform is noise free. See the discussion following Eq. (3–30).

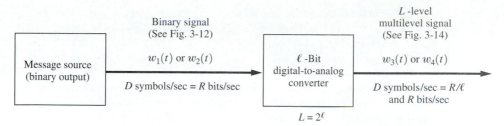

Figure 3–13   Binary-to-multilevel signal conversion.

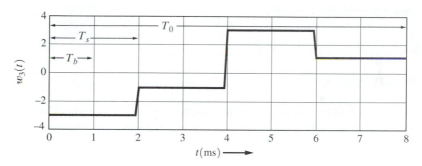

(a) Rectangular Pulse Shape $T_b = 1$ ms

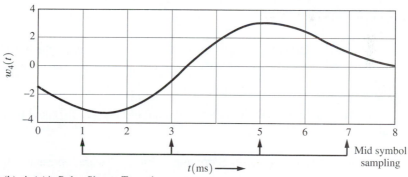

(b) sin(x)/x Pulse Shape, $T_b = 1$ ms

Figure 3–14 $L = 4$-level signaling (computed).

For these $L = 4$-level signals, the equivalent bit interval is $T_b = 1$ ms because each symbol carries $\ell = 2$ bits of data (i.e., one of $L = 4$-levels as shown in Table 3–3). The bit rate is $R = n/T_0 = \ell/T_s = 1$ kbit/s (same as for Example 3–3), and the baud rate is $D = N/T_0 = 1/T_s = 0.5$ kbaud (different from that of Example 3–3). The bit rate and the baud rate are related by

$$R = \ell D \qquad\qquad (3\text{--}34)$$

where $\ell = \log_2(L)$ is the number of bits read in by the DAC on each clock cycle.

The null bandwidth of the rectangular-pulse multilevel waveform, Fig. 3–14a, is $B = 1/T_s = D = 500$ Hz. From Eq. (3–32), the absolute bandwidth of the $\sin(x)/x$-pulse multilevel waveform, Fig. 3–14b, is $B = N/(2T_0) = 1/(2T_s) = D/2 = 250$ Hz. Thus, each of these $L = 4$-level waveforms has half the bandwidth of the corresponding binary signal (with the same pulse shape). In general, an L-level multilevel signal would have $1/\ell$ the bandwidth of the corresponding binary signal, where $\ell = \log_2(L)$. This bandwidth reduction is achieved because the symbol duration of the multilevel signal is ℓ times that of the binary signal. The bit rate R of the binary signal is ℓ times the symbol rate of the multilevel signal.

In the next section, exact formulas are obtained for the power spectral density of binary and multilevel signals.

3–5 LINE CODES AND SPECTRA

Binary Line Coding

Binary 1's and 0's, such as in PCM signaling, may be represented in various serial-bit signaling formats called *line codes*. Some of the more popular line codes are shown in Fig. 3–15.[†] There are two major categories: *return-to-zero* (RZ) and *nonreturn-to-zero* (NRZ). With RZ coding, the waveform returns to a zero-volt level for a portion (usually one-half) of the bit interval. The waveforms for the line code may be further classified according to the rule that is used to assign voltage levels to represent the binary data. Some examples follow.

Unipolar Signaling.   In positive-logic unipolar signaling, the binary 1 is represented by a high level ($+A$ volts) and a binary 0 by a zero level. This type of signaling is also called *on–off keying*.

Polar Signaling.   Binary 1's and 0's are represented by equal positive and negative levels.

Bipolar (Pseudoternary) Signaling.   Binary 1's are represented by alternately positive or negative values. The binary 0 is represented by a zero level. The term *pseudoternary* refers to the use of three encoded signal levels to represent two-level (binary) data. This is also called *alternate mark inversion* (AMI) signaling.

Manchester Signaling.   Each binary 1 is represented by a positive half-bit period pulse followed by a negative half-bit period pulse. Similarly, a binary 0 is represented by a negative half-bit period pulse followed by a positive half-bit period pulse. This type of signaling is also called *split-phase encoding*.

[†] Strictly speaking, punched paper tape is a storage medium, not a line code. However, it is included for historical purposes to illustrate the origin of the terms *mark* and *space*. In punched paper tape, the binary 1 corresponds to a hole (mark), and a binary 0 corresponds to no hole (space).

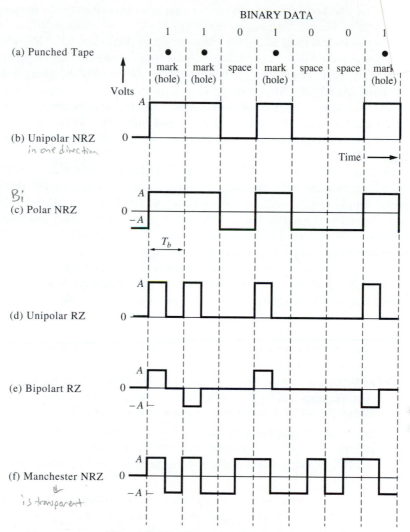

Figure 3–15   Binary signaling formats.

Later in this book, we will often use shortened notations. *Unipolar NRZ will be denoted simply by unipolar, polar NRZ by polar, and bipolar RZ by bipolar.* In this regard, unfortunately, the term *bipolar* has two different conflicting definitions. The meaning is usually made clear by the context in which it is used: (1) In the space communication industry, polar NRZ is sometimes called bipolar NRZ, or simply bipolar (this meaning will not be used in this book); and (2) in the telephone industry, the term *bipolar* denotes pseudoternary signaling (this is the meaning we use in this book), as in the T1 bipolar RZ signaling described in Sec. 3–9.

The line codes shown in Fig. 3–15 are also known by other names [Deffeback and Frost, 1971; Sklar, 1988]. For example, polar NRZ is also called NRZ-L, where L denotes the normal logical level assignment. Bipolar RZ is also called RZ-AMI, where AMI denotes alternate mark (binary 1) inversion. Bipolar NRZ is called NRZ-M, where M denotes inversion on mark. Negative logic bipolar NRZ is called NRZ-S, where S denotes inversion on space. Manchester NRZ is called Bi-φ-L, for biphase with normal logic level.

Other line codes, too numerous to list here, can also be found in the literature. [Bylanski and Ingram, 1976; Bic, Duponteil, and Imbeaux, 1991]. For example, the bipolar (pseudoternary) type may be extended into several subclasses as briefly discussed following Eq. (3–45).

Each of the line codes shown in Fig. 3–15 has advantages and disadvantages. For example, the unipolar NRZ line code has the advantage of using circuits that require only one power supply (e.g., a single +5-V power supply for TTL circuits), but it has the disadvantage of requiring channels that are dc coupled (i.e., with frequency response down to $f = 0$), because the waveform has a nonzero dc value. The polar NRZ line code does not require a dc-coupled channel, *provided that* the data toggles between binary 1's and 0's often and that equal numbers of binary 1's and 0's are sent. However, the circuitry that produces the polar NRZ signal requires a negative-voltage power supply, as well as a positive-voltage power supply. The Manchester NRZ line code has the advantage of always having a 0-dc value, regardless of the data sequence, but it has twice the bandwidth of the unipolar NRZ or polar NRZ code because the pulses are half the width. (See Fig. 3–15.)

The following are some of the desirable properties of a line code:

- *Self-synchronization.* There is enough timing information built into the code so that bit synchronizers can be designed to extract the timing or clock signal. A long series of binary 1's and 0's should not cause a problem in time recovery.

- *Low probability of bit error.* Receivers can be designed that will recover the binary data with a low probability of bit error when the input data signal is corrupted by noise or ISI. The ISI problem is discussed in Sec. 3–6, and the effect of noise is covered in Chapter 7.

- *A spectrum that is suitable for the channel.* For example, if the channel is ac coupled, the PSD of the line code signal should be negligible at frequencies near zero. In addition, the signal bandwidth needs to be sufficiently small compared to the channel bandwidth, so that ISI will not be a problem.

- *Transmission bandwidth.* This should be as small as possible.

- *Error detection capability.* It should be possible to implement this feature easily by the addition of channel encoders and decoders, or the feature should be incorporated into the line code.

- *Transparency.* The data protocol and line code are designed so that *every possible sequence of data* is faithfully and *transparently* received.

A protocol is not transparent if certain words are reserved for control sequences so that, for example, a certain word instructs the receiver to send all data that follow that code word to the printer. This feature causes a problem when a random data file (such as a machine language file) is transferred over the link, since some of the words in the file might be control character sequences. These sequences would be intercepted by the receiving system, and the defined action would be carried out, instead of passing the word on to the intended destination. In addition, a code is not transparent if some sequence will result in a loss of clocking signal (out of the bit synchronizer at the receiver). Because a string of zeros will result in a loss of the clocking signal, the bipolar format is not transparent.

The particular type of waveform selected for digital signaling depends on the application. The advantages and disadvantages of each signal format are discussed further after their spectra have been derived.

Power Spectra for Binary Line Codes

The PSD can be evaluated by using either a deterministic or a stochastic technique. This was first discussed in Chapter 1 and later illustrated in Example 2–18. To evaluate the PSD using the deterministic technique, the waveform for a line code that results from a particular data sequence is used. The approximate PSD is then evaluated by using Eq. (2–66) or, if the line code is periodic, Eq. (2–126). (Work Prob. 3–21 to apply this deterministic approach.) Alternatively, the PSD may be evaluated by using the stochastic approach that is developed in Chapter 6. The stochastic approach will be used to obtain the PSD of line codes that are shown in Fig. 3–15, because it gives the PSD for the line code with a random data sequence (instead of that for a particular data sequence).

As discussed and illustrated in Section 3–4, a digital signal (or line code) can be represented by

$$s(t) = \sum_{n=-\infty}^{\infty} a_n f(t - nT_s) \tag{3–35}$$

where $f(t)$ is the symbol pulse shape and T_s is the duration of one symbol. For binary signaling, $T_s = T_b$, where T_b is the time that it takes to send 1 bit. For multilevel signaling, $T_s = \ell T_b$. The set $\{a_n\}$ is the set of random data. For example, for the unipolar NRZ line code, $f(t) = \Pi\left(\dfrac{t}{T_b}\right)$, and $a_n = +A$ V when a binary 1 is sent and $a_n = 0$ V when a binary 0 is sent.

As is proven in Sec. 6–2, from Eq. (6–70), the general expression for the PSD of a digital signal is

$$\mathcal{P}_s(f) = \frac{|F(f)|^2}{T_s} \sum_{k=-\infty}^{\infty} R(k)e^{j2\pi kfT_s} \quad \text{\scriptsize Power spectra} \tag{3–36a}$$

where $F(f)$ is the Fourier transform of the pulse shape $f(t)$ and $R(k)$ is the autocorrelation of the data. This autocorrelation is given by

$$R(k) = \sum_{i=1}^{I} (a_n a_{n+k})_i P_i \tag{3–36b}$$

where a_n and $a_{n + k}$ are the (voltage) levels of the data pulses at the nth and $(n + k)$th symbol positions, respectively, and P_i is the probability of having the ith $a_n a_{n + k}$ product. Note that Eq. (3–36a) shows that the spectrum of the digital signal depends on two things: (1) the pulse shape used and (2) statistical properties of the data.

Using Eq. (3–36), which is the stochastic approach, we can evaluate the PSD for the various line codes shown in Fig. 3-15.

Unipolar NRZ Signaling. For unipolar signaling, the possible levels for the a's are $+A$ and 0 V. Assume that these values are equally likely to occur and that the data are independent. Now, evaluate $R(k)$ as defined by Eq. (3–36b). For $k = 0$, the possible products of $a_n a_n$ are $A \times A = A^2$ and $0 \times 0 = 0$, and consequently, $I = 2$. For random data, the probability of having A^2 is $\frac{1}{2}$ and the probability of having 0 is $\frac{1}{2}$, so that

$$R(0) = \sum_{i=1}^{2} (a_n a_n)_i P_i = A^2 \cdot \left(\tfrac{1}{2}\right) + 0 \cdot \tfrac{1}{2} = \tfrac{1}{2} A^2$$

For $k \neq 0$, there are $I = 4$ possibilities for the product values: $A \times A$, $A \times 0$, and $0 \times A$, 0×0. They all occur with a probability of $\frac{1}{4}$. Thus, for $k \neq 0$,

$$R(k) = \sum_{i=1}^{4} (a_n a_{n + k}) P_i = A^2 \left(\tfrac{1}{4}\right) + 0 \cdot \tfrac{1}{4} + 0 \cdot \tfrac{1}{4} + 0 \cdot \tfrac{1}{4} = \tfrac{1}{4} A^2$$

Hence,

$$R_{\text{unipolar}}(k) = \begin{cases} \dfrac{1}{2} A^2, & k = 0 \\[2ex] \dfrac{1}{4} A^2, & k \neq 0 \end{cases} \qquad (3\text{–}37a)$$

For rectangular NRZ pulse shapes, the Fourier transform pair is

$$f(t) = \Pi\left(\frac{t}{T_b}\right) \leftrightarrow F(f) = T_b \frac{\sin \pi f T_b}{\pi f T_b}. \qquad (3\text{–}37b)$$

Using Eq. (3–36a) with $T_s = T_b$, we find that the PSD for the unipolar NRZ line code is

$$\mathcal{P}_{\text{unipolar NRZ}}(f) = \frac{A^2 T_b}{4} \left(\frac{\sin \pi f T_b}{\pi f T_b}\right)^2 \left[1 + \sum_{k=-\infty}^{\infty} e^{j2\pi k f T_b}\right]$$

But[†]

$$\sum_{k=-\infty}^{\infty} e^{jk2\pi f T_b} = \frac{1}{T_b} \sum_{n=-\infty}^{\infty} \delta\left(f - \frac{n}{T_b}\right) \qquad (3\text{–}38)$$

Thus,

$$\mathcal{P}_{\text{unipolar NRZ}}(f) = \frac{A^2 T_b}{4} \left(\frac{\sin \pi f T_b}{\pi f T_b}\right)^2 \left[1 + \frac{1}{T_b} \sum_{n=-\infty}^{\infty} \delta\left(f - \frac{n}{T_b}\right)\right] \qquad (3\text{–}39a)$$

[†] Equation (3–38) is known as the *Poisson sum formula,* as derived in Eq. (2–115).

But because $[\sin{(\pi f T_b)}/(\pi f T_b)] = 0$ at $f = n/T_b$, for $n \neq 0$, this reduces to

$$\mathscr{P}_{\text{unipolar NRZ}}(f) = \frac{A^2 T_b}{4} \left(\frac{\sin{\pi f T_b}}{\pi f T_b}\right)^2 \left[1 + \frac{1}{T_b}\,\delta(f)\right] \tag{3–39b}$$

If A is selected so that the normalized average power of the unipolar NRZ signal is unity, then[†] $A = \sqrt{2}$. This PSD is plotted in Fig. 3–16a, where $1/T_b = R$, the bit rate of the line code. The disadvantage of unipolar NRZ is the waste of power due to the dc level and the fact that the spectrum is not approaching zero near dc. Consequently, dc coupled circuits are needed. The advantages of unipolar signaling are that it is easy to generate using TTL and CMOS circuits and requires the use of only one power supply.

Polar NRZ Signaling.   For polar NRZ signaling, the possible levels for the a's are $+A$ and $-A$ V. For equally likely occurrences of $+A$ and $-A$, and assuming that the data are independent from bit to bit, we get

$$R(0) = \sum_{i=1}^{2} (a_n a_n)_i P_i = A^2 \frac{1}{2} + (-A)^2 \frac{1}{2} = A^2$$

For $k \neq 0$,

$$R(k) = \sum_{i=1}^{4} (a_n a_{n+k}) P_i = A^2 \frac{1}{4} + (-A)(A) \frac{1}{4} + (A)(-A) \frac{1}{4} + (-A)^2 \frac{1}{4} = 0$$

Thus,

$$R_{\text{polar}}(k) = \begin{cases} A^2, & k = 0 \\ 0, & k \neq 0 \end{cases} \tag{3–40}$$

Then, substituting Eqs. (3–40) and (3–37a) into Eq. (3–36a), we obtain the PSD for the polar NRZ signal:

$$\mathscr{P}_{\text{polar NRZ}}(f) = A^2 T_b \left(\frac{\sin{\pi f T_b}}{\pi f T_b}\right)^2 \tag{3–41}$$

If A is selected so that the normalized average power of the polar NRZ signal is unity, then $A = 1$, and the resulting PSD is shown in Fig. 3–16b, where the bit rate is $R = 1/T_b$.

The polar signal has the disadvantage of having a large PSD near dc. On the other hand, polar signals are relatively easy to generate, although positive and negative power supplies are required, unless special-purpose integrated circuits are used that generate dual supply voltages from a single supply. The probability of bit error performance is superior to that of other signaling methods. (See Fig. 7–14.)

Unipolar RZ.   The autocorrelation for unipolar data was calculated previously and is given by Eq. (3–37a). For RZ signaling, the pulse duration is $T_b/2$ instead of T_b, as used in NRZ signaling. That is, for RZ, $R(o) = \sum_{i=1}^{64}$ $R(k) = \sum_{i=1}^{4 \times 64}$

[†] This can be demonstrated easily by using a line code with periodic data 10101010. Then, setting $P_s = \dfrac{1}{T_0} \displaystyle\int_0^{T_0} s^2(t)\,dt = \dfrac{1}{2T_b} A^2 T_b$ equal to unity, we get $A = \sqrt{2}$.

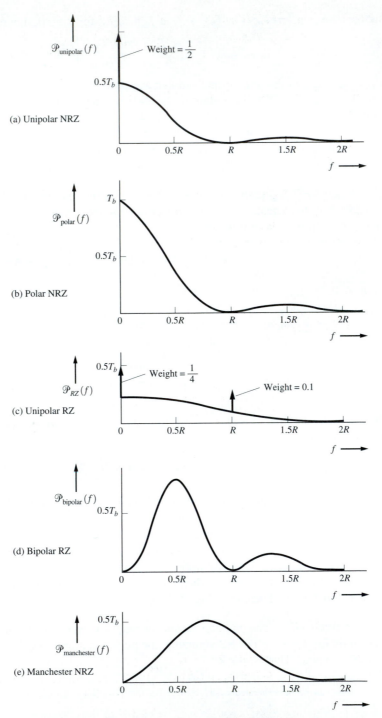

Figure 3–16   PSD for line codes (positive frequencies shown).

$$F(f) = \frac{T_b}{2} \left(\frac{\sin(\pi f T_b/2)}{\pi f T_b/2} \right) \tag{3-42}$$

Then, referring to Eqs. (3-37b) and (3-39a), we get the PSD for the unipolar RZ line code:

$$\mathcal{P}_{\text{unipolar RZ}}(f) = \frac{A^2 T_b}{16} \left(\frac{\sin(\pi f T_b/2)}{\pi f T_b/2} \right)^2 \left[1 + \frac{1}{T_b} \sum_{n=-\infty}^{\infty} \delta\left(f - \frac{n}{T_b} \right) \right] \tag{3-43}$$

If A is selected so that the normalized average power of the unipolar RZ signal is unity, then $A = 2$. The PSD for this unity-power case is shown in Fig. 3-16c, where $R = 1/T_b$.

As expected, the first null bandwidth is twice that for unipolar or polar signaling, since the pulse width is half as wide. There is a discrete (impulse) term at $f = R$. Consequently, this periodic component can be used for recovery of the clock signal. One disadvantage of this scheme is that it requires 3 dB more signal power than polar signaling for the same probability of bit error. (See Chapter 7.) Moreover, the spectrum is not negligible for frequencies near dc.

Bipolar RZ Signaling. The PSD for a bipolar signal can also be obtained using Eq. (3-36a). The permitted values for a_n are $+A$, $-A$, and 0, where binary 1's are represented by alternating $+A$ and $-A$ values and a binary 0 is represented by $a_n = 0$. For $k = 0$, the products $a_n a_n$ are A^2 and 0, where each of these products occurs with a probability of $\frac{1}{2}$. Thus,

$$R(0) = \frac{A^2}{2}$$

For $k = 1$ (the adjacent-bit case) and the data sequences $(1, 1)$, $(1, 0)$, $(0, 1)$, and $(0, 0)$, the possible $a_n a_{n+1}$ products are $-A^2$, 0, 0, and 0. Each of these sequences occurs with a probability of $\frac{1}{4}$. Consequently,

$$R(1) = \sum_{i=1}^{4} (a_n a_{n+1})_i P_i = -\frac{A^2}{4}$$

For $k > 1$, the bits being considered are not adjacent, and the $a_n a_{n+k}$ products are $\pm A^2$, 0, 0, and 0; these occur with a probability of $\frac{1}{4}$. Then

$$R(k > 1) = \sum_{i=1}^{5} (a_n a_{n+k})_i P_i = A^2 \cdot \frac{1}{8} - A^2 \cdot \frac{1}{8} = 0$$

Thus,

$$R_{\text{bipolar}}(k) = \begin{cases} \dfrac{A^2}{2}, & k = 0 \\ -\dfrac{A^2}{4}, & |k| = 1 \\ 0, & |k| > 1 \end{cases} \tag{3-44}$$

Using Eqs. (3–44) and (3–42) in Eq. (3–36a), where $T_s = T_b$, we find that the PSD for the bipolar RZ line code is

$$\mathscr{P}_{\text{bipolar RZ}}(f) = \frac{A^2 T_b}{8} \left(\frac{\sin(\pi f T_b/2)}{\pi f T_b/2} \right)^2 (1 - \cos(2\pi f T_b))$$

or

$$\mathscr{P}_{\text{bipolar RZ}}(f) = \frac{A^2 T_b}{4} \left(\frac{\sin(\pi f T_b/2)}{\pi f T_b/2} \right)^2 \sin^2(\pi f T_b) \qquad (3\text{–}45)$$

where $A = 2$ if the normalized average power is unity. This PSD is plotted in Fig. 3–16d. Bipolar signaling has a spectral null at dc, so ac coupled circuits may be used in the transmission path.

The clock signal can easily be extracted from the bipolar waveform by converting the bipolar format to a unipolar RZ format by the use of full-wave rectification. The resulting (unipolar) RZ signal has a periodic component at the clock frequency. (See Fig. 3–16c.) Bipolar signals are not transparent. That is, a string of zeros will cause a loss in the clocking signal. This difficulty can be prevented by using *high-density bipolar n* (HDB*n*) signaling, in which a string of more than *n* consecutive zeros is replaced by a "filling" sequence that contains some pulses.[†] The calculation of the PSD for HDB*n* codes is difficult, because the $R(k)$'s have to be individually evaluated for large values of k [Davis and Barber, 1973].

Bipolar signals also have single-error detection capabilities built in, since a single error will cause a violation of the bipolar line code rule. Any violations can easily be detected by receiver logic.

Two disadvantages of bipolar signals are that the receiver has to distinguish between three levels ($+A$, $-A$, and 0), instead of just two levels in the other signaling formats previously discussed. Also, the bipolar signal requires approximately 3 dB more signal power than a polar signal for the same probability of bit error. [It has $\frac{3}{2}$ the error of unipolar signaling as described by Eq. (7–28).]

Manchester NRZ Signaling.   The Manchester signal uses the pulse shape

$$f(t) = \Pi\left(\frac{t + T_b/4}{T_b/2} \right) - \Pi\left(\frac{t - T_b/4}{T_b/2} \right) \qquad (3\text{–}46a)$$

and the resulting pulse spectrum is

$$F(f) = \frac{T_b}{2} \left[\frac{\sin(\pi f T_b/2)}{\pi f T_b/2} \right] e^{j\omega T_b/4} - \frac{T_b}{2} \left[\frac{\sin(\pi f T_b/2)}{\pi f T_b/2} \right] e^{-j\omega T_b/4}$$

or

$$F(f) = jT_b \left[\frac{\sin(\pi f T_b/2)}{\pi f T_b/2} \right] \sin\left(\frac{\omega T_b}{4} \right) \qquad (3\text{–}46b)$$

[†] For example, for HDB3, the filling sequences used to replace $n + 1 = 4$ zeros are the alternating sequences 000V and 100V, where the 1 bit is encoded according to the bipolar rule and the V is a 1 pulse of such polarity as to violate the bipolar rule. The alternating filling sequences are designed so that consecutive V pulses alternate in sign. Consequently, there will be a 0 dc value in the line code, and the PSD will have a null at $f = 0$. To decode the HDB3 code, the bipolar decoder has to detect the bipolar violations and count the number of zeros preceding each violation so that the substituted 1's can be deleted.

Substituting this and Eq. (3–40) into Eq. (3–36a), the PSD for Manchester NRZ becomes

$$\mathscr{P}_{\text{Manchester NRZ}}(f) = A^2 T_b \left(\frac{\sin(\pi f T_b/2)}{\pi f T_b/2} \right)^2 \sin^2(\pi f T_b/2) \tag{3–46c}$$

where $A = 1$ if the normalized average power is unity.

This spectrum is plotted in Fig. 3–16e. The null bandwidth of the Manchester format is twice that of the bipolar bandwidth. However, the Manchester code has a zero dc level on a bit-by-bit basis. Moreover, a string of zeros will not cause a loss of the clocking signal.

In reviewing our study of PSD for digital signals, it should be emphasized that the spectrum is a function of the bit pattern (via the bit autocorrelation) as well as the pulse shape. The general result for the PSD, Eq. (3–36), is valid for multilevel as well as binary signaling.

Differential Coding

When serial data are passed through many circuits along a communication channel, the waveform is often unintentionally inverted (i.e., data complemented). This result can occur in a twisted-pair transmission line channel just by switching the two leads at a connection point when a polar line code is used. (Note that such switching would not affect the data on a bipolar signal.) To ameliorate the problem, *differential coding*, as illustrated in Fig. 3–17, is often employed. The encoded differential data are generated by

$$e_n = d_n \oplus e_{n-1} \tag{3–47}$$

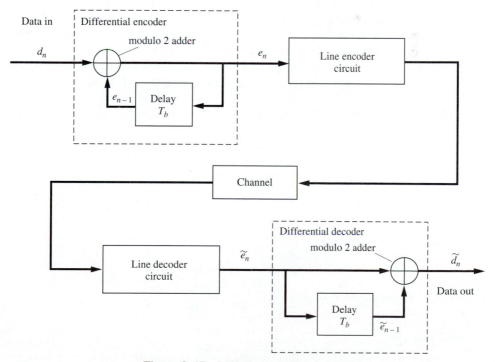

Figure 3–17   Differential coding system.

where $\oplus$ is a modulo 2 adder or an exclusive-OR gate (XOR) operation. The received encoded data are decoded by

$$\tilde{d}_n = \tilde{e}_n \oplus \tilde{e}_{n-1} \tag{3–48}$$

where the tilde denotes receiving-end data.

Each digit in the encoded sequence is obtained by comparing the present input bit with the past encoded bit. A binary 1 is encoded if the present input bit and the past encoded bit are of opposite state, and a binary 0 is encoded if the states are the same. This is equivalent to the truth table of an XOR (exclusive-OR) gate or a modulo 2 adder. An example of an encoded sequence is shown in Table 3–4, where the beginning reference digit is a binary 1. At the receiver, the encoded signal is decoded by comparing the state of adjacent bits. If the present received encoded bit has the same state as the past encoded bit, a binary 0 is the decoded output. Similarly, a binary 1 is decoded for opposite states. As shown in the table, the polarity of the differentially encoded waveform may be inverted without affecting the decoded data. This is a great advantage when the waveform is passed through thousands of circuits in a communication system and the positive sense of the output is lost or changes occasionally as the network changes, such as sometimes occurs during switching between several data paths.

Eye Patterns

The effect of channel filtering and channel noise can be seen by observing the received line code on an analog oscilloscope. The left side of Fig. 3–18 shows received corrupted polar NRZ waveforms for the cases of (a) ideal channel filtering, (b) filtering that produces intersymbol interference (ISI), and (c) noise plus ISI. (ISI is described in Sec. 3–6.) On the right side of the figure, corresponding oscilloscope presentations of the corrupted signal are shown with multiple sweeps, where each sweep is triggered by a clock signal and the sweep width

TABLE 3–4 EXAMPLE OF DIFFERENTIAL CODING

Encoding									
Input sequence	d_n		1	1	0	1	0	0	1
Encoded sequence	e_n	1	0	1	1	0	0	0	1
Reference digit ———		↑							
Decoding (with correct channel polarity)									
Received sequence (correct polarity)	$\tilde{e}_n$	1	0	1	1	0	0	0	1
Decoded sequence	$\tilde{d}_n$		1	1	0	1	0	0	1
Decoding (with inverted channel polarity)									
Received sequence (inverted polarity)	$\tilde{e}_n$	0	1	0	0	1	1	1	0
Decoded sequence	$\tilde{d}_n$		1	1	0	1	0	0	1

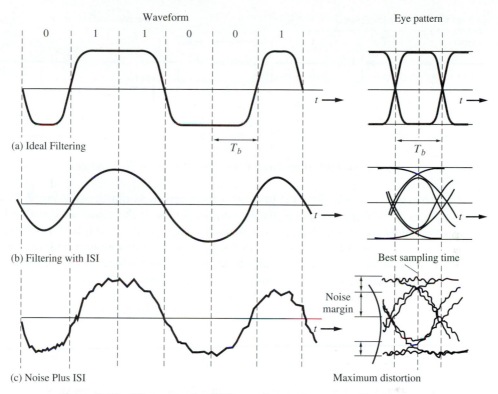

Figure 3–18   Distorted polar NRZ waveform and corresponding eye pattern.

is slightly larger than T_b. These displays on the right are called *eye patterns*, because they resemble the picture of a human eye. Under normal operating conditions (i.e., for no detected bit errors), the eye will be open. If there is a great deal of noise or ISI, the eye will close; this indicates that bit errors will be produced at the receiver output. The eye pattern provides an excellent way of assessing the quality of the received line code and the ability of the receiver to combat bit errors. As shown in the figure, the eye pattern provides the following information:

- The *timing error* allowed on the sampler at the receiver is given by the width inside the eye, called the *eye opening*. Of course, the preferred time for sampling is at the point where the vertical opening of the eye is the largest.
- The *sensitivity* to timing error is given by the slope of the open eye (evaluated at, or near, the zero-crossing point).
- The *noise margin* of the system is given by the height of the eye opening.

Regenerative Repeaters

When a line code digital signal (such as PCM) is transmitted over a hardwire channel (such as a twisted-pair telephone line), it is attenuated, filtered, and corrupted by noise. Consequently, for long lines, the data cannot be recovered at the receiving end unless repeaters

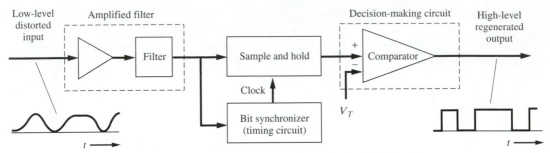

Figure 3–19   Regenerative repeater for unipolar NRZ signaling.

are placed in cascade along the line and at the receiver, as illustrated in Fig. 3–7. These repeaters amplify and "clean up" the signal periodically. If the signal were analog instead of digital, only linear amplifiers with appropriate filters could be used, since relative amplitude values would need to be preserved. In this case, in-band distortion would accumulate from linear repeater to linear repeater. This is one of the disadvantages of analog signaling. However, with digital signaling, nonlinear processing can be used to regenerate a "noise-free" digital signal. This type of nonlinear processing is called a *regenerative repeater*. A simplified block diagram of a regenerative repeater for unipolar NRZ signaling is shown in Fig. 3–19. The amplifying filter increases the amplitude of the low-level input signal to a level that is compatible with the remaining circuitry and filters the signal in such a way as to minimize the effects of channel noise and ISI. (The filter that reduces ISI is called an *equalizing filter* and is discussed in Sec. 3–6.) The bit synchronizer generates a clocking signal at the bit rate that is synchronized, so that the amplified distorted signal is sampled at a point where the eye opening is a maximum. (Bit synchronizers are discussed in detail in the next section.) For each clock pulse, the sample-and-hold circuit produces the sample value that is held (for T_b, a 1-bit interval) until the next clock pulse occurs. The comparator produces a high-level output only when the sample value is larger than the threshold level V_T. The latter is usually selected to be one-half the expected peak-to-peak variation of the sample values.[†] If the input noise is small and there is negligible ISI, the comparator output will be high only when there is a binary 1 (i.e., a high level) on the corrupted unipolar NRZ line code at the input to the repeater. The comparator—a threshold apparatus—acts as a decision-making device. Thus, the unipolar NRZ line code is regenerated "noise free," except for bit errors that are caused when the input noise and ISI alter the sample values sufficiently so that the sample values occur on the wrong side of V_T. Chapter 7 shows how the probability of bit error is influenced by the SNR at the input to the repeater, by the filter that is used, and by the value of V_T that is selected.[‡]

   In long-distance digital communication systems, many repeaters may be used in cascade, as shown in Fig. 3–7. Of course, the spacing between the repeaters is governed by the path loss of the transmission medium and the amount of noise that is added. A repeater is required when the SNR at a point along the channel becomes lower than the value that is needed to maintain the overall probability-of-bit-error specification. Assume that the

---

[†] This is the optimum V_T when the binary 1's and 0's are equally likely.

[‡] For minimum probability of bit error, the sample-and-hold circuit of Fig. 3–19 is replaced by an optimum sample-detection circuit called a *matched filter* (MF), described in Sec. 6–8.

repeaters are spaced so that each of them has the same probability of bit error, P_e, and that there are m repeaters in the system, including the end receiver. Then, for m repeaters in cascade, the overall probability of bit error, P_{me}, can be evaluated by the use of the binomial distribution from Appendix B. Using Eq. (B–33), we see that the probability of i out of the m repeaters producing a bit error is

$$P_i = \binom{m}{i} P_e^i (1 - P_e)^{m - i} \tag{3–49}$$

However, there is a bit error at the system output only when each of an *odd* number of the cascaded repeaters produces a bit error. Consequently, the overall probability of a bit error for m cascaded repeaters is

$$P_{me} = \sum_{\substack{i=1 \\ i=\text{odd}}}^{m} P_i = \sum_{\substack{i=1 \\ i=\text{odd}}}^{m} \binom{m}{i} P_e^i (1 - P_e)^{m - i}$$

$$= mP_e(1 - P_e)^{m - 1} + \frac{m(m - 1)(m - 2)}{3!} P_e^3(1 - P_e)^{m - 3} + \cdots \tag{3–50a}$$

Under useful operating conditions, $P_e \ll 1$, so only the first term of this series is significant. Thus, the overall probability of bit error is approximated by

$$P_{me} \approx mP_e \tag{3–50b}$$

where P_e is the probability of bit error for a single repeater.

Bit Synchronization

Synchronization signals are clock-type signals that are necessary within a receiver (or repeater) for detection (or regeneration) of the data from the corrupted input signal. These clock signals have a precise frequency and phase relationship with respect to the *received* input signal, and they are delayed compared to the clock signals at the transmitter, since there is propagation delay through the channel.

Digital communications usually need at least three types of synchronization signals: (1) *bit sync*, to distinguish one bit interval from another, as discussed in this section; (2) *frame sync*, to distinguish groups of data, as discussed in Sec. 3–9 in regard to time-division multiplexing; and (3) *carrier sync*, for bandpass signaling with coherent detection, as discussed in Chapters 4, 5, and 7. Systems are designed so that the sync is derived either directly from the corrupted signal or from a separate channel that is used only to transmit the sync information.

We will concentrate on systems with bit synchronizers that derive the sync directly from the corrupted signal, because it is often not economically feasible to send sync over a separate channel. The complexity of the bit synchronizer circuit depends on the sync properties of the line code. For example, the bit synchronizer for a unipolar RZ code with a sufficient number of alternating binary 1's and 0's is almost trivial, since the PSD of that code has a delta function at the bit rate, $f = R$, as shown in Fig. 3–16c. Consequently, the bit sync clock signal can be obtained by passing the received unipolar RZ waveform through a narrowband bandpass filter that is tuned to $f_0 = R = 1/T_b$. This is illustrated by Fig. 3–20 if

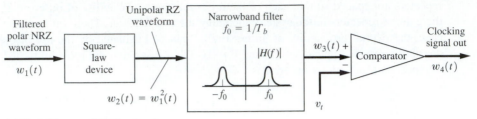

(a) Block Diagram of Bit Synchronizer

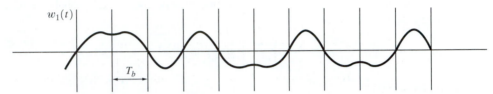

(b) Filltered Polar NRZ Input Waveform

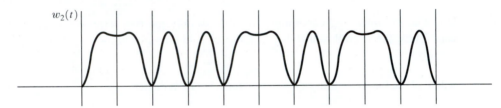

(c) Output of Square-law Device (Unipolar RZ)

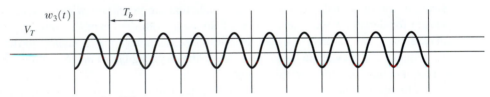

(d) Output of Narrowband Filter

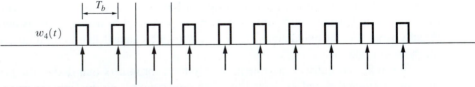

(e) Clocking Output Signal   Sampling times

Figure 3–20   Square-law bit synchronizer for polar NRZ signaling.

the square-law device is deleted. Alternatively, as described in Sec. 4–3, a phase-locked loop (PLL) can be used to extract the sync signal from the unipolar RZ line code by locking the PLL to the discrete spectral line at $f = R$. For a polar NRZ line code, the bit synchronizer is slightly more complicated, as shown in Fig. 3–20. Here the filtered polar NRZ waveform (Fig. 3–20b) is converted to a unipolar RZ waveform (Fig. 3–20c) by using a square-law (or, alternatively, a full-wave rectifier) circuit. The clock signal is easily recovered by using a filter or a PLL, since the unipolar RZ code has a delta function in its spectrum at $f = R$. All of the bit synchronizers discussed thus far use some technique for detecting the spectral line at $f = R$.

Another technique utilizes the symmetry property of the line code itself [Carlson, 1986]. Referring to Fig. 3–18, which illustrates the eye pattern for a polar NRZ code, we realize that a properly filtered line code has a pulse shape that is symmetrical about the optimum clocking (sampling) time, provided that the data are alternating between 1's and 0's. From Fig. 3–21, let $w_1(t)$ denote the filtered polar NRZ line code, and let $w_1(\tau_0 + nT_b)$ denote a sample value of the line code at the maximum (positive or negative) of the eye opening, where n is an integer, $R = 1/T_b$ is the bit rate, τ is the relative clocking time (i.e., clock phase), and τ_0 is the optimum value corresponding to samples at the maximum of the eye opening. Because the pulse shape of the line code is approximately symmetrical about the optimum clocking time for alternating data,

$$|w_1(\tau_0 + nT_b - \Delta)| \approx |w_1(\tau_0 + nT_b + \Delta)|$$

where τ_0 is the optimum clocking phase and $0 < \Delta < \frac{1}{2} T_b$. The quantity $w_1(\tau + nT_b - \Delta)$ is called the *early sample*, and $w_1(\tau + nT_b + \Delta)$ is called the *late sample*. These samples can be used to derive the optimum clocking signal, as illustrated in the *early–late bit synchronizer* shown in Fig. 3–21. The control voltage $w_3(t)$ for the voltage-controlled clock (VCC) is a smoothed (averaged) version of $w_2(t)$. That is,

$$w_3(t) = \langle w_2(t) \rangle \tag{3–51a}$$

where

$$w_2(t) = |w_1(\tau + nT_b - \Delta)| - |w_1(\tau + nT_b + \Delta)| \tag{3–51b}$$

(The averaging operation is needed so that the bit synchronizer will remain synchronized even if the data do not alternate for every bit interval.) If the VCC is producing clocking pulses with the optimum relative clocking time $\tau = \tau_0$ so that samples are taken at the maximum of the eye opening, Eq. (3–51) demonstrates that the control voltage $w_3(t)$ will be zero. If τ is late, $w_3(t)$ will be a positive correction voltage, and if τ is early, $w_3(t)$ will be negative. A positive (negative) control voltage will increase (decrease) the frequency of the VCC. Thus, the bit synchronizer will produce an output clock signal that is synchronized to the input data stream. Then $w_4(t)$ will be a pulse train with narrow clock pulses occurring at the time $t = \tau + nT_b$, where n is any integer and τ approximates τ_0, the optimum clock phase that corresponds to sampling at the maximum of the eye opening. It is interesting to realize that the early-late bit synchronizer of Fig. 3–21 has the same canonical form as the Costas carrier synchronization loop of Fig. 5–3.

Unipolar, polar, and bipolar bit synchronizers will work only when there are a sufficient number of alternating 1's and 0's in the data. The loss of synchronization because of long strings of all 1's or all 0's can be prevented by adopting one of two possible alternatives. One alternative, as discussed in Chapter 1, is to use bit interleaving (i.e., scrambling). In this case, the source data with strings of 1's or 0's are scrambled to produce data with

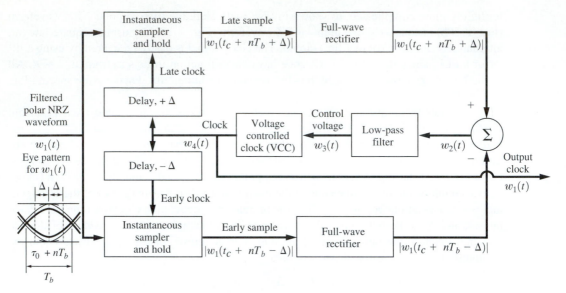

Figure 3–21   Early–late bit synchronizer for polar NRZ signaling.

alternating l's and 0's, which are transmitted over the channel by using a unipolar, polar, or bipolar line code. At the receiving end, scrambled data are first recovered by using the usual receiving techniques with bit synchronizers as just described; then the scrambled data are unscrambled. The other alternative is to use a completely different type of line code that does not require alternating data for bit synchronization. For example, Manchester NRZ encoding can be used, but it will require a channel with twice the bandwidth of that needed for a polar NRZ code.

Power Spectra for Multilevel Polar NRZ Signals

Multilevel signaling provides reduced bandwidth compared with binary signaling. The concept was introduced in Sec. 3–4. Here this concept will be extended and a formula for the PSD of a multilevel polar NRZ signal will be obtained. To reduce the signaling bandwidth, Fig. 3–22 shows how a binary signal is converted to a multilevel polar NRZ signal, where an ℓ-bit DAC is used to convert the binary signal with data rate R bits/sec to an $L = 2^{\ell}$-level multilevel polar NRZ signal.

For example, assume that an $\ell = 3$-bit DAC is used, so that $L = 2^3 = 8$ levels. Fig. 3–22b illustrates a typical input waveform, and Fig. 3–22c shows the corresponding eight-level multilevel output waveform, where T_s is the time it takes to send one multilevel symbol. To obtain this waveform, the code shown in Table 3–5 was used. From the figure, we see that $D = 1/T_s = 1/(3T_b) = R/3$, or, in general, the baud rate is

$$D = \frac{R}{\ell} \tag{3-52}$$

(a) ℓ Bit Digital-to-Analog Converter

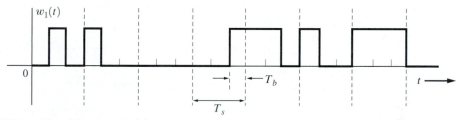

(b) Input Binary Waveform, $w_1(t)$

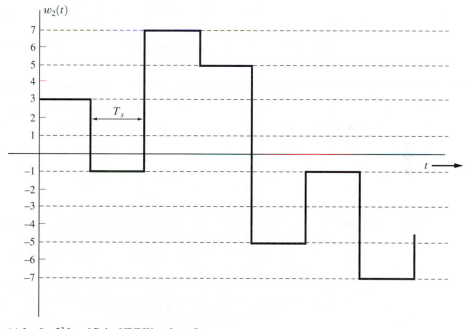

(c) $L = 8 = 2^3$ Level Polar NRZ Waveform Out

Figure 3–22   Binary-to-multilevel polar NRZ signal conversion.

The relationship between the output baud rate D and the associated bit rate R is identical to that discussed in conjunction with the dimensionality theorem in Sec. 3–4, where it was found that the bandwidth was constrained by $B \geq D/2$.

The PSD for the multilevel polar NRZ waveform (Fig. 3–22c) can be obtained by the use of Eq. (3–36a). Evaluating $R(k)$ for the case of equally likely levels a_n, as shown in Table 3–5, we have, for the case of $k = 0$,

TABLE 3–5   THREE-BIT DAC CODE

Digital Word	Output Level, $(a_n)_i$
000	+7
001	+5
010	+3
011	+1
100	−1
101	−3
110	−5
111	−7

$$R(0) = \sum_{i=1}^{8} (a_n)_i^2 P_i = 21$$

where $P_i = \frac{1}{8}$ for all of the eight possible values. For $k \neq 0$, $R(k) = 0$. Then, from Eq. (3–36a), the PSD for $w_2(t)$ is

$$\mathcal{P}_{w_2}(f) = \frac{|F(f)|^2}{T_s}(21 + 0)$$

where the pulse width (or symbol width) is now $T_s = 3T_b$. For the rectangular pulse shape of width $3T_b$, this becomes

$$\mathcal{P}_{w_2}(f) = 63T_b \left(\frac{\sin 3\pi f T_b}{3\pi f T_b} \right)^2$$

for the case where $\ell = 3$. Consequently, the first null bandwidth for this multilevel polar NRZ signal is $B_{\text{null}} = 1/(3T_b) = R/3$, or one-third the bandwidth of the input binary signal.

In general, for the case of $L = 2^\ell$ levels, the PSD of a multilevel polar NRZ signal with rectangular pulse shapes is

$$\mathcal{P}_{\text{multilevel NRZ}}(f) = K \left(\frac{\sin \ell \pi f T_b}{\ell \pi f T_b} \right)^2 \qquad (3\text{–}53)$$

where K is a constant and the null bandwidth is

$$B_{\text{null}} = R/\ell \qquad (3\text{–}54)$$

In summary, multilevel signaling, where $L > 2$, is used to reduce the bandwidth of a digital signal compared with the bandwidth required for binary signaling. In practice, filtered multilevel signals are often used to modulate a carrier for the transmission of digital information over a communication channel. This provides a relatively narrowband digital signal.

Spectral Efficiency

DEFINITION.   The *spectral efficiency* of a digital signal is given by the number of bits per second of data that can be supported by each hertz of bandwidth. That is,

$$\eta = \frac{R}{B} \text{ (bits/s)/Hz} \tag{3–55}$$

where R is the data rate and B is the bandwidth.

In applications in which the bandwidth is limited by physical and regulatory constraints, the job of the communication engineer is to choose a signaling technique that gives the highest spectral efficiency while achieving given cost constraints and meeting specifications for a low probability of bit error at the system output. Moreover, the maximum possible spectral efficiency is limited by the channel noise if the error is to be small. This maximum spectral efficiency is given by Shannon's channel capacity formula, Eq. (1–10),

$$\eta_{max} = \frac{C}{B} = \log_2\left(1 + \frac{S}{N}\right) \tag{3–56}$$

Shannon's theory does not tell us how to achieve a system with the maximum theoretical spectral efficiency; however, practical systems that approach this spectral efficiency usually incorporate error correction coding and multilevel signaling.

The spectral efficiency for multilevel polar NRZ signaling is obtained by substituting Eq. (3–54) into Eq. (3–55). We obtain

$$\eta = \ell\text{(bit/s)/Hz} \qquad \text{(multilevel polar NRZ signaling)} \tag{3–57}$$

where ℓ is the number of bits used in the DAC. Of course, ℓ cannot be increased without limit to an infinite efficiency, because it is limited by the signal-to-noise ratio as given in Eq. (3–56).

The spectral efficiencies for all the line codes studied in the previous sections can be easily evaluated from their PSDs. The results are shown in Table 3–6. Unipolar NRZ, polar NRZ, and bipolar RZ are twice as efficient as unipolar RZ or Manchester NRZ.

TABLE 3–6   SPECTRAL EFFICIENCIES OF LINE CODES

Code Type	First Null Bandwidth (Hz)	Spectral Efficiency $\eta = R/B$ [(bits/s)/Hz]
Unipolar NRZ	R	1
Polar NRZ	R	1
Unipolar RZ	$2R$	$\frac{1}{2}$
Bipolar RZ	R	1
Manchester NRZ	$2R$	$\frac{1}{2}$
Multilevel polar NRZ	R/ℓ	ℓ

All of these binary line codes have $\eta \leq 1$. Multilevel signaling can be used to achieve much greater spectral efficiency, but multilevel circuits are more costly. In practice, multilevel polar NRZ signaling is used in the T1G digital telephone lines, as described in Sec. 3–9.

3–6 INTERSYMBOL INTERFERENCE

The absolute bandwidth of rectangular multilevel pulses is infinity. If these pulses are filtered improperly as they pass through a communication system, they will spread in time, and the pulse for each symbol may be smeared into adjacent time slots and cause *intersymbol interference* (ISI), as illustrated in Fig. 3–23. Now, how can we restrict the bandwidth and still not introduce ISI? Of course, with a restricted bandwidth, the pulses would have rounded tops (instead of flat ones). This problem was first studied by Nyquist [1928]. He discovered three different methods for pulse shaping that could be used to eliminate ISI. Each of these methods will be studied in the sections that follow.

Consider a digital signaling system as shown in Fig. 3–24, in which the flat-topped multilevel signal at the input is

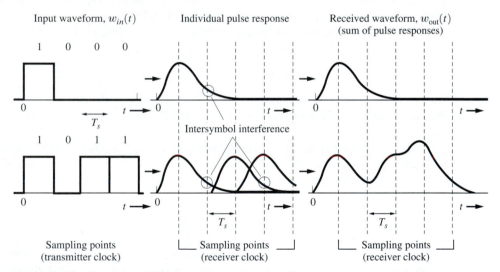

Figure 3–23   Examples of ISI on received pulses in a binary communication system.

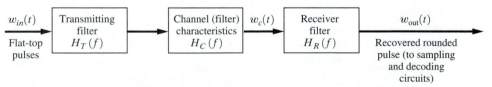

Figure 3–24   Baseband pulse-transmission system.

$$w_{\text{in}}(t) = \sum_n a_n h(t - nT_s) \tag{3–58}$$

where $h(t) = \Pi(t/T_s)$ and a_n may take on any of the allowed L multilevels ($L = 2$ for binary signaling). The symbol rate is $D = 1/T_s$ pulses/s. Then Eq. (3–58) may be written as

$$w_{\text{in}}(t) = \sum_n a_n h(t) * \delta(t - nT_s)$$
$$= \left[\sum_n a_n \delta(t - nT_s) \right] * h(t) \tag{3–59}$$

The output of the linear system of Fig. 3–24 would be just the input impulse train convolved with the equivalent impulse response of the overall system; that is,

$$w_{\text{out}}(t) = \left[\sum_n a_n \delta(t - nT_s) \right] * h_e(t) \tag{3–60}$$

where the equivalent impulse response is

$$h_e(t) = h(t) * h_T(t) * h_C(t) * h_R(t) \tag{3–61}$$

Note that $h_e(t)$ is also the pulse shape that will appear at the output of the receiver filter when a single flat-top pulse is fed into the transmitting filter (Fig. 3–24).

The equivalent system transfer function is

$$H_e(f) = H(f) H_T(f) H_C(f) H_R(f) \tag{3–62}$$

where

$$H(f) = \mathcal{F}\left[\Pi\left(\frac{t}{T_s}\right)\right] = T_s\left(\frac{\sin \pi T_s f}{\pi T_s f}\right) \tag{3–63}$$

Equation (3–63) is used so that flat-top pulses will be present at the input to the transmitting filter. The receiving filter is given by

$$H_R(f) = \frac{H_e(f)}{H(f) H_T(f) H_C(f)} \tag{3–64}$$

where $H_e(f)$ is the overall filtering characteristic.

When $H_e(f)$ is chosen to minimize the ISI, $H_R(f)$, obtained from Eq. (3–64), is called an *equalizing filter*. The equalizing filter characteristic depends on $H_C(f)$, the channel frequency response, as well as on the required $H_e(f)$. When the channel consists of dial-up telephone lines, the channel transfer function changes from call to call and the equalizing filter may need to be an *adaptive filter*. In this case, the equalizing filter adjusts itself to minimize the ISI. In some adaptive schemes, each communication session is preceded by a test bit pattern that is used to adapt the filter electronically for the maximum eye opening (i.e., minimum ISI). Such sequences are called *learning* or *training sequences and preambles*.

We can rewrite Eq. (3–60) so that the rounded pulse train at the output of the receiving filter is

$$w_{\text{out}}(t) = \sum_n a_n h_e(t - nT_s) \qquad (3\text{--}65)$$

The output pulse shape is affected by the input pulse shape (flat-topped in this case), the transmitter filter, the channel filter, and the receiving filter. Because, in practice, the channel filter is already specified, the problem is to determine the transmitting filter and the receiving filter that will minimize the ISI on the rounded pulse at the output of the receiving filter.

From an applications point of view, when the required transmitter filter and receiving filter transfer functions are found, they can each be multiplied by $Ke^{-j\omega T_d}$, where K is a convenient gain factor and T_d is a convenient time delay. These parameters are chosen to make the filters easier to build. The $Ke^{-j\omega T_d}$ factor(s) would not affect the zero ISI result, but, of course, would modify the level and delay of the output waveform.

Nyquist's First Method (Zero ISI)

Nyquist's first method for eliminating ISI is to use an equivalent transfer function, $H_e(f)$, such that the impulse response satisfies the condition

$$h_e(kT_s + \tau) = \begin{cases} C, & k = 0 \\ 0, & k \neq 0 \end{cases} \qquad (3\text{--}66)$$

where k is an integer, T_s is the symbol (sample) clocking period, τ is the offset in the receiver sampling clock times compared with the clock times of the input symbols, and C is a nonzero constant. That is, for a single flat-top pulse of level a present at the input to the transmitting filter at $t = 0$, the received pulse would be $ah_e(t)$. It would have a value of aC at $t = \tau$ but would not cause interference at other sampling times because $h_e(kT_s + \tau) = 0$ for $k \neq 0$.

Now suppose that we choose a $(\sin x)/x$ function for $h_e(t)$. In particular, let $\tau = 0$, and choose

$$h_e(t) = \frac{\sin \pi f_s t}{\pi f_s t} \qquad (3\text{--}67)$$

where $f_s = 1/T_s$. This impulse response satisfies Nyquist's first criterion for zero ISI, Eq. (3–66). Consequently, if the transmit and receive filters are designed so that the overall transfer function is

$$H_e(f) = \frac{1}{f_s} \Pi\left(\frac{f}{f_s}\right) \qquad (3\text{--}68)$$

there will be no ISI. Furthermore, the absolute bandwidth of this transfer function is $B = f_s/2$. From our study of the sampling theorem and the dimensionality theorem in Chapter 2 and Sec. 3–4, we realize that this is the optimum filtering to produce a minimum-bandwidth system. It will allow signaling at a baud rate of $D = 1/T_s = 2B$ pulses/s, where B is the absolute bandwidth of the system. However, the $(\sin x)/x$ type of overall pulse shape has two practical difficulties:

- The overall amplitude transfer characteristic $H_e(f)$ has to be flat over $-B < f < B$ and zero elsewhere. This is physically unrealizable (i.e., the impulse response would be noncausal and of infinite duration). $H_e(f)$ is difficult to approximate because of the steep skirts in the filter transfer function at $f = \pm B$.

- The synchronization of the clock in the decoding sampling circuit has to be almost perfect, since the $(\sin x)/x$ pulse decays only as $1/x$ and is zero in adjacent time slots only when t is at the exactly correct sampling time. Thus, inaccurate sync will cause ISI.

Because of these difficulties, we are forced to consider other pulse shapes that have a slightly wider bandwidth. The idea is to find pulse shapes that go through zero at adjacent sampling points and yet have an envelope that decays much faster than $1/x$ so that clock jitter in the sampling times does not cause appreciable ISI. One solution for the equivalent transfer function, which has many desirable features, is the raised cosine-rolloff Nyquist filter.

Raised Cosine-Rolloff Nyquist Filtering

DEFINITION.   The *raised cosine-rolloff Nyquist filter* has the transfer function

$$H_e(f) = \begin{cases} 1, & |f| < f_1 \\ \frac{1}{2}\left\{1 + \cos\left[\dfrac{\pi(|f| - f_1)}{2f_\Delta}\right]\right\}, & f_1 < |f| < B \\ 0, & |f| > B \end{cases} \qquad (3\text{–}69)$$

where B is the *absolute bandwidth* and the parameters

$$f_\Delta = B - f_0 \qquad (3\text{–}70)$$

and

$$f_1 \triangleq f_0 - f_\Delta \qquad (3\text{–}71)$$

f_0 is the *6-dB bandwidth* of the filter. The *rolloff factor* is defined to be

$$r = \frac{f_\Delta}{f_0} \qquad (3\text{–}72)$$

This filter characteristic is illustrated in Fig. 3–25. The corresponding impulse response is

$$h_e(t) = \mathcal{F}^{-1}[H_e(f)] = 2f_0\left(\frac{\sin 2\pi f_0 t}{2\pi f_0 t}\right)\left[\frac{\cos 2\pi f_\Delta t}{1 - (4f_\Delta t)^2}\right] \qquad (3\text{–}73)$$

Plots of the frequency response and the impulse response are shown in Fig. 3–26 for rolloff factors $r = 0$, $r = 0.5$, and $r = 1.0$. The $r = 0$ characteristic is the minimum-bandwidth case, where $f_0 = B$ and the impulse response is the $(\sin x)/x$ pulse shape. From this figure, it is seen that *as the absolute bandwidth is increased* (e.g., $r = 0.5$ or $r = 1.0$), (1) *the filtering requirements are relaxed*, although $h_e(t)$ is still noncausal, and (2) *the clock timing requirements are relaxed* also, since the envelope of the impulse response decays faster than $1/|t|$ (on the order of $1/|t|^3$ for large values of t).

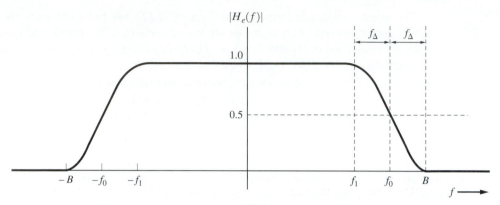

Figure 3–25   Raised cosine-rolloff Nyquist filter characteristics.

Let us now develop a formula which gives the baud rate that the raised cosine-rollof system can support without ISI. From Fig. 3–26b, the zeros in the system impulse response occur at $t = n/2f_0$, where $n \neq 0$. Therefore, data pulses may be inserted at each of these zero points without causing ISI. That is, referring to Eq. (3–66) with $\tau = 0$, we see that the raised cosime-rolloff filter satisfies Nyquist's first criterion for the absence of ISI if we select the symbol clock period to be $T_s = 1/(2f_0)$. The corresponding baud rate is $D = 1/T_s = 2f_0$ symbols/s. That is, *the 6-dB bandwidth of the raised cosine-rolloff filter, f_0, is designed to be half the symbol (baud) rate.* Using Eqs. (3–70) and (3–72), we see that the baud rate which can be supported by the system is

$$D = \frac{2B}{1 + r} \tag{3–74}$$

where B is the absolute bandwidth of the system and r is the system rolloff factor.

Example 3–1 (CONTINUED)

Assume that a binary digital signal, with polar NRZ signaling, is passed through a communication system with a raised cosine-rolloff filtering characteristic. Let the rolloff factor be 0.25. The bit rate of the digital signal is 64 kbits/s. Determine the absolute bandwidth of the filtered digital signal.

From Eq. (3–74), the absolute bandwidth is $B = 40$ kHz. This is less than the unfiltered digital signal null bandwidth of 64 kHz.

The raised cosine filter is also called a *Nyquist fitter*. It is only one of a more general class of filters that satisfy Nyquist's first criterion. This general class is described by the following theorem:

THEOREM.   *A filter is said to be a Nyquist filter if the effective transfer function is*

$$H_e(f) = \begin{cases} \Pi\left(\dfrac{f}{2f_0}\right) + Y(f), & |f| < 2f_0 \\ 0, & f \text{ elsewhere} \end{cases} \tag{3–75}$$

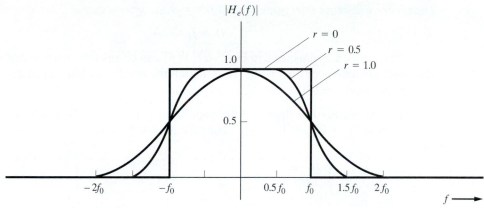

(a) Magnitude Frequency Response

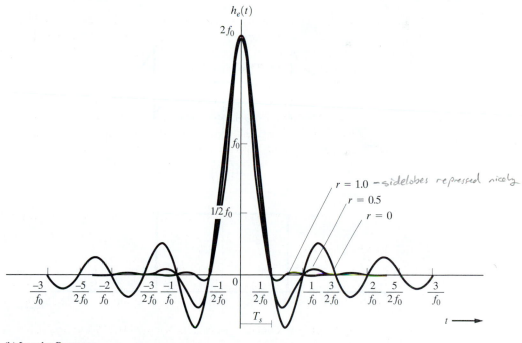

(b) Impulse Response

Figure 3–26   Frequency and time response for different rolloff factors.

where $Y(f)$ is a real function that is even symmetric about $f = 0$; that is,

$$Y(-f) = Y(f), \qquad |f| < 2f_0 \qquad\qquad (3\text{–}76a)$$

and Y is odd symmetric about $f = f_0$; that is,

$$Y(-f + f_0) = -Y(f + f_0), \qquad |f| < f_0 \qquad\qquad (3\text{–}76b)$$

Then there will be no intersymbol interference at the system output if the symbol rate is

$$D = f_s = 2f_0 \tag{3-77}$$

This theorem is illustrated in Fig. 3–27. $Y(f)$ can be any real function that satisfies the symmetry conditions of Eq. (3–76). Thus, an infinite number of filter characteristics can be used to produce zero ISI.

Proof. We need to show that the impulse response of this filter is 0 at $t = nT_s$, for $n \neq 0$, where $T_s = 1/f_s = 1/(2f_0)$. Taking the inverse Fourier transform of Eq. (3–75), we have

$$h_e(t) = \int_{-2f_0}^{-f_0} Y(f)e^{j\omega t}\, df + \int_{-f_0}^{f_0} [1 + Y(f)]e^{j\omega t}\, df + \int_{-f_0}^{2f_0} Y(f)e^{j\omega t}\, df$$

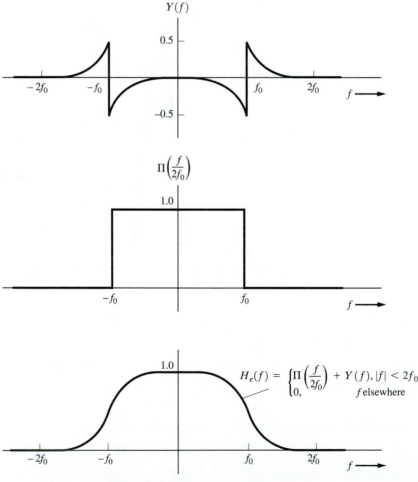

Figure 3–27 Nyquist filter characteristic.

or

$$h_e(t) = \int_{-f_0}^{f_0} e^{j\omega t}\, df + \int_{-2f_0}^{2f_0} Y(f)e^{j\omega t}\, df$$

$$= 2f_0 \left(\frac{\sin \omega_0 t}{\omega_0 t}\right) + \int_{-2f_0}^{0} Y(f)e^{j\omega t}\, df + \int_{0}^{2f_0} Y(f)e^{j\omega t}\, df$$

Letting $f_1 = f + f_0$ in the first integral and $f_1 = f - f_0$ in the second integral, we obtain

$$h_e(t) = 2f_0 \left(\frac{\sin \omega_0 t}{\omega_0 t}\right) + e^{-j\omega_0 t} \int_{-f_0}^{f_0} Y(f_1 - f_0)e^{j\omega_1 t}\, df_1$$

$$+ e^{-j\omega_0 t} \int_{-f_0}^{f_0} Y(f_1 + f_0)e^{j\omega_1 t}\, df_1$$

From Eqs. (3–76a) and (3–76b), we know that $Y(f_1 - f_0) = -Y(f_1 + f_0)$; thus, we have

$$h_e(t) = 2f_0 \left(\frac{\sin \omega_0 t}{\omega_0 t}\right) + j2 \sin \omega_0 t \int_{-f_0}^{f_0} Y(f_1 + f_0)e^{j\omega_1 t}\, df_1$$

This impulse response is real because $H_e(-f) = H_e^*(f)$, and it satisfies Nyquist's first criterion because $h_e(t)$ is zero at $t = n/(2f_0)$, $n \neq 0$, and $\tau = 0$. Thus, if we sample at $t = n/(2f_0)$, there will be no ISI.

However, the filter is noncausal. Of course, we could use a filter with a linear phase characteristic $H_e(f)e^{-j\omega T_d}$, and there would be no ISI if we delayed the clocking by T_d sec, since the $e^{-j\omega T_d}$ factor is the transfer function of an ideal delay line. This would move the peak of the impulse response to the right (along the time axis), and then the filter would be approximately causal.

At the digital receiver, in addition to minimizing the ISI, we would like to minimize the effect of channel noise by proper filtering. As will be shown in Chapter 6, the filter that minimizes the effect of channel noise is the *matched filter*. Unfortunately, if a matched filter is used for $H_R(f)$ at the receiver, the overall filter characteristic, $H_e(f)$, will usually not satisfy the Nyquist characteristic for minimum ISI. However, it can be shown that for the case of Gaussian noise into the receiver, the effects of both ISI and channel noise are minimized if the transmitter and receiver filters are designed so that [Shanmugan, 1979; Sunde, 1969; Ziemer and Peterson, 1985]

$$|H_T(f)| = \frac{\sqrt{|H_e(f)|}\;[\mathscr{P}_n(f)]^{1/4}}{\alpha|H(f)|\;\sqrt{|H_c(f)|}} \tag{3–78a}$$

and

$$|H_R(f)| = \frac{\alpha\sqrt{|H_e(f)|}}{\sqrt{H_c(f)|}\;[\mathscr{P}_n(f)]^{1/4}} \tag{3–78b}$$

where $\mathcal{P}_n(f)$ is the PSD for the noise at the receiver input and α is an arbitrary positive constant (e.g., choose $\alpha = 1$ for convenience). $H_e(f)$ is selected from any appropriate frequency response characteristic that satisfies Nyquist's first criterion as discussed previously, and $H(f)$ is given by Eq. (3–63). Any appropriate phase response can be used for $H_T(f)$ and $H_R(f)$, as long as the overall system phase response is linear. This results in a constant time delay versus frequency. The transmit and receive filters given by Eq. (3–78) become *square-root raised cosine-rolloff filters* for the case of a flat channel transfer function, flat noise, and a raised cosine-rolloff equivalent filter.

Nyquist's Second and Third Methods for Control of ISI

Nyquist's *second method* of ISI control allows some ISI to be introduced in a *controlled* way so that it can be canceled out at the receiver and the data can be recovered without error if no noise is present [Couch, 1993]. This technique also allows for the possibility of doubling the bit rate or, alternatively, halving the channel bandwidth. This phenomenon was observed by telegraphers in the 1900s and is known as "doubling the dotting speed" [Bennett and Davey, 1965].

In *Nyquist's third method* of ISI control, the effect of ISI is eliminated by choosing $h_e(t)$ so that the area under the $h_e(t)$ pulse within the desired symbol interval, T_s, is not zero, but the areas under $h_e(t)$ in adjacent symbol intervals are zero. For data detection, the receiver evaluates the area under the receiver waveform over each T_s interval. Pulses have been found that satisfy Nyquist's third criterion, but their performance in the presence of noise is inferior to the examples that were discussed previously [Sunde, 1969].

3–7 DIFFERENTIAL PULSE CODE MODULATION

When audio or video signals are sampled, it is usually found that adjacent samples are close to the same value. This means that there is a lot of redundancy in the signal samples and, consequently, that the bandwidth and the dynamic range of a PCM system are wasted when redundant sample values are retransmitted. One way to minimize redundant transmission and reduce the bandwidth is to transmit PCM signals corresponding to the difference in adjacent sample values. This, crudely speaking, is *differential pulse code modulation* (DPCM). At the receiver, the present sample value is regenerated by using the past value plus the update differential value that is received over the differential system.

Moreover, the present value can be estimated from the past values by using a *prediction filter*. Such a filter may be realized by using a tapped delay line (a bucket brigade device) to form a *transversal filter*, as shown in Fig. 3–28. When the tap gains $\{a_l\}$ are set so that the filter output will predict the present value from past values, the filter is said to be a *linear prediction filter* [Spilker, 1977]. The optimum tap gains are a function of the correlation properties of the audio or video signal [Jayant and Noll, 1984]. The output samples are

$$z(nT_s) = \sum_{l=1}^{K} a_l y(nT_s - lT_s) \qquad (3\text{–}79a)$$

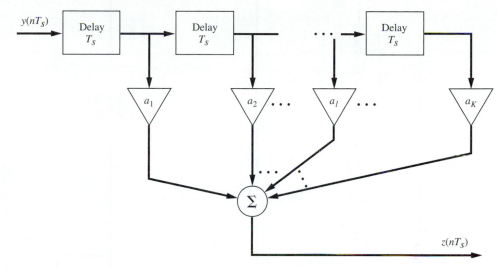

Figure 3–28   Transversal filter.

or, in simplified notation,

$$z_n = \sum_{l=1}^{K} a_l y_{n-l} \tag{3–79b}$$

where y_{n-l} denotes the sample value at the filter input at time $t = (n - l)T_s$ and there are K delay devices in the transversal filter.

The linear prediction filter may be used in a differential configuration to produce DPCM. Two possible configurations will be examined.

The first DPCM configuration, shown in Fig. 3–29, uses the predictor to obtain a differential pulse amplitude-modulated (DPAM) signal that is quantized and encoded to produce the DPCM signal. The recovered analog signal at the receiver output will be the same as that at the system input, plus *accumulated* quantizing noise. We may eliminate the accumulation effect by using the transmitter configuration of Fig. 3–30.

In the second DPCM configuration, shown in Fig. 3–30, the predictor operates on quantized values at the transmitter as well as at the receiver in order to minimize the quantization noise on the recovered analog signal. The analog output at the receiver is the same as the input analog signal at the transmitter, except for quantizing noise; furthermore, the quantizing noise does not accumulate, as was the case in the first configuration.

It can be shown that DPCM, like PCM, follows the 6-dB rule [Jayant and Noll, 1984]

$$\left(\frac{S}{N}\right)_{dB} = 6.02n + \alpha \tag{3–80a}$$

where

$$-3 < \alpha < 15 \qquad \text{for DPCM speech} \tag{3–80b}$$

and n is the number of quantizing bits ($M = 2^n$). Unlike companded PCM, the α for DPCM varies over a wide range, depending on the properties of the input analog signal. Equa-

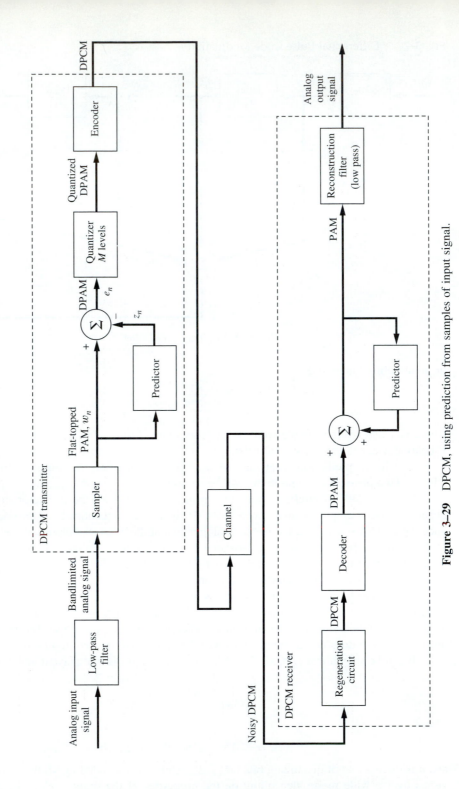

Figure 3–29 DPCM, using prediction from samples of input signal.

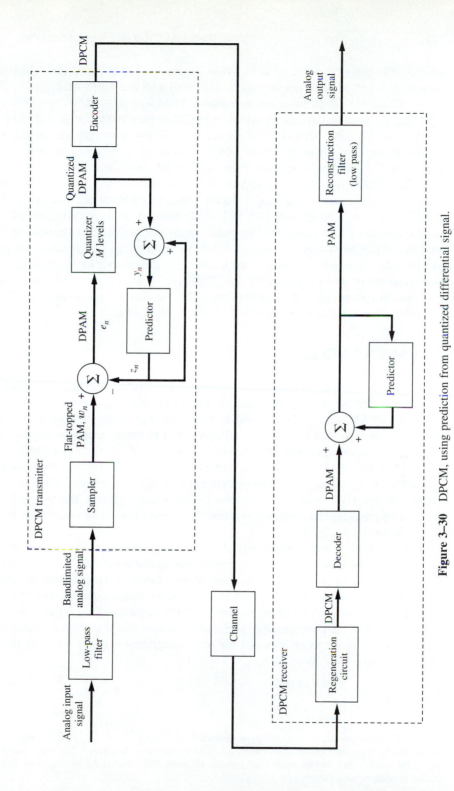

Figure 3–30 DPCM, using prediction from quantized differential signal.

191

tion (3–80b) gives the range of α for voice-frequency (300 to 3,400 Hz) telephone-quality speech. This DPCM performance may be compared with that for PCM. Equation (3–26b) indicates that $\alpha = -10$ dB for μ-law companded PCM with $\mu = 255$. Thus, there may be an SNR improvement as large as 25 dB when DPCM is used instead of $\mu = 255$ PCM. Alternatively, for the same SNR, DPCM could require 3 or 4 fewer bits per sample than companded PCM. This is why telephone DPCM systems often operate at a bit rate of $R = 32$ kbits/s or $R = 24$ kbits/s, instead of the standard 64 kbits/s needed for companded PCM.

The CCITT has adopted a 32-kbits/s DPCM standard that uses 4-bit quantization at an 8-ksample/s rate for encoding 3.2-kHz bandwidth VF signals [Decina and Modena, 1988]. Moreover, a 64-kbits/s DPCM CCITT standard (4-bit quantization and 16 ksamples/s) has been adopted for encoding audio signals that have a 7-kHz bandwidth. A detailed analysis of DPCM systems is difficult and depends on the type of input signal present, the sample rate, the number of quantizing levels used, the number of stages in the prediction filter, and the predictor gain coefficients. This type of analysis is beyond the scope of this text, but for further study, the reader is referred to published work on the topic [Flanagan et al., 1979; Jayant, 1974; Jayant and Noll, 1984; O'Neal, 1966b].

3–8 DELTA MODULATION

From a block diagram point of view, *delta modulation* (DM) is a special case of DPCM in which there are two quantizing levels. As shown in Fig. 3–30, for the case of $M = 2$, the quantized DPAM signal is *binary*, and the encoder is not needed because the function of the encoder is to convert the multilevel DPAM signal to binary code words. For the case of $M = 2$, the DPAM signal is a DPCM signal where the code words are one bit long. The cost of a DM system is less than that of a DPCM system ($M > 2$) because the analog-to-digital converter (ADC) and digital-to-analog converter (DAC) are not needed. This is the main attraction of the DM scheme—it is relatively inexpensive. In fact, the cost may be further reduced by replacing the predictor by a low-cost integration circuit (such as an RC low-pass filter), as shown in Fig. 3–31.

In the DM circuit shown in Fig. 3–31, the operations of the subtractor and two-level quantizer are implemented by using a comparator so that the output is $\pm V_c$ (binary). In this case, the DM signal is a polar signal. A set of waveforms associated with the delta modulator is shown in Fig. 3–32. In Fig. 3–32a, an assumed analog input waveform is illustrated. If the instantaneous samples of the flat-topped PAM signal are taken at the beginning of each sampling period, the corresponding accumulator output signal is shown.[†] Here the integrator is assumed to act as an accumulator (e.g., integrating impulses) so that the integrator output at time $t = nT_s$ is given by

$$z_n = \frac{1}{V_c} \sum_{i=0}^{n} \delta y_i \qquad (3\text{–}81)$$

[†] The sampling frequency, $f_s = 1/T_s$, is selected to be within the range $2B_{\text{in}} < f_s < 2B_{\text{channel}}$, where B_{in} is the bandwidth of the input analog signal and B_{channel} is the bandwidth of the channel. The lower limit prevents aliasing of the analog signal, and the upper limit prevents ISI in the DM signal at the receiver. (See Example 3–5 for further restrictions on the selection of f_s.)

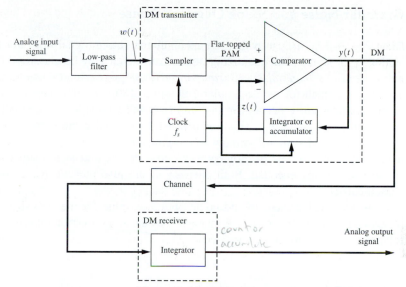

Figure 3–31   DM system.

where $y_i = y(iT_s)$ and δ is the accumulator gain or step size. The corresponding DM output waveform is shown in Fig. 3–32b.

At the receiver, the DM signal may be converted back to an analog signal approximation to the analog signal at the system input. This is accomplished by using an integrator for the receiver that produces a smoothed waveform corresponding to a smoothed version of the accumulator output waveform that is present in the transmitter (Fig. 3–32a).

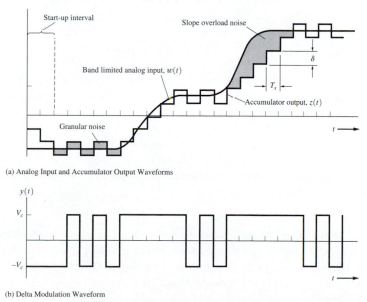

(a) Analog Input and Accumulator Output Waveforms

(b) Delta Modulation Waveform

Figure 3–32   DM system waveforms.

Granular Noise and Slope Overload Noise

From Fig. 3–32a, it is seen that the accumulator output signal does not always track the analog input signal. The quantizing noise error signal may be classified into two types of noise: *slope overload noise* and *granular noise*. Slope overload noise occurs when the step size δ is too small for the accumulator output to follow quick changes in the input waveform. Granular noise occurs for any step size, but is smaller for a small step size. Thus, we would like to have δ as small as possible to minimize the granular noise. The granular noise in a DM system is similar to the granular noise in a PCM system, whereas slope overload noise is a new phenomenon due to a differential signal (instead of the original signal itself) being encoded. Both phenomena are also present in the DPCM system discussed earlier.

It is clear that there should be an optimum value for the step size δ, because if δ is increased, the granular noise will increase, but the slope overload noise will decrease. This relationship is illustrated in Fig. 3–33.

Example 3–5 DESIGN OF A DM SYSTEM

Find the step size δ required to prevent slope overload noise for the case when the input signal is a sine wave.

The maximum slope that can be generated by the accumulator output is

$$\frac{\delta}{T_s} = \delta f_s \tag{3–82}$$

For the case of the sine-wave input, where $w(t) = A \sin \omega_a t$, the slope is

$$\frac{dw(t)}{dt} = A\omega_a \cos \omega_a t \tag{3–83}$$

and the maximum slope of the input signal is $A\omega_a$. Consequently, to avoid slope overload, we require that $\delta f_s > A\omega_a$, or

$$\delta > \frac{2\pi f_a A}{f_s} \tag{3–84}$$

However, we do not want to make δ too much larger than this value, or the granular noise will become too large.

The resulting detected signal-to-noise ratio can also be calculated. It has been determined experimentally that the spectrum of the granular noise is uniformly distributed over the frequency band $|f| \leq f_s$. It can also be shown that the total granular quantizing noise

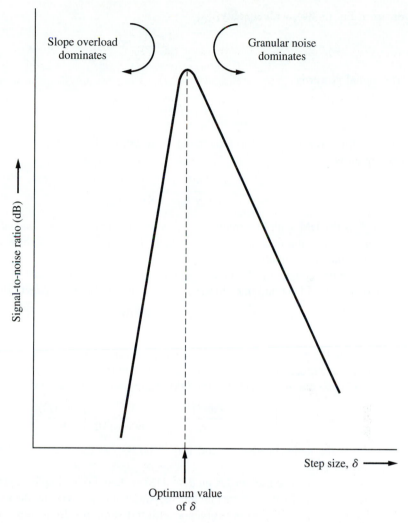

Figure 3–33   Signal-to-noise ratio out of a DM system as a function of step size.

is $\delta^2/3$. (See Sec. 7–7, where $\delta/2$ of PCM is replaced by δ for DM.) Thus, the PSD for the noise is $\mathcal{P}_n(f) = \delta^2/(6f_s)$. The granular noise power in the analog signal band, $|f| \leq B$, is

$$N = \langle n^2 \rangle = \int_{-B}^{B} \mathcal{P}_n(f)\, df = \frac{\delta^2 B}{3f_s} \qquad (3\text{–}85)$$

or, from Eq. (3–84), with equality,

$$N = \frac{4\pi^2 A^2 f_a^2 B}{3f_s^3}$$

The signal power is

$$S = \langle w^2(t)\rangle = \frac{A^2}{2} \tag{3-86}$$

The resulting average signal-to-quantizing noise ratio out of a DM system with a sine-wave test signal is

$$\left(\frac{S}{N}\right)_{\text{out}} = \frac{3}{8\pi^2}\frac{f_s^3}{f_a^2 B} \tag{3-87}$$

where f_s is the DM sampling frequency, f_a is the frequency of the sinusoidal input, and B is the bandwidth of the receiving system. Recall that Eq. (3–87) was shown to be valid only for sinusoidal-type signals.

For voice-frequency (VF) audio signals, it has been shown that Eq. (3–84) is too restrictive if $f_a = 4$ kHz and that slope overload is negligible if [deJager, 1952]

$$\delta \geq \frac{2\pi 800 W_p}{f_s} \tag{3-88}$$

where W_p is the peak value of the input audio waveform $w(t)$. (This is due to the fact that the midrange frequencies around 800 Hz dominate in the VF signal.) Combining Eqs. (3–88) and (3–85), we obtain the S/N for the DM system with a VF-type signal, viz.,

$$\left(\frac{S}{N}\right)_{\text{out}} = \frac{\langle w^2(t)\rangle}{N} = \frac{3f_s^3}{(1{,}600\pi)^2 B}\left(\frac{\langle w^2(t)\rangle}{W_p^2}\right) \tag{3-89}$$

where B is the audio bandwidth and $(\langle w^2(t)\rangle/W^2)$ is the average-audio-power to peak-audio-power ratio.

This result can be used to design a VF DM system. For example, suppose that we desire an SNR of at least 30 dB. Assume that the VF bandwidth is 4 kHz and the average-to-peak audio power is $\frac{1}{2}$. Then Eq. (3–89) gives a required sampling frequency of 40.7 kbits/s, or $f_s = 10.2B$. It is also interesting to compare this DM system with a PCM system that has the same bandwidth (i.e., bit rate). The number of bits, n, required for each PCM word is determined by $R = (2B)n = 10.2B$ or $n \approx 5$. Then the average-signal to quantizing-noise ratio of the comparable PCM system is 30.1 dB. (See Table 3–2.) Thus, under these conditions, the PCM system with a comparable bandwidth has about the same SNR performance as the DM system. Furthermore, repeating the preceding procedure, it can be shown that if an SNR larger than 30 dB were desired, the PCM system would have a larger SNR than that of a DM system with comparable bandwidth; on the other hand, if an SNR *less* than 30 dB were sufficient, the DM system would outperform (i.e., have a larger SNR than) the PCM system of the same bandwidth. Note that the SNR for DM increases as f_s^3, or a 9-dB-per-octave increase in f_s.

It is also possible to improve the SNR performance of a DM system by using double integration instead of single integration, as was studied here. With a double-integration system, the SNR increases as f_s^5, or 15 dB per octave [Jayant and Noll, 1984].

Adaptive Delta Modulation and
Continuously Variable Slope Delta Modulation

To minimize the slope overload noise while holding the granular noise at a reasonable value, *adaptive delta modulation* (ADM) is used. Here the step size is varied as a function of time as the input waveform changes. The step size is kept small to minimize the granular noise until the slope overload noise begins to dominate. Then the step size is increased to reduce the slope overload noise. The step size may be adapted, for example, by examining the DM pulses at the transmitter output. When the DM pulses consist of a string of pulses with the same polarity, the step size is increased (see Fig. 3-32) until the DM pulses begin to alternate in polarity, then the step size is decreased, and so on. One possible algorithm for varying the step size is shown in Table 3-7, where the step size changes with discrete variation. Here the step size is normally set to a value δ when the ADM signal consists of data with alternating 1's and 0's or when two successive binary 1's or 0's occur. However, if three successive binary 1's or three successive binary 0's occur, the step size is increased to 2δ, and 4δ for four successive binary 1's or 0's. Figure 3-34 gives a block diagram for this ADM system.

Papers have been published with demonstration records that illustrate the quality of ADM and other digital techniques when voice signals are sent over digital systems and recovered [Flanagan et al., 1979; Jayant, 1974]. Another variation of ADM is *continuously variable slope delta modulation* (CVSD). Here an integrator (instead of an accumulator) is used, so that $z(t)$ is made continuously variable instead of stepped in discrete increments as shown in Fig. 3-32a.

The Motorola MC34115 is a CVSD integrated circuit that varies the slope (step size) as a function of the last 3 bits of the encoded data sequence; the MC3418 has a similar 4-bit algorithm. Both of these CVSD codec chips can be connected either as modulators (encoders) or as demodulators (decoders) and may be operated with bit rates ranging from 9.6 to 64 kbits/s. A better SNR is obtained at the decoder output if a higher bit rate is used. If MC3418s are used in a 37.7-kbit/s system, the output SNR will be 30 dB. This performance is similar to that obtained for the DM and PCM systems discussed after Eq. (3-89).

The following question might be asked: Which is better, PCM or DM? The answer, of course, depends on the criterion used for comparison and the type of message. If the objective is to have a relatively simple, low-cost system, delta modulation may be the best.

TABLE 3-7 STEP-SIZE ALGORITHM

Data Sequence[a]				Number of Successive Binary 1's or 0's	Step-Size Algorithm, $f(d)$
×	×	0	1	1	δ
×	0	1	1	2	δ
0	1	1	1	3	2δ
1	1	1	1	4	4δ

[a] ×, do not care.

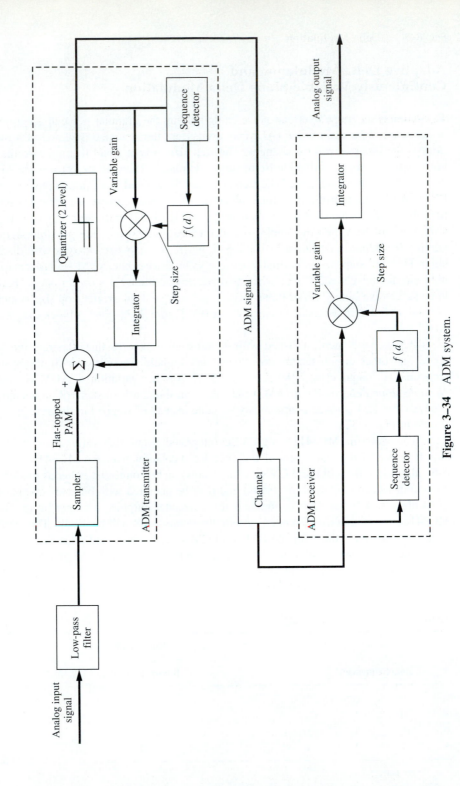

Figure 3–34 ADM system.

However, the cost of ADCs is dropping, so this may not be a factor. If high output SNR is the criterion, PCM is probably the best [O'Neal, 1966a]. If one is interfacing to existing equipment, compatibility would be a prime consideration. In this regard, PCM has the advantage, because it was adopted first and is widely used.

Speech Coding

Digital speech coders can be classified into two categories: waveform coders and vocoders. *Waveform coders* use algorithms to encode and decode speech so that the system output is an approximation to the input waveform. *Vocoders* encode speech by extracting a set of parameters that are digitized and transmitted to the receiver, where they are used to set values for parameters in function generators and filters, which, in turn, synthesize the output speech sound. Usually, the vocoder output waveform does not approximate the input waveform and may have an artificial, unnatural sound. Although the words of the speaker may be clearly understandable, the speaker may not be identifiable. With waveform encoders (e.g., PCM, DPCM, DM, and CVSD), it was demonstrated that VF-quality speech may be encoded at bit rates as low as 24 kbits/s. More advanced techniques reduce the required bit rate to 8 kbits/s, and speech coding is possible even at 2 kbits/s. Some techniques that are available to achieve coders at low bit rates are linear prediction (already discussed as applied to DPCM), adaptive subband coding, and vector quantization. Adaptive subband coding allocates bits according to the input speech spectrum and the properties of hearing. With vector quantization, whole blocks of samples are encoded at a time, instead of encoding on a sample-by-sample basis. Examples are code excited linear prediction (CELP) and vector-sum excited linear prediction (VSELP), used in digital cellular telephones, as described in Sec. 8–8. These coders employ linear-prediction-based *analysis-by-synthesis* (LPAS) techniques, in which the talker's speech signal is partitioned into 20-ms segments for analysis and synthesis. The encoder sequences through the possible codebook excitation patterns and the possible values for the filter parameters to find a synthesized waveform that gives the best match to a speech segment. The encoder parameters that specify this best match are then transmitted to the receiver via digital data. The received data establish the parameters for the receiver speech synthesizer so that the voice signal is reproduced for the listener.

One strives to use lower bit-rate codecs in order to reduce data transmission costs. However, lower bit-rate codecs require greater computation complexity, produce longer delays of the reproduced speech at the system output, and have poorer speech quality. (For more details about speech coders and the trade-off involved, see the literature [e.g., Gershio, 1994; Spanias, 1994; Budagavi and Gibson, 1998].)

3–9 TIME-DIVISION MULTIPLEXING†

> **DEFINITION.**   *Time-division multiplexing* (TDM) is the time interleaving of samples from several sources so that the information from these sources can be transmitted serially over a single communication channel.

---

† In this method, a common channel or system is shared by many users. Other methods for sharing a common communication system are discussed under the topic of multiple access techniques in Sec. 8–5.

Figure 3–35 illustrates the TDM concept as applied to three analog sources that are multiplexed over a PCM system. For convenience, natural sampling is shown together with the corresponding gated TDM PAM waveform. In practice, an electronic switch is used for the commutation (sampler). In this example, the pulse width of the TDM PAM signal is $T_s/3 = 1/(3f_s)$, and the pulse width of the TDM PCM signal is $T_s/(3n)$, where n is the number of bits used in the PCM word. Here $f_s = 1/T_s$ denotes the frequency of rotation for the commutator, and f_s satisfies the Nyquist rate for the analog source with the largest bandwidth. In some applications in which the bandwidth of the sources is markedly different, the larger bandwidth sources may be connected to several switch positions on the sampler so that they will be sampled more often than the smaller bandwidth sources.

At the receiver, the decommutator (sampler) has to be synchronized with the incoming waveform so that the PAM samples corresponding to source 1, for example, will appear on the channel 1 output. This is called *frame synchronization*. Low-pass filters are used to reconstruct the analog signals from the PAM samples. ISI resulting from poor channel filtering would cause PCM samples from one channel to appear on another channel, even though perfect bit and frame synchronization were maintained. Feedthrough of one channel's signal into another channel is called *crosstalk*.

Frame Synchronization

Frame synchronization is needed at the TDM receiver so that the received multiplexed data can be sorted and directed to the appropriate output channel. The frame sync can be provided to the receiver demultiplexer (demux) circuit either by sending a frame sync signal from the transmitter over a separate channel or by deriving the frame sync from the TDM signal itself. Because the implementation of the first approach is obvious, we will concentrate on that of the latter approach, which is usually more economical, since a separate sync channel is not needed. As illustrated in Fig. 3–36, frame sync may be multiplexed along with the information words in an N-channel TDM system by transmitting a unique K-bit sync word at the beginning of each frame. As illustrated in Fig. 3–37, the frame sync is recovered from the corrupted TDM signal by using a frame synchronizer circuit that cross-correlates the regenerated TDM signal with the expected unique sync word $\mathbf{s} = (s_1, s_2, \ldots, s_K)$. The elements of the unique sync word vector $\mathbf{s}$, denoted by $s_1, s_2 \ldots s_j, \ldots s_k$, are binary 1's or 0's (which, for TTL logic would represent +5V or 0 V, respectively). The current bit of the regenerated TDM signal is clocked into the first stage of the shift register and then shifted to the next stage on the next clock pulse so that the most immediate K bits are always stored in the shift register. The s_j's within the triangles below the shift register denote the presence or absence of an inverter. That is, if s_j is a binary 0, then there is an inverter in the jth leg. If s_j is a binary 1, there is no inverter. The coincident detector is a K-input AND gate.

If the unique sync word happens to be present in the shift register, all the inputs to the coincident detector will be binary 1's, and the output of the coincident detector will be a binary 1 (i.e., a high level). Otherwise, the output of the coincident detector is a binary 0 (i.e., a low level). Consequently, the coincident detector output will go high only during the T_b-s interval when the sync word is perfectly aligned in the shift register. Thus, the frame synchronizer recovers the frame sync signal.

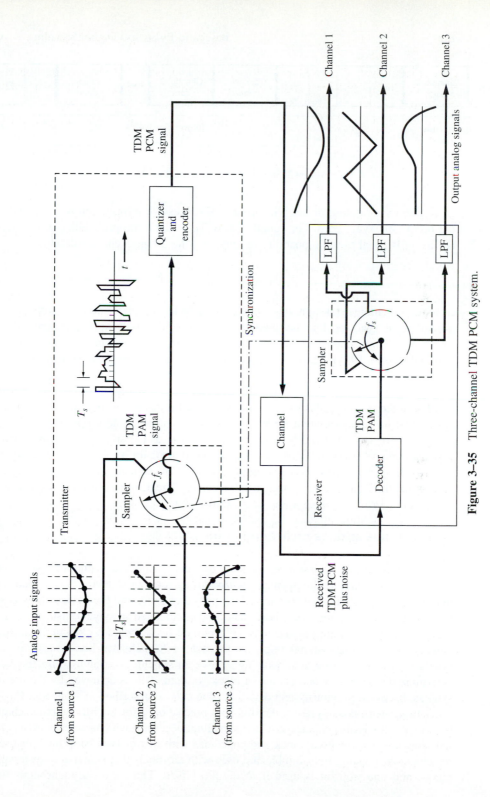

Figure 3–35 Three-channel TDM PCM system.

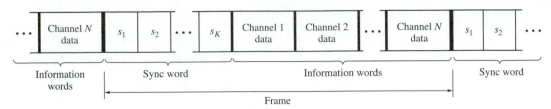

Figure 3–36   TDM frame sync format.

False sync output pulses will occur if K successive information bits happen to match the bits in the sync word. For equally likely TDM data, the probability of this false sync occurring is equal to the probability of obtaining the unique sync word, which is

$$P_f = \left(\frac{1}{2}\right)^K = 2^{-K} \tag{3–90}$$

In frame synchronizer design, this equation may be used to determine the number of bits, K, needed in the sync word so that the false lock probability will meet specifications. Alternatively, more sophisticated techniques such as aperture windows can be used to suppress false lock pulses [Ha, 1986]. The information words may also be encoded so that they are not allowed to have the bit strings that match the unique sync word.

Since the output of the coincident detector is a digitized crosscorrelation of the sync word with the passing K-bit word stored in the shift register, the sync word needs to be chosen so that its autocorrelation function, $R_s(k)$, has the desirable properties: $R_s(0) = 1$ and $R(k) \approx 0$ for $k \neq 0$. The PN codes (studied in Sec. 5–13) are almost ideal in this regard. For example, if $P_f = 4 \times 10^{-5}$ is the allowed probability of false sync, then, from Eq. (3–90), a $(K = 15)$-bit sync word is required. Consequently, a 15-stage shift register is needed for the frame synchronizer in the receiver. The 15-bit PN sync word can be generated at the transmitter using a four-stage shift register.

Synchronous and Asynchronous Lines

For bit sync, data transmission systems are designed to operate with either synchronous or asynchronous serial data lines. In a *synchronous* system, each device is designed so that its internal clock is relatively stable for a long period of time, and the device clock is synchronized to a system master clock. Each bit of data is clocked in synchronism with the master clock. The synchronizing signal may be provided by a separate clocking line or may be embedded in the data signal (e.g., by the use of Manchester line codes). In addition, synchronous transmission requires a higher level of synchronization to allow the receiver to determine the beginning and ending of blocks of data. This is achieved by the use of frame sync as discussed previously and data link protocols as described in Appendix C.

In *asynchronous* systems, the timing is precise only for the bits within each character (or word). This is also called *start–stop* signaling, because each character consists of a "start bit" that starts the receiver clock and concludes with one or two "stop bits" that terminate the clocking. Usually, two stop bits are used with terminals that signal at rates less than 300 bits/s, and one stop bit is used if $R \geq 300$ bits/s. Thus, with asynchronous lines, the

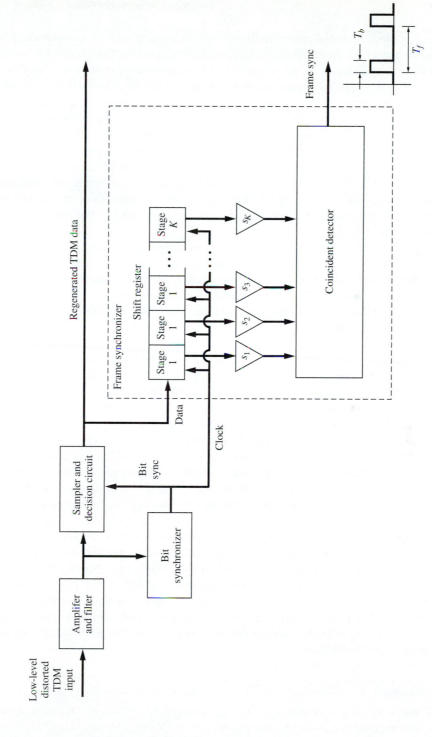

Figure 3–37 Frame synchronizer with TDM receiver front end.

receiver clock is started aperiodically and no synchronization with a master clock is required. The frequency of the receiver clock is sufficiently accurate that correct bit timing is maintained for the duration of one word. This aperiodic mode of operation is ideal for keyboard terminals, at which the typist does not type at an even pace and whose input rate is much slower than that of the data communication system. These asynchronous terminals often use a 7-bit ASCII code (see Appendix C), and the complete character consists of one start bit, 7 bits of the ASCII code, one parity bit, and one stop bit (for $R \geq 300$ bits/s). This gives a total character length of 10 bits. In TDM of the asynchronous type, the different sources are multiplexed on a *character-interleaved* (i.e., character-by-character) basis instead of interleaving on a bit-by-bit basis. The synchronous transmission system is more efficient, because start and stop bits are not required. However, the synchronous mode of transmission requires that the clocking signal be passed along with the data and that the receiver synchronize to the clocking signal.

"Intelligent" TDMs may be used to *concentrate* data arriving from many different terminals or sources. These TDMs are capable of providing speed, code, and protocol conversion. At the input to a large mainframe computer, they are called *front-end processors*. The hardware in the intelligent TDM consists of microprocessors or minicomputers. Usually, they connect "on the fly" to the input data lines that have data present and momentarily disconnect the lines that do not have data present. For example, a keyboard terminal is disconnected from the system while it is inactive (although to the user, it does not appear to be disconnected) and is connected as each character or block of data is sent. Thus the output data rate of the multiplexer is much less than the sum of the data capacities of the input lines. This technique is called *statistical* multiplexing; it allows many more terminals to be connected on line to the system.

Multiplexers can also be classified into three general types. The first TDM type consists of those that connect to synchronous lines. The second TDM type consists of those that connect to quasi-synchronous lines. In this case, the individual clocks of the input data sources are not exactly synchronized in frequency. Consequently, there will be some variation in the bit rates between the data arriving from different sources. In addition, in some applications the clock rates of the input data streams are not related by a rational number. In these cases, the TDM output signal will have to be clocked at an increased rate above the nominal value to accommodate those inputs that are not synchronous. When a new input bit is not available at the multiplexer clocking time (due to nonsynchronization), *stuff bits*, which are dummy bits, are inserted in the TDM output data stream. This strategy is illustrated by the bit-interleaved multiplexer shown in Fig. 3–38. The stuff bits may be binary 1's, 0's, or some alternating pattern, depending on the choice of the system designer. The third TDM type consists of those that operate with asynchronous sources and produce a high-speed asynchronous output (no stuff bits required) or high-speed synchronous output (stuff bits required).

Example 3–6 DESIGN OF A TIME-DIVISION MULTIPLEXER

Design a time-division multiplexer that will accommodate 11 sources. Assume that the sources have the following specifications:

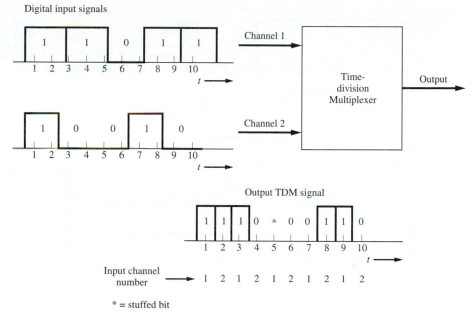

Figure 3–38   Two-channel bit-interleaved TDM with pulse stuffing.

Source 1.   Analog, 2-kHz bandwidth.

Source 2.   Analog, 4-kHz bandwidth.

Source 3.   Analog, 2-kHz bandwidth.

Sources 4–11.   Digital, synchronous at 7,200 bits/s.

Now suppose that the analog sources will be converted into 4-bit PCM words and, for simplicity, that frame sync will be provided via a separate channel and synchronous TDM lines are used. To satisfy the Nyquist rate for the analog sources, sources 1, 2, and 3 need to be sampled at 4, 8, and 4 kHz, respectively. As shown in Fig. 3–39, this can be accomplished by rotating the first commutator at $f_1 = 4$ kHz and sampling source 2 twice on each revolution. This produces a 16-kilosamples/s TDM PAM signal on the commutator output. Each of the analog sample values is converted into a 4-bit PCM word, so that the rate of the TDM PCM signal on the ADC output is 64 kbits/s. The digital data on the ADC output may be merged with the data from the digital sources by using a second commutator rotating at $f_2 = 8$ kHz and wired so that the 64-kbits/s PCM signal is present on 8 of 16 terminals. This arrangement provides an effective sampling rate of 64 kbits/s. On the other eight terminals, the digital sources are connected to provide a data transfer rate of 8 kbits/s for each source. Since the digital sources are supplying a 7.2-kbit/s data stream, pulse stuffing is used to raise the source rate to 8 kbits/s.

The preceding example illustrates the main advantage of TDM: It can easily accommodate both analog and digital sources. Unfortunately, when analog signals are converted to digital signals without redundancy reduction, they consume a great deal of digital system capacity.

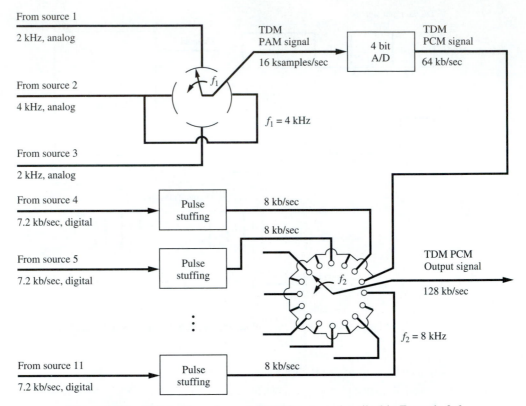

Figure 3–39   TDM with analog and digital inputs as described in Example 3-6.

TDM Hierarchy

In practice, TDMs may be grouped into two categories. The first category consists of mul-tiplexers used in conjunction with digital computer systems to merge digital signals from several sources for TDM transmission over a high-speed line to a digital computer. The out-put rate of these multiplexers has been standardized to 1.2, 2.4, 3.6, 4.8, 7.2, 9.6, 14.4, 19.2, and 28.8 kbits/s and to 10 and 100 Mbits/s.

The second category of TDMs is used by common carriers, such as the American Tele-phone and Telegraph Company (AT&T), to combine different sources into a high-speed dig-ital TDM signal for transmission over toll networks. Unfortunately, the standards adopted by North America and Japan are different from those that have been adopted in other parts of the world. The North America–Japan standards were first adopted by AT&T, and another set of standards has been adopted by CCITT under the auspices of ITU. The North American TDM hierarchy is shown in Fig. 3–40 [James and Muench, 1972].[†] The telephone industry

---

[†] The Japanese TDM hierarchy is the same as that for North America for multiplex levels 1 and 2, but dif-fers for levels 3, 4, and 5. For level 3, the Japanese standard is 32.064 Mbits/s (480 VF), level 4 is 97.728 Mbits/s (1440 VF), and level 5 is 397.200 Mbits/s (5760 VF). Dissimilarities between standards are briefly discussed and summarized by Jacobs [1986].

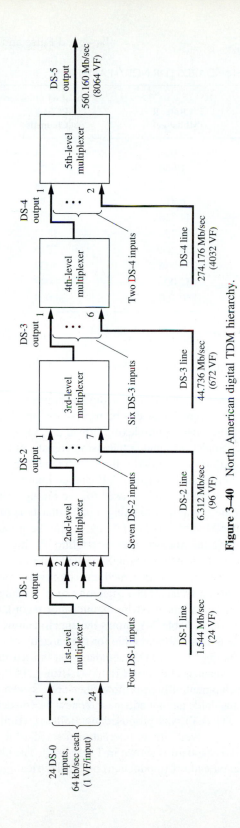

Figure 3-40 North American digital TDM hierarchy.

TABLE 3–8   TDM STANDARDS FOR NORTH AMERICA

Digital Signal Number	Bit Rate, R (Mbits/sec)	No. of 64 kbits/sec PCM VF Channels	Transmission Media Used
DS-0	0.064	1	Wire pairs
DS-1	1.544	24	Wire pairs
DS-1C	3.152	48	Wire pairs
DS-2	6.312	96	Wire pairs, fiber
DS-3	44.736	672	Coax., radio, fiber
DS-3C	90.254	1344	Radio, fiber
DS-4E	139.264	2016	Radio, fiber, coax.
DS-4	274.176	4032	Coax., fiber
DS-432	432.00	6048	Fiber
DS-5	560.160	8064	Coax., fiber

has standardized the bit rates to 1.544 Mbits/s, 6.312 Mbits/s, etc., and designates them as DS-1 for digital signal, type 1; DS-2 for digital signal, type 2; etc. as listed in Table 3–8. In Fig. 3–40, all input lines are assumed to be digital (binary) streams, and the number of voice-frequency (VF) analog signals that can be represented by these digital signals is shown in parentheses. The higher level multiplexing inputs are not always derived from lower level multiplexers. For example, one analog television signal can be converted directly to a DS-3 data stream (44.73 Mbits/s). Similarly, the DS streams can carry a mixture of information from a variety of sources such as video, VF, and computers.

The transmission medium that is used for the multiplex levels depends on the DS level involved and on the economics of using a particular type of medium at a particular location (Table 3–8). For example, higher DS levels may be transmitted over coaxial cables or fiber-optic cable or via microwave radio or satellite. A single DS-1 signal is usually transmitted over one pair of twisted wires. (One pair is used for each direction.) This type of DS-1 transmission over a twisted-pair medium is known (from its development in 1962 by AT&T) as the *T1 carrier system*. DS-1 signaling over a T1 system is very popular because of its relatively low cost and its excellent maintenance record. (T1 will be discussed in more detail in the next section.) Table 3–9 shows the specifications for the T-carrier digital baseband systems. Table 8–2 is a similar table for the capacity of common-carrier bandpass systems. The corresponding CCITT TDM standard that is used throughout the world except in North America and Japan is shown in Fig. 3–41 [Irmer, 1975].

With the development of high-bit-rate fiber-optic systems, it has become apparent that the original TDM standards are not adequate. A new TDM standard called the (Synchronous Optical Network (SONET) was proposed by Bellcore (Bell Communications Research) around 1985 and has evolved into an international standard that was adopted by the CCITT in 1989. This SONET standard is shown in Table 3–10. The OC-1 signal is an optical (light) signal that is turned on and off (modulated) by an electrical binary signal that has a line rate

TABLE 3-9 SPECIFICATIONS FOR T-CARRIER BASEBAND DIGITAL TRANSMISSION SYSTEMS

| System | Rate (Mbits/s) | System Capacity | | | Line Code | Repeater Spacing (miles) | Maximum System Length (miles) | System Error Rate |
		Digital Signal No.	Voice Channels	Medium				
T1	1.544	DS-1	24	Wire pair	Bipolar RZ	1	50	10^{-6}
T1C	3.152	DS-1C	48	Wire pair	Bipolar RZ	1	—	10^{-6}
T1D	3.152	DS-1C	48	Wire pair	Duobinary NRZ	1	—	10^{-6}
T1G	6.443	DS-2	96	Wire pair	4-level NRZ	1	200	10^{-6}
T2	6.312	DS-2	96	Wire pair[a]	B6ZS[b] RZ	2.3	500	10^{-7}
T3	44.736	DS-3	672	Coax.	B3ZS[b] RZ	c	c	c
T4	274.176	DS-4	4032	Coax.	Polar NRZ	1	500	10^{-6}
T5	560.160	DS-5	8064	Coax.	Polar NRZ	1	500	4×10^{-7}

[a] Special two-wire cable is required for 12,000-ft repeater spacing. Because T2 cannot use standard exchange cables, it is not as popular as T1.

[b] BnZS denotes *binary n-zero substitution*, where a string of *n* zeros in the bipolar line code is replaced with a special three-level code word so that synchronization can be maintained [Fike and Friend, 1984; Bic, Duponteil, and Imbeaux, 1991].

[c] Used in central telephone office for building multiplex levels; not used for transmission from office to office.

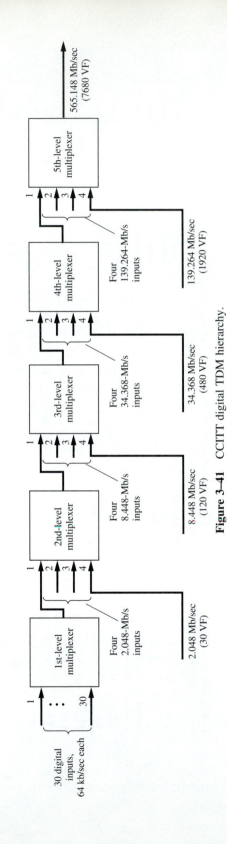

Figure 3–41 CCITT digital TDM hierarchy.

TABLE 3–10   SONET SIGNAL HIERARCHY

OC Level	Line Rate (M Bits/s)	Equivalent Number of		
		DS-3s	DS-1s	DS-0s
OC-1	51.84	1	28	672
OC-3	155.52	3	84	2,016
OC-9	466.56	9	252	6,048
OC-12	622.08	12	336	8,064
OC-18	933.12	18	504	12,096
OC-24	1,244.16	24	672	16,128
OC-36	1,866.24	36	1,008	24,192
OC-48	2,488.32	48	1,344	32,256
OC-192	9,953.28	192	5,376	129,024

of 51.84 Mbits/s. The electrical signal is called the Synchronous Transport Signal—level 1 (STS-1 signal). Other OC-N signals have line rates of exactly N times the OC-1 rate and are formed by modulating a light signal with an STS-N electrical signal. The STS-N signal is obtained by byte-interleaving (scrambling) N STS-1 signals. (More details about fiber-optic systems are given in Sec. 8–7.)

The telephone industry can also provide an all-digital network that integrates voice and data over a single telephone line from each user to the telephone company equipment. One approach is called the *integrated service digital network* (ISDN). Another approach is a digital subscriber line (DSL) technique called G.Lite. This provides an "always-on" 1.5-Mb/s data path (for Internet access) plus a standard VF telephone signal over a single twisted-pair line. (For details on these techniques, see Section 8–3.)

The T1 PCM System

For telephone voice service, the first-level TDM multiplexer in Fig. 3–40 is replaced by a TDM PCM system, which will convert 24-VF analog telephone signals to a DS-1 (1.544-Mbits/s) data stream. In AT&T terminology, this is either a D-type channel bank or a digital carrier trunk (DCT) unit. A T1 line span is a twisted-pair telephone line that is used to carry the DS-1 (1.544 Mbit/s) data stream. Two lines, one for transmiting and one for receiving, are used in the system. If the T1 line is connecting telephone equipment at different sites, repeaters are required about every mile.

The T1 system was developed by Bell Laboratories for short-haul digital communication of VF traffic up to 50 mi. The sampling rate used on each of the 24-VF analog signals is 8 kHz, which means that one frame length is $1/(8 \text{ kHz}) = 125 \ \mu\text{s}$, as shown in Fig. 3–42. Currently, each analog sample is nominally encoded into an 8-bit PCM word, so that there are $8 \times 24 = 192$ bits of data, plus one bit that is added for frame synchronization, yielding a total of 193 bits per frame. The T1 data rate is then $(193 \text{ bits/frame})(8{,}000 \text{ frames/s}) = 1.544$ Mbits/s, and the corresponding duration of each bit is 0.6477, μs. The

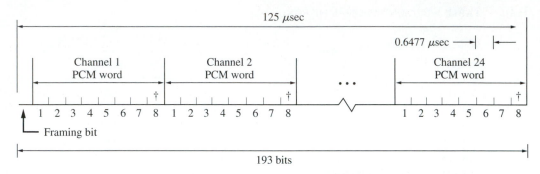

Figure 3–42   T1 TDM format for one frame.

signaling is incorporated into the T1 format by replacing the eighth bit (the least significant bit) in each of the 24 channels of the T1 signal by a signaling bit in every sixth frame. Thus, the signaling data rate for each of the 24 input channels is (1 bit/6 frames)(8,000 frames/s) = 1.333 kbits/s. The framing bit used in the even-numbered frames follows the sequence 001110, and in the odd-numbered frames it follows the sequence 101010, so that the frames with the signaling information in the eighth bit position (for each channel) may be identified. The digital signaling on the T1 line is represented by a bipolar RZ waveform format (see Fig. 3–15) with peak levels of ± 3 V across a 100-Ω load. Consequently, there is no dc component on the T1 line regardless of the data pattern. In encoding the VF PAM samples, a μ = 255-type compression characteristic is used, as described earlier in this chapter. Because the T1 data rate is 1.544 Mbits/s and the line code is bipolar, the first zero-crossing bandwidth is 1.544 MHz and the spectrum peaks at 772 kHz, as shown in Fig. 3–16d. If the channel filtering transfer function was of the raised cosine-rolloff type with $r = 1$, the absolute channel bandwidth would be 1.544 MHz. In the past, when twisted-pair lines were used for analog transmission only, *loading* coils (inductors) were used to improve the frequency (amplitude) response. However, they cause the line to have a phase response that is not linear with frequency, resulting in ISI. Thus, they must be removed when the line is used for T1 service.

The T1G carrier is described in Table 3–9. Instead of binary levels, it uses $M = 4$ (quaternary) multilevel polar NRZ signaling, where +3 V represents the two binary bits 11, +1 V represents 01, −1 V represents 00, and −3 V represents 10 [Azaret et al., 1985]. Thus, a data rate of 6.443 Mbits/s is achieved via 3.221-Mbaud signaling, giving a 3.221-MHz zero-crossing bandwidth, which is close to the 3.152-MHz zero-crossing bandwidth of the T1C system. This bandwidth can be supported by standard twisted-pair wires (one pair for each direction) with repeaters spaced at one-mile intervals.

Fiber-optic cable systems have phenomenal bandwidths and are relatively inexpensive on a per-channel basis. For example, the FT-2000 fiber-optic TDM system (see Tables 8–2 and 8–6) has a capacity of 32,256 VF channels, and the WaveStar system has a capacity of 6.25 million VF channels.

3–10 PACKET TRANSMISSION SYSTEM

TDM is a *synchronous transfer mode* (STM) technology. That is, a data source is assigned a specific time slot with a constant (fixed) data rate. In a number of applications, this fixed data rate assignment is not cost effective, since stuff bits are inserted to match the assigned data rate when the source does not have any data to send. For example, a user might send a large file (for which a large data rate is needed) and then type in some parameters (for which a low data rate is needed). This type of bursty data source is not efficiently accommodated by an STM system, but can be efficiently accommodated by a packet system.

A *packet transmission system* partitions the source data into data packets, each of which contains a destination address header. Many users share the high-speed channel by merging their packets into a high-speed data stream. Routers along the network read the header information on the packets and route the packet to the appropriate destination. To accommodate high-speed sources, more packets are sent from these sources over a given time interval, compared with only a few packets that are merged onto the network from low-speed sources.

Appendix C gives examples of packet transmission systems. Some of these are the Internet TCP/IP technology and the TELCO (telephone company) *asynchronous transfer mode* (ATM) technology.

A packet network efficiently assigns network resources when the sources have bursty data. However, there is network overhead caused by the transmission of the packet header information. STM networks are more efficient when the sources have a fixed data rate (i.e., when they are not bursty).

3–11 PULSE TIME MODULATION: PULSE WIDTH MODULATION AND PULSE POSITION MODULATION

Pulse time modulation (PTM) is a class of signaling techniques that encodes the sample values of an analog signal onto the *time axis* of a digital signal. PTM is analogous to angle modulation, which is described in Chapter 5. (As we have seen, PAM, PCM, and DM techniques encode the sample values into the amplitude characteristics of the digital signal.)

The two main types of PTM are *pulse width modulation* (PWM) and *pulse position modulation* (PPM). (See Fig. 3–43.) In PWM, which is also called *pulse duration modulation* (PDM), sample values of the analog waveform are used to determine the *width* of the pulse signal. Either *instantaneous* sampling or *natural* sampling can be used. Figure 3–44 shows a technique for generating PWM signals with instantaneous sampling, and Fig. 3–45 displays PWM with natural sampling. In PPM, the analog sample value determines the *position* of a narrow pulse relative to the clocking time. Techniques for generating PPM are also shown in the figures, and it is seen that PPM is easily obtained from PWM by using a monostable multivibrator circuit. In the literature on PTM signals, the comparator level V_r is often called the *slicing level*.

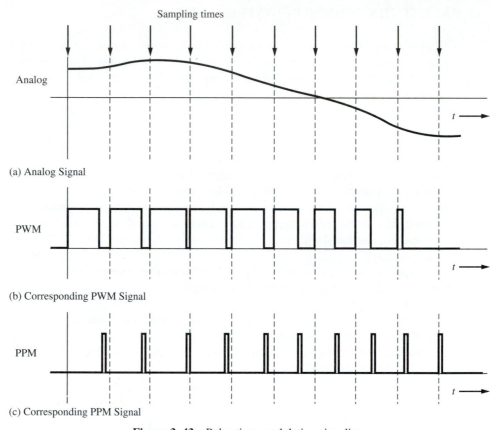

Figure 3–43 Pulse time modulation signaling.

PWM or PPM signals may be converted back to the corresponding analog signal by a receiving system (Fig. 3–46). For PWM detection, the PWM signal is used to start and stop the integration of an integrator; that is, the integrator is reset to zero, and integration is begun when the PWM pulse goes from a low level to a high level and the integrator integrates until the PWM pulse goes low. If the integrator input is connected to a constant voltage source, the output will be a truncated ramp. After the PWM signal goes low, the amplitude of the truncated ramp signal will be equal to the corresponding PAM sample value. At clocking time, the output of the integrator is gated to a PAM output line (slightly before the integrator is reset to zero), using a sample-and-hold circuit. The PAM signal is then converted to the analog signal by low-pass filtering. In a similar way, PPM may be converted to PAM by using the clock pulse to reset the integrator to zero and start the integration. The PPM pulse is then used to stop the integration. The final value of the ramp is the PAM sample that is used to regenerate the analog signal.

Pulse time modulation signaling is not widely used to communicate across channels, because a relatively wide bandwidth channel is needed, especially for PPM. However, PTM signals may be found internally in digital communications terminal equipment. The spectra of PTM signals are quite difficult to evaluate because of the nonlinear nature of the modu-

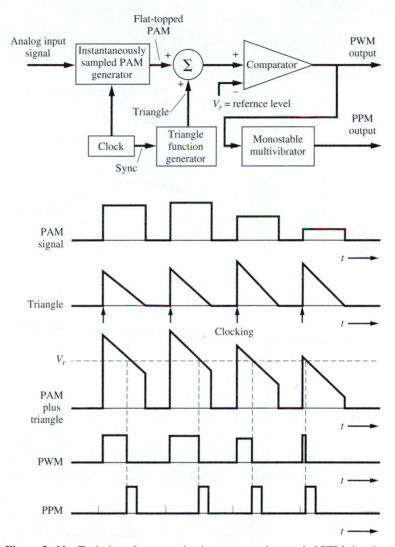

Figure 3–44   Technique for generating instantaneously sampled PTM signals.

lation [Rowe, 1965]. The main advantage of PTM signals is that they have a great immunity to additive noise compared to PAM signaling, and they are easier to generate and detect than PCM, which requires ADCs.

3–12 SUMMARY

In this study of baseband digital signaling, we concentrated on four major topics: (1) how the information in analog waveforms can be represented by digital signaling, (2) how to compute the spectra for line codes, (3) how filtering of the digital signal, due to the communication channel, affects our ability to recover the digital information at the receiver

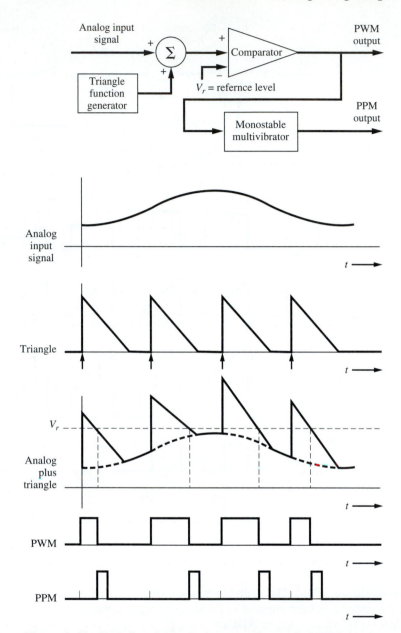

Figure 3–45   Technique for generating naturally sampled PTM signals.

[i.e., the intersymbol interference (ISI) problem], and (4) how we can merge the information from several sources into one digital signal by using time-division multiplexing (TDM). U.S. and worldwide standards for TDM telecommunications systems were given.

   PCM is an analog-to-digital conversion scheme that involves three basic operations: (1) *sampling* a bandlimited analog signal, (2) *quantizing* the analog samples into M discrete

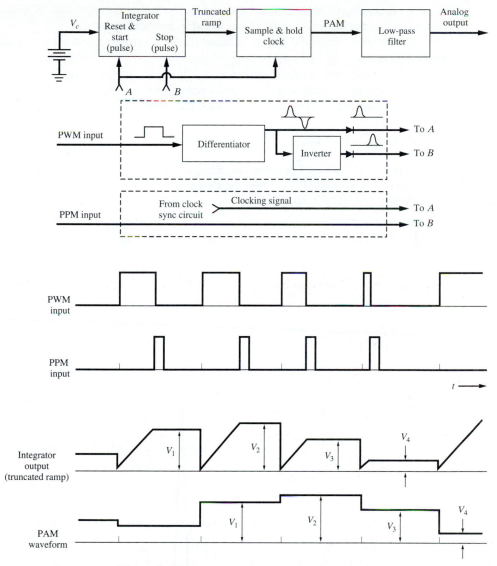

Figure 3–46   Detection of PWM and PPM signals.

values and (3) *encoding* each sample value into an *n*-bit word where $M = 2^n$. There are two sources of noise in the signal that is recovered at the receiver output: (1) quantizing noise due to the approximation of the sample values using the M allowed values and (2) noise due to receiver bit detection errors caused by channel noise or by ISI that arises because of improper channel frequency response. If the original analog signal is not strictly bandlimited, there will be a third noise component on the receiver output due to aliasing.

In studying the effect of improper channel filtering in producing ISI, the raised cosine-rolloff Nyquist filter was examined. Here it was found that the minimum bandwidth required to pass a digital signal without ISI was equal to one-half of the baud rate. A channel bandwidth equal to the baud rate ($r = 1$) was found to be more realistic.

The channel bandwidth may be reduced if multilevel signal techniques are used (for a given data rate R).

This chapter has focused on *baseband* signaling. In the next chapter, we will be concerned with modulating baseband signals onto a carrier so that the spectrum will be concentrated about some desired frequency called the *carrier frequency*.

3–13 STUDY-AID EXAMPLES

SA3–1 PAM Signal Spectrum and Bandwidth   An analog waveform, $w(t)$ is converted into a flat-topped PAM signal by using a sampling rate of 8 kHz and a pulse width of 100 μs. Assume that $W(f) = 2\Lambda(f/B)$, where $B = 3$ kHz.

(a) Find and sketch the magnitude spectrum for the PAM signal.

(b) Find a numerical value for the first null bandwidth of the PAM signal.

Solution.

(a) Using $W(f) = 2\Lambda(f/B)$ in Eq. (3–10), MATLAB computes and plots the spectrum shown in Fig. 3–47. The plot shows how $W(f)$ is repeated at harmonics of the sampling frequency and weighted by the $\tau\left(\dfrac{\sin \pi \tau f}{\pi \tau f}\right)$ function (caused by the rectangular pulse shape).

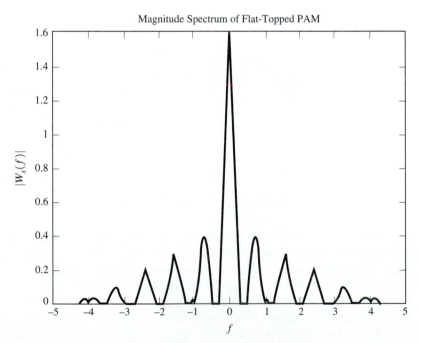

Figure 3–47   Solution for SA3-1.

(b) The spectrum first goes to zero at $B = 3$ kHz. For this spectrum, 3 kHz is not a good measure of bandwidth, because the spectral magnitude becomes large again at higher frequencies. In examples like this, engineers use the *envelope* of the spectrum to specify the null bandwidth. Thus, the first null bandwidth of the spectral envelope, $\tau \left| \dfrac{\sin \pi \tau f}{\pi \tau f} \right|$, is $B_{\text{null}} = 1/\tau = 1/100 \ \mu s = 10$ kHz.

SA3–2 PCM Signal Bandwidth and SNR

In a communications-quality audio system, an analog voice-frequency (VF) signal with a bandwidth of 3,200 Hz is converted into a PCM signal by sampling at 7,000 samples/s and by using a uniform quantizer with 64 steps. The PCM binary data are transmitted over a noisy channel to a receiver that has a bit error rate (BER) of 10^{-4}.

(a) What is the null bandwidth of the PCM signal if a polar line code is used?

(b) What is the average SNR of the recovered analog signal at the receiving end?

Solution.

(a) $M = 64$ quantizing steps generate 6-bit PCM words because $M = 2^n$. Using Eq. (3–15b), we find what the null bandwidth is

$$B_{\text{null}} = nf_s = 6(7,000) = 42 \text{ kHz}$$

Note: If $\sin x / x$ pulse shapes were used, the bandwidth would be

$$B_{\text{null}} = \tfrac{1}{2} nf_s = 21 \text{ kHz}$$

(b) Using Eq. (3–16b) with $M = 64$ and $P_e = 10^{-4}$ yields

$$\left(\frac{S}{N} \right) = \frac{M^2}{1 + 4(M^2 - 1)P_e} = \frac{4,096}{1 + 1.64} = 1,552 = 31.9 \text{ dB}$$

(*Note:* The 1 in the denominator represents quantization noise, and the 1.64 represents noise in the recovered analog signal caused by bit errors at the receiver. In this example, both noise effects contribute almost equally. For the case of $M = 64$, if the BER was less than 10^{-5}, the quantizing noise would dominate, or if the BER was larger than 10^{-3}, noise resulting from receiver bit errors would dominate.)

SA3–3 Properties of NRZ Line Codes

A unipolar NRZ line code is converted to a multilevel signal for transmission over a channel as illustrated in Fig. 3–13. The number of possible values in the multilevel signal is 32, and the signal consists of rectangular pulses that have a pulse width of 0.3472 ms. For the multilevel signal,

(a) What is the baud rate?

(b) What is the equivalent bit rate?

(c) What is the null bandwidth?

(d) Repeat (a) to (c) for the unipolar NRZ line code.

Solution.

(a) Using Eq. (3–28) where $N = 1$ pulse occurs in $T_0 = 0.3452$ ms, we get

$$D = N/T_0 = 1/0.3472 \text{ ms} = 2,880 \text{ baud}$$

(b) Because $L = 32 = 2^\ell$, $\ell = 5$. Using Eq. (3–34),

$$R = \ell D = 5(2880) = 14{,}400 \text{ bits/s}$$

(c) Using Eq. (3–54), we find that the null bandwidth is

$$B_{\text{null}} = R/\ell = D = 2{,}880 \text{ Hz}$$

(d) For the unipolar NRZ line code, there are $N = 5$ pulses in $T_0 = 0.3472$ ms, or

$$D = 5/0.3472 \text{ ms} = 14{,}400 \text{ baud}$$

$R = D$ because the unipolar NRZ line code is binary (i.e., $L = 2^\ell$ or $\ell = 1$). Thus, $R = 14{,}400$ bits/s. The null bandwidth is

$$B_{\text{null}} = R/\ell = D = 14{,}400 \text{ Hz}$$

SA3–4 Bandwidth of RS-232 Signals   The RS-232 serial port on a personal computer is transmitting data at a rate of 38,400 bits/s using a polar NRZ line code. Assume that binary 1's and 0's are equally likely to occur. Compute and plot the PSD for this RS-232 signal. Use a dB scale with the PSD being normalized so that 0 dB occurs at the peak of the PSD plot. Discuss the bandwidth requirements for this signal.

Solution.   Referring to Eq. (3–41), set $A^2 T_b$ equal to 1 so that 0 dB occurs at the peak. Then the PSD, in dB units, is

$$\mathcal{P}_{\text{dB}}(f) = 10 \log\left[\left(\frac{\sin \pi f T_b}{\pi f T_b}\right)^2\right]$$

where $T_b = 1/R$ and $R = 38{,}400$ bits/s.

This result is plotted in Fig. 3–48 using a dB scale. The plot reveals that the spectrum is broad for this case of digital signaling with rectangular pulse shapes. Although the null bandwidth is 38,000 Hz ($B_{\text{null}} = R$), it gives a false sense that the spectrum is relatively narrow, because the first sidelobe peak (at $f = 57{,}600$ Hz $= 1.5R$) is down by only 13.5 dB from the main lobe, and the second sidelobe peak (at $f = 96{,}000$ Hz $= 2.5R$) is down by only 17.9 dB. The power spectrum is falling off as $1/f^2$, which is only 6 dB per octave. Referring to Fig. 3–48 and knowing that the envelope of the PSD is described by $(1/\pi f T_b)^2$, we find that a bandwidth of $f = 386$ kHz $= 10.1R$ (i.e., 10 times the data rate) is needed to pass the frequency components that are not attenuated by more than 30 dB. Thus, the spectrum is broad when rectangular sampling pulses are used. (This was first illustrated in Fig. 2–24.)

Consequently, for applications of digital signaling over bandlimited channels, some sort of filtered pulse shape is required to provide good out-of-band spectral attenuation and yet not introduce ISI. For example, from Eq. (3–74), a filtered pulse corresponding to an $r = 0.5$ raised cosine characteristic would have infinite attenuation at the frequency (absolute bandwidth) of $B = \frac{1}{2}(1 + r)D = (0.5)(1.5)(38{,}400) = 28{,}800$ Hz $= 0.75R$. Referring to Fig. 3–26a and using Eq. (3–69), we see that the $r = 0.5$ raised cosine spectrum would be down by 30 dB at $f = 20{,}070$ Hz $= 0.523R$ and 100 dB at $f = 22{,}217$ Hz $= 0.579R$. For the 30-dB-down bandwidth criterion, this is a bandwidth (savings) of 19 times smaller than that required for rectangular pulse signaling.

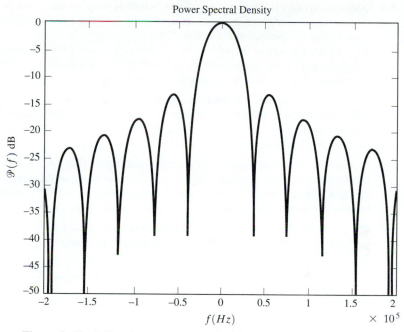

Figure 3–48   PSD of an RS-232 signal with a data rate of 38,400 bits/s.

PROBLEMS

3–1 Demonstrate that the Fourier series coefficients for the switching waveform shown in Fig. 3–1b are given by Eq. (3–5b).

3–2 **(a)** Sketch the naturally sampled PAM waveform that results from sampling a 1-kHz sine wave at a 4-kHz rate.

(b) Repeat part (a) for the case of a flat-topped PAM waveform.

3–3 The spectrum of an analog audio signal is shown in Fig. P3–3. The waveform is to be sampled at a 10-kHz rate with a pulse width of $\tau = 50$ μs.

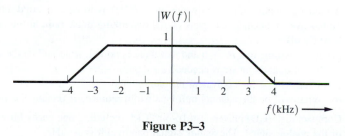

Figure P3–3

(a) Find an expression for the spectrum of the naturally sampled PAM waveform. Sketch your result.

(b) Find an expression for the spectrum of the flat-topped PAM waveform. Sketch your result.

3–4 **(a)** Show that an analog output waveform (which is proportional to the original input analog waveform) may be recovered from a naturally sampled PAM waveform by using the demodulation technique shown in Fig. 3–4.

   (b) Find the constant of proportionality, C, that is obtained with this demodulation technique, where $w(t)$ is the original waveform and $Cw(t)$ is the recovered waveform. Note that C is a function of n, where the oscillator frequency is nf_s.

3–5 Figure 3–4 illustrates how a naturally sampled PAM signal can be demodulated to recover the analog waveform by the use of a product detector. Show that the product detector can also be used to recover $w(t)$ from an instantaneously sampled PAM signal, provided that the appropriate filter $H(f)$ is used. Find the required $H(f)$ characteristic.

3–6 Assume that an analog signal with the spectrum shown in Fig. P3–3 is to be transmitted over a PAM system that has ac coupling. Thus, a PAM pulse shape of the Manchester type, as given by Eq. (3–46a), is used. The PAM sampling frequency is 10 kHz. Find the spectrum for the Manchester encoded flat-topped PAM waveform. Sketch your result.

3–7 In a binary PCM system, if the quantizing noise is not to exceed $\pm P$ percent of the peak-to-peak analog level, show that the number of bits in each PCM word needs to be

$$n \geq [\log_2 10] \left[\log_{10}\left(\frac{50}{P} \right) \right] = 3.32 \, \log_{10}\left(\frac{50}{P} \right)$$

(*Hint:* Look at Fig. 3–8c.)

3–8 The information in an analog voltage waveform is to be transmitted over a PCM system with a $\pm 0.1\%$ accuracy (full scale). The analog waveform has an absolute bandwidth of 100 Hz and an amplitude range of -10 to $+10$ V.

   (a) Determine the minimum sampling rate needed.
   (b) Determine the number of bits needed in each PCM word.
   (c) Determine the minimum bit rate required in the PCM signal.
   (d) Determine the minimum absolute channel bandwidth required for transmission of this PCM signal.

3–9 An 850-Mbyte hard disk is used to store PCM data. Suppose that a voice-frequency (VF) signal is sampled at 8 ksamples/s and the encoded PCM is to have an average SNR of at least 30 dB. How many minutes of VF conversation (i.e., PCM data) can be stored on the hard disk?

3–10 An analog signal with a bandwidth of 4.2 MHz is to be converted into binary PCM and transmitted over a channel. The peak-signal quantizing noise ratio at the receiver output must be at least 55 dB.

   (a) If we assume that $P_e = 0$ and that there is no ISI, what will be the word length and the number of quantizing steps needed?
   (b) What will be the equivalent bit rate?
   (c) What will be the channel null bandwidth required if rectangular pulse shapes are used?

3–11 Compact disk (CD) players use 16-bit PCM, including one parity bit with 8 times oversampling of the analog signal. The analog signal bandwidth is 20 kHz.

   (a) What is the null bandwidth of this PCM signal?
   (b) Using Eq. (3–18), find the peak (SNR) in decibels.

3–12 Given an audio signal with spectral components in the frequency band 300 to 3,000 Hz, assume that a sampling rate of 7 kHz will be used to generate a PCM signal. Design an appropriate PCM system as follows:

(a) Draw a block diagram of the PCM system, including the transmitter, channel, and receiver.

(b) Specify the number of uniform quantization steps needed and the channel null bandwidth required, assuming that the peak signal-to-noise ratio at the receiver output needs to be at least 30 dB and that polar NRZ signaling is used.

(c) Discuss how nonuniform quantization can be used to improve the performance of the system.

3–13 The SNRs, as given by Eqs. (3–17a) and (3–17b), assume no ISI and no bit errors due to channel noise (i.e., $P_e = 0$). How large can P_e become before Eqs. (3–17a) and (3–17b) are in error by 0.1% if $M = 4, 8$, or 16.

3–14 In a PCM system, the bit error rate due to channel noise is 10^{-4}. Assume that the peak signal-to-noise ratio on the recovered analog signal needs to be at least 30 dB.

(a) Find the minimum number of quantizing steps that can be used to encode the analog signal into a PCM signal.

(b) If the original analog signal had an absolute bandwidth of 2.7 kHz, what is the null bandwidth of the PCM signal for the polar NRZ signaling case?

3–15 Referring to Fig. 3–20 for a bit synchronizer using a square-law device, draw some typical waveforms that will appear in the bit synchronizer if a Manchester encoded PCM signal is present at the input. Discuss whether you would expect this bit synchronizer to work better for the Manchester encoded PCM signal or for a polar NRZ encoded PCM signal.

3–16 (a) Sketch the complete $\mu = 10$ compressor characteristic that will handle input voltages over the range -5 to $+5$ V.

(b) Plot the corresponding expandor characteristic.

(c) Draw a 16-level nonuniform quantizer characteristic that corresponds to the $\mu = 10$ compression characteristic.

3–17 For a 4-bit PCM system, calculate and sketch a plot of the output SNR (in decibels) as a function of the relative input level, $20 \log (x_{rms}/V)$ for

(a) A PCM system that uses $\mu = 10$ law companding.

(b) A PCM system that uses uniform quantization (no companding).

Which of these systems is better to use in practice? Why?

3–18 The performance of a $\mu = 255$ law companded PCM system is to be examined when the input consists of a sine wave having a peak value of V volts. Assume that $M = 256$.

(a) Find an expression that describes the output SNR for this companded PCM system.

(b) Plot $(S/N)_{out}$ (in decibels) as a function of the relative input level, $20 \log (x_{rms}/V)$. Compare this result with that shown in Fig. 3–10.

3–19 A multilevel digital communication system sends one of 16 possible levels over the channel every 0.8 ms.

(a) What is the number of bits corresponding to each level?

(b) What is the baud rate?

(c) What is the bit rate?

3–20 A multilevel digital communication system is to operate at a data rate of 9,600 bits/s.

(a) If 4-bit words are encoded into each level for transmission over the channel, what is the minimum required bandwidth for the channel?

(b) Repeat part (a) for the case of 8-bit encoding into each level.

3–21 Consider a deterministic test pattern consisting of alternating binary 1's and binary 0's. Determine the magnitude spectra (not the PSD) for the following types of signaling formats as a function of T_b, the time needed to send one bit of data:

(a) Unipolar NRZ signaling.

(b) Unipolar RZ signaling where the pulse width is $\tau = \frac{3}{4}T_b$.

How would each of these magnitude spectra change if the test pattern was changed to an alternating sequence of four binary 1's followed by four binary 0's?

3–22 Consider a random data pattern consisting of binary 1's and 0's, where the probability of obtaining either a binary 1 or a binary 0 is $\frac{1}{2}$. Calculate the PSD for the following types of signaling formats as a function of T_b, the time needed to send one bit of data:

(a) Unipolar NRZ signaling.

(b) Unipolar RZ signaling where the pulse width is $\tau = \frac{3}{4}T_b$.

How do these PSDs for the random data cases compare to the magnitude spectra for the deterministic case of Prob. 3–21? What is the spectral efficiency for each of these cases?

3–23 Consider a deterministic data pattern consisting of alternating binary 1's and 0's. Determine the magnitude spectra (not the PSD) for the following types of signaling formats as a function of T_b, the time needed to send one bit of data:

(a) Polar NRZ signaling.

(b) Manchester NRZ signaling.

How would each of these magnitude spectra change if the test pattern was changed to an alternating sequence of four binary 1's followed by two binary 0's?

3–24 Consider a random data pattern consisting of binary 1's and 0's, where the probability of obtaining either a binary 1 or a binary 0 is $\frac{1}{2}$. Calculate the PSD for the following types of signaling formats as a function of T_b, the time needed to send 1 bit of data:

(a) Polar RZ signaling where the pulse width is $\tau = \frac{1}{2}T_b$.

(b) Manchester RZ signaling where the pulse width is $\tau = \frac{1}{4}T_b$. What is the first null bandwidth of these signals? What is the spectral efficiency for each of these signaling cases?

3–25 Obtain the equations for the PSD of the bipolar NRZ and bipolar RZ (pulse width $\frac{1}{2}T_b$) line codes assuming peak values of ± 3 V. Plot these PSD results for the case of $R = 1.544$ Mbits/s.

3–26 In Fig. 3–16, the PSDs for several line codes are shown. These PSDs were derived assuming unity power for each signal so that the PSDs could be compared on an equal transmission power basis. Rederive the PSDs for these line codes, assuming that the peak level is unity (i.e., $A = 1$). Plot the PSDs so that the spectra can be compared on an equal peak-signal-level basis.

3–27 Using Eq. (3–36), determine the conditions required so that there are delta functions in the PSD for line codes. Discuss how this affects the design of bit synchronizers for these line codes. [*Hint:* Examine Eq. (3–43) and (6–70d).]

3–28 Consider a random data pattern consisting of binary 1's and 0's, where the probability of obtaining either a binary 1 or a binary 0 is $\frac{1}{2}$. Assume that these data are encoded into a polar-type waveform such that the pulse shape of each bit is given by

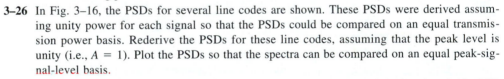

$$f(t) = \begin{cases} \cos\left(\dfrac{\pi t}{T_b}\right), & |t| < T_b/2 \\ 0, & |t| \text{ elsewhere} \end{cases}$$

where T_b is the time needed to send one bit.

(a) Sketch a typical example of this waveform.

(b) Find the expression for the PSD of this waveform and sketch it.

(c) What is the spectral efficiency of this type of binary signal?

3–29 The data stream 01101000101 appears at the input of a differential encoder. Depending on the initial start-up condition of the encoder, find two possible differentially encoded data streams that can appear at the output.

3–30 Create a practical block diagram for a differential encoding and decoding system. Explain how the system works by showing the encoding and decoding for the sequence 001111010001. Assume that the reference digit is a binary 1. Show that error propagation cannot occur.

3–31 Design a regenerative repeater with its associated bit synchronizer for a polar RZ line code. Explain how your design works. (*Hint:* See Fig. 3–19 and the discussion of bit synchronizers.)

3–32 Design a bit synchronizer for a Manchester NRZ line code by completing the following steps:
(a) Give a simplified block diagram.
(b) Explain how the synchronizer works.
(c) Specify the synchronizer's filter requirements.
(d) Explain the advantages and disadvantages of using this design for the Manchester NRZ line code compared with using a polar NRZ line code and its associated bit synchronizer.

3–33 Figure 3–22c illustrates an eight-level multilevel signal. Assume that this line code is passed through a channel that filters the signal and adds some noise.
(a) Draw a picture of the eye pattern for the received waveform.
(b) Design a possible receiver with its associated symbol synchronizer for this line code.
(c) Explain how your receiver works.

3–34 The information in an analog waveform is first encoded into binary PCM and then converted to a multilevel signal for transmission over the channel. The number of multilevels is eight. Assume that the analog signal has a bandwidth of 2,700 Hz and is to be reproduced at the receiver output with an accuracy of ±1% (full scale).
(a) Determine the minimum bit rate of the PCM signal.
(b) Determine the minimum baud rate of the multilevel signal.
(c) Determine the minimum absolute channel bandwidth required for transmission of this PCM signal.

3–35 A binary waveform of 9,600 bits/s is converted into an octal (multilevel) waveform that is passed through a channel with a raised cosine-rolloff Nyquist filter characteristic. The channel has a conditioned (equalized) phase response out to 2.4 kHz.
(a) What is the baud rate of the multilevel signal?
(b) What is the rolloff factor of the filter characteristic?

3–36 Assume that the spectral properties of an $L = 64$-level waveform with rectangular RZ-type pulse shapes are to be examined. The pulse shape is given by

$$f(t) = \Pi\left(\frac{2t}{T_s}\right)$$

where T_s is the time needed to send one of the multilevel symbols.
(a) Determine the expression for the PSD for the case of equally likely levels where the peak signal levels for this multilevel waveform are +10 V.
(b) What is the null bandwidth?
(c) What is the spectral efficiency?

3–37 A binary communication system uses polar signaling. The overall impulse response is designed to be of the $(\sin x)/x$ type, as given by Eq. (3–67), so that there will be no ISI. The bit rate is $R = f_s = 300$ bits/s.
(a) What is the bandwidth of the polar signal?
(b) Plot the waveform of the polar signal at the system output when the input binary data is 01100101. Can you discern the data by looking at this polar waveform?

3–38 Equation (3–67) gives one possible noncausal impulse response for a communication system that will have no ISI. For a causal approximation, select

$$h_e(t) = \frac{\sin \pi f_s(t - 4 \times 10^{-3})}{\pi f_s(t - 4 \times 10^{-3})} \; \Pi \left(\frac{t - 4 \times 10^{-3}}{8 \times 10^{-3}} \right)$$

where $f_s = 1{,}000$.

(a) Using a PC, calculate $H_e(f)$ by the use of the Fourier transform integral, and plot $|H_e(f)|$.

(b) What is the bandwidth of this causal approximation, and how does it compare with the bandwidth of the noncausal filter described by Eqs. (3–67) and (3–68)?

3–39 Starting with Eq. (3–69), prove that the impulse response of the raised cosine rolloff filter is given by Eq. (3–73).

3–40 Consider the raised cosine-rolloff Nyquist filter given by Eqs. (3–69) and (3–73).

(a) Plot $|H_e(f)|$ for the case of $r = 0.75$, indicating f_1, f_0, and B on your sketch in a manner similar to Fig. 3–25.

(b) Plot $h_e(t)$ for the case of $r = 0.75$ in terms of $1/f_0$. Your plot should be similar to Fig. 3–26.

3–41 Find the PSD of the waveform out of an $r = 0.5$ raised cosine-rolloff channel when the input is a polar NRZ signal. Assume that equally likely binary signaling is used and the channel bandwidth is just large enough to prevent ISI.

3–42 Equation (3–66) gives the condition for the absence of ISI (Nyquist's first method). Using that equation with $C = 1$ and $\tau = 0$, show that Nyquist's first method for eliminating ISI is also satisfied if

$$\sum_{k=-\infty}^{\infty} H_e\left(f + \frac{k}{T_s}\right) = T_s \qquad \text{for } |f| \le \frac{1}{2T_s}$$

3–43 Using the results of Prob. 3–42, demonstrate that the following filter characteristics do or do not satisfy Nyquist's criterion for eliminating ISI ($f_s = 2f_0 = 2/T_0$).

(a) $H_e(f) = \dfrac{T_0}{2} \; \Pi \left(\dfrac{1}{2} fT_0 \right).$

(b) $H_e(f) = \dfrac{T_0}{2} \; \Pi \left(\dfrac{2}{3} fT_0 \right).$

3–44 Assume that a pulse transmission system has the overall raised cosine-rolloff Nyquist filter characteristic described by Eq. (3–69).

(a) Find the $Y(f)$ Nyquist function of Eq. (3–75) corresponding to the raised cosine-rolloff Nyquist filter characteristic.

(b) Sketch $Y(f)$ for the case of $r = 0.75$.

(c) Sketch another $Y(f)$ that is not of the raised cosine-rolloff type, and determine the absolute bandwidth of the resulting Nyquist filter characteristic.

3–45 An analog signal is to be converted into a PCM signal that is a binary polar NRZ line code. The signal is transmitted over a channel that is absolutely bandlimited to 4 kHz. Assume that the PCM quantizer has 16 steps and that the overall equivalent system transfer function is of the raised cosine-rolloff type with $r = 0.5$.

(a) Find the maximum PCM bit rate that can be supported by this system without introducing ISI.

(b) Find the maximum bandwidth that can be permitted for the analog signal.

3–46 Rework Prob. 3–45 for the case of a multilevel polar NRZ line code when the number of levels is four.

3–47 Multilevel data with an equivalent bit rate of 2,400 bits/s is sent over a channel using a four-level line code that has a rectangular pulse shape at the output of the transmitter. The overall transmission system (i.e., the transmitter, channel, and receiver) has an $r = 0.5$ raised cosine-rolloff Nyquist filter characteristic.
  (a) Find the baud rate of the received signal.
  (b) Find the 6-dB bandwidth for this transmission system.
  (c) Find the absolute bandwidth for the system.

3–48 Assume that a PCM-type system is to be designed such that an audio signal can be delivered at the receiver output. This audio signal is to have a bandwidth of 3,400 Hz and an SNR of at least 40 dB. Determine the bit rate requirements for a design that uses
  (a) $\mu = 255$ companded PCM signaling.
  (b) DPCM signaling.
  Discuss which of the preceding systems would be used in your design and why.

3–49 Refer to Fig. 3–32, which shows typical DM waveforms. Draw an analog waveform that is different from the one shown in the figure. Draw the corresponding DM and integrator output waveforms. Denote the regions where slope overload noise dominates and where granular noise dominates.

3–50 A DM system is tested with a 10-kHz sinusoidal signal, 1 V peak to peak, at the input. The signal is sampled at 10 times the Nyquist rate.
  (a) What is the step size required to prevent slope overload and to minimize granular noise?
  (b) What is the PSD for the granular noise?
  (c) If the receiver input is bandlimited to 200 kHz, what is the average signal–quantizing noise power ratio?

3–51 Assume that the input to a DM is $0.1t^8 - 5t + 2$. The step size of the DM is 1 V, and the sampler operates at 10 samples/s. Sketch the input waveform, the delta modulator output, and the integrator output over a time interval of 0 to 2 s. Denote the granular noise and slope overload regions.

Use Matlab

3–52 Rework Prob. 3–51 for the case of an adaptive delta modulator where the step size is selected according to the number of successive binary 1's or 0's on the DM output. Assume that the step size is 1.5 V when there are four or more binary digits of the same sign, 1 V for the case of three successive digits, and 0.5 V for the case of two or fewer successive digits.

3–53 A delta modulator is to be designed to transmit the information of an analog waveform that has a peak-to-peak level of 1 V and a bandwidth of 3.4 kHz. Assume that the waveform is to be transmitted over a channel whose frequency response is extremely poor above 1 MHz.
  (a) Select the appropriate step size and sampling rate for a sine-wave test signal, and discuss the performance of the system, using the parameter values you have selected.
  (b) If the DM system is to be used to transmit the information of a voice (analog) signal, select the appropriate step size when the sampling rate is 25 kHz. Discuss the performance of the system under these conditions.

3–54 One analog waveform $w_1(t)$ is bandlimited to 3 kHz, and another, $w_2(t)$, is bandlimited to 9 kHz. These two signals are to be sent by TDM over a PAM-type system.
  (a) Determine the minimum sampling frequency for each signal, and design a TDM commutator and decommutator to accommodate the signals.
  (b) Draw some typical waveforms for $w_1(t)$ and $w_2(t)$, and sketch the corresponding TDM PAM waveform.

3–55 Three waveforms are time-division multiplexed over a channel using instantaneously sampled PAM. Assume that the PAM pulse width is very narrow and that each of the analog waveforms is sampled every 0.15 s. Plot the (composite) TDM waveform when the input analog waveforms are

$$w_1(t) = 3\sin(2\pi t)$$

$$w_2(t) = \Pi\left(\frac{t-1}{2}\right)$$

and

$$w_3(t) = -\Lambda(t-1)$$

3–56 Twenty-three analog signals, each with a bandwidth of 3.4 kHz, are sampled at an 8-kHz rate and multiplexed together with a synchronization channel (8 kHz) into a TDM PAM signal. This TDM signal is passed through a channel with an overall raised cosine-rolloff Nyquist filter characteristic of $r = 0.75$.

(a) Draw a block diagram for the system, indicating the f_s of the commutator and the overall pulse rate of the TDM PAM signal.

(b) Evaluate the absolute bandwidth required for the channel.

3–57 Two flat-topped PAM signals are time-division multiplexed together to produce a composite TDM PAM signal that is transmitted over a channel. The first PAM signal is obtained from an analog signal that has a rectangular spectrum, $W_1(f) = \Pi(f/2B)$. The second PAM signal is obtained from an analog signal that has a triangular spectrum $W_2(f) = \Lambda(f/B)$, where $B = 3$ kHz.

(a) Determine the minimum sampling frequency for each signal, and design a TDM commutator and decommutator to accommodate these signals.

(b) Calculate and sketch the magnitude spectrum for the composite TDM PAM signal.

3–58 Rework Prob. 3–56 for a TDM pulse code modulation system in which an 8-bit quantizer is used to generate the PCM words for each of the analog inputs and an 8-bit synchronization word is used in the synchronization channel.

3–59 Design a TDM PCM system that will accommodate four 300-bit/s (synchronous) digital inputs and one analog input that has a bandwidth of 500 Hz. Assume that the analog samples will be encoded into 4-bit PCM words. Draw a block diagram for your design, analogous to Fig. 3–39, indicating the data rates at the various points on the diagram. Explain how your design works.

3–60 Design a TDM system that will accommodate two 2,400-bit/s synchronous digital inputs and an analog input that has a bandwidth of 2,700 Hz. Assume that the analog input is sampled at 1.11111 times the Nyquist rate and converted into 4-bit PCM words. Draw a block diagram for your design, and indicate the data rate at various points on your diagram. Explain how your TDM scheme works.

3–61 Find the number of the following devices that could be accommodated by a T1-type TDM line if 1% of the line capacity were reserved for synchronization purposes.

(a) 110-bit/s teleprinter terminals.

(b) 8,000-bit/s speech codecs.

(c) 9,600-bit/s computer output ports.

(d) 64-kbit/s PCM VF lines.

(e) 144-kbit/s ISDN terminals.

How would these numbers change if each of the sources was operational an average of 10% of the time?

3–62 Assume that a sine wave is sampled at four times the Nyquist rate using instantaneous sampling.
   (a) Sketch the corresponding PWM signal.
   (b) Sketch the corresponding PPM signal.

3–63 Discuss why a PPM system requires a synchronizing signal, whereas PAM and PWM can be detected without the need for a synchronizing signal.

3–64 Compare the bandwidth required to send a message by using PPM and PCM signaling. Assume that the digital source sends out 8 bits/character, so that it can send 256 different messages (characters). Assume that the source rate is 10 characters/s. Use the dimensionality theorem, $N/T_0 = 2B$, to determine the minimum bandwidth B required.
   (a) Determine the minimum bandwidth required for a PCM signal that will encode the source information.
   (b) Determine the minimum bandwidth required for a PPM signal that will encode the source information.

CHAPTER 4

BANDPASS SIGNALING PRINCIPLES AND CIRCUITS

This chapter is concerned with *bandpass* signaling techniques. As indicated in Chapter 1, the bandpass communication signal is obtained by modulating a baseband analog or digital signal onto a carrier. This is an exciting chapter because the basic principles of bandpass signaling are revealed. The complex envelope is used, since it can represent any type of bandpass signal. This is the basis for understanding digital and analog communication systems that are described in more detail in Chapters 5 and 8. This chapter also describes practical aspects of the building blocks used in communication systems. These building blocks are filters, linear and nonlinear amplifiers, mixers, up and down converters, modulators, detectors, and phase-locked loops. The chapter concludes with descriptions of transmitters, receivers, and software radios.

4–1 COMPLEX ENVELOPE REPRESENTATION OF BANDPASS WAVEFORMS

What is a general representation for bandpass digital and analog signals? How do we represent a modulated signal? How do we represent bandpass noise? These are some of the questions that are answered in this section.

230

Definitions: Baseband, Bandpass, and Modulation

DEFINITION.   A *baseband* waveform has a spectral magnitude that is nonzero for frequencies in the vicinity of the origin (i.e., $f = 0$) and negligible elsewhere.

DEFINITION.   A *bandpass* waveform has a spectral magnitude that is nonzero for frequencies in some band concentrated about a frequency $f = \pm f_c$, where $f_c \gg 0$. The spectral magnitude is negligible elsewhere. f_c is called the *carrier frequency*.

For bandpass waveforms, the value of f_c may be arbitrarily assigned for mathematical convenience in *some* problems. In others, namely, modulation problems, f_c is the frequency of an oscillatory signal in the transmitter circuit and is the assigned frequency of the transmitter, such as, for example, 850 kHz for an AM broadcasting station.

In communication problems, the information source signal is usually a baseband signal—for example, a transistor–transistor logic (TTL) waveform from a digital circuit or an audio (analog) signal from a microphone. The communication engineer has the job of building a system that will transfer the information in the source signal $m(t)$ to the desired destination. As shown in Fig. 4–1, this usually requires the use of a bandpass signal, $s(t)$, which has a bandpass spectrum that is concentrated at $\pm f_c$, where f_c is selected so that $s(t)$ will propagate across the communication channel (either a wire or a wireless channel).

DEFINITION.   *Modulation* is the process of imparting the source information onto a bandpass signal with a carrier frequency f_c by the introduction of amplitude or phase perturbations or both. This bandpass signal is called the *modulated* signal $s(t)$, and the baseband source signal is called the *modulating* signal $m(t)$.

Examples of exactly how modulation is accomplished are given later in this chapter. This definition indicates that modulation may be visualized as a mapping operation that maps the source information onto the bandpass signal, $s(t)$. The bandpass signal will be transmitted over the channel.

As the modulated signal passes through the channel, noise corrupts it. The result is a bandpass signal-plus-noise waveform that is available at the receiver input, $r(t)$. (See Fig. 4–1.) The receiver has the job of trying to recover the information that was sent from the source; $\tilde{m}$ denotes the corrupted version of m.

Complex Envelope Representation

All bandpass waveforms, whether they arise from a modulated signal, interfering signals, or noise, may be represented in a convenient form given by the theorem that follows. $v(t)$ will be used to denote the bandpass waveform canonically; specifically, $v(t)$ can represent the

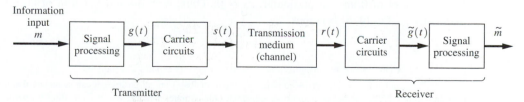

Figure 4–1   Communication system.

signal when $s(t) \equiv v(t)$, the noise when $n(t) \equiv v(t)$, the filtered signal plus noise at the channel output when $r(t) \equiv v(t)$, or any other type of bandpass waveform.[†]

THEOREM. *Any physical bandpass waveform can be represented by*

$$v(t) = \mathrm{Re}\{g(t)e^{j\omega_c t}\} \tag{4-1a}$$

Here, $\mathrm{Re}\{\cdot\}$ denotes the real part of $\{\cdot\}$, $g(t)$ is called the complex envelope of $v(t)$, and f_c is the associated carrier frequency (in hertz) where $\omega_c = 2\pi f_c$. Furthermore, two other equivalent representations are

$$v(t) = R(t)\cos[\omega_c t + \theta(t)] \tag{4-1b}$$

and

$$v(t) = x(t)\cos\omega_c t - y(t)\sin\omega_c t \tag{4-1c}$$

where

$$g(t) = x(t) + jy(t) = |g(t)|e^{j\underline{/g(t)}} \equiv R(t)e^{j\theta(t)} \tag{4-2}$$

$$x(t) = \mathrm{Re}\{g(t)\} \equiv R(t)\cos\theta(t) \tag{4-3a}$$

$$y(t) = \mathrm{Im}\{g(t)\} \equiv R(t)\sin\theta(t) \tag{4-3b}$$

$$R(t) \triangleq |g(t)| \equiv \sqrt{x^2(t) + y^2(t)} \tag{4-4a}$$

and

$$\theta(t) \triangleq \underline{/g(t)} = \tan^{-1}\left(\frac{y(t)}{x(t)}\right) \tag{4-4b}$$

Proof. Any physical waveform (it does not have to be periodic) may be represented over all time, $T_0 \to \infty$, by the complex Fourier series:

$$v(t) = \sum_{n=-\infty}^{n=\infty} c_n e^{jn\omega_0 t}, \quad \omega_0 = 2\pi/T_0 \tag{4-5}$$

Furthermore, because the physical waveform is real, $c_{-n} = c_n^*$, and, using $\mathrm{Re}\{\cdot\} = \frac{1}{2}\{\cdot\} + \frac{1}{2}\{\cdot\}^*$, we obtain

$$v(t) = \mathrm{Re}\left\{c_0 + 2\sum_{n=1}^{\infty} c_n e^{jn\omega_0 t}\right\} \tag{4-6}$$

Furthermore, because $v(t)$ is a bandpass waveform, the c_n have negligible magnitudes for n in the vicinity of 0 and, in particular, $c_0 = 0$. Thus, with the introduction of an arbitrary parameter f_c, Eq. (4-6) becomes[†]

[†] The symbol $\equiv$ denotes an equivalence, and the symbol $\triangleq$ denotes a definition.

[†] Because the frequencies involved in the argument of $\mathrm{Re}\{\cdot\}$ are all positive, it can be shown that the complex function $2\sum_{n=1}^{\infty} c_n e^{jn\omega_0 t}$ is analytic in the upper-half complex t plane. Many interesting properties result because this function is an analytic function of a complex variable.

$$v(t) = \text{Re}\left\{\left(2\sum_{n=1}^{n=\infty} c_n e^{j(n\omega_0 - \omega_c)t}\right) e^{j\omega_c t}\right\} \tag{4-7}$$

so that Eq. (4–1a) follows, where

$$g(t) \equiv 2\sum_{n=1}^{\infty} c_n e^{j(n\omega_0 - \omega_c)t} \tag{4-8}$$

Because $v(t)$ is a bandpass waveform with nonzero spectrum concentrated near $f = f_c$, the Fourier coefficients c_n are nonzero only for values of n in the range $\pm nf_0 \approx f_c$. Therefore, from Eq. (4–8), $g(t)$ has a spectrum that is concentrated near $f = 0$. That is, $g(t)$ is a *baseband* waveform.

The waveforms $g(t)$, (and consequently) $x(t)$, $y(t)$, $R(t)$, and $\theta(t)$ are all baseband waveforms, and, except for $g(t)$, they are all real waveforms. $R(t)$ is a nonnegative real waveform. Equation (4–1) is a low-pass-to-bandpass transformation. The $e^{j\omega_c t}$ factor in Eq. (4–1a) shifts (i.e., translates) the spectrum of the baseband signal $g(t)$ from baseband up to the carrier frequency f_c. In communications terminology, the frequencies in the baseband signal $g(t)$ are said to be *heterodyned* up to f_c. The complex envelope, $g(t)$, is usually a complex function of time, and it is the generalization of the phasor concept. That is, if $g(t)$ happens to be a complex constant, then $v(t)$ is a pure sinusoidal waveshape of frequency f_c, and this complex constant is the phasor representing the sinusoid. If $g(t)$ is not a constant, then $v(t)$ is not a pure sinusoid, because the amplitude and phase of $v(t)$ vary with time, caused by the variations in $g(t)$.

Representing the complex envelope in terms of two real functions in Cartesian coordinates, we have

$$g(x) \equiv x(t) + jy(t)$$

where $x(t) = \text{Re}\{g(t)\}$ and $y(t) = \text{Im}\{g(t)\}$. $x(t)$ is said to be the *in-phase modulation* associated with $v(t)$, and $y(t)$ is said to be the *quadrature modulation* associated with $v(t)$. Alternatively, the polar form of $g(t)$, represented by $R(t)$ and $\theta(t)$, is given by Eq. (4–2), where the identities between Cartesian and polar coordinates are given by Eqs. (4–3) and (4–4). $R(t)$ and $\theta(t)$ are real waveforms, and in addition, $R(t)$ is always nonnegative. $R(t)$ is said to be the *amplitude modulation (AM)* on $v(t)$, $\theta(t)$ is said to be the *phase modulation* (PM) on $v(t)$.

The usefulness of the complex envelope representation for bandpass waveforms cannot be overemphasized. In modern communication systems, the bandpass signal is often partitioned into two channels, one for $x(t)$ called the I (in-phase) channel and one for $y(t)$ called the Q (quadrature-phase) channel. In digital computer simulations of bandpass signals, the sampling rate used in the simulation can be minimized by working with the complex envelope $g(t)$, instead of with the bandpass signal $v(t)$, because $g(t)$ is the baseband equivalent of the bandpass signal.

4–2 REPRESENTATION OF MODULATED SIGNALS

Modulation is the process of encoding the source information $m(t)$ (modulating signal) into a bandpass signal $s(t)$ (modulated signal). Consequently, the modulated signal is just a special application of the bandpass representation. The *modulated signal* is given by

$$s(t) = \text{Re}\{g(t)e^{j\omega_c t}\} \tag{4-9}$$

where $\omega_c = 2\pi f_c$, in which f_c is the carrier frequency. The complex envelope $g(t)$ is a function of the modulating signal $m(t)$. That is,

$$g(t) = g[m(t)] \tag{4-10}$$

Thus, $g[\cdot]$ performs a mapping operation on $m(t)$. This was shown in Fig. 4–1.

Table 4–1 gives the "big picture" of the modulation problem. Examples of the mapping function $g[m]$ are given for amplitude modulation (AM), double-sideband suppressed carrier (DSB-SC), phase modulation (PM), frequency modulation (FM), single-sideband AM suppressed carrier (SSB-AM- SC), single-sideband PM (SSB-PM), single-sideband FM (SSB-FM), single-sideband envelope detectable (SSB-EV), single-sideband square-law detectable (SSB-SQ), and quadrature modulation (QM). Digital and analog modulated signals are discussed in detail in Chapter 5. Digitally modulated bandpass signals are obtained when $m(t)$ is a digital baseband signal—for example, the output of a transistor–transistor logic (TTL) circuit.

Obviously, it is possible to use other $g[m]$ functions that are not listed in Table 4–1. The question is; Are they useful? $g[m]$ functions that are easy to implement and that will give desirable spectral properties are sought. Furthermore, in the receiver, the inverse function $m[g]$ is required. The inverse should be single valued over the range used and should be easily implemented. The mapping should suppress as much noise as possible so that $m(t)$ can be recovered with little corruption.

4–3 SPECTRUM OF BANDPASS SIGNALS

The spectrum of a bandpass signal is directly related to the spectrum of its complex envelope.

THEOREM. *If a bandpass waveform is represented by*

$$v(t) = \text{Re}\{g(t)e^{j\omega_c t}\} \tag{4-11}$$

then the spectrum of the bandpass waveform is

$$V(f) = \tfrac{1}{2}[G(f - f_c) + G^*(-f - f_c)] \tag{4-12}$$

and the PSD of the waveform is

$$\mathcal{P}_v(f) = \tfrac{1}{4}[\mathcal{P}_g(f - f_c) + \mathcal{P}_g(-f - f_c)] \tag{4-13}$$

where $G(f) = \mathcal{F}[g(t)]$ and $\mathcal{P}_g(f)$ is the PSD of $g(t)$.

Proof.

$$v(t) = \text{Re}\{g(t)e^{j\omega_c t}\} = \tfrac{1}{2}g(t)e^{j\omega_c t} + \tfrac{1}{2}g^*(t)e^{-j\omega_c t}$$

Thus,

$$V(f) = \mathcal{F}[v(t)] = \tfrac{1}{2}\,\mathcal{F}[g(t)e^{j\omega_c t}] + \tfrac{1}{2}\mathcal{F}[g^*(t)e^{-j\omega_c t}] \tag{4-14}$$

TABLE 4-1 COMPLEX ENVELOPE FUNCTIONS FOR VARIOUS TYPES OF MODULATION[a]

Type of Modulation	Mapping Functions $g(m)$	Corresponding Quadrature Modulation	
		$x(t)$	$y(t)$
AM	$A_c[1+m(t)]$	$A_c[1+m(t)]$	0
DSB-SC	$A_c m(t)$	$A_c m(t)$	0
PM	$A_c e^{jD_p m(t)}$	$A_c \cos[D_p m(t)]$	$A_c \sin[D_p m(t)]$
FM	$A_c e^{jD_f \int_{-\infty}^{t} m(\sigma)\, d\sigma}$	$A_c \cos\left[D_f \int_{-\infty}^{t} m(\sigma)\, d\sigma\right]$	$A_c \sin\left[D_f \int_{-\infty}^{t} m(\sigma)\, d\sigma\right]$
SSB-AM-SC[b]	$A_c[m(t) \pm j\hat{m}(t)]$	$A_c m(t)$	$\pm A_c \hat{m}(t)$
SSB-PM[b]	$A_c e^{jD_p[m(t) \pm j\hat{m}(t)]}$	$A_c e^{\mp D_p \hat{m}(t)} \cos[D_p m(t)]$	$A_c e^{\mp D_p \hat{m}(t)} \sin[D_p m(t)]$
SSB-FM[b]	$A_c e^{jD_f \int_{-\infty}^{t}[m(\sigma) \pm j\hat{m}(\sigma)]d\sigma}$	$A_c e^{\mp D_f \int_{-\infty}^{t} \hat{m}(\sigma) d\sigma} \cos\left[D_f \int_{-\infty}^{t} m(\sigma)\, d\sigma\right]$	$A_c e^{\mp D_f \int_{-\infty}^{t} \hat{m}(\sigma) d\sigma} \sin\left[D_f \int_{-\infty}^{t} m(\sigma)\, d\sigma\right]$
SSB-EV[b]	$A_c e^{\{\ln[1+m(t)] \pm j\hat{\ln}\,[1+m(t)]\}}$	$A_c[1+m(t)] \cos\{\hat{\ln}[1+m(t)]\}$	$\pm A_c[1+m(t)] \sin\{\hat{\ln}[1+m(t)]\}$
SSB-SQ[b]	$A_c e^{(1/2)\{\ln[1+m(t)] \pm j\hat{\ln}\,[1+m(t)]\}}$	$A_c\sqrt{1+m(t)} \cos\{\tfrac{1}{2}\hat{\ln}[1+m(t)]\}$	$\pm A_c\sqrt{1+m(t)} \sin\{\tfrac{1}{2}\hat{\ln}[1+m(t)]\}$
QM	$A_c[m_1(t) + jm_2(t)]$	$A_c m_1(t)$	$A_c m_2(t)$

235

TABLE 4–1 COMPLEX ENVELOPE FUNCTIONS FOR VARIOUS TYPES OF MODULATION (*cont.*)

Type of Modulation	Corresponding Amplitude and Phase Modulation		Linearity	Remarks
	$R(t)$	$\theta(t)$		
AM	$A_c\lvert 1 + m(t)\rvert$	$\begin{cases}0, & m(t) > -1 \\ 180°, & m(t) < -1\end{cases}$	L^c	$m(t) > -1$ required for envelope detection
DSB-SC	$A_c\lvert m(t)\rvert$	$\begin{cases}0, & m(t) > 0 \\ 180°, & m(t) < 0\end{cases}$	L	Coherent detection required
PM	A_c	$D_p m(t)$	NL *nonlinear*	D_p is the phase deviation constant (rad/volt)
FM	A_c	$D_f \displaystyle\int_{-\infty}^{t} m(\sigma)\,d\sigma$	NL	D_f is the frequency deviation constant (rad/volt-sec)
SSB-AM-SC[b]	$A_c\sqrt{[m(t)]^2 + [\hat{m}(t)]^2}$	$\tan^{-1}[\pm\hat{m}(t)/m(t)]$	L	Coherent detection required
SSB-PM[b]	$A_c e^{\pm D_p \hat{m}(t)}$	$D_p m(t)$	NL	
SSB-FM[b]	$A_c e^{\pm D_f \int_{-\infty}^{t}\hat{m}(\sigma)\,d\sigma}$	$D_f \displaystyle\int_{-\infty}^{t} m(\sigma)\,d\sigma$	NL	
SSB-EV[b]	$A_c\lvert 1 + m(t)\rvert$	$\pm\widehat{\ln}[1 + m(t)]$	NL	$m(t) > -1$ is required so that $\ln(\cdot)$ will have a real value
SSB-SQ[b]	$A_c\sqrt{1 + m(t)}$	$\pm\tfrac{1}{2}\widehat{\ln}[1 + m(t)]$	NL	$m(t) > -1$ is required so that $\ln(\cdot)$ will have a real value
QM	$A_c\sqrt{m_1^2(t) + m_2^2(t)}$	$\tan^{-1}[m_2(t)/m_1(t)]$	L	Used in NTSC color television; requires coherent detection

[a] $A_c > 0$ is a constant that sets the power level of the signal as evaluated by the use of Eq. (4–17); L, linear; NL, nonlinear; $[\hat{\ }]$ is the Hilbert transform (i.e., the $-90°$ phase-shifted version) of $[\cdot]$. (See Sec. 5–5 and Sec. A–7, Appendix A.)

[b] Use upper signs for upper sideband signals and lower signs for lower sideband signals.

[c] In the strict sense, AM signals are not linear, because the carrier term does not satisfy the linearity (superposition) condition.

If we use $\mathcal{F}[g^*(t)] = G^*(-f)$ from Table 2–1 and the frequency translation property of Fourier transforms from Table 2–1, this equation becomes

$$V(f) = \tfrac{1}{2}\{G(f - f_c) + G^*[-(f + f_c)]\} \qquad (4\text{–}15)$$

which reduces to Eq. (4–12).

The PSD for $v(t)$ is obtained by first evaluating the autocorrelation for $v(t)$:

$$R_v(\tau) = \langle v(t)v(t + \tau)\rangle = \langle \mathrm{Re}\{g(t)e^{j\omega_c t}\}\,\mathrm{Re}\{g(t + \tau)\,e^{j\omega_c(t+\tau)}\}\rangle$$

Using the identity (see Prob. 2–66)

$$\mathrm{Re}(c_2)\,\mathrm{Re}(c_1) = \tfrac{1}{2}\,\mathrm{Re}(c_2^*c_1) + \tfrac{1}{2}\,\mathrm{Re}(c_2 c_1)$$

where $c_2 = g(t)e^{j\omega_c t}$ and $c_1 = g(t + \tau)\,e^{j\omega_c(t+\tau)}$, we get

$$R_v(\tau) = \langle \tfrac{1}{2}\,\mathrm{Re}\{g^*(t)g(t + \tau)\,e^{-j\omega_c t}\,e^{j\omega_c(t+\tau)}\}\rangle + \langle \tfrac{1}{2}\,\mathrm{Re}\{g(t)g(t + \tau)\,e^{j\omega_c t}\,e^{j\omega_c(t+\tau)}\}\rangle$$

Realizing that both $\langle\ \rangle$ and $\mathrm{Re}\{\ \}$ are linear operators, we may exchange the order of the operators without affecting the result, and the autocorrelation becomes

$$R_v(\tau) = \tfrac{1}{2}\,\mathrm{Re}\{\langle g^*(t)g(t + \tau)\,e^{j\omega_c \tau}\rangle\} + \tfrac{1}{2}\,\mathrm{Re}\{\langle g(t)g(t + \tau)\,e^{j2\omega_c t}\,e^{j\omega_c \tau}\rangle\}$$

or

$$R_v(\tau) = \tfrac{1}{2}\,\mathrm{Re}\{\langle g^*(t)g(t + \tau)\rangle\,e^{j\omega_c \tau}\} + \tfrac{1}{2}\,\mathrm{Re}\{\langle g(t)g(t + \tau)\,e^{j2\omega_c t}\rangle e^{j\omega_c \tau}\}$$

But $\langle g^*(t)g(t + \tau)\rangle = R_g(\tau)$. The second term on the right is negligible because $e^{j2\omega_c t} = \cos 2\omega_c t + j \sin 2\omega_c t$ oscillates much faster than variations in $g(t)g(t + \tau)$. In other words, f_c is much larger than the frequencies in $g(t)$, so the integral is negligible. This is an application of the Riemann–Lebesque lemma from integral calculus [Olmsted, 1961]. Thus, the autocorrelation reduces to

$$R_v(\tau) = \tfrac{1}{2}\,\mathrm{Re}\{R_g(\tau)e^{j\omega_c \tau}\} \qquad (4\text{–}16)$$

The PSD is obtained by taking the Fourier transform of Eq. (4–16) (i.e., applying the Wiener–Khintchine theorem). Note that Eq. (4–16) has the same mathematical form as Eq. (4–11) when t is replaced by τ, so the Fourier transform has the same form as Eq. (4–12). Thus,

$$\mathcal{P}_v(f) = \mathcal{F}[R_v(\tau)] = \tfrac{1}{4}\,[\mathcal{P}_g(f - f_c) + \mathcal{P}_g^*(-f - f_c)]$$

But $\mathcal{P}_g^*(f) = \mathcal{P}_g(f)$, since the PSD is a real function. Hence, the PSD is given by Eq. (4–13).

4–4 EVALUATION OF POWER

THEOREM.   *The total average normalized power of a bandpass waveform $v(t)$ is*

$$P_v = \langle v^2(t)\rangle = \int_{-\infty}^{\infty} \mathcal{P}_v(f)\,df = R_v(0) = \tfrac{1}{2}\langle |g(t)|^2\rangle \qquad (4\text{–}17)$$

where "normalized" implies that the load is equivalent to one ohm.

Proof. Substituting $v(t)$ into Eq. (2–67), we get

$$P_v = \langle v^2(t) \rangle = \int_{-\infty}^{\infty} \mathcal{P}_v(f)\, df$$

But $R_v(\tau) = \mathcal{F}^{-1}[\mathcal{P}_v(f)] = \int_{-\infty}^{\infty} \mathcal{P}_v(f) e^{j2\pi f\tau}\, df$, so

$$R_v(0) = \int_{-\infty}^{\infty} \mathcal{P}_v(f)\, df$$

Also, from Eq. (4–16),

$$R_v(0) = \tfrac{1}{2}\operatorname{Re}\{R_g(0)\} = \tfrac{1}{2}\operatorname{Re}\{\langle g^*(t)g(t+0)\rangle\}$$

or

$$R_v(0) = \tfrac{1}{2}\operatorname{Re}\{\langle |g(t)|^2 \rangle\}$$

But $|g(t)|$ is always real, so

$$R_v(0) = \tfrac{1}{2}\langle |g(t)|^2 \rangle$$

Another type of power rating, called the *peak envelope power* (PEP), is useful for transmitter specifications.

DEFINITION. The *peak envelope power* (PEP) is the average power that would be obtained if $|g(t)|$ were to be held constant at its peak value.

This is equivalent to evaluating the average power in an unmodulated RF sinusoid that has a peak value of $A_p = \max[v(t)]$, as is readily seen from Fig. 5–1b.

THEOREM. *The normalized PEP is given by*

$$P_{\text{PEP}} = \tfrac{1}{2}[\max|g(t)|]^2 \tag{4–18}$$

A proof of this theorem follows by applying the definition to Eq. (4–17). As described later in Chapters 5 and 8, the PEP is useful for specifying the power capability of AM, SSB, and television transmitters.

Example 4–1 AMPLITUDE-MODULATED SIGNAL

Evaluate the magnitude spectrum for an amplitude-modulated (AM) signal. From Table 4–1, the complex envelope of an AM signal is

$$g(t) = A_c[1 + m(t)]$$

so that the spectrum of the complex envelope is

$$G(f) = A_c\delta(f) + A_c M(f) \tag{4–19}$$

Using Eq. (4–9), we obtain the AM signal waveform

$$s(t) = A_c[1 + m(t)]\cos \omega_c t$$

and, using Eq. (4–12), we get the AM spectrum

$$S(f) = \tfrac{1}{2} A_c[\delta(f - f_c) + M(f - f_c) + \delta(f + f_c) + M(f + f_c)] \qquad (4\text{–}20a)$$

where, because $m(t)$ is real, $M^*(f) = M(-f)$ and $\delta(f) = \delta(-f)$ (the delta function was defined to be even) were used. Suppose that the magnitude spectrum of the modulation happens to be a triangular function, as shown in Fig. 4–2a. This spectrum might arise from an analog audio source in which the bass frequencies are emphasized. The resulting AM spectrum, using Eq. (4–20a), is shown in Fig. 4–2b. Note that because $G(f - f_c)$ and $G^*(-f - f_c)$ do not overlap, the magnitude spectrum is

$$|S(f)| = \begin{cases} \tfrac{1}{2} A_c \delta(f - f_c) + \tfrac{1}{2} A_c |M(f - f_c)|, & f > 0 \\ \tfrac{1}{2} A_c \delta(f + f_c) + \tfrac{1}{2} A_c |M(-f - f_c)|, & f < 0 \end{cases} \qquad (4\text{–}20b)$$

The 1 in $g(t) = A_c[1 + m(t)]$ causes delta functions to occur in the spectrum at $f = \pm f_c$, where f_c is the assigned carrier frequency. Using Eq. (4–17), we obtain the total average signal power

$$P_s = \tfrac{1}{2} A_c^2 \langle |1 + m(t)|^2 \rangle = \tfrac{1}{2} A_c^2 \langle 1 + 2m(t) + m^2(t) \rangle$$

$$= \tfrac{1}{2} A_c^2 [1 + 2\langle m(t) \rangle + \langle m^2(t) \rangle]$$

If we assume that the dc value of the modulation is zero, as shown in Fig. 4–2a, the average signal power becomes

$$P_s = \tfrac{1}{2} A_c^2 [1 + P_m] \qquad (4\text{–}21)$$

where $P_m = \langle m^2(t) \rangle$ is the power in the modulation $m(t)$, $\tfrac{1}{2} A_c^2$ is the carrier power, and $\tfrac{1}{2} A_c^2 P_m$ is the power in the sidebands of $s(t)$.

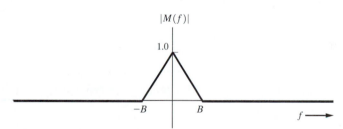

(a) Magnitude Spectrum of Modulation

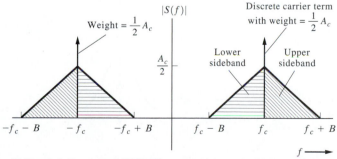

(b) Magnitude Spectrum of AM Signal

Figure 4–2   Spectrum of AM signal.

4–5 BANDPASS FILTERING AND LINEAR DISTORTION

Equivalent Low-Pass Filter

In Sec. 2–6, the general transfer function technique was described for the treatment of linear filter problems. Now a shortcut technique will be developed for modeling a bandpass filter by using an equivalent low-pass filter that has a complex-valued impulse response. (See Fig. 4–3a.) $v_1(t)$ and $v_2(t)$ are the input and output bandpass waveforms, with the corresponding complex envelopes $g_1(t)$ and $g_2(t)$. The impulse response of the bandpass filter, $h(t)$, can also be represented by its corresponding complex envelope $k(t)$. In addition, as

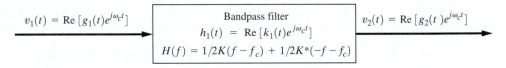

$$v_1(t) = \text{Re}\left[g_1(t)e^{j\omega_c t}\right]$$

Bandpass filter
$$h_1(t) = \text{Re}\left[k_1(t)e^{j\omega_c t}\right]$$
$$H(f) = 1/2K(f - f_c) + 1/2K^*(-f - f_c)$$

$$v_2(t) = \text{Re}\left[g_2(t)e^{j\omega_c t}\right]$$

(a) Bandpass Filter

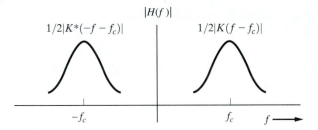

$|H(f)|$

$1/2|K^*(-f - f_c)|$ $1/2|K(f - f_c)|$

$-f_c$ f_c $f \longrightarrow$

(b) Typical Bandpass Filter Frequency Response

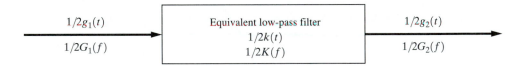

$1/2g_1(t)$

$1/2G_1(f)$

Equivalent low-pass filter
$1/2k(t)$
$1/2K(f)$

$1/2g_2(t)$

$1/2G_2(f)$

(c) Equivalent (Complex Impulse Response) Low-pass Filter

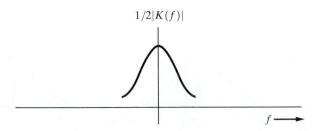

$1/2|K(f)|$

$f \longrightarrow$

(d) Typical Equivalent Low-pass Filter Frequency Response

Figure 4–3   Bandpass filtering.

shown in Fig. 4–3a, the frequency domain description, $H(f)$, can be expressed in terms of $K(f)$ with the help of Eqs. (4–11) and (4–12). Figure 4–3b shows a typical bandpass frequency response characteristic $|H(f)|$.

THEOREM.   *The complex envelopes for the input, output, and impulse response of a bandpass filter are related by*

$$\tfrac{1}{2}g_2(t) = \tfrac{1}{2}g_1(t) * \tfrac{1}{2}k(t) \qquad (4\text{–}22)$$

where $g_1(t)$ is the complex envelope of the input and $k(t)$ is the complex envelope of the impulse response. It also follows that

$$\tfrac{1}{2}G_2(f) = \tfrac{1}{2}G_1(f)\tfrac{1}{2}K(f) \qquad (4\text{–}23)$$

Proof.   We know that the spectrum of the output is

$$V_2(f) = V_1(f)\,H(f) \qquad (4\text{–}24)$$

Because $v_1(t)$, $v_2(t)$, and $h(t)$ are all bandpass waveforms, the spectra of these waveforms are related to the spectra of their complex envelopes by Eq. (4–15); thus, Eq. (4–24) becomes

$$\tfrac{1}{2}\big[G_2(f - f_c) + G_2^*(-f - f_c)\big]$$

$$= \tfrac{1}{2}\big[G_1(f - f_c) + G_1^*(-f - f_c)\big]\tfrac{1}{2}\big[K(f - f_c) + K^*(-f - f_c)\big] \qquad (4\text{–}25)$$

$$= \tfrac{1}{4}\big[G_1(f - f_c)K(f - f_c) + G_1(f - f_c)K^*(-f - f_c)$$

$$+ G_1^*(-f - f_c)K(f - f_c) + G_1^*(-f - f_c)K^*(-f - f_c)\big]$$

But $G_1(f - f_c)\,K^*(-f - f_c) = 0$, because the spectrum of $G_1(f - f_c)$ is zero in the region of frequencies around $-f_c$, where $K^*(-f - f_c)$ is nonzero. That is, there is no spectral overlap of $G_1(f - f_c)$ and $K^*(-f - f_c)$, because $G_1(f)$ and $K(f)$ have nonzero spectra around only $f = 0$ (i.e., baseband, as illustrated in Fig. 4–3d). Similarly, $G_1^*(-f - f_c)\,K(f - f_c) = 0$. Consequently, Eq. (4–25) becomes

$$\big[\tfrac{1}{2}G_2(f - f_c)\big] + \big[\tfrac{1}{2}G_2^*(-f - f_c)\big]$$

$$= \big[\tfrac{1}{2}G_1(f - f_c)\tfrac{1}{2}K(f - f_c)\big] + \big[\tfrac{1}{2}G_1^*(-f - f_c)\tfrac{1}{2}K^*(-f - f_c)\big] \qquad (4\text{–}26)$$

Thus, $\tfrac{1}{2}G_2(f) = \tfrac{1}{2}G_1(f)\tfrac{1}{2}K(f)$, which is identical to Eq. (4–23). Taking the inverse Fourier transform of both sides of Eq. (4–23), Eq. (4–22) is obtained.

This theorem indicates that any bandpass filter system may be described and analyzed by using an *equivalent low-pass filter* as shown in Fig. 4–3c. A typical equivalent low-pass frequency response characteristic is shown in Fig. 4–3d. Equations for equivalent low-pass filters are usually much less complicated than those for bandpass filters, so the equivalent low-pass filter system model is very useful. Because the highest frequency is much smaller in the equivalent low-pass filter, it is the basis for computer programs that use sampling to simulate bandpass communication systems (discussed in Sec. 4–6). Also, as shown in Prob. 4–12 and Fig. P4–12, the equivalent low-pass filter with complex impulse response may be realized by using four low-pass filters with real impulse response; however, if the frequency response of the bandpass filter is Hermitian symmetric about $f = f_c$, only two low-pass filters with real impulse response are required.

A linear bandpass filter can cause variations in the phase modulation at the output, $\theta_2(t) = \underline{/g_2(t)}$, as a function of the amplitude modulation on the input complex envelope, $R_1(t) = |g_1(t)|$. This is called *AM-to-PM conversion*. Similarly, the filter can cause variations in the amplitude modulation at the output, $R_2(t)$, because of the PM on the input, $\theta_1(t)$. This is called *PM-to-AM conversion*.

Because $h(t)$ represents a linear filter, $g_2(t)$ will be a linear filtered version of $g_1(t)$; however, $\theta_2(t)$ and $R_2(t)$—the PM and AM components, respectively, of $g_2(t)$—will be a *nonlinear* filtered version of $g_1(t)$, since $\theta_2(t)$ and $R_2(t)$ are nonlinear functions of $g_2(t)$. The analysis of the nonlinear distortion is very complicated. Although many analysis techniques have been published in the literature, none has been entirely satisfactory. Panter [1965] gives a three-chapter summary of some of these techniques, and a classical paper is also recommended [Bedrosian and Rice, 1968]. Furthermore, nonlinearities that occur in a practical system will also cause nonlinear distortion and AM-to-PM conversion effects. Nonlinear effects can be analyzed by several techniques, including power-series analysis; this is discussed in the section on amplifiers that follows later in this chapter. If a nonlinear effect in a *bandpass* system is to be analyzed, a Fourier series technique that uses the Chebyshev transform has been found to be useful [Spilker, 1977].

Linear Distortion

In Sec. 2–6, the general conditions were found for distortionless transmission. For linear *bandpass filters* (channels), a less restrictive set of conditions will now be shown to be satisfactory. For distortionless transmission of bandpass signals, the channel transfer function, $H(f) = |H(f)|e^{j\theta(f)}$, needs to satisfy the following requirements:

- The amplitude response is constant. That is,

$$|H(f)| = A \tag{4–27a}$$

  where A is a positive (real) constant.
- The derivative of the phase response is a constant. That is,

$$-\frac{1}{2\pi}\frac{d\theta(f)}{df} = T_g \tag{4–27b}$$

  where T_g is a constant called the complex *envelope delay* or, more concisely, the *group delay* and $\theta(f) = \underline{/H(f)}$.

This is illustrated in Fig. 4–4. Note that Eq. (4–27a) is identical to the general requirement of Eq. (2–150a), but Eq. (4–27b) is less restrictive than Eq. (2–150b). That is, if Eq. (2-150b) is satisfied, Eq. (4–27b) is satisfied, where $T_d = T_g$; however, if Eq. (4–27b) is satisfied, Eq. (2–150b) is not necessarily satisfied, because the integral of Eq. (4–27b) is

$$\theta(f) = -2\pi f T_g + \theta_0 \tag{4–28}$$

where θ_0 is a phase-shift constant, as shown in Fig. 4–4b. If θ_0 happens to be nonzero, Eq. (2–150b) is not satisfied.

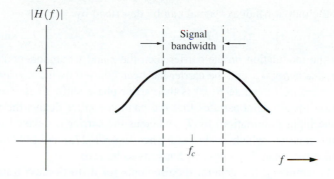

(a) Magnitude Response

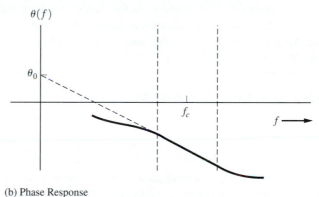

(b) Phase Response

Figure 4–4   Transfer characteristics of a distortionless bandpass channel.

Now it will be shown that Eqs. (4–27a) and (4–27b) are sufficient requirements for distortionless transmission of bandpass signals. From Eqs. (4–27a) and (4–28), the channel (or filter) transfer function is

$$H(f) = Ae^{j(-2\pi f T_g + \theta_0)} = (Ae^{j\theta_0})e^{-j2\pi f T_g} \tag{4-29}$$

over the bandpass of the signal. If the input to the bandpass channel is represented by

$$v_1(t) = x(t) \cos \omega_c t - y(t) \sin \omega_c t$$

then, using Eq. (4–29) and realizing that $e^{-j2\pi f T_g}$ causes a delay of T_g, we find that the output of the channel is

$$v_2(t) = Ax(t - T_g) \cos[\omega_c(t - T_g) + \theta_0] - Ay(t - T_g) \sin[\omega_c(t - T_g) + \theta_0]$$

Using Eq. (4–28), we obtain

$$v_2(t) = Ax(t - T_g) \cos[\omega_c t + \theta(f_c)] - Ay(t - T_g) \sin[\omega_c t + \theta(f_c)]$$

where, by the use of Eq. (2–150b) evaluated at $f = f_c$,

$$\theta(f_c) = -\omega_c T_g + \theta_0 = -2\pi f_c T_d$$

Thus, the output bandpass signal can be described by

$$v_2(t) = Ax(t - T_g) \cos[\omega_c(t - T_d)] - Ay(t - T_g) \sin[\omega_c(t - T_d)] \qquad (4\text{--}30)$$

where the modulation on the carrier (i.e., the x and y components) has been delayed by the *group time delay*, T_g, and the carrier has been delayed by the *carrier time delay*, T_d. Because $\theta(f_c) = -2\pi f_c T_d$, where $\theta(f_c)$ is the carrier phase shift, T_d is also called the *phase delay*. Equation (4–30) demonstrates that the bandpass filter delays the input complex envelope (i.e., the input information) by T_g, whereas the carrier is delayed by T_d. This is distortionless transmission, which is obtained when Eqs. (4–27a) and (4–27b) are satisfied. Note that T_g will differ from T_d, unless θ_0 happens to be zero.

In summary, the general requirements for distortionless transmission of either baseband or bandpass signals are given by Eqs. (2–150a) and (2–150b). However, for the bandpass case, Eq. (2–150b) is overly restrictive and may be replaced by Eq. (4–27b). In this case, $T_d \neq T_g$ unless $\theta_0 = 0$ where T_d is the *carrier or phase delay* and T_g is the *envelope or group delay*. For distortionless *bandpass* transmission, it is only necessary to have a transfer function with a *constant amplitude* and a *constant phase derivative* over the bandwidth of the signal.

4–6 BANDPASS SAMPLING THEOREM

Sampling is used in software radios and for simulation of communication systems. If the sampling is carried out at the Nyquist rate or larger ($f_s \geq 2B$, where B is the highest frequency involved in the spectrum of the RF signal), the sampling rate can be ridiculous. For example, consider a satellite communication system with a carrier frequency of $f_c = 6$ GHz. The sampling rate required can be at least 12 GHz. Fortunately, for signals of this type (bandpass signals), it can be shown that the sampling rate depends only on the *bandwidth* of the signal, not on the absolute frequencies involved. This is equivalent to saying that we can reproduce the signal from samples of the complex envelope.

THEOREM. BANDPASS SAMPLING THEOREM: *If a (real) bandpass waveform has a nonzero spectrum only over the frequency interval $f_1 < |f| < f_2$, where the transmission bandwidth B_T is taken to be the absolute bandwidth $B_T = f_2 - f_1$, then the waveform may be reproduced from sample values if the sampling rate is*

$$f_s \geq 2B_T \qquad (4\text{--}31)$$

For example, Eq. (4–31) indicates that if the 6-GHz bandpass signal previously discussed had a bandwidth of 10 MHz, a sampling frequency of only 20 MHz would be required instead of 12GHz. This is a savings of three orders of magnitude.

The bandpass sampling theorem of Eq. (4–31) can be proved by using the Nyquist sampling theorem of Eqs. (2–158) and (2–160) in the quadrature bandpass representation, which is

$$v(t) = x(t) \cos \omega_c t - y(t) \sin \omega_c t \qquad (4\text{--}32)$$

Let f_c be the center of the bandpass, so that $f_c = (f_2 + f_1)/2$. Then, from Eq. (4–8), both $x(t)$ and $y(t)$ are baseband signals and are absolutely bandlimited to $B = B_T/2$. From

(2–160), the sampling rate required to represent the baseband signal is $f_b \geq 2B = B_T$. Equation (4–32) becomes

$$v(t) = \sum_{n=-\infty}^{n=\infty} \left[x\left(\frac{n}{f_b}\right) \cos \omega_c t - y\left(\frac{n}{f_b}\right) \sin \omega_c t \right] \left[\frac{\sin\{\pi f_b[t - (n/f_b)]\}}{\pi f_b[t - (n/f_b)]} \right] \quad (4\text{–}33)$$

For the general case, where the $x(n/f_b)$ and $y(n/f_b)$ samples are independent, two real samples are obtained for each value of n, so that the overall sampling rate for $v(t)$ is $f_s = 2f_b \geq 2B_T$. This is the bandpass sampling frequency requirement of Eq. (4–31). The x and y samples can be obtained by sampling $v(t)$ at $t \approx (n/f_b)$, but adjusting t slightly, so that $\cos \omega_c t = 1$ and $\sin \omega_c t = -1$ at the *exact* sampling time for x and y, respectively. That is, for $t \approx n/f_s$, $v(n/f_b) = x(n/f_b)$ when $\cos \omega_c t = 1$ (i.e., $\sin \omega_c t = 0$), and $v(n/f_b) = y(n/f_b)$ when $\sin \omega_c t = -1$ (i.e., $\cos \omega_c t = 0$). Alternatively $x(t)$ and $y(t)$ can first be obtained by the use of two quadrature product detectors, as described by Eq. (4–76). The $x(t)$ and $y(t)$ baseband signals can then be individually sampled at a rate of f_b, and the overall equivalent sampling rate is still $f_s = 2f_b \geq 2B_T$.

In the application of this theorem, it is assumed that the bandpass signal $v(t)$ is reconstructed by the use of Eq. (4–33). This implies that *nonuniformly* spaced synchronized samples of $v(t)$ are used, since the samples are taken in pairs (for the x and y components) instead of being uniformly spaced T_s apart. *Uniformly* spaced samples of $v(t)$ itself can be used with a minimum sampling frequency of $2B_T$, *provided* that either f_1 or f_2 is a harmonic of f_s, [Hsu, 1999; Taub and Schilling, 1986]. Otherwise, a minimum sampling frequency larger than $2B_T$, but not larger than $4B_T$ is required [Hsu, 1999; Taub and Schilling, 1986]. This phenomenon occurs with impulse sampling [Eq. (2–173)] because f_s needs to be selected so that there is no spectral overlap in the $f_1 < f < f_2$ band when the bandpass spectrum is translated to harmonics of f_s.

THEOREM.   BANDPASS DIMENSIONALITY THEOREM: *Assume that a bandpass waveform has a nonzero spectrum only over the frequency interval $f_1 < |f| < f_2$, where the transmission bandwidth B_T is taken to be the absolute bandwidth given by $B_T = f_2 - f_1$ and $B_T \ll f_1$. The waveform may be completely specified over a T_0-second interval by*

$$N = 2B_T T_0 \quad (4\text{–}34)$$

independent pieces of information. N is said to be the number of dimensions required to specify the waveform.

Computer simulation is often used to analyze communication systems. The bandpass dimensionality theorem tells us that a bandpass signal B_T Hz wide can be represented over a T_0-s interval, provided that at least $N = 2B_T T_0$ samples are used. More details about the bandpass sampling theorem are discussed in study-aid Prob. SA4–5.

4–7 RECEIVED SIGNAL PLUS NOISE

Using the representation of bandpass signals and including the effects of channel filtering, we can obtain a model for the received signal plus noise. Referring to Fig. 4–1, the signal out of the transmitter is

$$s(t) = \text{Re}[g(t)e^{j\omega_c t}]$$

where $g(t)$ is the complex envelope for the particular type of modulation used. (See Table 4–1.) If the channel is linear and time invariant, the received signal plus noise is

$$r(t) = s(t) * h(t) + n(t) \qquad (4\text{--}35)$$

where $h(t)$ is the impulse response of the channel and $n(t)$ is the noise at the receiver input. Furthermore, if the channel is distortionless, its transfer function is given by Eq. (4–29), and consequently, the signal plus noise at the receiver input is

$$r(t) = \text{Re}[Ag(t - T_g)e^{j(\omega_c t + \theta(f_c))} + n(t)] \qquad (4\text{--}36)$$

where A is the gain of the channel (a positive number usually less than 1), T_g is the channel group delay, and $\theta(f_c)$ is the carrier phase shift caused by the channel. In practice, the values for T_g and $\theta(f_c)$ are often not known, so that if values for T_g and $\theta(f_c)$ are needed by the receiver to detect the information that was transmitted, receiver circuits estimate the received carrier phase $\theta(f_c)$ and the group delay (e.g., a bit synchronizer in the case of digital signaling). We will assume that the receiver circuits are designed to make errors due to these effects negligible; therefore, we can consider the signal plus noise at the receiver input to be

$$r(t) = \text{Re}[g(t)e^{j\omega_c t}] + n(t) \qquad (4\text{--}37)$$

where the effects of channel filtering, if any, are included by some modification of the complex envelope $g(t)$ and the constant A_c that is implicit within $g(t)$ (see Table 4–1) is adjusted to reflect the effect of channel attenuation. Details of this approach are worked out in Sec. 8-6.

4–8 CLASSIFICATION OF FILTERS AND AMPLIFIERS

Filters

Filters are devices that take an input waveshape and modify the frequency spectrum to produce the output waveshape. Filters may be classified in several ways. One is by the type of construction used, such as LC elements or quartz crystal elements. Another is by the type of transfer function that is realized, such as the Butterworth or Chebyshev response (defined subsequently). These two classifications are discussed in this section.

Filters use energy storage elements to obtain frequency discrimination. In any physical filter, the energy storage elements are imperfect. For example, a physical inductor has some series resistance as well as inductance, and a physical capacitor has some shunt (leakage) resistance as well as capacitance. A natural question, then, is, What is the quality Q of a circuit element or filter? Unfortunately, two different measures of filter quality are used in the technical literature. The first definition is concerned with the efficiency of *energy storage* in a circuit [Ramo, Whinnery, and vanDuzer, 1967, 1984] and is

$$Q = \frac{2\pi(\text{maximum energy stored during one cycle})}{\text{energy dissipated per cycle}} \qquad (4\text{--}38)$$

A larger value for Q corresponds to a more perfect storage element. That is, a perfect L or C element would have infinite Q. The second definition is concerned with the *frequency selectivity* of a circuit and is

$$Q = \frac{f_0}{B} \qquad (4\text{–}39)$$

where f_0 is the resonant frequency and B is the 3-dB bandwidth. Here, the larger the value of Q, the better is the frequency selectivity, because, for a given f_0, the bandwidth would be smaller.

In general, the value of Q as evaluated using Eq. (4–38) is different from the value of Q obtained from Eq. (4–39). However, these two definitions give identical values for an *RLC* series resonant circuit driven by a voltage source or for an *RLC* parallel resonant circuit driven by a current source [Nilsson, 1990]. For bandpass filtering applications, frequency selectivity is the desired characteristic, so Eq. (4–39) is used. Also, Eq. (4–39) is easy to evaluate from laboratory measurements. If we are designing a passive filter (not necessarily a single-tuned circuit) of center frequency f_0 and 3-dB bandwidth B, the individual circuit elements will each need to have much larger Q's than f_0/B. Thus, for a practical filter design, we first need to answer the question. What are the Q's needed for the filter elements, and what kind of elements will give these values of Q? This question is answered in Table 4–2, which lists filters as classified by the type of energy storage elements used in their construction and gives typical values for the Q of the elements. Filters that use lumped[†] L and C elements become impractical to build above 300 MHz, because the parasitic capacitance and inductance of the leads significantly affect the frequency response at high frequencies. Active filters, which use operational amplifiers with *RC* circuit elements, are practical only below 500 kHz, because the op amps need to have a large open-loop gain over the operating band. For very low-frequency filters, *RC* active filters are usually preferred to *LC* passive filters because the size of the *LC* components becomes large and the Q of the inductors becomes small in this frequency range. Active filters are difficult to implement within integrated circuits because the resistors and capacitors take up a significant portion of the chip area. This difficulty is reduced by using a switched-capacitor design for IC implementation. In that case, resistors are replaced by an arrangement of electronic switches and capacitors that are controlled by a digital clock signal [Schaumann, et al., 1990].

Crystal filters are manufactured from quartz crystal elements, which act as a series resonant circuit in parallel with a shunt capacitance caused by the holder (mounting). Thus, a parallel resonant, as well as a series resonant, mode of operation is possible. Above 100 MHz the quartz element becomes physically too small to manufacture, and below 1 kHz the size of the element becomes prohibitively large. Crystal filters have excellent performance because of the inherently high Q of the elements, but they are more expensive than *LC* and ceramic filters.

Mechanical filters use the vibrations of a resonant mechanical system to obtain the filtering action. The mechanical system usually consists of a series of disks spaced along a

---

[†] A lumped element is a discrete *R*-, *L*-, or *C*-type element, compared with a continuously distributed *RLC* element, such as that found in a transmission line.

TABLE 4–2   FILTER CONSTRUCTION TECHNIQUES

Type of Construction	Description of Elements or Filter	Center Frequency Range	Unloaded Q (Typical)	Filter Application[a]
LC (passive)		dc–300 MHz	100	Audio, video, IF, and RF
Active and Switched Capacitor		dc–500 kHz	200[b]	Audio
Crystal	Quartz crystal	1 kHz–100 MHz	100,000	IF
Mechanical	Transducers / Rod / Disk	50–500 kHz	1,000	IF
Ceramic	Ceramic disk / Electrodes	10 kHz–10.7 MHz	1,000	IF
Surface acoustic waves (SAW)	One section / Fingers / Piezoelectric substrate / Finger overlap region	10–800 MHz	c	IF and RF
Transmission line	$\lambda/4$	UHF and microwave	1,000	RF
Cavity		Microwave	10,000	RF

[a] IF, intermediate frequency; RF, radio frequency. (See Sec. 4–16).

[b] Bandpass Q's.

[c] Depends on design: $N = f_0/B$, where N is the number of sections, f_0 is the center frequency, and B is the bandwidth. Loaded Q's of 18,000 have been.

rod. Transducers are mounted on each end of the rod to convert the electrical signals to me-chanical vibrations at the input, and vice versa at the output. Each disk is the mechanical equivalent of a high-Q electrical parallel resonant circuit. The mechanical filter usually has a high insertion loss, resulting from the inefficiency of the input and output transducers.

Ceramic filters are constructed from piezoelectric ceramic disks with plated electrode connections on opposite sides of the disk. The behavior of the ceramic element is similar to that of the crystal filter element, as discussed earlier, except that the Q of the ceramic ele-ment is much lower. The advantage of the ceramic filter is that it often provides adequate performance at a cost that is low compared with that of crystal filters.

Surface acoustic wave (SAW) filters utilize acoustic waves that are launched and trav-el on the surface of a piezoelectric substrate (slab). Metallic interleaved "fingers" have been deposited on the substrate. The voltage signal on the fingers is converted to an acoustic sig-nal (and vice versa) as the result of the piezoelectric effect. The geometry of the fingers de-termines the frequency response of the filter, as well as providing the input and output coupling [Dorf, 1993, pp. 1073–1074]. The insertion loss is somewhat larger than that for crystal or ceramic filters. However, the ease of shaping the transfer function and the wide bandwidth that can be obtained with controlled attenuation characteristics make the SAW filters very attractive. This technology is used to provide excellent IF amplifier characteris-tics in modern television sets.

SAW devices can also be tapped so that they are useful for transversal filter configu-rations (Fig. 3–28) operating in the RF range. At lower frequencies, charge transfer devices (CTDs) can be used to implement transversal filters [Gersho, 1975].

Transmission line filters utilize the resonant properties of open-circuited or short-cir-cuited transmission lines. These filters are useful at UHF and microwave frequencies, at which wavelengths are small enough so that filters of reasonable size can be constructed. Similarly, the resonant effect of cavities is useful in building filters for microwave frequen-cies at which the cavity size becomes small enough to be practical.

Filters are also characterized by the type of transfer function that is realized. The trans-fer function of a linear filter with lumped circuit elements may be written as the ratio of two polynomials,

$$H(f) = \frac{b_0 + b_1(j\omega) + b_2(j\omega)^2 + \cdots + b_k(j\omega)^k}{a_0 + a_1(j\omega) + a_2(j\omega)^2 + \cdots + a_n(j\omega)^n} \tag{4–40}$$

where the constants a_i and b_i are functions of the element values and $\omega = 2\pi f$. The pa-rameter n is said to be the *order* of the filter. By adjusting the constants to certain values, desirable transfer function characteristics can be obtained. Table 4–3 lists three different filter characteristics and the optimization criterion that defines each one. The Chebyshev fil-ter is used when a sharp attenuation characteristic is required for a minimum number of circuit elements. The Bessel filter is often used in data transmission when the pulse shape is to be preserved, since it attempts to maintain a linear phase response in the passband. The Butterworth filter is often used as a compromise between the Chebyshev and Bessel characteristics.

The topic of filters is immense, and not all aspects of filtering can be covered here. For example, with the advent of inexpensive microprocessors, *digital filtering* and *digital signal processing* are becoming very important [Oppenheim and Schafer, 1975, 1989].

TABLE 4–3 SOME FILTER CHARACTERISTICS

Type	Optimization Criterion	Transfer Characteristic for the Low-Pass Filter[a]
Butterworth	Maximally flat: as many derivatives of $|H(f)|$ as possible go to zero as $f \rightarrow 0$	$$|H(f)| = \frac{1}{\sqrt{1 + (f/f_b)^{2n}}}$$
Chebyshev	For a given peak-to-peak ripple in the passband of the $|H(f)|$ characteristic, the $|H(f)|$ attenuates the fastest for any filter of nth order	$$|H(f)| = \frac{1}{\sqrt{1 + \varepsilon^2 C_n^2(f/f_b)}}$$ ε = a design constant; $C_n(f)$ is the nth-order Chebyshev polynomial defined by the recursion relation $$C_n(x) = 2xC_{n-1}(x) - C_{n-2}(x),$$ where $C_0(x) = 1$ and $C_1(x) = x$
Bessel	Attempts to maintain linear phase in the passband	$$H(f) = \frac{K_n}{B_n(f/f_b)}$$ K_n is a constant chosen to make $H(0) = 1$, and the Bessel recursion relation is $$B_n(x) = (2n - 1)\, B_{n-1}(x) - x^2 B_{n-2}(x),$$ where $B_0(x) = 1$ and $B_1(x) = 1 + jx$

[a] f_b is the cutoff frequency of the filter.

For additional reading on analog filters with an emphasis on communication system applications, see Bowron and Stephenson [1979].

Amplifiers

For analysis purposes, electronic circuits and, more specifically, amplifiers can be classified into two main categories: Nonlinear and linear. Linearity was defined in Sec. 2–6. In practice, all circuits are nonlinear to some degree, even at low (voltage and current) signal levels, and become highly nonlinear for high signal levels. Linear circuit analysis is often used for the low signal levels, since it greatly simplifies the mathematics and gives accurate answers if the signal level is sufficiently small.

The main categories of nonlinear and linear amplifiers can be further classified into the subcategories of circuits with memory and circuits with no memory. Circuits with memory contain inductive and capacitive effects that cause the present output value to be a function of previous input values as well as the present input value. If a circuit has no memory, its present output value is a function only of its present input value.

In introductory electrical engineering courses, it is first assumed that circuits are linear with no memory (resistive circuits) and, later, linear with memory (*RLC* circuits). It follows that linear amplifiers with memory may be described by a transfer function that is the ratio of the Fourier transform of the output signal to the Fourier transform of the input signal.

As discussed in Sec. 2–6, the transfer function of a distortionless amplifier is given by $Ke^{-j\omega T_d}$, where K is the voltage gain of the amplifier and T_d is the delay between the output and input waveforms. If the transfer function of the linear amplifier is not of this form, the output signal will be a *linearly* distorted version of the input signal.

4–9 NONLINEAR DISTORTION

In addition to linear distortion, practical amplifiers produce nonlinear distortion. To examine the effects of nonlinearity and yet keep a mathematical model that is tractable, we will assume *no memory* in the following analysis. Thus, we will look at the present output as a function of the present input in the time domain. If the amplifier is linear, this relationship is

$$v_0(t) = Kv_i(t) \tag{4-41}$$

where K is the voltage gain of the amplifier. In practice, the output of the amplifier becomes saturated at some value as the amplitude of the input signal is increased. This is illustrated by the nonlinear output-to-input characteristic shown in Fig. 4–5. The output-to-input characteristic may be modeled by a Taylor's expansion about $v_i = 0$ (i.e., a Maclaurin series); that is,

$$v_0 = K_0 + K_1 v_i + K_2 v_i^2 + \cdots = \sum_{n=0}^{\infty} K_n v_i^n \tag{4-42}$$

where

$$K_n = \frac{1}{n!} \left(\frac{d^n v_0}{dv_i^n} \right) \Bigg|_{v_i=0} \tag{4-43}$$

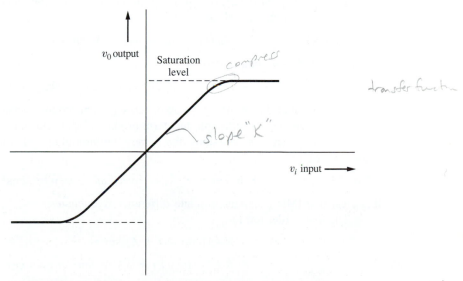

Figure 4–5   Nonlinear amplifier output-to-input characteristic.

There will be nonlinear distortion on the output signal if K_2, K_3, ... are not zero. K_0 is the output dc offset level, $K_1 v_i$ is the first-order (linear) term, $K_2 v_i^2$ is the second-order (square-law) term, and so on. Of course, K_1 will be larger than K_2, K_3, ... if the amplifier is anywhere near to being linear.

The *harmonic distortion* associated with the amplifier output is determined by applying a single sinusoidal test tone to the amplifier input. Let the input test tone be represented by

$$v_i(t) = A_0 \sin \omega_0 t \qquad (4\text{--}44)$$

Then the second-order output term is

$$K_2(A_0 \sin \omega_0 t)^2 = \frac{K_2 A_0^2}{2} (1 - \cos 2\omega_0 t) \qquad (4\text{--}45)$$

This indicates that the second-order distortion creates a dc level $K_2 A_0^2/2$ (in addition to any dc bias) and second harmonic distortion with amplitude $K_2 A_0^2/2$. In general, for a single-tone input, the output will be

$$\begin{aligned} v_{\text{out}}(t) = V_0 &+ V_1 \cos(\omega_0 t + \varphi_1) + V_2 \cos(2\omega_0 t + \varphi_2) \\ &+ V_3 \cos(3\omega_0 t + \varphi_3) + \cdots \end{aligned} \qquad (4\text{--}46)$$

where V_n is the peak value of the output at the frequency $n f_0$ hertz. Then, the percentage of *total harmonic distortion* (THD) is defined by

$$\text{THD } (\%) = \frac{\sqrt{\sum_{n=2}^{\infty} V_n^2}}{V_1} \times 100 \qquad (4\text{--}47)$$

The THD of an amplifier can be measured by using a distortion analyzer, or it can be evaluated by Eq. (4–47), with the V_n's obtained from a spectrum analyzer.

The *intermodulation distortion* (IMD) of the amplifier is obtained by using a two-tone test. If the input (tone) signals are

$$v_i(t) = A_1 \sin \omega_1 t + A_2 \sin \omega_2 t \qquad (4\text{--}48)$$

then the second-order output term is

$$K_2(A_1 \sin \omega_1 t + A_2 \sin \omega_2 t)^2 = K_2(A_1^2 \sin^2 \omega_1 t + 2A_1 A_2 \sin \omega_1 t \sin \omega_2 t + A_2^2 \sin^2 \omega_2 t)$$

The first and last terms on the right side of this equation produce harmonic distortion at frequencies $2f_1$ and $2f_2$. The cross-product term produces IMD. This term is present only when *both* input terms are present—thus, the name "intermodulation distortion." Then the second-order IMD is

$$2K_2 A_1 A_2 \sin \omega_1 t \sin \omega_2 t = K_2 A_1 A_2 \{\cos[(\omega_1 - \omega_2)t] - \cos[(\omega_1 + \omega_2)t]\}$$

It is clear that IMD generates sum and difference frequencies.

The third-order term is

$$\begin{aligned} K_3 v_i^3 &= K_3(A_1 \sin \omega_1 t + A_2 \sin \omega_2 t)^3 \\ &= K_3(A_1^3 \sin^3 \omega_1 t + 3A_1^2 A_2 \sin^2 \omega_1 t \sin \omega_2 t \\ &+ 3A_1 A_2^2 \sin \omega_1 t \sin^2 \omega_2 t + A_2^3 \sin^3 \omega_2 t) \end{aligned} \qquad (4\text{--}49)$$

The first and last terms on the right side of this equation will produce harmonic distortion, and the second term, a cross product, becomes

$$3K_3 A_1^2 A_2 \sin^2 \omega_1 t \sin \omega_2 t = \tfrac{3}{2} K_3 A_1^2 A_2 \sin \omega_2 t (1 - \cos 2\omega_1 t)$$

$$= \tfrac{3}{2} K_3 A_1^2 A_2 \{ \sin \omega_2 t - \tfrac{1}{2} [\sin(2\omega_1 + \omega_2)t$$

$$-\sin(2\omega_1 - \omega_2)t] \} \qquad (4\text{–}50)$$

Similarly, the third term of Eq. (4–49) is

$$3K_3 A_1 A_2^2 \sin \omega_1 t \sin^2 \omega_2 t$$

$$= \tfrac{3}{2} K_3 A_1 A_2^2 \{ \sin \omega_1 t - \tfrac{1}{2} [\sin(2\omega_2 + \omega_1)t - \sin(2\omega_2 - \omega_1)t] \} \quad (4\text{–}51)$$

The last two terms in Eqs. (4–50) and (4–51) are intermodulation terms at nonharmonic frequencies. For the case of *bandpass* amplifiers where f_1 and f_2 are within the bandpass with f_1 close to f_2 (i.e., $f_1 \approx f_2 \gg 0$), the distortion products at $2f_1 + f_2$ and $2f_2 + f_1$ will usually fall outside the passband and, consequently, may not be a problem. However, the terms at $2f_1 - f_2$ and $2f_2 - f_1$ will fall inside the passband and will be close to the desired frequencies f_1 and f_2. These will be the main distortion products for bandpass amplifiers, such as those used for RF amplification in transmitters and receivers.

As Eqs. (4–50) and (4–51) show, if either A_1 or A_2 is increased sufficiently, the IMD will become significant, since the desired output varies linearly with A_1 or A_2 and the IMD output varies as $A_1^2 A_2$ or $A_1 A_2^2$. Of course, the exact input level required for the intermodulation products to be a problem depends on the relative values of K_3 and K_1. The level may be specified by the amplifier third-order *intercept point*, which is evaluated by applying two equal amplitude test tones (i.e., $A_1 = A_2 = A$). The desired linearly amplified outputs will have amplitudes of $K_1 A$, and each of the third-order intermodulation products will have amplitudes of $3K_3 A^3/4$. The ratio of the desired output to the IMD output is then

$$R_{\text{IMD}} = \frac{4}{3} \left(\frac{K_1}{K_3 A^2} \right) \qquad (4\text{–}52)$$

The input intercept point, defined as the input level that causes R_{IMD} to be unity, is shown in Fig. 4–6. The solid curves are obtained by measurement, using two sinusoidal signal generators to generate the tones and measuring the level of the desired output (at f_1 or f_2) and the IMD products (at $2f_1 - f_2$ or $2f_2 - f_1$) with a spectrum analyzer. The intercept point is a fictitious point that is obtained by extrapolation of the linear portion (decibel plot) of the desired output and IMD curves until they intersect. The desired output (the output at either f_1 or f_2) actually becomes saturated when measurements are made, since the higher-order terms in the Taylor series have components at f_1 and f_2 that subtract from the linearly amplified output. For example, with K_3 being negative, the leading term in Eq. (4–51) occurs at f_1 and will subtract from the linearly amplified component at f_1, thus producing a saturated characteristic for the sinusoidal component at f_1. For an amplifier that happens to have the particular nonlinear characteristic shown in Fig. 4–6, the intercept point occurs for an RF input level of -10 dBm. Overload characteristics of receivers, such as those used in police walkie-talkies, are characterized by the third-order intercept-point specification. This dBm value is the RF signal level at the antenna input that corresponds to the intercept point. When

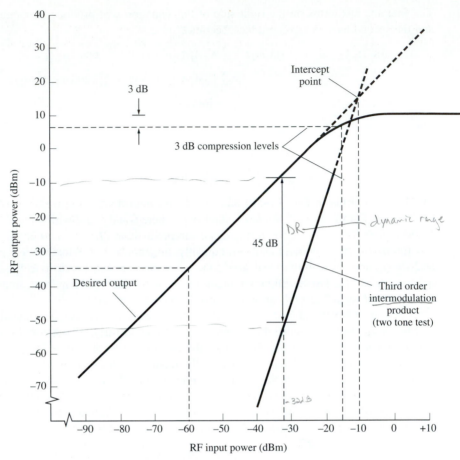

Figure 4–6   Amplifier output characteristics.

the receiver is deployed, input signal levels need to be much lower than that value in order to keep the undesired interfering intermodulation signals generated by the receiver circuits to an acceptable level. For transmitter applications, the intercept-point specification is the output signal level corresponding to the intercept point.

Other properties of an amplifier are also illustrated by Fig. 4–6. The gain of the amplifier is 25 dB in the linear region, because a −60 dBm input produces a −35 dBm output level. The desired output is compressed by 3 dB for an input level of −15 dBm. Consequently, the amplifier might be considered to be linear only if the input level is less than −15 dBm. Furthermore, if the third-order IMD products are to be down by at least 45 dBm, the input level will have to be kept lower than −32 dBm.

Another term in the distortion products at the output of a nonlinear amplifier is called *cross-modulation*. Cross-modulation terms are obtained when one examinies the third-order products resulting from a two-tone test. As shown in Eqs. (4–50) and (4–51), the terms $\frac{3}{2}K_3 A_1^2 A_2 \sin \omega_2 t$ and $\frac{3}{2}K_3 A_1 A_2^2 \sin \omega_1 t$ are cross-modulation terms. Let us examine the term

$\frac{3}{2}K_3 A_1^2 A_2 \sin \omega_2 t$. If we allow some amplitude variation in the input signal $A_1 \sin \omega_1 t$, so that it looks like an AM signal $A_1[1 + m_1(t)] \sin \omega_1 t$, where $m_1(t)$ is the modulating signal, a third-order distortion product becomes

$$\tfrac{3}{2}K_3 A_1^2 A_2 [1 + m_1(t)]^2 \sin \omega_2 t \qquad (4\text{–}53)$$

Thus, the AM on the signal at the carrier frequency f_1 will produce a signal at frequency f_2 with distorted modulation. That is, if two signals are passed through an amplifier having third-order distortion products in the output, and if either input signal has some AM, the amplified output of the *other* signal will be amplitude modulated to some extent by a distorted version of the modulation. This phenomenon is cross-modulation.

Passive as well as active circuits may have nonlinear characteristics and, consequently, will produce distortion products. For example, suppose that two AM broadcast stations have strong signals in the vicinity of a barn or house that has a metal roof with rusted joints. The roof may act as an antenna to receive and reradiate the RF energy, and the rusted joints may act as a diode (a nonlinear passive circuit). Signals at harmonics and intermodulation frequencies may be radiated and interfere with other communication signals. In addition, cross-modulation products may be radiated. That is, a distorted modulation of one station is heard on radios (located in the vicinity of the rusted roof) that are tuned to the other station's frequency.

When amplifiers are used to produce high-power signals, as in transmitters, it is desirable to have amplifiers with high efficiency in order to reduce the costs of power supplies, cooling equipment, and energy consumed. The efficiency is the ratio of the output signal power to the dc input power. Amplifiers may be grouped into several categories, depending on the biasing levels and circuit configurations used. Some of these are Class A, B, C, D, E, F, G, H, and S [Krauss, Bostian, and Raab, 1980; Smith, 1998]. For Class A operation, the bias on the amplifier stage is adjusted so that current flows during the complete cycle of an applied input test tone. For Class B operation, the amplifier is biased so that current flows for 180° of the applied signal cycle. Therefore, if a Class B amplifier is to be used for a baseband linear amplifier, such as an audio power amplifier in a hi-fi system, two devices are wired in push-pull configuration so that each one alternately conducts current over half of the input signal cycle. In *bandpass* Class B linear amplification, where the bandwidth is a small percentage of the operating frequency, only one active device is needed, since tuned circuits may be used to supply the output current over the other half of the signal cycle. For Class C operation, the bias is set so that (collector or plate) current flows in pulses, each one having a pulse width that is usually much less than half of the input cycle. Unfortunately, with Class C operation, it is not possible to have linear amplification, even if the amplifier is a bandpass RF amplifier with tuned circuits providing current over the nonconducting portion of the cycle. If one tries to amplify an AM signal with a Class C amplifier or other types of nonlinear amplifiers, the AM on the output will be distorted. However, RF signals with a constant real envelope, such as FM signals, may be amplified without distortion, because a nonlinear amplifier preserves the zero-crossings of the input signal.

The efficiency of a Class C amplifier is determined essentially by the conduction angle of the active device, since poor efficiency is caused by signal power being wasted in the device itself during the conduction time. The Class C amplifier is most efficient, having an efficiency factor of 100% in the ideal case. Class B amplifiers have an efficiency of

$\pi/4 \times 100 = 78.5\%$ or less, and Class A amplifiers have an efficiency of 50% or less [Krauss, Bostian, and Raab, 1980]. Because Class C amplifiers are the most efficient, they are generally used to amplify constant envelope signals, such as FM signals used in broadcasting. Class D, E, F, G, H, and S amplifiers usually employ switching techniques in specialized circuits to obtain high efficiency [Krauss, Bostian, and Raab, 1980; Smith, 1998].

Many types of microwave amplifiers, such as traveling-wave tubes (TWTs), operate on the velocity modulation principle. The input microwave signal is fed into a slow-wave structure. Here, the velocity of propagation of the microwave signal is reduced so that it is slightly below the velocity of the dc electron beam. This enables a transfer of kinetic energy from the electron beam to the microwave signal, thereby amplifying the signal. In this type of amplifier, the electron current is *not* turned on and off to provide the amplifying mechanism; thus, it is not classified in terms of Class B or C operation. The TWT is a linear amplifier when operated at the appropriate drive level. If the drive level is increased, the efficiency (RF output/dc input) is improved, but the amplifier becomes nonlinear. In this case, constant envelope signals, such as PSK or PM, need to be used so that the intermodulation distortion will not cause a problem. This is often the mode of operation of satellite transponders (transmitters in communication satellites), where solar cells are costly and have limited power output. The subject is discussed in more detail in Chapter 8 in the section on satellite communications.

4–10 LIMITERS

A *limiter* is a nonlinear circuit with an output saturation characteristic. A soft saturating limiter characteristic is shown in Fig. 4–5. Figure 4–7 shows a hard (ideal) limiter characteristic, together with an illustration of the unfiltered output waveform obtained for an input waveform. The ideal limiter transfer function is essentially identical to the output-to-input characteristic of an ideal comparator with a zero reference level. The waveforms shown in Fig. 4–7 illustrate how amplitude variations in the input signal are eliminated in the output signal. A *bandpass limiter* is a nonlinear circuit with a saturating characteristic followed by a bandpass filter. In the case of an ideal bandpass limiter, the filter output waveform would be sinusoidal, since the harmonics of the square wave would be filtered out. In general, any bandpass input (even a modulated signal plus noise) can be represented, using Eq. (4–1b), by

$$v_{in}(t) = R(t) \cos[\omega_c t + \theta(t)] \qquad (4\text{–}54)$$

where $R(t)$ is the equivalent real envelope and $\theta(t)$ is the equivalent phase function. The corresponding output of an ideal bandpass limiter becomes

$$v_{out}(t) = KV_L \cos[\omega_c t + \theta(t)] \qquad (4\text{–}55)$$

where K is the level of the fundamental component of the square wave, $4/\pi$, multiplied by the gain of the output (bandpass) filter. This equation indicates that any AM that was present on the limiter input does not appear on the limiter output, but that the phase function is preserved (i.e., the zero-crossings of the input are preserved on the limiter output). Limiters are often used in receiving systems designed for angle-modulated signaling—such as PSK,

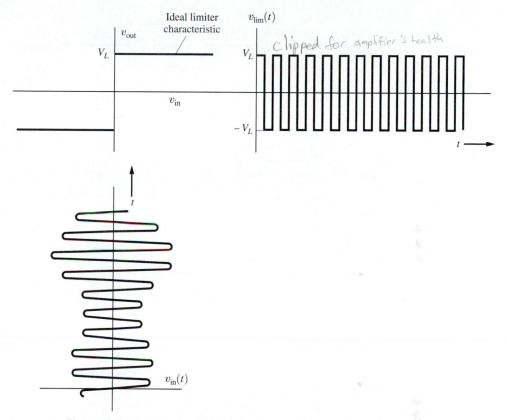

Figure 4–7   Ideal limiter characteristic with illustrative input and unfiltered output waveforms.

FSK, and analog FM—to eliminate any variations in the real envelope of the receiver input signal that are caused by channel noise or signal fading.

4–11 MIXERS, UP CONVERTERS, AND DOWN CONVERTERS

An *ideal mixer* is an electronic circuit that functions as a mathematical multiplier of two input signals. Usually, one of these signals is a sinusoidal waveform produced by a local oscillator, as illustrated in Fig. 4–8.

Mixers are used to obtain frequency translation of the input signal. Assume that the input signal is a bandpass signal that has a nonzero spectrum in a band around or near $f = f_c$. Then the signal is represented by

$$v_{in}(t) = \text{Re}\{g_{in}(t)e^{j\omega_c t}\} \tag{4–56}$$

where $g_{in}(t)$ is the complex envelope of the input signal. The signal out of the ideal mixer is then

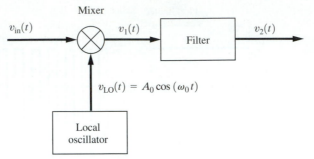

Figure 4–8 Mixer followed by a filter for either up or down conversion.

$$v_1(t) = [A_0 \operatorname{Re}\{g_{\text{in}}(t)e^{j\omega_c t}\}]\cos \omega_0 t$$

$$= \frac{A_0}{4}[g_{\text{in}}(t)e^{j\omega_c t} + g_{\text{in}}^*(t)e^{-j\omega_c t}](e^{j\omega_0 t} + e^{-j\omega_0 t})$$

$$= \frac{A_0}{4}[g_{\text{in}}(t)e^{j(\omega_c + \omega_0)t} + g_{\text{in}}^*(t)e^{-j(\omega_c + \omega_0)t} + g_{\text{in}}(t)e^{j(\omega_c - \omega_0)t} + g_{\text{in}}^*(t)e^{-j(\omega_c - \omega_0)t}]$$

or

$$v_1(t) = \frac{A_0}{2}\operatorname{Re}\{g_{\text{in}}(t)e^{j(\omega_c + \omega_0)t}\} + \frac{A_0}{2}\operatorname{Re}\{g_{\text{in}}(t)e^{j(\omega_c - \omega_0)t}\} \qquad (4\text{–}57)$$

Equation (4–57) illustrates that the input bandpass signal with a spectrum near $f = f_c$ has been converted (i.e., frequency translated) into two output bandpass signals, one at the up-conversion frequency band, where $f_u = f_c + f_0$, and one at the down-conversion band, where $f_d = f_c - f_0$. A filter, as illustrated in Fig. 4–8, may be used to select either the up-conversion component or the down-conversion component. This combination of a mixer plus a filter to remove one of the mixer output components is often called a *single-sideband mixer*. A *bandpass* filter is used to select the up-conversion component, but the down-conversion component is selected by either a *baseband* filter or a *bandpass* filter, depending on the location of $f_c - f_0$. For example, if $f_c - f_0 = 0$, a low-pass filter would be needed, and the resulting output spectrum would be a baseband spectrum. If $f_c - f_0 > 0$, where $f_c - f_0$ was larger than the bandwidth of $g_{\text{in}}(t)$, a bandpass filter would be used, and the filter output would be

$$v_2(t) = \operatorname{Re}\{g_2(t)e^{j(\omega_c - \omega_0)t}\} = \frac{A_0}{2}\operatorname{Re}\{g_{\text{in}}(t)e^{j(\omega_c - \omega_0)t}\} \qquad (4\text{–}58)$$

For this case of $f_c > f_0$, it is seen that the modulation on the mixer input signal $v_{\text{in}}(t)$ is preserved on the mixer up- or down-converted signals.

If $f_c < f_0$, we rewrite Eq. (4–57), obtaining

$$v_1(t) = \frac{A_0}{2}\operatorname{Re}\{g_{\text{in}}(t)e^{j(\omega_c + \omega_0)t}\} + \frac{A_0}{2}\operatorname{Re}\{g_{\text{in}}^*(t)e^{j(\omega_0 - \omega_c)t}\} \qquad (4\text{–}59)$$

because the frequency in the exponent of the bandpass signal representation needs to be positive for easy physical interpretation of the location of spectral components. For this case of $f_c < f_0$, the complex envelope of the down-converted signal has been conjugated compared to the complex envelope of the input signal. This is equivalent to saying that the sidebands have been exchanged; that is, the upper sideband of the input signal spectrum becomes the lower sideband of the down-converted output signal, and so on. This is demonstrated mathematically by looking at the spectrum of $g^*(t)$, which is

$$\mathcal{F}[g_{in}^*(t)] = \int_{-\infty}^{\infty} g_{in}^*(t)e^{-j\omega t}\,dt = \left[\int_{-\infty}^{\infty} g_{in}(t)e^{-j(-\omega)t}\,dt\right]^*$$

$$= G_{in}^*(-f) \tag{4–60}$$

The $-f$ indicates that the upper and lower sidebands have been exchanged, and the conjugate indicates that the phase spectrum has been inverted.

In summary, the complex envelope for the signal out of an *up converter* is

$$g_2(t) = \frac{A_0}{2}\,g_{in}(t) \tag{4–61a}$$

where $f_u = f_c + f_0 > 0$. Thus, the same modulation is on the output signal as was on the input signal, but the amplitude has been changed by the $A_0/2$ scale factor.

For the case of *down conversion*, there are two possibilities. For $f_d = f_c - f_0 > 0$, where $f_0 < f_c$,

$$g_2(t) = \frac{A_0}{2}\,g_{in}(t) \tag{4–61b}$$

This is called down conversion with *low-side injection*, because the LO frequency is below that of the incoming signal (i.e., $f_0 < f_c$). Here the output modulation is the same as that of the input, except for the $A_0/2$ scale factor. The other possibility is $f_d = f_0 - f_c > 0$, where $f_0 > f_c$, which produces the output complex envelope

$$g_2 = \frac{A_0}{2}\,g_{in}^*(t) \tag{4–61c}$$

This is down conversion with *high-side* injection, because $f_0 > f_c$. Here the sidebands on the down-converted output signal are reversed from those on the input (e.g., an LSSB input signal becomes a USSB output signal).

Ideal mixers act as linear time-varying circuit elements, since

$$v_1(t) = (A \cos \omega_0 t)v_{in}(t)$$

where $A \cos \omega_0 t$ is the time-varying gain of the linear circuit. It should also be recognized that mixers used in communication circuits are essentially mathematical multipliers. They should not be confused with the audio mixers that are used in radio and TV broadcasting studios. An *audio mixer* is a summing amplifier with multiple inputs so that several inputs from several sources—such as microphones, tape decks, and CD decks—can be "mixed" (added) to produce one output signal. Unfortunately, the term *mixer* means entirely different things, depending on the context used. As used in transmitters and receivers, it means a multiplying operation that produces a frequency translation of the input signal. In audio systems, it means a summing operation to combine several inputs into one output signal.

In practice, the multiplying operation needed for mixers may be realized by using one of the following:

1. A continuously variable transconductance device, such as a dual-gate FET.
2. A nonlinear device.
3. A linear device with a time-varying discrete gain.

In the first method, when a dual-gate FET is used to obtain multiplication, $v_{in}(t)$ is usually connected to gate 1 and the local oscillator is connected to gate 2. The resulting output is

$$v_1(t) = K v_{in}(t) v_{LO}(t) \tag{4–62}$$

over the operative region, where $v_{LO}(t)$ is the local oscillator voltage. The multiplier is said to be of a *single-quadrant* type if the multiplier action of Eq. (4–62) is obtained only when both input waveforms, v_{in} and $v_{LO}(t)$, have either nonnegative or nonpositive values [i.e., a plot of the values of $v_{in}(t)$ versus $v_{LO}(t)$ falls within a single quadrant]. The multiplier is of the *two-quadrant* type if multiplier action is obtained when either $v_{in}(t)$ or $v_{LO}(t)$ is nonnegative or nonpositive and the other is arbitrary. The multiplier is said to be of the *four-quadrant* type when multiplier action is obtained regardless of the signs of $v_{in}(t)$ and $v_{LO}(t)$.

In the second technique, a nonlinear device can be used to obtain multiplication by summing the two inputs as illustrated in Fig. 4–9. Looking at the square-law component at the output, we have

$$v_1(t) = K_2(v_{in} + v_{LO})^2 + \text{other terms}$$
$$= K_2(v_{in}^2 + 2v_{in}v_{LO} + v_{LO}^2) + \text{other terms} \tag{4–63}$$

The cross-product term gives the desired multiplier action:

$$2K_2 v_{in} v_{LO} = 2K_2 A_0 v_{in}(t) \cos \omega_0 t \tag{4–64}$$

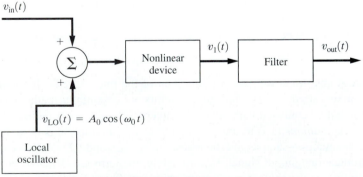

$v_{in}(t)$

$v_1(t)$

$v_{out}(t)$

Nonlinear device

Filter

Σ

$v_{LO}(t) = A_0 \cos(\omega_0 t)$

Local oscillator

Figure 4–9   Nonlinear device used as a mixer.

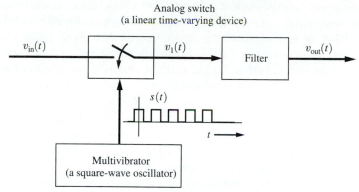

Figure 4–10   Linear time-varying device used as a mixer.

If we assume that $v_{in}(t)$ is a bandpass signal, the filter can be used to pass either the up- or down-conversion terms. However, some distortion products may also fall within the output passband if ω_c and ω_0 are not chosen carefully.

In the third method, a linear device with time-varying gain is used to obtain multiplier action. This is demonstrated in Fig. 4–10, in which the time-varying device is an analog switch (such as a CMOS 4016 integrated circuit) that is activated by a square-wave oscillator signal $v_0(t)$. The gain of the switch is either unity or zero. The waveform at the output of the analog switch is

$$v_1(t) = v_{in}(t)s(t) \tag{4–65}$$

where $s(t)$ is a unipolar switching square wave that has unity peak amplitude. (This is analogous to the PAM with natural sampling that was studied in Chapter 3.) Using the Fourier series for a rectangular wave, we find that Eq. (4–65) becomes

$$v_1(t) = v_{in}(t)\left[\frac{1}{2} + \sum_{n=1}^{\infty} \frac{2\sin(n\pi/2)}{n\pi} \cos n\omega_0 t\right] \tag{4–66}$$

The multiplying action is obtained from the $n = 1$ term, which is

$$\frac{2}{\pi} v_{in}(t) \cos \omega_0 t \tag{4–67}$$

This term would generate up- and down-conversion signals at $f_c + f_0$ and $f_c - f_0$ if $v_{in}(t)$ were a bandpass signal with a nonzero spectrum in the vicinity of $f = f_c$. However, Eq. (4–66) shows that other frequency bands are also present in the output signal, namely, at frequencies $f = |f_c \pm nf_0|$, $n = 3, 5, 7, \ldots$ and, in addition, there is the feed-through term $\frac{1}{2}v_{in}(t)$ appearing at the output. Of course, a filter may be used to pass either the up- or down-conversion component appearing in Eq. (4–66).

Mixers are often classified as being *unbalanced*, *single balanced*, or *double balanced*. That is, in general, we obtain

$$v_1(t) = C_1 v_{in}(t) + C_2 v_0(t) + C_3 v_{in}(t)v_0(t) + \text{other terms} \tag{4–68}$$

at the output of mixer circuits. When C_1 and C_2 are not zero, the mixer is said to be *unbalanced*, since $v_{in}(t)$ and $v_0(t)$ feed through to the output. An unbalanced mixer was illustrated in Fig. 4–9, in which a nonlinear device was used to obtain mixing action. In the Taylor's expansion of the nonlinear device output-to-input characteristics, the linear term would provide feed-through of both $v_{in}(t)$ and $v_0(t)$. A *single-balanced mixer* has feed-through for only one of the inputs; that is, either C_1 or C_2 of Eq. (4–68) is zero. An example of a single-balanced mixer is given in Fig. 4–10, which uses sampling to obtain mixer action. In this example, Eq. (4–66) demonstrates that $v_0(t)$ is balanced out (i.e., $C_2 = 0$) and $v_{in}(t)$ feeds through with a gain of $C_1 = \frac{1}{2}$. A *double-balanced mixer* has no feed-through from either input; that is, both C_1 and C_2 of Eq. (4–68) are zero. One kind of double-balanced mixer is discussed in the next paragraph.

Figure 4–11a shows the circuit for a double-balanced mixer. This circuit is popular because it is relatively inexpensive and has excellent performance. The third-order IMD is

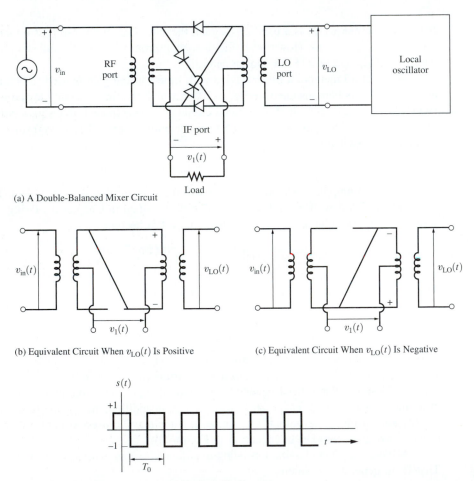

(a) A Double-Balanced Mixer Circuit

(b) Equivalent Circuit When $v_{LO}(t)$ Is Positive   (c) Equivalent Circuit When $v_{LO}(t)$ Is Negative

(d) Switching Waveform Due to the Local Oscillator Signal

Figure 4–11   Analysis of a double-balanced mixer circuit.

typically down at least 50 dB compared with the desired output components. This mixer is usually designed for source and load impedances of 50 Ω and has broadband input and output ports. The RF [i.e., $v_{in}(t)$] port and the LO (local oscillator) port are often usable over a frequency range of 1,000:1, say, 1 to 1,000 MHz; and the IF (intermediate frequency) output port, $v_1(t)$, is typically usable from dc to 600 MHz. The transformers are made by using small toroidal cores, and the diodes are matched hot carrier diodes. The input signal level at the RF port is relatively small, usually less than -5 dBm, and the local oscillator level at the LO port is relatively large, say, $+5$ dBm. The LO signal is large, and, in effect, turns the diodes on and off so that the diodes will act as switches. The LO provides the switching control signal. This circuit thus acts as a time-varying linear circuit (with respect to the RF input port), and its analysis is very similar to that used for the analog-switch mixer of Fig. 4–10. During the portion of the cycle when $v_{LO}(t)$ has a positive voltage, the output voltage is proportional to $+v_{in}(t)$, as seen from the equivalent circuit shown in Fig. 4–11b. When $v_{LO}(t)$ is negative, the output voltage is proportional to $-v_{in}(t)$, as seen from the equivalent circuit shown in Fig. 4–11c. Thus, the output of this double-balanced mixer is

$$v_1(t) = K v_{in}(t)\, s(t) \tag{4–69}$$

where $s(t)$ is a bipolar switching waveform, as shown in Fig. 4–11d. Since the switching waveform arises from the LO signal, its period is $T_0 = 1/f_0$. The switching waveform is described by

$$s(t) = 4 \sum_{n=1}^{\infty} \frac{\sin(n\pi/2)}{n\pi} \cos n\omega_0 t \tag{4–70}$$

so that the mixer output is

$$v_1(t) = [v_{in}(t)] \left[4K \sum_{n=1}^{\infty} \frac{\sin(n\pi/2)}{n\pi} \cos n\omega_0 t \right] \tag{4–71}$$

This equation shows that if the input is a bandpass signal with nonzero spectrum in the vicinity of f_c, the spectrum of the input will be translated to the frequencies $|f_c \pm nf_0|$, where $n = 1, 3, 5, \ldots$. In practice, the value K is such that the conversion gain (which is defined as the desired output level divided by the input level) at the frequency $|f_c \pm f_0|$ is about -6 dB. Of course, an output filter may be used to select the up-converted or down-converted frequency band.

In addition to up- or down-conversion applications, mixers (i.e., multipliers) may be used for amplitude modulators to translate a baseband signal to an RF frequency band, and mixers may be used as product detectors to translate RF signals to baseband. These applications will be discussed in later sections that deal with transmitters and receivers.

4–12 FREQUENCY MULTIPLIERS

Frequency multipliers consist of a nonlinear circuit followed by a tuned circuit, as illustrated in Fig. 4–12. If a bandpass signal is fed into a frequency multiplier, the output will appear in a frequency band at the nth harmonic of the input carrier frequency. Because the device

Frequency multiplier

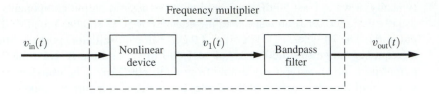

(a) Block Diagram of a Frequency Multiplier

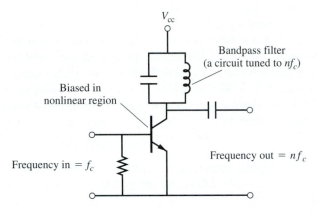

(b) Circuit Diagram of a Frequency Multiplier

Figure 4–12   Frequency multiplier.

is nonlinear, the bandwidth of the nth harmonic output is larger than that of the input signal. In general, the bandpass input signal is represented by

$$v_{in}(t) = R(t) \cos[\omega_c t + \theta(t)] \tag{4–72}$$

The transfer function of the nonlinear device may be expanded in a Taylor's series, so that the nth-order output term is

$$v_1(t) = K_n v_{in}^n(t) = K_n R^n(t) \cos^n[\omega_c t + \theta(t)]$$

or[†]

$$v_1(t) = CR^n(t) \cos[n\omega_c t + n\theta(t)] + \text{other terms}$$

Because the bandpass filter is designed to pass frequencies in the vicinity of nf_c, the output is

$$v_{out}(t) = CR^n(t) \cos[n\omega_c t + n\theta(t)] \tag{4–73}$$

[†] mth-order output terms, where $m > n$, may also contribute to the nth harmonic output, provided that K_m is sufficiently large with respect to K_n. This condition is illustrated by the trigonometric identity $8\cos^4 x = 3 + 4 \cos 2x + \cos 4x$, in which $m = 4$ and $n = 2$.

This illustrates that the input *amplitude* variation $R(t)$ *appears distorted* on the output signal because the real envelope on the output is $R^n(t)$. The waveshape of the *angle* variation, $\theta(t)$, is *not distorted* by the frequency multiplier, but the frequency multiplier does increase the magnitude of the angle variation by a factor of n. Thus, frequency multiplier circuits are not used on signals if AM is to be preserved; but as we will see, the frequency multiplier is very useful in PM and FM problems, since it effectively "amplifies" the angle variation waveform $\theta(t)$. The $n = 2$ multiplier is called a *doubler stage*, and the $n = 3$ frequency multiplier is said to be a *tripler stage*.

The frequency multiplier should not be confused with a mixer. The frequency multiplier acts as a nonlinear device. The mixer circuit (which uses a mathematical multiplier operation) acts as a linear circuit with time-varying gain (caused by the LO signal). The bandwidth of the signal at the output of a frequency multiplier is larger than that of the input signal, and it appears in a frequency band located at the nth harmonic of the input. The bandwidth of a signal at the output of a mixer is the same as that of the input, but the input spectrum has been translated either up or down, depending on the LO frequency and the bandpass of the output filter. A frequency multiplier is essentially a nonlinear amplifier followed by a bandpass filter that is designed to pass the nth harmonic.

4–13 DETECTOR CIRCUITS

As indicated in Fig. 4–1, the receiver contains carrier circuits that convert the input bandpass waveform into an output baseband waveform. These carrier circuits are called *detector circuits*. The sections that follow will show how detector circuits can be designed to produce $R(t)$, $\theta(t)$, $x(t)$, or $y(t)$ at their output for the corresponding bandpass signal that is fed into the detector input.

Envelope Detector

An *ideal envelope detector* is a circuit that produces a waveform at its output that is proportional to the real envelope $R(t)$ of its input. From Eq. (4–1b), the bandpass input may be represented by $R(t) \cos[\omega_c t + \theta(t)]$, where $R(t) \geq 0$; then the output of the ideal envelope detector is

$$v_{\text{out}}(t) = KR(t) \tag{4–74}$$

where K is the proportionality constant.

A simple diode detector circuit that approximates an ideal envelope detector is shown in Fig. 4–13a. The diode current occurs in pulses that are proportional to the positive part of the input waveform. The current pulses charge the capacitor to produce the output voltage waveform, as illustrated in Fig. 4–13b. The RC time constant is chosen so that the output signal will follow the real envelope $R(t)$ of the input signal. Consequently, the cutoff frequency of the low-pass filter needs to be much smaller than the carrier frequency f_c and much larger than the bandwidth of the (detected) modulation waveform B. That is,

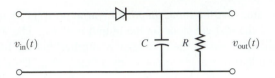

(a) A Diode Envelope Detector

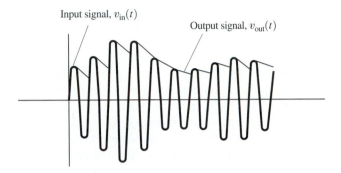

(b) Waveforms Associated with the Diode Envelope Detector

Figure 4–13   Envelope detector.

$$B \ll \frac{1}{2\pi RC} \ll f_c \qquad (4\text{–}75)$$

where RC is the time constant of the filter.

The envelope detector is typically used to detect the modulation on AM signals. In this case, $v_{in}(t)$ has the complex envelope $g(t) = A_c[1 + m(t)]$, where $A_c > 0$ represents the strength of the received AM signal and $m(t)$ is the modulation. If $|m(t)| < 1$, then

$$v_{out} = KR(t) = K|g(t)| = KA_c[1 + m(t)] = KA_c + KA_c m(t)$$

KA_c is a DC voltage that is used to provide *automatic gain control* (AGC) for the AM receiver. That is, for KA_c relatively small (a weak AM signal received), the receiver gain is increased and vice versa. $KA_c m(t)$ is the detected modulation. For the case of audio (not video) modulation, typical values for the components of the envelope detector are $R = 10$ kΩ and $C = 0.001$ μfd. This combination of values provides a low-pass filter cutoff frequency (3 dB down) of $f_{co} = 1/(2\pi RC) = 15.9$ kHz, much less than f_c and larger than the highest audio frequency, B, used in typical AM applications.

Product Detector

A *product detector* (Fig. 4–14) is a mixer circuit that down-converts the input (bandpass signal plus noise) to baseband. The output of the multiplier is

$$v_1(t) = R(t) \cos[\omega_c t + \theta(t)]A_0 \cos(\omega_c t + \theta_0)$$

$$= \tfrac{1}{2}A_0 R(t) \cos[\theta(t) - \theta_0] + \tfrac{1}{2}A_0 R(t) \cos[2\omega_c t + \theta(t) + \theta_0]$$

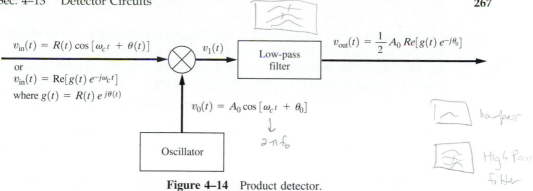

Figure 4–14   Product detector.

where the frequency of the oscillator is f_c and the phase is θ_0. The low-pass filter passes on-ly the down-conversion term, so that the output is

$$v_{out}(t) = \tfrac{1}{2}A_0 R(t) \cos[\theta(t) - \theta_0] = \tfrac{1}{2}A_0 \text{Re}\{g(t)e^{-j\theta_0}\} \qquad (4\text{–}76)$$

where the complex envelope of the input is denoted by

$$g(t) = R(t)e^{j\theta(t)} = x(t) + jy(t)$$

and $x(t)$ and $y(t)$ are the quadrature components. [See Eq. (4–2).] Because the frequency of the oscillator is the same as the carrier frequency of the incoming signal, the oscillator has been *frequency synchronized* with the input signal. Furthermore, if, in addition, $\theta_0 = 0$, the oscillator is said to be *phase synchronized* with the in-phase component, and the output becomes

$$v_{out}(t) = \tfrac{1}{2}A_0 x(t) \qquad (4\text{–}77a)$$

If $\theta_0 = 90°$,

$$v_{out} = \tfrac{1}{2}A_0 y(t) \qquad (4\text{–}77b)$$

Equation (4–76) also indicates that a product detector is sensitive to AM and PM. For example, if the input contains no angle modulation, so that $\theta(t) = 0$, and if the reference phase is set to zero (i.e., $\theta_0 = 0$), then

$$v_{out}(t) = \tfrac{1}{2}A_0 R(t) \qquad (4\text{–}78a)$$

which implies that $x(t) \geq 0$, and the real envelope is obtained on the product detector out-put, just as in the case of the envelope detector discussed previously. However, if an angle-modulated signal $A_c \cos[\omega_c t + \theta(t)]$ is present at the input and $\theta_0 = 90°$, the product detector output is

$$v_{out}(t) = \tfrac{1}{2}A_0 \text{Re}\{A_c e^{j[\theta(t)-90°]}\}$$

or

$$v_{out}(t) = \tfrac{1}{2}A_0 A_c \sin \theta(t) \qquad (4\text{–}78b)$$

In this case, the product detector acts like a *phase detector* with a *sinusoidal characteristic*, because the output voltage is proportional to the sine of the phase difference between the input signal and the oscillator signal. Phase detector circuits are also available that yield triangle and sawtooth characteristics [Krauss, Bostian, and Raab, 1980]. Referring to Eq. (4–78b) for the phase detector with a sinusoidal characteristic, and assuming that the phase difference is small [i.e., $|\theta(t)| \ll \pi/2$], we see that $\sin \theta(t) \approx \theta(t)$ and

$$v_{\text{out}}(t) \approx \tfrac{1}{2} A_0 A_c \, \theta(t) \tag{4–79}$$

which is a linear characteristic (for small angles). Thus, the output of this phase detector is directly proportional to the phase differences when the difference angle is small. (See Fig. 4–20a).

The product detector acts as a linear time-varying device with respect to the input $v_{\text{in}}(t)$, in contrast to the envelope detector, which is a nonlinear device. The property of being either linear or nonlinear significantly affects the results when two or more components, such as a signal plus noise, are applied to the input. This topic will be studied in Chapter 7.

Detectors may also be classified as being either coherent or noncoherent. A *coherent* detector has two inputs—one for a reference signal, such as the synchronized oscillator signal, and one for the modulated signal that is to be demodulated. The product detector is an example of a coherent detector. A *noncoherent* detector has only one input, namely, the modulated signal port. The envelope detector is an example of a noncoherent detector.

Frequency Modulation Detector

An *ideal frequency modulation (FM) detector* is a device that produces an output that is proportional to the instantaneous frequency of the input. That is, if the bandpass input is represented by $R(t) \cos[\omega_c t + \theta(t)]$, the output of the ideal FM detector is

$$v_{\text{out}}(t) = \frac{K d[\omega_c t + \theta(t)]}{dt} = K\left[\omega_c + \frac{d\theta(t)}{dt}\right] \tag{4–80}$$

when the input is not zero (i.e. $R(t) \neq 0$).

Usually, the FM detector is *balanced*. This means that the dc voltage $K\omega_c$ does not appear on the output if the detector is tuned to (or designed for) the carrier frequency f_c. In this case, the output is

$$v_{\text{out}}(t) = K \frac{d\theta(t)}{dt} \tag{4–81}$$

There are many ways to build FM detectors, but almost all of them are based on one of three principles:

- FM-to-AM conversion.
- Phase-shift or quadrature detection.
- Zero-crossing detection.

A *slope detector* is one example of the FM-to-AM conversion principle. A block diagram is shown in Fig. 4–15. A bandpass limiter is needed to suppress any amplitude variations on the input signal, since these would distort the desired output signal.

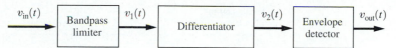

Figure 4–15   Frequency demodulation using slope detection.

The slope detector may be analyzed as follows. Suppose that the input is a fading signal with frequency modulation. From Table 4–1, this FM signal may be represented by

$$v_{in}(t) = A(t) \cos[\omega_c t + \theta(t)] \qquad (4\text{–}82)$$

where

$$\theta(t) = K_f \int_{-\infty}^{t} m(t_1) \, dt_1 \qquad (4\text{–}83)$$

$A(t)$ represents the envelope that is fading, and $m(t)$ is the modulation (e.g., audio) signal. It follows that the limiter output is proportional to

$$v_1(t) = V_L \cos[\omega_c t + \theta(t)] \qquad (4\text{–}84)$$

and the output of the differentiator becomes

$$v_2(t) = -V_L \left[\omega_c + \frac{d\theta(t)}{dt}\right] \sin[\omega_c t + \theta(t)] \qquad (4\text{–}85)$$

The output of the envelope detector is the magnitude of the complex envelope for $v_2(t)$:

$$v_{out}(t) = \left| -V_L \left[\omega_c + \frac{d\theta(t)}{dt}\right] \right|$$

Because $\omega_c \gg d\theta/dt$ in practice, this becomes

$$v_{out}(t) = V_L \left[\omega_c + \frac{d\theta(t)}{dt}\right]$$

Using Eq. (4–83), we obtain

$$v_{out}(t) = V_L \omega_c + V_L K_f m(t) \qquad (4\text{–}86)$$

which indicates that the output consists of a dc voltage $V_L \omega_c$, plus the ac voltage $V_L K_f m(t)$, which is proportional to the modulation on the FM signal. Of course, a capacitor could be placed in series with the output so that only the ac voltage would be passed to the load.

The differentiation operation can be obtained by any circuit that acts like a frequency-to-amplitude converter. For example, a single-tuned resonant circuit can be used as illustrated in Fig. 4–16, where the magnitude transfer function is $|H(f)| = K_1 f + K_2$ over the linear (useful) portion of the characteristic. A balanced FM detector, which is also called a *balanced discriminator*, is shown in Fig. 4–17. Two tuned circuits are used to balance out the dc when the input has a carrier frequency of f_c and to provide an extended linear frequency-to-voltage conversion characteristic.

Balanced discriminators can also be built that function because of the phase-shift properties of a double-tuned RF transformer circuit with primary and secondary windings [Stark,

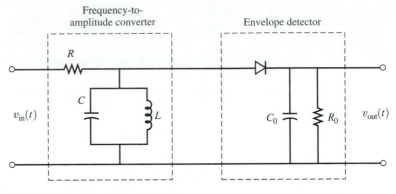

(a) Circuit Diagram of a Slope Detector

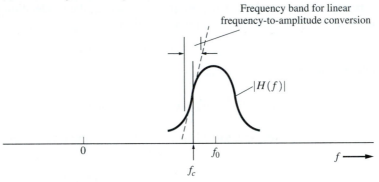

(b) Magnitude of Filter Transfer Function

Figure 4–16 Slope detection using a single-tuned circuit for frequency-to-am-plitude conversion.

Tuteur, and Anderson, 1988]. In practice, discriminator circuits have been replaced by integrated circuits that operate on the quadrature principle.

The *quadrature detector* is described as follows: A quadrature signal is first obtained from the FM signal; then, through the use of a product detector, the quadrature signal is multiplied with the FM signal to produce the demodulated signal $v_{out}(t)$. The quadrature signal can be produced by passing the FM signal through a capacitor (large) reactance that is connected in series with a parallel resonant circuit tuned to f_c. The quadrature signal voltage appears across the parallel resonant circuit. The series capacitance provides a 90° phase shift, and the resonant circuit provides an additional phase shift that is proportional to the instantaneous frequency deviation (from f_c) of the FM signal. From Eqs. (4–84) and (4–83), the FM signal is

$$v_{in}(t) = V_L \cos[\omega_c t + \theta(t)] \tag{4–87}$$

and the quadrature signal is

$$v_{quad}(t) = K_1 V_L \sin\left[\omega_c t + \theta(t) + K_2 \frac{d\theta(t)}{dt}\right] \tag{4–88}$$

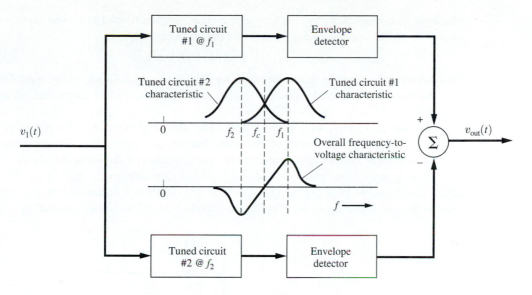

(a) Block Diagram

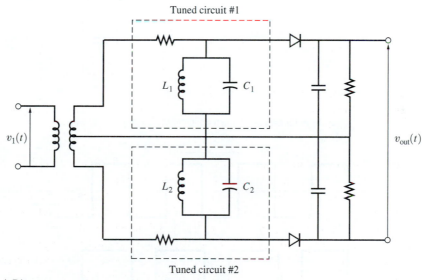

(b) Circuit Diagram

Figure 4–17   Balanced discriminator.

where K_1 and K_2 are constants that depend on component values used for the series capacitor and in the parallel resonant circuit. These two signals, Eqs. (4–87) and (4–88), are multiplied together by a product detector (e.g., see Fig. 4–14) to produce the output signal

$$v_{out}(t) = \tfrac{1}{2} K_1 V_L^2 \sin\left[K_2 \frac{d\theta(t)}{dt} \right] \tag{4–89}$$

where the sum-frequency term is eliminated by the low-pass filter. For K_2 sufficiently small, $\sin x \approx x$, and by the use of Eq. (4–83), the output becomes

$$v_{\text{out}}(t) = \tfrac{1}{2}K_1 K_2 V_L^2 K_f m(t) \tag{4–90}$$

This demonstrates that the quadrature detector detects the modulation on the input FM signal. The quadrature detector principle is also used by phase-locked loops that are configured to detect FM. [See Eq. (4–110).

As indicated by Eq. (4–80), the output of an ideal FM detector is directly proportional to the instantaneous frequency of the input. This linear frequency-to-voltage characteristic may be obtained directly by counting the zero-crossings of the input waveform. An FM detector utilizing this technique is called a *zero-crossing detector*. A hybrid circuit (i.e., a circuit consisting of both digital and analog devices) that is a balanced FM zero-crossing detector is shown in Fig. 4–18. The limited (square-wave) FM signal, denoted by $v_1(t)$, is

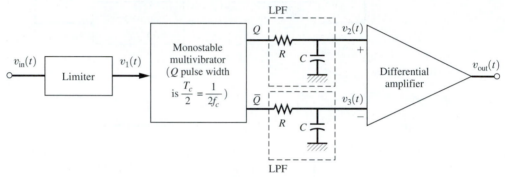

(a) Circuit

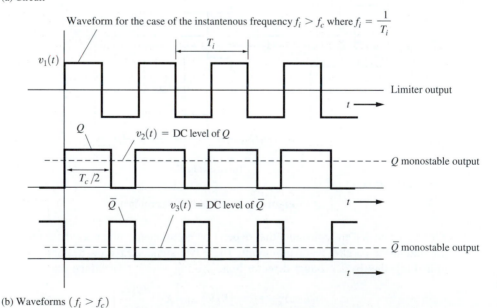

(b) Waveforms ($f_i > f_c$)

Figure 4–18 Balanced zero-crossing FM detector.

shown in Fig. 4–18b. For purposes of illustration, it is assumed that $v_1(t)$ is observed over that portion of the modulation cycle when the instantaneous frequency

$$f_i(t) = f_c + \frac{1}{2\pi}\frac{d\theta(t)}{dt} \tag{4–91}$$

is larger than the carrier frequency f_c. That is, $f_i > f_c$ in the illustration. Since the modulation voltage varies slowly with respect to the input FM signal oscillation, $v_1(t)$ appears (in the figure) to have a constant frequency, although it is actually varying in frequency according to $f_i(t)$. The monostable multivibrator (one-shot) is triggered on the positive slope zero-crossings of $v_1(t)$. For balanced FM detection, the pulse width of the Q output is set to $T_c/2 = 1/2f_c$, where f_c is the carrier frequency of the FM signal at the input. Thus, the differential amplifier output voltage is zero if $f_i = f_c$. For $f_i > f_c$ [as illustrated by the $v_1(t)$ waveform in the figure], the output voltage is positive, and for $f_i < f_c$, the output voltage will be negative. Hence, a linear frequency-to-voltage characteristic, $C[f_i(t) - f_c]$, is obtained where, for an FM signal at the input, $f_i(t) = f_c + (1/2\pi)K_f m(t)$.

Another circuit that can be used for FM demodulation, as well as for other purposes, is the phase-locked loop.

4–14 PHASE-LOCKED LOOPS AND FREQUENCY SYNTHESIZERS

A phase-locked loop (PLL) consists of three basic components: (1) a phase detector, (2) a low-pass filter, and (3) a voltage-controlled oscillator (VCO), as shown in Fig. 4–19. The VCO is an oscillator that produces a periodic waveform with a frequency that may be varied about some free-running frequency f_0, according to the value of the applied voltage $v_2(t)$. The free-running frequency f_0, is the frequency of the VCO output when the applied voltage $v_2(t)$ is zero. The phase detector produces an output signal $v_1(t)$ that is a function of the phase difference between the incoming signal $v_{in}(t)$ and the oscillator signal $v_0(t)$. The filtered signal $v_2(t)$ is the control signal that is used to change the frequency of the VCO output. The PLL configuration may be designed so that it acts as a narrowband tracking filter when the low-pass filter (LPF) is a narrowband filter. In this operating mode, the frequency of the VCO will become that of one of the line components of the input signal spectrum, so that, in effect, the VCO output signal is a periodic signal with a frequency equal

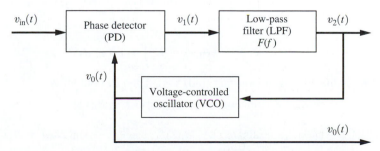

Figure 4–19 Basic PLL.

to the average frequency of this input signal component. Once the VCO has acquired the frequency component, the frequency of the VCO will track the input signal component if it changes slightly in frequency. In another mode of operation, the bandwidth of the LPF is wider so that the VCO can track the instantaneous frequency of the whole input signal. When the PLL tracks the input signal in either of these ways, the PLL is said to be "locked."

If the applied signal has an initial frequency of f_0, the PLL will acquire a lock and the VCO will track the input signal frequency over some range, provided that the input frequency changes slowly. However, the loop will remain locked only over some finite range of frequency shift. This range is called the *hold-in* (or *lock*) *range*. The hold-in range depends on the overall dc gain of the loop, which includes the dc gain of the LPF. On the other hand, if the applied signal has an initial frequency different from f_0, the loop may not acquire lock even though the input frequency is within the hold-in range. The frequency range over which the applied input will cause the loop to lock is called the *pull-in* (or *capture*) *range*. This range is determined primarily by the loop filter characteristics, and it is never greater than the hold-in range. (See Fig. 4–23.) Another important PLL specification is the *maximum locked sweep rate*, which is defined as the maximum rate of change of the input frequency for which the loop will remain locked. If the input frequency changes faster than this rate, the loop will drop out of lock.

If the PLL is built using analog circuits, it is said to be an *analog phase-locked loop* (APLL). Conversely, if digital circuits and signals are used, the PLL is said to be a *digital phase-locked loop* (DPLL). For example, the phase detection (PD) characteristic depends on the exact implementation used. Some PD characteristics are shown in Fig. 4–20. The sinusoidal characteristic is obtained if an (analog circuit) multiplier is used and the periodic signals are sinusoids. The multiplier may be implemented by using a double-balanced mixer. The triangle and sawtooth PD characteristics are obtained by using digital circuits. In addition to using digital VCO and PD circuits, the DPLL may incorporate a digital loop filter and signal-processing techniques that use microprocessors. Gupta [1975] published a fine tutorial paper on analog phase-locked loops in the *IEEE Proceedings*, and Lindsey and Chie [1981] followed with a survey paper on digital PLL techniques. In addition, there are excellent books available [Blanchard, 1976; Gardner, 1979; Best, 1999].

The PLL may be studied by examining the APLL, as shown in Fig. 4–21. In this figure, a multiplier (sinusoidal PD characteristic) is used. Assume that the input signal is

$$v_{\text{in}}(t) = A_i \sin[\omega_0 t + \theta_i(t)] \tag{4–92}$$

and that the VCO output signal is

$$v_0(t) = A_0 \cos[\omega_0 t + \theta_0(t)] \tag{4–93}$$

where

$$\theta_0(t) = K_v \int_{-\infty}^{t} v_2(\tau)\, d\tau \tag{4–94}$$

and K_v is the VCO gain constant (rad/V-s). Then PD output is

$$v_1(t) = K_m A_i A_0 \sin[\omega_0 t + \theta_i(t)] \cos[\omega_0 t + \theta_0(t)]$$

$$= \frac{K_m A_i A_0}{2} \sin[\theta_i(t) - \theta_0(t)] + \frac{K_m A_i A_0}{2} \sin[2\omega_0 t + \theta_i(t) + \theta_0(t)] \tag{4–95}$$

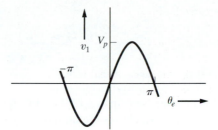

(a) Sinusoidal Characteristics

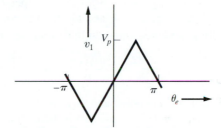

(b) Triangle Characteristics

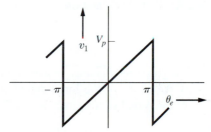

(c) Sawtooth Characteristics

Figure 4-20   Some phase detector characteristics.

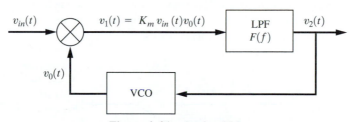

Figure 4-21   Analog PLL.

where K_m is the gain of the multiplier circuit. The sum frequency term does not pass through the LPF, so the LPF output is

$$v_2(t) = K_d[\sin \theta_e(t)] * f(t) \qquad (4\text{-}96)$$

where

$$\theta_e(t) \triangleq \theta_i(t) - \theta_0(t) \tag{4-97}$$

$$K_d = \frac{K_m A_i A_0}{2} \tag{4-98}$$

and $f(t)$ is the impulse response of the LPF. $\theta_e(t)$ is called the *phase error*; K_d is the equivalent PD constant, which, for the multiplier-type PD, depends on the levels of the input signal A_i and the level of the VCO signal A_0.

The overall equation describing the operation of the PLL may be obtained by taking the derivative of Eqs. (4–94) and (4–97) and combining the result by the use of Eq. (4–96). The resulting nonlinear equation that describes the PLL becomes

$$\frac{d\theta_e(t)}{dt} = \frac{d\theta_i(t)}{dt} - K_d K_v \int_0^t [\sin \theta_e(\lambda)] f(t - \lambda) \, d\lambda \tag{4-99}$$

where $\theta_e(t)$ is the unknown and $\theta_i(t)$ is the forcing function.

In general, this PLL equation is difficult to solve. However, it may be reduced to a linear equation if the gain K_d is large, so that the loop is locked and the error $\theta_e(t)$ is small. In this case, $\sin \theta_e(t) \approx \theta_e(t)$, and the resulting linear equation is

$$\frac{d\theta_e(t)}{dt} = \frac{d\theta_i(t)}{dt} - K_d K_v \theta_e(t) * f(t) \tag{4-100}$$

A block diagram based on this linear equation is shown in Fig. 4–22. In this linear PLL model (Fig. 4–22), the *phase* of the input signal and the *phase* of the VCO output signal are used instead of the actual signals themselves (Fig. 4–21). The closed-loop transfer function $\Theta_0(f)/\Theta_i(f)$ is

$$H(f) = \frac{\Theta_0(f)}{\Theta_i(f)} = \frac{K_d K_v F(f)}{j2\pi f + K_d K_v F(f)} \tag{4-101}$$

where $\Theta_0(f) = \mathcal{F}[\theta_0(t)]$ and $\Theta_i(f) = \mathcal{F}[\theta_i(t)]$. Of course, the design and analysis techniques used to evaluate linear feedback control systems, such as Bode plots, which will indicate phase gain and phase margins, are applicable. In fact, they are extremely useful in describing the performance of *locked* PLLs.

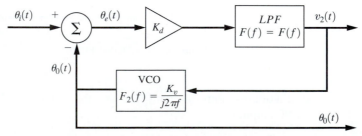

Figure 4–22   Linear model of the analog PLL.

The equation for the *hold-in range* may be obtained by examining the nonlinear behavior of the PLL. From Eqs. (4–94) and (4–96), the instantaneous frequency deviation of the VCO from ω_0 is

$$\frac{d\theta_0(t)}{dt} = K_v v_2(t) = K_v K_d[\sin\,\theta_e(t)] * f(t) \tag{4–102}$$

To obtain the hold-in range, the input frequency is changed very slowly from f_0. Here the dc gain of the filter is the controlling parameter, and Eq. (4–102) becomes

$$\Delta\omega = K_v K_d F(0)\sin\,\theta_e \tag{4–103}$$

The maximum and minimum values of $\Delta\omega$ give the hold-in range, and these are obtained when $\sin\,\theta_e = \pm 1$. Thus, the maximum hold-in range (the case with no noise) is

$$\Delta f_h = \frac{1}{2\pi}\,K_v K_d F(0) \tag{4–104}$$

A typical lock-in characteristic is illustrated in Fig. 4–23. The solid curve shows the VCO control signal $v_2(t)$ as the sinusoidal testing signal is swept from a low frequency to a high frequency (with the free-running frequency of the VCO, f_0, being within the swept band). The dashed curve shows the result when sweeping from high to low. The hold-in range Δf_h is related to the dc gain of the PLL as described by Eq. (4–104).

The *pull-in range* Δf_p is determined primarily by the loop-filter characteristics. For example, assume that the loop has not acquired lock and that the testing signal is swept slowly toward f_0. Then, the PD output, there will be a beat (oscillatory) signal, and its frequency $|f_{in} - f_0|$ will vary from a large value to a small value as the test signal frequency sweeps toward f_0. As the testing signal frequency comes closer to f_0, the beat-frequency waveform will become nonsymmetrical, in which case it will have a nonzero dc value. This dc value tends to change the frequency of the VCO to that of the input signal frequency, so that the loop will tend to lock. The pull-in range, Δf_p, where the loop acquires lock will depend on

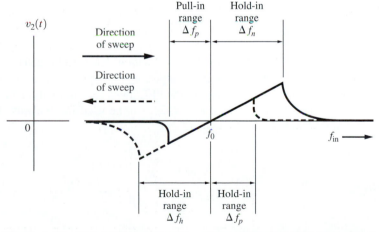

Figure 4–23   PLL VCO control voltage for a swept sinusoidal input signal.

exactly how the loop filter $F(f)$ processes the PD output to produce the VCO control signal. Furthermore, even if the input signal is within the pull-in range, it may take a fair amount of time for the loop to acquire lock, since the LPF acts as an integrator and it takes some time for the control voltage (filter output) to build up to a value large enough for locking to occur. The analysis of the pull-in phenomenon is complicated. It is actually statistical in nature, because it depends on the initial phase relationship of the input and VCO signals and on noise that is present in the circuit. Consequently, in the measurement of Δf_p, several repeated trials may be needed to obtain a typical value.

The locking phenomenon is not peculiar to PLL circuits, but occurs in other types of circuits as well. For example, if an external signal is injected into the output port of an oscillator (i.e., a plain oscillator, not a VCO), the oscillator signal will tend to change frequency and will eventually lock onto the frequency of the external signal if the latter is within the pull-in range of the oscillator. This phenomenon is called *injection locking* or *synchronization* of an oscillator and may be modeled by a PLL model [Couch, 1971].

The PLL has numerous applications in communication systems, including (1) FM detection, (2) the generation of highly stable FM signals, (3) coherent AM detection, (4) frequency multiplication, (5) frequency synthesis, and (6) use as a building block within complicated digital systems to provide bit synchronization and data detection.

Let us now find what conditions are required for the PLL to become an FM detector. Referring to Fig. 4–21, let the PLL input signal be an FM signal. That is,

$$v_{\text{in}}(t) = A_i \sin\left[\omega_c t + D_f \int_{-\infty}^{t} m(\lambda)\, d\lambda\right] \tag{4-105a}$$

where

$$\theta_i(t) = D_f \int_{-\infty}^{t} m(\lambda)\, d\lambda \tag{4-105b}$$

or

$$\Theta_i(f) = \frac{D_f}{j 2\pi f}\, M(f) \tag{4-105c}$$

and $m(t)$ is the baseband (e.g., audio) modulation that is to be detected. We would like to find the conditions such that the PLL output, $v_2(t)$, is proportional to $m(t)$. Assume that f_c is within the capture (pull-in) range of the PLL; thus, for simplicity, let $f_0 = f_c$. Then the linearized PLL model, as shown in Fig. 4–22, can be used for analysis. Working in the frequency domain, we obtain the output

$$V_2(f) = \frac{\left(j\, \dfrac{2\pi f}{K_v}\right) F_1(f)}{F_1(f) + j\left(\dfrac{2\pi f}{K_v K_d}\right)}\, \Theta_i(f)$$

which, by the use of Eq. (4–105c), becomes

$$V_2(f) = \frac{\dfrac{D_f}{K_v}F_1(f)}{F_1(f) + j\left(\dfrac{2\pi f}{K_v K_d}\right)} M(f) \tag{4–106}$$

Now we find the conditions such that $V_2(f)$ is proportional to $M(f)$. Assume that the bandwidth of the modulation is B hertz, and let $F_1(f)$ be a low-pass filter. Thus,

$$F(f) = F_1(f) = 1, \qquad |f| < B \tag{4–107}$$

Also, let

$$\frac{K_v K_d}{2\pi} \gg B \tag{4–108}$$

Then Eq. (4–106) becomes

$$V_2(f) = \frac{D_f}{K_v} M(f) \tag{4–109}$$

or

$$v_2(t) = Cm(t) \tag{4–110}$$

where the constant of proportionality is $C = D_f/K_v$. Hence, the PLL circuit of Fig. 4–21 will become an FM detector circuit, where $v_2(t)$ is the detected FM output when the conditions of Eqs. (4–107) and (4–108) are satisfied.

In another application, the PLL may be used to supply the coherent oscillator signal for product detection of an AM signal (Fig. 4–24). Recall from Eqs. (4–92) and (4–93) that the VCO of a PLL locks 90° out of phase with respect to the incoming signal.[†] Then $v_0(t)$ needs to be shifted by −90° so that it will be in phase with the carrier of the input AM signal, the requirement for coherent detection of AM, as given by Eq. (4–77). In this application, the bandwidth of the LPF needs to be just wide enough to provide the necessary pull-in range in order for the VCO to be able to lock onto the carrier frequency f_c.

Figure 4–25 illustrates the use of a PLL in a frequency synthesizer. The synthesizer generates a periodic signal of frequency

$$f_{\text{out}} = \left(\frac{N}{M}\right) f_x \tag{4–111}$$

where f_x is the frequency of the stable oscillator and N and M are the frequency-divider parameters. This result is verified by recalling that when the loop is locked, the dc control signal $v_3(t)$ shifts the frequency of the VCO so that $v_2(t)$ will have the same frequency as $v_{\text{in}}(t)$. Thus,

---

[†] This results from the characteristic of the phase detector circuit. The statement is correct for a PD that produces a zero dc output voltage when the two PD input signals are 90° out of phase (i.e., a multiplier-type PD). However, if the PD circuit produced a zero dc output when the two PD inputs were in phase, the VCO of the PLL would lock in phase with the incoming PLL signal.

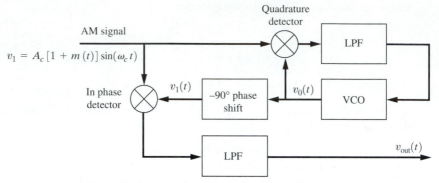

Figure 4–24   PLL used for coherent detection of AM.

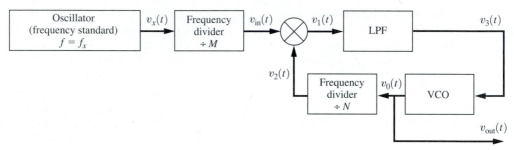

Figure 4–25   PLL used in a frequency synthesizer.

$$\frac{f_x}{M} = \frac{f_{\text{out}}}{N} \qquad\qquad (4\text{--}112)$$

which is equivalent to Eq. (4–111).

Classical frequency dividers use *integer values* for M and N. Furthermore, if programmable dividers are used, the synthesizer output frequency may be changed by using software that programs a microprocessor to select the appropriate values of M and N, according to Eq. (4–111). This technique is used in frequency synthesizers that are built into modern receivers with digital tuning. (See study aid Prob. SA4–6 for an example of frequency synthesizer design). For the case of $M = 1$, the frequency synthesizer acts as a frequency multiplier.

Equivalent *noninteger values* for N can be obtained by periodically changing the divider count over a set of similar integer values. This produces an average N value that is noninteger and is called the *fractional-N* technique. With fractional-N synthesizers, the instantaneous value of N changes with time, and this can modulate the VCO output signal to produce unwanted (spurious) sidebands in the spectrum. By careful design the sideband noise can be reduced to a low level [Conkling, 1998]. More complicated PLL synthesizer configurations can be built that incorporate mixers and additional oscillators.

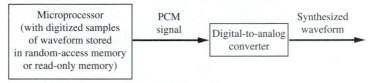

Figure 4–26   Direct digital synthesis (DDS).

4–15 DIRECT DIGITAL SYNTHESIS

Direct digital synthesis (DDS) is a method for generating a desired waveform (such as a sine wave) by using the computer technique described in Fig. 4–26. To configure the DDS system to generate a waveform, samples of the desired waveform are converted into PCM words and stored in the memory (random-access memory [RAM] or read-only memory [ROM]) of the microprocessor system. The DDS system can then generate the desired waveform by "playing back" the stored words into the digital-to-analog converter.

This DDS technique has many attributes. For example, if the waveform is periodic, such as a sine wave, only one cycle of samples needs to be stored in memory. The continuous sine wave can be generated by repeatedly cycling through the memory. The frequency of the generated sine wave is determined by the rate at which the memory is read out. If desired, the microprocessor can be programmed to generate a certain frequency during a certain time interval and then switch to a different frequency (or another waveshape) during a different time interval. Also, simultaneous sine and cosine (two-phase) outputs can be generated by adding another DAC. The signal-to-quantizing noise can be designed to be as large as desired by selecting the appropriate number of bits that are stored for each PCM word, as described by Eq. (3–18).

The DDS technique is replacing analog circuits in many applications. For example, in higher-priced communications receivers, the DDS technique is used as a frequency synthesizer to generate local oscillator signals that tune the radio. (See Sec. 4–16.) In electronic pipe organs and music synthesizers, DDS can be used to generate authentic as well as weird sounds. Instrument manufacturers are using DDS to generate the output waveforms for function generators and arbitrary waveform generators (AWG). Telephone companies are using DDS to generate dial tones and busy signals. (See Chapter 8.)

4–16 TRANSMITTERS AND RECEIVERS

Generalized Transmitters

Transmitters generate the modulated signal at the carrier frequency f_c from the modulating signal $m(t)$. In Secs. 4–1 and 4–2, it was demonstrated that any type of modulated signal could be represented by

$$v(t) = \text{Re}\{g(t)\ e^{j\omega_c t}\} \tag{4–113}$$

or, equivalently,

$$v(t) = R(t) \cos[\omega_c t + \theta(t)] \tag{4–114}$$

and

$$v(t) = x(t) \cos \omega_c t - y(t) \sin \omega_c t \tag{4-115}$$

where the complex envelope

$$g(t) = R(t) e^{j\theta(t)} = x(t) + jy(t) \tag{4-116}$$

is a function of the modulating signal $m(t)$. The particular relationship that is chosen for $g(t)$ in terms of $m(t)$ defines the type of modulation that is used, such as AM, SSB, or FM. (See Table 4–1.) A generalized approach may be taken to obtain universal transmitter models that may be reduced to those used for a particular type of modulation. We will also see that there are equivalent models that correspond to different circuit configurations, yet they may be used to produce the same type of modulated signal at their outputs. It is up to the designer to select an implementation method that will maximize performance, yet minimize cost, based on the state of the art in circuit development.

There are two canonical forms for the generalized transmitter, as indicated by Eqs. (4–114) and (4–115). Equation (4–114) describes an AM–PM type of circuit, as shown in Fig. 4–27. The baseband signal-processing circuit generates $R(t)$ and $\theta(t)$ from $m(t)$. The R and θ are functions of the modulating signal $m(t)$, as given in Table 4–1 for the particular type of modulation desired. The signal processing may be implemented by using either nonlinear analog circuits or a digital computer that incorporates the R and θ algorithms under software program control. In the implementation using a digital computer, one ADC will be needed at the input and two DACs will be needed at the output. The remainder of the AM–PM canonical form requires RF circuits, as indicated in the figure.

Figure 4–28 illustrates the second canonical form for the generalized transmitter. This uses in-phase and quadrature-phase (IQ) processing. Similarly, the formulas relating $x(t)$ and $y(t)$ to $m(t)$ are shown in Table 4–1, and the baseband signal processing may be implemented by using either analog hardware or digital hardware with software. The remainder of the canonical form uses RF circuits as indicated.

Once again, it is stressed that any type of signal modulation (AM, FM, SSB, QPSK, etc.) may be generated by using either of these two canonical forms. Both of these forms conveniently separate baseband processing from RF processing. Digital techniques are

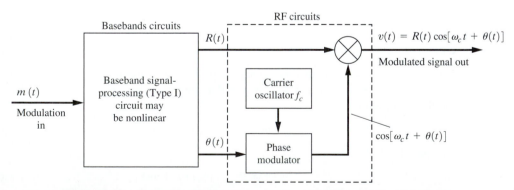

Figure 4–27   Generalized transmitter using the AM–PM generation technique.

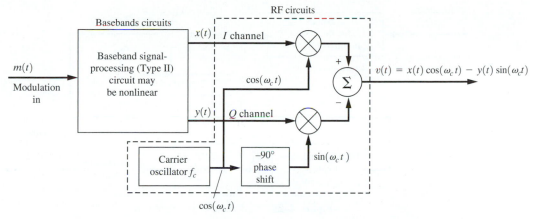

Figure 4–28   Generalized transmitter using the quadrature generation technique.

especially useful to realize the baseband-processing portion. Furthermore, if digital computing circuits are used, any desired type of modulation can be obtained by selecting the appropriate software algorithm.

Most of the practical transmitters in use today are special variations on these canonical forms. Practical transmitters may perform the RF operations at some convenient lower RF frequency and then up-convert to the desired operating frequency. In the case of RF signals that contain no AM, frequency multipliers may be used to arrive at the operating frequency. Of course, power amplifiers are usually required to bring the output power level up to the specified value. If the RF signal contains no amplitude variations, Class C amplifiers (which have relatively high efficiency) may be used; otherwise, Class B amplifiers are used.

Generalized Receiver: The Superheterodyne Receiver

The receiver has the job of extracting the source information from the received modulated signal that may be corrupted by noise. Often, it is desired that the receiver output be a replica of the modulating signal that was present at the transmitter input. There are two main classes of receivers: the *tuned radio-frequency* (TRF) receiver and the *superheterodyne* receiver.

The TRF receiver consists of a number of cascaded high-gain RF bandpass stages that are tuned to the carrier frequency f_c, followed by an appropriate detector circuit (an envelope detector, a product detector, an FM detector, etc.). The TRF is not very popular, because it is difficult to design tunable RF stages so that the desired station can be selected and yet have a narrow bandwidth so that adjacent channel stations are rejected. In addition, it is difficult to obtain high gain at radio frequencies and to have sufficiently small stray coupling between the output and input of the RF amplifying chain so that the chain will not become an oscillator at f_c. The "crystal set" that is built by the Cub Scouts is an example of a single–RF-stage TRF receiver that has no gain in the RF stage. TRF receivers are often used to measure time-dispersive (multipath) characteristics of radio channels [Rappaport, 1989].

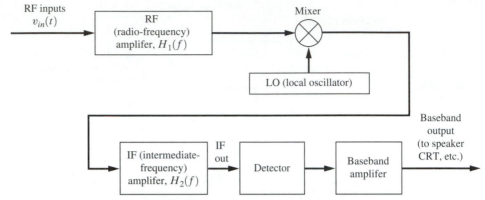

Figure 4–29   Superheterodyne receiver.

Most receivers employ the *superheterodyne* receiving technique as shown in Fig. 4–29. The technique consists of either down-converting or up-converting the input signal to some convenient frequency band, called the *intermediate frequency* (IF) band, and then extracting the information (or modulation) by using the appropriate detector.[†] This basic receiver structure is used for the reception of all types of bandpass signals, such as television, FM, AM, satellite, cellular, and radar signals. The RF amplifier has a bandpass characteristic that passes the desired signal and provides amplification to override additional noise that is generated in the mixer stage. The RF filter characteristic also provides some rejection of adjacent channel signals and noise, but the main adjacent channel rejection is accomplished by the IF filter.

The IF filter is a bandpass filter that selects either the up-conversion or down-conversion component (whichever is chosen by the receiver's designer). When up conversion is selected, the complex envelope of the IF (bandpass) filter output is the same as the complex envelope for the RF input, except for RF filtering, $H_1(f)$, and IF filtering, $H_2(f)$. However, if down conversion is used with $f_{LO} > f_c$, the complex envelope at the IF output will be the conjugate of that for the RF input. [See Eq. (4–61c).] This means that the sidebands of the IF output will be inverted (i.e., the upper sideband on the RF input will become the lower sideband, etc., on the IF output). If $f_{LO} < f_c$, the sidebands are not inverted.

The center frequency selected for the IF amplifier is chosen on the basis of three considerations:

- The IF frequency should be such that a stable high-gain IF amplifier can be economically attained.

- The IF frequency needs to be low enough so that, with practical circuit elements in the IF filters, values of Q can be attained that will provide a steep attenuation characteristic outside the bandwidth of the IF signal. This decreases the noise and minimizes the interference from adjacent channels.

[†] *Dual-conversion* superheterodyne receivers can also be built, in which a second mixer and a second IF stage follow the first IF stage shown in Fig. 4–29.

- The IF frequency needs to be high enough so that the receiver image response can be made acceptably small.

The *image response* is the reception of an unwanted signal located at the image frequency due to insufficient attenuation of the image signal by the RF amplifier filter. The image response is best illustrated by an example.

Example 4–2 AM BROADCAST SUPERHETERODYNE RECEIVER

Assume that an AM broadcast band radio is tuned to receive a station at 850 kHz and that the LO frequency is on the high side of the carrier frequency. If the IF frequency is 455 kHz, the LO frequency will be 850 + 455 = 1,305 kHz. (See Fig. 4–30.) Furthermore, assume that other signals are present at the RF input of the radio and, particularly, that there is a signal at 1,760 kHz; this signal will be down-converted by the mixer to 1,760 − 1,305 = 455 kHz. That is, the undesired (1,760-kHz) signal will be translated to 455 kHz and will be added at the mixer output to the desired (850-kHz) signal, which was also down-converted to 455 kHz. This *undesired* signal that has been converted to the IF band is called the *image signal*. If the gain of the RF amplifier is down by, say, 25 dB at 1,760 kHz compared to the gain at 850 kHz, and if the undesired signal is 25 dB stronger at the receiver input than the desired signal, both signals will have the same level when translated to the IF. In this case, the undesired signal will definitely interfere with the desired signal in the detection process.

For down converters (i.e., $f_{IF} = |f_c - f_{LO}|$), the image frequency is

$$f_{image} = \begin{cases} f_c + 2f_{IF}, & \text{if } f_{LO} > f_c \quad \text{(high-side injection)} \\ f_c - 2f_{IF}, & \text{if } f_{LO} < f_c \quad \text{(low-side injection)} \end{cases} \qquad (4\text{–}117a)$$

where f_c is the desired RF frequency, f_{IF} is the IF frequency, and f_{LO} is the local oscillator frequency. For up converters (i.e., $f_{IF} = f_c + f_{LO}$), the image frequency is

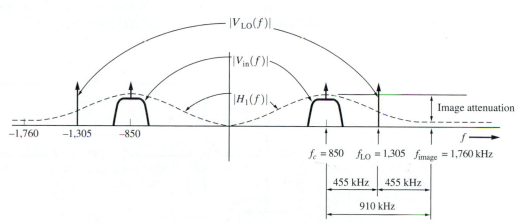

Figure 4–30 Spectra of signals and transfer function of an RF amplifier in a superheterodyne receiver.

TABLE 4–4   SOME POPULAR IF FREQUENCIES IN THE UNITED STATES.

IF Frequency	Application
262.5 kHz	AM broadcast radios (in automobiles)
455 kHz	AM broadcast radios
10.7 MHz	FM broadcast radios
21.4 MHz	FM two-way radios
30 MHz	Radar receivers
43.75 MHz (video carrier)	TV sets
60 MHz	Radar receivers
70 MHz	Satellite receivers

$$f_{\text{image}} = f_c + 2f_{\text{LO}} \tag{4–117b}$$

From Fig. 4–30, it is seen that the image response will usually be reduced if the IF frequency is increased, since f_{image} will occur farther away from the main peak (or lobe) of the RF filter characteristic, $|H_1(f)|$.

Recalling our earlier discussion on mixers, we also realize that other spurious responses (in addition to the image response) will occur in practical mixer circuits. These must also be taken into account in good receiver design.

Table 4–4 illustrates some typical IF frequencies that have become de facto standards. For the intended application, the IF frequency is low enough that the IF filter will provide good adjacent channel signal rejection when circuit elements with a realizable Q are used; yet the IF frequency is large enough to provide adequate image-signal rejection by the RF amplifier filter.

The type of detector selected for use in the superheterodyne receiver depends on the intended application. For example, a product detector may be used in a PSK (digital) system, and an envelope detector is used in AM broadcast receivers. If the complex envelope $g(t)$ is desired for generalized signal detection or for optimum reception in digital systems, the $x(t)$ and $y(t)$ quadrature components, where $x(t) + jy(t) = g(t)$, may be obtained by using quadrature product detectors, as illustrated in Fig. 4–31. $x(t)$ and $y(t)$ could be fed into a signal

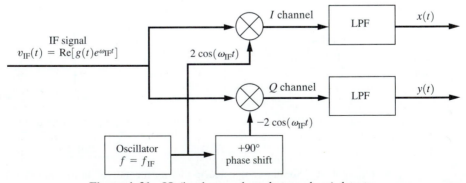

Figure 4–31   IQ (in-phase and quadrature-phase) detector.

processor to extract the modulation information. Disregarding the effects of noise, the signal processor could recover $m(t)$ from $x(t)$ and $y(t)$ (and, consequently, demodulate the IF signal) by using the inverse of the complex envelope generation functions given in Table 4–1.

The superheterodyne receiver has many advantages and some disadvantages. The main advantage is that extraordinarily high gain can be obtained without instability (self-oscillation). The stray coupling between the output of the receiver and the input does not cause oscillation because the gain is obtained in disjoint frequency bands—RF, IF, and baseband. The receiver is easily tunable to another frequency by changing the frequency of the LO signal (which may be supplied by a frequency synthesizer) and by tuning the bandpass of the RF amplifier to the desired frequency. Furthermore, high-Q elements—which are needed (to produce steep filter skirts) for adjacent channel rejection—are needed only in the fixed tuned IF amplifier. The main disadvantage of the superheterodyne receiver is the response to spurious signals that will occur if one is not careful with the design.

Zero-IF Receivers

When the LO frequency of a superheterodyne receiver is selected to be the carrier frequency ($f_{LO} = f_c$) then $f_{IF} = 0$, and the superheterodyne receiver becomes a *zero-IF* or *direct conversion* receiver.[†] In this case, the IF filter becomes a low-pass filter (LPF). This mixer–LPF combination functions as a product detector (and the detector stage of Figure 4–29 is not needed). A quadrature down converter can also be added so that the $x(t)$ and $y(t)$ components of the complex envelope can be recovered. In this case the zero-IF receiver has a block diagram as shown in Fig. 4–31, where the input signal is at f_c and ω_c replaces ω_{IF} in the figure. The components $x(t)$ and $y(t)$ may be sampled and digitized with ADC so that the complex envelope, $g(t) = x(t) + jy(t)$, may be processed digitally with DSP hardware, which is discussed in Sec. 4–17. The analog LPF acts as an antialiasing filter for the sampler and the DSP hardware. The zero-IF receiver is also similar to a TRF receiver with product detection.

Zero-IF receivers have several advantages. They have no image response. The same zero-IF receiver hardware can be used in many different applications for manufacturing economy. Since DSP hardware is used, the effective RF bandpass characteristics and the detector characteristics are determined by DSP software. (See next section.) The software can be changed easily to match the desired application. The same zero-IF hardware can be used for receivers in different VHF and UHF bands by selecting the appropriate LO frequency ($F_{LO} = f_c$) and tuning the front-end filter (usually a single-tuned circuit) to f_c.

The zero-IF receiver has the disadvantage of possibly leaking LO radiation out of the antenna input port due to feed-through from the mixer. Also, there will be a *dc* offset on the mixer output, if there is LO leakage into the antenna input since a sine wave (LO signal) multiplied by itself produces a dc term (plus an out-of-band second harmonic). The use of a high-quality balance mixer and LO shielding will minimize these problems. The receiver can also have a poor noise figure, since the front end usually is not a high-gain, low-noise stage. As in any receiver, the hardware has to be carefully designed so that there is sufficient dynamic range to prevent strong signals from overloading the receiver (producing spurious

---

[†] A direct-conversion receiver is also called a *homodyne* or *synchrodyne receiver*.

signals due to nonlinearities) and yet sufficient gain for detecting weak signals. In spite of these difficulties, the zero-IF receiver provides an economical, high-performance solution for many applications. A practical zero-IF receiver with excellent selectivity provided by DSP filtering is described in QST [Frohne, 1998].

Interference

A discussion of receivers would not be complete without considering some of the causes of *interference*. Often the owner of the receiver thinks that a certain signal, such as an amateur radio signal, is causing the difficulty. This may or may not be the case. The origin of the interference may be at any of three locations:

- At the interfering signal *source*, a transmitter may generate out-of-band signal components (such as harmonics) that fall in the band of the desired signal.
- At the *receiver* itself, the front end may overload or produce spurious responses. Front-end overload occurs when the RF or mixer stage of the receiver is driven into the nonlinear range by the interfering signal and the nonlinearity causes cross-modulation on the desired signal at the output of the receiver RF amplifier.
- In the *channel*, a nonlinearity in the transmission medium may cause undesired signal components in the band of the desired signal.

For more discussion of receiver design and examples of practical receiver circuits, the reader is referred to the *ARRL Handbook* [ARRL, 1999].

4–17 SOFTWARE RADIOS

Software radios use DSP hardware, microprocessors, specialized digital ICs, and software to produce modulated signals for transmission (see Table 4–1 and Fig. 4–28) and to demodulate signals at the receiver. Ultimately, the ideal software receiver would sample and digitize received signals at the antenna with analog-to-digital conversion (ADC) and process the signal with digital signal-processing (DSP) hardware. Software would be used to compute the receiver output. The difficulty with this approach is that it is almost impossible to build ADC/DSP hardware that operates fast enough to directly process wideband modulated signals with gigahertz carrier frequencies [Baines, 1995]. However, the complex envelope of these signals may be obtained by using a superheterodyne receiver with quadrature detectors (Fig. 4–31). For sufficiently modest bandpass bandwidth (say, 25 MHz), the I and Q components, $x(t)$ and $y(t)$, of the complex envelope can be sampled and processed with practical DSP hardware so that software programming can be used.

In another approach, a high-speed ADC can be used to provide samples of the IF signal that are passed to a *digital down-converter* (DDC) integrated circuit (e.g., the Harris, now Intersil, HSP50016) [Chester, 1999]. The DDC multiplies the IF samples with samples of cosine and sine LO signals. This down-converts the IF samples to baseband I and Q samples. The DDC uses ROM lookup tables to obtain the LO cosine and sine samples, a method similar to the direct digital synthesis (DDS) technique discussed in Sec. 4–15. To simultaneously receive multiple adjacent channel signals, multiple DDC ICs can be used in parallel with the LO of each DDC tuned to the appropriate frequency to down convert the signal to baseband I and Q samples for that signal. (For more details, see the Intersil Web site at http://www.intersil.com.)

The I and Q samples of the complex envelope, $g(t) = x(t) + jy(t)$, can be filtered to provide equivalent bandpass IF filtering (as described in Sec. 4–5). The filtering can provide excellent equivalent IF filter characteristics with tight skirts for superb adjacent channel interference rejection. The filter characteristic may be changed easily by changing the software. Raised cosine-rolloff filtering is often used to reduce the transmission bandwidth of digital signals without introducing ISI. For minimization of bit errors due to channel noise, as well as elimination of ISI, a square-root raised-cosine filter is used at both the transmitter and the receiver [as shown by Eq. (3–78) of Sec. 3–6].

AM and PM detection is accomplished by using the filtered I and Q components to compute the magnitude and phase of the complex envelope, as shown by Eqs. (4–4a) and (4–4b), respectively. FM detection is obtained by computing the derivative of the phase, as shown by Eq. (4–8).

The Fourier transform can also be used in software radios, since the FFT can be computed efficiently with DSP ICs. For example, the FFT spectrum can be used to determine the presence or absence of adjacent channel signals. Then, appropriate software processing can either enhance or reject a particular signal (as desired for a particular application). The FFT can also be used to simultaneously detect the data on a large number of modulated carriers that are closely spaced together. (For details, see Sec. 5–12 on OFDM.)

The software radio concept has many advantages. Two of these are that the same hardware may be used for many different types of radios, since the software distinguishes one type from another, and that, after software radios are sold, they can be updated in the field to include the latest protocols and features by downloading revised software. The software radio concept is becoming more economical and practical each day. It is the "way of the future."

4–18 SUMMARY

The basic techniques used for bandpass signaling have been studied in this chapter. The complex-envelope technique for representing bandpass signals and filters was found to be very useful. A description of communication circuits with output analysis was presented for filters, amplifiers, limiters, mixers, frequency multipliers, phase-locked loops, and detector circuits. Nonlinear as well as linear circuit analysis techniques were used. The superheterodyne receiving circuit was found to be fundamental in communication receiver design. Generalized transmitters, receivers, and software radios were studied. Practical aspects of their design, such as techniques for evaluating spurious signals, were examined.

4–19 STUDY-AID EXAMPLES

SA4–1 Voltage Spectrum for an AM Signal   An AM voltage signal $s(t)$ with a carrier frequency of 1,150 kHz has a complex envelope $g(t) = A_c[1 + m(t)]$. $A_c = 500$ V, and the modulation is a 1-kHz sinusoidal test tone described by $m(t) = 0.8 \sin(2\pi 1{,}000t)$. Evaluate the voltage spectrum for this AM signal.

Solution. Using the definition of a sine wave from Sec. A–1,

$$m(t) = \frac{0.8}{j2} \left[e^{j2\pi1000t} - e^{-j2\pi1000t} \right] \tag{4-118}$$

Using Eq. (2–26) with the help of Sec. A–5, we find that the Fourier transform of $m(t)$ is[†]

$$M(f) = -j\,0.4\,\delta(f - 1{,}000) + j\,0.4\,\delta(f + 1{,}000) \tag{4-119}$$

Substituting this into Eq. (4–20a) yields the voltage spectrum of the AM signal:

$$S(f) = 250\,\delta(f - f_c) - j100\,\delta(f - f_c - 1{,}000) + j100\,\delta(f - f_c + 1{,}000)$$
$$+\ 250\,\delta(f + f_c) - j100\,\delta(f + f_c - 1{,}000) + j100\,\delta(f + f_c + 1{,}000) \tag{4-120}$$

SA4–2 PSD for an AM Signal   Compute the PSD for the AM signal that is described in SA4–1.

Solution. Using Eq. (2–71), we obtain the autocorrelation for the sinusoidal modulation $m(t)$ namely,

$$R_m(\tau) = \frac{A^2}{2} \cos \omega_0 \tau = \frac{A^2}{4} \left[e^{j\omega_0\tau} + e^{-j\omega_0\tau} \right] \tag{4-121}$$

where $A = 0.8$ and $\omega_0 = 2\pi1{,}000$. Taking the Fourier transform by the use of Eq. (2–26) we obtain the PSD of $m(t)$:[‡]

$$\mathcal{P}_m(f) = \frac{A^2}{4} \left[\delta(f - f_0) + \delta(f + f_0) \right]$$

or

$$\mathcal{P}_m(f) = 0.16 \left[\delta(f - 1{,}000) + \delta(f + 1{,}000) \right] \tag{4-122}$$

The autocorrelation for the complex envelope of the AM signal is

$$R_g(\tau) = \langle g^*(t)\,g(t + \tau) \rangle = A_c^2 \,\langle [1 + m(t)][1 + m(t + \tau)] \rangle$$
$$= A_c^2 \left[\langle 1 \rangle + \langle m(t) \rangle + \langle m(t + \tau) \rangle + \langle m(t)\,m(t + \tau) \rangle \right]$$

But $\langle 1 \rangle = 1$, $\langle m(t) \rangle = 0$, $\langle m(t + \tau) \rangle = 0$, and $\langle m(t)\,m(t + \tau) \rangle = R_m(\tau)$. Thus,

$$R_g(\tau) = A_c^2 + A_c^2 R_m(\tau) \tag{4-123}$$

Taking the Fourier transform of both sides of Eq. (4–123), we get

$$\mathcal{P}_g(f) = A_c^2 \,\delta(f) + A_c^2 \,\mathcal{P}_m(f) \tag{4-124}$$

Substituting Eq. (4–124) into Eq. (4–13), with the aid of Eq. (4–122), we obtain the PSD for the AM signal:

[†] Because $m(t)$ is periodic, an alternative method for evaluating $M(f)$ is given by Eq. (2–109), where $c_{-1} = j0.4$, $c_1 = -j0.4$, and the other c_n's are zero.

[‡] Because $m(t)$ is periodic, Eq. (2–126) can be used as an alternative method of evaluating $\mathcal{P}_m(f)$. That is, by using Eq. (2–126) with $c_1 = c_{-1}^* = A/(2j) = -j0.8/2 = -j0.4$ (and the other c_n's are zero), Eq. (4–122) is obtained.

$$\mathcal{P}_s(f) = 62{,}500 \, \delta(f - f_c) + 10{,}000 \, \delta(f - f_c - 1{,}000)$$

$$+ \, 10{,}000 \, \delta(f - f_c + 1{,}000) + 62{,}500 \, \delta(f + f_c)$$

$$+ \, 10{,}000 \, \delta(f + f_c - 1{,}000) + 10{,}000 \, \delta(f + f_c + 1{,}000) \qquad (4\text{--}125)$$

(*Note:* We realize that this bandpass PSD for $s(t)$ is found by translating (i.e., moving) the baseband PSD of $g(t)$ up to f_c and down to $-f_c$. Furthermore, for the case of AM, the PSD of $g(t)$ consists of the PSD for $m(t)$ plus the superposition of a delta function at $f = 0$).

SA4–3 Average Power for an AM Signal   Assume that the AM voltage signal $s(t)$, as described in SA4–1, appears across a 50-Ω resistive load. Compute the actual average power dissipated in the load.

Solution.   From Eq. (4–21), the normalized average power is

$$(P_s)_{\text{norm}} = (V_s)^2_{\text{rms}} = \tfrac{1}{2} A_c^2 \left[1 + (V_m)^2_{\text{rms}} \right]$$

$$= \tfrac{1}{2} (500)^2 \left[1 + \left(\frac{0.8}{\sqrt{2}} \right) \right]^2 = 165 \text{ kW} \qquad (4\text{--}126a)$$

Note: An alternative method of computing $(P_s)_{\text{norm}}$ is to calculate the area under the PDF for $s(t)$. That is, by using Eq. (4–125),

$$(P_s)_{\text{norm}} = (V_s)^2_{\text{rms}} = \int_{-\infty}^{\infty} \mathcal{P}_s(f) \, df = 165 \text{ kW} \qquad (4\text{--}126b)$$

Using Eq. (4–126a) or Eq. (4–126b), we obtain the actual average power dissipated in the 50-Ω load:[†]

$$(P_s)_{\text{actual}} = \frac{(V_s)^2_{\text{rms}}}{R_L} = \frac{1.65 \times 10^5}{50} = 3.3 \text{ kW} \qquad (4\text{--}127)$$

SA4–4 PEP for an AM Signal   If the AM voltage signal of SA4–1 appears across a 50-Ω resistive load, compute the actual peak envelope power (PEP).

Solution.   Using Eq. (4–18), we get the normalized PEP:

$$(P_{\text{PEP}})_{\text{norm}} = \tfrac{1}{2} \left[\max |g(t)| \right]^2 = \tfrac{1}{2} A_c^2 [1 + \max m(t)]^2$$

$$= \tfrac{1}{2} (500)^2 [1 + 0.8]^2 = 405 \text{ kW} \qquad (4\text{--}128)$$

Then the actual PEP for this AM voltage signal with a 50-Ω load is

$$(P_{\text{PEP}})_{\text{actual}} = \frac{(P_{\text{PEP}})_{\text{norm}}}{R_L} = \frac{4.05 \times 10^5}{50} = 8.1 \text{ kW} \qquad (4\text{--}129)$$

SA4–5 Sampling Methods for Bandpass Signals   Suppose that a bandpass signal $s(t)$ is to be sampled and that the samples are to be stored for processing at a later time. As shown in Fig. 4–32a, this bandpass signal has a bandwidth of B_T centered about f_c, where $f_c \gg B_T$ and $B_T > 0$. The signal $s(t)$ is to be sampled by using any one of three methods shown in

---

[†] If $s(t)$ is a current signal (instead of a voltage signal), then $(P_s)_{\text{actual}} = (I_s)^2_{\text{rms}} R_L$.

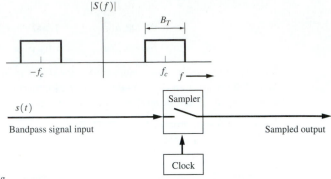

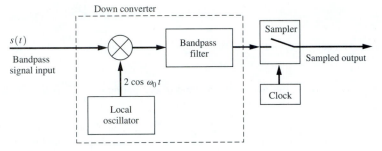

(a) Method I—Direct sampling

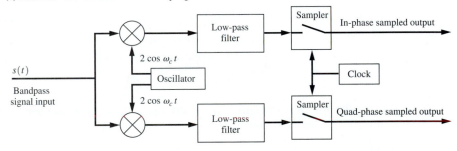

(b) Method II—Down conversion and sampling

(c) Method III—IQ (in-phase and quad-phase) sampling

Figure 4–32   Three methods for sampling bandpass signals.

Fig. 4–32.[‡] For each of these sampling methods, determine the minimum sampling frequency (i.e., minimum clock frequency) required, and discuss the advantages and disadvantages of each method.

Solution.

Method I   Referring to Fig. 4–32a, we see that Method I uses direct sampling as described in Chapter 2. From Eq. (2–168), the minimum sampling frequency is $(f_s)_{min} = 2B$, where B is

‡ Thanks to Professor Christopher S. Anderson, Department of Electrical and Computer Engineering, University of Florida, for suggesting Method II.

the highest frequency in the signal. For this bandpass signal, the highest frequency is $B = f_c + B_T/2$. Thus, for Method I, the minimum sampling frequency is

$$(f_s)_{\min} = 2f_c + B_T \qquad \text{Method I} \qquad (4\text{--}130)$$

For example, if $f_c = 100$ MHz and $B_T = 1$ MHz, a minimum sampling frequency of $(f_s)_{\min} = 201$ MHz would be required.

Method II   Referring to Fig. 4–32b, we see that Method II down converts the bandpass signal to an IF, so that the highest frequency that is to be sampled is drastically reduced. For maximum reduction of the highest frequency, we choose[†] the local oscillator frequency to be $f_0 = f_c - B_T/2$. The highest frequency in the down-converted signal (at the sampler input) is $B = (f_c + B_T/2) - f_0 = f_c + B_T/2 - f_c + B_T/2 = B_T$, and the lowest frequency (in the positive-frequency part of the down-converted signal) is $(f_c - B_T/2) - f_0 = f_c - B_T/2 - f_c + B_T/2 = 0$. Using Eq. (2–168), we find that the minimum sampling frequency is

$$(f_s)_{\min} = 2B_T \qquad \text{Method II} \qquad (4\text{--}131)$$

when the frequency of the LO is chosen to be $f_0 = f_c - B_T/2$. For this choice of LO frequency, the bandpass filter becomes a low-pass filter with a cutoff frequency of B_T. Note that Method II gives a drastic reduction in the sampling frequency (an advantage) compared with Method I. For example, if $f_c = 100$ MHz and $B_T = 1$ MHz, then the minimum sampling frequency is now $(f_s)_{\min} = 2$ MHz, instead of the 201 MHz required in Method I. However, Method II requires the use of a down converter (a disadvantage). Note also that $(f_s)_{\min}$ of Method II, as specified by Eq. (4–131), satisfies the $(f_s)_{\min}$ given by the *bandpass sampling theorem*, as described by Eq. (4–31).

Method II is one of the most efficient ways to obtain samples for a bandpass signal. When the bandpass signal is reconstructed from the sample values with the use of Eq. (2–158) and (2–160), the *down-converted* bandpass signal is obtained. To obtain the original bandpass signal $s(t)$, an up-converter is needed to convert the down-converted signal back to the original bandpass region of the spectrum.

Method II can also be used to obtain samples of the quadrature (i.e., I and Q) components of the complex envelope. From Fig. 4–32b, the IF signal at the input to the sampler is

$$v_{\text{IF}}(t) = x(t)\cos\omega_{\text{IF}}t - y(t)\sin\omega_{\text{IF}}t$$

where $f_{\text{IF}}(t) = B_T/2$. Samples of $x(t)$ can be obtained if $v_{\text{IF}}(t)$ is sampled at the times corresponding to $\cos\omega_{\text{IF}}t = \pm1$ (and $\sin\omega_{\text{IF}}t = 0$). This produces B_T samples of $x(t)$ per second. Likewise, samples of $y(t)$ are obtained at the times when $\sin\omega_{\text{IF}}t = \pm1$ (and $\cos\omega_{\text{IF}}t = 0$). This produces B_T samples of $y(t)$ per second. The composite sampling rate for the clock is $f_s = 2B_T$. Thus, the sampler output contains the following sequence of I and Q values: $x, -y, -x, y, x, -y, \ldots$. The sampling clock can be synchronized to the IF phase by using carrier synchronization circuits. Method III uses a similar approach.

Method III   From Fig. 4–32c, Method III uses in-phase (I) and quadrature-phase (Q) product detectors to produce the $x(t)$ and $y(t)$ quadrature components of $s(t)$. (This was discussed in Sec. 4–16 and illustrated in Fig. 4–31.) The highest frequencies in $x(t)$ and $y(t)$ are $B = B_T/2$. Thus, using Eq. (2–168), we find that the minimum sampling frequency for the clock of the I and Q samplers is

---

[†] Low-side LO injection is used so that any asymmetry in the two sidebands of $s(t)$ will be preserved in the same way in the down-converted signal.

$$(f_s)_{\min} = B_T \quad \text{(each sampler)} \qquad \text{Method III} \qquad (4\text{--}132)$$

Because there are two samplers, the combined sampling rate is $(f_s)_{\min \text{ overall}} = 2B_T$. This also satisfies the minimum sampling rate allowed for bandpass signals as described by Eq. (4–31). Thus, Method III (like Method II) gives one of the most efficient ways to obtain samples of bandpass signals. For the case of $f_c = 100$ MHz and $B_T = 1$ MHz, an overall sampling rate of 2 MHz is required for Method III, which is the same as that obtained by Method II. Because IQ samples have been obtained, they may be processed by using DSP algorithms to perform equivalent bandpass filtering, as described by Sec. 4–5, or equivalent modulation of another type, as described by Sec. 4–2. If desired, the original bandpass signal may be reconstructed with the use of Eq. (4–32).

SA4–6 Frequency Synthesizer Design for a Receiver LO. Design a frequency synthesizer for use as the local oscillator in an AM superheterodyne radio. The radio has a 455-kHz IF and can be tuned across the AM band from 530 kHz to 1,710 kHz in 10-kHz steps. The synthesizer uses a 1-MHz reference oscillator and generates a high-side LO injection signal.

Solution. Referring to Eq. (4–59) and Fig. 4–29 for the case of down conversion and high-side injection, we find that the required frequency for the LO is $f_0 = f_c + f_{\text{IF}}$. If $f_c = 530$ kHz and $f_{\text{IF}} = 455$ kHz, the desired synthesizer output frequency is $f_0 = 985$ kHz. Referring to the block diagram for the frequency synthesizer (Fig. 4–25), we select the frequency of the mixer input signal, $v_{\text{in}}(t)$, to be 5 kHz, which is one-half the desired 10-kHz step. Then $M = f_x/f_{\text{in}} = 1,000$ kHz$/5$ kHz $= 200$, and an integer value can be found for N to give the needed LO frequency. Using Eg. (4–112), we obtain $N = f_0/f_{\text{in}}$. For $f_0 = 985$ kHz and $f_{\text{in}} = 5$ kHz, we get $N = 197$. Thus, to tune the radio to $f_c = 530$ kHz, the required values of M and N are $M = 200$ and $N = 197$. In a similar way, other values for N can be obtained to tune the radio to 540, 550, ... , 1,710 kHz. (M remains at 200.) Table 4–6 lists the results. The selected values for M and N are kept small in order to minimize the spurious sideband noise on the synthesized LO signal. M and N are minimized by making the size of the step frequency, f_{in}, as large as possible.

The spectral sideband noise on the synthesizer output signal is minimized by using a low-noise reference oscillator and a low-noise VCO and by choosing a small value of N for the reduction in the number of intermodulation noise components on the synthesized signal. The bandwidth of the loop filter is also minimized, but if it is too small, the pull-in range will not

TABLE 4–5 SOLUTION TO SA4–6 WITH DIVIDER RATIOS M AND N FOR AN AM RADIO-FREQUENCY SYNTHESIZER

Reference frequency = 1,000 Hz; IF frequency = 455 kHz

Received frequency, f_c (kHz)	Local oscillator frequency, f_0 (kHz)	M	N
530	985	200	197
540	995	200	199
550	1,005	200	201
⋮	⋮	⋮	⋮
1,700	2,155	200	431
1,710	2,165	200	433

be sufficient for reliable locking of the synthesizer PLL when it is turned on. In this example, N can be reduced by a factor of about $\frac{1}{2}$ the IF frequency is chosen to be 450 kHz instead of 455 kHz. For example, for $f_{\text{IF}} = 450$ kHz and $f_c = 530$ kHz, we need $f_0 = 980$ kHz. This LO frequency is attained if $M = 100$ (for a step size of $f_{\text{in}} = 10$ kHz) and $N = 98$, compared with (see Table 4–6) $M = 200$ and $N = 197$, which was needed for the case when $f_{\text{IF}} = 455$ kHz.

PROBLEMS

4–1 Show that if $v(t) = \text{Re}\{g(t)e^{j\omega_c t}\}$, Eqs. (4–1b) and (4–1c) are correct, where $g(t) = x(t) + jy(t) = R(t)e^{j\theta(t)}$.

4–2 A double-sideband suppressed carrier (DSB-SC) signal $s(t)$ with a carrier frequency of 3.8 MHz has a complex envelope $g(t) = A_c m(t)$. $A_c = 50$ V, and the modulation is a 1-kHz sinusoidal test tone described by $m(t) = 2 \sin(2\pi 1,000t)$. Evaluate the voltage spectrum for this DSB-SC signal.

4–3 Assume that the DSB-SC voltage signal $s(t)$, as described in Prob. 4–2 appears across a 50-Ω resistive load.
(a) Compute the actual average power dissipated in the load.
(b) Compute the actual PEP.

4–4 A bandpass filter is shown in Fig. P4–4.

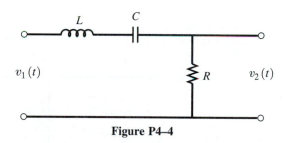

Figure P4–4

(a) Find the mathematical expression for the transfer function of this filter, $H(f) = V_2(f)/V_1(f)$, as a function of R, L, and C. Sketch the magnitude transfer function $|H(f)|$.
(b) Find the expression for the equivalent low-pass filter transfer function, and sketch the corresponding low-pass magnitude transfer function.

4–5 Let the transfer function of an ideal bandpass filter be given by

$$H(f) = \begin{cases} 1, & |f + f_c| < B_T/2 \\ 1, & |f - f_c| < B_T/2 \\ 0, & f \text{ elsewhere} \end{cases}$$

where B_T is the absolute bandwidth of the filter.
(a) Sketch the magnitude transfer function $|H(f)|$.
(b) Find an expression for the waveform at the output, $v_2(t)$, if the input consists of the pulsed carrier

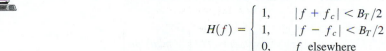

(c) Sketch the output waveform $v_2(t)$ for the case when $B_T = 4/T$ and $f_c \gg B_T$.
(*Hint:* Use the complex-envelope technique, and express the answer as a function of the sine integral, defined by

$$\text{Si}(u) = \int_0^u \frac{\sin \lambda}{\lambda} \, d\lambda$$

The sketch can be obtained by looking up values for the sine integral from published tables [Abramowitz and Stegun, 1964] or by numerically evaluating $\text{Si}(u)$.)

4–6 Examine the distortion properties of an *RC* low-pass filter (shown in Fig. 2–15). Assume that the filter input consists of a bandpass signal that has a bandwidth of 1 kHz and a carrier frequency of 15 kHz. Let the time constant of the filter be $\tau_0 = RC = 10^{-5}$ s.
  (a) Find the phase delay for the output carrier.
  (b) Determine the group delay at the carrier frequency.
  (c) Evaluate the group delay for frequencies around and within the frequency band of the signal. Plot this delay as a function of frequency.
  (d) Using the results of (a) through (c), explain why the filter does or does not distort the bandpass signal.

4–7 A bandpass filter as shown in Fig. P4–7 has the transfer function

$$H(s) = \frac{Ks}{s^2 + (\omega_0/Q)s + \omega_0^2}$$

where $Q = R\sqrt{C/L}$, the resonant frequency is $f_0 = 1/(2\pi\sqrt{LC})$, $\omega_0 = 2\pi f_0$, K is a constant, and values for R, L, and C are given in the figure. Assume that a bandpass signal with $f_c = 4$ kHz and a bandwidth of 200 Hz passes through the filter, where $f_0 = f_c$.

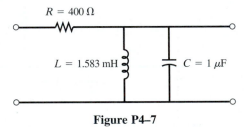

Figure P4–7

  (a) Using Eq. (4–39), find the bandwidth of the filter.
  (b) Plot the carrier delay as a function of f about f_0.
  (c) Plot the group delay as a function of f about f_0.
  (d) Explain why the filter does or does not distort the signal.

4–8 An FM signal is of the form

$$s(t) = \cos \left[\omega_c t + D_f \int_{-\infty}^{t} m(\sigma) \, d\sigma \right]$$

where $m(t)$ is the modulating signal and $\omega_c = 2\pi f_c$, in which f_c is the carrier frequency. Show that the functions $g(t)$, $x(t)$, $y(t)$, $R(t)$, and $\theta(t)$, as given for FM in Table 4–1, are correct.

4–9 Let a modulated signal,

$$s(t) = 100 \sin(\omega_c + \omega_a)t + 500 \cos \omega_c t - 100 \sin(\omega_c - \omega_a)t$$

where the unmodulated carrier is $500 \cos \omega_c t$.

(a) Find the complex envelope for the modulated signal. What type of modulation is involved? What is the modulating signal?

(b) Find the quadrature modulation components $x(t)$ and $y(t)$ for this modulated signal.

(c) Find the magnitude and PM components $R(t)$ and $\theta(t)$ for this modulated signal.

(d) Find the total average power, where $s(t)$ is a voltage waveform that is applied across a 50-Ω load.

4–10 Find the spectrum of the modulated signal given in Prob. 4–9 by two methods:

(a) By direct evaluation using the Fourier transform of $s(t)$.

(b) By the use of Eq. (4–12).

4–11 Given a pulse-modulated signal of the form

$$s(t) = e^{-at} \cos[(\omega_c + \Delta\omega)t]u(t)$$

where a, ω_c, and $\Delta\omega$ are positive constants and the carrier frequency, $\omega_c \gg \Delta\omega$,

(a) Find the complex envelope.

(b) Find the spectrum $S(f)$.

(c) Sketch the magnitude and phase spectra $|S(f)|$ and $\theta(f) = \underline{/S(f)}$.

4–12 In a digital computer simulation of a bandpass filter, the complex envelope of the impulse response is used, where $h(t) = \operatorname{Re}[k(t)\,e^{j\omega_c t}]$, as shown in Fig. 4–3. The complex impulse response can be expressed in terms of quadrature components as $k(t) = 2h_x(t) + j2h_y(t)$, where $h_x(t) = \frac{1}{2}\operatorname{Re}[k(t)]$ and $h_y(t) = \frac{1}{2}\operatorname{Im}[k(t)]$. The complex envelopes of the input and output are denoted, respectively, by $g_1(t) = x_1(t) + jy_1(t)$ and $g_2(t) = x_2(t) + jy_2(t)$. The bandpass filter simulation can be carried out by using four *real baseband* filters (i.e., filters having real impulse responses), as shown in Fig. P4–12. Note that although there are four filters, there are only two different impulse responses: $h_x(t)$ and $h_y(t)$.

(a) Using Eq. (4–22), show that Fig. P4–12 is correct.

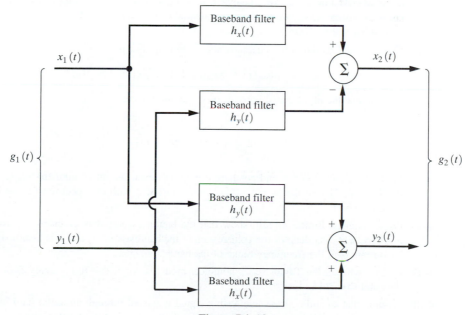

Figure P4–12

(b) Show that $h_y(t) \equiv 0$ (i.e., no filter is needed) if the bandpass filter has a transfer function with Hermitian symmetry about f_c,—that is, if $H(-\Delta f + f_c) = H^*(\Delta f + f_c)$, where $|\Delta f| < B_T/2$ and B_T is the bounded spectral bandwidth of the bandpass filter. This Hermitian symmetry implies that the magnitude frequency response of the bandpass filter is even about f_c and the phase response is odd about f_c.

4–13 Evaluate and sketch the magnitude transfer function for **(a)** Butterworth, **(b)** Chebyshev, and **(c)** Bessel low-pass filters. Assume that $f_b = 10$ Hz and $\varepsilon = 1$.

4–14 Plot the amplitude response, the phase response, and the phase delay as a function of frequency for the following low-pass filters, where $B = 100$ Hz:

(a) Butterworth filter, second order:

$$H(f) = \frac{1}{1 + \sqrt{2}\,(jf/B) + (jf/B)^2}$$

(b) Butterworth filter, fourth order:

$$H(f) = \frac{1}{[1 + 0.765(jf/B) + (jf/B)^2][1 + 1.848(jf/B) + (jf/b)^2]}$$

Compare your results for the two filters.

4–15 Assume that the output-to-input characteristic of a bandpass amplifier is described by Eq. (4–42) and that the linearity of the amplifier is being evaluated by using a two-tone test.

(a) Find the frequencies of the fifth-order intermodulation products that fall within the amplifier bandpass.

(b) Evaluate the levels for the fifth-order intermodulation products in terms of A_1, A_2, and the K's.

4–16 An amplifier is tested for total harmonic distortion (THD) by using a single-tone test. The output is observed on a spectrum analyzer. It is found that the peak values of the three measured harmonics decrease according to an exponential recursion relation $V_{n+1} = V_n e^{-n}$, where $n = 1$, 2, 3. What is the THD?

4–17 The nonlinear output–input characteristic of an amplifier is

$$v_{\text{out}}(t) = 5v_{\text{in}}(t) + 1.5v_{\text{in}}^2(t) + 1.5v_{\text{in}}^3(t)$$

Assume that the input signal consists of seven components:

$$v_{\text{in}}(t) = \frac{1}{2} + \frac{4}{\pi^2} \sum_{k=1}^{6} \frac{1}{(2k-1)^2} \cos[(2k-1)\pi t]$$

(a) Plot the output signal and compare it with the linear output component $5v_{\text{in}}(t)$.

(b) Take the FFT of the output $v_{\text{out}}(t)$, and compare it with the spectrum for the linear output component.

4–18 For a bandpass limiter circuit, show that the bandpass output is given by Eq. (4–55), where $K = (4/\pi)A_0$. A_0 denotes the voltage gain of the bandpass filter, and it is assumed that the gain is constant over the frequency range of the bandpass signal.

4–19 Discuss whether the Taylor series nonlinear model is applicable to the analysis of **(a)** soft limiters and **(b)** hard limiters.

4–20 Assume that an audio sine-wave testing signal is passed through an audio hard limiter circuit. Evaluate the total harmonic distortion (THD) on the signal at the limiter output.

4–21 Using the mathematical definition of linearity given in Chapter 2, show that the analog switch multiplier of Fig. 4–10 is a linear device.

4–22 An audio signal with a bandwidth of 10 kHz is transmitted over an AM transmitter with a carrier frequency of 1.0 MHz. The AM signal is received on a superheterodyne receiver with an envelope detector. What is the constraint on the RC time constant for the envelope detector?

4–23 Assume that an AM receiver with an envelope detector is tuned to an SSB-AM signal that has a modulation waveform given by $m(t)$. Find the mathematical expression for the audio signal that appears at the receiver output in terms of $m(t)$. Is the audio output distorted?

4–24 Evaluate the sensitivity of the zero-crossing FM detector shown in Fig. 4–18. Assume that the differential amplifier is described by $v_{out}(t) = A[v_2(t) - v_3(t)]$, where A is the voltage gain of the amplifier. In particular, show that $v_{out} = Kf_d$, where $f_d = f_i - f_c$, and find the value of the sensitivity constant K in terms of A, R, and C. Assume that the peak levels of the monostable outputs Q and $\bar{Q}$ are 4 V (TTL circuit levels).

4–25 **(a)** Using Eq. (4–100), show that the linearized block diagram model for a PLL is given by Fig. 4–22.

(b) Show that Eq. (4–101) describes the linear PLL model as given in Fig. 4–22.

4–26 Using the Laplace transform and the final value theorem, find an expression for the steady-state phase error, $\lim_{t \to \infty} \theta_e(t)$, for a PLL as described by Eq. (4–100). [*Hint:* The final value theorem is $\lim_{t \to \infty} f(t) = \lim_{s \to 0} sF(s)$.]

4–27 Assume that the loop filter of a PLL is a low-pass filter, as shown in Fig. P4–27.

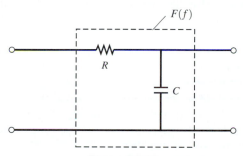

Figure P4–27

(a) Evaluate the closed-loop transfer function $H(f) = \dfrac{\Theta_0(f)}{\Theta_i(f)}$ for a linearized PLL.

(b) Sketch the Bode plot $[|H(f)|]_{dB} \triangleq 20 \log |H(f)|$ for this PLL.

4–28 Assume that the phase noise characteristic of a PLL is being examined. The internal phase noise of the VCO is modeled by the input $\theta_n(t)$, as shown in Fig. P4–28.

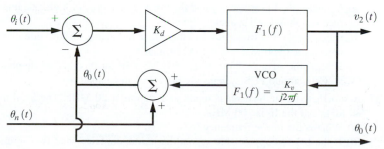

Figure P4–28

(a) Find an expression for the closed-loop transfer function $\Theta_0(f)/\Theta_n(f)$, where $\theta_i(t) = 0$.

(b) If $F_1(f)$ is the low-pass filter given in Fig. P4–27, sketch the Bode plot $[|\Theta_0(f)/\Theta_n(f)|]_{dB}$ for the phase noise transfer function.

4–29 The input to a PLL is $v_{in}(t) = A \sin(\omega_0 t + \theta_i)$. The LPF has a transfer function $F(s) = (s + a)/s$.

(a) What is the steady-state phase error?

(b) What is the maximum hold-in range for the noiseless case?

4–30 (a) Refer to Fig. 4–25 for a PLL frequency synthesizer. Design a synthesizer that will cover a range of 144 to 148 MHz in 5-kHz steps, starting at 144.000 MHz. Assume that the frequency standard operates at 5 MHz, that the M divider is fixed at some value, and that the N divider is programmable so that the synthesizer will cover the desired range. Sketch a diagram of your design, indicating the frequencies present at various points of the diagram.

(b) Modify your design so that the output signal can be frequency modulated with an audio-frequency input such that the peak deviation of the RF output is 5 kHz.

4–31 Assume that an SSB-AM transmitter is to be realized using the AM–PM generation technique, as shown in Fig. 4–27.

(a) Sketch a block diagram for the baseband signal-processing circuit.

(b) Find expressions for $R(t)$, $\theta(t)$, and $v(t)$ when the modulation is $m(t) = A_1 \cos \omega_1 t + A_2 \cos \omega_2 t$.

4–32 Rework Prob. 4–31 for the case of generating an FM signal.

4–33 Assume that an SSB-AM transmitter is to be realized using the quadrature generation technique, as shown in Fig. 4–28.

(a) Sketch a block diagram for the baseband signal-processing circuit.

(b) Find expressions for $x(t)$, $y(t)$, and $v(t)$ when the modulation is $m(t) = A_1 \cos \omega_1 t + A_2 \cos \omega_2 t$.

4–34 Rework Prob. 4–33 for the case of generating an FM signal.

4–35 An FM radio is tuned to receive an FM broadcasting station of frequency 96.9 MHz. The radio is of the superheterodyne type with the LO operating on the high side of the 96.9-MHz input and using a 10.7-MHz IF amplifier.

(a) Determine the LO frequency.

(b) If the FM signal has a bandwidth of 180 kHz, give the requirements for the RF and IF filters.

(c) Calculate the frequency of the image response.

4–36 A dual-mode cellular phone is designed to operate with either analog cellular phone service in the 900-MHz band or digital PCS in the 1,900-MHz band. The phone uses a superheterodyne receiver with a 500-MHz IF for both modes.

(a) Calculate the LO frequency and the image frequency for high-side injection when the phone receives an 880-MHz analog signal.

(b) Calculate the LO frequency and the image frequency for low-side injection when the phone receives a 1,960-MHz digital PCS signal.

(c) Discuss the advantage of using a 500-MHz IF for this dual-mode phone.

(*Note*: Cellular and PCS systems are described in Chapter 8.)

4–37 A superheterodyne receiver is tuned to a station at 20 MHz. The local oscillator frequency is 80 MHz and the IF is 100 MHz.

(a) What is the image frequency?

(b) If the LO has appreciable second-harmonic content, what two additional frequencies are received?

filter, one pole   6dB per octet × # of poles in filter
                                                    50

(c) If the RF amplifier contains a single-tuned parallel resonant circuit with $Q = 50$ tuned to 20 MHz, what will be the image attentuation in dB?

4–38 An SSB-AM receiver is tuned to receive a 7.225-MHz lower SSB (LSSB) signal. The LSSB signal is modulated by an audio signal that has a 3-kHz bandwidth. Assume that the receiver uses a superheterodyne circuit with an SSB IF filter. The IF amplifier is centered on 3.395 MHz. The LO frequency is on the high (frequency) side of the input LSSB signal.
  (a) Draw a block diagram of the single-conversion superheterodyne receiver, indicating frequencies present and typical spectra of the signals at various points within the receiver.
  (b) Determine the required RF and IF filter specifications, assuming that the image frequency is to be attenuated by 40 dB.

4–39 (a) Draw a block diagram of a superheterodyne FM receiver that is designed to receive FM signals over a band from 144 to 148 MHz. Assume that the receiver is of the dual-conversion type (i.e., a mixer and an IF amplifier, followed by another mixer and a second IF amplifier), where the first IF is 10.7 MHz and the second is 455 kHz. Indicate the frequencies of the signals at different points on the diagram, and, in particular, show the frequencies involved when a signal at 146.82 MHz is being received.
  (b) Replace the first oscillator by a frequency synthesizer such that the receiver can be tuned in 5-kHz steps from 144.000 to 148.000 MHz. Show the diagram of your synthesizer design and the frequencies involved.

4–40 An AM broadcast-band radio is tuned to receive a 1,080-kHz AM signal and uses high-side LO injection. The IF is 455 kHz.
  (a) Sketch the frequency response for the RF and IF filters.
  (b) What is the image frequency?

4–41 Commercial AM broadcast stations operate in the 540- to 1,700-kHz band, with a transmission bandwidth limited to 10 kHz.
  (a) What are the maximum number of stations that can be accommodated?
  (b) If stations are not assigned to adjacent channels (in order to reduce interference on receivers that have poor IF characteristics), how many stations can be accommodated?
  (c) For 455-kHz IF receivers, what is the band of image frequencies for the AM receiver that uses a down-converter with high-side injection?

CHAPTER 5

Chapter Objectives

- Amplitude modulation and single sideband

- Frequency and phase modulation

- Digitally modulated signals (OOK, BPSK, FSK, MSK, MPSK, QAM, QPSK, $\pi/4$QPSK, and OFDM)

- Spread spectrum and CDMA systems

AM, FM, AND DIGITAL MODULATED SYSTEMS

This chapter is concerned with the bandpass techniques of amplitude modulation (AM), single-sideband (SSB), phase modulation (PM), and frequency modulation (FM); and with the digital modulation techniques of on–off keying (OOK), binary phase-shift keying (BPSK), frequency-shift keying (FSK), quadrature phase-shift keying (QPSK), quadrature amplitude modulation (QAM) and orthogonal frequency-division multiplexing (OFDM). All of these bandpass signaling techniques consist of modulating an analog or digital baseband signal onto a carrier. This approach was first introduced in Sec. 4–2. In particular, the modulated bandpass signal can be described by

$$s(t) = \text{Re}\{g(t)^{j\omega_c t}\} \tag{5–1}$$

where $\omega_c = 2\pi f_c$ and f_c is the carrier frequency. The desired type of modulated signal, $s(t)$, is obtained by selecting the appropriate modulation mapping function $g[m(t)]$ of Table 4–1, where $m(t)$ is the analog or digital baseband signal.

The voltage (or current) spectrum of the bandpass signal is

$$S(f) = \tfrac{1}{2}\left[G(f - f_0) + G^*(-f - f_c)\right] \tag{5–2a}$$

and the PSD is

$$\mathcal{P}_s(f) = \tfrac{1}{4}\left[\mathcal{P}_g(f - f_c) + \mathcal{P}_g(-f - f_c)\right] \tag{5–2b}$$

where $G(f) = \mathcal{F}[g(t)]$ and $\mathcal{P}_g(f)$ is the PSD of the complex envelope $g(t)$.

General results that apply to both digital and analog modulating waveforms are developed in the first half of the chapter (Secs. 5–1 to 5–8). Digital modulated signals are emphasized in the second half (Secs. 5–9 to 5–13).

The goals of this chapter are to

- Study $g(t)$ and $s(t)$ for various types of analog and digital modulations.
- Evaluate the spectrum for various types of analog and digital modulations.
- Examine some transmitter and receiver structures.
- Study some adopted standards.
- Learn about spread spectrum systems.

5–1 AMPLITUDE MODULATION

From Table 4–1, the complex envelope of an AM signal is given by

$$g(t) = A_c[1 + m(t)] \tag{5–3}$$

where the constant A_c has been included to specify the power level and $m(t)$ is the modulating signal (which may be analog or digital). These equations reduce to the following representation for the AM signal:

$$s(t) = A_c[1 + m(t)] \cos \omega_c t \tag{5–4}$$

A waveform illustrating the AM signal, as seen on an oscilloscope, is shown in Fig. 5–1. For convenience, it is assumed that the modulating signal $m(t)$ is a sinusoid. $A_c[1 + m(t)]$ corresponds to the in-phase component $x(t)$ of the complex envelope; it also corresponds to the real envelope $|g(t)|$ when $m(t) \geq -1$ (the usual case).

If $m(t)$ has a peak positive value of $+1$ and a peak negative value of -1, the AM signal is said to be 100% modulated.

DEFINITION.   The *percentage of positive modulation* on an AM signal is

$$\% \text{ positive modulation} = \frac{A_{\max} - A_c}{A_c} \times 100 = \max[m(t)] \times 100 \tag{5–5a}$$

and the *percentage of negative modulation* is

$$\% \text{ negative modulation} = \frac{A_c - A_{\min}}{A_c} \times 100 = -\min[m(t)] \times 100 \tag{5–5b}$$

The overall modulation percentage is

$$\% \text{ modulation} = \frac{A_{\max} - A_{\min}}{2A_c} \times 100 = \frac{\max[m(t)] - \min[m(t)]}{2} \times 100 \tag{5–6}$$

(a) Sinusoidal Modulating Wave

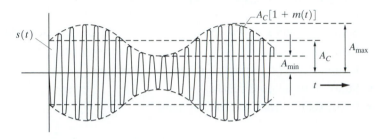

(b) Resulting AM Signal

Figure 5–1   AM signal waveform.

where A_{max} is the maximum value of $A_c[1 + m(t)]$, A_{min} is the minimum value, and A_c is the level of the AM envelope in the absence of modulation [i.e., $m(t) = 0$].

Equation (5–6) may be obtained by averaging the positive and negative modulation as given by Eqs. (5–5a) and (5–5b). A_{max}, A_{min}, and A_c are illustrated in Fig. 5–1b, where, in this example, $A_{max} = 1.5A_c$ and $A_{min} = 0.5A_c$, so that the percentages of positive and negative modulation are both 50% and the overall modulation is 50%.

The percentage of modulation can be over 100% (A_{min} will have a negative value), provided that a four-quadrant multiplier is used to generate the product of $A_c[1 + m(t)]$ and $\cos \omega_c t$ so that the true AM waveform, as given by Eq. (5–4), is obtained.[†] However, if the transmitter uses a two-quadrant multiplier that produces a zero output when $A_c[1 + m(t)]$ is negative, the output signal will be

$$s(t) = \begin{cases} A_c[1 + m(t)] \cos \omega_c t, & \text{if } m(t) \geq -1 \\ 0, & \text{if } m(t) < -1 \end{cases} \qquad (5\text{–}7)$$

which is a distorted AM signal. The bandwidth of this signal is much wider than that of the undistorted AM signal, as is easily demonstrated by spectral analysis. This is the overmodulated condition that the FCC does not allow. An AM transmitter that uses plate modulation is an example of a circuit that acts as a two-quadrant multiplier. Here, for high-power AM signal generation, the unmodulated carrier signal is applied to the grid of the tube, and the dc plate voltage is varied proportionally to $A_c[1 + m(t)]$, where $A_c[1 + m(t)] \geq 0$. This produces the product $A_c[1 + m(t)] \cos \omega_c t$, provided that $m(t) \geq -1$, but produces no output when $m(t) < -1$.

[†] If the percentage of modulation becomes very large (approaching infinity), the AM signal becomes the double-sideband suppressed carrier signal that is described in the next section.

If the percentage of negative modulation is less than 100%, an envelope detector may be used to recover the modulation without distortion, since the envelope, $|g(t)| = |A_c[1 + m(t)]|$, is identical to $A_c[1 + m(t)]$. If the percentage of negative modulation is over 100%, undistorted modulation can still be recovered provided that the proper type of detector—a product detector—is used. This is seen from Eq. (4–76) with $\theta_0 = 0$. Furthermore, the product detector may be used for any percentage of modulation. In Chapter 7, we will see that a product detector is superior to an envelope detector when the input signal-to-noise ratio is small.

From Eq. (4–17), the *normalized average power* of the AM signal is

$$\langle s^2(t) \rangle = \tfrac{1}{2} \langle |g(t)|^2 \rangle = \tfrac{1}{2} A_c^2 \langle [1 + m(t)]^2 \rangle$$
$$= \tfrac{1}{2} A_c^2 \langle 1 + 2m(t) + m^2(t) \rangle$$
$$= \tfrac{1}{2} A_c^2 + A_c^2 \langle m(t) \rangle + \tfrac{1}{2} A_c^2 \langle m^2(t) \rangle \qquad (5\text{–}8)$$

If the modulation contains no dc level, then $\langle m(t) \rangle = 0$ and the *normalized* power of the AM signal is

$$\langle s^2(t) \rangle = \underbrace{\tfrac{1}{2} A_c^2}_{\substack{\text{discrete} \\ \text{carrier power}}} + \underbrace{\tfrac{1}{2} A_c^2 \langle m^2(t) \rangle}_{\text{sideband power}} \qquad (5\text{–}9)$$

DEFINITION.   The *modulation efficiency* is the percentage of the total power of the modulated signal that conveys information.

In AM signaling, only the sideband components convey information, so the modulation efficiency is

$$E = \frac{\langle m^2(t) \rangle}{1 + \langle m^2(t) \rangle} \times 100\% \qquad (5\text{–}10)$$

The highest efficiency that can be attained for a 100% AM signal would be 50%, (for the case when square-wave modulation is used).

Using Eq. (4–18), we obtain the normalized peak envelope power (PEP) of the AM signal:

$$P_{\text{PEP}} = \frac{A_c^2}{2} \{1 + \max[m(t)]\}^2 \qquad (5\text{–}11)$$

The voltage spectrum of the AM signal is given by Eq. (4–20a) of Example 4–1 and is

$$S(f) = \frac{A_c}{2} [\delta(f - f_c) + M(f - f_c) + \delta(f + f_c) + M(f + f_c)] \qquad (5\text{–}12)$$

The AM spectrum is just a translated version of the modulation spectrum plus delta functions that give the carrier line spectral component. The *bandwidth* is *twice* that of the modulation. As shown is Sec. 5–6, the spectrum for an FM signal is much more complicated, since the modulation mapping function $g(m)$ is nonlinear.

Example 5–1 POWER OF AN AM SIGNAL

The FCC rates AM broadcast band transmitters by their average *carrier* power; this rating system is common in other AM audio applications as well. Suppose that a 5,000-W AM transmitter is connected to a 50-Ω load; then the constant A_c is given by $\tfrac{1}{2} A_c^2/50 = 5,000$. Thus, the peak voltage across the load will be $A_c = 707$ V during the times when there is no modulation. If the transmitter is then 100% modulated by a 1,000-Hz test tone, the total (carrier plus side-band) average power will be, from Eq. (5–9),

$$1.5 \left[\frac{1}{2} \left(\frac{A_c^2}{50} \right) \right] = (1.5) \times (5,000) = 7,500 \text{ W}$$

because $\langle m^2(t) \rangle = \frac{1}{2}$ for 100% sinusoidal modulation. Note that 7,500 W is the actual power, not the normalized power. The peak voltage (100% modulation) is $(2)(707) = 1414$ V across the 50-Ω load. From Eq. (5–11), the PEP is

$$4 \left[\frac{1}{2} \left(\frac{A_c^2}{50} \right) \right] = (4)(5,000) = 20,000 \text{ W}$$

The modulation efficiency would be 33%, since $\langle m^2(t) \rangle = \frac{1}{2}$.

There are many ways of building AM transmitters. One might first consider generating the AM signal at a low power level (by using a multiplier) and then amplifying it. This, however, requires the use of linear amplifiers (such as Class A or B amplifiers, discussed in Sec. 4–9) so that the AM will not be distorted. Because these linear amplifiers are not very efficient in converting the power-supply power to an RF signal, much of the energy is wasted in heat.[†] Consequently, high-power AM broadcast transmitters are built by amplifying the carrier oscillator signal to a high power level with efficient Class C or Class D amplifiers and then amplitude modulating the last high-power stage. This is called *high-level* modulation. One example is shown in Fig. 5–2a, in which a pulse width modulation (PWM) technique is used to achieve the AM with high conversion efficiency [DeAngelo, 1982]. The audio input is converted to a PWM signal that is used to control a high-power switch (tube or transistor) circuit. The output of this switch circuit consists of a high-level PWM signal that is filtered by a low-pass filter to produce the "dc" component that is used as the power supply for the power amplifier (PA) stage. The PWM switching frequency is usually chosen in the range of 70 to 80 kHz so that the fundamental and harmonic components of the PWM signal can be easily suppressed by the low-pass filter, and yet the "dc" can vary at an audio rate as high as 12 or 15 kHz for good AM audio frequency response. This technique provides excellent frequency response and low distortion, since no high-power audio transformers are needed, but vacuum tubes are often used in the PA and electronic switch circuits, as transistors do not have sufficiently larger dissipation.

Another technique allows an all solid-state high-power transmitter to be built. It uses digital processing to generate AM. A 50-kW AM transmitter can be built that uses 50 to 100 transistor PA modules, each of which produces either 100 W, 300 W, 500 W, or 1,000 W. See http://www.broadcast.harris.com. Each module generates a constant-amplitude square wave at the carrier frequency (which is filtered to produce the sine wave fundamental). To synthesize the AM signal, the analog audio signal is sampled and digitized via an ADC. The samples are used to determine (compute) the combination of modules that need

---

[†] Do not confuse this conversion efficiency with modulation efficiency, which was defined by Eq. (5–10).

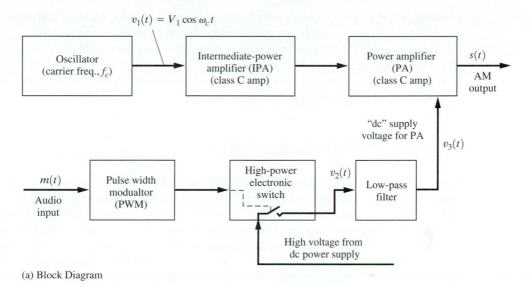

$$v_1(t) = V_1 \cos \omega_c t$$

(a) Block Diagram

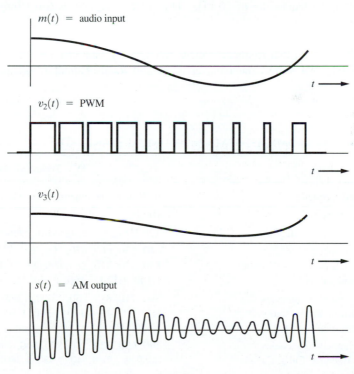

(b) Waveforms

Figure 5–2   Generation of high-power AM by the use of PWM.

to be turned on (from sample to sample) in order to generate the required amplitude on the combined (summed) signal. If one module fails, another module (or a combination of modules) is substituted for it, ensuring excellent on-the-air transmitter reliability, since the transmitter will continue to function with failed modules. Any failed modules can be replaced or repaired later at a convenient time. This AM transmitter has an AC-power-to-RF conversion efficiency of 86% and excellent audio fidelity.

5–2 AM BROADCAST TECHNICAL STANDARDS

Some of the FCC technical standards for AM broadcast stations are shown in Table 5–1. Since the channel bandwidth is 10 kHz, the highest audio frequency is limited to 5 kHz if the resulting AM signal is not to interfere with the stations assigned to the adjacent channels. This low fidelity is not an inherent property of AM, but occurs because the channel bandwidth was limited by the 10 kHz standard the FCC chose instead of, say, 30 kHz, so that three times the number of channels could be accommodated in the AM broadcast band. In practice, the FCC allows stations to have an audio bandwidth of 10 kHz, which produces an AM signal bandwidth of 20 kHz. This, of course, causes some interference to adjacent channel stations.

In the United States, the carrier frequencies are designated according to the intended coverage area for that frequency: clear-channel, regional channel, or local channel frequencies. Table 5–1 shows the clear-channel and local-channel frequencies. The others are

Table 5–1 AM BROADCAST STATION TECHNICAL STANDARDS

Item	FCC Technical Standard
Assigned frequency, f_c	In 10-kHz increments from 540 to 1,700 kHz
Channel bandwidth	10 kHz
Carrier frequency stability	± 20 Hz of the assigned frequency
Clear-channel frequencies (One Class A, 50-kW station) (Nondirectional)	640, 650, 660, 670, 700, 720, 750, 760, 770, 780, 820, 830, 840, 870, 880, 890, 1,020, 1,030, 1,040, 1,070, 1,100, 1,120, 1,160, 1,180, 1,200, and 1,210 kHz
Clear-channel frequencies (Multiple 50-kW stations) (Directional night)	680, 710, 810, 850, 1,000, 1,060, 1,080, 1,090, 1,110, 1,130, 1,140, 1,170, 1,190, 1,500, 1,510, 1,520, and 1,530 kHz
Clear-channel frequencies (For Bahama, Cuba, Canada, or Mexico)	540, 690, 730, 740, 800, 860, 900, 940, 990, 1,010, 1,050, 1,220, 1,540, 1,550, 1,560, 1,570, and 1,580 kHz
Local channel frequencies (1-kW stations)	1,230, 1,240, 1,340, 1,400, 1,450, and 1,490 kHz
Maximum power licensed	50 kW

regional. Class A clear-channel stations operate full time (day and night), and most have a power of 50 kW. These stations are intended to cover large areas. Moreover, to accommodate as many stations as possible, nonclear-channel stations may be assigned to operate on clear-channel frequencies when they can be implemented without interfering with the dominant clear-channel station. Often, to prevent interference, these secondary stations have to operate with directional antenna patterns such that there is a null in the direction of the dominant station. This is especially true for nighttime operation, when sky-wave propagation allows a clear-channel station to cover half of the United States.

Class B stations operate full time to cover a regional area. Most Class B stations have a power of 5 kW, although some operate with power as large as 50 kW. Class C stations cover a local area full time, and most operate with a power of 1 kW. Hundreds of Class C stations are assigned to each local-channel frequency (see Table 5–1), so the nighttime sky-wave interference is large on these frequencies. Because of this interference, the night time coverage radius of a Class C station may be as small as 5 miles from the transmitter site. Class D stations operate daytime, some with powers as large as 50 kW, and, if allowed to operate at night, with nighttime power of 250 watts or less.

International broadcast AM stations, which operate in the shortwave bands (3 to 30 MHz), generally do so with high power levels. Some of these feed 500 kW of carrier power into directional antennas that produce effective radiated power levels in the megawatt range (i.e., when the gain of the directional antenna is included).

Many AM broadcasters transmit stereo audio simultaneously with compatible monaural audio using the Motorola C-QUAM[†] system. The system uses quadrature modulation (QM). (See Table 4–1.) The monaural audio, consisting of the sum of the left and right channels, is used to modulate a cosine carrier. That is, $m_1(t) = V_0 + m_L(t) + m_R(t)$, where V_0 is the dc offset used to produce the discrete carrier term in the AM spectrum, $m_L(t)$ is the left-channel audio, and $m_R(t)$ is the right-channel audio. The quadrature carrier is modulated by $m_2(t) = m_L(t) - m_R(t)$. The resulting C-QUAM signal is compatible for mono reception using a conventional AM receiver with an envelope detector [Mennie, 1978].

Stereo AM receivers use an envelope detector to recover $m_L(t) + m_R(t)$ and a quadrature product detector to recover $m_L(t) - m_R(t)$. Sum and difference networks can then be used to obtain the left- and right-channel audio, $m_L(t)$ and $m_R(t)$, respectively.

5–3 DOUBLE-SIDEBAND SUPPRESSED CARRIER

A *double-sideband suppressed carrier* (DSB-SC) signal is an AM signal that has a suppressed discrete carrier. The DSB-SC signal is given by

$$s(t) = A_c m(t) \cos \omega_c t \tag{5–13}$$

where $m(t)$ is assumed to have a zero dc level for the suppressed carrier case. The spectrum is identical to that for AM given by Eq. (5–12), except that the delta functions at $\pm f_c$ are missing. That is, the spectrum for DSB-SC is

---

[†] C-QUAM is the acronym used by Motorola for compatible *q*uadrature *a*mplitude *m*odulation.

$$S(f) = \frac{A_c}{2} [M(f - f_c) + M(f + f_c)] \qquad (5\text{–}14)$$

Compared with an AM signal, the percentage of modulation on a DSB-SC signal is infinite, because there is no carrier line component. Furthermore, the modulation efficiency of a DSB-SC signal is 100%, since no power is wasted in a discrete carrier. However, a product detector (which is more expensive than an envelope detector) is required for demodulation of the DSB-SC signal. If transmitting circuitry restricts the modulated signal to a certain peak value, say, A_p, it can be demonstrated (see Prob. 5–8) that the sideband power of a DSB-SC signal is *four* times that of a comparable AM signal with the same peak level. In this sense, the DSB-SC signal has a fourfold power advantage over that of an AM signal.

If $m(t)$ is a polar binary data signal (instead of an audio signal), then Eq. (5–13) is a BPSK signal, first described in Example 2–18. Details of BPSK signaling will be covered in Sec. 5–9. As shown in Table 4–1, a QM signal can be generated by adding two DSB signals where there are two signals, $m_1(t)$ and $m_2(t)$, modulating cosine and sine carriers, respectively.

5–4 COSTAS LOOP AND SQUARING LOOP

The coherent reference for product detection of DSB-SC cannot be obtained by the use of an ordinary phase-locked tracking loop, because there are no spectral line components at $\pm f_c$. However, since the DSB-SC signal has a spectrum that is symmetrical with respect to the (suppressed) carrier frequency, either one of the two types of carrier recovery loops shown in Fig. 5–3 may be used to demodulate the DSB-SC signal. Figure 5–3a shows the *Costas PLL* and Fig. 5–3b shows the *squaring loop*. The noise performances of these two loops are equivalent [Ziemer and Peterson, 1985], so the choice of which loop to implement depends on the relative cost of the loop components and the accuracy that can be realized when each is built.

As shown in Fig. 5–3a, the Costas PLL is analyzed by assuming that the VCO is locked to the input suppressed carrier frequency, f_c, with a constant phase error of θ_e. Then the voltages $v_1(t)$ and $v_2(t)$ are obtained at the output of the baseband low-pass filters as shown. Since θ_e is small, the amplitude of $v_1(t)$ is relatively large compared to that of $v_2(t)$ (i.e., $\cos \theta_e \gg \sin \theta_e$). Furthermore, $v_1(t)$ is proportional to $m(t)$, so it is the demodulated (product detector) output. The product voltage $v_3(t)$ is

$$v_3(t) = \tfrac{1}{2} (\tfrac{1}{2}A_0A_c)^2 m^2(t) \sin 2\theta_e$$

The voltage $v_3(t)$ is filtered with an LPF that has a cutoff frequency near dc, so that this filter acts as an integrator to produce the dc VCO control voltage

$$v_4(t) = K \sin 2\theta_e$$

where $K = \tfrac{1}{2} (\tfrac{1}{2}A_0A_c)^2 \langle m^2(t) \rangle$ and $\langle m^2(t) \rangle$ is the dc level of $m^2(t)$. This dc control voltage is sufficient to keep the VCO locked to f_c with a small phase error θ_e.

The squaring loop, shown in Fig. 5–3b, is analyzed by evaluating the expression for the signal out of each component block, as illustrated in the figure. Either the Costas PLL

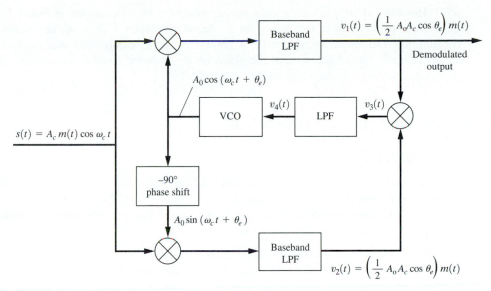

(a) Costas Phase-Locked Loop

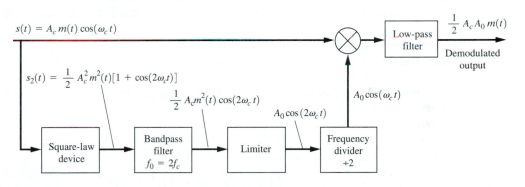

(b) Squaring Loop

Figure 5–3 Carrier recovery loops for DSB-SC signals.

or the squaring loop can be used to demodulate a DSB-SC signal, because, for each case, the output is $Cm(t)$, where C is a constant. Furthermore, either of these loops can be used to recover (i.e., demodulate) a BPSK signal, since the BPSK signal has the same mathematical form as a DSB-SC signal, where $m(t)$ is a polar NRZ data signal as given in Fig. 3-15c.

Both the Costas PLL and the squaring loop have one major disadvantage—a 180° phase ambiguity. For example, suppose that the input is $-A_c m(t) \cos \omega_c t$ instead of $+A_c m(t) \cos \omega_c t$. Retracing the steps in the preceding analysis, we see that the output would be described by *exactly* the same equation that was obtained before. Then, whenever the loop is energized, it is just as likely to phase lock such that the demodulated output is proportional to $-m(t)$ as it is to $m(t)$. Thus, we cannot be sure of the polarity of the output. This is no problem if $m(t)$ is a monaural audio signal, because $-m(t)$ sounds the same to our ears as

$m(t)$. However, if $m(t)$ is a polar data signal, binary 1's might come out as binary 0's after the circuit is energized, or vice versa. As we found in Chapter 3, there are two ways of abrogating this 180° phase ambiguity: (1) A known test signal can be sent over the system after the loop is turned on, so that the sense of the polarity can be determined, and (2) differential coding and decoding may be used.

5–5 ASYMMETRIC SIDEBAND SIGNALS

Single Sideband

> **DEFINITION.** An *upper single sideband* (USSB) signal has a zero-valued spectrum for $|f| < f_c$, where f_c is the carrier frequency.
>
> A *lower single sideband* (LSSB) signal has a zero-valued spectrum for $|f| > f_c$, where f_c is the carrier frequency.

There are numerous ways in which the modulation $m(t)$ may be mapped into the complex envelope $g[m]$ such that an SSB signal will be obtained. Table 4–1 lists some of these methods. SSB-AM is by far the most popular type. It is widely used by the military and by radio amateurs in high-frequency (HF) communication systems. It is popular because the bandwidth is the same as that of the modulating signal (which is half the bandwidth of an AM or DSB-SC signal). For these reasons, we will concentrate on this type of SSB signal. In the usual application, the term *SSB* refers to the SSB-AM type of signal, unless otherwise denoted.

> **THEOREM.** *An SSB signal (i.e., SSB-AM type) is obtained by using the complex envelope*

$$g(t) = A_c[m(t) \pm j\hat{m}(t)] \tag{5–15}$$

which results in the SSB signal waveform

$$s(t) = A_c[m(t) \cos \omega_c t \pm \hat{m}(t) \sin \omega_c t] \tag{5–16}$$

where the upper $(-)$ sign is used for USSB and the lower $(+)$ sign is used for LSSB. $\hat{m}(t)$ denotes the Hilbert transform of $m(t)$, which is given by[†]

$$\hat{m}(t) \triangleq m(t) * h(t) \tag{5–17}$$

where

$$h(t) = \frac{1}{\pi t} \tag{5–18}$$

and $H(f) = \mathcal{F}[h(t)]$ corresponds to a $-90°$ phase-shift network:

$$H(f) = \begin{cases} -j, & f > 0 \\ j, & f < 0 \end{cases} \tag{5–19}$$

[†] A table of Hilbert transform pairs is given in Sec. A–7 (Appendix A).

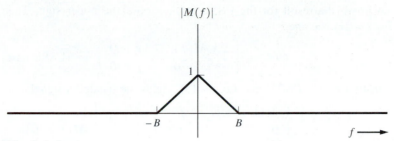

(a) Baseband Magnitude Spectrum

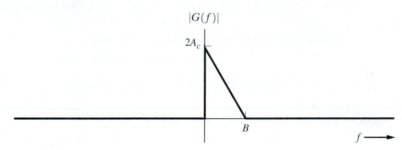

(b) Magnitude of Corresponding Spectrum of the Complex Envelope for USSB

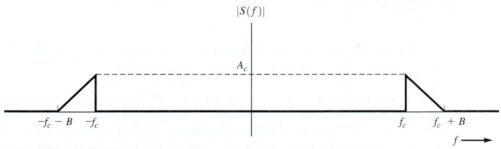

(c) Magnitude of Corresponding Spectrum of the USSB Signal

Figure 5–4   Spectrum for a USSB signal.

Figure 5–4 illustrates this theorem. Assume that $m(t)$ has a magnitude spectrum that is of triangular shape, as shown in Fig. 5–4a. Then, for the case of USSB (upper signs), the spectrum of $g(t)$ is zero for negative frequencies, as illustrated in Fig. 5–4b, and $s(t)$ has the USSB spectrum shown in Fig. 5–4c. This result is proved as follows:

Proof. We need to show that the spectrum of $s(t)$ is zero on the appropriate sideband (depending on the sign chosen). Taking the Fourier transform of Eq. (5–15), we get

$$G(f) = A_c\{M(f) \pm j\mathcal{F}[\hat{m}(t)]\} \tag{5–20}$$

and, using Eq. (5–17), we find that the equation becomes

$$G(f) = A_c M(f)[1 \pm jH(f)] \tag{5–21}$$

To prove the result for the USSB case, choose the upper sign. Then, from Eq. (5–19), Eq. (5–21) becomes

$$G(f) = \begin{Bmatrix} 2A_c M(f), & f > 0 \\ 0, & f < 0 \end{Bmatrix} \qquad \text{(USSB case)} \qquad (5\text{–}22)$$

Substituting Eq. (5–22) into Eq. (4–15), yields the bandpass signal:

$$S(f) = A_c \begin{Bmatrix} M(f - f_c), & f > f_c \\ 0, & f < f_c \end{Bmatrix} + A_c \begin{Bmatrix} 0, & f > -f_c \\ M(f + f_c), & f < -f_c \end{Bmatrix} (5\text{–}23)$$

This is indeed a USSB signal (see Fig. 5–4).

If the lower signs of Eq. (5–21) were chosen, an LSSB signal would have been obtained.

The *normalized average power* of the SSB signal is

$$\langle s^2(t) \rangle = \tfrac{1}{2} \langle |g(t)|^2 \rangle = \tfrac{1}{2} A_c^2 \langle m^2(t) + [\hat{m}(t)]^2 \rangle \qquad (5\text{–}24)$$

As shown in study-aid Example SA5–1, $\langle \hat{m}(t)^2 \rangle = \langle m^2(t) \rangle$, so that the SSB signal power is

$$\langle s^2(t) \rangle = A_c^2 \langle m^2(t) \rangle \qquad (5\text{–}25)$$

which is the power of the modulating signal $\langle m^2(t) \rangle$ multiplied by the power gain factor A_c^2.

The *normalized peak envelope power (PEP)* is

$$\tfrac{1}{2} \max\{|g(t)|^2\} = \tfrac{1}{2} A_c^2 \max\{m^2(t) + [\hat{m}(t)]^2\} \qquad (5\text{–}26)$$

Figure 5–5 illustrates two techniques for generating the SSB signal. The *phasing method* is identical to the IQ canonical form discussed earlier (Fig. 4–28) as applied to SSB signal generation. The *filtering method* is a special case in which RF processing (with a sideband filter) is used to form the equivalent $g(t)$, instead of using baseband processing to generate $g[m]$ directly. The filter method is the most popular method because excellent sideband suppression can be obtained when a crystal filter is used for the sideband filter.[†] Crystal filters are relatively inexpensive when produced in quantity at standard IF frequencies. In addition to these two techniques for generating SSB, there is a third technique, called *Weaver's method* [Weaver, 1956]. This is described by Fig. P5–12 and Prob. 5–12. A practical SSB transmitter incorporates an up converter to translate the SSB signal to the desired operating frequency and uses a Class B amplifier to amplify the signal to the desired power level.

SSB signals have *both* AM and PM. Using Eq. (5–15), we have, for the AM component (real envelope),

$$R(t) = |g(t)| = A_c \sqrt{m^2(t) + [\hat{m}(t)]^2} \qquad (5\text{–}27)$$

and for the PM component,

$$\theta(t) = \underline{/g(t)} = \tan^{-1}\left[\frac{\pm \hat{m}(t)}{m(t)} \right] \qquad (5\text{–}28)$$

[†] Excellent sideband suppression is possible because communications-quality audio has negligible spectral content below 300 Hz. Thus, the sideband filter can be designed to provide the required sideband attenuation over a $2 \times 300 = 600$-Hz transition band.

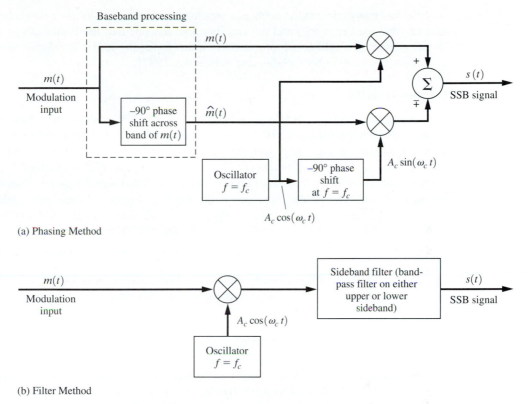

(a) Phasing Method

(b) Filter Method

Figure 5–5   Generation of SSB.

SSB signals may be received by using a superheterodyne receiver that incorporates a product detector with $\theta_0 = 0$. Thus, the receiver output is

$$v_{out}(t) = K \, Re\{g(t)e^{-j\theta_0}\} = KA_c m(t) \tag{5–29}$$

where K depends on the gain of the receiver and the loss in the channel. In detecting SSB signals with audio modulation, the reference phase θ_0 does not have to be zero, because the same intelligence is heard regardless of the value of the phase used [although the $v_{out}(t)$ waveform will be drastically different, depending on the value of θ_0]. For digital modulation, the phase has to be exactly correct so that the digital waveshape is preserved. Furthermore, SSB is a poor modulation technique to use *if* the modulating data signal consists of a line code with a *rectangular* pulse shape. The rectangular shape (zero rise time) causes the value of the SSB-AM waveform to be infinite adjacent to the switching times of the data because of the Hilbert transform operation. (This result will be demonstrated in a homework problem.) Thus, an SSB signal with this type of modulation cannot be generated by any practical device, since a device can produce only finite peak value signals. However, if rolled-off pulse shapes are used in the line code, such as $(\sin x)/x$ pulses, the SSB signal will have a reasonable peak value, and digital data transmission can then be accommodated via SSB.

SSB has many advantages, such as a superior detected signal-to-noise ratio compared to that of AM (see Chapter 7) and the fact that SSB has one-half the bandwidth of AM or DSB-SC signals. (For additional information on this topic, the reader is referred to a book that is devoted wholly to SSB [Sabin and Schoenike, 1987]).

Vestigial Sideband

In certain applications (such as television broadcasting), a DSB modulation technique takes too much bandwidth for the (television) channel, and an SSB technique is too expensive to implement, although it takes only half the bandwidth. In this case, a compromise between DSB and SSB, called *vestigial sideband* (VSB), is often chosen. VSB is obtained by partial suppression of one of the sidebands of a DSB signal. The DSB signal may be either an AM signal or a DSB-SC signal. This approach is illustrated in Fig. 5–6, where one sideband of the DSB signal is attenuated by using a bandpass filter, called a vestigial sideband filter, that has an asymmetrical frequency response about $\pm f_c$. The VSB signal is given by

$$s_{\text{VSB}}(t) = s(t) * h_v(t) \tag{5–30}$$

where $s(t)$ is a DSB signal described by either Eq. (5–4) with carrier or Eq. (5–13) with suppressed carrier and $h_v(t)$ is the impulse response of the VSB filter. The spectrum of the VSB signal is

$$S_{\text{VSB}}(f) = S(f)H_v(f) \tag{5–31}$$

as illustrated in Fig. 5–6d.

The modulation on the VSB signal can be recovered by a receiver that uses product detection or, if a large carrier is present, by the use of envelope detection. For recovery of undistorted modulation, the transfer function for the VSB filter must satisfy the constraint

$$H_v(f - f_c) + H_v(f + f_c) = C, \quad |f| \le B \tag{5–32}$$

where C is a constant and B is the bandwidth of the modulation. An application of this constraint is shown in Fig. 5–6e, where it is seen that the condition specified by Eq. (5–32) is satisfied for the VSB filter characteristic shown in Fig. 5–6c.

The need for the constraint of Eq. (5–32) will now be proven. Assume that $s(t)$ is a DSB-SC signal. Then, by the use of Eqs. (5–14) and (5–31), the spectrum of the VSB signal is

$$S_{\text{VSB}}(f) = \frac{A_c}{2} \left[M(f - f_c)H_v(f) + M(f + f_c)H_v(f) \right]$$

Referring to Fig. 4-14, we see that the product detector output is

$$v_{\text{out}}(t) = [A_0 s_{\text{VSB}}(t) \cos \omega_c t] * h(t)$$

where $h(t)$ is the impulse response of the low-pass filter having a bandwidth of B hertz. In the frequency domain, this equation becomes

$$V_{\text{out}}(f) = A_0 \left\{ S_{\text{VSB}}(f) * \left[\tfrac{1}{2}\delta(f - f_c) + \tfrac{1}{2}\delta(f + f_c) \right] \right\} H(f)$$

where $H(f) = 1$ for $|f| < B$ and 0 for f elsewhere. Substituting for $S_{\text{VSB}}(f)$ and using the convolutional property $x(f) * \delta(f - a) = x(f - a)$ yields

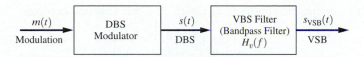

(a) Generation of VSB Signal

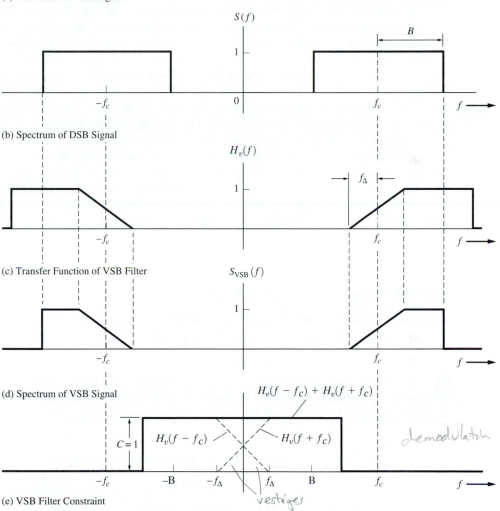

(b) Spectrum of DSB Signal

(c) Transfer Function of VSB Filter

(d) Spectrum of VSB Signal

(e) VSB Filter Constraint

Figure 5–6   VSB transmitter and spectra.

$$V_{\text{out}}(f) = \frac{A_c A_0}{4} \left[M(f) H_v(f - f_c) + M(f) H_v(f + f_c) \right], \quad |f| < B$$

or

$$V_{\text{out}}(f) = \frac{A_c A_0}{4} M(f) \left[H_v(f - f_c) + H_v(f + f_c) \right], \quad |f| < B$$

If $H_v(f)$ satisfies the constraint of Eq. (5–32), this equation becomes

$$V_{out}(f) = KM(f)$$

or $V_{out}(t) = Km(t)$, where $K = A_cA_0/4$, which shows that the output of the product detector is undistorted when Eq. (5–32) is satisfied.

As discussed in Chapter 8, both analog and digital broadcast television use VSB to reduce the required channel bandwidth to 6 MHz. For analog TV, as shown in Fig. 8–31c, the frequency response of the visual TV transmitter is flat over the upper sideband out to 4.2 MHz above the visual carrier frequency and is flat over the lower sideband out to 0.75 MHz below the carrier frequency. The IF filter in the TV receiver has the VSB filter characteristic shown in Fig. 5–6c, where $f_\Delta = 0.75$ MHz. This gives an overall frequency response characteristic that satisfies the constraint of Eq. (5–32), so that the video modulation on the visual VSB TV signal can be recovered at the receiver without distortion.

Digital TV (in the United States; see Sec. 8–9) uses an $r = 0.0575$ square-root raised cosine-rolloff filter at the transmitter to obtain VSB as shown in Fig. 8–39. In practice, a square-root raised cosine-rolloff filter is used in both the digital TV transmitter and the receiver to give the overall raised cosine-rolloff characteristic, thereby minimizing the effect of channel noise as well as ISI. (See Sec. 3–6.) This overall response also satisfies Eq. (5–32), so that the digital TV signal (8-level serial data) is recovered without distortion or ISI.

5–6 PHASE MODULATION AND FREQUENCY MODULATION

Representation of PM and FM Signals

Phase modulation (PM) and *frequency modulation* (FM) are special cases of angle-modulated signaling. In this kind of signaling the complex envelope is

$$g(t) = A_c e^{j\theta(t)} \tag{5–33}$$

Here the real envelope, $R(t) = |g(t)| = A_c$, is a constant, and the phase $\theta(t)$ is a linear function of the modulating signal $m(t)$. However, $g(t)$ is a *nonlinear* function of the modulation. Using Eq. (5–33), we find the resulting *angle-modulated signal* to be

$$s(t) = A_c \cos[\omega_c t + \theta(t)] \tag{5–34}$$

For PM, the phase is directly proportional to the modulating signal; that is,

$$\theta(t) = D_p m(t) \tag{5–35}$$

where the proportionality constant D_p is the phase sensitivity of the phase modulator, having units of radians per volt [assuming that $m(t)$ is a voltage waveform]. For FM, the phase is proportional to the integral of $m(t)$, so that

$$\theta(t) = D_f \int_{-\infty}^{t} m(\sigma) \, d\sigma \tag{5–36}$$

where the frequency deviation constant D_f has units of radians/volt-second.

Comparing the last two equations, we see that if we have a PM signal modulated by $m_p(t)$, there is *also* FM on the signal, corresponding to a *different* modulation waveshape that is given by

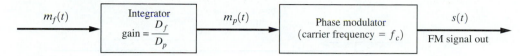

(a) Generation of FM Using a Phase Modulator

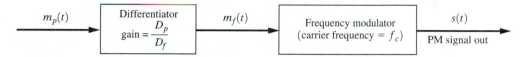

(b) Generation of PM Using a Frequency Modulator

Figure 5–7   Generation of FM from PM, and vice versa.

$$m_f(t) = \frac{D_p}{D_f}\left[\frac{dm_p(t)}{dt}\right] \tag{5–37}$$

where the subscripts f and p denote frequency and phase, respectively. Similarly, if we have an FM signal modulated by $m_f(t)$, the corresponding phase modulation on this signal is

$$m_p(t) = \frac{D_f}{D_p}\int_{-\infty}^{t} m_f(\sigma)\,d\sigma \tag{5–38}$$

From Eq. (5–38), a PM circuit may be used to synthesize an FM circuit by inserting an integrator in cascade with the phase modulator input. (See Fig. 5–7a.)

Direct PM circuits are realized by passing an unmodulated sinusoidal signal through a time-varying circuit which introduces a phase shift that varies with the applied modulating voltage. (See Fig. 5–8a.) D_p is the gain of the PM circuit (rad/V). Similarly, a direct FM circuit is obtained by varying the tuning of an oscillator tank (resonant) circuit according to the modulation voltage. This is shown in Fig. 5–8b, where D_f is the gain of the modulator circuit (which has units of radians per volt-second).

DEFINITION.   If a bandpass signal is represented by

$$s(t) = R(t)\cos\psi(t)$$

where $\psi(t) = \omega_c t + \theta(t)$, then the *instantaneous* frequency (hertz) of $s(t)$ is [Boashash, 1992]

$$f_i(t) = \frac{1}{2\pi}\,\omega_i(t) = \frac{1}{2\pi}\left[\frac{d\psi(t)}{dt}\right]$$

or

$$f_i(t) = f_c + \frac{1}{2\pi}\left[\frac{d\theta(t)}{dt}\right] \tag{5–39}$$

For the case of FM, using Eq. (5–36), we obtain the instantaneous frequency:

$$f_i(t) = f_c + \frac{1}{2\pi}D_f m(t) \tag{5–40}$$

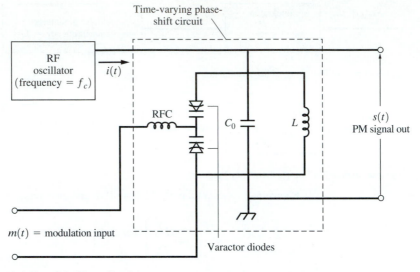

(a) A Phase Modulator Circuit

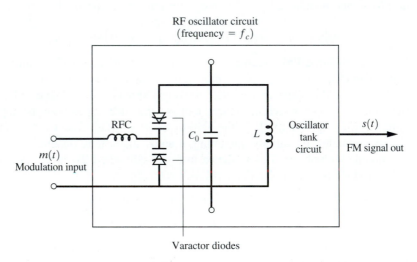

(b) A Frequency Modulator Circuit

Figure 5–8   Angle modulator circuits. RFC = radio-frequency choke.

Of course, this is the reason for calling this type of signaling *frequency modulation*—the instantaneous frequency varies about the assigned carrier frequency f_c in a manner that is directly proportional to the modulating signal $m(t)$. Figure 5–9b shows how the instantaneous frequency varies when a sinusoidal modulation (for illustrative purposes) is used. The resulting FM signal is shown in Fig. 5–9c.

The instantaneous frequency should not be confused with the term *frequency* as used in the spectrum of the FM signal. The spectrum is given by the Fourier transform of $s(t)$ and

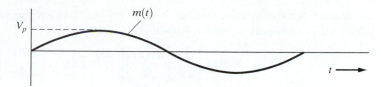

(a) Sinusoidal Modulating Signal

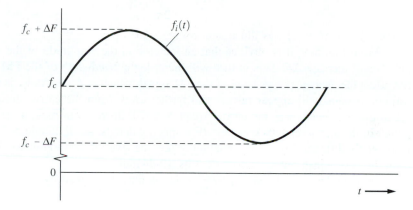

(b) Instantaneous Frequency of the Corresponding FM Signal

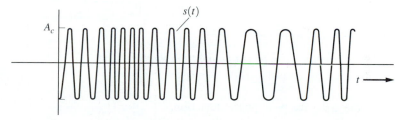

(c) Corresponding FM Signal

Figure 5–9   FM with a sinusoidal baseband modulating signal.

is evaluated by looking at $s(t)$ over the infinite time interval $(-\infty < t < \infty)$. Thus, the spectrum tells us what frequencies are present in the signal (on the average) *over all time*. The instantaneous frequency is the frequency that is present at a *particular* instant of time.

The *frequency deviation* from the carrier frequency is

$$f_d(t) = f_i(t) - f_c = \frac{1}{2\pi}\left[\frac{d\theta(t)}{dt}\right] \qquad (5\text{--}41)$$

and the *peak frequency deviation* is

$$\Delta F = \max\left\{\frac{1}{2\pi}\left[\frac{d\theta(t)}{dt}\right]\right\} \qquad (5\text{--}42)$$

Note that ΔF is a nonnegative number. In some applications, such as (unipolar) digital modulation, the peak-to-peak deviation is used. This is defined by

$$\Delta F_{pp} = \max\left\{\frac{1}{2\pi}\left[\frac{d\theta(t)}{dt}\right]\right\} - \min\left\{\frac{1}{2\pi}\left[\frac{d\theta(t)}{dt}\right]\right\} \tag{5-43}$$

For FM signaling, the peak frequency deviation is related to the peak modulating voltage by

$$\Delta F = \frac{1}{2\pi} D_f V_p \tag{5-44}$$

where $V_p = \max[m(t)]$, as illustrated in Fig. 5–9a.

From Fig. 5–9, it is obvious that an increase in the amplitude of the modulation signal V_p will increase ΔF. This in turn will increase the bandwidth of the FM signal, but will not affect the average power level of the FM signal, which is $A_c^2/2$. As V_p is increased, spectral components will appear farther and farther away from the carrier frequency, and the spectral components near the carrier frequency will decrease in magnitude, since the total power in the signal remains constant. (For specific details, see Example 5–2.) This situation is distinctly different from AM signaling, where the level of the modulation affects the power in the AM signal, but does not affect its bandwidth.

In a similar way, the *peak phase deviation* may be defined by

$$\Delta\theta = \max[\theta(t)] \tag{5-45}$$

which, for PM, is related to the peak modulating voltage by

$$\Delta\theta = D_p V_p \tag{5-46}$$

where $V_p = \max[m(t)]$.

DEFINITION. [†]   The *phase modulation index* is given by

$$\beta_p = \Delta\theta \tag{5-47}$$

where $\Delta\theta$ is the peak phase deviation.

The *frequency modulation index* is given by

$$\beta_f = \frac{\Delta F}{B} \tag{5-48}$$

where ΔF is the peak frequency deviation and B is the bandwidth of the modulating signal, which, for the case of sinusoidal modulation, is f_m, the frequency of the sinusoid. [‡]

For the case of PM or FM signaling with *sinusoidal* modulation such that the PM and FM signals have the same peak frequency deviation, β_p is identical to β_f.

[†] For digital signals, an alternative definition of modulation index is sometimes used and is denoted by h in the literature. This digital modulation index is $h = 2\Delta\theta/\pi$, where $2\Delta\theta$ is the maximum peak-to-peak phase deviation change during the time that it takes to send one symbol, T_s. [See Eq. (5–82).]

[‡] Strictly speaking, the FM index is defined only for the case of single-tone (i.e., sinusoidal) modulation. However, it is often used for other waveshapes, where B is chosen to be the highest frequency or the dominant frequency in the modulating waveform.

Spectra of Angle-Modulated Signals

Using Eq. (4–12), we find that the spectrum of an angle modulated signal is given by

$$S(f) = \tfrac{1}{2}[G(f - f_c) + G^*(-f - f_c)] \tag{5–49}$$

where

$$G(f) = \mathcal{F}[g(t)] = \mathcal{F}[A_c e^{j\theta(t)}] \tag{5–50}$$

When the spectra for AM, DSB-SC, and SSB were evaluated, we were able to obtain relatively simple formulas relating $S(f)$ to $M(f)$. For angle modulation signaling, this is not the case, because $g(t)$ is a nonlinear function of $m(t)$. Thus, a general formula relating $G(f)$ to $M(f)$ cannot be obtained. This, is unfortunate, but it is a fact of life. That is, to evaluate the spectrum for an angle-modulated signal, Eq. (5–50) must be evaluated on a case-by-case basis for the particular modulating waveshape of interest. Furthermore, since $g(t)$ is a nonlinear function of $m(t)$, superposition does not hold, and the FM spectrum for the sum of two modulating waveshapes is not the same as summing the FM spectra that were obtained when the individual waveshapes were used.

One example of the spectra obtained for an angle-modulated signal is given in Chapter 2. (See Example 2–18.) There, a carrier was phase modulated by a square wave where the peak-to-peak phase deviation was 180°. In that example, the spectrum was easy to evaluate because this was the very special case where the PM signal reduces to a DSB-SC signal. In general, of course, the evaluation of Eq. (5–50) into a closed form is not easy, and one often has to use numerical techniques to approximate the Fourier transform integral. An example for the case of a sinusoidal modulating waveshape will now be worked out.

Example 5–2 Spectrum of a PM or FM Signal with Sinusoidal Modulation

Assume that the modulation on the PM signal is

$$\beta = \begin{cases} \beta_p \\ \text{or} \\ \beta_f \end{cases}$$

$$m_p(t) = A_m \sin \omega_m t \tag{5–51}$$

Then

$$\theta(t) = \beta \sin \omega_m t \tag{5–52}$$

where $\beta_p = D_p A_m = \beta$ is the phase modulation index.

The same phase function $\theta(t)$, as given by Eq. (5–52), could also be obtained if FM were used, where

$$m_f(t) = A_m \cos \omega_m t \tag{5–53}$$

and $\beta = \beta_f = D_f A_m / \omega_m$. The peak frequency deviation would be $\Delta F = D_f A_m / 2\pi$.

The complex envelope is

$$g(t) = A_c e^{j\theta(t)} = A_c e^{j\beta \sin \omega_m t} \tag{5–54}$$

which is periodic with period $T_m = 1/f_m$. Consequently, $g(t)$ could be represented by a Fourier series that is valid over all time ($-\infty < t < \infty$); namely,

$$g(t) = \sum_{n=-\infty}^{n=\infty} c_n\, e^{jn\omega_m t} \tag{5-55}$$

where

$$c_n = \frac{A_c}{T_m} \int_{-T_m/2}^{T_m/2} \left(e^{j\beta\,\sin\,\omega_m t}\right) e^{-jn\omega_m t}\, dt \tag{5-56}$$

which reduces to

$$c_n = A_c \left[\frac{1}{2\pi} \int_{-\pi}^{\pi} e^{j\,(\beta\,\sin\,\theta - n\theta)}\, d\theta\right] = A_c J_n(\beta) \tag{5-57}$$

This integral—known as the *Bessel function of the first kind of the nth order, $J_n(\beta)$*—cannot be evaluated in closed form, but it has been evaluated numerically. Some tabulated values for $J_n(\beta)$ are given in Table 5–2. Extensive tables are available [Abramowitz and Stegun, 1964] or can be computed using MATLAB. The Bessel functions are invoked by standard function calls in mathematical personal computer programs such as MATLAB. Examination of the integral shows that (by making a change in variable)

$$J_{-n}(\beta) = (-1)^n\, J_n(\beta) \tag{5-58}$$

A plot of the Bessel functions for various orders n as a function of β is shown in Fig. 5–10.
    Taking the Fourier transform of Eq. (5–55), we obtain

$$G(f) = \sum_{n=-\infty}^{n=\infty} c_n\, \delta(f - nf_m) \tag{5-59}$$

or

$$G(f) = A_c \sum_{n=-\infty}^{n=\infty} J_n(\beta)\delta(f - nf_m) \tag{5-60}$$

Using this result in Eq. (5–49), we get the spectrum of the angle-modulated signal. The magnitude spectrum for $f > 0$ is shown in Fig. 5–11 for the cases of $\beta = 0.2, 1.0, 2.0, 5.0,$ and 8.0. Note that the discrete carrier term (at $f = f_c$) is proportional to $|J_0(\beta)|$; consequently, the level (magnitude) of the discrete carrier depends on the modulation index. It will be zero if $J_0(\beta) = 0$, which occurs if $\beta = 2.40, 5.52,$ and so on, as shown in Table 5–3.

    Figure 5–11 also shows that the bandwidth of the angle-modulated signal depends on β and f_m. In fact, it can be shown that 98% of the total power is contained in the bandwidth

$$B_T = 2(\beta + 1)B \tag{5-61}$$

where β is either the phase modulation index or the frequency modulation index and B is the bandwidth of the modulating signal (which is f_m for sinusoidal modulation).[†] This formula gives a rule-of-thumb expression for evaluating the transmission bandwidth of PM and FM signals; it is called *Carson's rule*. B_T is shown in Fig. 5–11 for various values of β. Carson's rule is very important because it gives an easy way to compute the bandwidth of angle-

---

[†] For the case of FM (not PM) with $2 < B < 10$, Carson's rule, Eq. (5–61), actually underestimates B_T somewhat. In this case, a better approximation is $B_T = 2(\beta + 2)B$. Also, if the modulation signal contains discontinuities, such as square-wave modulation, both formulas may not be too accurate, and the B_T should be evaluated by examining the spectrum of the angle-modulated signal. However, to avoid confusion in computing B_T, we will assume that Eq. (5–61) is approximately correct for all cases.

Table 5–2 FOUR-PLACE VALUES OF THE BESSEL FUNCTIONS $J_n(\beta)$

β: \ n	0.5	1	2	3	4	5	6	7	8	9	10
0	0.9385	0.7652	0.2239	−0.2601	−0.3971	−0.1776	0.1506	0.3001	0.1717	−0.09033	−0.2459
1	0.2423	0.4401	0.5767	0.3391	−0.06604	−0.3276	−0.2767	−0.004683	0.2346	0.2453	0.04347
2	0.03060	0.1149	0.3528	0.4861	0.3641	0.04657	−0.2429	−0.3014	−0.1130	0.1448	0.2546
3	0.002564	0.01956	0.1289	0.3091	0.4302	0.3648	0.1148	−0.1676	−0.2911	−0.1809	0.05838
4		0.002477	0.03400	0.1320	0.2811	0.3912	0.3576	0.1578	−0.1054	−0.2655	−0.2196
5			0.007040	0.04303	0.1321	0.2611	0.3621	0.3479	0.1858	−0.05504	−0.2341
6			0.001202	0.01139	0.04909	0.1310	0.2458	0.3392	0.3376	0.2043	−0.01446
7				0.002547	0.01518	0.05338	0.1296	0.2336	0.3206	0.3275	0.2167
8					0.004029	0.01841	0.05653	0.1280	0.2235	0.3051	0.3179
9						0.005520	0.02117	0.05892	0.1263	0.2149	0.2919
10						0.001468	0.006964	0.02354	0.06077	0.1247	0.2075
11							0.002048	0.008335	0.02560	0.06222	0.1231
12								0.002656	0.009624	0.02739	0.06337
13									0.003275	0.01083	0.02897
14									0.001019	0.003895	0.01196
15										0.001286	0.004508
16											0.001567

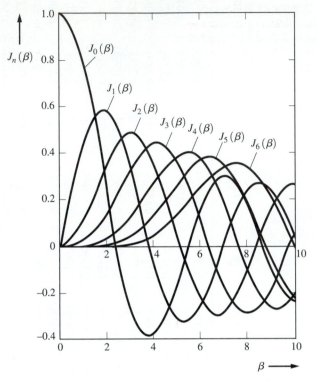

Figure 5–10   Bessel functions for $n = 0$ to $n = 6$.

Table 5–3   ZEROS OF BESSEL FUNCTIONS: VALUES FOR β WHEN $J_n(\beta) = 0$

	Order of Bessel Function, n						
	0	**1**	**2**	**3**	**4**	**5**	**6**
β for 1st zero	2.40	3.83	5.14	6.38	7.59	8.77	9.93
β for 2nd zero	5.52	7.02	8.42	9.76	11.06	12.34	13.59
β for 3rd zero	8.65	10.17	11.62	13.02	14.37	15.70	17.00
β for 4th zero	11.79	13.32	14.80	16.22	17.62	18.98	20.32
β for 5th zero	14.93	16.47	17.96	19.41	20.83	22.21	23.59
β for 6th zero	18.07	19.61	21.12	22.58	24.02	25.43	26.82
β for 7th zero	21.21	22.76	24.27	25.75	27.20	28.63	30.03
β for 8th zero	24.35	25.90	27.42	28.91	30.37	31.81	33.23

modulated signals. Computation of the bandwidth using other definitions, such as the 3-dB bandwidth, can be very difficult, since the spectrum of the FM or PM signal must first be evaluated. This is a nontrivial task, except for simple cases such as single-tone (sinusoidal) modulation or unless a digital computer is used to compute the approximate spectrum.

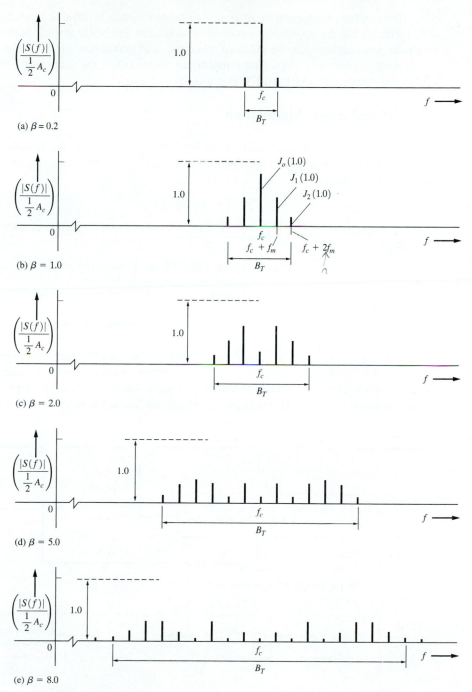

Figure 5–11   Magnitude spectra for FM or PM with sinusoidal modulation for various modulation indexes.

Because the exact spectrum of angle-modulated signals is difficult to evaluate in general, formulas for the approximation of the spectra are extremely useful. Some relatively simple approximations may be obtained when the peak phase deviation is small and when the modulation index is large. These topics are discussed in the sections on narrowband angle modulation and wideband FM that follow.

Narrowband Angle Modulation

When $\theta(t)$ is restricted to a small value, say, $|\theta(t)| < 0.2$ rad, the complex envelope $g(t) = A_c e^{j\theta}$ may be approximated by a Taylor's series in which only the first two terms are used. Thus, because $e^x \approx 1 + x$ for $|x| \ll 1$,

$$g(t) \approx A_c[1 + j\theta(t)] \qquad (5\text{--}62)$$

Using this approximation in Eq. (4–9) or Eq. (5–1), we obtain the expression for a *narrowband angle-modulated signal:*

$$s(t) = \underbrace{A_c \cos \omega_c t}_{\substack{\text{discrete} \\ \text{carrier term}}} - \underbrace{A_c\,\theta(t) \sin \omega_c t}_{\text{sideband term}} \qquad (5\text{--}63)$$

This result indicates that a narrowband angle-modulated signal consists of two terms: a discrete carrier component (which does not change with the modulating signal) and a sideband term. This signal is similar to an AM-type signal, *except* that the sideband term is 90° out of phase with the discrete carrier term. The narrowband signal can be generated by using a balanced modulator (multiplier), as shown in Fig. 5–12a for the case of narrowband frequency modulation (NBFM). Furthermore, wideband frequency modulation (WBFM) may

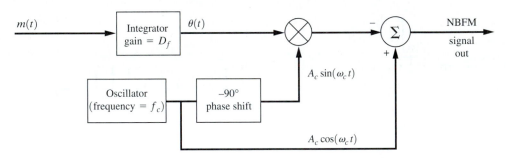

(a) Generation of NBFM Using a Balanced Modulator

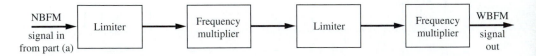

(b) Generation of WBFM From a NBFM Signal

Figure 5–12   Indirect method of generating WBFM (Armstrong method).

be generated from the NBFM signal by using frequency multiplication, as shown in Fig. 5–12b. Limiter circuits are needed to suppress the incidental AM [which is $\sqrt{1 + \theta^2(t)}$ as the result of the approximation of Eq. (5–62)] that is present in the NBFM signal. This method of generating WBFM is called the *Armstrong method* or *indirect method*.

From Eqs. (5–62) and (5–49), the spectrum of the narrowband angle-modulated signal is

$$S(f) = \frac{A_c}{2} \left\{ [\delta(f - f_c) + \delta(f + f_c)] + j[\Theta(f - f_c) - \Theta(f + f_c)] \right\} \quad (5\text{–}64)$$

where

$$\Theta(f) = \mathscr{F}[\theta(t)] = \begin{cases} D_p M(f), & \text{for PM signaling} \\[2mm] \dfrac{D_f}{j2\pi f} \, M(f), & \text{for FM signaling} \end{cases} \quad (5\text{–}65)$$

Wideband Frequency Modulation

A *direct method* of generating wideband frequency modulation (WBFM) is to use a voltage-controlled oscillator (VCO), as illustrated in Fig. 5–8b. However, for VCOs that are designed for wide frequency deviation (ΔF large), the stability of the carrier frequency $f_c = f_0$ is not very good, so the VCO is incorporated into a PLL arrangement wherein the PLL is locked to a stable frequency source, such as a crystal oscillator. (See Fig. 5–13.) The frequency divider is needed to lower the modulation index of the WBFM signal to produce an NBFM signal ($\beta \approx 0.2$) so that a large discrete carrier term will always be present at frequency f_c/N to beat with the crystal oscillator signal and produce the dc control voltage. [See Fig. 5–11a and Eqs. (5–63) and (5–64).] This dc control voltage holds the VCO on the assigned frequency with a tolerance determined by the crystal oscillator circuit.

The power spectral density (PSD) of a WBFM signal may be *approximated* by using the probability density function (PDF) of the modulating signal. This is reasonable from an intuitive viewpoint, since the instantaneous frequency varies directly with the modulating

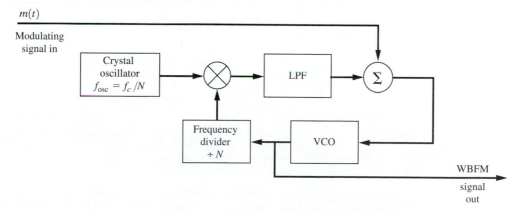

Figure 5–13   Direct method of generating WBFM.

signal voltage for the case of FM [$D_f/(2\pi)$ being the proportionality constant]. If the modulating signal spends more time at one voltage value than another, the instantaneous frequency will dwell at the corresponding frequency, and the power spectrum will have a peak at this frequency. A discussion of the approximation involved—called the *quasi-static approximation*—is well documented [Rowe, 1965]. This result is stated in the following theorem.

THEOREM. *For WBFM signaling, where*

$$s(t) = A_c \cos\left[\omega_c t + D_f \int_{-\infty}^{t} m(\sigma)\,d\sigma\right]$$

$$\beta_f = \frac{D_f \max[m(t)]}{2\pi B} > 1$$

and B is the bandwidth of m(t), the normalized PSD of the WBFM signal is approximated by

$$\mathscr{P}(f) = \frac{\pi A_c^2}{2D_f}\left[f_m\left(\frac{2\pi}{D_f}(f - f_c)\right) + f_m\left(\frac{2\pi}{D_f}(-f - f_c)\right)\right] \qquad (5\text{--}66)$$

where $f_m(\cdot)$ is the PDF of the modulating signal.[†]

This theorem is proved in Prob. 6–52.

Example 5–3 SPECTRUM FOR WBFM WITH TRIANGULAR MODULATION

The spectrum for a WBFM signal with a triangular modulating signal (Fig. 5–14a) will be evaluated. The associated PDF for triangular modulation is shown in Fig. 5–14b. The PDF is described by

$$f_m(m) = \begin{cases} \dfrac{1}{2V_p}, & |m| < V_p \\ 0, & m \text{ otherwise} \end{cases} \qquad (5\text{--}67)$$

where V_p is the peak voltage of the triangular waveform. Substituting this equation into Eq. (5–66), we obtain

$$\mathscr{P}(f) = \frac{\pi A_c^2}{2D_f}\left[\left\{\begin{matrix} \dfrac{1}{2V_p}, & \left|\dfrac{2\pi}{D_f}(f - f_c)\right| < V_p \\ 0, & f \text{ otherwise} \end{matrix}\right\} \right.$$

$$\left. + \left\{\begin{matrix} \dfrac{1}{2V_p}, & \left|\dfrac{2\pi}{D_f}(f + f_c)\right| < V_p \\ 0, & f \text{ otherwise} \end{matrix}\right\}\right]$$

[†] See Appendix B for the definition of PDF and examples of PDFs for various waveforms. This topic may be deleted if the reader is not sufficiently familiar with PDFs. Do not confuse the PDF of the modulation, $f_m(\cdot)$, with the frequency variable f.

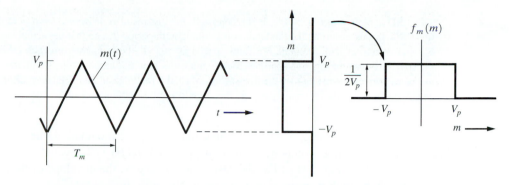

(a) Triangle Modulating Waveform   (b) PDF of Triangle Modulation

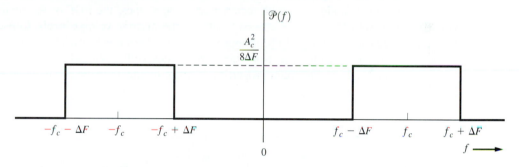

(c) PSD of WBFM Signal with Triangle Modulation, $\Delta F = D_f V_p/2\pi$

Figure 5–14   Approximate spectrum of a WBFM signal with triangle modulation.

The PSD of the WBFM signal becomes

$$
\mathcal{P}(f) = \begin{cases} \dfrac{A_c^2}{8\Delta F}, & (f_c - \Delta F) < f < (f_c + \Delta F) \\[2ex] 0, & f \text{ otherwise} \end{cases}
$$

$$
+ \begin{cases} \dfrac{A_c^2}{8\Delta F}, & (-f_c - \Delta F) < f < (-f_c + \Delta F) \\[2ex] 0, & f \text{ otherwise} \end{cases} \qquad (5\text{–}68)
$$

where the peak frequency deviation is

$$
\Delta F = \frac{D_f V_p}{2\pi} \qquad (5\text{–}69)
$$

This PSD is plotted in Fig. 5–14c. It is recognized that this result is an *approximation* to the actual PSD. In this example, the modulation is periodic with period T_m, so that the actual spectrum is a line spectrum with the delta functions spaced $f_m = 1/T_m$ hertz apart from each other (like the spacing shown in Fig. 5–11 for the case of sinusoidal modulation). This approximation gives us the *approximate envelope* of the line spectrum. (If the modulation had been nonperiodic, the exact spectrum would be continuous and thus would not be a line spectrum.)

In the other example, which uses sinusoidal (periodic) modulation, the exact PSD is known to contain line components with weights $[A_c J_n(\beta)]^2/2$ located at frequencies $f = f_c + nf_m$. (See Fig. 5–11.) For the case of a high index, the envelope of these weights approximates the PDF of a sinusoid as given in Appendix B.

The approximate spectra for *wideband digital PM* signals can also be evaluated by Eq. (5–66). For digital modulation with rectangular pulse shapes, the PDF of the modulation voltage, $m(t)$, consists of delta functions located at the discrete voltage levels. Consequently, the PSD for wideband digital FM is approximated by delta functions. This approximation is illustrated in Fig. 5–15 for wideband binary frequency shift keying (FSK).

In summary, some important properties of angle-modulated signals are as follows:

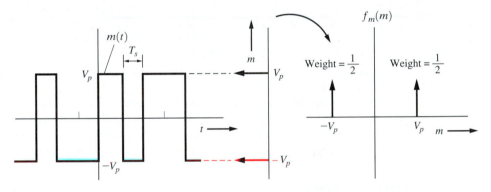

(a) Rectangular Modulating Waveform
(equally likely binary 1's and 0's)

(b) PDF of Digital Modulation

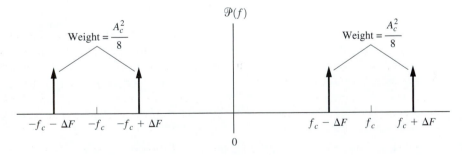

(c) PSD of Wideband FSK with Binary Modulation, $\Delta F = D_f V_p/2\pi$

Figure 5–15   Approximate spectrum of wideband binary FSK.

- An angle-modulated signal is a nonlinear function of the modulation, and consequently, the bandwidth of the signal increases as the modulation index increases.
- The discrete carrier level changes, depending on the modulating signal, and is zero for certain types of modulating waveforms.
- The bandwidth of a narrowband angle-modulated signal is twice the modulating signal bandwidth (the same as that for AM signaling).
- The real envelope of an angle-modulated signal, $R(t) = A_c$, is constant and, consequently, does not depend on the level of the modulating signal.

Preemphasis and Deemphasis in Angle-Modulated Systems

In angle-modulated systems, the signal-to-noise ratio at the output of the receiver can be improved if the level of the modulation (at the transmitter) is boosted at the top end of the (e.g., audio) spectrum—this is called *preemphasis*—and attenuated at high frequencies on the receiver output—called *deemphasis*. This gives an overall baseband frequency response that is flat, while improving the signal-to-noise ratio at the receiver output. (See Fig. 5–16.) In the preemphasis characteristic, the second corner frequency f_2 occurs much above the baseband spectrum of the modulating signal (say, 25 kHz for audio modulation). In FM broadcasting, the time constant τ_1 is usually 75 μs, so that f_1 occurs at 2.12 kHz. The resulting overall system frequency response obtained using preemphasis at the transmitter and deemphasis at the receiver is flat over the band of the modulating signal. In FM broadcasting, with 75-μs preemphasis, the signal that is transmitted is an FM signal for modulating frequencies up to 2.1 kHz, but a *phase-modulated* signal for audio frequencies above 2.1 kHz, because the preemphasis network acts as a differentiator for frequencies between f_1 and f_2. Hence, *preemphasized FM is actually a combination of FM and PM* and combines the advantages of both with respect to noise performance. In Chapter 7, we demonstrated that preemphasis–deemphasis improves the signal-to-noise ratio at the receiver output.

5–7 FREQUENCY-DIVISION MULTIPLEXING AND FM STEREO

Frequency-division multiplexing (FDM) is a technique for transmitting multiple messages simultaneously over a wideband channel by first modulating the message signals onto several subcarriers and forming a composite baseband signal that consists of the sum of these modulated subcarriers. This composite signal may then be modulated onto the main carrier, as shown in Fig. 5–17. Any type of modulation, such as AM, DSB, SSB, PM, FM, and so on, can be used. The types of modulation used on the subcarriers, as well as the type used on the main carrier, may be different. However, as shown in Fig. 5–17b, the composite signal spectrum must consist of modulated signals that do not have overlapping spectra; otherwise, crosstalk will occur between the message signals at the receiver output. The composite baseband signal then modulates a main transmitter to produce the FDM signal that is transmitted over the wideband channel.

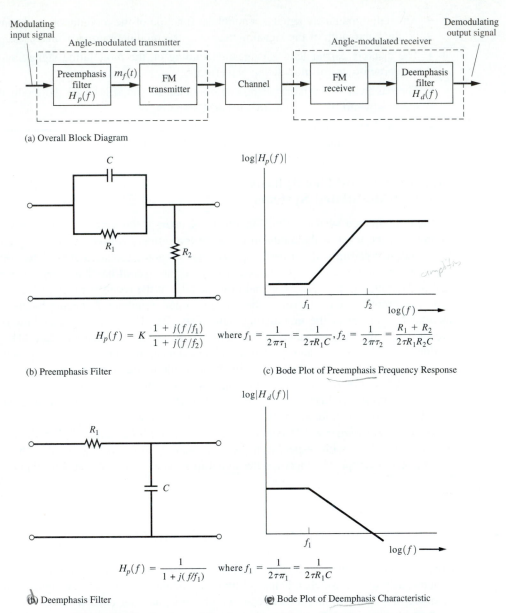

(a) Overall Block Diagram

$$H_p(f) = K \frac{1 + j(f/f_1)}{1 + j(f/f_2)} \quad \text{where } f_1 = \frac{1}{2\pi\tau_1} = \frac{1}{2\pi R_1 C}, f_2 = \frac{1}{2\pi\tau_2} = \frac{R_1 + R_2}{2\pi R_1 R_2 C}$$

(b) Preemphasis Filter   (c) Bode Plot of Preemphasis Frequency Response

$$H_p(f) = \frac{1}{1 + j(f/f_1)} \quad \text{where } f_1 = \frac{1}{2\pi\tau_1} = \frac{1}{2\pi R_1 C}$$

(d) Deemphasis Filter   (e) Bode Plot of Deemphasis Characteristic

Figure 5–16   Angle-modulated system with preemphasis and deemphasis.

The received FDM signal is first demodulated to reproduce the composite baseband signal that is passed through filters to separate the individual modulated subcarriers. Then the subcarriers are demodulated to reproduce the message signals $m_1(t)$, $m_2(t)$, and so on.

The FM stereo broadcasting system that has been adopted in the United States is an example of an FDM system. Furthermore, it is compatible with the monaural FM system that

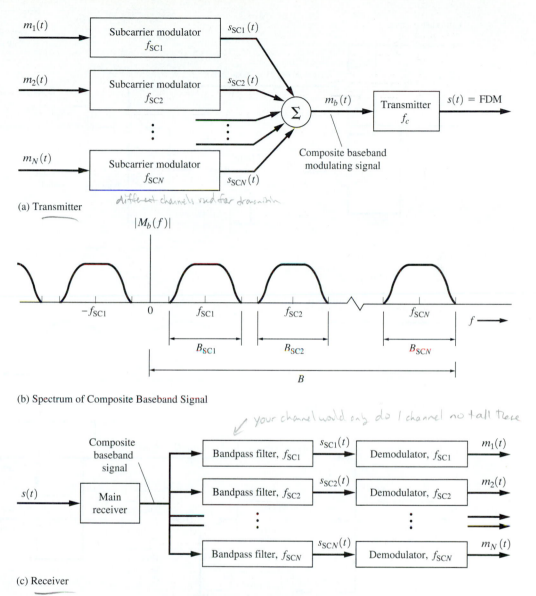

(a) Transmitter   *different channels used for transmission*

(b) Spectrum of Composite Baseband Signal

✓ your channel would only do 1 channel not all there

(c) Receiver

Figure 5–17   FDM system.

has existed since the 1940s. That is, a listener with a conventional monaural FM receiver will hear the monaural audio (which consists of the left- plus the right-channel audio), while a listener with a stereo receiver will receive the left-channel audio on the left speaker and the right-channel audio on the right speaker (Fig. 5–18). To obtain the compatibility feature, the left- and right-channel audios are combined (summed) to produce the monaural signal, and the difference audio is used to modulate a 38-kHz DSB-SC signal. A 19-kHz pilot

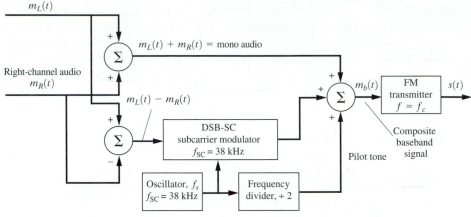

(a) FM Stereo Transmitter

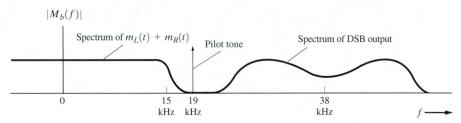

(b) Spectrum of Composite Baseband Signal

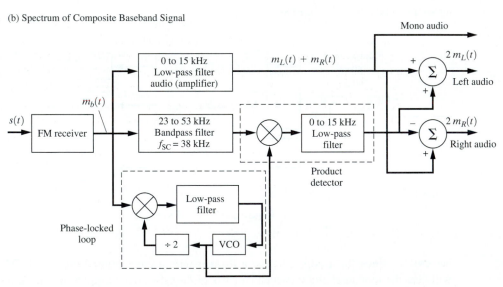

(c) FM Stereo Receiver

Figure 5–18   FM stereo system.

tone is added to the composite baseband signal $m_b(t)$ to provide a reference signal for coherent (product) subcarrier demodulation in the receiver. As seen from Fig. 5–18c, this system is compatible with existing FM monaural receivers. In Prob. 5–44 we will find that a relatively simple switching (sampling) technique may be used to implement the demodulation of the subcarrier and the separation of the left and right signals in one operation.

The FM station may also be given *subsidiary communications authorization* (SCA) by the FCC. This allows the station to add an FM subcarrier to permit the transmission of a second analog-audio program or background music to business subscribers for use in their stores or offices. The SCA FM subcarrier frequency is usually 67 kHz, although this frequency is not specified by FCC rules. Moreover, up to four SCA subcarriers are permitted by the FCC, and each may carry either data or analog audio material. Using a data rate just under 1,200 b/s on a 57-kHz subcarrier, the *radio broadcast data system* (RBDC) provides auxiliary text information—such as station call letters, the titles of programs, the names of musical artists, and auto traffic congestion reports.

5–8 FM BROADCAST TECHNICAL STANDARDS

Table 5–4 gives some of the FCC technical standards that have been adopted for FM systems. In the United States, FM stations are classified into one of three major categories, depending on their intended coverage area. Class A stations are local stations. They have a maximum effective radiated power (ERP) of 6 kW and a maximum antenna height of 300 ft above average terrain. The ERP is the average transmitter output power multiplied by the power gains of both the transmission line (a number less than unity) and the antenna. (See Sec. 8-9 for some TV ERP calculations.) Class B stations have a maximum ERP of 50 kW, with a maximum antenna height of 500 ft above average terrain. Class B stations are assigned to the northeastern part of the United States, southern California, Puerto Rico, and the Virgin Islands. Class C stations are assigned to the remainder of the United States. They have a maximum ERP of 100 kW and a maximum antenna height of 2,000 ft above average terrain. As shown in the table, FM stations are further classified as commercial or noncommercial. Noncommercial stations operate in the 88.1 to 91.9 MHz segment of the FM band and provide educational programs with no commercials. In the commercial segment of the FM band, 92.1 to 107.9 MHz, certain frequencies are reserved for Class A stations, and the remainder are for Class B or Class C station assignment. A listing of these frequencies and the specific station assignments for each city is available [Broadcasting, 1995].

5–9 BINARY MODULATED BANDPASS SIGNALING

Digitally modulated bandpass signals are generated by using the complex envelopes for AM, PM, FM, or QM (quadrature modulation) signaling that were first shown in Table 4–1 and then studied in previous sections. For digital modulated signals, the modulating signal $m(t)$ is a digital signal given by the binary or multilevel line codes that were developed in Chapter 3. In this section, details of binary modulated signals are given. In Sections 5–10 and 5–11, multilevel and minimum-shift–keyed (MSK) digitally modulated signals are described.

Table 5–4   FCC FM STANDARDS

Class of Service	Item	FCC Standard
FM broadcasting	Assigned frequency, f_c	In 200-kHz increments from 88.1 MHz (FM Channel 201) to 107.9 MHz (FM Channel 300)
	Channel bandwidth	200 kHz
	Noncommercial stations	88.1 MHz (Channel 201) to 91.9 MHz (Channel 220)
	Commercial stations	92.1 MHz (Channel 221) to 107.9 MHz (Channel 300)
	Carrier frequency stability	$\pm 2{,}000$ Hz of the assigned frequency
	100% modulation [a]	$\Delta F = 75$ kHz
	Audio frequency response [b]	50 Hz to 15 kHz, following a 75-μs preemphasis curve
	Modulation index	5 (for $\Delta F = 75$ kHz and $B = 15$ kHz)
	% harmonic distortion [b]	$<3.5\%$ (50–100 Hz)
		$<2.5\%$ (100–7500 Hz)
		$<3.0\%$ (7500–15,000 Hz)
	FM noise	At least 60 dB below 100% modulation at 400 Hz
	AM noise	50 dB below the level corresponding to 100% AM in a band 50 Hz–15 kHz
	Maximum power licensed	100 kW in horizontal polarized plane plus 100 kW in vertical polarized plane
Two-way FM mobile radio	100% modulation	$\Delta F = 5$ kHz
	Modulation index	1 (for $\Delta F = 5$ kHz and $B = 5$ kHz)
	Carrier frequencies are within the frequency bands	32–50 MHz (low VHF band)
		144–148 MHz (2-m amateur band)
		148–174 MHz (high VHF band) [c]
		420–450 MHz ($\frac{3}{4}$-m amateur band)
		450–470 MHz (UHF band)
		470–512 MHz (UHF, T band)
		806–928 MHz (900-MHz band)
Analog TV aural (FM) signal	100% modulation	$\Delta F = 25$ kHz
	Modulation index	1.67 (for $\Delta F = 25$ kHz and $B = 15$ kHz)

[a] For stereo transmission, the 19-kHz pilot tone may contribute as much as 10% of the total allowed 75-kHz peak deviation. If SCA is used, each SCA subcarrier may also contribute up to 10%, and the total peak deviation may be 110% of 75 kHz.

[b] Under the new FCC deregulation policy, these requirements are deleted from the FCC rules, although broadcasters still use them as guidelines for minimum acceptable performance.

[c] Amplitude-compressed SSB is also permitted in the 150- to 170-MHz band in 5-kHz bandwidth channels.

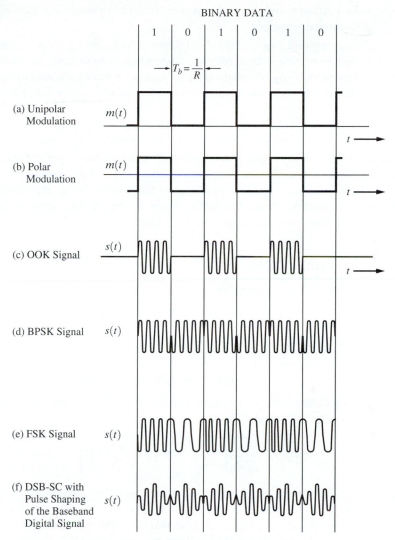

Figure 5–19   Bandpass digitally modulated signals.

The most common binary bandpass signaling techniques, illustrated in Fig. 5–19, are as follows:

- *On–off keying* (OOK), also called *amplitude shift keying* (ASK), which consists of keying (switching) a carrier sinusoid on and off with a unipolar binary signal. OOK is identical to *unipolar* binary modulation on a DSB-SC signal [Eq. (5–13)]. Morse code radio transmission is an example of this technique. OOK was one of the first modulation techniques to be used and precedes analog communication systems.
- *Binary phase-shift keying* (BPSK), which consists of shifting the phase of a sinusoidal carrier $0°$ or $180°$ with a unipolar binary signal. BPSK is equivalent to PM signaling

with a digital waveform and is also equivalent to modulating a DSB-SC signal with a polar digital waveform.

- *Frequency-shift keying* (FSK), which consists of shifting the frequency of a sinusoidal carrier from a mark frequency (corresponding, for example, to sending a binary 1) to a space frequency (corresponding to sending a binary 0), according to the baseband digital signal. FSK is identical to modulating an FM carrier with a binary digital signal.

As indicated in Sec. 3–6, the bandwidth of the digital signal needs to be minimized to achieve spectral conservation. This may be accomplished by using a premodulation raised cosine-rolloff filter to minimize the bandwidth of the digital signal and yet not introduce ISI. The shaping of the baseband digital signal produces an analog baseband waveform that modulates the transmitter. Figure 5–19f illustrates the resulting DSB-SC signal when a premodulation filter is used. Thus, the BPSK signal of Fig. 5–19d becomes a DSB-SC signal (Fig. 5–19f) when premodulation filtering is used.

On–Off Keying (OOK)

The OOK signal is represented by

$$s(t) = A_c m(t) \cos \omega_c t \qquad (5\text{–}70)$$

where $m(t)$ is a unipolar baseband data signal, as shown in Fig. 5–19a. Consequently, the complex envelope is simply

$$g(t) = A_c m(t) \qquad \text{for OOK} \qquad (5\text{–}71)$$

and the PSD of this complex envelope is proportional to that for the unipolar signal. Using Eq. (3–39b), we find that this PSD is

$$\mathcal{P}_g(f) = \frac{A_c^2}{2}\left[\delta(f) + T_b \left(\frac{\sin \pi f T_b}{\pi f T_b} \right)^2 \right] \qquad \text{(for OOK)} \qquad (5\text{–}72)$$

where $m(t)$ has a peak value of $A = \sqrt{2}$ so that $s(t)$ has an average normalized power of $A_c^2/2$. The PSD for the corresponding OOK signal is then obtained by substituting Eq. (5–72) into Eq. (5–2b). The result is shown for positive frequencies in Fig. 5–20a, where $R = 1/T_b$ is the bit rate. The null-to-null bandwidth is $B_T = 2R$, and the absolute bandwidth is $B_T = \infty$. Also, the transmission bandwidth of the OOK signal is $B_T = 2B$, where B is the baseband bandwidth, since OOK is AM-type signaling.

If raised cosine-rolloff filtering is used (to conserve bandwidth), the absolute bandwidth of the filtered binary signal is related to the bit rate R by Eq. (3–74), where $D = R$ for binary digital signaling. Thus, the absolute baseband bandwidth is

$$B = \tfrac{1}{2}(1 + r)R \qquad (5\text{–}73)$$

where r is the rolloff factor of the filter. This gives an absolute transmission bandwidth of

$$B_T = (1 + r)R \qquad (5\text{–}74)$$

for OOK signaling with raised cosine-rolloff filtering.

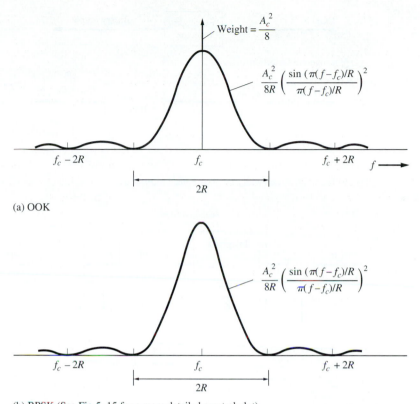

(a) OOK

(b) BPSK (See Fig 5–15 for a more detailed spectral plot)

Figure 5–20   PSD of bandpass digital signals (positive frequencies shown).

OOK may be detected by using either an envelope detector (noncoherent detection) or a product detector (coherent detection), because it is a form of AM signaling. (In radio-frequency receiver applications in which the input RF signal is small, a superheterodyne receiver circuit of Fig. 4–29 is used, and one of the detector circuits is placed after the IF output stage.) These detectors are shown in Figs. 5–21a and 5–21b. For product detection, the carrier reference, $\cos(\omega_c t)$, must be provided. This is usually obtained from a PLL circuit (studied in Sec. 4–14), where the PLL is locked onto a discrete carrier term (see Fig. 5–20a) of the OOK signal.

For optimum detection of OOK—that is, to obtain the lowest BER when the input OOK signal is corrupted by additive white Gaussian noise (AWGN)—product detection with matched filter processing is required. This is shown in Fig. 5–21c, where waveforms at various points of the circuit are illustrated for the case of receiving an OOK signal that corresponds to the binary data stream 1101. Details about the operation, the performance, and the realization of the matched filter are given in Sec. 6–8. Note that the matched filter also requires a clocking signal that is used to reset the integrator at the beginning of each bit interval and to clock the sample-and-hold circuit at the end of each bit interval. This clock signal is provided by a bit synchronizer circuit (studied in Chapter 4).

(a) Noncoherent Detection

(b) Coherent Detection with Low-Pass Filter Processing

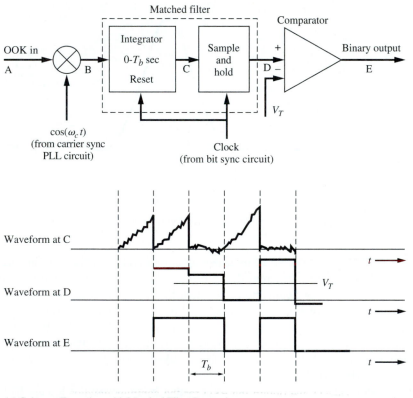

(c) Coherent Detection with Matched Filter Processing

Figure 5–21   Detection of OOK.

The optimum coherent OOK detector of Fig. 5–21c is more costly to implement than the noncoherent OOK detector of Fig. 5–21a. If the input noise is small, the noncoherent receiver may be the best solution, considering both cost and noise performance. The trade-off in BER performance between optimum coherent detection and nonoptimun noncoherent detection is studied in Sec. 7–6.

Binary Phase-Shift Keying (BPSK)

The BPSK signal is represented by

$$s(t) = A_c \cos[\omega_c t + D_p m(t)] \qquad (5\text{--}75a)$$

where $m(t)$ is a polar baseband data signal. For convenience, let $m(t)$ have peak values of ± 1 and a rectangular pulse shape.

We now show that BPSK is also a form of AM-type signaling. Expanding Eq. (5–75a), we get

$$s(t) = A_c \cos(D_p m(t)) \cos \omega_c t - A_c \sin(D_p m(t)) \sin \omega_c t$$

Recalling that $m(t)$ has values of ± 1 and that $\cos(x)$ and $\sin(x)$ are even and odd functions of x, we see that the representation of the BPSK signal reduces to

$$s(t) = \underbrace{(A_c \cos D_p) \cos \omega_c t}_{\text{pilot carrier term}} - \underbrace{(A_c \sin D_p) m(t) \sin \omega_c t}_{\text{data term}} \qquad (5\text{--}75b)$$

The level of the pilot carrier term is set by the value of the peak deviation, $\Delta\theta = D_p$.

For digital angle-modulated signals, the *digital modulation index h* is defined by

$$h = \frac{2\Delta\theta}{\pi} \qquad (5\text{--}76)$$

where $2\Delta\theta$ is the maximum peak-to-peak phase deviation (radians) during the time required to send one symbol, T_s. For binary signaling, the symbol time is equal to the bit time $(T_s = T_b)$.

The level of the pilot carrier term is set by the value of the peak deviation, which is $\Delta\theta = D_p$ for $m(t) = \pm 1$. If D_p is small, the pilot carrier term has a relatively large amplitude compared to the data term; consequently, there is very little power in the data term (which contains the source information). To maximize the signaling efficiency (so that there is a low probability of error), the power in the data term needs to be maximized. This is accomplished by letting $\Delta\theta = D_p = 90° = \pi/2$ radians, which corresponds to a digital modulation index of $h = 1$. For this optimum case of $h = 1$, the BPSK signal becomes

$$s(t) = -A_c m(t) \sin \omega_c t \qquad (5\text{--}77)$$

Throughout the text, we assume that $\Delta\theta = 90°$, $h = 1$, is used for BPSK signaling (unless otherwise stated). Equation (5–77) shows that BPSK is equivalent to DSB-SC signaling with a polar baseband data waveform. The complex envelope for this BPSK signal is

$$g(t) = jA_c m(t) \qquad \text{for BPSK} \qquad (5\text{--}78)$$

Using Eq. (3–41), we obtain the PSD for the complex envelope, viz.,

$$\mathcal{P}_g(f) = A_c^2 \, T_b \left(\frac{\sin \pi f T_b}{\pi f T_b}\right)^2 \qquad \text{(for BPSK)} \qquad (5\text{--}79)$$

where $m(t)$ has values of ± 1, so that $s(t)$ has an average normalized power of $A_c^2/2$. The PSD for the corresponding BPSK signal is readily evaluated by translating the baseband spectrum to the carrier frequency, as specified by substituting Eq. (5–79) into Eq. (5–2b). The resulting BPSK spectrum is shown in Fig. 5–20b. The null-to-null bandwidth for BPSK is also $2R$, the same as that found for OOK.

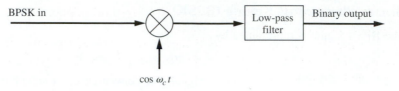

(a) Detection of BPSK (Coherent Detection)

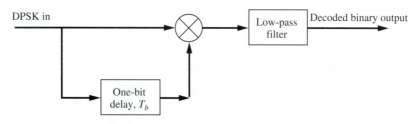

(b) Detection of DPSK (Partially Coherent Detection)

Figure 5–22   Detection of BPSK and DPSK.

To detect BPSK, synchronous detection must be used, as illustrated in Fig. 5–22a. Since there is no discrete carrier term in the BPSK signal, a PLL may be used to extract the carrier reference *only* if a low-level pilot carrier is transmitted together with the BPSK signal. Otherwise, a Costas loop or a squaring loop (Fig. 5–3) may be used to synthesize the carrier reference from this DSB-SC (i.e., BPSK) signal and to provide coherent detection. However, the 180° phase ambiguity must be resolved, as discussed in Sec. 5–4. This can be accomplished by using differential coding at the transmitter input and differential decoding at the receiver output, as illustrated previously in Fig. 3–17.

For optimum detection of BPSK (i.e., the lowest possible BER for the case of AWGN), the low-pass filter in Fig. 5–22a is replaced by an integrate-and-dump matched filter processing that was illustrated in Fig. 5–21c, where V_T is set to 0 V for the case of BPSK. The resulting probability of bit error is given in Sec. 7–3.

Differential Phase-Shift Keying (DPSK)

Phase-shift–keyed signals cannot be detected incoherently. However, a partially coherent technique can be used whereby the phase reference for the present signaling interval is provided by a delayed version of the signal that occurred during the previous signaling interval. This is illustrated by the receiver shown in Fig. 5–22b, where differential decoding is provided by the (one-bit) delay and the multiplier. Consequently, if the data on the BPSK signal are differentially encoded (e.g., see the illustration in Table 3–4), the *decoded* data sequence will be recovered at the output of the receiver. This signaling technique consisting of transmitting a differentially encoded BPSK signal is known as DPSK.

For optimum detection of DPSK the low-pass filter of Fig. 5–22 is replaced by an integrate-and-dump matched filter, and the DPSK input signal needs to be prefiltered by a bandpass filter that has an impulse response of $h(t) = \Pi\left[(t - 0.5T_b)/T_b\right]\cos(\omega_c t)$. (For more details about this optimum receiver, see Fig. 7–12.) The resulting BER is given by Eqs. (7–66) and (7–67).

In practice, DPSK is often used instead of BPSK, because the DPSK receiver does not require a carrier synchronizer circuit. An example is the Bell 212A modem (1,200 bits/s) described in Appendix C, Table C-6.

Frequency-Shift Keying (FSK)

The FSK signal can be characterized as one of two different types, depending on the method used to generate it. One type is generated by switching the transmitter output line between two different oscillators, as shown in Fig. 5–23a. This type generates an output waveform that is discontinuous at the switching times. It is called *discontinuous-phase FSK,* because $\theta(t)$ is discontinuous at the switching times. The discontinuous-phase FSK signal is represented by

$$s(t) = A_c \cos[\omega_c t + \theta(t)] = \begin{cases} A_c \cos(\omega_1 t + \theta_1), & \text{for } t \text{ in the time interval when} \\ & \text{a binary 1 is being sent} \\ A_c \cos(\omega_2 t + \theta_2), & \text{for } t \text{ in the time interval when} \\ & \text{a binary 0 is being sent} \end{cases} \quad (5\text{–}80)$$

where f_1 is called the *mark* (binary 1) frequency and f_2 is called the *space* (binary 0) frequency. θ_1 and θ_2 are the start-up phases of the two oscillators. The discontinuous phase function is

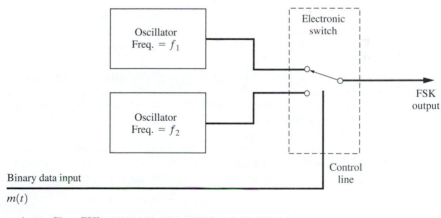

(a) Discontinuous-Phase FSK

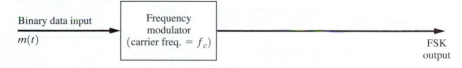

(b) Continuous-Phase FSK

Figure 5–23   Generation of FSK.

$$\theta(t) = \begin{cases} \omega_1 t + \theta_1 - \omega_c t, & \text{for } t \text{ during the binary 1 interval} \\ \omega_2 t + \theta_2 - \omega_c t, & \text{for } t \text{ during the binary 0 interval} \end{cases}$$

Since FSK transmitters are not usually built this way, we will turn to the second type, shown in Fig. 5–23b.

The *continuous-phase FSK* signal is generated by feeding the data signal into a frequency modulator, as shown in Fig. 5–23b. This FSK signal is represented (see Sec. 5–6) by

$$s(t) = A_c \cos\left[\omega_c t + D_f \int_{-\infty}^{t} m(\lambda)\, d\lambda \right]$$

or

$$s(t) = \text{Re}\{g(t) e^{j\omega_c t}\} \tag{5–81a}$$

where

$$g(t) = A_c e^{j\theta(t)} \tag{5–81b}$$

$$\theta(t) = D_f \int_{-\infty}^{t} m(\lambda)\, d\lambda \quad \text{for FSK} \tag{5–81c}$$

and $m(t)$ is a baseband digital signal. Although $m(t)$ is discontinuous at the switching time, the phase function $\theta(t)$ is continuous because $\theta(t)$ is proportional to the integral of $m(t)$. If the serial data input waveform is binary, such as a polar baseband signal, the resulting FSK signal is called a *binary FSK* signal. Of course, a multilevel input signal would produce a multilevel FSK signal. We will assume that the input is a binary signal in this section and examine the properties of binary FSK signals.

In general, the spectra of FSK signals are difficult to evaluate since the complex envelope, $g(t)$, is a nonlinear function of $m(t)$. However, the techniques that were developed in Sec. 5–6 are applicable, as we show in the next example.

Example 5–4 SPECTRUM OF THE BELL-TYPE 103 FSK MODEM

Personal computers are often connected to remote computers via modems and analog dial-up telephone lines. These telephone lines have a passband over the VF range of 300 to 3,300 Hz. Because baseband digital signals (such as polar line code signals) do not have dominant frequencies in this band, they are usually modulated onto a carrier to produce a bandpass signal that has dominant spectral components within the VF range. To accomplish this, modems (i.e., a modulator and a demodulator) are connected to the phone line at each end. (See Fig. 5–24.)

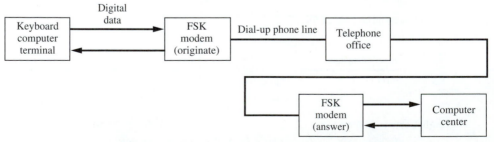

Figure 5–24   Computer communication using FSK signaling.

Popular 28.8-kb/s V.34 modems use QAM signaling as developed in Sec. 5–10 and discussed in Sec. C–8, while 56-kb/s V.90 modems use a PCM technique (see Sec. 3–3). Historically, FSK signaling was used, and it continues to be used for telephone caller ID signaling (Bell 202 standard; see Tables 8–1 and C–5) and in many wireless applications, because it is cost effective.

The evaluation of spectra for FSK signals will be demonstrated by using the 300-b/s Bell System 103 modem standard as an example. Referring to Fig. 5–24, we find that each modem contains both an FSK transmitter and an FSK receiver, so that the computer terminal can both "talk" and "listen." Two FSK frequency bands are used (one around 1 kHz and another around 2 kHz), making it possible to talk and listen simultaneously. This approach is called *full-duplex capability.* (In *half-duplex,* one cannot listen while talking, and vice versa; in *simplex,* one can only talk or only listen.) The standard mark and space frequencies for the two bands are shown in Table 5–5. From this table, it is seen that the peak-to-peak deviation is $2\Delta F = 200$ Hz.

The spectrum for a Bell-type 103 modem will now be evaluated for the case of the widest-bandwidth FSK signal, obtained when the input data signal consists of a deterministic (periodic) square wave corresponding to an alternating data pattern (i.e., 10101010). [†] This is illustrated in Fig. 5–25a, where T_b is the time interval for one bit and $T_0 = 2T_b$ is the period of the data modulation. The spectrum is obtained by using the Fourier series technique developed in Sec. 5–6 and Example 5–2. Since the modulation is periodic, we would expect the FSK spectrum to be a line spectrum (i.e., delta functions). Using Eqs. (5–81c) and (5–42), we find that the peak frequency deviation is $\Delta F = D_f /(2\pi)$ for $m(t)$ having values of ± 1. This results in the triangular phase function shown in Fig. 5–25b. From the figure, the digital modulation index is

$$h = \frac{2\Delta\theta}{\pi} = \Delta FT_0 = \frac{2\Delta F}{R} \tag{5–82}$$

where the bit rate is $R = 1/T_b = 2/T_0$. In this application, note that the digital modulation index, as given by Eq. (5–82), is identical to the FM modulation index defined by Eq. (5–48), since

$$h = \frac{\Delta F}{1/T_0} = \frac{\Delta F}{B} = \beta_f$$

provided that the bandwidth of $m(t)$ is defined as $B = 1/T_0$.

The Fourier series for the complex envelope is

$$g(t) = \sum_{-\infty}^{\infty} c_n \, e^{jn\omega_0 t} \tag{5–83}$$

where $f_0 = 1/T_0 = R/2$,

$$c_n = \frac{A_c}{T_0} \int_{-T_0/2}^{T_0/2} e^{j\theta(t)} e^{-jn\omega_0 t} \, dt$$

$$= \frac{A_c}{T_0} \left[\int_{-T_0/4}^{T_0/4} e^{j\Delta\omega t - jn\omega_0 t} \, dt + \int_{T_0/4}^{3T_0/4} e^{-j\Delta\omega(t-(T_0/2))} e^{-jn\omega_0 t} \, dt \right] \tag{5–84}$$

[†] For the case of random data, the PSD for $g(t)$ is given by the $h = 0.7 \approx 0.67$ curve of Fig. 5–27.

Table 5–5   MARK AND SPACE FREQUENCIES FOR THE BELL-TYPE 103 MODEM

	Originate Modem (Hz)	Answer Modem (Hz)
Transmit frequencies		
Mark (binary 1)	$f_1 = 1{,}270$	$f_1 = 2{,}225$
Space (binary 0)	$f_2 = 1{,}070$	$f_2 = 2{,}025$
Receive frequencies		
Mark (binary 1)	$f_1 = 2{,}225$	$f_1 = 1{,}270$
Space (binary 0)	$f_2 = 2{,}025$	$f_2 = 1{,}070$

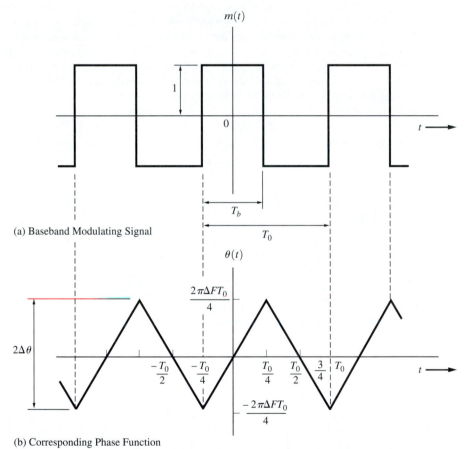

(a) Baseband Modulating Signal

(b) Corresponding Phase Function

Figure 5–25   Input data signal and FSK signal phase function.

and $\Delta\omega = 2\pi\Delta F = 2\pi h/T_0$. Equation (5–84) reduces to

$$c_n = \frac{A_c}{2}\left[\left(\frac{\sin\left[(\pi/2)(h-n)\right]}{(\pi/2)(h-n)}\right) + (-1)^n\left(\frac{\sin\left[(\pi/2)(h+n)\right]}{(\pi/2)(h+n)}\right)\right] \qquad (5\text{--}85)$$

where the digital modulation index is $h = 2\Delta F/R$, in which $2\Delta F$ is the peak-to-peak frequency shift and R is the bit rate. Using Eqs. (5–49) and (5–59), we see that the spectrum of this FSK signal with alternating data is

$$S(f) = \tfrac{1}{2}\left[G(f - f_c) + G^*(-f - f_c)\right] \tag{5–86a}$$

where

$$G(f) = \sum_{-\infty}^{\infty} c_n\, \delta(f - nf_0) = \sum_{-\infty}^{\infty} c_n\, \delta\!\left(f - \frac{nR}{2}\right) \tag{5–86b}$$

and c_n is given by Eq. (5–85).

FSK spectra can be evaluated easily for cases of different frequency shifts ΔF and bit rates R if a personal computer is used. A summary of three computer runs using different sets of parameters is shown in Fig. 5–26. Figure 5–26a gives the FSK spectrum for the Bell 103. For this case, where the Bell 103 parameters are used, the digital modulation index is $h = 0.67$ and there are no spectral lines at the mark and space frequencies, f_1 and f_2, respectively. Figures 5–26b and 5–26c give the FSK spectra for $h = 1.82$ and $h = 3.33$. Note that as the modulation index is increased, the spectrum concentrates about f_1 and f_2. This is the same result predicted by the PSD theorem for wideband FM [Eq. (5–66)], since the PDF of the binary modulation consists of two delta functions. (See Fig. 5–15).

The approximate transmission bandwidth B_T for the FSK signal is given by Carson's rule, $B_T = 2(\beta + 1)B$, where $\beta = \Delta F/B$. This is equivalent to

$$B_T = 2\Delta F + 2B \tag{5–87}$$

where B is the bandwidth of the digital (e.g., square-wave) modulation waveform. In our example of an alternating binary 1 and 0 test pattern waveform, the bandwidth of this square-wave modulating waveform (assuming that the first null type of bandwidth is used) is $B = R$, and, using Eq. (5–87), we find that the FSK transmission bandwidth becomes

$$B_T = 2(\Delta F + R) \tag{5–88}$$

This result is illustrated in Fig. 5–26. If a raised cosine-rolloff premodulation filter is used, the transmission bandwidth of the FSK signal becomes

$$B_T = 2\Delta F + (1 + r)R \tag{5–89}$$

For wideband FSK, where $\beta \gg 1$, ΔF dominates in these equations, and we have $B_T = 2\Delta F$. For narrowband FSK, the transmission bandwidth is $B_T = 2B$.

The exact PSD for continuous-phase FSK signals is difficult to evaluate for the case of random data modulation. However, it can be done by using some elegant statistical techniques [Proakis, 1995, pp. 209–215; Anderson and Salz, 1965; Bennett and Rice, 1963]. The resulting PSD for the complex envelope of the FSK signal is[†]

$$\mathcal{P}_g(f) = \frac{A_c^2 T_b}{2}$$

$$\times \left(A_1^2(f)[1 + B_{11}(f)] + A_2^2(f)[1 + B_{22}(f)] + 2B_{12}(f)A_1(f)A_2(f)\right) \tag{5–90a}$$

[†] It is assumed that $h \neq 0, 1, 2, \ldots$. When $h = 0, 1, 2, \ldots$, there are also discrete terms (delta functions) in the spectrum.

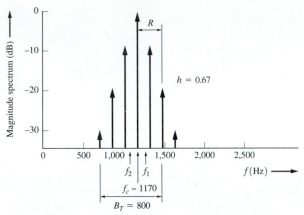

(a) FSK Spectrum with f_2 = 1,070 Hz, f_1 = 1,270 Hz, and R = 300 bits/sec
(Bell 103 Parameters, Originate mode) for h = 0.67

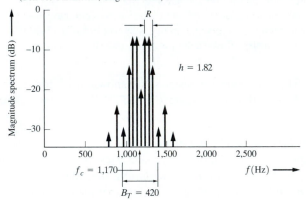

(b) FSK Spectrum with f_2 = 1,070 Hz, f_1 = 1,270 Hz, and R = 110 bits/sec
for h = 1.82

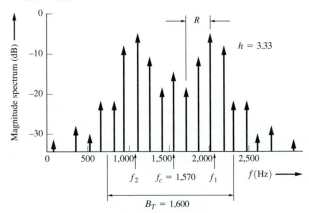

(c) FSK Spectrum with f_2 = 1,070 Hz, f_1 = 2,070 Hz, and R = 300 bits/sec
for h = 3.33

Figure 5–26   FSK spectra for alternating data modulation (positive frequencies shown with one-sided magnitude values).

where

$$A_n(f) = \frac{\sin[\pi T_b(f - \Delta F(2n - 3))]}{\pi T_b(f - \Delta F(2n - 3))} \tag{5–90b}$$

and

$$B_{nm}(f) =$$

$$\frac{\cos[2\pi f T_b - 2\pi\Delta F T_b(n + m - 3)] - \cos(2\pi\Delta F T_b)\cos[2\pi\Delta F T_b(n + m - 3)]}{1 + \cos^2(2\pi\Delta F T_b) - 2\cos(2\pi\Delta F T_b)\cos(2\pi f T_b)} \tag{5–90c}$$

in which ΔF is the peak frequency deviation, $R = 1/T_b$ is the bit rate, and the digital modulation index is $h = 2\Delta F/R$. The PSD is plotted in Fig. 5–27 for several values of the digital modulation index. The curves in this figure were obtained by using the MATLAB computation. The curve for $h = 0.7 \approx 0.67$ corresponds to the PSD of $g(t)$ for the 300-bits/s Bell 103 FSK modem of Example 5–4.

FSK can be detected by using either a frequency (noncoherent) detector or two product detectors (coherent detection), as shown in Fig. 5–28. A detailed study of coherent and noncoherent detection of FSK is given in Sections 7–3 and 7–4, where the BER is evaluated. To obtain the lowest BER when the FSK signal is corrupted by AWGN, coherent detection with matched filter processing and a threshold device (comparator) is required. (See Fig. 7–8.)

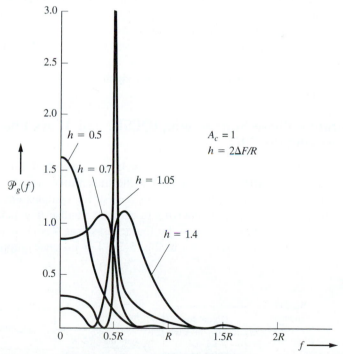

Figure 5–27   PSD for the complex envelope of FSK (positive frequencies shown).

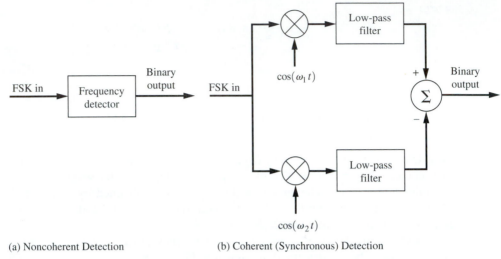

(a) Noncoherent Detection   (b) Coherent (Synchronous) Detection

Figure 5–28   Detection of FSK.

5–10 MULTILEVEL MODULATED BANDPASS SIGNALING

With multilevel signaling, digital inputs with more than two levels are allowed on the transmitter input. This technique is illustrated in Fig. 5–29, which shows how multilevel signals can be generated from a serial binary input stream by using a digital-to-analog converter (DAC). For example, suppose that an $\ell = 2$-bit DAC is used. Then the number of levels in the multilevel signal is $M = 2^\ell = 2^2 = 4$, as illustrated in Fig. 3–14a for rectangular pulses. The symbol rate (baud) of the multilevel signal is $D = R/\ell = \frac{1}{2}R$, where the bit rate is $R = 1/T_b$ bits/s.

Quadrature Phase-Shift Keying (QPSK) and *M*-ary Phase-Shift Keying (MPSK)

If the transmitter is a PM transmitter with an $M = 4$-level digital modulation signal, *M-ary phase-shift keying* (MPSK) is generated at the transmitter output. Assuming rectangular-shaped data pulses, a plot of the permitted values of the complex envelope, $g(t) = A_c e^{j\theta(t)}$, would contain four points, one value of g (a complex number in general) for each of the four multilevel values, corresponding to the four phases that θ is permitted to have. A plot of two possible sets of $g(t)$ is shown in Fig. 5–30. For instance, suppose that the permitted

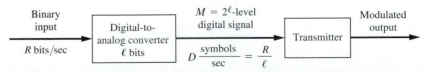

Figure 5–29   Multilevel digital transmission system.

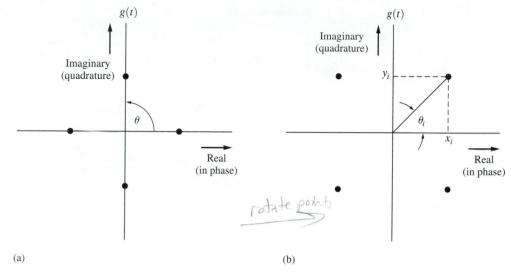

Figure 5–30   QPSK and $\pi/4$ QPSK signal constellations (permitted values of the complex envelope).

multilevel values at the DAC are -3, -1, $+1$, and $+3$ V; then, in Fig. 5–30a, these multi-level values might correspond to PSK phases of 0, 90, 180, and 270°, respectively. In Fig. 5–30b, those levels would correspond to carrier phases of 45, 135, 225, and 315°, respec-tively. These two signal constellations are essentially the same, except for a shift in the car-rier-phase reference.[†] This example of *M*-ary PSK where $M = 4$ is called *quadrature phase-shift-keyed* (QPSK) signaling.

MPSK can also be generated by using two quadrature carriers modulated by the x and y components of the complex envelope (instead of using a phase modulator); in that case,

$$g(t) = A_c e^{j\theta(t)} = x(t) + jy(t) \tag{5–91}$$

where the permitted values of x and y are

$$x_i = A_c \cos \theta_i \tag{5–92}$$

and

$$y_i = A_c \sin \theta_i \tag{5–93}$$

for the permitted phase angles θ_i, $i = 1, 2, \ldots, M$, of the MPSK signal. This situation is il-lustrated in Fig. 5–31, where the signal processing circuit implements Eqs. (5–92) and (5–93). Figure 5–30 gives the relationship between the permitted phase angles θ_i and the (x_i, y_i) components for two QPSK signal constellations. This is identical to the quadrature method of generating modulated signals presented in Fig. 4–28.

[†] A constellation is an *N*-dimensional plot of the possible signal vectors corresponding to the possible dig-ital signals. (See Sec. 3–4.)

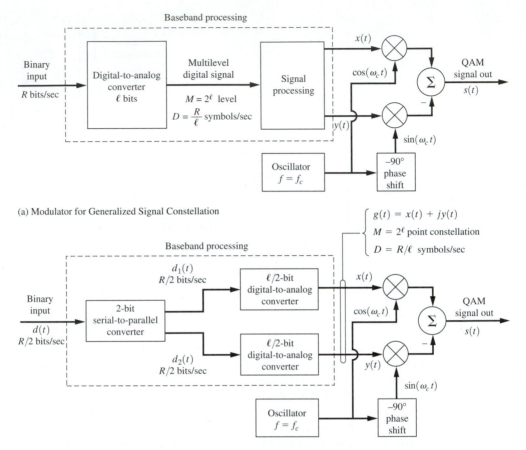

(a) Modulator for Generalized Signal Constellation

$$\begin{cases} g(t) = x(t) + jy(t) \\ M = 2^\ell \text{ point constellation} \\ D = R/\ell \text{ symbols/sec} \end{cases}$$

(b) Modulator for Rectangular Signal Constellation

Figure 5–31   Generation of QAM signals.

For rectangular-shaped data pulses, the envelope of the QPSK signal is constant. That is, there is no AM on the signal, even during the data transition times, when there is a 180° phase shift, since the data switches values (say, from +1 to −1) instantaneously. The rectangular-shaped data produce a $(\sin x/x)^2$-type power spectrum for the QPSK signal that has large undesirable spectral sidelobes. (See Fig. 5–33.) These undesirable sidelobes can be eliminated if the data pulses are filtered to a pulse shape corresponding to a raised cosine-rolloff filter. Unfortunately, this produces AM on the resulting QPSK signal, because the filtered data waveform cannot change instantaneously from one peak to another when 180° phase transitions occur. Although filtering solves the problem of poor spectral sidelobes, it creates another one: AM on the QPSK signal. Due to this AM, low-efficiency linear (Class A or Class B) amplifiers, instead of high-efficiency nonlinear (Class C) amplifiers, are required for amplifying the QPSK signal without distortion. In portable communication applications, these amplifiers increase the battery capacity requirements by as much as 50%. A possible solution to the dilemma is to use offset QPSK (OQPSK) or $\pi/4$ QPSK, each of which has a lower amount of AM. (OQPSK and $\pi/4$ QPSK are described after the next section.)

Quadrature Amplitude Modulation (QAM)

Quadrature carrier signaling, as shown in Fig. 5–31, is called *quadrature amplitude modulation* (QAM). In general, QAM signal constellations are not restricted to having permitted signaling points only on a circle (of radius A_c, as was the case for MPSK). The general QAM signal is

$$s(t) = x(t) \cos \omega_c t - y(t) \sin \omega_c t \qquad (5\text{–}94)$$

where

$$g(t) = x(t) + jy(t) = R(t)e^{j\theta(t)} \qquad (5\text{–}95)$$

For example, a popular 16-symbol ($M = 16$ levels) QAM constellation is shown in Fig. 5–32, where the relationship between (R_i, θ_i) and (x_i, y_i) can readily be evaluated for each of the 16 signal values permitted. This type of signaling is used by 2,400-bit/s V.22 bis computer modems. (See Table C–6.) Here, x_i and y_i are each permitted to have four levels per dimension. This 16-symbol QAM signal may be generated by using two ($\ell/2 = 2$)-bit digital-to-analog converters and quadrature-balanced modulators as shown in Fig. 5–31b. The waveforms of I and Q components are represented by

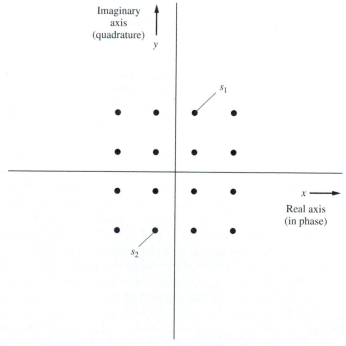

Figure 5–32   16-symbol QAM constellation (four levels per dimension).

$$x(t) = \sum_n x_n h_1 \left(t - \frac{n}{D} \right) \qquad (5\text{--}96)$$

and

$$y(t) = \sum_n y_n h_1 \left(t - \frac{n}{D} \right) \qquad (5\text{--}97)$$

where $D = R/\ell$ and (x_n, y_n) denotes one of the permitted (x_i, y_i) values during the symbol time that is centered on $t = nT_s = n/D$ s. (It takes T_s s to send each symbol.) $h_1(t)$ is the pulse shape that is used for each symbol. If the bandwidth of the QAM signal is not to be restricted, the pulse shape will be rectangular and of T_s s duration. In some applications, the timing between the $x(t)$ and $y(t)$ components is offset by $T_s/2 = 1/(2D)$ s. That is, $x(t)$ would be described by Eq. (5–96), and the offset would be described by

$$y(t) = \sum_n y_n h_1 \left(t - \frac{n}{D} - \frac{1}{2D} \right) \qquad (5\text{--}98)$$

One popular type of offset signaling is *offset QPSK (OQPSK)*, which is identical to offset QAM when $M = 4$. A special case of OQPSK when $h_1(t)$ has a sinusoidal type of pulse shape is called *minimum-shift keying* (MSK). This type of signaling will be studied in a later section. Furthermore, a QPSK signal is said to be *unbalanced* if the $x(t)$ and $y(t)$ components have unequal powers or unequal data rates (or both).

OQPSK and $\pi/4$ QPSK

Offset quadrature phase-shift keying (OQPSK) is $M = 4$ PSK in which the allowed data transition times for the I and Q components are offset by a $\frac{1}{2}$ symbol (i.e., by 1 bit) interval. The technique is described by Eqs. (5–96) and (5–98), where the offset is $1/(2D)$. This offset provides an advantage when nonrectangular (i.e., filtered) data pulses are used, because the offset greatly reduces the AM on the OQPSK signal compared with the AM on the corresponding QPSK signal. The AM is reduced because a maximum phase transition of only $90°$ occurs for OQPSK signaling (as opposed to $180°$ for QPSK), since the I and Q data cannot change simultaneously, because the data are offset.

A $\pi/4$ *quadrature phase-shift keying* ($\pi/4$ QPSK) signal is generated by alternating between two QPSK constellations that are rotated by $\pi/4 = 45°$ with respect to each other. The two QPSK constellations, shown in Fig. 5–30, are used alternately as follows: Given a point on one of the signal constellations that corresponds to two bits of input data, two new bits are read to determine the next point that is selected from the *other* constellation. That is, the two new input bits cause a phase shift of $\pm45°$ or $\pm135°$, depending on their value. For example, an 11 could correspond to a phase shift of $\Delta\theta = 45°$, a 10 to $\Delta\theta = -45°$, a 01 to $\Delta\theta = 135°$, and a 00 to $\Delta\theta = -135°$. Since this uses a form of differential encoding, it is called $\pi/4$ *differential quadrature phase-shift keying* ($\pi/4$ DQPSK).

At the receiver, the data on the $\pi/4$ QPSK signal can be easily detected by using an FM detector, followed by a resettable integrator that integrates over a symbol (2-bit) period. The FM detector produces the derivative of the phase, and the integrator evaluates the phase shift that occurs over the symbol interval. The result is one of the four possible phase shifts, $\pm45°$ and $\pm135°$. For example, if the detected phase shift at the integrator output is $-45°$, the corresponding detected data are 10. Data on the $\pi/4$ QPSK signal can also be detected by using baseband IQ processing or by using differential detection at the IF [Rappaport, 1996]. Computer simulations indicate that all three of these receiver structures have almost the same BER error performance [Anvari and Woo, 1991]. In an AWGN channel, the BER performance of the three differential (noncoherent) detectors is about 3 dB inferior to that of QPSK, but coherently detected $\pi/4$ QPSK has the same BER as QPSK (shown in Fig. 7–14). For the case of nonrectangular data pulses, the AM on $\pi/4$ QPSK is less than that on QPSK, since the maximum phase shift for $\pi/4$ QPSK is 135°, compared with 180° for QPSK. But the AM on OQPSK is even less as it has a maximum phase shift of 90°. However, $\pi/4$ QPSK is easy to detect and has been adopted for use in TDMA cellular telephone systems. (See Chapter 8.)

The power spectra for these signals are described in the next section.

PSD for MPSK, QAM, QPSK, OQPSK, and $\pi/4$ QPSK

The PSD for MPSK and QAM signals is relatively easy to evaluate for the case of rectangular bit-shape signaling. In this case, the PSD has the same spectral shape that was obtained for BPSK, *provided* that proper frequency scaling is used.

The PSD for the complex envelope, $g(t)$, of the MPSK or QAM signal can be obtained by using Eq. (6–70d). We know that

$$g(t) = \sum_{-\infty}^{\infty} c_n f(t - nT_s) \tag{5–99}$$

where c_n is a complex-valued random variable representing the multilevel value during the nth symbol pulse. $f(t) = \Pi(t/T_s)$ is the rectangular symbol pulse with symbol duration T_s. $D = 1/T_s$ is the symbol (or baud) rate. The rectangular pulse has the Fourier transform

$$F(f) = T_s\left(\frac{\sin \pi f T_s}{\pi f T_s}\right) = \ell T_b\left(\frac{\sin \ell \, \pi f T_b}{\ell \, \pi f T_b}\right) \tag{5–100}$$

where $T_s = \ell T_b$. That is, there are ℓ bits representing each allowed multilevel value. For symmetrical (polar type) signaling—for example, as illustrated in Fig. 5–32 for the case of $M = 16$—with equally likely multilevels, the mean value of c_n is

$$m_c = \overline{c_n} = 0 \tag{5–101a}$$

and the variance is

$$\sigma_c^2 = \overline{c_n c_n^*} = \overline{|c_n|^2} = C \tag{5–101b}$$

where C is a real positive constant. Substituting Eqs. (5–100) and (5–101) into Eq. (6–70d), we find that the PSD for the complex envelope of MPSK or QAM signals with data modulation of rectangular bit shape is

$$\mathscr{P}_g(f) = K\left(\frac{\sin \pi f \ell T_b}{\pi f \ell T_b}\right)^2, \quad \text{for MPSK and QAM} \tag{5–102}$$

where $K = C\ell T_b$, $M = 2^\ell$ is the number of points in the signal constellation, and the bit rate is $R = 1/T_b$. For a total transmitted power of P watts, the value of K is $K = 2P\ell T_b$, since

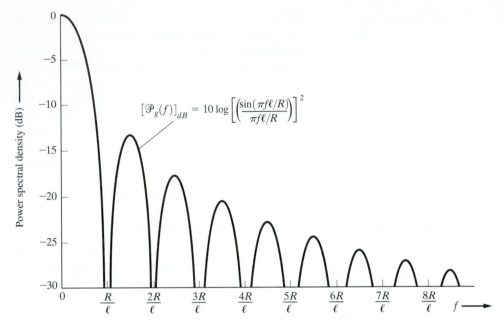

Figure 5–33 PSD for the complex envelope of MPSK and QAM with rectangular data pulses, where $M = 2^\ell$, R is the bit rate, and $R/\ell = D$ is the baud rate (positive frequencies shown). Use $\ell = 2$ for PSD of QPSK, OQPSK, and $\pi/4$ QPSK complex envelope.

$\int_{-\infty}^{\infty} \mathcal{P}_s(f)\, df = P$. This PSD for the complex envelope is plotted in Fig. 5–33. The PSD of the MPSK or QAM signal is obtained by simply translating the PSD of Fig. 5–33 to the carrier frequency, as described by Eq. (5–2b). For $\ell = 1$, the figure gives the PSD for BPSK (i.e., compare Fig. 5–33, $\ell = 1$, with Fig. 5–20b). It is also realized that the PSD for the complex envelope of bandpass multilevel signals, as described by Eq. (5–102), is essentially the same as the PSD for baseband multilevel signals that was obtained in Eq. (3–53).

Equation (5–102) and Fig. 5–33 also describe the PSD for QPSK, OQPSK, and $\pi/4$ QPSK for rectangular-shaped data pulses when $\ell = 2(M = 4)$ is used. For signaling with nonrectangular data pulses, the PSD formula can also be obtained by following the same procedure that gave Eq. (5–102), provided that the appropriate pulse transfer function is used to replace Eq. (5–100). For example, for raised cosine-rolloff filtering, where $f_0 = 1/(2\ell T_b)$, the PSD of Fig. 5–33 would become $\mathcal{P}_g(f) = 20 \log[|H_e(f)|]$, where $H_e(f)$ is the raised cosine transfer function of Eq. (3–69) and Fig. 3–26a.

From Fig. 5–33, we see that the null-to-null transmission bandwidth of MPSK or QAM is

$$B_T = 2R/\ell \qquad\qquad (5\text{--}103)$$

when rectangular data pulses are used.

Substituting Eq. (5–103) into Eq. (3–55), we find that the spectral efficiency of MPSK or QAM signaling with rectangular data pulses is

$$\eta = \frac{R}{B_T} = \frac{\ell}{2} \ \frac{\text{bits/s}}{\text{Hz}} \qquad (5\text{--}104)$$

where $M = 2^{\ell}$ is the number of points in the signal constellation. For an $M = 16$ QAM signal, the bandwidth efficiency is $\eta = 2$ bits/s per hertz of bandwidth.

Spectral Efficiency for MPSK, QAM, QPSK, OQPSK, and $\pi/4$ QPSK with Raised Cosine Filtering

The spectrum shown in Fig. 5–33 was obtained for the case of rectangular symbol pulses, and the spectral sidelobes are terrible. The first sidelobe is attenuated by only 13.4 dB. The sidelobes can be eliminated if raised cosine filtering is used (since the raised cosine filter has an absolutely bandlimited frequency response). Referring to Sec. 3–6, we select the 6-dB bandwidth of the raised cosine filter to be half the symbol (baud) rate in order for there to be no ISI. That is, $f_0 = \frac{1}{2}(R/\ell)$. The raised cosine filter has the disadvantage of introducing AM on MPSK signals (and modifying the AM on QAM signals). In practice, a square-root raised cosine (SRRC frequency response characteristic is often used at the transmitter, along with another SRRC filter at the receiver, in order to simultaneously prevent ISI on the received filtered pulses and to minimize the bit errors due to channel noise. However, the SRRC filter also introduces AM on the transmitted signal. If the overall pulse shape satisfies the raised cosine-rolloff filter characteristic, then, by the use of Eq. (3–74), the absolute bandwidth of the M-level modulating signal is

$$B = \tfrac{1}{2}(1 + r)D \qquad (5\text{--}105)$$

where $D = R/\ell$ and r is the rolloff factor of the filter characteristic. Furthermore, from our study of AM (and DSB-SC, to be specific), we know that the transmission bandwidth is related to the modulation bandwidth by $B_T = 2B$, so that the overall absolute transmission bandwidth of the QAM signal with raised cosine filtered pulses is

$$B_T = \left(\frac{1 + r}{\ell}\right)R \qquad (5\text{--}106)$$

This compares to an absolute bandwidth of infinity (and a null bandwidth of $B_T = 2R/\ell$) for the case of QAM with rectangular data pulses (as shown in Fig. 5–33).

Because $M = 2^{\ell}$, which implies that $\ell = \log_2 M = (\ln M)/(\ln 2)$, the spectral efficiency of QAM-type signaling with raised cosine filtering is

$$\eta = \frac{R}{B_T} = \frac{\ln M}{(1 + r)\ln 2} \ \frac{\text{bit/s}}{\text{Hz}} \qquad (5\text{--}107)$$

This result is important because it tells us how fast we can signal for a prescribed bandwidth. The result also holds for MPSK, since it is a special case of QAM. Equation (5–107) is used to generate Table 5–6, which illustrates the allowable bit rate per hertz of transmission bandwidth for QAM signaling. For example, suppose that we want to signal over a communications satellite that has an available bandwidth of 2.4 MHz. If we used BPSK ($M = 2$) with a 50% rolloff factor, we could signal at a rate of $B_T \times \eta = 2.4 \times 0.677 = 1.60$ Mbits/s; but if we used QPSK ($M = 4$) with a 25% rolloff factor, we could signal at a rate of $2.4 \times 1.6 = 3.84$ Mbits/s.

Table 5–6 SPECTRAL EFFICIENCY FOR QAM SIGNALING WITH RAISED COSINE-ROLLOFF
PULSE SHAPING (Use $M = 4$ for QPSK, OQPSK, and $\pi/4$ QPSK signaling)

Number of Levels, M (symbols)	Size of DAC, ℓ (bits)	$\eta = \dfrac{R}{B_T}\left(\dfrac{\text{bit/s}}{\text{Hz}}\right)$					
		$r = 0.0$	$r = 0.1$	$r = 0.25$	$r = 0.5$	$r = 0.75$	$r = 1.0$
2	1	1.00	0.909	0.800	0.667	0.571	0.500
4	2	2.00	1.82	1.60	1.33	1.14	1.00
8	3	3.00	2.73	2.40	2.00	1.71	1.50
16	4	4.00	3.64	3.20	2.67	2.29	2.00
32	5	5.00	4.55	4.0	3.33	2.86	2.50

To conserve bandwidth, the number of levels M in Eq. (5–107) cannot be increased too much, since, for a given peak envelope power (PEP), the spacing between the signal points in the signal constellation will decrease and noise on the received signal will cause errors. (Noise moves the received signal vector to a new location that might correspond to a different signal level.) However, we know that R certainly has to be less than C, the channel capacity (Sec. 1–9), if the errors are to be kept small. Consequently, using Eq. (1–10), we require that

$$\eta < \eta_{max} \qquad (5\text{–}108\text{a})$$

where

$$\eta_{max} = \log_2\left(1 + \frac{S}{N}\right) \qquad (5\text{–}108\text{b})$$

5–11 MINIMUM-SHIFT KEYING (MSK) AND GMSK

Minimum-shift keying is another bandwidth conservation technique that has been developed. It has the advantage of producing a constant-amplitude signal and, consequently, can be amplified with Class C amplifiers without distortion. As we will see, MSK is equivalent to OQPSK with sinusoidal pulse shaping [for $h_i(t)$].

DEFINITION. *Minimum-shift keying* (MSK) is continuous-phase FSK with a minimum modulation index ($h = 0.5$) that will produce orthogonal signaling.

First, let us show that $h = 0.5$ is the minimum index allowed for orthogonal continuous-phase FSK. For a binary 1 to be transmitted over the bit interval $0 < t < T_b$, the FSK signal would be $s_1(t) = A_c \cos(\omega_1 t + \theta_1)$, and for a binary 0 to be transmitted, the FSK signal would be $s_2(t) = A_c \cos(\omega_2 t + \theta_2)$, where $\theta_1 = \theta_2$ for the continuous-phase condition at the switching time $t = 0$. For orthogonal signaling, from Eq. (2–77), we require the integral of the product of the two signals over the bit interval to be zero. Thus, we require that

$$\int_0^{T_b} s_1(t)s_2(t)\,dt = \int_0^{T_b} A_c^2 \cos(\omega_1 t + \theta_1)\cos(\omega_2 t + \theta_2)\,dt = 0 \qquad (5\text{–}109a)$$

This reduces to the requirement that

$$\frac{A_c^2}{2}\left[\frac{\sin[(\omega_1 - \omega_2)T_b + (\theta_1 - \theta_2)] - \sin(\theta_1 - \theta_2)}{\omega_1 - \omega_2}\right]$$

$$+ \frac{A_c^2}{2}\left[\frac{\sin[(\omega_1 + \omega_2)T_b + (\theta_1 + \theta_2)] - \sin(\theta_1 + \theta_2)}{\omega_1 + \omega_2}\right] = 0 \qquad (5\text{–}109b)$$

The second term is negligible, because $\omega_1 + \omega_2$ is large,[†] so the requirement is that

$$\frac{\sin[2\pi h + (\theta_1 - \theta_2)] - \sin(\theta_1 - \theta_2)}{2\pi h} = 0 \qquad (5\text{–}110)$$

where $(\omega_1 - \omega_2)T_b = 2\pi(2\Delta F)T_b$ and, from Eq. (5–82), $h = 2\Delta F T_b$. For the *continuous-phase* case, $\theta_1 = \theta_2$; and Eq. (5–110) is satisfied for a minimum value of $h = 0.5$, or a peak frequency deviation of

$$\Delta F = \frac{1}{4T_b} = \frac{1}{4}R, \quad \text{for MSK} \qquad (5\text{–}111a)$$

For *discontinuous-phase* FSK, $\theta_1 \neq \theta_2$; and the minimum value for orthogonality is $h = 1.0$ or

$$\Delta F = \frac{1}{2T_b} = \frac{1}{2}R, \quad \text{for discontinuous-phase, FSK} \qquad (5\text{–}111b)$$

Now we will demonstrate that the MSK signal (which is $h = 0.5$ continuous-phase FSK) is a form of OQPSK signaling with sinusoidal pulse shaping. First, consider the FSK signal over the signaling interval $(0, T_b)$. When we use Eq. (5–81), the complex envelope is

$$g(t) = A_c e^{j\theta(t)} = A_c e^{j2\pi\Delta F \int_0^t m(\lambda)d\lambda}$$

where $m(t) = \pm 1,\ 0 < t < T_b$. Using Eq. (5–111), we find that the complex envelope becomes

$$g(t) = A_c e^{\pm j\pi t/(2T_b)} = x(t) + jy(t), \quad 0 < t < T_b$$

where the $\pm$ signs denote the possible data during the $(0, T_b)$ interval. Thus,

$$x(t) = A_c \cos\left(\pm 1\ \frac{\pi t}{2T_b}\right), \quad 0 < t < T_b \qquad (5\text{–}112a)$$

$$y(t) = A_c \sin\left(\pm 1\ \frac{\pi t}{2T_b}\right), \quad 0 < t < T_b \qquad (5\text{–}112b)$$

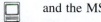

and the MSK signal is

$$s(t) = x(t)\cos\omega_c t - y(t)\sin\omega_c t \qquad (5\text{–}112c)$$

[†] If $\omega_1 + \omega_2$ is not sufficiently large to make the second term negligible, choose $f_c = \frac{1}{2}m/T_b = \frac{1}{2}mR$, where m is a positive integer. This will make the second term zero ($f_1 = f_c - \Delta F$ and $f_2 = f_c + \Delta F$).

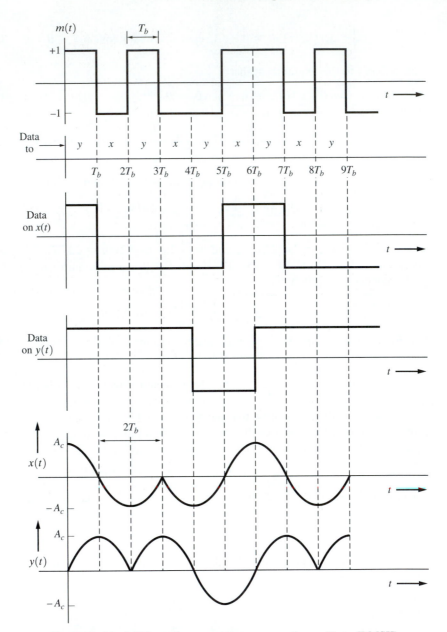

Figure 5–34   MSK quadrature component waveforms (Type II MSK).

A typical input data waveform $m(t)$ and the resulting $x(t)$ and $y(t)$ quadrature modulation waveforms are shown in Fig. 5–34. From Eqs. (5–112a) and (5–112b), and realizing that $\cos[\pm\pi t/(2T_b)] = \cos[\pi t/(2T_b)]$ and $\sin[\pm\pi t/(2T_b)] = \pm\sin[\pi t/(2T_b)]$, we see that the ±1 sign of $m(t)$ during the $(0, T_b)$ interval affects only $y(t)$, and not $x(t)$, over the signaling interval of $(0, 2T_b)$. We also realize that the $\sin[\pi t/(2T_b)]$ pulse of $y(t)$ is $2T_b$ s wide. Similarly, it can be seen that the ± 1 sign of $m(t)$ over the $(T_b, 2T_b)$ interval affects only $x(t)$, and not $y(t)$, over the interval $(T_b, 3T_b)$. In other words, the binary data of $m(t)$ *alternately* modulate the $x(t)$ and $y(t)$ components, and the pulse shape for the $x(t)$ and $y(t)$ symbols (which are $2T_b$ wide instead of T_b) is a sinusoid, as shown in the figure. Thus, MSK is equivalent to OQPSK with sinusoidal pulse shaping.

The $x(t)$ and $y(t)$ waveforms, as shown in Fig. 5–34, illustrate the so-called Type II MSK [Bhargava, Haccoun, Matyas, and Nuspl, 1981], in which the basic pulse shape is always a positive half-cosinusoid. For Type I MSK, the pulse shape [for both $x(t)$ and $y(t)$] alternates between a positive and a negative half-cosinusoid. For both Type I and Type II MSK, it can be shown that there is *no* one-to-one relationship between the input data $m(t)$ and the resulting mark and space frequencies, f_1 and f_2, respectively, in the MSK signal. This can be demonstrated by evaluating the instantaneous frequency f_i as a function of the data presented by $m(t)$ during the different bit intervals. The instantaneous frequency is $f_i = f_c + (1/2\pi)[d\theta(t)/dt] = f_c \pm \Delta F$, where $\theta(t) = \tan^{-1}[y(t)/x(t)]$. The $\pm$ sign is determined by the encoding technique (Type I or II) that is used to obtain the $x(t)$ and $y(t)$ waveforms in each bit interval T_b, as well as by the sign of the data on $m(t)$. To get a one-to-one frequency relationship between a Type I MSK signal and the corresponding $h = 0.5$ FSK signal, called a *fast frequency-shift keyed* (FFSK) signal, the data input to the Type I FSK modulator are first differentially encoded. Examples of the waveforms for Type I and II MSK and for FFSK can be found in the MATLAB solutions to Probs. 5–69, 5–70, and 5–71. Regardless of the differences noted, FFSK, Type I MSK, and Type II MSK are all constant-amplitude, continuous-phase FSK signals with a digital modulation index of $h = 0.5$.

The PSD for (Type I and Type II) MSK can be evaluated as follows: Because $x(t)$ and $y(t)$ have independent data and their dc value is zero, and since $g(t) = x(t) + jy(t)$, the PSD for the complex envelope is

$$\mathcal{P}_g(f) = \mathcal{P}_x(f) + \mathcal{P}_y(f) = 2\mathcal{P}_x(f)$$

where $\mathcal{P}_x(x) = \mathcal{P}_y(f)$ because $x(t)$ and $y(t)$ have the same type of pulse shape. When we use Eq. (3–40) in Eq. (3–36a) with a pulse width of $2T_b$, this PSD becomes

$$\mathcal{P}_g(f) = \frac{2}{2T_b}\, |F(f)|^2 \qquad\qquad (5\text{–}113)$$

where $F(f) = \mathcal{F}[f(t)]$ and $f(t)$ is the pulse shape. For the MSK half-cosinusoidal pulse shape, we have

$$f(t) = \begin{cases} A_c\, \cos\left(\dfrac{\pi t}{2T_b}\right), & |t| < T_b \\[2mm] 0, & t \text{ elsewhere} \end{cases} \qquad (5\text{–}114a)$$

and the Fourier transform is

$$F(f) = \frac{4A_c T_b\, \cos\, 2\pi T_b f}{\pi[1 - (4T_b f)^2]} \qquad\qquad (5\text{–}114b)$$

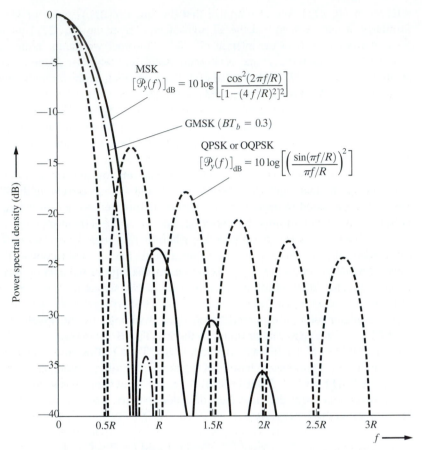

Figure 5–35 PSD for complex envelope of MSK, GMSK, QPSK, and OQPSK, where R is the bit rate (positive frequencies shown).

Thus, the PSD for the complex envelope of an MSK signal is

$$\mathcal{P}_g(f) = \frac{16A_c^2 T_b}{\pi^2}\left(\frac{\cos^2 2\pi T_b f}{[1 - (4T_b f)^2]^2}\right) \tag{5-115}$$

where the normalized power in the MSK signal is $A_c^2/2$. The PSD for MSK is readily obtained by translating this spectrum up to the carrier frequency, as described by Eq. (5–2b). This complex-envelope PSD for MSK is shown by the solid-line curve in Fig. 5–35.

Another form of MSK is *Gaussian-filtered MSK* (GMSK). For GMSK, the data (rectangular-shaped pulses) are filtered by a filter having a Gaussian-shaped frequency response characteristic before the data are frequency modulated onto the carrier. The transfer function of the Gausian low-pass filter is

$$H(f) = e^{-[(f/B)^2\,(\ln 2/2)]} \tag{5-116}$$

where B is the 3-dB bandwidth of the filter. This filter reduces the spectral sidelobes on the transmitted MSK signal. The PSD for GMSK is difficult to evaluate analytically, but can be obtained via computer simulation [Muroto, 1981]. The result is shown in Fig. 5–35 for the case when the 3-dB bandwidth is 0.3 of the bit rate (i.e., $BT_b = 0.3$). For smaller values of BT_b, the spectral sidelobes are reduced further, but the ISI increases. $BT_b = 0.3$ gives a good compromise for relatively low sidelobes and tolerable ISI that is below the noise floor for cellular telephone applications. For $BT_b = 0.3$, GMSK has lower spectral sidelobes than those for MSK, QPSK, or OQPSK (with rectangular-shaped data pulses). In addition, GMSK has a constant envelope, since it is a form of FM. Consequently, GMSK can be amplified without distortion by high-efficiency Class C amplifiers. GMSK and MSK can also be detected either coherently or incoherently. (See Sec. 7–5.) As discussed in Chapt. 8, GMSK with $BT_b = 0.3$ is the modulation format used in GSM cellular telephone systems.

Other digital modulation techniques, such as *tamed frequency modulation* (TFM), have even better spectral characteristics than MSK [DeJager and Dekker, 1978; Pettit, 1982; Taub and Schilling, 1986], and the optimum pulse shape for minimum spectral occupancy of FSK-type signals has been found [Campanella, LoFaso, and Mamola, 1984].

MSK signals can be generated by using any one of several methods, as illustrated in Fig. 5–36. Figure 5–36a shows the generation of FFSK (which is equivalent to Type I MSK with differential encoding of the input data). Here, a simple FM-type modulator having a peak deviation of $\Delta F = 1/(4T_b) = (1/4)R$ is used. Figure 5–36b shows an MSK Type I modulator that is a realization of Eq. (5–112). This is called the *parallel* method of generating MSK, since parallel in-phase (I) and quadrature-phase (Q) channels are used. Figure 5–36c shows the *serial* method of generating MSK. In this approach, BPSK is first generated at a carrier frequency of $f_2 = f_c - \Delta F$, and the bandpass is filtered about $f_1 = f_c + \Delta F$ to produce an MSK signal with a carrier frequency of f_c. (See Prob. 5–72 to demonstrate that this technique is correct.) More properties of MSK are given in Leib and Pasupathy [1993].

Sections 5–9, 5–10, and 5–11, on digital bandpass signaling techniques, are summarized in Table 5–7. The spectral efficiencies of various types of digital signals are shown for the case when rectangular-shaped data pulses are used and for two different bandwidth criteria—the null-to-null bandwidth and the 30-dB bandwidth. A larger value of η indicates a better spectral efficiency. Of course, as shown in Table 5–6, raised cosine-rolloff filtering of the rectangular pulses could be used to reduce the bandwidth and increase η. Alternatively, Gaussian filtering could be used, but it introduces some ISI.

When designing a communication system, one is concerned with the cost and the error performance, as well as the spectral occupancy of the signal. The topic of error performance is covered in Chapter 7.

5–12 ORTHOGONAL FREQUENCY DIVISION MULTIPLEXING (OFDM)

Orthogonal frequency division multiplexing (OFDM) is a technique for transmitting data in parallel by using a large number of modulated carriers with sufficient frequency spacing so that the carriers are orthogonal. As we shall see, OFDM provides resistance to data errors caused by multipath channels.

Over a T-sec interval, the complex envelope for the OFDM signal is

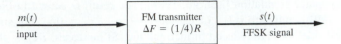

(a) Generation of Fast Frequency-Shift Keying (FFSK)

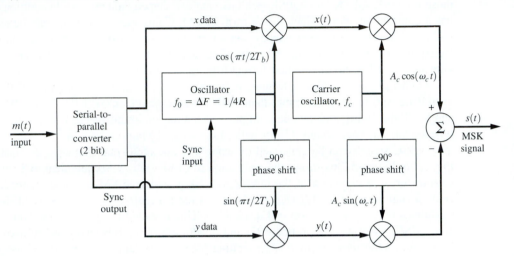

(b) Parallel Generation of Type I MSK (This will generate FFSK if a differential encoder is inserted at the input.)

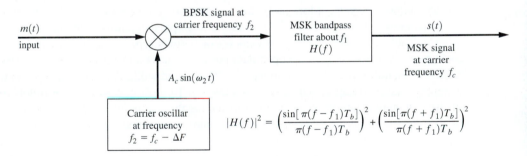

(c) Serial Generation of MSK

Figure 5–36   Generation of MSK signals.

$$g(t) = A_c \sum_{n=0}^{N-1} w_n \varphi_n(t), \quad 0 > t > T \qquad (5\text{--}117a)$$

where A_c is the carrier amplitude, w_n is the element of the N-element parallel data vector $\mathbf{w} = [w_0, w_1, ..., w_{N-1}]$, and the orthogonal carriers are

$$\varphi_n(t) = e^{j2\pi f_n t} \quad \text{where} \quad f_n = \frac{1}{T}\left(n - \frac{N-1}{2}\right) \qquad (5\text{--}117b)$$

Table 5–7   SPECTRAL EFFICIENCY OF DIGITAL SIGNALS

| Type of Signal | Spectral Efficiency, $\eta = \dfrac{R}{B_T} \left(\dfrac{\text{bits/s}}{\text{Hz}}\right)$ | |
	Null-to-Null Bandwidth	30-dB Bandwidth
OOK and BPSK	0.500	0.052
QPSK, OQPSK, and $\pi/4$ QPSK	1.00	0.104
MSK	0.667	0.438
16 QAM	2.00	0.208
64 QAM	3.00	0.313

The duration of the data symbol on each carrier is T seconds, and the carriers are spaced $1/T$ Hz apart. This assures that the carriers are orthogonal, since the $\varphi_n(t)$ satisfy the orthogonality condition of Eq. (2–77) over the T-sec interval (as shown in Example 2–11). Because the carriers are orthogonal, data can be detected on each of these closely spaced carriers without interference from the other carriers.

A key advantage of OFDM is that it can be generated by using FFT digital signal-processing techniques. For example, if we suppress the frequency offset $(N - 1)/2T$ of Eq. (5–117b) and substitute Eq. (5–117b) into Eq. (5–117a), where $t = kT/N$, then the elements of the IFFT vector, as defined by Eq. (2–177), are obtained. Thus, the OFDM signal may be generated by using the IFFT algorithm as shown in Fig. 5–37. In that figure, the complex envelope, $g(t)$, is described by the I and Q components $x(t)$ and $y(t)$, where $g(t) = x(t)$ and $jy(t)$. (This is an application of the generalized transmitter of Fig. 4–28b.)

Referring to Fig. 5–37, let the input serial data symbols have a duration of T_s sec each. These data can be binary (± 1) to produce BPSK modulated carriers or can be multilevel complex-valued serial data to produce (as appropriate) QPSK, MPSK, or QAM carriers. $D_s = 1/T_s$, is the input symbol (baud) rate. The serial-to-parallel converter reads in N input serial symbols at a time and holds their values (elements of $\mathbf{w}$) on the parallel output lines for $T = NT_s$ seconds, where T is the time span of the IFFT. The IFFT uses $\mathbf{w}$ to evaluate output IFFT vector $\mathbf{g}$, which contains elements representing samples of the complex envelope. The parallel-to-serial converter shifts out the element values of $\mathbf{g}$. These are the samples of the complex envelope for the OFDM signal described by Eq. (5–117), where $x(t)$ and $y(t)$ are the I and Q components of the complex envelope. The OFDM signal is produced by the IQ modulators as shown in the figure.

At the receiver, the serial data are recovered from the received OFDM signal by (1) demodulating the signal to produce serial I and Q data, (2) converting the serial data to parallel data, (3) evaluating the FTT, and (4) converting the FFT vector (parallel data) to serial output data.

The length-of-the-FFT vector determines the resistance of OFDM to errors caused by multipath channels. N is chosen so that $T = NT_s$ is much larger than the maximum delay time of echo components in the received multipath signal.

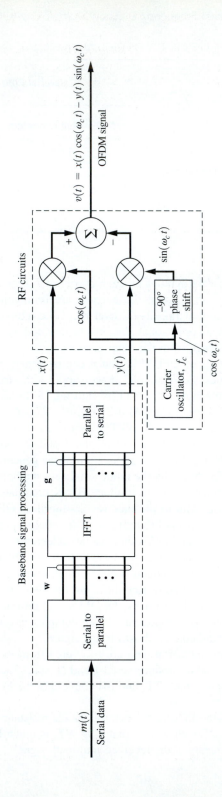

Figure 5–37 OFDM transmitter.

The PSD of the OFDM signal can be obtained relatively easily, since the OFDM signal of Eq. (5–117) consists of orthogonal carriers modulated by data with rectangular pulse shapes that have a duration of T sec. Consequently, the PSD of each carrier is of the form $|\text{Sa}[\pi(f - f_n)T]|^2$, and the overall PSD for the complex envelope of the OFDM signal is

$$\mathcal{P}_g(f) = C \sum_{n=0}^{N-1} \left| \frac{\sin[\pi(f - f_n)T]}{\pi(f - f_n)T} \right|^2 \tag{5–118}$$

where $C = A_c^2\, \overline{|w_n|^2}\, T$ and $\overline{w_n} = 0$. The spectrum is shown in Fig. 5–38 for the case of $N = 32$. Since the spacing between the carriers is $1/T$ Hz and there are N carriers, the null bandwidth of the OFDM signal is

$$B_T = \frac{N+1}{T} = \frac{N+1}{NT_s} \approx \frac{1}{T_s} = D_s \text{ Hz} \tag{5–119}$$

where the approximation is reasonable for $N > 10$ and D_s is the symbol (baud) rate of the input serial data for the OFDM transmitter. In more advanced OFDM systems, rounded (non-rectangular) pulse shapes can be used to reduce the PSD sidelobes outside the $B_T = D_s$ band.

In a generalization of OFDM called *discrete multitone* (DMT) signaling, the data rate on each carrier can be varied, depending on the received SNR on each. That is, the data rate

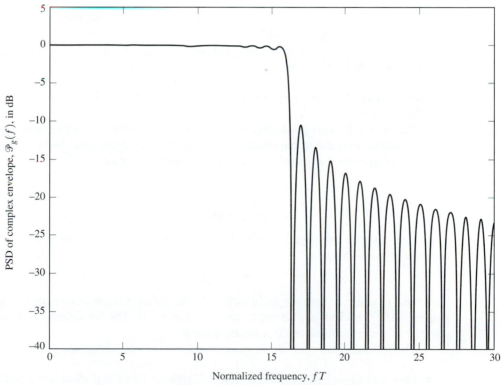

Figure 5–38   PSD for the complex envelope of OFDM with $N = 32$.

is reduced (or the carrier is turned off) on carrier frequencies that have low SNR. The data rates can be adapted as the fading conditions vary, resulting in almost error-free transmission over fading channels.

OFDM is used in the European digital broadcast systems (radio and television) and in some ASDL telephone line modems. (See Chapter 8.)

5–13 SPREAD SPECTRUM SYSTEMS

In our study of communication systems, we have been concerned primarily with the performance of communication systems in terms of bandwidth efficiency and energy efficiency (i.e., detected SNR or probability of bit error) with respect to natural noise. However, in some applications, we also need to consider multiple-access capability, antijam capability, interference rejection, and covert operation, or low-probability of intercept (LPI) capability. (The latter considerations are especially important in military applications.) These performance objectives can be optimized by using spread spectrum techniques.

Multiple-access capability is needed in cellular telephone and personal communication applications, where many users share a band of frequencies, because there is not enough available bandwidth to assign a permanent frequency channel to each user. Spread spectrum techniques can provide the simultaneous use of a wide frequency band via *code-division multiple-access* (CDMA) techniques, an alternative to band sharing. Two other approaches, time-division multiple access (TDMA) and frequency-division multiple access (FDMA), are studied in Sections 3–9, 5–7, and 8–5.

There are many types of *spread spectrum* (SS) systems. To be considered an SS system, a system must satisfy two criteria:

1. The bandwidth of the transmitted signal $s(t)$ needs to be much greater than that of the message $m(t)$.
2. The relatively wide bandwidth of $s(t)$ must be caused by an independent modulating waveform $c(t)$ called the *spreading signal*, and this signal must be known by the receiver in order for the message signal $m(t)$ to be detected.

The SS signal is

$$s(t) = \mathrm{Re}\{g(t)\, e^{j\omega_c t}\} \tag{5-120a}$$

where the complex envelope is a function of both $m(t)$ and $c(t)$. In most cases, a product function is used, so that

$$g(t) = g_m(t)g_c(t) \tag{5-120b}$$

where $g_m(t)$ and $g_c(t)$ are the usual types of modulation complex-envelope functions that generate AM, PM, FM, and so on, as given in Table 4–1. The SS signals are classified by the type of mapping functions that are used for $g_c(t)$.

The following are some of the most common types of SS signals:

- *Direct Sequence* (DS). Here, a DSB-SC type of spreading modulation is used [i.e., $g_c(t) = c(t)$], and $c(t)$ is a polar NRZ waveform.

- *Frequency Hopping* (FH). Here $g_c(t)$ is of the FM type where there are $M = 2^k$ hop frequencies determined by the k-bit words obtained from the spreading code waveform $c(t)$.
- *Hybrid* techniques that include both DS and FH.

We illustrate exactly how DS and FH systems work in the remaining sections of the chapter.

Direct Sequence

Assume that the information waveform $m(t)$ comes from a digital source and is a polar waveform having values of ± 1. Furthermore, let us examine the case of BPSK modulation, where $g_m(t) = A_c m(t)$. Thus, for DS, where $g_c(t) = c(t)$ is used in Eq. (5–120b), the complex envelope for the SS signal becomes

$$g(t) = A_c m(t) c(t) \tag{5–121}$$

The resulting $s(t) = \text{Re}\{g(t)e^{j\omega_c t}\}$ is called a *binary phase-shift keyed data, direct sequence spreading, spread spectrum signal* (BPSK-DS-SS), and $c(t)$ is a polar spreading signal. Furthermore, let this spreading waveform be generated by using a *pseudonoise* (PN) code generator, as shown in Fig. 5–39b, where the values of $c(t)$ are ± 1. The pulse width of $c(t)$ is denoted by T_c and is called a *chip* interval (as contrasted with a bit interval). The code generator uses a modulo-2 adder and r shift register stages that are clocked every T_c sec. It can be shown that $c(t)$ is periodic. Furthermore, feedback taps from the stage of the shift registers and modulo-2 adders are arranged so that the $c(t)$ waveform has a maximum period of N chips, where $N = 2^r - 1$. This type of PN code generator is said to generate a *maximum-length sequence,* or *m-sequence,* waveform.

Properties of Maximum-Length Sequences. The following are some properties of *m*-sequences [Peterson, Ziemer, and Borth, 1995]:

Property 1.   In one period, the number of 1's is always one more than the number of 0's.

Property 2.   The modulo-2 sum of any *m*-sequence, when summed chip by chip with a shifted version of the same sequence, produces another shifted version of the same sequence.

Property 3.   If a window of width r (where r is the number of stages in the shift register) is slid along the sequence for N shifts, then all possible r-bit words will appear exactly once, except for the all 0 r-bit word.

Property 4.   If the 0's and 1's are represented by -1 and $+1$ V, the autocorrelation of the sequence is

$$R_c(k) = \begin{cases} 1, & k = \ell N \\ -\dfrac{1}{N}, & k \neq \ell N \end{cases} \tag{5–122}$$

where $R_c(k) \triangleq (1/N) \sum_{n=0}^{N-1} c_n c_{n+k}$ and $c_n = \pm 1$.

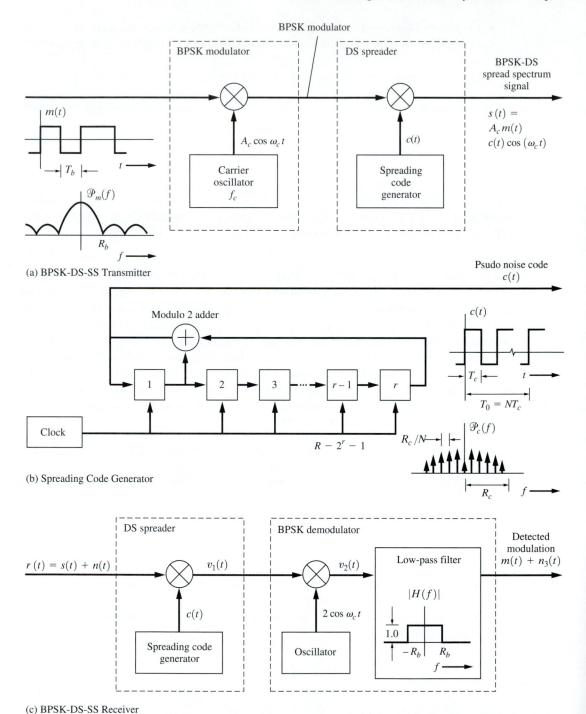

(a) BPSK-DS-SS Transmitter

(b) Spreading Code Generator

(c) BPSK-DS-SS Receiver

Figure 5-39   Direct sequence spread spectrum system (DS-SS).

The autocorrelation of the *waveform* $c(t)$ is

$$R_c(\tau) = \left(1 - \frac{\tau_\varepsilon}{T_c}\right) R_c(k) + \frac{\tau_\varepsilon}{T_c} R_c(k+1) \tag{5–123}$$

where $R_c(\tau) = \langle c(t)c(t+\tau)\rangle$ and τ_ε is defined by

$$\tau = kT_c + \tau_\varepsilon, \qquad \text{with} \qquad 0 \le \tau_\varepsilon < T_c \tag{5–124}$$

Equation (5–123) reduces to

$$R_c(\tau) = \left[\sum_{\ell=-\infty}^{\ell=\infty} \left(1 + \frac{1}{N}\right) \Lambda \left(\frac{\tau - \ell N T_c}{T_c}\right)\right] - \frac{1}{N} \tag{5–125}$$

This equation is plotted in Fig. 5–40a, where it is apparent that the autocorrelation function for the PN waveform is periodic with triangular pulses of width $2T_c$ repeated every NT_c

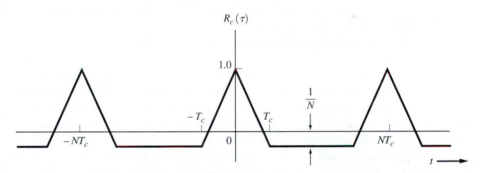

(a) Autocorrelation Function

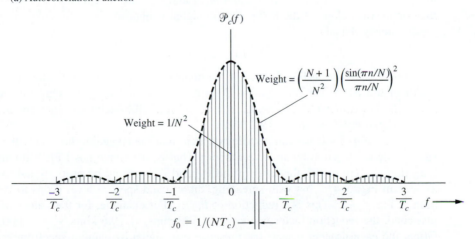

(b) Power Spectral Density (PSD)

Figure 5–40   Autocorrelation and PSD for an *m*-sequence PN waveform.

seconds and that a correlation level of $-1/N$ occurs between these triangular pulses. Furthermore, since the autocorrelation function is periodic, the associated PSD is a line spectrum. That is, the autocorrelation is expressed as the Fourier series

$$R_c(\tau) = \sum_{n=-\infty}^{\infty} r_n e^{j2\pi n f_0 t} \tag{5-126}$$

where $f_0 = 1/(NT_c)$ and $\{r_n\}$ is the set of Fourier series coefficients. Thus, using Eq. (2–109) yields

$$\mathcal{P}_c(f) = \mathcal{F}[R_c(\tau)] = \sum_{n=-\infty}^{\infty} r_n \delta(f - nf_0) \tag{5-127}$$

where the Fourier series coefficients are evaluated and found to be

$$r_n = \begin{cases} \dfrac{1}{N^2}, & n = 0 \\[2ex] \left(\dfrac{N+1}{N^2}\right)\left(\dfrac{\sin(\pi n/N)}{\pi n/N}\right)^2, & n \neq 0 \end{cases} \tag{5-128}$$

This PSD is plotted in Fig. 5–40b.

Now let us demonstrate that the bandwidth of the SS signal is relatively large compared to the data rate R_b and is determined primarily by the spreading waveform $c(t)$, and not by the data modulation $m(t)$. Referring to Fig. 5–39, we see that the PSDs of both $m(t)$ and $c(t)$ are of the $[(\sin x)/x]^2$ type, where the bandwidth of $c(t)$ is much larger than that of $m(t)$ because it is assumed that the chip rate $R_c = 1/T_c$ is much larger than the data rate $R_b = 1/T_b$. That is, $R_c \gg R_b$. To simplify the mathematics, approximate these PSDs by rectangular spectra, as shown in Figs. 5–41a and 5–41b, where the heights of the PSD are selected so that the areas under the curves are unity because the powers of $m(t)$ and $c(t)$ are unity. (They both have only ± 1 values.) From Eq. (5–121), $g(t)$ is obtained by multiplying $m(t)$ and $c(t)$ in the time domain, and $m(t)$ and $c(t)$ are independent. Thus, the PSD for the complex envelope of the BPSK-DS-SS signal is obtained by a convolution operation in the frequency domain:

$$\mathcal{P}_g(f) = A_c^2 \, \mathcal{P}_m(f) * \mathcal{P}_c(f) \tag{5-129}$$

This result is shown in Fig. 5–41c for the approximate PSDs of $m(t)$ and $c(t)$. The bandwidth of the BPSK-DS-SS signal is determined essentially by the chip rate R_c, because $R_c \gg R_b$. For example, let $R_b = 9.6$ kbits/s and $R_c = 9.6$ Mchips/s. Then the bandwidth of the SS signal is $B_T \approx 2R_c = 19.2$ MHz.

From Fig. 5–41, we can also demonstrate that the spreading has made the signal less susceptible to detection by an eavesdropper. That is, the signal has LPI. Without spreading [i.e., if $c(t)$ were unity], the level of the in-band PSD would be proportional to $A_c^2/(2R_b)$, as seen in Fig. 5–41a, but with spreading, the in-band spectral level drops to $A_c^2/(2R_c)$, as seen in Fig. 5–41c. This is a reduction of R_c/R_b. For example, for the values of R_b and R_c just cited, the reduction factor would be (9.6 Mchips/s)/(9.6 kbits/s) = 1,000, or 30 dB. Often, the eavesdropper detects the presence of a signal by using a spectrum analyzer, but when SS is used, the level will drop by 30 dB. This is often below the noise floor of the potential eavesdropper, and thus, the SS signal will escape detection by the eavesdropper.

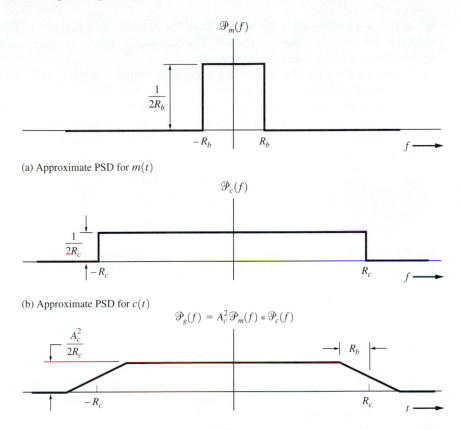

(a) Approximate PSD for $m(t)$

(b) Approximate PSD for $c(t)$

$$\mathcal{P}_g(f) = A_c^2 \mathcal{P}_m(f) * \mathcal{P}_c(f)$$

(c) Approximate PSD for Complex Envelope of the SS Signal

Figure 5–41   Approximate PSD of the BPSK-DS-SS signal.

Figure 5–39c shows a receiver that recovers the modulation on the SS signal. The receiver has a despreading circuit that is driven by a PN code generator in synchronism with the transmitter spreading code. Assume that the input to the receiver consists of the SS signal plus a narrowband (sine wave) jammer signal. Then

$$r(t) = s(t) + n(t) = A_c m(t) c(t) \cos \omega_c t + n_j(t), \qquad (5\text{–}130)$$

where the jamming signal is

$$n_J(t) = A_J \cos \omega_c t \qquad (5\text{–}131)$$

Here, it is assumed that the jamming power is $A_J^2/2$ relative to the signal power of $A_c^2/2$ and that the jamming frequency is set to f_c for the worst-case jamming effect. Referring to Fig. 5–39c, we find that the output of the despreader is

$$v_1(t) = A_c m(t) \cos \omega_c t + A_J c(t) \cos \omega_c t \qquad (5\text{–}132)$$

since $c^2(t) = (\pm 1)^2 = 1$. The BPSK-DS-SS signal has become simply a BPSK signal at the output of the despreader. That is, at the receiver input the SS signal has a bandwidth of $2R_c$, but at the despreader output the bandwidth of the resulting BPSK signal is $2R_b$, a 1,000:1 reduction for the figures previously cited. The data on the BPSK despread signal are recovered by using a BPSK detector circuit as shown.

Now we show that this SS receiver provides an antijam capability of 30 dB for the case of $R_b = 9.6$ kbits/s and $R_c = 9.6$ Mchips/s. From Eq. (5–132), it is seen that the narrowband jammer signal that was present at the receiver input has been *spread* by the despreader, since it has been multiplied by $c(t)$. It is this spreading effect on the jamming signal that produces the antijam capability. Using Eq. (5–132) and referring to Fig. 5–39c, we obtain an input to the LPF of

$$v_2(t) = A_c m(t) + n_2(t) \qquad (5–133)$$

where

$$n_2(t) = A_J c(t) \qquad (5–134)$$

and the terms about $f = 2f_c$ have been neglected because they do not pass through the LPF. Referring to Fig. 5–39c, we note that the jammer power at the receiver output is

$$P_{n_3} = \int_{-R_b}^{R_b} \mathcal{P}_{n_2}(f)\ df = \int_{-R_b}^{R_b} A_J^2\ \frac{1}{2R_c}\ df = \frac{A_J^2}{R_c/R_b} \qquad (5–135)$$

and the jammer power at the input to the LPF is A_J^2. $[\mathcal{P}_{n_2}(f) = A_J^2/(2R_c)$, as seen from Fig. 5–41b and Eq. (5–134).] For a conventional BPSK system (i.e., one without a spread spectrum), $c(t)$ would be unity and Eq. (5–134) would become $n_2(t) = A_J$, so the jamming power out of the LPF would be A_J^2 instead of $A_J^2/(R_c/R_b)$ for the case of an SS system. [The output signal should be $A_c m(t)$ for both cases.] Thus, the SS receiver has reduced the effect of narrowband jamming by a factor of R_c/R_b. This factor, R_c/R_b, is called the *processing gain* of the SS receiver.[†] For our example of $R_c = 9.6$ Mchips/s and $R_b = 9.6$ kbits/s, the processing gain is 30 dB, which means that the narrowband jammer would have to have 30 dB more power to have the same jamming effect on this SS system, compared with the conventional BPSK system (without SS). Thus, this SS technique provides 30 dB of antijam capability for the R_c/R_b ratio cited in the example.

SS techniques can also be used to provide multiple access, called *code division multiple access* (CDMA). Here, each user is assigned a spreading code such that the signals are orthogonal. The technique is used in CDMA cellular telephone systems. Thus, multiple SS signals can be transmitted simultaneously in the same frequency band, and yet the data on a particular SS signal can be decoded by a receiver, *provided that* the receiver uses a PN code that is identical to, and synchronized with, the particular SS signal that is to be decoded. CDMA links are designed to operate either in the synchronous mode or the asynchronous mode. For the *synchronous mode*, the symbol transition times of all users are aligned. This mode is often used for the forward link from the base station (BS) to the mobile station (MS). For the *asynchronous mode*, no effort is made to align the sequences. This mode is used for the reverse

---

[†] The processing gain is defined as the ratio of the noise power out without SS divided by the noise power out with SS. This is equivalent to the ratio $(S/N)_{\text{out}}/(S/N)_{\text{in}}$ when $(S/N)_{\text{in}}$ is the signal-to-noise power into the receiver and $(S/N)_{\text{out}}$ is the signal-to-noise power out of the LPF.

link from the MS to the BS. The asynchronous mode has more multiple access interference, but fewer design constraints.

To accommodate more users in frequency bands that are now saturated with conventional narrowband users (such as the two-way radio bands), it is possible to assign new SS stations. This is called *spread spectrum overlay*. The SS stations would operate with such a wide bandwidth that their PSD would appear to be negligible to narrowband receivers located sufficiently distant from the SS transmitters. On the other hand, to the SS receiver, the narrowband signals would have a minimal jamming effect because of the large coding gain of the SS receiver.

Frequency Hopping

As indicated previously, a frequency-hopped (FH) SS signal uses a $g_c(t)$ that is of the FM type, where there are $M = 2^k$ hop frequencies controlled by the spreading code, in which k chip words are taken to determine each hop frequency. An FH-SS transmitter is shown in Fig. 5–42a. The source information is modulated onto a carrier using conventional FSK or BPSK techniques to produce an FSK or a BPSK signal. The frequency hopping is accomplished by using a mixer circuit wherein the LO signal is provided by the output of a frequency synthesizer that is hopped by the PN spreading code. The serial-to-parallel converter reads k serial chips of the spreading code and outputs a k-chip parallel word to the programmable dividers in the frequency synthesizer. (See Fig. 4–25 and the related discussion of frequency synthesizers.) The k-chip word specifies one of the possible $M = 2^k$ hop frequencies, $\omega_1, \omega_2, \ldots, \omega_M$.

The FH signal is decoded as shown in Fig. 5–42b. Here, the receiver has the knowledge of the transmitter, $c(t)$, so that the frequency synthesizer in the receiver can be hopped in synchronism with that at the transmitter. This despreads the FH signal, and the source information is recovered from the dehopped signal with the use of a conventional FSK or BPSK demodulator, as appropriate.

In 1985, the FCC opened up three shared-frequency bands—902 to 928 MHz, 2,400 to 2,483.5 MHz, and 5,725 to 5,850 MHz—for commercial SS use with unlicensed 1-W transmitters. This has led to the production and use of SS equipment for telemetry systems, wireless local area networks for personal computers, and wireless fire and security systems. Some SS applications have advantages over other systems. For example, an SS cellular telephone system (i.e., CDMA) appears to be able to accommodate about 1,000 users per cell, compared to the 55 users per cell the U.S. analog cellular system accommodates [Schilling, Pickholtz, and Milstein, 1990]. (See Sec. 8–8 for CDMA cellular telephone standards.) For further study on SS systems, the reader is referred to books and papers that have been written on the this subject [Cooper and McGillem, 1986; Dixon, 1994; McGill, Natali, and Edwards, 1994; Peterson, Ziemer, and Borth, 1995; Rhee, 1998].

5–14 SUMMARY

In this chapter, a wide range of analog and digital modulation systems were examined on the basis of the theory developed in Chapters 1 through 4. AM, SSB, PM, and FM signaling techniques were considered in detail. Standards for AM and FM broadcasting signals were

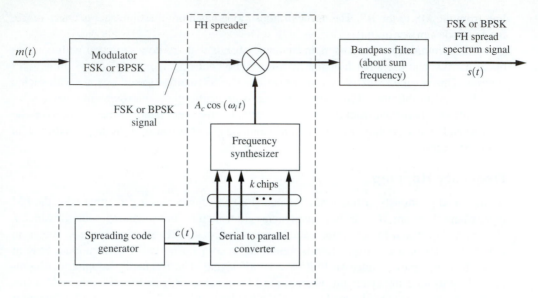

(a) Transmitter

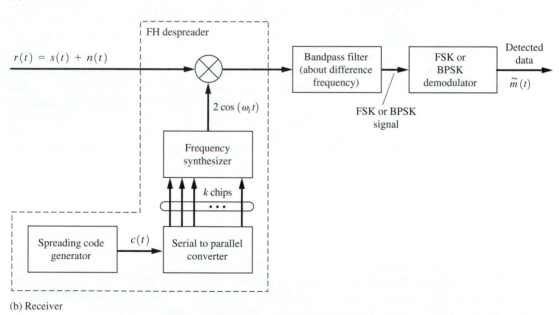

(b) Receiver

Figure 5–42   Frequency-hopped spread spectrum system (FH-SS).

given. Digital signaling techniques such as OOK, BPSK, FSK, MSK and OFDM were developed. The spectra for these digital signals were evaluated in terms of the bit rate of the digital information source. Multilevel digital signaling techniques such as QPSK, MPSK, and QAM, were also studied, and their spectra were evaluated.

Spread spectrum signaling was examined. This technique has multiple-access capability, antijam capability, interference rejection, and a low probability of intercept, properties that are applicable to personal communication systems and military systems.

5–15 STUDY-AID EXAMPLES

SA5–1 Formula for SSB Power   Prove that the normalized average power for an SSB signal is $\langle s^2(t) \rangle = A_c^2 \langle m^2(t) \rangle$, as given by Eq. (5–25).

Solution   For SSB, $g(t) = A_c[m(t) \pm j\hat{m}(t)]$. Using Eq. (4–17) yields

$$P_s = \langle s^2(t) \rangle = \tfrac{1}{2} \langle |g(t)|^2 \rangle = \tfrac{1}{2} A_c \langle m^2(t) + [\hat{m}(t)]^2 \rangle$$

or

$$P_s = \langle s^2(t) \rangle = \tfrac{1}{2} A_c \{ \langle m^2(t) \rangle + \langle [\hat{m}(t)]^2 \rangle \} \qquad (5\text{–}136)$$

But

$$\langle [\hat{m}(t)]^2 \rangle = \int_{-\infty}^{\infty} \mathcal{P}_{\hat{m}}(f)\, df = \int_{-\infty}^{\infty} |H(f)|^2\, \mathcal{P}_m(f)\, df$$

where $H(f)$ is the transfer function of the Hilbert transformer. Using Eq. (5–19), we find that $|H(f)| = 1$. Consequently,

$$\langle [\hat{m}(t)]^2 \rangle = \int_{-\infty}^{\infty} \mathcal{P}_{\hat{m}}(f)\, df = \langle m^2(t) \rangle \qquad (5\text{–}137)$$

Substituting Eq. (5–137) into Eq. (5–136), we get

$$P_s = \langle s^2(t) \rangle = A_c^2 \langle m^2(t) \rangle$$

SA5–2 Evaluation of SSB Power   An SSB transmitter with $A_c = 100$ is being tested by modulating it with a triangular waveform that is shown in Fig. 5–14a, where $V_p = 0.5$ V. The transmitter is connected to a 50-Ω resistive load. Calculate the actual power dissipated into the load.

Solution   Using Eq. (5–25) yields

$$P_{\text{actual}} = \frac{(V_s)_{\text{rms}}^2}{R_L} = \frac{\langle s^2(t) \rangle}{R_L} = \frac{A_c^2}{R_L} \langle m^2(t) \rangle \qquad (5\text{–}138)$$

For the waveform shown in Fig. 5–14a,

$$\langle m^2(t) \rangle = \frac{1}{T_m} \int_0^{T_m} m^2(t)\, dt = \frac{4}{T_m} \int_0^{T_m/4} \left(\frac{4V_p}{T_m} t - V_p \right)^2 dt$$

or

$$\langle m^2(t) \rangle = \frac{4V_p^2}{T_m} \left. \frac{\left(\dfrac{4}{T_m} t - 1 \right)^3}{3\left(\dfrac{4}{T_m} \right)} \right|_0^{T_m/4} = \frac{V_p^2}{3} \qquad (5\text{–}139)$$

Substituting Eq. (5–139) into Eq. (5–138), we get

$$P_{\text{actual}} = \frac{A_c^2 V_p^2}{3R_L} = \frac{(100)^2(0.5)^2}{3(50)} = 16.67 \text{ W}$$

SA5–3 FM Transmitter with Frequency Multipliers   As shown in Fig. 5–43, an FM transmitter consists of an FM exciter stage, a ×3 frequency multiplier, an up-converter (with a bandpass filter), a ×2 frequency multiplier, and a ×3 frequency multiplier. The oscillator has a frequency of 80.0150 MHz, and the bandpass filter is centered around the carrier frequency, which is located at approximately 143 MHz. The FM exciter has a carrier frequency of 20.9957 MHz and a peak deviation of 0.694 kHz when the audio input is applied. The bandwidth of the audio input is 3 kHz. Calculate the carrier frequency and the peak deviation for the FM signals at points B, C, D, E, and F. Also, calculate the bandwidth required for the bandpass filter and the exact center frequency for the bandpass filter.

Solution   As shown in Sec. 4–12, a frequency multiplier produces an output signal at the nth harmonic of the input, and it increases any PM or FM variation that is on the input signal by a factor of n. That is, if the input signal has an angle variation of $\theta(t)$, the output signal will have a variation of $n\theta(t)$, as shown by Eq. (4–73). Thus, the peak deviation at the output of a frequency multiplier is $(\Delta F)_{\text{out}} = n(\Delta F)_{\text{in}}$, because $\Delta F = (1/2\pi) \, d\theta(t)/dt$. The FM exciter output has a carrier frequency $(f_c)_A = 20.9957$ MHz and a peak deviation of $(\Delta F)_A = 0.694$ kHz. Thus, the FM signal at point B has the parameters

$$(f_c)_B = 3(f_c)_A = 62.9871 \text{ MHz} \quad \text{and} \quad (\Delta F)_B = 3(\Delta F)_A = 2.08 \text{ kHz}$$

The mixer (multiplier) produces two signals—a sum frequency term and a difference frequency term at point C—with the carrier frequencies

$$(f_c)_{C \text{ sum}} = f_0 + (f_c)_B = 143.0021 \text{ MHz}$$

and

$$(f_c)_{C \text{ diff}} = f_0 - (f_c)_B = 17.0279 \text{ MHz}$$

Because the mixer output signal has the same complex envelope as the complex envelope at its input (see Sec. 4–11), all the modulation output parameters at the mixer output are the same as those for the input. Thus, the sum and difference carrier frequencies are frequency modulated, and the peak deviation for each is $(\Delta F)_C = (\Delta F)_B = 2.08$ kHz. The bandpass filter passes the 143-MHz term. Consequently, the FM signal at point D has the parameters

$$(f_c)_D = (f_c)_{C \text{ sum}} = 143.0021 \text{ MHz} \quad \text{and} \quad (\Delta F)_D = (\Delta F)_C = 2.08 \text{ kHz}$$

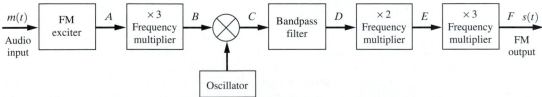

Figure 5–43   FM transmitter.

The FM signals at points E and F have the parameters

$$(f_c)_E = 2(f_c)_D = 286.0042 \text{ MHz and } (\Delta F)_E = 2(\Delta F)_D = 4.16 \text{ kHz}$$

$$(f_c)_F = 3(f_c)_E = 858.0126 \text{ MHz and } (\Delta F)_E = 3(\Delta F)_E = 12.49 \text{ kHz}$$

In summary, the circuit of Fig. 5–43 produces an FM signal at 858.0126 MHz that has a peak deviation of 12.49 kHz. The bandpass filter is centered at $(f_c)_{C \text{ sum}} = 143.0021$ MHz and has a bandwidth sufficient to pass the FM signal with the deviation $(\Delta F)_C = 2.08$ kHz. Using Carson's rule, [Eq. (5–61)], we find that the required bandwidth for the bandpass filter is

$$B_T = 2[\beta_C + 1]B = 2[(\Delta F)_C + B]$$

or

$$B_T = 2[2.08 + 3.0] = 10.16 \text{ kHz}$$

SA5–4 Using an SSB Transmitter to Translate Baseband Data to RF   Data are sent by an amateur radio operator on the 40 meter band by using an SSB transceiver. To accomplish this, a modem of the Bell 103 type (described in Example 5–4) is connected to the audio (microphone) input of the SSB transceiver. Assume that the modem is set to the answer mode and the transceiver is set to transmit a lower SSB signal on a suppressed carrier frequency of $(f_c)_{\text{SSB}} = 7.090$ MHz. Describe the type of digitally modulated signal that is emitted, and determine its carrier frequency. For alternating 101010 data, compute the spectrum of the transmitted signal.

Solution   Referring to Sec. 4–5, we note that an LSSB transmitter just translates the spectrum of the audio input signal up to the suppressed carrier frequency and deletes the upper sideband. From Table 5–5, the Bell 103 modem (answer mode) has a mark frequency of $f_1 = 2,225$ Hz, a space frequency of $f_2 = 2,025$ Hz, and a carrier frequency of $(f_c)_{\text{Bell 103}} = 2,125$ Hz. The LSSB transmitter translates these frequencies to a mark frequency (binary 1) of

$$(f_c)_{\text{SSB}} - f_1 = 7090 \text{ kHz} - 2.225 \text{ kHz} = 7087.775 \text{ kHz}$$

a space frequency (binary 0) of

$$(f_c)_{\text{SSB}} - f_2 = 7090 - 2.025 = 7087.975 \text{ kHz}$$

and a carrier frequency of

$$(f_c)_{\text{FSK}} = (f_c)_{\text{SSB}} - (f_c)_{\text{Bell 103}} = 7090 - 2.125 = 7087.875 \text{ kHz}.$$

Consequently, the SSB transceiver would produce an FSK digital signal with a carrier frequency of 7087.875 kHz.

For the case of alternating data, the spectrum of this FSK signal is given by Eq. (5–85) and (5–86), where $f_c = 7087.875$ kHz. The resulting spectral plot would be like that of Fig. 5–26a, where the spectrum is translated from $f_c = 1,170$ Hz to $f_c = 7087.875$ kHz. It is also realized that this spectrum appears on the lower sideband of the SSB carrier frequency $(f_c)_{\text{SSB}} = 7090$ kHz. If a DSB-SC transmitter had been used (instead of an LSSB transmitter), the spectrum would be replicated on the upper sideband as well as on the lower sideband, and two redundant FSK signals would be emitted.

For the case of random data, the PSD for the complex envelope is given by Eq. (5–90) and shown in Fig. 5–27 for the modulation index of $h = 0.7$. Using Eq. (5–2b), we find that the PSD for the FSK signal is the translation of the PSD for the complex envelope to the carrier frequency of 7087.875 kHz.

PROBLEMS

5–1 An AM broadcast transmitter is tested by feeding the RF output into a 50-Ω (dummy) load. Tone modulation is applied. The carrier frequency is 850 kHz and the FCC licensed power output is 5,000 W. The sinusoidal tone of 1,000 Hz is set for 90% modulation.
(a) Evaluate the FCC power in dBk (dB above 1 kW) units.
(b) Write an equation for the voltage that appears across the 50-Ω load, giving numerical values for all constants.
(c) Sketch the spectrum of this voltage as it would appear on a calibrated spectrum analyzer.
(d) What is the average power that is being dissipated in the dummy load?
(e) What is the peak envelope power?

5–2 An AM transmitter is modulated with an audio testing signal given by $m(t) = 0.2 \sin \omega_1 t + 0.5 \cos \omega_2 t$, where $f_1 = 500$ Hz, $f_2 = 500 \sqrt{2}$ Hz, and $A_c = 100$. Assume that the AM signal is fed into a 50-Ω load.
(a) Sketch the AM waveform.
(b) What is the modulation percentage?
(c) Evaluate and sketch the spectrum of the AM waveform.

5–3 For the AM signal given in Prob. 5–2,
(a) Evaluate the average power of the AM signal.
(b) Evaluate the PEP of the AM signal.

5–4 Assume that an AM transmitter is modulated with a video testing signal given by $m(t) = -0.2 + 0.6 \sin \omega_1 t$, where $f_1 = 3.57$ MHz. Let $A_c = 100$.
(a) Sketch the AM waveform.
(b) What are the percentages of positive and negative modulation?
(c) Evaluate and sketch the spectrum of the AM waveform about f_c.

5–5 A 50,000-W AM broadcast transmitter is being evaluated by means of a two-tone test. The transmitter is connected to a 50-Ω load, and $m(t) = A_1 \cos \omega_1 t + A_1 \cos 2\omega_1 t$, where $f_1 = 500$ Hz. Assume that a perfect AM signal is generated.
(a) Evaluate the complex envelope for the AM signal in terms of A_1 and ω_1.
(b) Determine the value of A_1 for 90% modulation.
(c) Find the values for the peak current and average current into the 50-Ω load for the 90% modulation case.

5–6 An AM transmitter uses a two-quadrant multiplier so that the transmitted signal is described by Eq. (5–7). Assume that the transmitter is modulated by $m(t) = A_m \cos \omega_m t$, where A_m is adjusted so that 120% positive modulation is obtained. Evaluate the spectrum of this AM signal in terms of A_c, f_c, and f_m. Sketch your result.

5–7 A DSB-SC signal is modulated by $m(t) = \cos\ \omega_1 t + 2\ \cos\ 2\omega_1 t$, where $\omega_1 = 2\pi f_1$, $f_1 = 500$ Hz, and $A_c = 1$.

(a) Write an expression for the DSB-SC signal and sketch a picture of this waveform.

(b) Evaluate and sketch the spectrum for this DSB-SC signal.

(c) Find the value of the average (normalized) power.

(d) Find the value of the PEP (normalized).

5–8 Assume that transmitting circuitry restricts the modulated output signal to a certain peak value, say, A_p, because of power-supply voltages that are used and because of the peak voltage and current ratings of the components. If a DSB-SC signal with a peak value of A_p is generated by this circuit, show that the sideband power of this DSB-SC signal is four times the sideband power of a comparable AM signal having the same peak value A_p that could also be generated by this circuit.

5–9 A DSB-SC signal can be generated from two AM signals as shown in Fig. P5–9. Using mathematics to describe signals at each point on the figure, prove that the output is a DSB-SC signal.

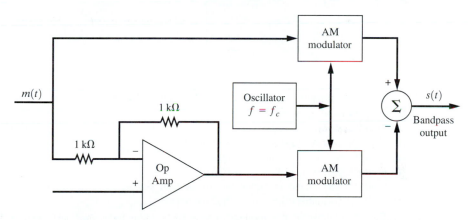

Figure P5–9

5–10 Show that the complex envelope $g(t) = m(t) - j\hat{m}(t)$ produces a lower SSB signal, provided that $m(t)$ is a real signal.

5–11 Show that the impulse response of a $-90°$ phase shift network (i.e., a Hilbert transformer) is $1/\pi t$. *Hint:*

$$H(f) = \lim_{\substack{\alpha \to 0 \\ \alpha > 0}} \begin{cases} -je^{-\alpha f}, & f > 0 \\ je^{\alpha f}, & f < 0 \end{cases}$$

5–12 SSB signals can be generated by the phasing method shown in Fig. 5–5a, by the filter method, of Fig. 5–5b, or by the use of Weaver's method [Weaver, 1956], as shown in Fig. P5–12. For Weaver's method (Fig. P5–12), where B is the bandwidth of $m(t)$,

(a) Find a mathematical expression that describes the waveform out of each block on the block diagram.

(b) Show that $s(t)$ is an SSB signal.

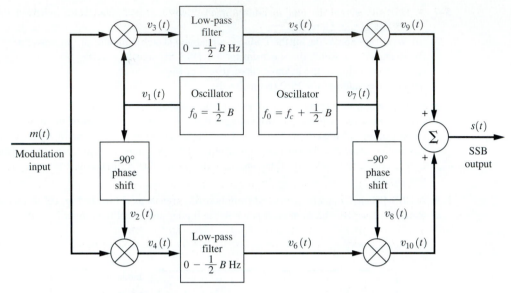

Figure P5–12   Weaver's method for generating SSB.

5–13 An SSB-AM transmitter is modulated with a sinusoid $m(t) = 5 \cos \omega_1 t$, where $\omega_1 = 2\pi f_1$, $f_1 = 500$ Hz, and $A_c = 1$.
 (a) Evaluate $\hat{m}(t)$.
 (b) Find the expression for a lower SSB signal.
 (c) Find the rms value of the SSB signal.
 (d) Find the peak value of the SSB signal.
 (e) Find the normalized average power of the SSB signal.
 (f) Find the normalized PEP of the SSB signal.
5–14 An SSB-AM transmitter is modulated by a rectangular pulse such that $m(t) = \Pi(t/T)$ and $A_c = 1$.
 (a) Prove that

$$\hat{m}(t) = \frac{1}{\pi} \ln \left| \frac{2t + T}{2t - T} \right|$$

 as given in Table A–7.
 (b) Find an expression for the SSB-AM signal $s(t)$, and sketch $s(t)$.
 (c) Find the peak value of $s(t)$.
5–15 For Prob. 5–14,
 (a) Find the expression for the spectrum of a USSB-AM signal.
 (b) Sketch the magnitude spectrum, $|S(f)|$.
5–16 A USSB transmitter is modulated with the pulse

$$m(t) = \frac{\sin \pi a t}{\pi a t}$$

 (a) Prove that

$$\hat{m}(t) = \frac{\sin^2[(\pi a/2)t]}{(\pi a/2)t}$$

(b) Plot the corresponding USSB signal waveform for the case of $A_c = 1$, $a = 2$, and $f_c = 20$ Hz.

5–17 A USSB-AM signal is modulated by a rectangular pulse train:

$$m(t) = \sum_{n=-\infty}^{\infty} \Pi[(t - nT_0)/T]$$

Here, $T_0 = 2T$.

(a) Find the expression for the spectrum of the SSB-AM signal.

(b) Sketch the magnitude spectrum, $|S(f)|$.

5–18 A phasing-type SSB-AM detector is shown in Fig. P5–18. This circuit is attached to the IF output of a conventional superheterodyne receiver to provide SSB reception.

(a) Determine whether the detector is sensitive to LSSB or USSB signals. How would the detector be changed to receive SSB signals with the opposite type of sidebands?

(b) Assume that the signal at point A is a USSB signal with $f_c = 455$ kHz. Find the mathematical expressions for the signals at points B through I.

(c) Repeat part (b) for the case of an LSSB-AM signal at point A.

(d) Discuss the IF and LP filter requirements if the SSB signal at point A has a 3-kHz bandwidth.

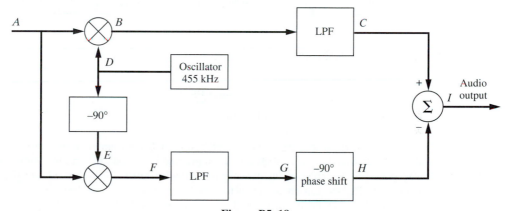

Figure P5–18

5–19 Can a Costas loop, shown in Fig. 5–3, be used to demodulate an SSB-AM signal? Use mathematics to demonstrate that your answer is correct.

5–20 A modulated signal is described by the equation

$$s(t) = 10 \cos[(2\pi \times 10^8)t + 10 \cos(2\pi \times 10^3 t)]$$

Find each of the following:

(a) Percentage of AM.

(b) Normalized power of the modulated signal.

(c) Maximum phase deviation.

(d) Maximum frequency deviation.

5–21 A sinusoidal signal $m(t) = \cos 2\pi f_m t$ is the input to an angle-modulated transmitter, where the carrier frequency is $f_c = 1$ Hz and $f_m = f_c/4$.

(a) Plot $m(t)$ and the corresponding PM signal, where $D_p = \pi$.

(b) Plot $m(t)$ and the corresponding FM signal, where $D_f = \pi$.

5–22 A sinusoidal modulating waveform of amplitude 4 V and frequency 1 kHz is applied to an FM exciter that has a modulator gain of 50 Hz/V.
(a) What is the peak frequency deviation?
(b) What is the modulation index?

5–23 An FM signal has sinusoidal modulation with a frequency of $f_m = 15$ kHz and modulation index of $\beta = 2.0$.
(a) Find the transmission bandwidth by using Carson's rule.
(b) What percentage of the total FM signal power lies within the Carson rule bandwidth?

5–24 An FM transmitter has the block diagram shown in Fig. P5–24. The audio frequency response is flat over the 20-Hz-to-15-kHz audio band. The FM output signal is to have a carrier frequency of 103.7 MHz and a peak deviation of 75 kHz.
(a) Find the bandwidth and center frequency required for the bandpass filter.
(b) Calculate the frequency f_0 of the oscillator.
(c) What is the required peak deviation capability of the FM exciter?

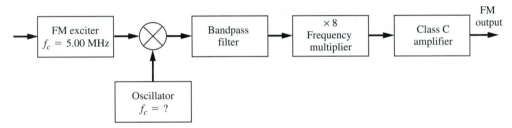

Figure P5–24

5–25 Analyze the performance of the FM circuit of Fig. 5–8b. Assume that the voltage appearing across the reverse-biased diodes, which provide the voltage variable capacitance, is $v(t) = 5 + 0.05m(t)$, where the modulating signal is a test tone, $m(t) = \cos \omega_1 t$, $\omega_1 = 2\pi f_1$, and $f_1 = 1$ kHz. The capacitance of each of the biased diodes is $C_d = 100/\sqrt{1 + 2v(t)}$ pF. Assume that $C_0 = 180$ pF and that L is chosen to resonate at 5 MHz.
(a) Find the value of L.
(b) Show that the resulting oscillator signal is an FM signal. For convenience, assume that the peak level of the oscillator signal is 10 V. Find the parameter D_f.

5–26 A modulated RF waveform is given by $500 \cos[\omega_c t + 20 \cos \omega_1 t]$, where $\omega_1 = 2\pi f_1$, $f_1 = 1$ kHz, and $f_c = 100$ MHz.
(a) If the phase deviation constant is 100 rad/V, find the mathematical expression for the corresponding phase modulation voltage $m(t)$. What is its peak value and its frequency?
(b) If the frequency deviation constant is 1×10^6 rad/V-s, find the mathematical expression for the corresponding FM voltage $m(t)$. What is its peak value and its frequency?
(c) If the RF waveform appears across a 50-Ω load, determine the average power and the PEP.

5–27 Consider the FM signal $s(t) = 10 \cos[\omega_c t + 100 \int_{-\infty}^{t} m(\sigma)\, d\sigma]$, where $m(t)$ is a polar square-wave signal with a duty cycle of 50%, a period of 1 s, and a peak value of 5 V.
(a) Sketch the instantaneous frequency waveform and the waveform of the corresponding FM signal. (See Fig. 5–9.)
(b) Plot the phase deviation $\theta(t)$ as a function of time.
(c) Evaluate the peak frequency deviation.

5–28 A carrier $s(t) = 100 \cos(2\pi \times 10^9 t)$ of an FM transmitter is modulated with a tone signal. For this transmitter, a 1-V (rms) tone produces a deviation of 30 kHz. Determine the amplitude and frequency of all FM signal components (spectral lines) that are greater than 1% of the unmodulated carrier amplitude for the following modulating signals:

(a) $m(t) = 2.5 \cos(3\pi \times 10^4 t)$.
(b) $m(t) = 1 \cos(6\pi \times 10^4 t)$.

5–29 Referring to Eq. (5–58), show that

$$J_{-n}(\beta) = (-1)^n J_n(\beta)$$

5–30 Consider an FM exciter with the output $s(t) = 100 \cos[2\pi\, 1{,}000t + \theta(t)]$. The modulation is $m(t) = 5 \cos(2\pi 8t)$, and the modulation gain of the exciter is 8 Hz/V. The FM output signal is passed through an ideal (brick-wall) bandpass filter that has a center frequency of 1,000 Hz, a bandwidth of 56 Hz, and a gain of unity. Determine the normalized average power

(a) At the bandpass filter input.
(b) At the bandpass filter output.

5–31 A 1-kHz sinusoidal signal phase modulates a carrier at 146.52 MHz with a peak phase deviation of 45°. Evaluate the exact magnitude spectra of the PM signal if $A_c = 1$. Sketch your result. Using Carson's rule, evaluate the approximate bandwidth of the PM signal, and see if it is a reasonable number when compared with your spectral plot.

5–32 A 1-kHz sinusoidal signal frequency modulates a carrier at 146.52 MHz with a peak deviation of 5 kHz. Evaluate the exact magnitude spectra of the FM signal if $A_c = 1$. Sketch your result. Using Carson's rule, evaluate the approximate bandwidth of the FM signal, and see if it is a reasonable number when compared with your spectral plot.

5–33 The calibration of a frequency deviation monitor is to be verified by using a Bessel function test. An FM test signal with a calculated frequency deviation is generated by frequency modulating a sine wave onto a carrier. Assume that the sine wave has a frequency of 2 kHz and that the amplitude of the sine wave is slowly increased from zero until the discrete carrier term (at f_c) of the FM signal vanishes, as observed on a spectrum analyzer. What is the peak frequency deviation of the FM test signal when the discrete carrier term is zero? Suppose that the amplitude of the sine wave is increased further until this discrete carrier term appears, reaches a maximum, and then disappears again. What is the peak frequency deviation of the FM test signal now?

5–34 A frequency modulator has a modulator gain of 10 Hz/V, and the modulating waveform is

$$m(t) = \begin{cases} 0, & t < 0 \\ 5, & 0 < t < 1 \\ 15, & 1 < t < 3 \\ 7, & 3 < t < 4 \\ 0, & 4 < t \end{cases}$$

(a) Plot the frequency deviation in hertz over the time interval $0 < t < 5$.
(b) Plot the phase deviation in radians over the time interval $0 < t < 5$.

5–35 A square-wave (digital) test signal of 50% duty cycle phase modulates a transmitter where $s(t) = 10 \cos[\omega_c t + \theta(t)]$. The carrier frequency is 60 MHz and the peak phase deviation is 45°. Assume that the test signal is of the unipolar NRZ type with a period of 1 ms and that it is symmetrical about $t = 0$. Find the exact spectrum of $s(t)$.

5–36 Two sinusoids, $m(t) = A_1 \cos \omega_1 t + A_2 \cos \omega_2 t$, phase modulate a transmitter. Derive a formula that gives the exact spectrum for the resulting PM signal in terms of the signal parameters $A_c, \omega_c, D_p, A_1, A_2,$ $\omega_1,$ and $\omega_2.$ [*Hint:* Use $e^{ja(t)} = (e^{ja_1(t)})(e^{ja_2(t)})$, where $a(t) = a_1(t) + a_2(t)$.]

5–37 Plot the magnitude spectrum centered on $f = f_c$ for an FM signal where the modulating signal is

$$m(t) = A_1 \cos 2\pi f_1 t + A_2 \cos 2\pi f_2 t$$

Assume that $f_1 = 10$ Hz and $f_2 = 17$ Hz, and that A_1 and A_2 are adjusted so that each tone contributes a peak deviation of 20 Hz.

5–38 For small values of β, $J_n(\beta)$ can be approximated by $J_n(\beta) = \beta^n/(2^n n!)$. Show that, for the case of FM with sinusoidal modulation, $\beta = 0.2$ is sufficiently small to give NBFM.

5–39 A polar square wave with a 50% duty cycle frequency modulates an NBFM transmitter such that the peak phase deviation is 10°. Assume that the square wave has a peak value of 5 V, a period of 10 ms, and a zero-crossing at $t = 0$ with a positive-going slope.
(a) Determine the peak frequency deviation of this NBFM signal.
(b) Evaluate and sketch the spectrum of the signal, using the narrowband analysis technique. Assume that the carrier frequency is 30 MHz.

5–40 Design a wideband FM transmitter that uses the indirect method for generating a WBFM signal. Assume that the carrier frequency of the WBFM signal is 96.9 MHz and that the transmitter is capable of producing a high-quality FM signal with a peak deviation of 75 kHz when modulated by a 1-V (rms) sinusoid of frequency 20 Hz. Show a complete block diagram of your design, indicating the frequencies and peak deviations of the signals at various points.

5–41 An FM signal, $100 \cos[\omega_c t + D_f \int_{-\infty}^{t} m(\sigma)\, d\sigma]$, is modulated by the waveform shown in Fig. P5–41. Let $f_c = 420$ MHz.
(a) Determine the value of D_f so that the peak-to-peak frequency deviation is 25 kHz.
(b) Evaluate and sketch the approximate PSD.
(c) Determine the bandwidth of this FM signal such that spectral components are down at least 40 dB from the unmodulated carrier level for frequencies outside that bandwidth.

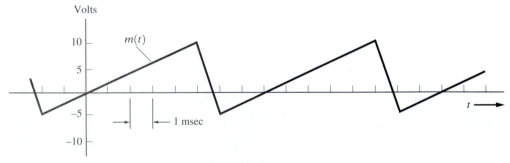

Figure P5–41

5–42 A periodic multilevel digital test signal as shown in Fig. P5–42 modulates a WBFM transmitter. Evaluate and sketch the approximate power spectrum of this WBFM signal if $A_c = 5$, $f_c = 2$ GHz, and $D_f = 10^5$.

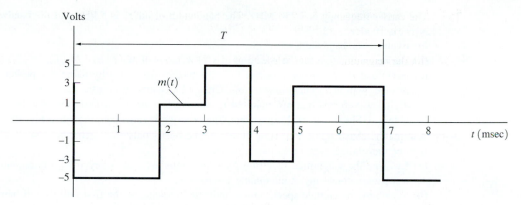

Figure P5–42

5–43 Refer to Fig. 5–16b, which displays the circuit diagram for a commonly used preemphasis filter network.

(a) Show that $f_1 = 1/(2\pi R_1 C)$.

(b) Show that $f_2 = (R_1 + R_2)/(2\pi R_1 R_2 C) \approx 1/(2\pi R_2 C)$.

(c) Evaluate K in terms of R_1, R_2, and C.

5–44 The composite baseband signal for FM stereo transmission is given by

$$m_b(t) = [m_L(t) + m_R(t)] + [m_L(t) - m_R(t)] \cos(\omega_{sc} t) + K \cos(\tfrac{1}{2}\omega_{sc} t)$$

A stereo FM receiver that uses a switching type of demultiplexer to recover $m_L(t)$ and $m_R(t)$ is shown in Fig. P5–44.

(a) Determine the switching waveforms at points C and D that activate the analog switches. Be sure to specify the correct phasing for each one. Sketch these waveforms.

(b) Draw a more detailed block diagram showing the blocks inside the PLL.

(c) Write equations for the waveforms at points A through F, and explain how this circuit works by sketching typical waveforms at each of these points.

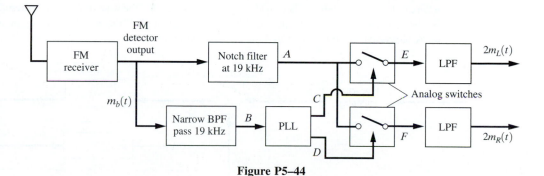

Figure P5–44

5–45 In a communication system, two baseband signals (they may be analog or digital) are transmitted simultaneously by generating the RF signal

$$s(t) = m_1(t) \cos \omega_c t + m_2(t) \sin \omega_c t$$

The carrier frequency is 7.250 MHz. The bandwidth of $m_1(t)$ is 5 kHz and the bandwidth of $m_2(t)$ is 10 kHz.

(a) Evaluate the bandwidth of $s(t)$.

(b) Derive an equation for the spectrum of $s(t)$ in terms of $M_1(f)$ and $M_2(f)$.

(c) $m_1(t)$ and $m_2(t)$ can be recovered (i.e., detected) from $s(t)$ by using a superheterodyne receiver with two switching detectors. Draw a block diagram for the receiver and sketch the digital waveforms that are required to operate the samplers. Describe how they can be obtained. Show that the Nyquist sampling criterion is satisfied. (*Hint:* See Fig. P5–44.)

5–46 A digital baseband signal consisting of rectangular binary pulses occurring at a rate of 24 kbits/s is to be transmitted over a bandpass channel.

(a) Evaluate the magnitude spectrum for OOK signaling that is keyed by a baseband digital test pattern consisting of alternating 1's and 0's.

(b) Sketch the magnitude spectrum and indicate the value of the first null-to-null bandwidth. Assume a carrier frequency of 150 MHz.

(c) For a random data pattern, find the PSD and plot the result. Compare this result with that obtained in parts (a) and (b) for alternating data.

5–47 Repeat Prob. 5–46 for the case of BPSK signaling.

5–48 A carrier is angle modulated with a polar baseband data signal to produce a BPSK signal $s(t) = 10 \cos[\omega_c t + D_p m(t)]$, where $m(t) = \pm 1$ corresponds to the binary data 10010110. $T_b = 0.0025$ sec and $\omega_c = 1,000\pi$. Using MATLAB, plot the BPSK signal waveform and its corresponding FFT spectrum for the following digital modulation indices:

(a) $h = 0.2$.

(b) $h = 0.5$.

(c) $h = 1.0$.

5–49 Evaluate the magnitude spectrum for an FSK signal with alternating 1 and 0 data. Assume that the mark frequency is 50 kHz, the space frequency is 55 kHz, and the bit rate is 2,400 bits/s. Find the first null-to-null bandwidth.

5–50 Assume that 4,800-bit/s random data are sent over a bandpass channel by BPSK signaling. Find the transmission bandwidth B_T such that the spectral envelope is down at least 35 dB outside this band.

5–51 As indicated in Fig. 5–22a, a BPSK signal can be demodulated by using a coherent detector wherein the carrier reference is provided by a Costas loop for the case of $h = 1.0$. Alternatively, the carrier reference can be provided by a squaring loop that uses a ×2 frequency multiplier. A block diagram for a squaring loop is shown in Fig. P5–51.

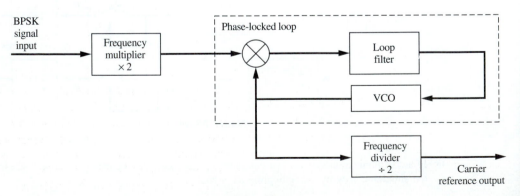

Figure P5–51

(a) Using the squaring loop, draw an overall block diagram for a BPSK receiver.

(b) By using mathematics to represent the waveforms, show how the squaring loop recovers the carrier reference.

(c) Demonstrate that the squaring loop does or does not have a 180° phase ambiguity problem.

5–52 A binary data signal is differentially encoded and modulates a PM transmitter to produce a differentially encoded phase-shift-keyed signal (DPSK). The peak-to-peak phase deviation is 180° and f_c is harmonically related to the bit rate R.

(a) Draw a block diagram for the transmitter, including the differential encoder.

(b) Show typical waveforms at various points on the block diagram if the input data sequence is 01011000101.

(c) Assume that the receiver consists of a superheterodyne circuit. The detector that is used is shown in Fig. P5–52, where $T = 1/R$. If the DPSK IF signal $v_1(t)$ has a peak value of A_c volts, determine the appropriate value for the threshold voltage setting V_T.

(d) Sketch the waveforms that appear at various points of this detector circuit for the data sequence in part (b).

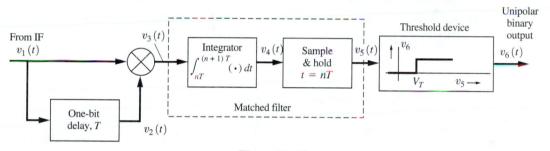

Figure P5–52

5–53 A binary baseband signal is passed through a raised cosine-rolloff filter with a 50% rolloff factor and is then modulated onto a carrier. The data rate is 64 kbits/s. Evaluate

(a) The absolute bandwidth of a resulting OOK signal.

(b) The approximate bandwidth of a resulting FSK signal when the mark frequency is 150 kHz and the space frequency is 155 kHz.

(*Note:* It is interesting to compare these bandwidths with those obtained in Probs. 5–46 and 5–49.)

5–54 Evaluate the exact magnitude spectrum of the FSK signal that is emitted by a Bell-type 103 modem operating in the answer mode at 300 bits/s. Assume that the data are alternating 1's and 0's.

5–55 Starting with Eq. (5–84), work out all of the mathematical steps to derive the result given in Eq. (5–85).

5–56 When FSK signals are detected using coherent detection as shown in Fig. 5–28b, it is assumed that cos $\omega_1 t$ and cos $\omega_2 t$ are orthogonal functions. This is approximately true if $f_1 - f_2 = 2 \Delta F$ is sufficiently large. Find the *exact* condition required for the mark and space FSK signals to be orthogonal. (*Hint:* The answer relates f_1, f_2, and R.)

5–57 Show that the approximate transmission bandwidth for FSK is given by $B_T = 2R(1 + h/2)$, where h is the digital modulation index and R is the bit rate.

5–58 Assume that a QPSK signal is used to send data at a rate of 30 Mbits/s over a satellite transponder. The transponder has a bandwidth of 24 MHz.

    (a) If the satellite signal is equalized to have an equivalent raised cosine filter characteristic, what is the rolloff factor r required?

    (b) Could a rolloff factor r be found so that a 50-Mbit/s data rate could be supported?

5-59 A QPSK signal is generated with nonrectangular data pulses on the I and Q channels. The data pulses have spectra corresponding to the transfer function of a square-root raised cosine-rolloff filter.

    (a) Find the formula for the PSD of the QPSK signal.

    (b) Plot the PSD for the complex envelope of the QPSK signal, where the rolloff factor is $r = 0.35$ and the data rate is normalized to $R = 1$ bit/s. Plot your result in dB versus frequency normalized to the bit rate, similar to that shown in Fig. 5–33.

5–60 Show that two BPSK systems can operate simultaneously over the same channel by using quadrature (sine and cosine) carriers. Give a block diagram for the transmitters and receivers. What is the overall aggregate data rate R for this quadrature carrier multiplex system as a function of the channel null bandwidth B_T? How does the aggregate data rate of the system compare to the data rate for a system that time-division multiplexes the two sources and then transmits the TDM data via a QPSK carrier?

5–61 An 18,000-ft twisted-pair telephone line has a usable bandwidth of 750 kHz. Find the maximum data rate tha can be supported on this line to produce a null-to-null bandwidth of 750 kHz if

    (a) QPSK signaling (rectangular pulses) with a single carrier is used.

    (b) OFDM signaling with QPSK carriers is used.

5–62 Assume that a telephone line channel is equalized to allow bandpass data transmission over a frequency range of 400 to 3,100 Hz so that the available channel bandwidth is 2,700 Hz and the midchannel frequency is 1,750 Hz. Design a 16-symbol QAM signaling scheme that will allow a data rate of 9,600 bits/s to be transferred over the channel. In your design, choose an appropriate rolloff factor r and indicate the absolute and 6-dB QAM signal bandwidth. Discuss why you selected the particular value of r that you used.

5–63 Assume that $R = 9,600$ bits/s. For rectangular data pulses, calculate the *second* null-to-null bandwidth of BPSK, QPSK, MSK, 64PSK, and 64QAM. Discuss the advantages and disadvantages of using each of these signaling methods.

5–64 Referring to Fig. 5–31b, sketch the waveforms that appear at the output of each block, assuming that the input is a TTL-level signal with data 110100101 and $\ell = 4$. Explain how this QAM transmitter works.

5–65 Design a receiver (i.e., determine the block diagram) that will detect the data on a QAM waveform having an $M = 16$-point signal constellation as shown in Fig. 5–32. Explain how your receiver works. (*Hint:* Study Fig. 5–31b.)

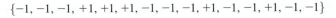

5–66 Using MATLAB, plot the QPSK and OQPSK in-phase and quadrature modulation waveforms for the data stream

$$\{-1, -1, -1, +1, +1, +1, -1, -1, -1, +1, -1, -1, +1, -1, -1\}$$

Use a rectangular pulse shape. For convenience, let $T_b = 1$.

5–67 For $\pi/4$ QPSK signaling,

    (a) Calculate the carrier phase shifts when the input data stream is 10110100101010, where the leftmost bits are first applied to the transmitter.

    (b) Find the absolute bandwidth of the signal if $r = 0.5$ raised cosine-rolloff filtering is used and the data rate is 1.5 Mbits/s.

5–68 **(a)** Fig. 5–34 shows the $x(t)$ and $y(t)$ waveforms for Type II MSK. Redraw these waveforms for the case of Type I MSK.
(b) Show that Eq. (5–114b) is the Fourier transform of Eq. (5–114a).

5–69 Using MATLAB, plot the MSK Type I modulation waveforms $x(t)$ and $y(t)$ and the MSK signal $s(t)$. Assume that the input data stream is

$$\{+1, -1, -1, +1, -1, -1, +1, -1, -1, +1, +1, -1, +1, +1, -1\}$$

Assume also that $T_b = 1$ and that f_c has a value such that a good plot of $s(t)$ is obtained in a reasonable amount of computer time.

5–70 Repeat Prob. 5–69, but use the data stream

$$\{-1, -1, -1, +1, +1, +1, -1, -1, -1, +1, -1, -1, +1, -1, -1\}$$

This data stream is the differentially encoded version of the data stream in Prob. 5–69. The generation of FFSK is equivalent to Type I MSK with differential encoding of the data input. FFSK has a one-to-one relationship between the input data and the mark and space frequencies.

5–71 Using MATLAB, plot the MSK Type II modulation waveforms $x(t)$ and $y(t)$ and the MSK signal $s(t)$. Assume that the input data stream is

$$\{-1, -1, -1, +1, +1, +1, -1, -1, -1, +1, -1, -1, +1, -1, -1\}$$

Assume also that $T_b = 1$ and f_c has a value such that a good plot of $s(t)$ is obtained in a reasonable amount of computer time.

5–72 Show that MSK can be generated by the serial method of Fig. 5–36c. That is, show that the PSD for the signal at the output of the MSK bandpass filter is the MSK spectrum as described by Eqs. (5–115) and (5–2b).

5–73 GMSK is generated by filtering rectangular-shaped data pulses with a Gaussian filter and applying the filtered signal to an MSK transmitter.
(a) Show that the Gaussian filtered data pulse is

$$p(t) = \left(\sqrt{\frac{2\pi}{\ln 2}}\right)(BT_b)\int_{\frac{t}{TB_b}-\frac{1}{2}}^{\frac{t}{TB_b}+\frac{1}{2}} e^{-\left[\frac{2\pi^2}{\ln 2}(BT_b)^2 x^2\right]} \, dx$$

[*Hint:* Evaluate $p(t) = h(t)*\Pi(t/T_b)$, where $h(t) = \mathfrak{F}^{-1}[H(f)]$ and $H(f)$ is described by Eq. (5–116).]
(b) Plot $p(t)$ for $BT_b = 0.3$ and T_b normalized to $T_b = 1$.

5–74 Recompute the spectral efficiencies for all of the signals shown in Table 5–7 by using a 40-dB bandwidth criterion.

5–75 Evaluate and plot the PSD for OFDM signaling with $N = 64$. Find the bandwidth of this OFDM signal if the input data rate is 10 Mbits/s and each carrier uses 16PSK modulation.

5–76 Prove that Eq. (5–123)—the autocorrelation for an m-sequence PN code—is correct. *Hint:* Use the definition of the autocorrelation function, $R_c(\tau) = \langle c(t)c(t + \tau)\rangle$, and Eq. (5–122), where

$$c(t) = \sum_{-\infty}^{\infty} c_n p(t - nT_c)$$

and

$$p(t) = \begin{cases} 1, & 0 < t < T_c \\ 0, & t \text{ elsewhere} \end{cases}$$

5–77 Find an expression for the PSD of an *m*-sequence PN code when the chip rate is 10 MHz and there are eight stages in the shift register. Sketch your result.

5–78 Referring to Fig. 5–40a, show that the complex Fourier series coefficients for the autocorrelation of an *m*-sequence PN waveform are given by Eq. (5–128).

5–79 Assume that the modulator and demodulator for the FH-SS system of Fig. 5–42 are of the FSK type.
(a) Find a mathematical expression for the FSK-FH-SS signal $s(t)$ at the transmitter output.
(b) Using your result for $s(t)$ from part (a) as the receiver input in Fig. 5–42b [i.e., $r(t) = s(t)$], show that the output of the receiver bandpass filter is an FSK signal.

CHAPTER 6

Chapter Objectives

- Random processes
- Power spectral density
- Characteristics of linear systems
- Gaussian random processes
- Matched filters

RANDOM PROCESSES AND SPECTRAL ANALYSIS

The mathematics used to describe random signals and noise will be examined in this chapter. Recall from Chapter 1 that random, or stochastic, signals (in contrast to deterministic signals) are used to convey information. Noise is also described in terms of statistics. Thus, knowledge of random signals and noise is fundamental to an understanding of communication systems.

This chapter is written with the assumption that the reader has a basic understanding of probability, random variables, and ensemble averages, all of which are developed in Appendix B. If the reader is not familiar with these ideas, Appendix B should be studied now just like a regular chapter. For the reader who has had courses in this area, Appendix B can be used as a quick review.

Random processes are extensions of the concepts associated with random variables when the time parameter is brought into the problem. As we will see, this enables frequency response to be incorporated into the statistical description.

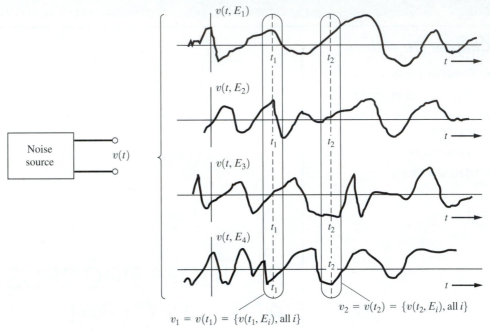

Figure 6–1   Random-noise source and some sample functions of the random-noise process.

6–1 SOME BASIC DEFINITIONS

Random Processes

DEFINITION.   A *real random process* (or *stochastic process*) is an indexed set of real functions of some parameter (usually time) that has certain statistical properties.

Consider voltage waveforms that might be emitted from a noise source. (See Fig. 6–1.) One possible waveform is $v(t, E_1)$. Another is $v(t, E_2)$. In general, $v(t, E_i)$ denotes the waveform that is obtained when the event E_i of the sample space occurs. $v(t, E_i)$ is said to be a *sample* function of the sample space. The set of all possible sample functions $\{v(t, E_i)\}$ is called the *ensemble* and defines the *random process* $v(t)$ that describes the noise source. That is, the events $\{E_i\}$ are mapped into a set of time functions $\{v(t, E_i)\}$. This collection of functions is the random process $v(t)$. When we observe the voltage waveform generated by the noise source, we see one of the sample functions.

Sample functions can be obtained by observing simultaneously the outputs of many identical noise sources. To obtain all of the sample functions in general, an infinite number of noise sources would be required.

The definition of a random process may be compared with that of a random variable. A random variable maps events into *constants*, whereas a random process maps events into *functions* of the parameter t.

THEOREM.   *A random process may be described by an indexed set of random variables.*

Referring to Fig. 6–1, define a set of random variables $v_1 = v(t_1)$, $v_2 = v(t_2)$, ..., where $v(t)$ is the random process. Here, the random variable $v_j = v(t_j)$ takes on the values described by the set of constants $\{v(t_j, E_i), \text{ for all } i\}$.

For example, suppose that the noise source has a Gaussian distribution. Then any of the random variables will be described by

$$f_{v_j}(v_j) = \frac{1}{\sqrt{2\pi}\sigma_j} \, e^{-(v_j - m_j)^2/(2\sigma j^2)} \tag{6–1}$$

where $v_j \triangleq v(t_j)$. We realize that, in general, the probability density function (PDF) depends implicitly on time, since m_j and σ_j respectively correspond to the mean value and standard deviation measured at the time $t = t_j$. The $N = 2$ joint distribution for the Gaussian source for $t = t_1$ and $t = t_2$ is the bivariate Gaussian PDF $f_v(v_1, v_2)$, as given by Eq. (B–97), where $v_1 = v(t_1)$ and $v_2 = v(t_2)$.

To describe a general random process $x(t)$ completely, an N-dimensional PDF, $f_x(\mathbf{x})$, is required, where $\mathbf{x} = (x_1, x_2, \ldots, x_j, \ldots, x_N)$, $x_j \triangleq x(t_j)$, and $N \to \infty$. Furthermore, the N-dimensional PDF is an implicit function of N time constants $t_1, t_2, \ldots, t_N$, since

$$f_x(\mathbf{x}) = f_x(x(t_1), x(t_2), \ldots, x(t_N)) \tag{6–2}$$

Random processes may be classified as continuous or discrete. A *continuous random process* consists of a random process with associated continuously distributed random variables $v_j = v(t_j)$. The Gaussian random process (described previously) is an example of a continuous random process. Noise in linear communication circuits is usually of the continuous type. (In many cases, noise in nonlinear circuits is also of the continuous type.) A *discrete random process* consists of random variables with discrete distributions. For example, the output of an ideal (hard) limiter is a binary (discrete with two levels) random process. Some sample functions of a binary random process are illustrated in Fig. 6–2.

Stationarity and Ergodicity

DEFINITION.   A random process $x(t)$ is said to be *stationary to the order N* if, for any $t_1, t_2, \ldots, t_N$,

$$\underset{\text{random variables}}{f_x(x(t_1), x(t_2), \ldots, x(t_N))} = f_x(x(t_1 + t_0), x(t_2 + t_0), \ldots, x(t_N + t_0)) \tag{6–3}$$

where t_0 is any arbitrary real constant. Furthermore, the process is said to be *strictly stationary* if it is stationary to the order $N \to \infty$.

This definition implies that if a stationary process of order N is translated in time, then the Nth-order statistics do not change. Furthermore, the N-dimensional PDF depends on $N - 1$ time differences $t_2 - t_1$, $t_3 - t_1$, ..., $t_N - t_1$ since t_0 could be chosen to be $-t_1$.

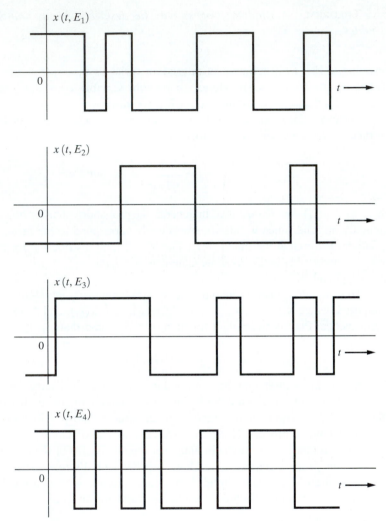

Figure 6–2   Sample functions of a binary random process.

Example 6–1 First-order Stationarity

Examine a random process $x(t)$ to determine whether it is first-order stationary. From Eq. (6–3), the requirement for first-order stationarity is that the first-order PDF not be a function of time. Let the random process be

$$x(t) = A \sin(\omega_0 t + \theta_0) \tag{6–4}$$

CASE 1: A STATIONARY RESULT.   First, assume that A and ω_0 are deterministic constants and θ_0 is a random variable. t is the time parameter. In addition, assume that θ_0 is uniformly distributed over $-\pi$ to π. Then $\psi \triangleq \theta_0 + \omega_0 t$ is a random variable uniformly distributed over the interval $\omega_0 t - \pi < \psi < \omega_0 t + \pi$. The first-order PDF for $x(t)$ can be obtained by the transformation technique developed in Sec. B–8 of Appendix B. This is essentially the same problem as the one worked out in Example B–5. From Eq. (B–71), the first-order PDF for $x(t)$ is

$$f(x) = \begin{cases} \dfrac{1}{\pi\sqrt{A^2 - x^2}}, & |x| \leq A \\ 0, & x \text{ elsewhere} \end{cases} \tag{6–5a}$$

Because this PDF is not a function of t, $x(t)$ is a first-order stationary process for the assumptions of Case 1, where θ_0 is a random variable. This result would be applicable to problems in which θ_0 is the random start-up phase of an unsynchronized oscillator.

CASE 2: A NONSTATIONARY RESULT.   Second, assume that A, ω_0, and θ_0 are deterministic constants. Then, at any time, the value of $x(t)$ is known with a probability of unity. Thus, the first-order PDF of $x(t)$ is

$$f(x) = \delta(x - A\sin(\omega_0 t + \theta_0)) \tag{6–5b}$$

This PDF is a function of t; consequently, $x(t)$ is not first-order stationary for the assumption of Case 2, where θ_0 is a deterministic constant. This result will be applicable to problems in which the oscillator is synchronized to some external source so that the oscillator start-up phase will have the known value θ_0.

DEFINITION.   A random process is said to be _ergodic_ if all time averages of any sample function are equal to the corresponding ensemble averages (expectations).

Two important averages in electrical engineering are dc and rms values. These values are defined in terms of time averages, but if the process is ergodic, they may be evaluated by the use of ensemble averages. The dc value of $x(t)$ is $x_{dc} \triangleq \langle x(t) \rangle$. When $x(t)$ is ergodic, the time average is equal to the ensemble average, so we obtain

$$x_{dc} \triangleq \langle x(t) \rangle \equiv \overline{[x(t)]} = m_x \tag{6–6a}$$

where the time average is

$$\langle [x(t)] \rangle = \lim_{T \to \infty} \frac{1}{T} \int_{-T/2}^{T/2} [x(t)]\,dt \tag{6–6b}$$

the ensemble average is

$$\overline{[x(t)]} = \int_{-\infty}^{\infty} [x]f_x(x)\,dx = m_x \tag{6–6c}$$

and m_x denotes the mean value. Similarly, the rms value is obtained as

$$X_{rms} \triangleq \sqrt{\langle x^2(t) \rangle} \equiv \sqrt{\overline{x^2}} = \sqrt{\sigma_x^2 + m_x^2} \tag{6–7}$$

where σ_x^2 is the variance of $x(t)$.

In summary, if a process is ergodic, all time and ensemble averages are interchangeable. Then the time average cannot be a function of time, since the time parameter has been averaged out. Furthermore, the ergodic process must be stationary, because otherwise the ensemble averages (such as moments) would be a function of time. However, not all stationary processes are ergodic.

Example 6–2 ERGODIC RANDOM PROCESS

Let a random process be given by

$$x(t) = A \cos(\omega_0 t + \theta) \qquad (6\text{–}8)$$

where A and ω_0 are constants and θ is a random variable that is uniformly distributed over $(0, 2\pi)$.

First, we evaluate some ensemble averages. The mean and the second moment are

$$\bar{x} = \int_{-\infty}^{\infty} [x(\theta)] f_\theta(\theta) \, d\theta = \int_0^{2\pi} [A \cos(\omega_0 t + \theta)] \frac{1}{2\pi} \, d\theta = 0 \qquad (6\text{–}9)$$

and

$$\overline{x^2(t)} = \int_0^{2\pi} [A \cos(\omega_0 t + \theta)]^2 \frac{1}{2\pi} \, d\theta = \frac{A^2}{2} \qquad (6\text{–}10)$$

In this example, the time parameter t disappeared when the ensemble averages were evaluated, which would not be the case unless $x(t)$ was stationary.

Second, we evaluate the corresponding time averages by using a typical sample function of the random process. One sample function is $x(t, E_1) = A \cos \omega_0 t$, which occurs when $\theta = 0$, corresponding to one of the events (outcomes). The time average for any of the sample functions can be evaluated by letting θ be the appropriate value between 0 and 2π. The time averages for the first and second moments are

$$\langle x(t) \rangle = \frac{1}{T_0} \int_0^{T_0} A \cos(\omega_0 t + \theta) \, dt = 0 \qquad (6\text{–}11)$$

and

$$\langle x^2(t) \rangle = \frac{1}{T_0} \int_0^{T_0} [A \cos(\omega_0 t + \theta)]^2 \, dt = \frac{A^2}{2} \qquad (6\text{–}12)$$

where $T_0 = 1/f_0$ and the time-averaging operator for a periodic function, Eq. (2–4), has been used. In this example, θ disappears when the time average is evaluated. This is a consequence of $x(t)$ being an ergodic process. However, in a nonergodic example, the time average would be a random variable.

Comparing Eq. (6–9) with Eq. (6–11) and Eq. (6–10) with Eq. (6–12), we see that the time average is equal to the ensemble average for the first and second moments. Consequently, we suspect that this process might be ergodic. However, we have not proven that the process is ergodic, because we have not evaluated all the possible time and ensemble averages or all the moments. However, it seems that the other time and ensemble averages would be equal, so we will *assume* that the process is ergodic. In general, it is difficult to prove that a process is ergodic, so we will assume that this is the case if the process appears to be stationary and some of the time averages are equal to the corresponding ensemble averages. An ergodic process has to be stationary, since the time averages cannot be functions of time. However, if a process is known to be stationary, it may or may not be ergodic.

At the end of the chapter in Prob. 6–2, it will be shown that the random process described by Eq. (6–8) would not be stationary (and, consequently, would not be ergodic) if θ were uniformly distributed over $(0, \pi/2)$ instead of $(0, 2\pi)$.

Correlation Functions and Wide-Sense Stationarity

DEFINITION.   The *autocorrelation function* of a real process $x(t)$ is

$$R_x(t_1, t_2) = \overline{x(t_1)x(t_2)} = \int_{-\infty}^{\infty} \int_{-\infty}^{\infty} x_1 x_2 f_x(x_1, x_2)\, dx_1\, dx_2 \qquad (6\text{-}13)$$

where $x_1 = x(t_1)$ and $x_2 = x(t_2)$. If the process is stationary to the second order, the autocorrelation function is a function only of the time difference $\tau = t_2 - t_1$.

That is,

$$R_x(\tau) = \overline{x(t)x(t + \tau)} \qquad (6\text{-}14)$$

if $x(t)$ is second-order stationary.[†]

DEFINITION.   A random process is said to be *wide-sense stationary* if

1. $\overline{x(t)}$ = constant and $\qquad\qquad\qquad\qquad\qquad$ (6-15a)
2. $R_x(t_1, t_2) = R_x(\tau)$ $\qquad\qquad\qquad\qquad\qquad\qquad$ (6-15b)

where $\tau = t_2 - t_1$.

A process that is stationary to order 2 or greater is wide-sense stationary. However, the converse is not necessarily true, because only *certain* ensemble averages, namely, those of Eq. (6-15), need to be satisfied for wide-sense stationarity.[‡] As indicated by Eq. (6-15), the mean and autocorrelation functions of a wide-sense stationary process do not change with a shift in the time origin. This implies that the associated circuit elements that generate the wide-sense stationary random process do not drift (age) with time.

The autocorrelation function gives us an idea of the frequency response that is associated with the random process. For example, if $R_x(\tau)$ remains relatively constant as τ is increased from zero to some positive number, then, on the average, sample values of x taken at $t = t_1$ and $t = t_1 + \tau$ are nearly the same. Consequently, $x(t)$ does not change rapidly with time (on the average), and we would expect $x(t)$ to have low frequencies present. On the other hand, if $R_x(\tau)$ decreases rapidly as τ is increased, we would expect $x(t)$ to change rapidly with time, and consequently, high frequencies would be present. We formulate this result rigorously in Sec. 6-2, where it will be shown that the PSD, which is a function of frequency, is the Fourier transform of the autocorrelation function.

Properties of the autocorrelation function of a real wide-sense stationary process are as follows:

1. $R_x(0) = \overline{x^2(t)}$ = second moment $\qquad\qquad\qquad$ (6-16)
2. $R_x(\tau) = R_x(-\tau)$ $\qquad\qquad\qquad\qquad\qquad\qquad$ (6-17)
3. $R_x(0) \geq |R_x(\tau)|$ $\qquad\qquad\qquad\qquad\qquad\qquad$ (6-18)

[†] The time average type of the autocorrelation function was defined in Chapter 2. The time average autocorrelation function, Eq. (2-68), is identical to the ensemble average autocorrelation function, Eq. (6-14), when the process is ergodic.

[‡] An exception occurs for the Gaussian random process, in which wide-sense stationarity does imply that the process is strict-sense stationary, since the $N \to \infty$-dimensional Gaussian PDF is completely specified by the mean, variance, and covariance of $x(t_1), x(t_2), \ldots, x(t_N)$.

The first two properties follow directly from the definition of $R_x(\tau)$, as given by Eq. (6–14). Furthermore, if $x(t)$ is ergodic, $R(0)$ is identical to the square of the rms value of $x(t)$. Property 3 is demonstrated as follows: $[x(t) \pm x(t + \tau)]^2$ is a nonnegative quantity, so that

$$\overline{[x(t) \pm x(t + \tau)]^2} \geq 0$$

or

$$\overline{x^2(t)} \pm \overline{2x(t)x(t + \tau)} + \overline{x^2(t + \tau)} \geq 0$$

This equation is equivalent to

$$R_x(0) \pm 2R_x(\tau) + R_x(0) \geq 0$$

which reduces to property 3.

The autocorrelation may be generalized to define an appropriate correlation function for two random processes.

DEFINITION.   The *cross-correlation function* for two real processes $x(t)$ and $y(t)$ is

$$R_{xy}(t_1, t_2) = \overline{x(t_1)y(t_2)} \qquad (6\text{–}19)$$

Furthermore, if $x(t)$ and $y(t)$ are jointly stationary,[†] the cross-correlation function becomes

$$R_{xy}(t_1, t_2) = R_{xy}(\tau)$$

where $\tau = t_2 - t_1$.

Some *properties* of cross-correlation functions of jointly stationary real processes are

$$1. \ R_{xy}(-\tau) = R_{yx}(\tau) \qquad (6\text{–}20)$$

$$2. \ |R_{xy}(\tau)| \leq \sqrt{R_x(0)R_y(0)} \qquad (6\text{–}21)$$

and

$$3. \ |R_{xy}(\tau)| \leq \tfrac{1}{2}[R_x(0) + R_y(0)] \qquad (6\text{–}22)$$

The first property follows directly from the definition, Eq. (6–19). Property 2 follows from the fact that

$$\overline{[x(t) + Ky(t + \tau)]^2} \geq 0 \qquad (6\text{–}23)$$

for any real constant K. Expanding Eq. (6–23), we obtain an equation that is a quadratic in K:

[†] Similar to the definition of Eq. (6–3), $x(t)$ and $y(t)$ are said to be jointly stationary when

$f_{xy}(x(t_1), x(t_2), \ldots, x(t_N), y(t_{N+1}), y(t_{N+2}), \ldots, y(t_{N+M}))$

$\qquad = f_{xy}(x(t_1 + t_0), x(t_2 + t_0), \ldots, x(t_N + t_0), y(t_{N+1} + t_0), y(t_{N+2} + t_0), \ldots, y(t_{N+M} + t_0))$

$x(t)$ and $y(t)$ are strict-sense jointly stationary if this equality holds for $N \to \infty$ and $M \to \infty$ and are wide-sense jointly stationary for $N = 2$ and $M = 2$.

$$[R_y(0)]K^2 + [2R_{xy}(\tau)]K + [R_x(0)] \geq 0 \tag{6-24}$$

For K to be real, it can be shown that the discriminant of Eq. (6–24) has to be nonpositive.[†] That is,

$$[2R_{xy}(\tau)]^2 - 4[R_y(0)][R_x(0)] \leq 0 \tag{6-25}$$

This is equivalent to property 2, as described by Eq. (6–21). Property 3 follows directly from Eq. (6–24), where $K = \pm 1$. Furthermore,

$$|R_{xy}(\tau)| \leq \sqrt{R_x(0)R_y(0)} \leq \tfrac{1}{2}[R_x(0) + R_y(0)] \tag{6-26}$$

because the geometric mean of two positive numbers $R_x(0)$ and $R_y(0)$ does not exceed their arithmetic mean.

Note that the cross-correlation function of two random processes $x(t)$ and $y(t)$ is a generalization of the concept of the joint mean of two random variables defined by Eq. (B–91). Here, x_1 is replaced by $x(t)$, and x_2 is replaced by $y(t + \tau)$. Thus, two random processes $x(t)$ and $y(t)$ are said to be *uncorrelated* if

$$R_{xy}(\tau) = [\overline{x(t)}][\overline{y(t + \tau)}] = m_x m_y \tag{6-27}$$

for all values of τ. Similarly, two random processes $x(t)$ and $y(t)$ are said to be *orthogonal* if

$$R_{xy}(\tau) = 0 \tag{6-28}$$

for all values of τ.

As mentioned previously, if $y(t) = x(t)$, the cross-correlation function becomes an autocorrelation function. In this sense, the autocorrelation function is a special case of the cross-correlation function. Of course, when $y(t) \equiv x(t)$, all the properties of the cross-correlation function reduce to those of the autocorrelation function.

If the random processes $x(t)$ and $y(t)$ are jointly ergodic, the time average may be used to replace the ensemble average. For correlation functions, this becomes

$$R_{xy}(\tau) \triangleq \overline{x(t)y(t + \tau)} \equiv \langle x(t)y(t + \tau) \rangle \tag{6-29}$$

where

$$\langle [\cdot] \rangle = \lim_{T \to \infty} \frac{1}{T} \int_{-T/2}^{T/2} [\cdot]\, dt \tag{6-30}$$

when $x(t)$ and $y(t)$ are jointly ergodic. In this case, cross-correlation functions and auto-correlation functions of voltage or current waveforms may be measured by using an electronic circuit that consists of a delay line, a multiplier, and an integrator. The measurement technique is illustrated in Fig. 6–3.

Complex Random Processes

In previous chapters, the complex envelope $g(t)$ was found to be extremely useful in describing bandpass waveforms. Bandpass *random* signals and noise can also be described in terms of the complex envelope, where $g(t)$ is a *complex* baseband *random process*.

[†] The parameter K is a root of this quadratic only when the quadratic is equal to zero. When the quadratic is positive, the roots have to be complex.

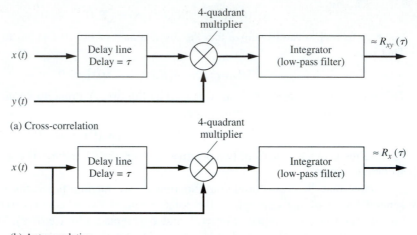

(a) Cross-correlation

(b) Autocorrelation

Figure 6–3   Measurement of correlation functions.

DEFINITION.   A *complex random process* is

$$g(t) \triangleq x(t) + jy(t) \tag{6-31}$$

where $x(t)$ and $y(t)$ are real random processes and $j = \sqrt{-1}$.

A complex process is strict-sense stationary if $x(t)$ and $y(t)$ are jointly strict-sense stationary; that is,

$$f_g(x(t_1), y(t_1'), x(t_2), y(t_2'), \ldots, x(t_N), y(t_N'))$$
$$= f_g(x(t_1 + t_0), y(t_1' + t_0), \ldots, x(t_N + t_0), y(t_N' + t_0)) \tag{6-32}$$

for any value of t_0 and any $N \to \infty$.

The definitions of the correlation functions can be generalized to cover *complex* random processes.

DEFINITION.   The *autocorrelation function* for a *complex* random process is

$$R_g(t_1, t_2) = \overline{g^*(t_1) g(t_2)} \tag{6-33}$$

where the asterisk denotes the complex conjugate.

Furthermore, the complex random process is stationary in the wide sense if $\overline{g(t)}$ is a complex constant and $R_g(t_1, t_2) = R_g(\tau)$, where $\tau = t_2 - t_1$. The autocorrelation of a wide-sense stationary complex process has the Hermitian symmetry property

$$R_g(-\tau) = R_g^*(\tau) \tag{6-34}$$

DEFINITION.   The *cross-correlation function* for two *complex* random processes $g_1(t)$ and $g_2(t)$ is

$$R_{g_1 g_2}(t_1, t_2) = \overline{g_1^*(t_1) g_2(t_2)} \tag{6-35}$$

When the complex random processes are jointly wide-sense stationary, the cross-correlation function becomes

$$R_{g_1 g_2}(t_1, t_2) = R_{g_1 g_2}(\tau)$$

where $\tau = t_2 - t_1$.

In Sec. 6–7, we use these definitions in the statistical description of bandpass random signals and noise.

6–2 POWER SPECTRAL DENSITY

Definition

A definition of the PSD $\mathcal{P}_x(f)$ was given in Chapter 2 for the case of deterministic waveforms by Eq. (2–66). Here we develop a more general definition that is applicable to the spectral analysis of random processes.

Suppose that $x(t, E_i)$ represents a sample function of a random process $x(t)$. A truncated version of this sample function can be defined according to the formula

$$x_T(t, E_i) = \begin{cases} x(t, E_i), & |t| < \tfrac{1}{2}T \\ 0, & t \text{ elsewhere} \end{cases} \tag{6–36}$$

where the subscript T denotes the truncated version. The corresponding Fourier transform is

$$\begin{aligned} X_T(f, E_i) &= \int_{-\infty}^{\infty} x_T(t, E_i) e^{-j2\pi ft}\, dt \\ &= \int_{-T/2}^{T/2} x(t, E_i) e^{-j2\pi ft}\, dt \end{aligned} \tag{6–37}$$

which indicates that X_T is itself a random process, since x_T is a random process. We will now simplify the notation and denote these functions simply by $X_T(f)$, $x_T(t)$, and $x(t)$, because it is clear that they are all random processes.

The normalized[†] energy in the time interval $(-T/2, T/2)$ is

$$E_T = \int_{-\infty}^{\infty} x_T^2(t)\, dt = \int_{-\infty}^{\infty} |X_T(f)|^2\, df \tag{6–38}$$

Here, Parseval's theorem was used to obtain the second integral. E_T is a random variable because $x(t)$ is a random process. Furthermore, the mean normalized energy is obtained by taking the ensemble average of Eq. (6–38):

$$\bar{E}_T = \int_{-T/2}^{T/2} \overline{x^2(t)}\, dt = \int_{-\infty}^{\infty} \overline{x_T^2(t)}\, dt = \int_{-\infty}^{\infty} \overline{|X_T(f)|^2}\, df \tag{6–39}$$

[†] If $x(t)$ is a voltage or current waveform, E_T is the energy on a per-ohm (i.e., $R = 1$) normalized basis.

The normalized average power is the energy expended per unit time, so the normalized average power is

$$P = \lim_{T \to \infty} \frac{1}{T} \int_{-T/2}^{T/2} \overline{x^2(t)} \, dt = \lim_{T \to \infty} \frac{1}{T} \int_{-\infty}^{\infty} \overline{x_T^2(t)} \, dt$$

or

$$P = \int_{-\infty}^{\infty} \left[\lim_{T \to \infty} \frac{1}{T} \overline{|X_T(f)|^2} \right] df = \left\langle \overline{x^2(t)} \right\rangle \tag{6-40}$$

In the evaluation of the limit in Eq. (6–40), it is important that the ensemble average be evaluated *before* the limit operation is carried out, because we want to ensure that $X_T(f)$ is finite. [Since $x(t)$ is a power signal, $X(f) = \lim_{T \to \infty} X_T(f)$ may not exist.] Note that Eq. (6–40) indicates that, for a random process, the average normalized power is given by the time average of the second moment. Of course, if $x(t)$ is wide-sense stationary, $\left\langle \overline{x^2(t)} \right\rangle = \overline{x^2(t)}$ because $\overline{x^2(t)}$ is a constant.

From the definition of the PSD in Chapter 2, we know that

$$P = \int_{-\infty}^{\infty} \mathcal{P}(f) \, df \tag{6-41}$$

Thus, we see that the following definition of the PSD is consistent with that given by Eq. (2–66) in Chapter 2.

DEFINITION.   The *power spectral density* (PSD) for a random process $x(t)$ is given by

$$\mathcal{P}_x(f) = \lim_{T \to \infty} \left(\frac{\overline{[\,|X_T(f)|^2\,]}}{T} \right) \tag{6-42}$$

where

$$X_T(f) = \int_{-T/2}^{T/2} x(t) e^{-j2\pi ft} \, dt \tag{6-43}$$

Wiener–Khintchine Theorem

Often, the PSD is evaluated from the autocorrelation function for the random process by using the following theorem.

WIENER–KHINTCHINE THEOREM[†]   *When $x(t)$ is a wide-sense stationary process, the PSD can be obtained from the Fourier transform of the autocorrelation function:*

$$\mathcal{P}_x(f) = \mathcal{F}[R_x(\tau)] = \int_{-\infty}^{\infty} R_x(\tau) e^{-j2\pi f\tau} \, d\tau \tag{6-44}$$

[†] Named after the American mathematician Norbert Wiener (1894−1964) and the German mathematician A. I. Khintchine (1894−1959). Other spellings of the German name are Khinchine and Khinchin.

Conversely,

$$R_x(\tau) = \mathcal{F}^{-1}[\mathcal{P}_x(f)] = \int_{-\infty}^{\infty} \mathcal{P}_x(f) e^{j2\pi f\tau} \, df \tag{6-45}$$

provided that $R(\tau)$ becomes sufficiently small for large values of τ, so that

$$\int_{-\infty}^{\infty} |\tau R(\tau)| \, d\tau < \infty \tag{6-46}$$

This theorem is also valid for a nonstationary process, *provided* that we replace $R_x(\tau)$ by $\langle R_x(t, t + \tau) \rangle$.

   Proof. From the definition of PSD,

$$\mathcal{P}_x(f) = \lim_{T \to \infty} \left(\frac{\overline{|X_T(f)|^2}}{T} \right)$$

where

$$\overline{|X_T(f)|^2} = \overline{\left| \int_{-T/2}^{T/2} x(t) e^{-j\omega t} \, dt \right|^2}$$

$$= \int_{-T/2}^{T/2} \int_{-T/2}^{T/2} \overline{x(t_1) x(t_2)} e^{-j\omega t_1} e^{j\omega t_2} \, dt_1 dt_2$$

and $x(t)$ is assumed to be real. But $\overline{x(t_1) x(t_2)} = R_x(t_1, t_2)$. Furthermore, let $\tau = t_2 - t_1$, and make a change in variable from t_2 to $\tau + t_1$. Then

$$\overline{|X_T(f)|^2} = \int_{t_1=-T/2}^{t_1=T/2} \underbrace{\left[\int_{\tau=-T/2-t_1}^{\tau=T/2-t_1} R_x(t_1, t_1 + \tau) e^{-j\omega\tau} \, d\tau \right] dt_1}_{\textcircled{1}} \tag{6-47}$$

The area of this two-dimensional integration is shown in Fig. 6–4. In the figure, \textcircled{1} denotes the area covered by the product of the inner integral and the differential width dt_1. To evaluate this two-dimensional integral easily, the order of integration will be exchanged. As seen in the figure, this is accomplished by covering the total area by using \textcircled{2} when $\tau < 0$ and \textcircled{3} when $\tau \geq 0$. Thus Eq. (6–47) becomes

$$\overline{|X_T(f)|^2} = \underbrace{\int_{-T}^{0} \left[\int_{t_1=-T/2-\tau}^{t_1=T/2} R_x(t_1, t_1 + \tau) e^{-j\omega\tau} \, dt_1 \right] d\tau}_{\textcircled{2}}$$

$$+ \underbrace{\int_{0}^{T} \left[\int_{t_1=-T/2}^{t_1=T/2-\tau} R_x(t_1, t_1 + \tau) e^{-j\omega\tau} \, dt_1 \right] d\tau}_{\textcircled{3}} \tag{6-48}$$

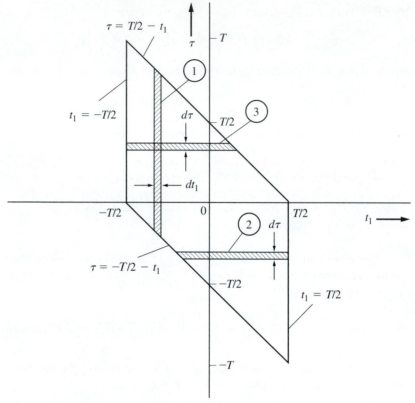

Figure 6–4 Region of integration for Eqs. (6–47) and (6–48).

Now, assume that $x(t)$ is stationary, so that $R_x(t_1, t_1 + \tau) = R_x(\tau)$, and factor $R_x(\tau)$ outside the inner integral. Then

$$\overline{|X_T(f)|^2} = \int_{-T}^{0} R_x(\tau) e^{-j\omega\tau} \left[t_1 \Big|_{-T/2-\tau}^{T/2} \right] d\tau + \int_{0}^{T} R(\tau) e^{-j\omega\tau} \left[t_1 \Big|_{-T/2}^{T/2-\tau} \right] d\tau$$

$$= \int_{-T}^{0} (T + \tau) R_x(\tau) e^{-j\omega\tau} \, d\tau + \int_{0}^{T} (T - \tau) R_x(\tau) e^{-j\omega\tau} \, d\tau$$

This equation can be written compactly as

$$\overline{|X_T(f)|^2} = \int_{-T}^{T} (T - |\tau|) R_x(\tau) e^{-j\omega\tau} \, d\tau \qquad (6\text{–}49)$$

By substituting Eq. (6–49) into Eq. (6–42), we obtain

$$\mathcal{P}_x(f) = \lim_{T \to \infty} \int_{-T}^{T} \left(\frac{T - |\tau|}{T} \right) R_x(\tau) e^{-j\omega\tau} \, d\tau \tag{6–50a}$$

or

$$\mathcal{P}_x(f) = \int_{-\infty}^{\infty} R_x(\tau) e^{j\omega\tau} \, d\tau - \lim_{T \to \infty} \int_{-T}^{T} \frac{|\tau|}{T} R_x(\tau) e^{-j\omega\tau} \, d\tau \tag{6–50b}$$

Using the assumption of Eq. (6–46), we observe that the right-hand integral is zero and Eq. (6–50b) reduces to Eq. (6–44). Thus, the theorem is proved. The converse relationship follows directly from the properties of Fourier transforms. Furthermore, if $x(t)$ is not stationary, Eq. (6–44) is still obtained if we replace $R_x(t_1, t_1 + \tau)$ into Eq. (6–48) by $\langle R_x(t_1, t_1 + \tau) \rangle = R_x(\tau)$.

Comparing the definition of the PSD with results of the Wiener–Khintchine theorem, we see that there are two different methods that may be used to evaluate the PSD of a random process:

1. The PSD is obtained using the *direct method* by evaluating the definition as given by Eq. (6–42).
2. The PSD is obtained using the *indirect method* by evaluating the Fourier transform of $R_x(\tau)$, where $R_x(\tau)$ has to be obtained first.

Both methods will be demonstrated in Example 6–3.

Properties of the PSD

Some properties of the PSD are:

1. $\mathcal{P}_x(f)$ is always real.   (6–51)
2. $\mathcal{P}_x(f) \geq 0$.   (6–52)
3. When $x(t)$ is real, $\mathcal{P}_x(-f) = \mathcal{P}_x(f)$.   (6–53)
4. $\int_{-\infty}^{\infty} \mathcal{P}_x(f) \, df = P = $ total normalized power.   (6–54a)
   When $x(t)$ is wide-sense stationary,

$$\int_{-\infty}^{\infty} \mathcal{P}_x(f) \, df = P = \overline{x^2} = R_x(0) \tag{6–54b}$$

5. $\mathcal{P}_x(0) = \int_{-\infty}^{\infty} R_x(\tau) \, d\tau.$ (6–55)

These properties follow directly from the definition of a PSD and the use of the Wiener–Khintchine theorem.

Example 6–3 EVALUATION OF THE PSD FOR A POLAR BASEBAND SIGNAL

Let $x(t)$ be a polar signal with random binary data. A sample function of this signal is illustrated in Fig. 6–5a. Assume that the data are independent from bit to bit and that the probability of obtaining a binary 1 during any bit interval is $\frac{1}{2}$. Find the PSD of $x(t)$.

The polar signal may be modeled by

$$x(t) = \sum_{n=-\infty}^{\infty} a_n f(t - nT_b) \tag{6-56}$$

where $f(t)$ is the signaling pulse shape, shown in Fig. 6–5b, and T_b is the duration of one bit. $\{a_n\}$ is a set of random variables that represent the binary data. It is given that the random variables are independent. Clearly, each one is discretely distributed at $a_n = \pm 1$ and $P(a_n = 1) = P(a_n = -1) = \frac{1}{2}$, as described in the statement of the problem.

The PSD for $x(t)$ will be evaluated first by using method 1, which requires that $X_T(f)$ be obtained. We can obtain $x_T(t)$ by truncating Eq. (6–56); that is,

$$x_T(t) = \sum_{n=-N}^{n=N} a_n f(t - nT_b)$$

where $T/2 = (N + \frac{1}{2})T_b$. Then

$$X_T(f) = \mathscr{F}[x_T(t)] = \sum_{n=-N}^{N} a_n \mathscr{F}[f(t - nT_b)] = \sum_{n=-N}^{N} a_n F(f) e^{-j\omega nT_b}$$

or

$$X_T(f) = F(f) \sum_{n=-N}^{N} a_n e^{-j\omega nT_b} \tag{6-57}$$

where $F(f) = \mathscr{F}[f(t)]$. When we substitute Eq. (6–57) into Eq. (6–42), we find that the PSD is

$$\mathscr{P}_x(f) = \lim_{T \to \infty} \left(\frac{1}{T} |F(f)|^2 \overline{\left| \sum_{n=-N}^{N} a_n e^{-j\omega nT_b} \right|^2} \right)$$

$$= |F(f)|^2 \lim_{T \to \infty} \left(\frac{1}{T} \sum_{n=-N}^{N} \sum_{m=-N}^{N} \overline{a_n a_m} e^{j(m-n)\omega T_b} \right) \tag{6-58}$$

The average $\overline{a_n a_m}$ now needs to be evaluated for the case of polar signaling ($a_n = \pm 1$). We have

$$\overline{a_n a_m} = \begin{cases} \overline{a_n^2}, & n = m \\ \overline{a_n}\,\overline{a_m}, & n \neq m \end{cases}$$

where $\overline{a_n a_m} = \overline{a_n}\,\overline{a_m}$ for $n \neq m$, since a_n and a_m are independent. Using the discrete distribution for a_n, we get

$$\overline{a_n} = (+1)\tfrac{1}{2} + (-1)\tfrac{1}{2} = 0$$

Similarly, $\overline{a_m} = 0$. Further,

$$\overline{a_n^2} = (+1)^2(\tfrac{1}{2}) + (-1)^2(\tfrac{1}{2}) = 1$$

Thus,

$$\overline{a_n a_m} = \begin{cases} 1, & n = m \\ 0, & n \neq m \end{cases} \tag{6-59}$$

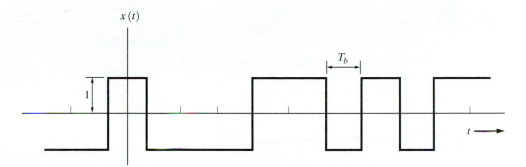

(a) Polar Signal

(b) Signaling Pulse Shape

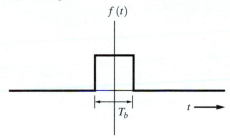

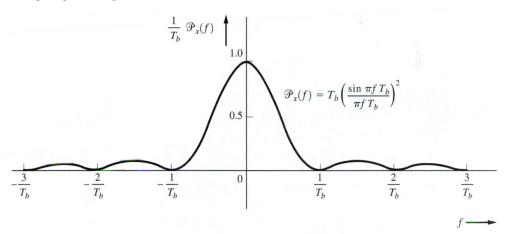

$$\mathcal{P}_x(f) = T_b \left(\frac{\sin \pi f T_b}{\pi f T_b} \right)^2$$

(c) Power Spectral Density of a Polar Signal

Figure 6–5   Random polar signal and its PSD.

With this result, Eq. (6–58) becomes

$$\mathcal{P}_x(f) = |F(f)|^2 \lim_{T \to \infty} \left(\frac{1}{T} \sum_{n=-N}^{N} 1 \right)$$

and, with $T = 2\left(N + \frac{1}{2}\right)T_b$,

$$\mathcal{P}_x(f) = |F(f)|^2 \lim_{N \to \infty}\left[\frac{2N + 1}{(2N + 1)T_b}\right]$$

or

$$\mathcal{P}_x(f) = \frac{1}{T_b}|F(f)|^2 \qquad \text{(polar signaling)} \qquad (6\text{--}60)$$

For the rectangular pulse shape shown in Fig. 6–5b,

$$F(f) = T_b\left(\frac{\sin \pi f T_b}{\pi f T_b}\right) \qquad\qquad (6\text{--}61)$$

Thus, the PSD for a polar signal with a rectangular pulse shape is

$$\mathcal{P}_x(f) = T_b\left(\frac{\sin \pi f T_b}{\pi f T_b}\right)^2 \qquad\qquad (6\text{--}62)$$

This PSD is plotted in Fig. 6–5c.[†] The null bandwidth is $B = 1/T_b = R$, where R is the bit rate. Note that Eq. (6–62) satisfies the properties for a PSD function listed previously.

Method 2 will now be used to evaluate the PSD of the polar signal. This involves calculating the autocorrelation function and then evaluating the Fourier transform of $R_x(\tau)$ to obtain the PSD. When we use Eq. (6–56), the autocorrelation is

$$R_x(t, t + \tau) = \overline{x(t)x(t + \tau)}$$

$$= \overline{\sum_{n=-\infty}^{\infty} a_n f(t - nT_b) \sum_{m=-\infty}^{\infty} a_m f(t + \tau - mT_b)}$$

$$= \sum_{n}\sum_{m}\overline{a_n a_m}f(t - nT_b)f(t + \tau - mT_b)$$

By using Eq. (6–59), this equation reduces to

$$R_x(t, t + \tau) = \sum_{n=-\infty}^{\infty} f(t - nT_b)f(t + \tau - nT_b) \qquad (6\text{--}63)$$

Obviously, $x(t)$ is *not* a wide-sense stationary process, since the autocorrelation function depends on absolute time t. To reduce Eq. (6–63), a particular type of pulse shape needs to be designated. Assume, once again, the rectangular pulse shape

$$f(t) = \begin{cases} 1, & |t| \leq T_b/2 \\ 0, & t \text{ elsewhere} \end{cases}$$

The pulse product then becomes

$$f(t - nT_b)f(t + \tau - nT_b) = \begin{cases} 1, & \text{if } |t - nT_b| \leq T_b/2 \text{ and } |t + \tau - nT_b| \leq T_b/2 \\ 0, & \text{otherwise} \end{cases}$$

Working with the inequalities, we get a unity product only when

$$\left(n - \tfrac{1}{2}\right)T_b \leq t \leq \left(n + \tfrac{1}{2}\right)T_b$$

[†] The PSD of this polar signal is purely continuous because the positive and negative pulses are assumed to be of equal amplitude.

and

$$(n - \tfrac{1}{2})T_b - \tau \le t \le (n + \tfrac{1}{2})T_b - \tau$$

Assume that $\tau \ge 0$; then we will have a unity product when

$$(n - \tfrac{1}{2})T_b \le t \le (n + \tfrac{1}{2})T_b - \tau$$

provided that $\tau \le T_b$. Thus, for $0 \le \tau \le T_b$,

$$R_x(t, t + \tau) = \sum_{n=-\infty}^{\infty} \begin{cases} 1, & (n - \tfrac{1}{2})T_b \le t \le (n + \tfrac{1}{2})T_b - \tau \\ 0, & \text{otherwise} \end{cases} \tag{6-64}$$

We know that the Wiener–Khintchine theorem is valid for nonstationary processes if we let $R_x(\tau) = \langle R_x(t, t + \tau) \rangle$. Using Eq. (6–64), we get

$$R_x(\tau) = \langle R_x(t, t + \tau) \rangle = \lim_{T \to \infty} \frac{1}{T} \int_{-T/2}^{T/2} \sum_{n=-\infty}^{\infty} \begin{cases} 1, & (n - \tfrac{1}{2})T_b \le t \le (n + \tfrac{1}{2})T_b - \tau \\ 0, & \text{elsewhere} \end{cases} dt$$

or

$$R_x(\tau) = \lim_{T \to \infty} \frac{1}{T} \sum_{n=-N}^{N} \left(\int_{(n+1/2)T_b}^{(n+1/2)T_b - \tau} 1 \ dt \right)$$

where $T/2 = (N - \tfrac{1}{2})T_b$. This reduces to

$$R_x(\tau) = \lim_{N \to \infty} \left[\frac{1}{(2N + 1)\,T_b}\,(2N + 1) \begin{cases} (T_b - \tau), & 0 \le \tau \le T_b \\ 0, & \tau > T_b \end{cases} \right]$$

or

$$R_x(\tau) = \begin{cases} \dfrac{T_b - \tau}{T_b}, & 0 \le \tau \le T_b \\[2mm] 0, & \tau > T_b \end{cases} \tag{6-65}$$

Similar results can be obtained for $\tau < 0$. However, we know that $R_x(-\tau) = R_x(\tau)$, so Eq. (6–65) can be generalized for all values of τ. Thus,

$$R_x(\tau) = \begin{cases} \dfrac{T_b - |\tau|}{T_b}, & |\tau| \le T_b \\[2mm] 0, & \text{otherwise} \end{cases} \tag{6-66}$$

which shows that $R_x(\tau)$ has a triangular shape. Evaluating the Fourier transform of Eq. (6–66), we obtain the PSD for the polar signal with a rectangular bit shape:

$$\mathcal{P}_x(f) = T_b \left(\frac{\sin \pi f T_b}{\pi f T_b} \right)^2 \tag{6-67}$$

This result, obtained by the use of method 2, is identical to Eq. (6–62), obtained by using method 1.

General Formula for the PSD of Digital Signals

We now derive a general formula for the PSD of digital signals. The formulas for the PSD in Example 6–3 are valid only for polar signaling with $a_n = \pm 1$ and no correlation between the bits. A more general result can be obtained in terms of the autocorrelation of the data, a_n, by starting with Eq. (6–56). As illustrated in Figs. 3–12 and 3–14, the data may be binary or multilevel. The duration (width) of the symbol pulse $f(t)$ is T_s. For binary data, $T_s = T_b$, where T_b is the duration of 1 bit. We define the autocorrelation of the data by

$$R(k) = \overline{a_n a_{n+k}} \tag{6–68}$$

Next, we make a change in the index in Eq. (6–58), letting $m = n + k$. Then, by using Eq. (6–68) and $T = (2N + 1)T_s$, Eq. (6–58) becomes

$$\mathcal{P}_x(f) = |F(f)|^2 \lim_{N \to \infty} \left[\frac{1}{(2N + 1)T_s} \sum_{n=-N}^{n=N} \sum_{k=-N-n}^{k=N-n} R(k)e^{jk\omega T_s} \right]$$

Replacing the outer sum over the index n by $2N + 1$, we obtain the following expression. [This procedure is not strictly correct, since the inner sum is also a function of n. The correct procedure would be to exchange the order of summation, in a manner similar to that used in Eq. (6–47) through (6–50), where the order of integration was exchanged. The result would be the same as that as obtained below when the limit is evaluated as $N \to \infty$.]

$$\mathcal{P}_x(f) = \frac{|F(f)|^2}{T_s} \lim_{N \to \infty} \left[\frac{(2N + 1)}{(2N + 1)} \sum_{k=-N-n}^{k=N-n} R(k)e^{jk\omega T_s} \right]$$

$$= \frac{|F(f)|^2}{T_s} \sum_{k=-\infty}^{k=\infty} R(k)e^{jk\omega T_s}$$

$$= \frac{|F(f)|^2}{T_s} \left[R(0) + \sum_{k=-\infty}^{-1} R(k)e^{jk\omega T_s} + \sum_{k=1}^{\infty} R(k)e^{jk\omega T_s} \right]$$

or

$$\mathcal{P}_x(f) = \frac{|F(f)|^2}{T_s} \left[R(0) + \sum_{k=1}^{\infty} R(-k)e^{-jk\omega T_s} + \sum_{k=1}^{\infty} R(k)e^{jk\omega T_s} \right] \tag{6–69}$$

But because $R(k)$ is an autocorrelation function, $R(-k) = R(k)$, and Eq. (6–69) becomes

$$\mathcal{P}_x(f) = \frac{|F(f)|^2}{T_s} \left[R(0) + \sum_{k=1}^{\infty} R(k)\left(e^{jk\omega T_s} + e^{-jk\omega T_s}\right) \right]$$

In summary, the general expression for the PSD of a digital signal is

$$\mathcal{P}_x(f) = \frac{|F(f)|^2}{T_s} \left[R(0) + 2\sum_{k=1}^{\infty} R(k)\,\cos(2\pi k f T_s) \right] \tag{6–70a}$$

An equivalent expression is

$$\boxed{\mathcal{P}_x(f) = \frac{|F(f)|^2}{T_s} \left[\sum_{k=-\infty}^{\infty} R(k)e^{jk\omega T_s} \right]} \tag{6–70b}$$

where the autocorrelation of the data is

$$R(k) = \overline{a_n a_{n+k}} = \sum_{i=1}^{I} (a_n a_{n+k})_i P_i \tag{6–70c}$$

in which P_i is the probability of getting the product $(a_n a_{n+k})_i$, of which there are I possible values. $F(f)$ is the spectrum of the pulse shape of the digital symbol.

Note that the quantity in brackets in Eq. (6–70b) is similar to the discrete Fourier transform (DFT) of the data autocorrelation function $R(k)$, except that the frequency variable ω is continuous. Thus, the PSD of the baseband digital signal is influenced by *both* the "spectrum" of the data and the spectrum of the pulse shape used for the line code. Furthermore, the spectrum *may* also contain delta functions if the mean value of the data, $\bar{a}_n$, is nonzero. To demonstrate this result, we first assume that the data symbols are uncorrelated; that is,

$$R(k) = \begin{cases} \overline{a_n^2}, & k = 0 \\ \bar{a}_n \overline{a}_{n+k}, & k \neq 0 \end{cases} = \begin{cases} \sigma_a^2 + m_a^2, & k = 0 \\ m_a^2, & k \neq 0 \end{cases}$$

where, as defined in Appendix B, the mean and the variance for the data are $m_a = \bar{a}_n$ and $\sigma_a^2 = \overline{(a_n - m_a)^2} = \overline{a_n^2} - m_a^2$. Substituting the preceding equation for $R(k)$ into Eq. (6–70b), we get

$$\mathcal{P}_x(f) = \frac{|F(f)|^2}{T_s} \left[\sigma_a^2 + m_a^2 \sum_{k=-\infty}^{\infty} e^{jk\omega T_s} \right]$$

From Eq. (2–115), the Poisson sum formula, we obtain

$$\sum_{k=-\infty}^{\infty} e^{\pm jk\omega T_s} = D \sum_{n=-\infty}^{\infty} \delta(f - nD)$$

where $D = 1/T_s$ is the baud rate. We see that the PSD then becomes

$$\mathcal{P}_x(f) = \frac{|F(f)|^2}{T_s} \left[\sigma_a^2 + m_a^2 D \sum_{n=-\infty}^{\infty} \delta(f - nD) \right]$$

Thus, for the case of uncorrelated data, the PSD of the digital signal is

$$\mathcal{P}_x(f) = \sigma_a^2 \, D |F(f)|^2 + (m_a D)^2 \sum_{n=-\infty}^{\infty} |F(nD)|^2 \delta(f - nD) \qquad (6\text{–}70\text{d})$$

$$\underbrace{\qquad\qquad}_{\text{Continuous spectrum}} \qquad \underbrace{\qquad\qquad\qquad\qquad\qquad}_{\text{Discrete spectrum}}$$

For the general case where there is correlation between the data, let the data autocorrelation function $R(k)$ be expressed in terms of the *normalized-data* autocorrelation function $\rho(k)$. That is, let $\tilde{a}_n$ represent the corresponding data that have been normalized to have unity variance and zero mean. Thus,

$$a_n = \sigma_a \tilde{a}_n + m_a$$

and, consequently,

$$R(k) = \sigma_a^2 \, \rho(k) + m_a^2$$

where

$$\rho(k) = \overline{\tilde{a}_n \tilde{a}_{n+k}}$$

Substituting this expression for $R(k)$ into Eq. (6–70b) and using $R(-k) = R(k)$, we get

$$\mathcal{P}_x(f) = \frac{|F(f)|^2}{T_s} \left[\sigma_a^2 \sum_{k=-\infty}^{\infty} \rho(k)e^{-jk\omega T_s} + m_a^2 \sum_{k=-\infty}^{\infty} e^{-jk\omega T_s} \right]$$

Thus, for the general case of correlated data, the PSD of the digital signal is

$$\mathcal{P}_x(f) = \sigma_a^2 \; D|F(f)|^2 \, \mathcal{W}_\rho(f) + (m_a D)^2 \sum_{n=-\infty}^{\infty} |F(nD)|^2 \delta(f - nD) \quad (6\text{–}70e)$$

$$\underbrace{\hspace{6cm}}_{\text{Continuous spectrum}} \qquad \underbrace{\hspace{7cm}}_{\text{Discrete spectrum}}$$

where

$$\mathcal{W}_\rho(f) \sum_{k=-\infty}^{\infty} \rho(k)e^{-j2\pi kfT_s} \qquad (6\text{–}70f)$$

is a spectral weight function obtained from the Fourier transform of the normalized auto-correlation impulse train

$$\sum_{k=-\infty}^{\infty} \rho(k)\,\delta(\tau - kT_s)$$

This demonstrates that the PSD of the digital signal consists of a continuous spectrum that depends on the pulse-shape spectrum $F(f)$ and the data correlation. Furthermore, if $m_a \neq 0$ and $F(nD) \neq 0$, the PSD will also contain spectral lines (delta functions) spaced at harmonics of the baud rate D.

Examples of the application of these results are given in Sec. 3–5, where the PSD for unipolar RZ, bipolar, and Manchester line codes are evaluated. (See Fig. 3–16 for a plot of the PSDs for these codes.) Examples of bandpass digital signaling, such as OOK, BPSK, QPSK, MPSK, and QAM, are given in Sections 5–9 and 5–10.

White-Noise Processes

DEFINITION. A random process $x(t)$ is said to be a *white-noise process* if the PSD is constant over all frequencies; that is,

$$\mathcal{P}_x(f) = \frac{N_0}{2} \qquad (6\text{–}71)$$

where N_0 is a positive constant.

The autocorrelation function for the white-noise process is obtained by taking the inverse Fourier transform of Eq. (6–71). The result is

$$R_x(\tau) = \frac{N_0}{2}\,\delta(\tau) \qquad (6\text{–}72a)$$

For example, the thermal-noise process described in Sec. 8–6 can be considered to be a white-noise process over the operating band where

$$N_0 = kT \tag{6–72b}$$

Thermal noise also happens to have a Gaussian distribution. Of course, it is also possible to have white noise with other distributions.

Measurement of PSD

The PSD may be measured by using analog or digital techniques. In either case, the measurement can only approximate the true PSD, because the measurement is carried out over a finite time interval instead of the infinite interval specified Eq. (6–42).

Analog Techniques.   Analog measurement techniques consist of using either a bank of parallel narrowband bandpass filters with contiguous bandpass characteristics or using a single bandpass filter with a center frequency that is tunable. In the case of the filter bank, the waveform is fed simultaneously into the inputs of all the filters, and the power at the output of each filter is evaluated. The output powers are divided (scaled) by the effective bandwidth of the corresponding filter so that an approximation for the PSD is obtained. That is, the PSD is evaluated for the frequency points corresponding to the center frequencies of the filters. Spectrum analyzers with this parallel type of analog processing are usually designed to cover the audio range of the spectrum, where it is economically feasible to build a bank of bandpass filters.

RF spectrum analyzers are usually built by using a single narrowband IF bandpass filter that is fed from the output of a mixer (up- or down-converter) circuit. The local oscillator (LO) of the mixer is swept slowly across an appropriate frequency band so that the RF spectrum analyzer is equivalent to a tunable narrowband bandpass filter wherein the center frequency is swept across the desired spectral range. Once again, the PSD is obtained by evaluating the scaled power output of the narrowband filter as a function of the sweep frequency.

Numerical Computation of the PSD.   The PSD is evaluated numerically in spectrum analyzers that use digital signal processing. One approximation for the PSD is

$$\mathcal{P}_T(f) = \frac{|X_T(f)|^2}{T} \tag{6–73a}$$

where the subscript T indicates that the approximation is obtained by viewing $x(t)$ over a T-s interval. T is called the *observation interval* or *observation length*. Of course, Eq. (6–73a) is an approximation to the true PSD, as defined in Eq. (6–42), because T is finite and because only a sample of the ensemble is used. In more sophisticated spectrum analyzers, $\mathcal{P}_T(f)$ is evaluated for several $x(t)$ records, and the average value of $\mathcal{P}_T(f)$ at each frequency is used to approximate the ensemble average required for a true $\mathcal{P}_x(f)$ of Eq. (6–42). In evaluating $\mathcal{P}_T(f)$, the DFT is generally used to approximate $X_T(f)$. This brings the pitfalls of DFT analysis into play, as described in Sec. 2–8.

It should be stressed that Eq. (6–73a) is an approximation or *estimate* of the PSD. This estimate is called a *periodogram*, because it was used historically to search for periodicities in data records that appear as delta functions in the PSD. [Delta functions are relatively easy to find in the PSD, and thus, the periodicities in $x(t)$ are easily determined.] It is desirable that the estimate have an ensemble average that gives the true PSD. If this is the case, the estimator is said to be *unbiased*. We can easily check Eq. (6–73a) to see if it is unbiased. We have

$$\overline{\mathscr{P}_T(f)} = \overline{\left[\frac{|X_T(f)|^2}{T}\right]}$$

and, using Eq. (6–50a) for T finite, we obtain

$$\overline{\mathscr{P}_T(f)} = \mathscr{F}\left[R_x(\tau) \; \Lambda \left(\frac{\tau}{T}\right)\right]$$

and then, referring to Table 2–2,

$$\overline{\mathscr{P}_T(f)} = T\mathscr{P}_x(f) * \left(\frac{\sin \pi f T}{\pi f T}\right)^2 \tag{6–73b}$$

Because $\overline{\mathscr{P}_T(f)} \neq \mathscr{P}_x(f)$, the periodogram is a *biased* estimate. The bias is caused by the triangular window function $\Lambda(\tau/T)$ that arises from the truncation of $x(t)$ to a T-second interval. Using Sec. A–8, where $a = \pi T$, we see that $\lim_{T \to \infty} [\overline{\mathscr{P}_T(f)}] = \mathscr{P}_x(f)$, so that the periodogram becomes unbiased as $T \to \infty$. Consequently, the periodogram is said to be *asymptotically unbiased*.

In addition to being unbiased, it is desirable that an estimator be *consistent*. This means that the variance of the estimator should become small as $T \to \infty$. In this regard, it can be shown that Eq. (6–73a) gives an *inconsistent* estimate of the PSD when $x(t)$ is Gaussian [Bendat and Piersol, 1971].

Of course, window functions other than the triangle can be used to obtain different PSD estimators [Blackman and Tukey, 1958; Jenkins and Watts, 1968]. More modern techniques assume a mathematical model (i.e., form) for the autocorrelation function and estimate parameters for this model. The model is then checked to see if it is consistent with the data. Examples are the *moving average* (MA), *autoregressive* (AR), and *autoregressive-moving average* (ARMA) models [Kay, 1986; Kay and Marple, 1981; Marple, 1986; Scharf, 1991; Shanmugan and Breipohl, 1988].

Spectrum analyzers that use microprocessor-based circuits often utilize numerical techniques to evaluate the PSD. These instruments can evaluate the PSD only for relatively low-frequency waveforms, say, over the audio or ultrasonic frequency range, since real-time digital signal-processing circuits cannot be built to process signals at RF rates.

6–3 DC AND RMS VALUES FOR ERGODIC RANDOM PROCESSES

In Chapter 2, the dc value, the rms value, and the average power were defined in terms of time average operations. For ergodic processes, the time averages are equivalent to ensemble averages. Thus, the dc value, the rms value, and the average power (which are all fundamental concepts in electrical engineering) can be related to the moments of an ergodic random process. In the following summary of these relationships, $x(t)$ is an ergodic random process that may correspond to either a voltage or a current waveform:

1. Dc value:

$$X_{dc} \triangleq \langle x(t) \rangle \equiv \bar{x} = m_x \tag{6-74}$$

2. Normalized dc power:

$$P_{dc} \triangleq [\langle x(t) \rangle]^2 \equiv (\bar{x})^2 \tag{6-75}$$

3. Rms value:

$$X_{rms} \triangleq \sqrt{\langle x^2(t) \rangle} \equiv \sqrt{\overline{(x^2)}} = \sqrt{R_x(0)} = \sqrt{\int_{-\infty}^{\infty} \mathcal{P}_x(f)\,df} \tag{6-76}$$

4. Rms value of the ac part:

$$(X_{rms})_{ac} \triangleq \sqrt{\langle (x(t) - X_{dc})^2 \rangle} \equiv \sqrt{\overline{(x - \bar{x})^2}}$$

$$= \sqrt{\overline{x^2} - (\bar{x})^2} = \sqrt{R_x(0) - (\bar{x})^2}$$

$$= \sqrt{\int_{-\infty}^{\infty} \mathcal{P}_x(f)\,df - (\bar{x})^2} = \sigma_x = \text{standard deviation} \tag{6-77}$$

5. Normalized total average power:

$$P \triangleq \langle x^2(t) \rangle \equiv \overline{x^2} = R_x(0) = \int_{-\infty}^{\infty} \mathcal{P}_x(f)\,df \tag{6-78}$$

6. Normalized average power of the ac part:

$$P_{ac} \triangleq \langle (x(t) - X_{dc})^2 \rangle \equiv \overline{(x - \bar{x})^2}$$

$$= \overline{x^2} - (\bar{x})^2 = R_x(0) - (\bar{x})^2$$

$$= \int_{-\infty}^{\infty} \mathcal{P}_x(f)\,df - (\bar{x})^2 = \sigma_x^2 = \text{variance} \tag{6-79}$$

Furthermore, commonly available laboratory equipment may be used to evaluate the mean, second moment, and variance of an ergodic process. For example, if $x(t)$ is a voltage waveform, $\bar{x}$ can be measured by using a dc voltmeter, and σ_x can be measured by using a "true rms" (ac coupled) voltmeter.[†] Employing the measurements, we easily obtain the second moment from $\overline{x^2} = \sigma_x^2 + (\bar{x})^2$. At higher frequencies (e.g., radio, microwave, and optical), $\overline{x^2}$ and σ_x^2 can be measured by using a calibrated power meter. That is, $\sigma_x^2 = \overline{x^2} = RP$, where R is the load resistance of the power meter (usually 50 Ω) and P is the power meter reading.

[†] Most "true rms" meters do not have a frequency response down to dc. Thus, they do not measure the true rms value $\sqrt{\langle x^2(t) \rangle} = \sqrt{\overline{x^2}}$; instead, they measure σ.

6–4 LINEAR SYSTEMS

Input–Output Relationships

As developed in Chapter 2, a linear time-invariant system may be described by its impulse response $h(t)$ or, equivalently, by its transfer function $H(f)$. This is illustrated in Fig. 6–6, in which $x(t)$ is the input and $y(t)$ is the output. The input–output relationship is

$$y(t) = h(t) * x(t) \tag{6-80}$$

The corresponding Fourier transform relationship is

$$Y(f) = H(f)X(f) \tag{6-81}$$

If $x(t)$ and $y(t)$ are random processes, these relationships are still valid (just as they were for the case of deterministic functions). In communication systems, $x(t)$ might be a random signal plus (random) noise, or $x(t)$ might be noise alone when the signal is absent. In the case of random processes, autocorrelation functions and PSD functions may be used to describe the frequencies involved. Consequently, we need to address the following question: What are the autocorrelation function and the PSD for the output process $y(t)$ when the autocorrelation and PSD for the input $x(t)$ are known?

THEOREM.   *If a wide-sense stationary random process $x(t)$ is applied to the input of a time-invariant linear network with impulse response $h(t)$, the output autocorrelation is*

$$R_y(\tau) = \int_{-\infty}^{\infty} \int_{-\infty}^{\infty} h(\lambda_1)h(\lambda_2) R_x(\tau - \lambda_2 + \lambda_1) \, d\lambda_1 d\lambda_2 \tag{6-82a}$$

or

$$R_y(\tau) = h(-\tau) * h(\tau) * R_x(\tau) \tag{6-82b}$$

The output PSD is

$$\mathcal{P}_y(f) = |H(f)|^2 \, \mathcal{P}_x(f) \tag{6-83}$$

where $H(f) = \mathcal{F}[h(t)]$.

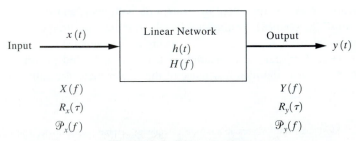

Figure 6–6   Linear system.

Equation (6–83) shows that the power transfer function of the network is

$$G_h(f) = \frac{\mathcal{P}_y(f)}{\mathcal{P}_x(f)} = |H(f)|^2 \tag{6–84}$$

as cited in Eq. (2–143).

Proof. From Eq. (6–80),

$$R_y(\tau) \triangleq \overline{y(t)y(t + \tau)}$$

$$= \overline{\left[\int_{-\infty}^{\infty} h(\lambda_1)x(t - \lambda_1)\, d\lambda_1 \right] \left[\int_{-\infty}^{\infty} h(\lambda_2)x(t + \tau - \lambda_2)\, d\lambda_2 \right]}$$

$$= \int_{-\infty}^{\infty} \int_{-\infty}^{\infty} h(\lambda_1)h(\lambda_2)\overline{x(t - \lambda_1)x(t + \tau - \lambda_2)}\, d\lambda_1 d\lambda_2 \tag{6–85}$$

But

$$\overline{x(t - \lambda_1)x(t + \tau - \lambda_2)} = R_x(t + \tau - \lambda_2 - t + \lambda_1) = R_x(\tau - \lambda_2 + \lambda_1)$$

so Eq. (6–85) is equivalent to Eq. (6–82a). Furthermore, Eq. (6–82a) may be written in terms of convolution operations as

$$R_y(\tau) = \int_{-\infty}^{\infty} h(\lambda_1) \left\{ \int_{-\infty}^{\infty} h(\lambda_2) R_x[(\tau + \lambda_1) - \lambda_2]\, d\lambda_2 \right\} d\lambda_1$$

$$= \int_{-\infty}^{\infty} h(\lambda_1) \{ h(\tau + \lambda_1) * R_x(\tau + \lambda_1) \}\, d\lambda_1$$

$$= \int_{-\infty}^{\infty} h(\lambda_1) \{ h[- ((-\tau) - \lambda_1)] * R_x[- ((-\tau) - \lambda_1)] \}\, d\lambda_1$$

$$= h(-\tau) * h[- (-\tau)] * R_x[- (-\tau)]$$

which is equivalent to the convolution notation of Eq. (6–82b).

The PSD of the output is obtained by taking the Fourier transform of Eq. (6–82b). We obtain

$$\mathcal{F}[R_y(\tau)] = \mathcal{F}[h(-\tau)]\mathcal{F}[h(\tau) * R_x(\tau)]$$

or

$$\mathcal{P}_y(f) = H^*(f)H(f)\mathcal{P}_x(f)$$

where $h(t)$ is assumed to be real. This equation is equivalent to Eq. (6–83).

This theorem may be applied to cascaded linear systems. For example, two cascaded networks are shown in Fig. 6–7.

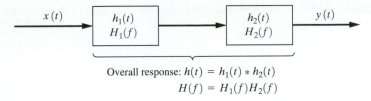

Overall response: $h(t) = h_1(t) * h_2(t)$
$$H(f) = H_1(f)H_2(f)$$

Figure 6–7 Two linear cascaded networks.

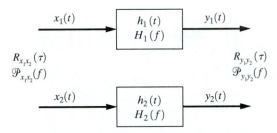

Figure 6–8 Two linear systems.

The theorem may also be generalized to obtain the cross-correlation or cross-spectrum of two linear systems, as illustrated in Fig. 6–8. $x_1(t)$ and $y_1(t)$ are the input and output processes of the first system, which has the impulse response $h_1(t)$. Similarly, $x_2(t)$ and $y_2(t)$ are the input and output processes for the second system.

THEOREM.   *Let $x_1(t)$ and $x_2(t)$ be wide-sense stationary inputs for two time-invariant linear systems, as shown in Fig. 6–8; then the output cross-correlation function is*

$$R_{y_1y_2}(\tau) = \int_{-\infty}^{\infty} \int_{-\infty}^{\infty} h_1(\lambda_1)h_2(\lambda_2) R_{x_1x_2}(\tau - \lambda_2 + \lambda_1) \, d\lambda_1 d\lambda_2 \qquad (6\text{–}86\text{a})$$

or

$$R_{y_1y_2}(\tau) = h_1(-\tau) * h_2(\tau) * R_{x_1x_2}(\tau) \qquad (6\text{–}86\text{b})$$

Furthermore, by definition, the output cross power spectral density is the Fourier transform of the cross-correlation function; thus,

$$\mathscr{P}_{y_1y_2}(f) = H_1^*(f)H_2(f)\mathscr{P}_{x_1x_2}(f) \qquad (6\text{–}87)$$

where $\mathscr{P}_{y_1y_2}(f) = \mathscr{F}[R_{y_1y_2}(\tau)]$, $\mathscr{P}_{x_1x_2}(f) = \mathscr{F}[R_{x_1x_2}(\tau)]$, $H_1(f) = \mathscr{F}[h_1(t)]$,   *and* $H_2(f) = \mathscr{F}[h_2(t)]$.

The proof of this theorem is similar to that for the preceding theorem and will be left to the reader as an exercise.

Example 6–4 OUTPUT AUTOCORRELATION AND PSD
            FOR AN *RC* LOW-PASS FILTER

An *RC* low-pass filter (LPF) is shown in Fig. 6–9. Assume that the input is an ergodic random process with a uniform PSD:

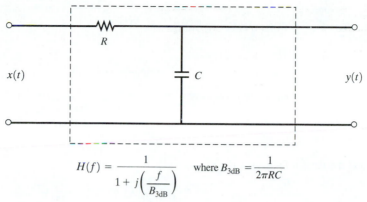

$$H(f) = \frac{1}{1 + j\left(\dfrac{f}{B_{3dB}}\right)} \quad \text{where } B_{3dB} = \frac{1}{2\pi RC}$$

Figure 6–9   *RC* LPF

$$\mathcal{P}_x(f) = \tfrac{1}{2}N_0$$

Then the PSD of the output is

$$\mathcal{P}_y(f) = |H(f)|^2 \mathcal{P}_x(f)$$

which becomes

$$\mathcal{P}_y(f) = \frac{\tfrac{1}{2}N_0}{1 + (f/B_{3dB})^2} \tag{6–88}$$

where $B_{3db} = 1/(2\pi RC)$. Note that $\mathcal{P}_y(B_{3dB})/\mathcal{P}_y(0)$ is $\tfrac{1}{2}$, so $B_{3dB} = 1/(2\pi RC)$ is indeed the 3-dB bandwidth. Taking the inverse Fourier transform of $\mathcal{P}_y(f)$, as described by Eq. (6–88), we obtain the output autocorrelation function for the RC filter.

$$R_y(\tau) = \frac{N_0}{4RC} e^{-|\tau|/(RC)} \tag{6–89}$$

The normalized output power, which is the second moment of the output, is

$$P_y = \overline{y^2} = R_y(0) = \frac{N_0}{4RC} \tag{6–90}$$

Furthermore, the dc value of the output, which is the mean value, is zero, since[†]

$$Y_{DC} = m_y = \sqrt{\lim_{\substack{\varepsilon \to 0 \\ \varepsilon > 0}} \int_{-\varepsilon}^{\varepsilon} \mathcal{P}_y(f)\, df} = 0 \tag{6–91}$$

The variance of the output is also given by Eq. (6–90) because $\sigma_y^2 = \overline{y^2} - m_y^2$, where $m_y = 0$.

Example 6–5 SIGNAL-TO-NOISE RATIO AT THE OUTPUT OF AN *RC* LOW-PASS FILTER

Refer to Fig. 6–9 again, and assume that $x(t)$ consists of a sine-wave (deterministic) signal plus white ergodic noise. Thus,

---

[†] When the integral is nonzero, the value obtained is equal to or larger than the square of the mean value.

$$x(t) = s_i(t) + n_i(t)$$

and

$$s_i(t) = A_0 \cos(\omega_0 t + \theta_0)$$

where A_0, ω_0, and θ_0 are known constants and $\mathcal{P}_{n_i}(f) = N_0/2$. The input signal power is

$$\langle s_i^2(t) \rangle = \frac{A_0^2}{2}$$

and the input noise power is

$$\langle n_i^2 \rangle = \overline{n_i^2} = \int_{-\infty}^{\infty} \mathcal{P}_{n_i}(f) \ df = \int_{-\infty}^{\infty} \frac{N_0}{2} \ df = \infty$$

Hence, the input signal-to-noise ratio (SNR) is

$$\left(\frac{S}{N}\right)_{\text{in}} = \frac{\langle s_i^2(t) \rangle}{\langle n_i^2(t) \rangle} = 0 \tag{6-92}$$

Because the system is linear, the output consists of the sum of the filtered input signal plus the filtered input noise:

$$y(t) = s_0(t) + n_0(t)$$

The output signal is

$$s_0(t) = s_i(t) * h(t)$$

$$= A_0 |H(f_0)| \cos[\omega_0 t + \theta_0 + \underline{/H(f_0)}]$$

and the output signal power is

$$\langle s_0^2(t) \rangle = \frac{A_0^2}{2} |H(f_0)|^2 \tag{6-93}$$

From Eq. (6–90) of Example 6–4, the output noise power is

$$\overline{n_0^2} = \frac{N_0}{4RC} \tag{6-94}$$

The output SNR is then

$$\left(\frac{S}{N}\right)_{\text{out}} = \frac{\langle s_0^2(t) \rangle}{\langle n_0^2(t) \rangle} = \frac{\langle s_0^2(t) \rangle}{\overline{n_0^2(t)}} = \frac{2A_0^2 |H(f_0)|^2 RC}{N_0} \tag{6-95}$$

or

$$\left(\frac{S}{N}\right)_{\text{out}} = \frac{2A_0^2 RC}{N_0[1 + (2\pi f_0 RC)^2]}$$

The reader is encouraged to use calculus to find the value of RC that maximizes the SNR. It is $RC = 1/(2\pi f_0)$. Thus, for maximum SNR, the filter is designed to have a 3-dB bandwidth equal to f_0.

6–5 BANDWIDTH MEASURES

Several bandwidth measures were defined in Sec. 2–9, namely, absolute bandwidth, 3-dB bandwidth, equivalent bandwidth, null-to-null (zero-crossing) bandwidth, bounded bandwidth, power bandwidth, and the FCC bandwidth parameter. These definitions can be applied to evaluate the bandwidth of a wide-sense stationary process $x(t)$, where $\mathcal{P}_x(f)$ replaces $|H(f)|^2$ in the definitions. In this section, we review the equivalent bandwidth and define a new bandwidth measure: rms bandwidth.

Equivalent Bandwidth

For a wide-sense stationary process $x(t)$, the equivalent bandwidth, as defined by Eq. (2–192), becomes

$$B_{\text{eq}} = \frac{1}{\mathcal{P}_x(f_0)} \int_0^\infty \mathcal{P}_x(f) \ df = \frac{R_x(0)}{2\mathcal{P}_x(f_0)} \tag{6–96}$$

where f_0 is the frequency at which $\mathcal{P}_x(f)$ is a maximum. This equation is valid for both bandpass and baseband processes. ($f_0 = 0$ is used for baseband processes.)

Rms Bandwidth

The rms bandwidth is the square root of the second moment of the frequency with respect to a properly normalized PSD. In this case, f, although not random, may be *treated* as a random variable that has the "density function" $\mathcal{P}_x(f)/\int_{-\infty}^{\infty} \mathcal{P}_x(\lambda) \ d\lambda$. This is a nonnegative function in which the denominator provides the correct normalization so that the integrated value of the ratio is unity. Thus, the function satisfies the properties of a PDF.

DEFINITION.   If $x(t)$ is a *low-pass* wide-sense stationary process, the *rms bandwidth* is

$$B_{\text{rms}} = \sqrt{\overline{f^2}} \tag{6–97}$$

where

$$\overline{f^2} = \int_{-\infty}^{\infty} f^2 \left[\frac{\mathcal{P}_x(f)}{\left[\int_{-\infty}^{\infty} \mathcal{P}_x(\lambda) \ d\lambda \right]} \right] df = \frac{\int_{-\infty}^{\infty} f^2 \ \mathcal{P}_x(f) \ df}{\int_{-\infty}^{\infty} \mathcal{P}_x(\lambda) \ d\lambda} \tag{6–98}$$

The rms measure of bandwidth is often used in theoretical comparisons of communication systems because the mathematics involved in the rms calculation is often easier to carry out than that for other bandwidth measures. However, the rms bandwidth is not easily measured with laboratory instruments.

THEOREM. *For a wide-sense stationary process* $x(t)$, *the mean-squared frequency is*

$$\overline{f^2} = \left[-\frac{1}{(2\pi)^2 R(0)} \right] \frac{d^2 R_x(\tau)}{d\tau^2} \bigg|_{\tau=0} \tag{6-99}$$

Proof. We know that

$$R_x(\tau) = \int_{-\infty}^{\infty} \mathcal{P}_x(f) e^{j2\pi f \tau} \, dt$$

Taking the second derivative of this equation with respect to τ, we obtain

$$\frac{d^2 R_x(\tau)}{d\tau^2} = \int_{-\infty}^{\infty} \mathcal{P}_x(f) e^{j2\pi f \tau} (j2\pi f)^2 \, df$$

Evaluating both sides of this equation at $\tau = 0$ yields

$$\frac{d^2 R_x(\tau)}{d\tau^2} \bigg|_{\tau=0} = (j2\pi)^2 \int_{-\infty}^{\infty} f^2 \mathcal{P}_x(f) \, df$$

Substituting this for the integral in Eq. (6–98), that equation becomes

$$\overline{f^2} = \frac{\displaystyle\int_{-\infty}^{\infty} f^2 \mathcal{P}_x(f) \, df}{\displaystyle\int_{-\infty}^{\infty} \mathcal{P}_x(\lambda) \, d\lambda} = \frac{-\dfrac{1}{(2\pi)^2} \left[\dfrac{d^2 R_x(\tau)}{d\tau^2} \right]_{\tau=0}}{R_x(0)}$$

which is identical to Eq. (6–99).

The rms bandwidth for a *bandpass* process can also be defined. Here we are interested in the square root of the second moment about the mean frequency of the *positive* frequency portion of the spectrum.

DEFINITION. If $x(t)$ is a *bandpass* wide-sense stationary process, the *rms bandwidth* is

$$B_{\text{rms}} = 2\sqrt{\overline{(f - f_0)^2}} \tag{6-100}$$

where

$$\overline{(f - f_0)^2} = \int_0^{\infty} (f - f_0)^2 \left(\frac{\mathcal{P}_x(f)}{\displaystyle\int_0^{\infty} \mathcal{P}_x(\lambda) \, d\lambda} \right) df \tag{6-101}$$

and

$$f_0 \triangleq \overline{f} = \int_0^{\infty} f \left(\frac{\mathcal{P}_x(f)}{\displaystyle\int_0^{\infty} \mathcal{P}_x(\lambda) \, d\lambda} \right) df \tag{6-102}$$

As a sketch of a typical bandpass PSD will verify, the quantity given by the radical in Eq. (6–100) is analogous to σ_f. Consequently, the factor of 2 is needed to give a reasonable definition for the bandpass bandwidth.

Example 6–6 EQUIVALENT BANDWIDTH AND RMS
BANDWIDTH FOR AN *RC* LPF

To evaluate the equivalent bandwidth and the rms bandwidth for a filter, white noise may be applied to the input. The bandwidth of the output PSD is the bandwidth of the filter, since the input PSD is a constant.

For the *RC* LPF (Fig. 6–9), the output PSD, for white noise at the input, is given by Eq. (6–88). The corresponding output autocorrelation function is given by Eq. (6–89). When we substitute these equations into Eq. (6–96), the equivalent bandwidth for the *RC* LPF is

$$B_{eq} = \frac{R_y(0)}{2\mathcal{P}_y(0)} = \frac{N_0/4RC}{2(\frac{1}{2}N_0)} = \frac{1}{4RC} \text{ hertz} \tag{6–103}$$

Consequently, for an *RC* LPF,

$$B_{eq} = \frac{\pi}{2}B_{3dB} \tag{6–104}$$

The rms bandwidth is obtained by substituting Eqs. (6–88) and (6–90) into Eq. (6–98). We obtain

$$B_{rms} = \sqrt{\frac{\int_{-\infty}^{\infty} f^2 \mathcal{P}_y(f)df}{R_y(0)}} = \sqrt{\frac{1}{2\pi^2 RC} \int_{-\infty}^{\infty} \frac{f^2}{(B_{3dB})^2 + f^2} df} \tag{6–105}$$

Examining the integral, we note that the integrand becomes *unity* as $f \to \pm\infty$, so that the value of the integral is infinity. Thus, $B_{rms} = \infty$ for an *RC* LPF. For the rms bandwidth to be finite, the PSD needs to decay faster than $1/|f|^2$ as the frequency becomes large. Consequently, for the RC LPF, the rms definition is not very useful.

6–6 THE GAUSSIAN RANDOM PROCESS

DEFINITION.   A random process $x(t)$ is said to be Gaussian if the random variables

$$x_1 = x(t_1), \, x_2 = x(t_2), \, \dots, \, x_N = x(t_N) \tag{6–106}$$

have an *N*-dimensional Gaussian PDF for any *N* and any $t_1, t_2, \dots, t_N$.

The *N*-dimensional Gaussian PDF can be written compactly by using matrix notation. Let **x** be the *column* vector denoting the *N* random variables:

$$\mathbf{x} = \begin{bmatrix} x_1 \\ x_2 \\ \vdots \\ x_N \end{bmatrix} = \begin{bmatrix} x(t_1) \\ x(t_2) \\ \vdots \\ x(t_N) \end{bmatrix} \tag{6–107}$$

The N-dimensional Gaussian PDF is

$$f_{\mathbf{x}}(\mathbf{x}) = \frac{1}{(2\pi)^{N/2}|\text{Det } \mathbf{C}|^{1/2}} \, e^{-(1/2)[(\mathbf{x}-\mathbf{m})^T \mathbf{C}^{-1}(\mathbf{x}-\mathbf{m})]} \tag{6-108}$$

where the mean vector is

$$\mathbf{m} = \bar{\mathbf{x}} = \begin{bmatrix} \bar{x}_1 \\ \bar{x}_2 \\ \vdots \\ \bar{x}_N \end{bmatrix} = \begin{bmatrix} m_1 \\ m_2 \\ \vdots \\ m_N \end{bmatrix} \tag{6-109}$$

and where $(\mathbf{x} - \mathbf{m})^T$ denotes the transpose of the column vector $(\mathbf{x} - \mathbf{m})$.

Det $\mathbf{C}$ is the determinant of the matrix $\mathbf{C}$, and $\mathbf{C}^{-1}$ is the inverse of the matrix $\mathbf{C}$. The covariance matrix is defined by

$$\mathbf{C} = \begin{bmatrix} c_{11} & c_{12} & \cdots & c_{1N} \\ c_{21} & c_{22} & \cdots & c_{2N} \\ \vdots & \vdots & & \vdots \\ c_{N1} & c_{N2} & \cdots & c_{NN} \end{bmatrix} \tag{6-110}$$

where the elements of the matrix are

$$c_{ij} = \overline{(x_i - m_i)(x_j - m_j)} = \overline{[x(t_i) - m_i][x(t_j) - m_j]} \tag{6-111}$$

For a wide-sense stationary process, $m_i = \overline{x(t_i)} = m_j = \overline{x(t_j)} = m$. The elements of the covariance matrix become

$$c_{ij} = R_x(t_j - t_i) - m^2 \tag{6-112}$$

If, in addition, the x_i happen to be uncorrelated, $\overline{x_i x_j} = \bar{x}_i \bar{x}_j$ for $i \neq j$, and the covariance matrix becomes

$$\mathbf{C} = \begin{bmatrix} \sigma^2 & 0 & \cdots & 0 \\ 0 & \sigma^2 & \cdots & 0 \\ \vdots & & \ddots & \vdots \\ 0 & 0 & \cdots & \sigma^2 \end{bmatrix} \tag{6-113}$$

where $\sigma^2 = \overline{x^2} - m^2 = R_x(0) - m^2$. That is, the covariance matrix becomes a diagonal matrix if the random variables are uncorrelated. Using Eq. (6–113) in Eq. (6–108), we can conclude that the Gaussian random variables are independent when they are uncorrelated.

Properties of Gaussian Processes

Some properties of Gaussian processes are:

1. $f_{\mathbf{x}}(\mathbf{x})$ depends only on $\mathbf{C}$ and $\mathbf{m}$, which is another way of saying that the N-dimensional Gaussian PDF is completely specified by the first- and second-order moments (i.e., means, variances, and covariances).
2. Since the $\{x_i = x(t_i)\}$ are jointly Gaussian, the $x_i = x(t_i)$ are individually Gaussian.

3. When **C** is a diagonal matrix, the random variables are uncorrelated. Furthermore, the Gaussian random variables are independent when they are uncorrelated.

4. A linear transformation of a set of Gaussian random variables produces another set of Gaussian random variables.

5. A wide-sense stationary Gaussian process is also strict-sense stationary[†] [Papoulis, 1984, p. 222; Shanmugan and Breipohl, 1988, p. 141].

Property 4 is very useful in the analysis of linear systems. This property, as well as the following theorem, will be proved subsequently.

THEOREM.   *If the input to a linear system is a Gaussian random process, the system output is also a Gaussian process.*

Proof. The output of a linear network having an impulse response $h(t)$ is

$$y(t) = h(t) * x(t) = \int_{-\infty}^{\infty} h(t - \lambda) x(\lambda) \, d\lambda$$

This can be approximated by

$$y(t) = \sum_{j=1}^{N} h(t - \lambda_j) x(\lambda_j) \, \Delta\lambda \tag{6–114}$$

which becomes exact as N gets large and $\Delta\lambda \to 0$.

The output random variables for the output random process are

$$y(t_1) = \sum_{j=1}^{N} [h(t_1 - \lambda_j) \, \Delta\lambda] x(\lambda_j)$$

$$y(t_2) = \sum_{j=1}^{N} [h(t_2 - \lambda_j) \, \Delta\lambda] x(\lambda_j)$$

$$\vdots$$

$$y(t_N) = \sum_{j=1}^{N} [h(t_N - \lambda_j) \, \Delta\lambda] x(\lambda_j)$$

In matrix notation, these equations become

$$\begin{bmatrix} y_1 \\ y_2 \\ \vdots \\ y_N \end{bmatrix} = \begin{bmatrix} h_{11} & h_{12} & \cdots & h_{1N} \\ h_{21} & h_{22} & \cdots & h_{2N} \\ \vdots & \vdots & \ddots & \vdots \\ h_{N1} & h_{N2} & \cdots & h_{NN} \end{bmatrix} \begin{bmatrix} x_1 \\ x_2 \\ \vdots \\ x_N \end{bmatrix}$$

or

$$\mathbf{y} = \mathbf{Hx} \tag{6–115}$$

[†] This follows directly from Eq. (6–112), because the N-dimensional Gaussian PDF is a function only of τ and not the absolute times.

where the elements of the $N \times N$ matrix $\mathbf{H}$ are related to the impulse response of the linear network by

$$h_{ij} = [h(t_i - \lambda_j)] \, \Delta\lambda \qquad (6\text{-}116)$$

We will now show that y is described by an N-dimensional Gaussian PDF when $\mathbf{x}$ is described by an N-dimensional Gaussian PDF. This may be accomplished by using the theory for a multivariate functional transformation, as given in Appendix B. From Eq. (B–99), the PDF of $\mathbf{y}$ is

$$f_\mathbf{y}(\mathbf{y}) = \frac{f_\mathbf{x}(\mathbf{x})}{|J(\mathbf{y}/\mathbf{x})|}\bigg|_{\mathbf{x}=H^{-1}\mathbf{y}} \qquad (6\text{-}117)$$

The Jacobian is

$$J\left(\frac{\mathbf{y}}{\mathbf{x}}\right) = \text{Det} \begin{bmatrix} \dfrac{dy_1}{dx_1} & \dfrac{dy_1}{dx_2} & \cdots & \dfrac{dy_1}{dx_N} \\[2mm] \dfrac{dy_2}{dx_1} & \dfrac{dy_2}{dx_2} & \cdots & \dfrac{dy_2}{dx_N} \\[2mm] \vdots & \vdots & & \vdots \\[2mm] \dfrac{dy_N}{dx_1} & \dfrac{dy_N}{dx_2} & \cdots & \dfrac{dy_N}{dx_N} \end{bmatrix} = \text{Det}[\mathbf{H}] \triangleq K \qquad (6\text{-}118)$$

where K is a constant. In this problem, $J(\mathbf{y}/\mathbf{x})$ is a *constant* (not a function of $\mathbf{x}$) because $\mathbf{y} = \mathbf{Hx}$ is a *linear* transformation. Thus,

$$f_\mathbf{y}(\mathbf{y}) = \frac{1}{|K|} f_\mathbf{x}(\mathbf{H}^{-1}\mathbf{y})$$

or

$$f_\mathbf{y}(\mathbf{y}) = \frac{1}{(2\pi)^{N/2}|K|\,|\text{Det } \mathbf{C}_x|^{1/2}} \, e^{-(1/2)[(\mathbf{H}^{-1}\mathbf{y}-\mathbf{m}_x)^T \mathbf{C}_x^{-1} \, (\mathbf{H}^{-1}\mathbf{y}-\mathbf{m}_x)]} \qquad (6\text{-}119)$$

where the subscript x has been appended to the quantities that are associated with $x(t)$. But we know that $\mathbf{m}_y = \mathbf{Hm}_x$, and from matrix theory, we have the property $[\mathbf{AB}]^T = \mathbf{B}^T\mathbf{A}^T$, so that the exponent of Eq. (6–119) becomes

$$-\tfrac{1}{2}[(\mathbf{y} - \mathbf{m}_y)^T(\mathbf{H}^{-1})^T]\mathbf{C}_x^{-1}[\mathbf{H}^{-1}(\mathbf{y} - \mathbf{m}_y)] = -\tfrac{1}{2}[(\mathbf{y} - \mathbf{m}_y)^T\mathbf{C}_y^{-1}(\mathbf{y} - \mathbf{m}_y)] \quad (6\text{-}120)$$

where

$$\mathbf{C}_y^{-1} = (\mathbf{H}^{-1})^T\mathbf{C}_x^{-1}\mathbf{H}^{-1} \qquad (6\text{-}121)$$

or

$$\mathbf{C}_y = \mathbf{HC}_x\mathbf{H}^T \qquad (6\text{-}122)$$

Thus, the PDF for **y** is

$$f_{\mathbf{y}}(\mathbf{y}) = \frac{1}{(2\pi)^{N/2}|K|\,|Det\ \mathbf{C_x}|^{1/2}}\ e^{-(1/2)(\mathbf{y}-\mathbf{m}_y)^T\,\mathbf{C}_y^{-1}\,(\mathbf{y}-\mathbf{m}_y)} \qquad (6\text{–}123)$$

which is an *N*-dimensional Gaussian PDF. This completes the proof of the theorem.

    If a linear system acts like an integrator or an LPF, the output random variables (of the output random process) tend to be proportional to the sum of the input random variables. Consequently, by applying the central limit theorem (see Appendix B), the output of the integrator or LPF will tend toward a Gaussian random process when the input random variables are independent with non-Gaussian PDFs.

Example 6–7 WHITE GAUSSIAN-NOISE PROCESS

Assume that a Gaussian random process $n(t)$ is given that has a PSD of

$$\mathcal{P}_n(f) = \begin{cases} \frac{1}{2}\,N_0, & |f| \leq B \\ 0, & f \quad \text{otherwise} \end{cases} \qquad (6\text{–}124)$$

where *B* is a positive constant. This describes a *bandlimited white Gaussian* process as long as *B* is finite, but becomes a completely *white* (all frequencies are present) *Gaussian process as* $B \to \infty$.

    The autocorrelation function for the bandlimited white process is

$$R_n(\tau) = BN_0 \left(\frac{\sin\ 2\pi B\tau}{2\pi B\tau} \right) \qquad (6\text{–}125)$$

The total average power is $P = R_n(0) = BN_0$. The mean value of $n(t)$ is zero, since there is no δ function in the PSD at $f = 0$. Furthermore, the autocorrelation function is zero for $\tau = k/(2B)$ when k is a nonzero integer. Therefore, the random variables $n_1 = n(t_1)$ and $n_2 = n(t_2)$ are uncorrelated when $t_2 - t_1 = \tau = k/(2B)$, $k \neq 0$. For other values of τ, the random variables are correlated. Since $n(t)$ is assumed to be Gaussian, n_1 and n_2 are jointly Gaussian random variables. Consequently, by property 3, the random variables are independent when $t_2 - t_1 = k/(2B)$. They are dependent for other values of t_2 and t_1. As $B \to \infty$, $R_n(\tau) \to \frac{1}{2}N_0\,\delta(\tau)$, and the random variables n_1 and n_2 become independent for all values of t_1 and t_2, provided that $t_1 \neq t_2$. Furthermore, as $B \to \infty$, the average power becomes infinite. Consequently, a white-noise process is not physically realizable. However, it is a very useful mathematical idealization for system analysis, just as a deterministic impulse is useful for obtaining the impulse response of a linear system, although the impulse itself is not physically realizable.

6–7 BANDPASS PROCESSES[†]

Bandpass Representations

In Sec. 4–1, it was demonstrated that any bandpass waveform could be represented by

$$v(t) = \text{Re}\{g(t)\,e^{j\omega_c t}\} \qquad (6\text{–}126a)$$

[†] In some other texts these are called *narrowband noise processes*, which is a misnomer because they may be wideband *or* narrowband.

or, equivalently, by

$$v(t) = x(t) \cos \omega_c t - y(t) \sin \omega_c t \qquad (6\text{--}126\text{b})$$

and

$$v(t) = R(t) \cos[\omega_c t + \theta(t)] \qquad (6\text{--}126\text{c})$$

where $g(t)$ is the complex envelope, $R(t)$ is the real envelope, $\theta(t)$ is the phase, and $x(t)$ and $y(t)$ are the quadrature components. Thus, the complex envelope is

$$g(t) = |g(t)| e^{j\underline{/g(t)}} = R(t) e^{j\,\theta(t)} = x(t) + jy(t) \qquad (6\text{--}127\text{a})$$

with the relationships

$$R(t) = |g(t)| = \sqrt{x^2(t) + y^2(t)} \qquad (6\text{--}127\text{b})$$

$$\theta(t) = \underline{/g(t)} = \tan^{-1}\left[\frac{y(t)}{x(t)}\right] \qquad (6\text{--}127\text{c})$$

$$x(t) = R(t) \cos \theta(t) \qquad (6\text{--}127\text{d})$$

and

$$y(t) = R(t) \sin \theta(t) \qquad (6\text{--}127\text{e})$$

Furthermore, the spectrum of $v(t)$ is related to that of $g(t)$ by Eq. (4–12), which is

$$V(f) = \tfrac{1}{2}[G(f - f_c) + G^*(-f - f_c)] \qquad (6\text{--}128)$$

In Chapters 4 and 5, the bandpass representation was used to analyze communication systems from a *deterministic* viewpoint. Here, we extend the representation to random processes, which may be random signals, noise, or signals corrupted by noise.

If $v(t)$ is a bandpass random process containing frequencies in the vicinity of $\pm f_c$, then $g(t)$, $x(t)$, $y(t)$, $R(t)$, and $\theta(t)$ will be baseband processes. In general, $g(t)$ is a complex process (as described in Sec. 6–1), and $x(t)$, $y(t)$, $R(t)$, and $\theta(t)$ are always real processes. This is readily seen from a Fourier series expansion of $v(t)$, as demonstrated by Eqs. (4–5) through (4–8), where the Fourier series coefficients form a set of random variables, since $v(t)$ is a random process. Furthermore, if $v(t)$ is a Gaussian random process, the Fourier series coefficients consist of a set of *Gaussian* random variables because they are obtained by linear operations on $v(t)$. Similarly, when $v(t)$ is a Gaussian process, $g(t)$, $x(t)$, and $y(t)$ are Gaussian processes, since they are linear functions of $v(t)$. However, $R(t)$ and $\theta(t)$ are not Gaussian, because they are nonlinear functions of $v(t)$. The first-order PDF for these processes will be evaluated in Example 6–10.

We now need to address the topic of stationarity as applied to the bandpass representation.

THEOREM. *If $x(t)$ and $y(t)$ are jointly wide-sense stationary (WSS) processes, the real bandpass process*

$$v(t) = \text{Re}\{g(t) e^{j\omega_c t}\} = x(t) \cos \omega_c t - y(t) \sin \omega_c t \qquad (6\text{--}129\text{a})$$

will be WSS if and only if

$$\textbf{1.}\ \ \overline{x(t)} = \overline{y(t)} = 0 \tag{6–129b}$$

$$\textbf{2.}\ \ R_x(\tau) = R_y(\tau) \tag{6–129c}$$

and

$$\textbf{3.}\ \ R_{xy}(\tau) = -R_{yx}(\tau) \tag{6–129d}$$

   Proof. The requirements for WSS are that $\overline{v(t)}$ be constant and $R_v(t, t + \tau)$ be a function only of τ. We see that $\overline{v(t)} = \overline{x(t)}\cos \omega_c t - \overline{y(t)}\sin \omega_c t$ is a constant for any value of t only if $\overline{x(t)} = \overline{y(t)} = 0$. Thus, condition 1 is required.
   The conditions required to make $R_v(t, t + \tau)$ only a function of τ are found as follows:

$$R_v(t, t + \tau)$$

$$= \overline{v(t)v(t + \tau)}$$

$$= \overline{[x(t)\ \cos\ \omega_c t - y(t)\ \sin\ \omega_c t][x(t + \tau)\ \cos\ \omega_c(t + \tau) - y(t + \tau)\ \sin\ \omega_c(t + \tau)]}$$

$$= \overline{x(t)x(t + \tau)}\ \cos\ \omega_c t\ \cos\ \omega_c(t + \tau) - \overline{x(t)y(t + \tau)}\ \cos\ \omega_c t\ \sin\ \omega_c(t + \tau)$$

$$- \overline{y(t)x(t + \tau)}\ \sin\ \omega_c t\ \cos\ \omega_c(t + \tau) + \overline{y(t)y(t + \tau)}\ \sin\ \omega_c t\ \sin\ \omega_c(t + \tau)$$

or

$$R_v(t, t + \tau) = R_x(\tau)\ \cos\ \omega_c t\ \cos\ \omega_c(t + \tau) - R_{xy}(\tau)\ \cos\ \omega_c t\ \sin\ \omega_c(t + \tau)$$

$$- R_{yx}(\tau)\ \sin\ \omega_c t\ \cos\ \omega_c(t + \tau) + R_y(\tau)\ \sin\ \omega_c t\ \sin\ \omega_c(t + \tau)$$

When we use trigonometric identities for products of sines and cosines, this equation reduces to

$$R_v(t, t + \tau) = \tfrac{1}{2}[R_x(\tau) + R_y(\tau)]\ \cos\ \omega_c \tau + \tfrac{1}{2}[R_x(\tau) - R_y(\tau)]\ \cos\ \omega_c(2t + \tau)$$

$$- \tfrac{1}{2}[R_{xy}(\tau) - R_{yx}(\tau)]\ \sin\ \omega_c \tau - \tfrac{1}{2}[R_{xy}(\tau) + R_{yx}(\tau)]\ \sin\ \omega_c(2t + \tau)$$

The autocorrelation for $v(t)$ can be made to be a function of τ only if the terms involving t are set equal to zero. That is, $[R_x(\tau) - R_y(\tau)] = 0$ and $[R_{xy}(\tau) + R_{yx}(\tau)] = 0$. Thus, conditions 2 and 3 are required.

   If conditions 1 through 3 of Eq. (6–129) are satisfied so that $v(t)$ is WSS, properties 1 through 5 of Eqs. (6–133a) through (6–133e) are valid. Furthermore, the $x(t)$ and $y(t)$ components of $v(t) = x(t)\cos \omega_c t - y(t)\sin \omega_c t$ satisfy properties 6 through 14, as described by Eqs. (6–133f) through (6–133n), when conditions 1 through 3 are satisfied. These properties are highly useful in analyzing the random processes at various points in communication systems.
   For a given bandpass waveform, the description of the complex envelope $g(t)$ is not unique. This is easily seen in Eq. (6–126), where the choice of the value for the parameter f_c is left to our discretion. Consequently, in the representation of a given bandpass waveform $v(t)$, the frequency components that are present in the corresponding complex envelope $g(t)$ depend on the value of f_c that is chosen in the model. Moreover, in representing random processes, one

is often interested in having a representation for a WSS bandpass process with certain PSD characteristics. In this case, it can be shown that $\text{Re}\{(-j)g(t)e^{j\omega_c t}\}$ gives the same PSD as $\text{Re}\{g(t)e^{j\omega_c t}\}$ when $v(t)$ is a WSS process [Papoulis, 1984, pp. 314–322]. Consequently, $g(t)$ is not unique, and it can be chosen to satisfy some additional desired condition. Yet, properties 1 through 14 will still be satisfied when the conditions of Eq. (6–129) are satisfied.

In some applications, conditions 1 through 3 of Eq. (6–129) are *not* satisfied. This will be the case, for example, when the $x(t)$ and $y(t)$ quadrature components do not have the same power, as in an unbalanced quadrature modulation problem. Another example is when $x(t)$ or $y(t)$ has a dc value. In these cases, the bandpass random process model described by Eq. (6–126) would be nonstationary. Consequently, one is faced with the following question: Can another bandpass model be found that models a WSS $v(t)$, but yet does not require conditions 1 through 3 of Eq. (6–129)? The answer is yes. Let the model of Eq. (6–126) be generalized to include a phase constant θ_c that is a random variable. Then we have the following theorem.

THEOREM. *If $x(t)$ and $y(t)$ are jointly WSS processes, the real bandpass process*

$$v(t) = \text{Re}\{g(t)e^{j(\omega_c t + \theta_c)}\} = x(t)\cos(\omega_c t + \theta_c) - y(t)\sin(\omega_c t + \theta_c) \quad (6\text{–}130)$$

will be WSS when θ_c is an independent random variable uniformly distributed over $(0, 2\pi)$.

This modification of the bandpass model should not worry us, because we can argue that it is actually a better model for physically obtainable bandpass processes. That is, the constant θ_c is often called the *random start-up phase*, since it depends on the "initial conditions" of the physical process. Any noise source or signal source starts up with a random-phase angle when it is turned on, unless it is synchronized by injecting some external signal.

Proof. Using Eq. (6–130) to model our bandpass process, we now demonstrate that this $v(t)$ is wide-sense stationary when $g(t)$ is wide-sense stationary, even though the conditions of Eq. (6–129) may be violated. To show that Eq. (6–130) is WSS, the first requirement is that $\overline{v(t)}$ be a constant:

$$\overline{v(t)} = \overline{\text{Re}\{g(t)e^{j(\omega_c t + \theta_c)}\}} = \text{Re}\{\overline{g(t)}e^{j\omega_c t}\,\overline{e^{j\theta_c}}\}$$

But $\overline{e^{j\theta_c}} = 0$, so we have $\overline{v(t)} = 0$, which is a constant. The second requirement is that $R_v(t, t + \tau)$ be only a function of τ:

$$R_v(t, t + \tau) = \overline{v(t)v(t + \tau)}$$

$$= \overline{\text{Re}\{g(t)e^{j(\omega_c t + \theta_c)}\}\ \text{Re}\{g(t + \tau)e^{j(\omega_c t + \omega_c \tau + \theta_c)}\}}$$

Using the identity $\text{Re}(c_1)\,\text{Re}(c_2) = \frac{1}{2}\text{Re}(c_1 c_2) + \frac{1}{2}\text{Re}(c_1^* c_2)$ and recalling that θ_c is an independent random variable, we obtain

$$R_v(t, t + \tau) = \tfrac{1}{2}\text{Re}\{\overline{g(t)g(t + \tau)}e^{j(2\omega_c t + \omega_c \tau)}\overline{e^{j2\theta_c}}\}$$
$$+ \tfrac{1}{2}\text{Re}\{\overline{g^*(t)g(t + \tau)}e^{j\omega_c \tau}\}$$

But $\overline{e^{j2\theta_c}} = 0$ and $R_g(\tau) = \overline{g^*(t)g(t + \tau)}$, since $g(t)$ is assumed to be wide-sense stationary. Thus,

$$R_v(t, t + \tau) = \tfrac{1}{2}\text{Re}\{R_g(\tau)e^{j\omega_c \tau}\} \quad (6\text{–}131)$$

The right-hand side of Eq. (6–131) is not a function of t, so $R_v(t, t + \tau) = R_v(\tau)$. Consequently, Eq. (6–130) gives a model for a wide-sense stationary bandpass process.

Furthermore, for this model as described by Eq. (6–130), properties 1 through 5 of Eq. (6–133a) through Eq. (6–133e) are valid, but all the properties 6 through 14 [Eqs. (6–133f) through (6–133h)] are not valid for the $x(t)$ and $y(t)$ components of $v(t) = x(t)$ cos $(\omega_c t + \theta_c) - y(t)$ sin $(\omega_c t + \theta_c)$, unless all the conditions of Eq. (6–129) are satisfied. However, as will be proved later, the *detected* $x(t)$ and $y(t)$ components at the output of quadrature product detectors (see Fig. 6–11) satisfy properties 6 through 14, provided that the start-up phase of the detectors, θ_0, is independent of $v(t)$. [Note that the $x(t)$ and $y(t)$ components associated with $v(t)$ at the input to the detector are not identical to the $x(t)$ and $y(t)$ quadrature output waveforms unless $\theta_c = \theta_0$; however, the PSDs may be identical.]

The complex-envelope representation of Eq. (6–130) is quite useful for evaluating the output of detector circuits. For example, if $v(t)$ is a signal-plus-noise process that is applied to a product detector, $x(t)$ is the output process if the reference is 2 cos $(\omega_c t + \theta_c)$ and $y(t)$ is the output process if the reference is -2 sin$(\omega_c t + \theta_c)$. (See Chapter 4.) Similarly, $R(t)$ is the output process for an envelope detector, and $\theta(t)$ is the output process for a phase detector.

Properties of WSS Bandpass Processes

Theorems giving the relationships between the autocorrelation functions and the PSD of $v(t)$, $g(t)$, $x(t)$, and $y(t)$ can be obtained. These and other theorems are listed subsequently as properties of bandpass random processes. The relationships assume that the bandpass process $v(t)$ is real and WSS.[†] The bandpass nature of $v(t)$ is described mathematically with the aid of Fig. 6–10a, in which

$$\mathcal{P}_v(f) = 0 \qquad \text{for } f_2 < |f| < f_1 \qquad (6\text{--}132)$$

where $0 < f_1 \le f_c \le f_2$. Furthermore, a positive constant B_0 is defined such that B_0 is the largest frequency interval between f_c and either band edge, as illustrated in Fig. 6–10a, and $B_0 < f_c$.

The properties are as follows:

1. $g(t)$ is a complex wide-sense-stationary baseband process. $\qquad$ (6–133a)
2. $x(t)$ and $y(t)$ are real jointly wide-sense stationary baseband processes. $\qquad$ (6–133b)
3. $R_v(\tau) = \frac{1}{2} \operatorname{Re}\{R_g(\tau) e^{j\omega_c \tau}\}$. $\qquad$ (6–133c)
4. $\mathcal{P}_v(f) = \frac{1}{4}[\mathcal{P}_g(f - f_c) + \mathcal{P}_g(-f - f_c)]$. $\qquad$ (6–133d)
5. $\overline{v^2} = \frac{1}{2} \overline{|g(t)|^2} = R_v(0) = \frac{1}{2} R_g(0)$. $\qquad$ (6–133e)

If the bandpass process $v(t)$ is WSS and if conditions 1 through 3 of Eq. (6–129) are satisfied, $v(t)$ can be represented by Eq. (6–129a), where the $x(t)$ and $y(t)$ components satisfy properties 6 through 14 listed below. On the other hand, if $v(t)$ is WSS, but does not satisfy

---

[†] $v(t)$ also has zero mean value, because it is a bandpass process.

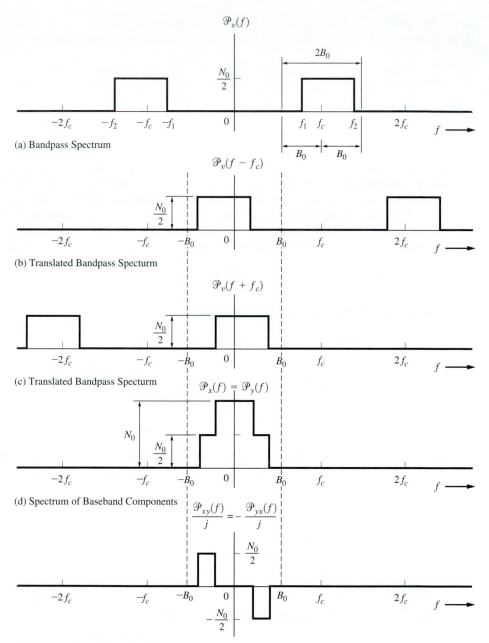

(a) Bandpass Spectrum

(b) Translated Bandpass Specturm

(c) Translated Bandpass Specturm

(d) Spectrum of Baseband Components

(e) Cross-spectrum of Baseband Components

Figure 6–10   Spectra of random process for Example 6–8.

all three conditions of Eq. (6–129), the representation of Eq. (6–129a) is not applicable, because it will not be a WSS model, and consequently, properties 6 through 14 are not applicable. However, for $v(t)$ WSS, the representation of Eq. (6–130) is always applicable, regardless of whether none, some, or all of conditions 1 through 3 of Eq. (6–129) are satisfied and, consequently, none, some, or all of the properties 6 through 14 are satisfied by the $x(t)$ and $y(t)$ components of Eq. (6–130). Furthermore, if the quadrature components of $v(t)$ are recovered by product detector circuits, as shown in Fig. 6–11, where θ_0 is a uniformly distributed random variable that is *independent* of $v(t)$, the *detected* $x(t)$ and $y(t)$ quadrature components satisfy properties 6 through 14. These properties are as follows:[†]

6. $\overline{x(t)} = \overline{y(t)} = 0.$ (6–133f)

7. $\overline{v^2(t)} = \overline{x^2(t)} = \overline{y^2(t)} = \frac{1}{2}\overline{|g(t)|^2}$

$\qquad\quad = R_v(0) = R_x(0) = R_y(0) = \frac{1}{2}R_g(0).$ (6–133g)

8. $R_x(\tau) = R_y(\tau) = 2\displaystyle\int_0^\infty \mathcal{P}_v(f)\,\cos\left[2\pi(f-f_c)\tau\right]df.$ (6–133h)

9. $R_{xy}(\tau) = 2\displaystyle\int_0^\infty \mathcal{P}_v(f)\,\sin\left[2\pi(f-f_c)\tau\right]df.$ (6–133i)

10. $R_{xy}(\tau) = -R_{xy}(-\tau) = -R_{yx}(\tau).$ (6–133j)

11. $R_{xy}(0) = 0.$ (6–133k)

12. $\mathcal{P}_x(f) = \mathcal{P}_y(f) = \begin{cases}[\mathcal{P}_v(f-f_c)+\mathcal{P}_v(f+f_c)], & |f| < B_0 \\ 0, & f \quad \text{elsewhere}\end{cases}.$ (6–133l)

13. $\mathcal{P}_{xy}(f) = \begin{cases}j[\mathcal{P}_v(f-f_c)-\mathcal{P}_v(f+f_c)], & |f| < B_0 \\ 0, & f \quad \text{elsewhere}\end{cases}.$ (6–133m)

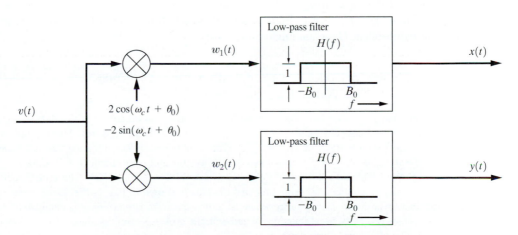

Figure 6–11   Recovery of $x(t)$ and $y(t)$ from $v(t)$

[†] Proofs of these properties will be given after Examples 6–8 and 6–9. Properties 6 through 14 also hold for the $x(t)$ and $y(t)$ components of Eq. (6–129a), provided that the conditions of Eqs. (6–129b, c, and d) are satisfied.

14. $\mathcal{P}_{xy}(f) = -\mathcal{P}_{xy}(-f) = -\mathcal{P}_{yx}(f)$. (6–133n)

Additional properties can be obtained when $v(t)$ is a wide-sense stationary *single-side-band (SSB) random process*. If $v(t)$ is SSB about $f = \pm f_c$, then, from Chapter 5, we have

$$g(t) = x(t) \pm j\hat{x}(t) \qquad (6\text{–}134)$$

where the upper sign is used for USSB and the lower sign is used for LSSB. $\hat{x}(t)$ is the Hilbert transform of $x(t)$. Using Eq. (6–134), we obtain some additional properties:

15. When $v(t)$ is an SSB process about $f = \pm f_c$,

$$R_g(\tau) = 2[R_x(\tau) \pm j\hat{R}_x(\tau)] \qquad (6\text{–}135a)$$

where $\hat{R}_x(\tau) = [1/(\pi\tau)] * R_x(\tau)$.

16. For USSB processes,

$$\mathcal{P}_g(f) = \begin{cases} 4\mathcal{P}_x(f), & f > 0 \\ 0, & f < 0 \end{cases} \qquad (6\text{–}135b)$$

17. For LSSB processes,

$$\mathcal{P}_g(f) = \begin{cases} 0, & f > 0 \\ 4\mathcal{P}_x(f), & f < 0 \end{cases} \qquad (6\text{–}135c)$$

From property 9, we see that if the PSD of $v(t)$ is even about $f = f_c$, $f > 0$, then $R_{xy}(\tau) \equiv 0$ for all τ. Consequently, $x(t)$ and $y(t)$ will be *orthogonal* processes when $\mathcal{P}_v(f)$ is even about $f = f_c$, $f > 0$. Furthermore, if, in addition, $v(t)$ is Gaussian, $x(t)$ and $y(t)$ will be independent Gaussian random processes.

Example 6–8 SPECTRA FOR THE QUADRATURE COMPONENTS
OF WHITE BANDPASS NOISE

Assume that $v(t)$ is an independent bandlimited white-noise process. Let the PSD of $v(t)$ be $N_0/2$ over the frequency band $f_1 \leq |f| \leq f_2$, as illustrated in Fig. 6–10a. Using property 12, we can evaluate the PSD for $x(t)$ and $y(t)$. This is obtained by summing the translated spectra $\mathcal{P}_v(f - f_c)$ and $\mathcal{P}_v(f + f_c)$, illustrated in Fig. 6–10b and c to obtain $\mathcal{P}_x(f)$ shown in Fig. 6–10d. Note that the spectrum $\mathcal{P}_x(f)$ is zero for $|f| > B_0$. Similarly, the cross-spectrum, $\mathcal{P}_{xy}(f)$, can be obtained by using property 13. This is shown in Fig. 6–10e. It is interesting to note that, over the frequency range where the cross-spectrum is nonzero, it is completely imaginary, since $\mathcal{P}_v(f)$ is a real function. In addition, the cross-spectrum is always an odd function.

The total normalized power is

$$P = \int_{-\infty}^{\infty} \mathcal{P}_v(f)\, df = N_0(f_2 - f_1)$$

The same result is obtained if the power is computed from $\mathcal{P}_x(f) = \mathcal{P}_y(f)$ by using property 7:

$$P = R_x(0) = R_y(0) = \int_{-\infty}^{\infty} \mathcal{P}_x(f)\, df = N_0(f_2 - f_1)$$

Example 6–9 PSD FOR A BPSK SIGNAL

The PSD for a BPSK signal that is modulated by random data will now be evaluated. In Chapters 2 and 5, it was demonstrated that the BPSK signal can be represented by

$$v(t) = x(t)\cos(\omega_c t + \theta_c) \tag{6–136}$$

where $x(t)$ represents the polar binary data (see Example 2–18) and θ_c is the random start-up phase.

The PSD of $v(t)$ is found by using property 4, where $g(t) = x(t) + j0$. Thus,

$$\mathcal{P}_v(f) = \tfrac{1}{4}[\mathcal{P}_x(f - f_c) + \mathcal{P}_x(-f - f_c)] \tag{6–137}$$

Now we need to find the PSD for the polar binary modulation $x(t)$. This was calculated in Example 6–3, where the PSD of a polar baseband signal with equally likely binary data was evaluated. Substituting Eq. (6–62) or, equivalently, Eq. (6–67) into Eq. (6–137), we obtain the PSD for the BPSK signal:

$$\mathcal{P}_v(f) = \frac{T_b}{4}\left\{\left[\frac{\sin\pi(f - f_c)\,T_b}{\pi(f - f_c)\,T_b}\right]^2 + \left[\frac{\sin\pi(f + f)\,T_b}{\pi(f + f_c)\,T_b}\right]^2\right\} \tag{6–138}$$

A sketch of this result is shown in Fig. 6–12. The result was cited earlier, in Eq. (2–200), where bandwidths were also evaluated.

Proofs of Some Properties

Giving proofs for all 17 properties listed previously would be a long task. We therefore present detailed proofs only for some of them. Proofs that involve similar mathematics will be left to the reader as exercise problems.

Properties 1 through 3 have already been proven in the discussion preceding Eq. (6–131). Property 4 follows readily from property 3 by taking the Fourier transform of

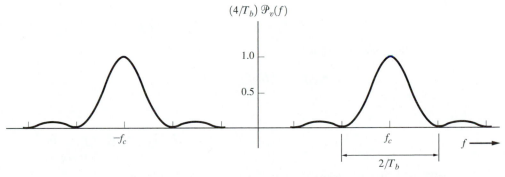

Figure 6–12   Power spectrum for a BPSK signal.

Eq. (6–133c). The mathematics is identical to that used in obtaining Eq. (4–25). Property 5 also follows directly from property 3. Property 6 will be shown subsequently. Property 7 follows from properties 3 and 8. As we will see later, properties 8 through 11 follow from properties 12 and 13.

Properties 6 and 12 are obtained with the aid of Fig. 6–11. That is, as shown by Eq. (4–77) in Sec. 4–13, $x(t)$ and $y(t)$ can be recovered by using product detectors. Thus,

$$x(t) = [2v(t) \cos(\omega_c t + \theta_0)] * h(t) \tag{6–139}$$

and

$$y(t) = -[2v(t) \sin(\omega_c t + \theta_0)] * h(t) \tag{6–140}$$

where $h(t)$ is the impulse response of an ideal LPF that is bandlimited to B_0 hertz and θ_0 is an independent random variable that is uniformly distributed over $(0, 2\pi)$ and corresponds to the random start-up phase of a phase-incoherent receiver oscillator. Property 6 follows from Eq. (6–139) by taking the ensemble average since $\overline{\cos(\omega_c t + \theta_0)} = 0$ and $\overline{\sin(\omega_c t + \theta_0)} = 0$. Property 12, which is the PSD for $x(t)$, can be evaluated by first evaluating the autocorrelation for $w_1(t)$ of Fig. 6–11:

$$w_1(t) = 2v(t) \cos(\omega_c t + \theta_0)$$
$$R_{w_1}(\tau) = \overline{w_1(t)w_1(t + \tau)}$$
$$= \overline{4v(t)v(t + \tau) \cos(\omega_c t + \theta_0) \cos(\omega_c(t + \tau) + \theta_0)}$$

But θ_0 is an independent random variable; hence, using a trigonometric identity, we have

$$R_{w_1}(\tau) = 4\overline{v(t)v(t + \tau)}[\tfrac{1}{2} \cos \omega_c \tau + \tfrac{1}{2} \overline{\cos(2\omega_c t + \omega_c \tau + 2\theta_0)}]$$

However, $\overline{\cos(2\omega_c t + \omega_c \tau + 2\theta_0)} = 0$, so

$$R_{w_1}(\tau) = 2R_v(\tau) \cos \omega_c \tau \tag{6–141}$$

The PSD of $w_1(t)$ is obtained by taking the Fourier transform of Eq. (6–141).

$$\mathcal{P}_{w_1}(f) = 2\mathcal{P}_v(f) * [\tfrac{1}{2}\delta(f - f_c) + \tfrac{1}{2}\delta(f + f_c)]$$

or

$$\mathcal{P}_{w_1}(f) = \mathcal{P}_v(f - f_c) + \mathcal{P}_v(f + f_c)$$

Finally, the PSD of $x(t)$ is

$$\mathcal{P}_x(f) = |H(f)|^2 \mathcal{P}_{w_1}(f)$$

or

$$\mathcal{P}_x(f) = \begin{cases} [\mathcal{P}_v(f - f_c) + \mathcal{P}_v(f + f_c)], & |f| < B_0 \\ 0, & \text{otherwise} \end{cases}$$

which is property 12.

Property 8 follows directly from property 12 by taking the inverse Fourier transform:

$$R_x(\tau) = \mathcal{F}^{-1}[\mathcal{P}_x(f)]$$

$$= \int_{-B_0}^{B_0} \mathcal{P}_v(f - f_c) e^{j2\pi f\tau}\, df + \int_{-B_0}^{B_0} \mathcal{P}_v(f + f_c) e^{j2\pi f\tau}\, df$$

Making changes in the variables, let $f_1 = -f + f_c$ in the first integral, and let $f_1 = f + f_c$ in the second integral. Then we have

$$R_x(\tau) = \int_{f_c - B_0}^{f_c + B_0} \mathcal{P}_v(-f_1)\; e^{j2\pi(f_c - f_1)\tau}\, df_1 + \int_{f_c - B_0}^{f_c + B_0} \mathcal{P}_v(f_1) e^{j2\pi(f_1 - f_c)\tau}\, df_1$$

But $\mathcal{P}_v(-f) = \mathcal{P}_v(f)$, since $v(t)$ is a real process. Furthermore, because $v(t)$ is bandlimited, the limits on the integrals may be changed to integrate over the interval $(0, \infty)$. Thus,

$$R_x(\tau) = 2\int_0^{\infty} \mathcal{P}_v(f_1) \left[\frac{e^{j2\pi(f_1 - f_c)\tau} + e^{-j2\pi(f_1 - f_c)\tau}}{2} \right] df_1$$

which is identical to property 8.

In a similar way, properties 13 and 9 can be shown to be valid. Properties 10 and 14 follow directly from property 9.

For SSB processes, $y(t) = \pm\hat{x}(t)$. Property 15 is then obtained as follows:

$$R_{gg}(\tau) = \overline{g^*(t)g(t + \tau)}$$

$$= \overline{[x(t) \mp j\hat{x}(t)][x(t + \tau) \pm j\,\hat{x}(t + \tau)]}$$

$$= \overline{[x(t)x(t + \tau) + \hat{x}(t)\hat{x}(t + \tau)]}$$

$$\pm j\overline{[-\hat{x}(t)x(t + \tau) + [x(t)\hat{x}(t + \tau)]} \tag{6-142}$$

Enploying the definition of a cross-correlation function and using property 10, we have

$$R_{x\hat{x}}(\tau) = \overline{x(t)\hat{x}(t + \tau)} = -R_{\hat{x}x}(\tau) = -\overline{\hat{x}(t)x(t + \tau)} \tag{6-143}$$

Furthermore, knowing that $\hat{x}(t)$ is the convolution of $x(t)$ with $1/(\pi t)$, we can demonstrate (see Prob. 6-40) that

$$R_{\hat{x}}(\tau) = R_x(\tau) \tag{6-144}$$

and

$$R_{x\hat{x}}(\tau) = \hat{R}_{xx}(\tau) \tag{6-145}$$

Thus, Eq. (6-142) reduces to property 15. Taking the Fourier transform of Eq. (6-135a), we obtain properties 16 and 17.

As demonstrated by property 6, the detected mean values $\overline{x(t)}$ and $\overline{y(t)}$ are zero when $v(t)$ is independent of θ_0. However, from Eqs. (6-139) and (6-140), we realize that a less restrictive condition is required. That is, $\overline{x(t)}$ or $\overline{y(t)}$ will be zero if $v(t)$ is orthogonal to $\cos(\omega_c t + \theta_0)$ or $\sin(\omega_c t + \theta_0)$, respectively; otherwise, they will be nonzero. For example, suppose that $v(t)$ is

$$v(t) = 5 \cos(\omega_c t + \theta_c) \qquad (6\text{--}146)$$

where θ_c is a random variable uniformly distributed over $(0, 2\pi)$. If the reference $2 \cos(\omega_c t + \theta_0)$ of Fig. 6–11 is phase coherent with $\cos(\omega_c t + \theta_c)$ (i.e., $\theta_0 \equiv \theta_c$), the output of the upper LPF will have a mean value of 5. On the other hand, if the random variables θ_0 and θ_c are independent, then $v(t)$ will be orthogonal to $\cos(\omega_c t + \theta_0)$, and $\bar{x}$ of the output of the upper LPF of Fig. 6–11 will be zero. The output dc value (i.e., the time average) will be $5 \cos(\theta_0 - \theta_c)$ in either case.

No properties have been given pertaining to the autocorrelation or the PSD of $R(t)$ and $\theta(t)$ as related to the autocorrelation and PSD of $v(t)$. In general, this is a difficult problem, because $R(t)$ and $\theta(t)$ are nonlinear functions of $v(t)$. The topic is discussed in more detail after Example 6–10.

As indicated previously, $x(t)$, $y(t)$, and $g(t)$ are Gaussian processes when $v(t)$ is Gaussian. However, as we will demonstrate, $R(t)$ and $\theta(t)$ are not Gaussian processes when $v(t)$ is Gaussian.

Example 6–10 PDF FOR THE ENVELOPE AND PHASE
FUNCTIONS OF A GAUSSIAN BANDPASS PROCESS

Assume that $v(t)$ is a wide-sense stationary *Gaussian* process with finite PSD that is symmetrical about $f = \pm f_c$. We want to find the one-dimensional PDF for the envelope process $R(t)$. Of course, this is identical to the process that appears at the output of an envelope detector when the input is a Gaussian process, such as Gaussian noise. Similarly, the PDF for the phase $\theta(t)$ (the output of a phase detector) will also be obtained.

The problem is solved by evaluating the two-dimensional random-variable transformation of $x = x(t)$ and $y = y(t)$ into $R = R(t)$ and $\theta = \theta(t)$, which is illustrated in Fig. 6–13. Because $v(t)$ is Gaussian, we know that x and y are jointly *Gaussian*. For $v(t)$ having a finite PSD that is symmetrical about $f = \pm f_c$, the mean values for x and y are both zero and the variances of both are

$$\sigma^2 = \sigma_x^2 = \sigma_y^2 = R_v(0) \qquad (6\text{--}147)$$

Furthermore, x and y are independent, because they are uncorrelated Gaussian random variables [since the PSD of $v(t)$ is symmetrical about $f = \pm f_c$]. Therefore, the joint PDF of x and y is

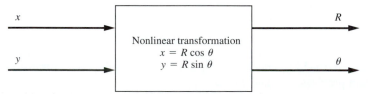

Figure 6–13   Nonlinear (polar) transformation of two Gaussian random variables.

$$f_{xy}(x, y) = \frac{1}{2\pi\sigma^2}\, e^{-(x^2+y^2)/(2\sigma^2)} \tag{6-148}$$

The joint PDF for R and θ is obtained by the two-dimensional transformation of x and y into R and θ.

$$f_{R\theta}(R,\theta) = \left. \frac{f_{xy}(x, y)}{|J[(R, \theta)/(x, y)]|}\right|_{\substack{x=R\,\cos\theta \\ y=R\,\sin\theta}}$$

$$= \left. f_{xy}(x, y)\left|J\left(\frac{(x, y)}{(R, \theta)}\right)\right|\right|_{\substack{x=R\,\cos\theta \\ y=R\,\sin\theta}} \tag{6-149}$$

We will work with $J[(x, y)/(R, \theta)]$ instead of $J[(R, \theta)/(x, y)]$, because in this problem the partial derivatives in the former are easier to evaluate than those in the latter. We have

$$J\left(\frac{(x, y)}{(R, \theta)}\right) = \mathrm{Det}\begin{bmatrix} \dfrac{\partial x}{\partial R} & \dfrac{\partial x}{\partial \theta} \\[2ex] \dfrac{\partial y}{\partial R} & \dfrac{\partial y}{\partial \theta} \end{bmatrix}$$

where x and y are related to R and θ as shown in Fig. 6–13. Of course, $R \geq 0$, and θ falls in the interval $(0, 2\pi)$, so that

$$J\left(\frac{(x, y)}{(R, \theta)}\right) = \mathrm{Det}\begin{bmatrix} \cos\theta & -R\,\sin\theta \\ \sin\theta & R\,\cos\theta \end{bmatrix}$$

$$= R[\cos^2\theta + \sin^2\theta] = R \tag{6-150}$$

Substituting Eqs. (6–148) and (6–150) into Eq. (6–149), the joint PDF of R and θ is

$$f_{R\theta}(R,\theta) = \begin{cases} \dfrac{R}{2\pi\sigma^2}\, e^{-R^2/2\sigma^2}, & R \geq 0 \text{ and } 0 \leq \theta \leq 2\pi \\[2ex] 0, & R \text{ and } \theta \text{ otherwise} \end{cases} \tag{6-151}$$

The PDF for the envelope is obtained by calculating the marginal PDF:

$$f_R(R) = \int_{-\infty}^{\infty} f_R(R, \theta)\, d\theta = \int_0^{2\pi} \frac{R}{2\pi\sigma^2}e^{-R/(2\sigma^2)}\, d\theta, \quad R \geq 0$$

or

$$f_R(R) = \begin{cases} \dfrac{R}{\sigma^2}\, e^{-R^2/(2\sigma^2)}, & R \geq 0 \\[2ex] 0, & R \text{ otherwise} \end{cases} \tag{6-152}$$

This is called a *Rayleigh* PDF. Similarly, the PDF of θ is obtained by integrating out R in the joint PDF:

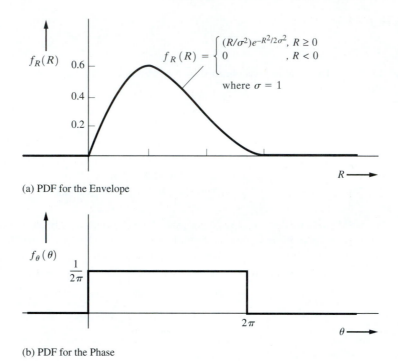

$$f_R(R) = \begin{cases} (R/\sigma^2)e^{-R^2/2\sigma^2}, & R \geq 0 \\ 0, & R < 0 \end{cases}$$

where $\sigma = 1$

(a) PDF for the Envelope

(b) PDF for the Phase

Figure 6–14 PDF for the envelope and phase of a Gaussian process.

$$f_\theta(\theta) = \begin{cases} \dfrac{1}{2\pi}, & 0 \leq \theta \leq 2\pi \\ 0, & \text{otherwise} \end{cases} \tag{6-153}$$

This is called a *uniform* PDF. Sketches of these PDFs are shown in Fig. 6–14.

The random variables $R = R(t_1)$ and $\theta = \theta(t_1)$ are independent, since $f_R(R, \theta) = f_R(R) f(\theta)$. However, $R(t)$ and $\theta(t)$ are *not* independent random processes because the random variables $R = R(t_1)$ and $\theta = \theta(t_1 + \tau)$ are not independent for all values of τ. To verify this statement, a four-dimensional transformation problem consisting of transforming the random variables $(x(t_1), x(t_2), y(t_1), y(t_2))$ into $(R(t_1), R(t_2), \theta(t_1), \theta(t_2))$ needs to be worked out, where $t_2 - t_1 = \tau$ (Davenport and Root, 1958).

The evaluation of the autocorrelation function for the envelope $R(t)$ generally requires that the two-dimensional density function of $R(t)$ be known, since $R_R(\tau) = \overline{R(t)R(t + \tau)}$. However, to obtain this joint density function for $R(t)$, the four-dimensional density function of $(R(t_1), R(t_2), \theta(t_1), \theta(t_2))$ must first be obtained by a four-dimensional transformation, as discussed in the preceding paragraph. A similar problem is worked to evaluate the autocorrelation function for the phase $\theta(t)$. These difficulties in evaluating the autocorrelation function for $R(t)$ and $\theta(t)$ arise because they are nonlinear functions of $v(t)$. The PSDs for $R(t)$ and $\theta(t)$ are obtained by taking the Fourier transform of the autocorrelation function.

6–8 MATCHED FILTERS

General Results

In preceding sections of this chapter, we developed techniques for describing random processes and *analyzing* the effect of linear systems on these processes. In this section, we develop a technique for *designing* a linear filter to minimize the effect of noise while maximizing the signal.

A general representation for a matched filter is illustrated in Fig. 6–15. The input signal is denoted by $s(t)$ and the output signal by $s_0(t)$. Similar notation is used for the noise. This filter is used in applications where the signal may or may not be present, but when the signal is present, *its waveshape is known.* (It will become clear in Examples 6–11 and 6–12, how the filter can be applied to digital signaling and radar problems.) The signal is assumed to be (absolutely) time limited to the interval $(0, T)$ and is zero otherwise. The PSD, $\mathcal{P}_n(f)$, of the additive input noise $n(t)$ is also known. We wish to determine the filter characteristic such that the instantaneous output signal power is maximized at a sampling time t_0, compared with the average output noise power. That is, we want to find $h(t)$ or, equivalently, $H(f)$, so that

$$\left(\frac{S}{N}\right)_{\text{out}} = \frac{s_0^2(t)}{n_0^2(t)} \tag{6–154}$$

is a maximum at $t = t_0$. This is the matched-filter design criterion.

The matched filter does *not* preserve the input signal waveshape, as that is not its objective. Rather, the objective is to distort the input signal waveshape and filter the noise so that at the sampling time t_0, the output signal level will be as large as possible with respect to the rms (output) noise level. In Chapter 7 we demonstrate that, under certain conditions, the filter minimizes the probability of error when receiving digital signals.

THEOREM.   *The **matched filter** is the linear filter that maximizes* $(S/N)_{\text{out}} = s_0^2(t_0)/\overline{n_0^2(t)}$ *of Fig. 6–15 and that has a transfer function given by*[†]

$$H(f) = K\frac{S^*(f)}{\mathcal{P}_n(f)}\, e^{-j\omega t_0} \tag{6–155}$$

where $S(f) = \mathcal{F}[s(t)]$ is the Fourier transform of the known input signal $s(t)$ of duration T sec. $\mathcal{P}_n(f)$ is the PSD of the input noise, t_0 is the sampling time when $(S/N)_{\text{out}}$ is evaluated, and K is an arbitrary real nonzero constant.

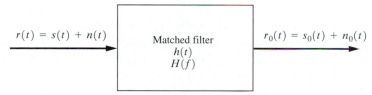

Figure 6–15   Matched filter.

[†] It appears that this formulation of the matched filter was first discovered independently by B. M. Dwork and T. S. George in 1950; the result for the white-noise case was shown first by D. O. North in 1943 [Root, 1987].

Proof. The output signal at time t_0 is

$$s_0(t_0) = \int_{-\infty}^{\infty} H(f)S(f)e^{j\omega t_0} \, df$$

The average power of the output noise is

$$\overline{n_0^2(t)} = R_{n_0}(0) = \int_{-\infty}^{\infty} |H(f)|^2 \mathcal{P}_n(f) \, df$$

Substituting these equations into Eq. (6–154), we get

$$\left(\frac{S}{N}\right)_{\text{out}} = \frac{\left| \int_{-\infty}^{\infty} H(f)S(f)e^{j\omega t_0} \, df \right|^2}{\int_{-\infty}^{\infty} |H(f)|^2 \mathcal{P}_n(f) \, df} \tag{6–156}$$

We wish to find the particular $H(f)$ that maximizes $(S/N)_{\text{out}}$. This can be obtained with the aid of the Schwarz inequality,[†] which is

$$\left| \int_{-\infty}^{\infty} A(f)B(f) \, df \right|^2 \le \int_{-\infty}^{\infty} |A(f)|^2 \, df \int_{-\infty}^{\infty} |B(f)|^2 \, df \tag{6–157}$$

where $A(f)$ and $B(f)$ may be complex functions of the real variable f. Furthermore, equality is obtained only when

$$A(f) = KB^*(f) \tag{6–158}$$

where K is any arbitrary real constant. The Schwarz inequality may be used to replace the numerator on the right-hand side of Eq. (6–156) by letting

$$A(f) = H(f) \sqrt{\mathcal{P}_n(f)}$$

and

$$B(f) = \frac{S(f)e^{j\omega t_0}}{\sqrt{\mathcal{P}_n(f)}}$$

Then Eq. (6–156) becomes

$$\left(\frac{S}{N}\right)_{\text{out}} \le \frac{\int_{-\infty}^{\infty} |H(f)|^2 \mathcal{P}_n(f) \, df \int_{-\infty}^{\infty} \frac{|S(f)|^2}{\mathcal{P}_n(f)} \, df}{\int_{-\infty}^{\infty} |H(f)|^2 \mathcal{P}_n(f) \, df}$$

[†] A proof for the Schwarz inequality is given in the appendix (Sec. 6–10) at the end of this chapter.

where it is realized that $\mathcal{P}_n(f)$ is a nonnegative real function. Thus,

$$\left(\frac{S}{N}\right)_{\text{out}} \leq \int_{-\infty}^{\infty} \frac{|S(f)|^2}{\mathcal{P}_n(f)}\, df \tag{6-159}$$

The maximum $(S/N)_{\text{out}}$ is obtained when $H(f)$ is chosen such that equality is attained. This occurs when $A(f) = KB^*(f)$, or

$$H(f)\, \sqrt{\mathcal{P}_n(f)} = \frac{KS^*(f)e^{-j\omega t_0}}{\sqrt{\mathcal{P}_n(f)}}$$

which reduces to Eq. (6–155) of the theorem.

From a practical viewpoint, it is realized that the constant K is arbitrary, since both the input signal and the input noise would be multiplied by K, and K cancels out when $(S/N)_{\text{out}}$ is evaluated. However, both the output signal and the noise levels depend on the value of the constant.

In this proof, no constraint was applied to assure that $h(t)$ would be causal. Thus, the filter specified by Eq. (6–155) may not be realizable (i.e., causal). However, the transfer function given by that equation can often be approximated by a realizable (causal) filter. If the causal constraint is included (in solving for the matched filter), the problem becomes more difficult, and a linear integral equation must be solved to obtain the unknown function $h(t)$ [Thomas, 1969].

Results for White Noise

For the case of white noise, the description of the matched filter is simplified as follows: For white noise, $\mathcal{P}_n(f) = N_0/2$. Thus, Eq. (6–155) becomes

$$H(f) = \frac{2K}{N_0}\, S^*(f)e^{-j\omega t_0}$$

From this equation, we obtain the following theorem.

THEOREM. *When the input noise is white, the impulse response of the matched filter becomes*

$$h(t) = Cs(t_0 - t) \tag{6-160}$$

where C is an arbitrary real positive constant, t_0 is the time of the peak signal output, and s(t) is the known input-signal waveshape.

Proof. We have

$$h(t) = \mathcal{F}^{-1}[H(f)] = \frac{2K}{N_0}\int_{-\infty}^{\infty} S^*(f)\ e^{-j\omega t_0}e^{j\omega t}\, df$$

$$= \frac{2K}{N_0}\left[\int_{-\infty}^{\infty} S(f)e^{j2\pi f(t_0 - t)}\, df\right]^*$$

$$= \frac{2K}{N_0}\left[s(t_0 - t)\right]^*$$

But $s(t)$ is a real signal; hence, let $C = 2K/N_0$, so that the impulse response is equivalent to Eq. (6–160).

Equation (6–160) shows that the impulse response of the matched filter (white-noise case) is simply the known signal waveshape that is "played backward" and translated by an amount t_0 (as illustrated in Example 6–11). Thus, the filter is said to be "matched" to the signal.

An important property is the actual value of $(S/N)_{out}$ that is obtained from the matched filter. From Eq. (6–159), using Parseval's theorem, as given by Eq. (2–41), we obtain

$$\left(\frac{S}{N}\right)_{out} = \int_{-\infty}^{\infty} \frac{|S(f)|^2}{N_0/2} \, df = \frac{2}{N_0} \int_{-\infty}^{\infty} s^2(t) \, dt$$

But $\int_{-\infty}^{\infty} s^2(t) \, dt = E_s$ is the energy in the (finite-duration) input signal. Hence,

$$\left(\frac{S}{N}\right)_{out} = \frac{2E_s}{N_0} \tag{6–161}$$

This is a very interesting result. It states that $(S/N)_{out}$ depends on the signal *energy* and PSD level of the noise, and not on the particular signal waveshape that is used. Of course, the signal energy can be increased to improve $(S/N)_{out}$ by increasing the signal amplitude, the signal duration, or both.

Equation (6–161) can also be written in terms of a time–bandwidth product and the ratio of the input average signal power (over T seconds) to the average noise power. Assume that the input noise power is measured in a band that is W hertz wide. We also know that the signal has a duration of T seconds. Then, from Eq. (6–161),

$$\left(\frac{S}{N}\right)_{out} = 2TW \frac{(E_s/T)}{(N_0 W)} = 2(TW) \left(\frac{S}{N}\right)_{in} \tag{6–162}$$

where $(S/N)_{in} = (E_s/T)/(N_0 W)$. From Eq. (6–162), we see that an increase in the time–bandwidth product (TW) does not change the output SNR, because the input SNR decreases correspondingly. In radar applications, increased TW provides increased ability to resolve (distinguish) targets, instead of presenting merged targets. Equation (6–161) clearly shows that it is the input-signal energy with respect to N_0 that actually determines the $(S/N)_{out}$ that is attained [Turin, 1976].

Example 6–11 INTEGRATE-AND-DUMP (MATCHED) FILTER

Suppose that the known signal is the rectangular pulse, as shown in Fig. 6–16a:

$$s(t) = \begin{cases} 1, & t_1 \leq t \leq t_2 \\ 0, & t \quad \text{otherwise} \end{cases} \tag{6–163}$$

The signal duration is $T = t_2 - t_1$. Then, for the case of white noise, the impulse response required for the matched filter is

$$h(t) = s(t_0 - t) = s(-(t - t_0)) \tag{6–164}$$

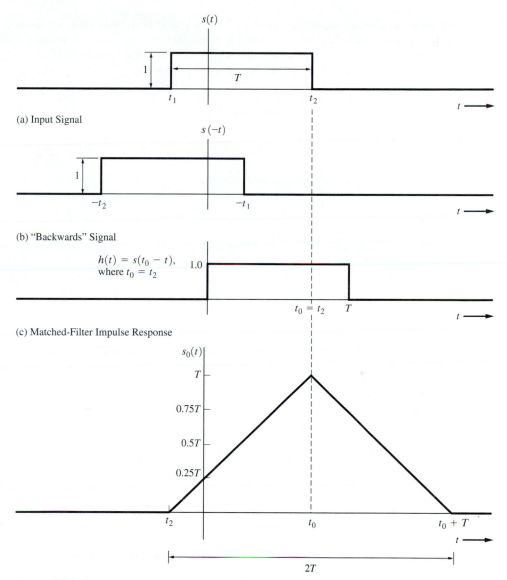

$s(t)$

(a) Input Signal

$s(-t)$

(b) "Backwards" Signal

$h(t) = s(t_0 - t)$,
where $t_0 = t_2$

(c) Matched-Filter Impulse Response

$s_0(t)$

(d) Signal Out of Matched Filter

Figure 6–16   Waveforms associated with the matched filter of Example 6–11.

C was chosen to be unity for convenience, and $s(-t)$ is shown in Fig. 6–16b. From this figure, it is obvious that, for the impulse response to be causal, we require that

$$t_0 \geq t_2 \qquad\qquad (6\text{--}165)$$

We will use $t_0 = t_2$ because this is the smallest allowed value that satisfies the causal condition and we would like to minimize the time that we have to wait before the maximum signal level occurs at the filter output (i.e., $t = t_0$). A sketch of $h(t)$ for $t_0 = t_2$ is shown in Fig. 6–16c. The resulting output signal is shown in Fig. 6–16d. Note that the peak output signal level does indeed occur at $t = t_0$ and that the input signal waveshape has been distorted by the filter in order to peak up the output signal at $t = t_0$.

In applications to digital signaling with a rectangular bit shape, this matched filter is equivalent to an *integrate-and-dump filter*, as we now illustrate. Assume that we signal with one rectangular pulse and are interested in sampling the filter output when the signal level is maximum. Then the filter output at $t = t_0$ is

$$r_0(t_0) = r(t_0) * h(t_0) = \int_{-\infty}^{\infty} r(\lambda) h(t_0 - \lambda) \, d\lambda$$

When we substitute for the matched-filter impulse response shown in Fig. 6–16c, this equation becomes

$$r_0(t_0) = \int_{t_0-T}^{t_0} r(\lambda) \, d\lambda \tag{6–166}$$

Thus, we need to integrate the digital input signal plus noise over one symbol period T (which is the bit period for binary signaling) and "dump" the integrator output at the end of the symbol period. This is illustrated in Fig. 6–17 for binary signaling. Note that for proper operation of this optimum filter, an external clocking signal called *bit sync* is required. (See Chapter 3 for a discussion of bit synchronizers.) In addition, the output signal is not binary, since the output sample values are still corrupted by noise (although the noise has been minimized by the matched filter). The output could be converted into a binary signal by feeding it into a comparator, which is exactly what is done in digital receivers, as described in Chapter 7.

Correlation Processing

THEOREM.   *For the case of white noise, the matched filter may be realized by correlating the input with $s(t)$; that is,*

$$r_0(t_0) = \int_{t_0-T}^{t_0} r(t) s(t) \, dt \tag{6–167}$$

where $s(t)$ is the known signal waveshape and $r(t)$ is the processor input, as illustrated in Fig. 6–18.

Proof.   The output of the matched filter at time t_0 is

$$r_0(t_0) = r(t_0) * h(t_0) = \int_{-\infty}^{t_0} r(\lambda) h(t_0 - \lambda) \, d\lambda$$

But from Eq. (6–160),

$$h(t) = \begin{cases} s(t_0 - t), & 0 \le t \le T \\ 0, & \text{elsewhere} \end{cases}$$

so

$$r_0(t_0) = \int_{t_0-T}^{t_0} r(\lambda) s[t_0 - (t_0 - \lambda)] \, d\lambda$$

which is identical to Eq. (6–167).

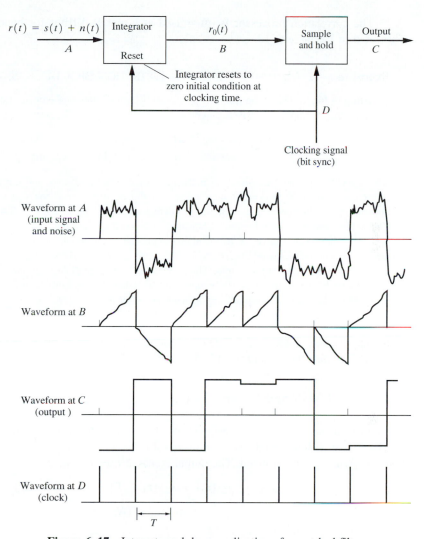

Figure 6–17   Integrate-and-dump realization of a matched filter.

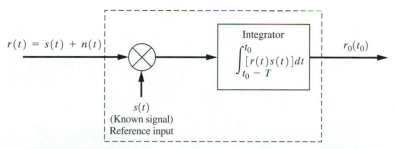

Figure 6–18   Matched-filter realization by correlation processing.

The correlation processor is often used as a matched filter for bandpass signals, as illustrated in Example 6–12.

Example 6–12 MATCHED FILTER FOR DETECTION OF A BPSK SIGNAL

Referring to Fig. 6–18, let the filter input be a BPSK signal plus white noise. For example, this might be the IF output of a BPSK receiver. The BPSK signal can be written as

$$s(t) = \begin{cases} +A \cos \omega_c t, & nT < t \le (n+1)T \quad \text{for a binary 1} \\ -A \cos \omega_c t, & nT < t \le (n+1)T \quad \text{for a binary 0} \end{cases}$$

where f_c is the IF center frequency, T is the duration of one bit of data, and n is an integer. The reference input to the correlation processor should be either $+A \cos \omega_c t$ or $-A \cos \omega_c t$, depending on whether we are attempting to detect a binary 1 or a binary 0. Since these wave-shapes are identical except for the ± 1 constants, we could just use $\cos \omega_c t$ for the reference and recognize that when a binary 1 BPSK signal is present at the input (no noise), a positive volt-age $\frac{1}{2}AT$ would be produced at the output. Similarly, a binary 0 BPSK signal would produce a negative voltage $-\frac{1}{2}AT$ at the output. Thus, for BPSK signaling plus white noise, we obtain the matched filter shown in Fig. 6–19. Notice that this looks like the familiar product detector, ex-cept that the LPF has been replaced by a gated integrator that is controlled by the bit-sync clock. With this type of postdetection processing, the product detector becomes a matched filter. How-ever, to implement such an optimum detector, both bit sync and carrier sync are needed. The technique, shown in Fig. 6–19, could be classified as a more general form of an integrate-and-dump filter (first shown in Fig. 6–17).

Transversal Matched Filter

A transversal filter can also be designed to satisfy the matched-filter criterion. Referring to Fig. 6–20, we wish to find the set of transversal filter coefficients $\{a_i; i = 1, 2, \ldots, N\}$ such that $s_0^2(t_0)/\overline{n_0^2(t)}$ is maximized. The output signal at time $t = t_0$ is

$$s_0(t_0) = a_1 s(t_0) + a_2 s(t_0 - T) + a_3 s(t_0 - 2T)$$

$$+ \ldots + a_N s(t_0 - (N-1)T)$$

or

$$s_0(t_0) = \sum_{k=1}^{N} a_k s(t_0 - (k-1)T) \qquad (6\text{–}168)$$

Similarly, for the output noise,

$$n_0(t) = \sum_{k=1}^{N} a_k n(t - (k-1)T) \qquad (6\text{–}169)$$

The average noise power is

$$\overline{n_0^2(t)} = \sum_{k=1}^{N} \sum_{l=1}^{N} a_k a_l \overline{n(t - (k-1)T)n(t - (l-1)T)}$$

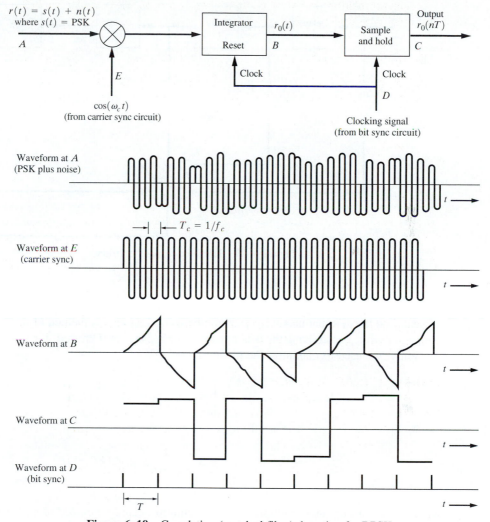

Figure 6–19   Correlation (matched-filter) detection for BPSK.

or

$$\overline{n_0^2(t)} = \sum_{k=1}^{N} \sum_{l=1}^{N} a_k a_l R_n(kT - lT) \tag{6–170}$$

where $R_n(\tau)$ is the autocorrelation of the input noise. Thus, the output-peak-signal-to-average-noise-power ratio is

$$\frac{s_0^2(t_0)}{\overline{n_0^2(t)}} = \frac{\left[\displaystyle\sum_{k=1}^{N} a_k s(t_0 - (k-1)T)\right]^2}{\displaystyle\sum_{k=1}^{N} \sum_{l=1}^{N} a_k a_l R_n(kT - lT)} \tag{6–171}$$

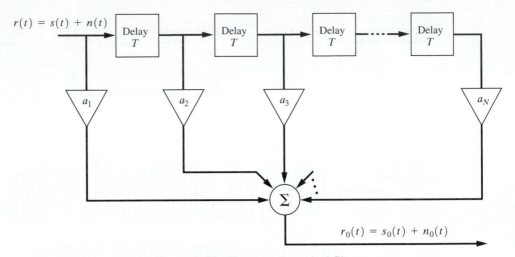

$r(t) = s(t) + n(t)$

$r_0(t) = s_0(t) + n_0(t)$

Figure 6–20   Transversal matched filter.

We can find the a_k's that maximize this ratio by using Lagrange's method of maximizing the numerator while constraining the denominator to be a constant [Olmsted, 1961, p. 518]. Consequently, we need to maximize the function

$$M(a_1, a_2, \ldots, a_N) = \left[\sum_{k=1}^{N} a_k s(t_0 - (k-1)T) \right]^2$$

$$- \lambda \sum_{k=1}^{N} \sum_{l=1}^{N} a_k a_l R_n(kT - lT) \tag{6-172}$$

where λ is the Lagrange multiplier. The maximum occurs when $\partial M / \partial a_i = 0$ for all the $i = 1, 2, \ldots, N$. Thus,

$$\frac{\partial M}{\partial a_i} = 0 = 2 \left[\sum_{k=1}^{N} a_k s(t_0 - (k-1)T) \right] s(t_0 - (i-1)T)$$

$$- 2\lambda \sum_{k=1}^{N} a_k R_n(kT - iT) \tag{6-173}$$

for $i = 1, 2, \ldots, N$. But $\sum_{k=1}^{N} a_k s(t_0 - (k-1)T) = s_0(t_0)$, which is a constant. Furthermore, let $\lambda = s_0(t_0)$. Then we obtain the required condition,

$$s(t_0 - (i-1)T) = \sum_{k=1}^{N} a_k R_n(kT - iT) \tag{6-174}$$

for $i = 1, 2, \ldots, N$. This is a set of N simultaneous linear equations that must be solved to obtain the a's. We can obtain these coefficients conveniently by writing Eq. (6–174) in matrix form. We define the elements

$$s_i \triangleq s[t_0 - (i - 1)T], \qquad i = 1, 2, \ldots, N \qquad (6\text{–}175)$$

and

$$r_{ik} = R_n(kT - iT), \qquad i = 1, \ldots, N \qquad (6\text{–}176)$$

In matrix notation, Eq. (6–174) becomes

$$\mathbf{s} = \mathbf{Ra} \qquad (6\text{–}177)$$

where the known signal vector is

$$\mathbf{s} = \begin{bmatrix} s_1 \\ s_2 \\ \vdots \\ s_N \end{bmatrix} \qquad (6\text{–}178)$$

the known autocorrelation matrix for the input noise is

$$\mathbf{R} = \begin{bmatrix} r_{11} & r_{12} & \cdots & r_{1N} \\ r_{21} & r_{22} & \cdots & r_{2N} \\ \vdots & \vdots & & \vdots \\ r_{N1} & r_{N2} & \cdots & r_{NN} \end{bmatrix} \qquad (6\text{–}179)$$

and the unknown transversal filter coefficient vector is

$$\mathbf{a} = \begin{bmatrix} a_1 \\ a_2 \\ \vdots \\ a_N \end{bmatrix} \qquad (6\text{–}180)$$

The coefficients for the transversal matched filter are then given by

$$\mathbf{a} = \mathbf{R}^{-1}\mathbf{s} \qquad (6\text{–}181)$$

where $\mathbf{R}^{-1}$ is the inverse of the autocorrelation matrix for the noise and $\mathbf{s}$ is the (known) signal vector.

6–9 SUMMARY

A *random process* is the extension of the concept of a random variable to random waveforms. A random process $x(t)$ is described by an N-dimensional PDF, where the random variables are $x_1 = x(t_1)$, $x_2 = x(t_2)$, $\ldots$, $x_N = x(t_N)$. If this PDF is invariant for a shift in the time origin, as $N \to \infty$, the process is said to be *strict-sense stationary*.

The *autocorrelation function* of a random process $x(t)$ is

$$R_x(t_1, t_2) = \overline{x(t_1)x(t_2)} = \int_{-\infty}^{\infty} \int_{-\infty}^{\infty} x_1 x_2 f_x(x_1, x_2)\, dx_1\, dx_2$$

In general, a two-dimensional PDF of $x(t)$ is required to evaluate $R_x(t_1, t_2)$. If the process is stationary,

$$R_x(t_1, t_2) = \overline{x(t_1)x(t_1 + \tau)} = R_x(\tau)$$

where $\tau = t_2 - t_1$.

If $\overline{x(t)}$ is a constant and if $R_x(t_1, t_2) = R_x(\tau)$, the process is said to be *wide-sense stationary*. If a process is strict-sense stationary, it is wide-sense stationary, but the converse is not generally true.

A process is *ergodic* when the time averages are equal to the corresponding ensemble averages. If a process is ergodic, it is also stationary, but the converse is not generally true. For ergodic processes, the dc value (a time average) is also $X_{dc} = \overline{x(t)}$ and the rms value (a time average) is also $X_{rms} = \sqrt{\overline{x^2(t)}}$.

The *power spectral density* (PSD), $\mathscr{P}_x(f)$, is the Fourier transform of the autocorrelation function $R_x(\tau)$ (Wiener–Khintchine theorem):

$$\mathscr{P}_x(f) = \mathscr{F}[R_x(\tau)]$$

The PSD is a nonnegative real function and is even about $f = 0$ for real processes. The PSD can also be evaluated by the ensemble average of a function of the Fourier transforms of the truncated sample functions.

The autocorrelation function of a wide-sense stationary real random process is a real function, and it is even about $\tau = 0$. Furthermore, $R_x(0)$ gives the total normalized average power, and this is the maximum value of $R_x(\tau)$.

For *white noise*, the PSD is a constant and the autocorrelation function is a Dirac delta function located at $\tau = 0$. White noise is not physically realizable, because it has infinite power, but it is a very useful approximation for many problems.

The *cross-correlation* function of two jointly stationary real processes $x(t)$ and $y(t)$ is

$$R_{xy}(\tau) = \overline{x(t)y(t + \tau)}$$

and the cross-PSD is

$$\mathscr{P}_{xy}(f) = \mathscr{F}[R_{xy}(\tau)]$$

The two processes are said to be *uncorrelated* if

$$R_{xy}(\tau) = [\overline{x(t)}][\overline{y(t)}]$$

for all τ. They are said to be orthogonal if

$$R_{xy}(\tau) = 0$$

for all τ.

Let $y(t)$ be the output process of a *linear system* and $x(t)$ the input process, where $y(t) = x(t) * h(t)$. Then

$$R_y(\tau) = h(-\tau) * h(t) * R_x(\tau)$$

and

$$\mathscr{P}_y(f) = |H(f)|^2 \mathscr{P}_x(f)$$

where $H(f) = \mathscr{F}[h(t)]$.

The *equivalent bandwidth* of a linear system is defined by

$$B = \frac{1}{|H(f_0)|^2} \int_0^\infty |H(f)|^2 \, df$$

where $H(f)$ is the transfer function of the system and f_0 is usually taken to be the frequency where $|H(f)|$ is a maximum. Similarly, the equivalent bandwidth of a random process $x(t)$ is

$$B = \frac{1}{\mathcal{P}_x(f_0)} \int_0^\infty \mathcal{P}_x(f) \, df = \frac{R_x(0)}{2\mathcal{P}_x(f_0)}$$

If the input to a linear system is a Gaussian process, the output is another *Gaussian* process.

A real stationary *bandpass random process* can be represented by

$$v(t) = \text{Re}\{g(t)e^{j(\omega_c t + \theta_c)}\}$$

where the complex envelope $g(t)$ is related to the quadrature processes $x(t)$ and $y(t)$. Numerous properties of these random processes can be obtained and are listed in Sec. 6–7. For example, $x(t)$ and $y(t)$ are independent Gaussian processes when the PSD of $v(t)$ is symmetrical about $f = f_c$, $f > 0$, and $v(t)$ is Gaussian. Properties for SSB bandpass processes are also obtained.

The *matched filter* is a linear filter that maximizes the instantaneous output signal power to the average output noise power for a given input signal waveshape. For the case of white noise, the impulse response of the matched filter is

$$h(t) = Cs(t_0 - t)$$

where $s(t)$ is the known signal waveshape, C is a real constant, and t_0 is the time that the output signal power is a maximum. The matched filter can be realized in many forms, such as the integrate-and-dump, the correlator, and the transversal filter configurations.

6–10 APPENDIX: PROOF OF SCHWARZ'S INEQUALITY

Schwarz's inequality is

$$\left| \int_{-\infty}^\infty f(t)g(t) \, dt \right|^2 \le \int_{-\infty}^\infty |f(t)|^2 \, dt \int_{-\infty}^\infty |g(t)|^2 \, dt \tag{6–182}$$

and becomes an *equality* if and only if

$$f(t) = Kg^*(t) \tag{6–183}$$

where K is an arbitrary real constant. $f(t)$ and $g(t)$ may be complex valued. It is assumed that both $f(t)$ and $g(t)$ have finite energy. That is,

$$\int_{-\infty}^\infty |f(t)|^2 \, dt < \infty \qquad \text{and} \qquad \int_{-\infty}^\infty |g(t)|^2 \, dt < \infty \tag{6–184}$$

Proof. Schwarz's inequality is equivalent to the inequality

$$\left| \int_{-\infty}^\infty f(t)g(t) \, dt \right| \le \sqrt{\int_{-\infty}^\infty |f(t)|^2 \, dt} \sqrt{\int_{-\infty}^\infty |g(t)|^2 \, dt} \tag{6–185}$$

Furthermore,

$$\left| \int_{-\infty}^{\infty} f(t)g(t) \ dt \right| \leq \int_{-\infty}^{\infty} |f(t)g(t)| \ dt = \int_{-\infty}^{\infty} |f(t)||g(t)| \ dt \qquad (6-186)$$

and equality holds if Eq. (6–183) is satisfied. Thus, if we can prove that

$$\int_{-\infty}^{\infty} |f(t)||g(t)| \ dt \leq \sqrt{\int_{-\infty}^{\infty} |f(t)|^2 \ dt} \ \sqrt{\int_{-\infty}^{\infty} |g(t)|^2 \ dt} \qquad (6-187)$$

then we have proved Schwarz's inequality. To simplify the notation, we replace $|f(t)|$ and $|g(t)|$ by the real-valued functions $a(t)$ and $b(t)$ where

$$a(t) = |f(t)| \qquad (6-188a)$$

and

$$b(t) = |g(t)| \qquad (6-188b)$$

Then we need to show that

$$\int_{a}^{b} a(t)b(t) \ dt < \sqrt{\int_{-\infty}^{\infty} a^2(t) \ dt} \ \sqrt{\int_{-\infty}^{\infty} b^2(t) \ dt} \qquad (6-189)$$

This can easily be shown by using an orthonormal functional series to represent $a(t)$ and $b(t)$. Let

$$a(t) = a_1 \varphi_1(t) + a_2 \varphi_2(t) \qquad (6-190a)$$

and

$$b(t) = b_1 \varphi_1(t) + b_2 \varphi_2(t) \qquad (6-190b)$$

where, as described in Chapter 2, $\mathbf{a} = (a_1,a_2)$ and $\mathbf{b} = (b_1,b_2)$ represent $a(t)$ and $b(t)$, respectively. These relationships are illustrated in Fig. 6–21.

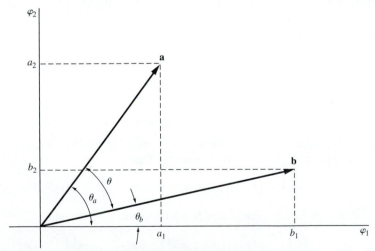

Figure 6–21 Vector representations of $a(t)$ and $b(t)$.

Then, using the figure, we obtain

$$\cos \theta = \cos(\theta_a - \theta_b) = \cos \theta_a \cos \theta_b + \sin \theta_a \sin \theta_b = \frac{a_1}{|\mathbf{a}|} \frac{b_1}{|\mathbf{b}|} + \frac{a_2}{|\mathbf{a}|} \frac{b_2}{|\mathbf{b}|}$$

or

$$\cos \theta = \frac{\mathbf{a} \cdot \mathbf{b}}{|\mathbf{a}| \ |\mathbf{b}|} \qquad (6\text{–}191)$$

Furthermore, the dot product is equivalent to the inner product:

$$\mathbf{a} \cdot \mathbf{b} = \int_{-\infty}^{\infty} a(t) b(t) \, dt \qquad (6\text{–}192a)$$

[This may be demonstrated by substituting $a(t) = \Sigma a_j \varphi_j(t)$ and $b(t) = \Sigma b_k \varphi_k(t)$ into the integral, where $\varphi_j(t)$ and $\varphi_k(t)$ are real orthonormal functions.]

From the Pythagorean theorem, the lengths (or norms) of the vectors $\mathbf{a}$ and $\mathbf{b}$ are

$$|\mathbf{a}| = \sqrt{a_1^2 + a_2^2} = \sqrt{\mathbf{a} \cdot \mathbf{a}} = \sqrt{\int_{-\infty}^{\infty} a^2(t) \, dt} \qquad (6\text{–}192b)$$

and

$$|\mathbf{b}| = \sqrt{b_1^2 + b_2^2} = \sqrt{\mathbf{b} \cdot \mathbf{b}} = \sqrt{\int_{-\infty}^{\infty} b^2(t) \, dt} \qquad (6\text{–}192c)$$

Because $|\cos \theta| \leq 1$, we have

$$\left| \int_{-\infty}^{\infty} a(t) b(t) \, dt \right| \leq \sqrt{\int_{-\infty}^{\infty} a^2(t) \, dt} \ \sqrt{\int_{-\infty}^{\infty} b^2(t) \, dt} \qquad (6\text{–}193)$$

where equality is obtained when we substitute

$$\mathbf{a} = K\mathbf{b} \qquad (6\text{–}194)$$

Using Eq. (6–188), when $f(t) = Kg^*(t)$, we find that Eq. (6–194) is also satisfied, so that equality is obtained. Thus, the proof of Eq. (6–193) proves Schwarz's inequality.

6–11 STUDY-AID EXAMPLES

SA6–1 PDF for a Linearly Transformed Gaussian RV   Let $y(t) = A\cos(10\pi t)$ be a random process, where A is a Gaussian random variable process with zero mean and variance σ_A^2. Find the PDF for $y(t)$. Is $y(t)$ WSS?

Solution $y(t)$ is a linear function of A. Thus, $y(t)$ is Gaussian because A is Gaussian. (See Sec. 6–6.) Since $\bar{A} = 0$,

$$m_y = \overline{y(t)} = \bar{A}\cos(10\pi t) = 0 \qquad (6\text{–}195a)$$

and

$$\sigma_y^2 = \overline{y^2} - m_y^2 = \overline{A^2} \cos^2(10\pi t) = \sigma_A^2 \cos^2(10\pi t) \qquad (6\text{–}195b)$$

Thus, the PDF for $y(t)$ is

$$f(y) = \frac{1}{\sqrt{2\pi}\,\sigma_y^2}\, e^{-y^2/2\sigma_y^2} \qquad (6\text{--}195c)$$

where σ_y^2 is given by Eq. (6–195b). Furthermore, $y(t)$ is not WSS, because

$$R_y(0) = \sigma_y^2 + m_x^2 = \sigma_A^2 \cos^2(10\pi t)$$

is a function of t.

SA6–2 Mean Value Out of a Linear Filter   Derive an expression for the mean value of the output of a linear time-invariant filter if the input $x(t)$ is a WSS random process.

Solution   The output of the filter is

$$y(t) = h(t) * x(t)$$

where $h(t)$ is the impulse response of the filter. Taking the expected value, we get

$$m_y = \overline{y(t)} = \overline{h(t) * x(t)} = h(t) * \overline{x(t)} = h(t) * m_x$$

$$= \int_{-\infty}^{\infty} h(\lambda) m_x\, d\lambda = m_x \int_{-\infty}^{\infty} h(\lambda)\, d\lambda \qquad (6\text{--}196)$$

However, the transfer function of the filter is

$$H(f) = \mathcal{F}[h(t)] = \int_{-\infty}^{\infty} h(t)\, e^{-j2\pi ft}\, dt$$

so that $H(0) = \int_{-\infty}^{\infty} h(t)\, dt$. Thus, Eq. (6–196) reduces to

$$m_y = m_x H(0) \qquad (6\text{--}197)$$

SA6–3 Average Power Out of a Differentiator   Let $y(t) = dn(t)/dt$, where $n(t)$ is a random-noise process that has a PSD of $\mathcal{P}_n(f) = N_0/2 = 10^{-6}$ W/Hz. Evaluate the normalized power of $y(t)$ over a low-pass frequency band of $B = 10$ Hz.

Solution $\mathcal{P}_y(f) = |H(f)|^2 \mathcal{P}_n(f)$, where, from Table 2–1, $H(f) = j2\pi f$ for a differentiator. Thus,

$$P_y = \int_{-B}^{B} \mathcal{P}_y(f)\, df = \int_{-B}^{B} (2\pi f)^2 \frac{N_0}{2}\, df = \frac{8\pi^2}{3}\left(\frac{N_0}{2}\right) B^3$$

or

$$P_y = \frac{8\pi^2}{3}\,(10^{-6})\,(10^3) = 0.0263 \text{ W} \qquad (6\text{--}198)$$

SA6–4 PSD for a Bandpass Process   A bandpass process is described by

$$v(t) = x(t) \cos(\omega_c t + \theta_c) - y(t) \sin(\omega_c t + \theta_c)$$

where $y(t) = x(t)$ is a WSS process with a PSD as shown in Fig. 6–22a. θ_c is an independent random variable uniformly distributed over $(0, 2\pi)$. Find the PSD for $v(t)$.

Solution   From Eq. (6–130), we know that $v(t)$ is a WSS bandpass process, with $\mathcal{P}_v(f)$ given by Eq. (6–133d). Thus, $\mathcal{P}_g(f)$ needs to be evaluated. Also,

$$v(t) = \text{Re}\{g(t) e^{j(\omega_c t + \theta_c)}\}$$

where $g(t) = x(t) + jy(t) = x(t) + jx(t) = (1 + j)x(t)$. We have, then,

$$R_g(\tau) = \overline{g^*(t)g(t+\tau)} = \overline{(1-j)(1+j)x(t)x(t+\tau)}$$

$$= (1+1)R_x(\tau) = 2R_x(\tau)$$

Thus,

$$\mathcal{P}_g(f) = 2\mathcal{P}_x(f) \tag{6–199}$$

Substituting Eq. (6–199) into Eq. (6–133d), we get

$$\mathcal{P}_v(f) = \frac{1}{2}[\mathcal{P}_x(f - f_c) + \mathcal{P}_x(-f - f_c)] \tag{6–200}$$

$\mathcal{P}_v(f)$ is plotted in Fig. 6–22b for the $\mathcal{P}_x(f)$ of Fig. 6–22a.

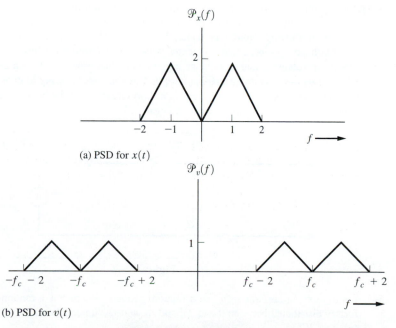

(a) PSD for $x(t)$

(b) PSD for $v(t)$

Figure 6–22   PSD for SA6–4.

PROBLEMS

6–1 Let a random process $x(t)$ be defined by

$$x(t) = At + B$$

(a) If B is a constant and A is uniformly distributed between -1 and $+1$, sketch a few sample functions.

(b) If A is a constant and B is uniformly distributed between 0 and 2, sketch a few sample functions.

6–2 Let a random process be given by

$$x(t) = A \cos(\omega_0 t + \theta)$$

where A and ω_0 are constants and θ is a random variable. Let

$$f(\theta) = \begin{cases} \dfrac{2}{\pi}, & 0 \le \theta \le \dfrac{\pi}{2} \\ 0, & \text{elsewhere} \end{cases}$$

(a) Evaluate $\overline{x(t)}$.

(b) From the result of part (a), what can be said about the stationarity of the process?

6–3 Using the random process described in Prob. 6–2,

(a) Evaluate $\langle x^2(t) \rangle$.

(b) Evaluate $\overline{x^2(t)}$.

(c) Using the results of parts (a) and (b), determine whether the process is ergodic for these averages.

6–4 A conventional average-reading ac voltmeter (volt-ohm multimeter) has a schematic diagram as shown in Fig. P6–4. The needle of the meter movement deflects proportionally to the average current flowing through the meter. The meter scale is marked to give the rms value of sine-wave voltages. Suppose that this meter is used to determine the rms value of a noise voltage. The noise voltage is known to be an ergodic Gaussian process having a zero mean value. What is the value of the constant that is multiplied by the meter reading to give the true rms value of the Gaussian noise? (*Hint:* The diode is a short circuit when the input voltage is positive and an open circuit when the input voltage is negative.)

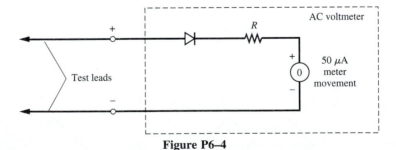

Figure P6–4

6–5 Let $x(t) = A_0 \sin(\omega_0 t + \theta)$ be a random process, where θ is a random variable that is uniformly distributed between 0 and 2π and A_0 and ω_0 are constants.

(a) Find $R_x(\tau)$.

(b) Show that $x(t)$ is wide-sense stationary.

(c) Verify that $R_x(\tau)$ satisfies the appropriate properties.

6–6 Let $r(t) = A_0 \cos \omega_0 t + n(t)$, where A_0 and ω_0 are constants. Assume that $n(t)$ is a wide-sense stationary random noise process with a zero mean value and an autocorrelation of $R_n(\tau)$.

(a) Find $\overline{r(t)}$ and determine whether $r(t)$ is wide-sense stationary.

(b) Find $R_r(t_1, t_2)$.

(c) Evaluate $\langle R_r(t, t + \tau) \rangle$, where $t_1 = t$ and $t_2 = t + \tau$.

6–7 Let an additive signal-plus-noise process be described by the equation $r(t) = s(t) + n(t)$.

(a) Show that $R_r(\tau) = R_s(\tau) + R_n(\tau) + R_{sn}(\tau) + R_{ns}(\tau)$.

(b) Simplify the result for part (a) for the case when $s(t)$ and $n(t)$ are independent and the noise has a zero mean value.

6–8 Consider the sum of two ergodic noise voltages:

$$n(t) = n_1(t) + n_2(t)$$

The power of $n_1(t)$ is 5 W, the power of $n_2(t)$ is 10 W, the dc value of $n_1(t)$ is -2 V, and the dc value n_2 is $+1$ V. Find the power of $n(t)$ if

(a) $n_1(t)$ and $n_2(t)$ are orthogonal.

(b) $n_1(t)$ and $n_2(t)$ are uncorrelated.

(c) The cross-correlation of $n_1(t)$ and $n_2(t)$ is 2 for $\tau = 0$.

6–9 Assume that $x(t)$ is ergodic, and let $x(t) = m_x + y(t)$, where $m_x = \overline{x(t)}$ is the dc value of $x(t)$ and $y(t)$ is the ac component of $x(t)$. Show that

(a) $R_x(\tau) = m_x^2 + R_y(\tau)$.

(b) $\lim_{\tau \to \infty} R_x(\tau) = m_x^2$.

(c) Can the dc value of $x(t)$ be determined from $R_x(\tau)$?

6–10 Determine whether the following functions satisfy the properties of autocorrelation functions:

(a) $\sin \omega_0 \tau$.

(b) $(\sin \omega_0 \tau)/(\omega_0 \tau)$.

(c) $\cos \omega_0 \tau + \delta(\tau)$.

(d) $e^{-a|\tau|}$, where $a < 0$.

(*Note*: $\mathscr{F}[R(\tau)]$ must also be a nonnegative function.)

6–11 A random process $x(t)$ has an autocorrelation function given by $R_x(\tau) = 5 + 8e^{-3|\tau|}$. Find

(a) The rms value for $x(t)$.

(b) The PSD for $x(t)$.

6–12 The autocorrelation of a random process is $R_x(\tau) = 4e^{-\tau^2} + 3$. Plot the PSD for $x(t)$ and evaluate the rms bandwidth for $x(t)$.

6–13 Show that two random processes $x(t)$ and $y(t)$ are uncorrelated (i.e., $R_{xy}(\tau) = m_x m_y$) if the processes are independent.

6–14 If $x(t)$ contains periodic components, show that

(a) $R_x(\tau)$ contains periodic components.

(b) $\mathscr{P}_x(f)$ contains delta functions.

6–15 Find the PSD for the random process described in Prob. 6–2.

6–16 Determine whether the following functions can be valid PSD functions for a real process:

(a) $2e^{-2\pi|f-45|}$.

(b) $4e^{-2\pi[f^2-16]}$.

(c) $25 + \delta(f - 16)$.

(d) $10 + \delta(f)$.

6–17 The PSD of an ergodic random process $x(t)$ is

$$\mathcal{P}_x(f) = \begin{cases} \dfrac{1}{B}\,(B - |f|), & |f| \le B \\ 0, & f \text{ otherwise} \end{cases}$$

where $B > 0$. Find
(a) The rms value of $x(t)$.
(b) $R_x(\tau)$.

6–18 Referring to the techniques described in Example 6–3, evaluate the PSD for a PCM signal that uses Manchester NRZ encoding. (See Fig. 3-15.) Assume that the data have values of $a_n = \pm 1$, which are equally likely, and that the data are independent from bit to bit.

6–19 The *magnitude* frequency response of a linear time-invariant network is to be determined from a laboratory setup as shown in Fig. P6–19. Discuss how $|H(f)|$ is evaluated from the measurements.

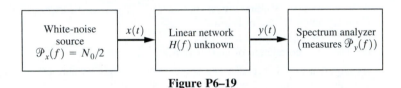

Figure P6–19

6–20 A linear time-invariant network with an unknown $H(f)$ is shown in Fig. P6–20.
(a) Find a formula for evaluating $h(t)$ in terms of $R_{xy}(\tau)$ and N_0.
(b) Find a formula for evaluating $H(f)$ in terms of $\mathcal{P}_{xy}(f)$ and N_0.

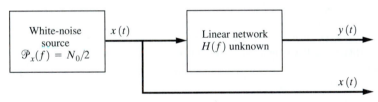

Figure P6–20

6–21 The output of a linear system is related to the input by $y(t) = h(t) * x(t)$, where $x(t)$ and $y(t)$ are jointly wide-sense stationary. Show that
(a) $R_{xy}(\tau) = h(\tau) * R_x(\tau)$.
(b) $\mathcal{P}_{xy}(f) = H(f)\mathcal{P}_x(f)$.
(c) $R_{yx}(\tau) = h(-\tau) * R_x(\tau)$.
(d) $\mathcal{P}_{yx}(f) = H^*(f)\mathcal{P}_x(f)$.
[*Hint:* Use Eqs. (6–86) and (6–87).]

6–22 Ergodic white noise with a PSD of $\mathcal{P}_n(f) = N_0/2$ is applied to the input of an ideal integrator with a gain of K (a real number) such that $H(f) = K/(j2\pi f)$.
(a) Find the PSD for the output.
(b) Find the rms value of the output noise.

6–23 A linear system has a power transfer function $|H(f)|^2$ as shown in Fig. P6–23. The input $x(t)$ is a Gaussian random process with a PSD given by

$$\mathscr{P}_x(f) = \begin{cases} \frac{1}{2}N_0, & |f| \leq 2B \\ 0, & f \text{ elsewhere} \end{cases}$$

(a) Find the autocorrelation function for the output $y(t)$.

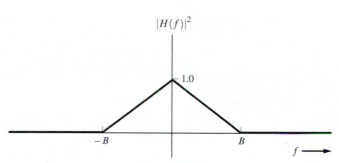

Figure P6–23

(b) Find the PDF for $y(t)$.
(c) When are the two random variables $y_1 = y(t_1)$ and $y_2 = y(t_2)$ independent?

6–24 A linear filter evaluates the T-second moving average of an input waveform, where the filter output is

$$y(t) = \frac{1}{T} \int_{t-(T/2)}^{t+(T/2)} x(u)\, du$$

and $x(t)$ is the input.
(a) Show that the impulse response is $h(t) = (1/T)\, \Pi(t/T)$.
(b) Show that

$$R_y(\tau) = \frac{1}{T} \int_{-T}^{T} \left(1 - \frac{|u|}{T}\right) R_x(\tau - u)\, du$$

(c) If $R_x(\tau) = e^{-|\tau|}$ and $T = 1$ sec, plot $R_y(\tau)$, and compare it with $R_x(\tau)$.

6–25 As shown in Example 6–5, the output signal-to-noise ratio of an RC LPF is given by Eq. (6–95) when the input is a sinusoidal signal plus white noise. Derive the value of the RC product such that the output signal-to-noise ratio will be a maximum.

6–26 Assume that a sine wave of peak amplitude A_0 and frequency f_0, plus white noise with $\mathscr{P}_n(f) = N_0/2$, is applied to a linear filter. The transfer function of the filter is

$$H(f) = \begin{cases} \dfrac{1}{B}\,(B - |f|), & |f| < B \\ 0, & f \text{ elsewhere} \end{cases}$$

where B is the absolute bandwidth of the filter. Find the signal-to-noise power ratio for the filter output.

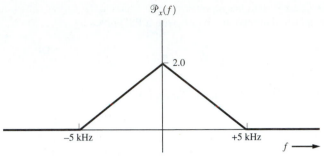

Figure P6–27

6–27 For the random process $x(t)$ with the PSD shown in Fig. P6–27, determine
(a) The equivalent bandwidth.
(b) The rms bandwidth.

6–28 If $x(t)$ is a real bandpass random process that is wide-sense stationary, show that the definition of the rms bandwidth, Eq. (6–100), is equivalent to

$$B_{\text{rms}} = 2\sqrt{\overline{f^2} - (f_0)^2}$$

where $\overline{f^2}$ is given by Eq. (6–98) or Eq. (6–99) and f_0 is given by Eq. (6–102).

6–29 In the definition for the rms bandwidth of a *bandpass* random process, f_0 is used. Show that

$$f_0 = \frac{1}{2\pi R_x(0)} \left(\frac{d\hat{R}_x(\tau)}{d\tau} \right)\Bigg|_{\tau=0}$$

where $\hat{R}_x(\tau)$ is the Hilbert transform of $R_x(\tau)$.

6–30 Two identical RC LPFs are coupled in cascade by an isolation amplifier that has a voltage gain of 10.
(a) Find the overall transfer function of the network as a function of R and C.
(b) Find the 3-dB bandwidth in terms of R and C.

6–31 Let $x(t)$ be a Gaussian process in which two random variables are $x_1 = x(t_1)$ and $x_2 = x(t_2)$. The random variables have variances of σ_1^2 and σ_2^2 and means of m_1 and m_2. The correlation coefficient is

$$\rho = \overline{(x_1 - m_1)(x_2 - m_2)}/(\sigma_1\sigma_2)$$

Using matrix notation for the $N = 2$-dimensional PDF, show that the equation for the PDF of $\mathbf{x}$ reduces to the bivariate Gaussian PDF as given by Eq. (B–97).

6–32 A bandlimited white Gaussian random process has an autocorrelation function that is specified by Eq. (6–125). Show that as $B \to \infty$, the autocorrelation function becomes $R_n(\tau) = \frac{1}{2}N_0\delta(\tau)$.

6–33 Let two random processes $x(t)$ and $y(t)$ be jointly Gaussian with zero-mean values. That is, $(x_1, x_2, \ldots, x_N, y_1, y_2, \ldots, y_M)$ is described by an $(N + M)$-dimensional Gaussian PDF. The crosscorrelation is

$$R_{xy}(\tau) = \overline{x(t_1)y(t_2)} = 10\sin(2\pi\tau)$$

(a) When are the random variables $x_1 = x(t_1)$ and $y_2 = y(t_2)$ independent?
(b) Show that $x(t)$ and $y(t)$ are or are not independent random processes.

6–34 Starting with Eq. (6–121), show that

$$\mathbf{C}_y = \mathbf{H}\mathbf{C}_x\mathbf{H}^T$$

(*Hint:* Use the identity matrix property, $\mathbf{A}\mathbf{A}^{-1} = \mathbf{A}^{-1}\mathbf{A} = \mathbf{I}$, where $\mathbf{I}$ is the identity matrix.)

6–35 Consider the random process

$$x(t) = A_0 \cos(\omega_c t + \theta)$$

where A_0 and ω_0 are constants and θ is a random variable that is uniformly distributed over the interval $(0, \pi/2)$.
(a) Determine whether $x(t)$ is wide-sense stationary.
(b) Find the PSD for $x(t)$.
(c) If θ is uniformly distributed over $(0, 2\pi)$, is $x(t)$ wide-sense stationary?

6–36 A bandpass WSS random process $v(t)$ is represented by Eq. (6–129a), where the conditions of Eq. (6–129) are satisfied. The PSD of $v(t)$ is shown in Fig. P6–36, where $f_c = 1$ MHz. Using MATLAB or MathCAD,
(a) Plot $\mathcal{P}_x(f)$.
(b) Plot $\mathcal{P}_{xy}(f)$.

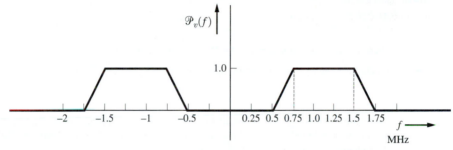

Figure P6–36

6–37 The PSD of a bandpass WSS process $v(t)$ is shown in Fig. P6–37. $v(t)$ is the input to a product detector, and the oscillator signal (i.e., the second input to the multiplier) is $5 \cos(\omega_c t + \theta_0)$, where $f_c = 1$ MHz and θ_0 is an independent random variable with a uniform distribution over $(0, 2\pi)$. Using MATLAB or MathCAD, plot the PSD for the output of the product detector.

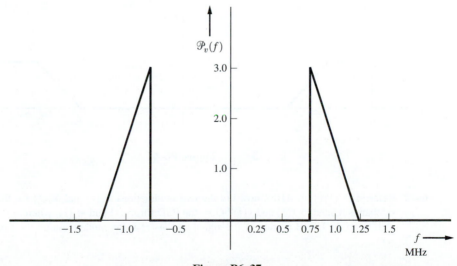

Figure P6–37

6–38 A WSS bandpass process $v(t)$ is applied to a product detector as shown in Fig. 6–11, where $\theta_c = 0$.

(a) Derive an expression for the autocorrelation of $w_1(t)$ in terms of $R_v(\tau)$. Is $w_1(t)$ WSS?

(b) Use $R_{w_1}(\tau)$ obtained in part (a) to find an expression for $\mathcal{P}_{w_1}(f)$. (*Hint:* Use the Wiener–Khintchine theorem.)

6–39 A USSB signal is

$$v(t) = 10\, \mathrm{Re}\{[x(t) + j\hat{x}(t)]e^{j(\omega_c t + \theta_c)}\}$$

where θ_c is a random variable that is uniformly distributed over $(0, 2\pi)$. The PSD for $x(t)$ is given in Fig. P6–27. Find

(a) The PSD for $v(t)$.

(b) The total power of $v(t)$.

6–40 Show that

(a) $R_{\hat{x}}(\tau) = R_x(\tau)$ and

(b) $R_{x\hat{x}}(\tau) = \hat{R}_x(\tau)$,

where the caret symbol denotes the Hilbert transform.

6–41 A bandpass random signal can be represented by

$$s(t) = x(t)\cos(\omega_c t + \theta_c) - y(t)\sin(\omega_c t + \theta_c)$$

where the PSD of $s(t)$ is shown in Fig. P6–41. θ_c is an independent random variable that is uniformly distributed over $(0, 2\pi)$. Assume that $f_3 - f_2 = f_2 - f_1$. Find the PSD for $x(t)$ and $y(t)$ when

(a) $f_c = f_1$. This is USSB signaling, where $y(t) = \hat{x}(t)$.

(b) $f_c = f_2$. This represents independent USSB and LSSB signaling with two different modulations.

(c) $f_1 < f_c < f_2$. This is vestigial sideband signaling.

(d) For which, if any, of these cases are $x(t)$ and $y(t)$ orthogonal?

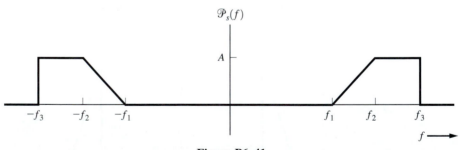

Figure P6–41

6–42 Referring to Prob. 6–41(b), how are the two modulations $m_1(t)$ and $m_2(t)$ for the independent sidebands related to $x(t)$ and $y(t)$? Give the PSD for $m_1(t)$ and $m_2(t)$, where $m_1(t)$ is the modulation on the USSB portion of the signal and $m_2(t)$ is the modulation on the LSSB portion of $s(t)$.

6–43 For the bandpass random process, show that
(a) Equation (6–133m) is valid (property 13).
(b) Equation (6–133i) is valid (property 9).

6–44 Referring to Example 6–9, find the PSD for a BPSK signal with Manchester-encoded data. (See Fig. 3-15.) Assume that the data have values of $a_n = \pm 1$ which are equally likely and that the data are independent from bit to bit.

6–45 The input to an envelope detector is an ergodic bandpass Gaussian noise process. The rms value of the input is 2 V and the mean value is 0 V. The envelope detector has a voltage gain of 10. Find
(a) The dc value of the output voltage.
(b) The rms value of the output voltage.

6–46 A narrowband-signal-plus-Gaussian noise process is represented by the equation

$$r(t) = A \cos(\omega_c t + \theta_c) + x(t) \cos(\omega_c t + \theta_c) - y(t) \sin(\omega_c t + \theta_c)$$

where $A \cos(\omega_c t + \theta_c)$ is a sinusoidal carrier and the remaining terms are the bandpass noise with independent Gaussian I and Q components. Let the noise have an rms value of σ and a mean of 0. The signal-plus-noise process appears at the input to an envelope detector. Show that the PDF for the output of the envelope detector is

$$f(R) = \begin{cases} \dfrac{R}{\sigma^2}\, e^{-[(R^2+A^2)/(2\sigma^2)]}\, I_0\left(\dfrac{RA}{\sigma^2}\right), & R \ge 0 \\ 0, & R < 0 \end{cases}$$

where

$$I_0(z) \triangleq \frac{1}{2\pi} \int_0^{2\pi} e^{z \cos\theta}\, d\theta$$

is the modified Bessel function of the first kind of zero order. $f(R)$ is known as a *Rician* PDF in honor of the late S. O. Rice, who was an outstanding engineer at Bell Telephone Laboratories.

6–47 Assume that

$$s(t) = \begin{cases} \dfrac{A}{T}\, t \, \cos\omega_c t, & 0 \le t \le T \\ 0, & t \text{ otherwise} \end{cases}$$

is a known signal. This signal plus white noise is present at the input to a matched filter.
(a) Design a matched filter for $s(t)$. Sketch waveforms analogous to those shown in Fig. 6–16.
(b) Sketch the waveforms for the correlation processor shown in Fig. 6–18.

6–48 A baseband digital communication system uses polar signaling with a bit rate of $R = 2{,}000$ bits/s. The transmitted pulses are rectangular and the frequency response of channel filter is

$$H_c(f) = \frac{B}{B + jf}$$

where $B = 6{,}000$ Hz. The filtered pulses are the input to a receiver that uses integrate-and-dump processing, as illustrated in Fig. 6–17. Examine the integrator output for ISI. In particular,

(a) Plot the integrator output when a binary "1" is sent.

(b) Plot the integrator output for an all-pass channel, and compare this result with that obtained in part (a).

6–49 Refer to Fig. 6–19 for the detection of a BPSK signal. Suppose that the BPSK signal at the input is

$$r(t) = s(t) = \sum_{n=0}^{7} d_n p(t - nT)$$

where

$$p(t) = \begin{cases} e^{-t}\cos(\omega_c t), & 0 < t < T \\ 0, & t \text{ elsewhere} \end{cases}$$

and the binary data d_n is the 8-bit string $\{+1, -1, -1, +1, +1, -1, +1, -1\}$. Use MATLAB or MathCAD to

(a) Plot the input waveform $r(t)$.

(b) Plot the integrator output waveform $r_0(t)$.

6–50 A matched filter is described in Fig. P6–50.

(a) Find the impulse response of the matched filter.

(b) Find the pulse shape to which this filter is matched (the white-noise case).

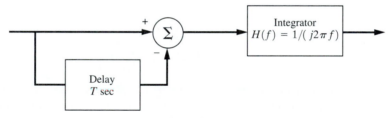

Figure P6–50

6–51 An FSK signal $s(t)$ is applied to a correlation receiver circuit that is shown in Fig. P6–51. The FSK signal is

$$s(t) = \begin{cases} A\cos(\omega_1 t), & \text{when a binary 1 is sent} \\ A\cos(\omega_2 t), & \text{when a binary 0 is sent} \end{cases}$$

where $f_1 = f_c + \Delta F$ and $f_2 = f_c - \Delta F$. Let ΔF be chosen to satisfy the MSK condition, which is $\Delta F = 1/(4T)$. T is the time that it takes to send one bit, and the integrator is reset every T seconds. Assume that $A = 1, f_c = 1{,}000$ Hz, and $\Delta F = 50$ Hz.

(a) If a binary 1 is sent, plot $v_1(t)$, $v_2(t)$, and $r_0(t)$ over a T-second interval.

(b) Refer to Fig. 6–16, and find an expression that describes the output of a filter that is matched to the FSK signal when a binary 1 is sent. Plot the filter output.

(c) Discuss how the plots obtained for parts (a) and (b) agree or disagree.

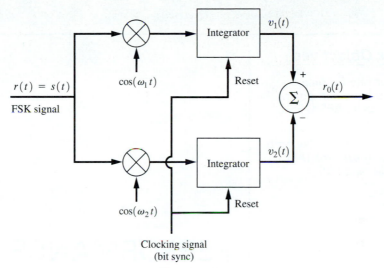

Figure P6–51

6–52 Let

$$s(t) = \text{Re}\{g(t)e^{j(\omega_c t + \theta_c)}\}$$

be a wideband FM signal, where

$$g(t) = A_c e^{jD_f \int_{-\infty}^{t} m(\lambda)\, d\lambda}$$

and $m(t)$ is a random modulation process.

(a) Show that

$$R_g(\tau) = A_c^2 \overline{[e^{jD_f \tau m(t)}]}$$

when the integral $\int_t^{t+\tau} m(\lambda)\, d\lambda$ is approximated by $\tau m(t)$.

(b) Using the results of part (a), prove that the PSD of the wideband FM signal is given by Eq. (5–66). That is, show that

$$\mathcal{P}_s(f) = \frac{\pi A_c^2}{2 D_f} \left[f_m\!\left(\frac{2\pi}{D_f}\,(f - f_c) \right) + f_m\!\left(\frac{2\pi}{D_f}\,(-f - f_c) \right) \right]$$

CHAPTER 7

Chapter Objectives

- Bit error rate for binary systems (unipolar, polar, bipolar, OOK, BPSK, FSK, and MSK)

- Output signal-to-noise ratio for analog systems (AM, SSB, PM, and FM)

PERFORMANCE OF COMMUNICATION SYSTEMS CORRUPTED BY NOISE

As discussed in Chapter 1, the two primary considerations in the design of a communication system are as follows:

1. The *performance* of the system when it is corrupted by noise. The performance measure for a digital system is the probability of error of the output signal. For analog systems, the performance measure is the output signal-to-noise ratio.

2. The channel *bandwidth* that is required for transmission of the communication signal. This bandwidth was evaluated for various types of digital and analog signals in the previous chapters.

There are numerous ways in which the information can be demodulated (recovered) from the received signal that has been corrupted by noise. Some receivers provide optimum performance, but most do not. Often a suboptimum receiver will be used in order to lower the cost. In addition, some suboptimum receivers perform almost as well as optimum ones for all practical purposes. Here we will *analyze* the performance of some suboptimum as well as some optimum receivers.

7–1 ERROR PROBABILITIES FOR BINARY SIGNALING

General Results

Figure 7–1 shows a general block diagram for a binary communication system. The receiver input $r(t)$ consists of the transmitted signal $s(t)$ plus channel noise $n(t)$. For *baseband* signaling, the processing circuits in the receiver consist of low-pass filtering with appropriate amplification. For *bandpass* signaling, such as OOK, BPSK, and FSK, the processing circuits normally consist of a superheterodyne receiver containing a mixer, an IF amplifier, and a detector. These circuits produce a baseband analog output $r_0(t)$. (For example, when BPSK signaling is used, the detector might consist of a product detector and an integrator as described in Sec. 6–8 and illustrated in Fig. 6–19.)

The analog baseband waveform $r_0(t)$ is sampled at the clocking time $t = t_0 + nT$ to produce the samples $r_0(t_0 + nT)$, which are fed into a threshold device (a comparator). The threshold device produces the binary serial-data waveform $\tilde{m}(t)$.

In this subsection, we develop a *general technique* for evaluating the probability of encountering a bit error, also called the *bit-error rate* (BER), for binary signaling. In later sections, this technique will be used to obtain specific expressions for the BER of various binary signaling schemes, such as OOK, BPSK, and FSK.

To develop a general formula for the BER of a detected binary signal, let T be the duration of time it takes to transmit one bit of data. The transmitted signal over a bit interval $(0,T)$ is

$$s(t) = \begin{cases} s_1(t), & 0 < t \le T, & \text{for a binary 1} \\ s_2(t), & 0 < t \le T, & \text{for a binary 0} \end{cases} \tag{7-1}$$

where $s_1(t)$ is the waveform that is used if a binary 1 is transmitted and $s_2(t)$ is the waveform that is used if a binary 0 is transmitted. If $s_1(t) = -s_2(t)$, $s(t)$ is called an *antipodal*

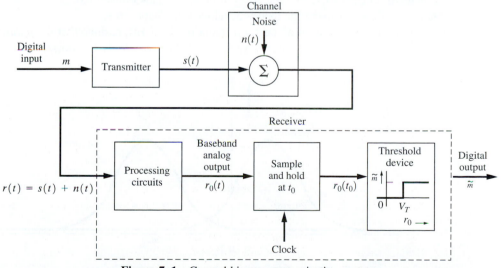

Figure 7–1   General binary communication system.

signal. The binary signal plus noise at the receiver input produces a baseband analog wave-form at the output of the processing circuits that is denoted by

$$r_0(t) = \begin{cases} r_{01}(t), & 0 < t \le T, \quad \text{for a binary 1 sent} \\ r_{02}(t), & 0 < t \le T, \quad \text{for a binary 0 sent} \end{cases} \tag{7–2}$$

where $r_{01}(t)$ is the output signal that is corrupted by noise for a binary 1 transmission and $r_{02}(t)$ is the output for the binary 0 transmission. (Note that if the receiver uses nonlinear processing circuits, such as an envelope detector, the superposition of the signal plus noise outputs are not valid operations.) This analog voltage waveform $r_0(t)$ is sampled at some time t_0 during the bit interval. That is, $0 < t_0 \le T$. For matched-filter processing circuits, t_0 is usually T. The resulting sample is

$$r_0(t_0) = \begin{cases} r_{01}(t_0), & \text{for a binary 1 sent} \\ r_{02}(t_0), & \text{for a binary 0 sent} \end{cases} \tag{7–3}$$

It is realized that $r_0(t_0)$ is a *random variable* that has a *continuous* distribution because the channel noise has corrupted the signal. To shorten the notation, we will denote $r_0(t_0)$ simply by r_0. That is,

$$r_0 = r_0(t_0) = \begin{cases} r_{01}, & \text{for a binary 1 sent} \\ r_{02}, & \text{for a binary 0 sent} \end{cases} \tag{7–4}$$

We call r_0 the *test statistic*.

For the moment, let us assume that we can evaluate the PDFs for the two random variables $r_0 = r_{01}$ and $r_0 = r_{02}$. These PDFs are actually *conditional PDFs,* since they depend, respectively, on a binary 1 or a binary 0 being transmitted. That is, when $r_0 = r_{01}$, the PDF is $f(r_0|s_1$ sent), and when $r_0 = r_{02}$, the PDF is $f(r_0|s_2$ sent). These conditional PDFs are shown in Fig. 7–2. For illustrative purposes, Gaussian shapes are shown. The actual shapes of the PDFs depend on the characteristics of the channel noise, the specific types of filter and detector circuits used, and the types of binary signals transmitted. (In later sections, we obtain specific PDFs using the theory developed in Chapter 6.)

In our development of a general formula for the BER, assume that the *polarity of the processing circuits* of the receiver is such that if the *signal only* (no noise) were present at

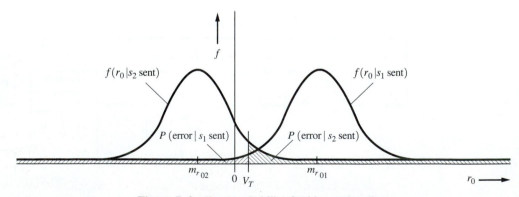

Figure 7–2　Error probability for binary signaling.

the receiver input, $r_0 > V_T$ when a *binary 1 is sent* and $r_0 < V_T$ when a *binary 0 is sent*; V_T is the threshold (voltage) setting of the comparator (threshold device).

When signal plus noise is present at the receiver input, errors can occur in two ways. An error occurs when $r_0 < V_T$ if a binary 1 is sent:

$$P(\text{error}|s_1 \text{ sent}) = \int_{-\infty}^{V_T} f(r_0|s_1)\, dr_0 \tag{7–5}$$

This is illustrated by a shaded area to the left of V_T in Fig. 7–2. Similarly, an error occurs when $r_0 > V_T$ if a binary 0 is sent:

$$P(\text{error}|s_2 \text{ sent}) = \int_{V_T}^{\infty} f(r_0|s_2)\, dr_0 \tag{7–6}$$

The BER is then

$$P_e = P(\text{error}|s_1 \text{ sent})P(s_1\text{sent}) + P(\text{error}|s_2 \text{ sent})P(s_2 \text{ sent}) \tag{7–7}$$

This follows from probability theory (see Appendix B), where the probability of an event that consists of joint events is

$$P(E) = \sum_{i=1}^{2} P(E, s_i) = \sum_{i=1}^{2} P(E|s_i)P(s_i)$$

When we combine Eqs. (7–5), (7–6), and (7–7), the general expression for the BER of any binary communication system is

$$P_e = P(s_1 \text{ sent}) \int_{-\infty}^{V_T} f(r_0|s_1)\, dr_0 + P(s_2 \text{ sent}) \int_{V_T}^{\infty} f(r_0|s_2)\, dr_0 \tag{7–8}$$

$P(s_1 \text{ sent})$ and $P(s_2 \text{ sent})$ are known as the *source statistics* or *a priori statistics*. In most applications, the source statistics are considered to be equally likely. That is,

$$P(\text{binary 1 sent}) = P(s_1 \text{ sent}) = \tfrac{1}{2} \tag{7–9a}$$

$$P(\text{binary 0 sent}) = P(s_2 \text{ sent}) = \tfrac{1}{2} \tag{7–9b}$$

In the results that we obtain throughout the remainder of this chapter, we will assume that the source statistics are equally likely. The conditional PDFs depend on the signaling waveshapes involved, the channel noise, and the receiver processing circuits used. These are obtained subsequently for the case of Gaussian channel noise and linear processing circuits.

Results for Gaussian Noise

Assume that the channel noise is a zero-mean wide-sense stationary Gaussian process and that the receiver processing circuits, except for the threshold device, are linear. Then we know (see Chapter 6) that for a Gaussian process at the input, the output of the linear processor will also be a Gaussian process. For baseband signaling, the processing circuits would consist of linear filters with some gain. For bandpass signaling, as we demonstrated in Chapter 4, a superheterodyne circuit (consisting of a mixer, IF stage, and product detector) is a linear circuit. However, if automatic gain control (AGC) or limiters are used, the receiver will be nonlinear, and the results of this section will not be applicable. In addition, if a non-linear detector such as an envelope detector is used, the output noise will not be Gaussian. For the case of a linear-processing receiver circuit with a binary signal plus noise at the input, the sampled output is

$$r_0 = s_0 + n_0 \tag{7–10}$$

Here the shortened notation $r_0(t_0) = r_0$ is used. $n_0(t_0) = n_0$ is a zero-mean Gaussian random variable, and $s_0(t_0) = s_0$ is a constant that depends on the signal being sent. That is,

$$s_0 = \begin{cases} s_{01}, & \text{for a binary 1 sent} \\ s_{02}, & \text{for a binary 0 sent} \end{cases} \tag{7–11}$$

where s_{01} and s_{02} are known constants for a given type of receiver with known input signaling waveshapes $s_1(t)$ and $s_2(t)$. Since the output noise sample n_0 is a zero-mean Gaussian random variable, the total output sample r_0 is a Gaussian random variable with a mean value of either s_{01} or s_{02}, depending on whether a binary 1 or a binary 0 was sent. This is illustrated in Fig. 7–2, where the mean value of r_0 is $m_{r_{01}} = s_{01}$ when a binary 1 is sent and the mean value of r_0 is $m_{r_{02}} = s_{02}$ when a binary 0 is sent. Thus, the two conditional PDFs are

$$f(r_0|s_1) = \frac{1}{\sqrt{2\pi}\,\sigma_0}\,e^{-(r_0-s_{01})^2/(2\sigma_0^2)} \tag{7–12}$$

and

$$f(r_0|s_2) = \frac{1}{\sqrt{2\pi}\,\sigma_0}\,e^{-(r_0-s_{02})^2/(2\sigma_0^2)} \tag{7–13}$$

$\sigma_0^2 = \overline{n_0^2} = \overline{n_0^2(t_0)} = \overline{n_0^2(t)}$ is the average power of the *output* noise from the receiver processing circuit where the output noise process is wide-sense stationary.

Using equally likely source statistics and substituting Eqs. (7–12) and (7–13) into Eq. (7–8), we find that the BER becomes

$$P_e = \frac{1}{2}\int_{-\infty}^{V_T} \frac{1}{\sqrt{2\pi}\,\sigma_0}\,e^{-(r_0-s_{01})^2/(2\sigma_0^2)}\,dr_0 + \frac{1}{2}\int_{V_T}^{\infty} \frac{1}{\sqrt{2\pi}\,\sigma_0}\,e^{-(r_0-s_{02})^2/(2\sigma_0^2)}\,dr_0 \tag{7–14}$$

This can be reduced to the $Q(z)$ functions defined in Sec. B–7 (Appendix B) and tabulated in Sec. A–10 (Appendix A). Let $\lambda = -(r_0 - s_{01})/\sigma_0$ in the first integral and $\lambda = (r_0 - s_{02})/\sigma_0$ in the second integral; then

$$P_e = \frac{1}{2}\int_{-(V_T-s_{01})/\sigma_0}^{\infty} \frac{1}{\sqrt{2\pi}}\,e^{-\lambda^2/2}\,d\lambda + \frac{1}{2}\int_{(V_T-s_{02})/\sigma_0}^{\infty} \frac{1}{\sqrt{2\pi}}\,e^{-\lambda^2/2}\,d\lambda$$

or

$$P_e = \frac{1}{2}\,Q\left(\frac{-V_T + s_{01}}{\sigma_0}\right) + \frac{1}{2}\,Q\left(\frac{V_T - s_{02}}{\sigma_0}\right) \tag{7–15}$$

By using the appropriate value for the comparator threshold, V_T, this probability of error can be minimized. To find the V_T that minimizes P_e, we need to solve $dP_e/dV_T = 0$. Using Leibniz's rule, Sec. A-2, for differentiating the integrals of Eq. (7–14), we obtain

$$\frac{dP_e}{dV_T} = \frac{1}{2}\,\frac{1}{\sqrt{2\pi}\,\sigma_0}\,e^{-(V_T-s_{01})^2/(2\sigma_0^2)} - \frac{1}{2}\,\frac{1}{\sqrt{2\pi}\,\sigma_0}\,e^{-(V_T-s_{02})^2/(2\sigma_0^2)} = 0$$

or

$$e^{-(V_T-s_{01})^2/(2\sigma_0^2)} = e^{-(V_T-s_{02})^2/(2\sigma_0^2)}$$

which implies the condition

$$(V_T - s_{01})^2 = (V_T - s_{02})^2$$

Consequently, for minimum P_e, the threshold setting of the comparator needs to be

$$V_T = \frac{s_{01} + s_{02}}{2} \tag{7–16}$$

Substituting Eq. (7–16) into Eq. (7–15), we obtain the expression for the minimum P_e. Thus, for binary signaling in Gaussian noise and with the optimum threshold setting as specified by Eq. (7–16), the BER is

$$P_e = Q\left(\frac{s_{01} - s_{02}}{2\sigma_0}\right) = Q\left(\sqrt{\frac{(s_{01} - s_{02})^2}{4\sigma_0^2}}\right) \tag{7–17}$$

where it is assumed that $s_{01} > V_T > s_{02}$.[†] So far, we have optimized only the threshold level, not the filters in the processing circuits.

Results for White Gaussian Noise and Matched-Filter Reception

If the receiving filter (in the processing circuits of Fig. 7–1) is optimized, the BER as given by Eq. (7–17) can be reduced. To *minimize* P_e, we need to *maximize* the argument of Q, as is readily seen from Fig. B–7. Thus, we need to find the linear filter that maximizes

$$\frac{[s_{01}(t_0) - s_{02}(t_0)]^2}{\sigma_0^2} = \frac{[s_d(t_0)]^2}{\sigma_0^2}$$

$s_d(t_0) \triangleq s_{01}(t_0) - s_{02}(t_0)$ is the *difference signal* sample value that is obtained by subtracting the sample s_{02} from s_{01}. The corresponding instantaneous power of the difference output signal at $t = t_0$ is $s_d^2(t_0)$. As derived in Sec. 6–8, the linear filter that maximizes the

---

[†] If $s_{01} < V_T < s_{02}$, the result is

$$P_e = Q\left(\frac{s_{02} - s_{01}}{2\sigma_0}\right) = Q\left(\sqrt{\frac{(s_{02} - s_{01})^2}{4\sigma_0^2}}\right)$$

where the characteristic of the threshold device in Fig. 7–1 is altered so that a binary 0 is chosen when $r_0 > V_T$ and a binary 1 is chosen when $r_0 < V_T$.

instantaneous output signal power at the sampling time $t = t_0$ when compared with the average output noise power $\sigma_0^2 = n_0^2(t)$ is the *matched filter*. For the case of *white noise* at the receiver input, *the matched filter needs to be matched to the difference signal* $s_d(t) = s_1(t) - s_2(t)$. Thus, the impulse response of the matched filter for binary signaling is

$$h(t) = C[s_1(t_0 - t) - s_2(t_0 - t)] \qquad (7\text{–}18)$$

where $s_1(t)$ is the signal (only) that appears at the receiver input when a binary 1 is sent, $s_2(t)$ is the signal that is received when a binary 0 is sent, and C is a real constant. Furthermore, by using Eq. (6–161), the output peak signal to average noise ratio that is obtained from the matched filter is

$$\frac{[s_d(t_0)]^2}{\sigma_0^2} = \frac{2E_d}{N_0}$$

$N_0/2$ is the PSD of the noise at the *receiver input*, and E_d is the difference signal energy at the *receiver input*, where

$$E_d = \int_0^T [s_1(t) - s_2(t)]^2 \, dt \qquad (7\text{–}19)$$

Thus, for binary signaling corrupted by *white* Gaussian noise, *matched-filter* reception, and by using the optimum threshold setting, the BER is

$$P_e = Q\left(\sqrt{\frac{E_d}{2N_0}}\right) \qquad (7\text{–}20)$$

We will use this result to evaluate the P_e for various types of binary signaling schemes where matched-filter reception is used.

Results for Colored Gaussian Noise and Matched-Filter Reception

The technique that we just used to obtain the BER for binary signaling in white noise can be modified to evaluate the BER for the colored noise case. The modification used is illustrated in Fig. 7–3. Here a *prewhitening filter* is inserted ahead of the receiver processing circuits. The transfer function of the prewhitening filter is

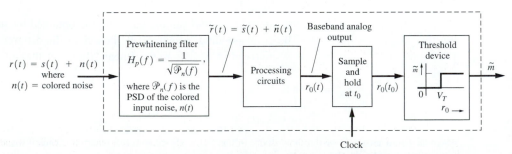

Figure 7–3   Matched-filter receiver for colored noise.

$$H_p(f) = \frac{1}{\sqrt{\mathcal{P}_n(f)}} \qquad (7\text{–}21)$$

so that the noise that appears at the filter output, $\tilde{n}(t)$, is white. We have now converted the colored noise problem into a white noise problem, so that the design techniques presented in the preceding section are applicable. The matched filter in the processing circuits is now matched to the *filtered waveshapes*,

$$\tilde{s}(t) = \tilde{s}_1(t) = s_1(t) * h_p(t) \qquad \text{(binary 1)} \qquad (7\text{–}22a)$$

$$\tilde{s}(t) = \tilde{s}_2(t) = s_2(t) * h_p(t) \qquad \text{(binary 0)} \qquad (7\text{–}22b)$$

where $h_p(t) = \mathcal{F}^{-1}[H_p(f)]$. Since the prewhitening will produce signals $\tilde{s}_1(t)$ and $\tilde{s}_2(t)$, which are spread beyond the T-second signaling interval, two types of degradation will result:

- The signal energy of the filtered signal that occurs beyond the T-second interval will not be used by the matched filter in maximizing the output signal.
- The portions of signals from previous signaling intervals that occur in the present signaling interval will produce ISI. (See Chapter 3.)

Both of these effects can be reduced if the duration of the original signal is made less than the T-second signaling interval, so that almost all of the spread signal will occur within that interval.

7–2 PERFORMANCE OF BASEBAND BINARY SYSTEMS

Unipolar Signaling

As illustrated in Fig. 7–4b, the two baseband signaling waveforms are

$$s_1(t) = +A, \qquad 0 < t \le T \qquad \text{(binary 1)} \qquad (7\text{–}23a)$$

$$s_2(t) = 0, \qquad 0 < t \le T \qquad \text{(binary 0)} \qquad (7\text{–}23b)$$

where $A > 0$. This unipolar signal plus white Gaussian noise is present at the receiver input.

First, evaluate the performance of a receiver that uses an LPF, $H(f)$, with unity gain. Choose the equivalent bandwidth of this LPF to be $B > 2/T$, so that the unipolar signaling waveshape is preserved (approximately) at the filter output, yet the noise will be reduced by the filter.[†] Thus, $s_{01}(t_0) \approx A$ and $s_{02}(t_0) \approx 0$. The noise power at the output of the filter is

---

[†] From Eq. (3–39b), the PSD of a unipolar signal (rectangular pulse shape) is proportional to $[\sin(\pi f T)/(\pi f T)]^2$, so that the second null bandwidth is $2/T$. Referring to study-aid examples SA7–1 and SA7–2, it is shown that if the equivalent bandwidth of the LPF is greater than $2/T$, the filtered signal will consist of pulses that are almost rectangular in shape and the peak values of the pulses are approximately equal to A.

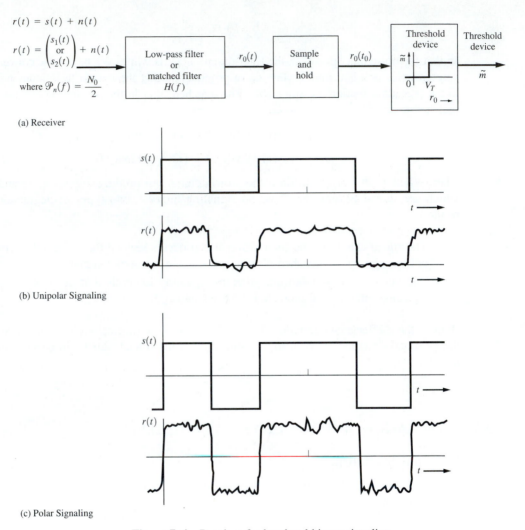

(a) Receiver

(b) Unipolar Signaling

(c) Polar Signaling

Figure 7–4   Receiver for baseband binary signaling.

$\sigma_0^2 = (N_0/2)(2B)$, where B is the equivalent bandwidth of the filter. The optimum threshold setting is then $V_T = \frac{1}{2}A$. When we use Eq. (7–17), the BER is

$$P_e = Q\left(\sqrt{\frac{A^2}{4N_0B}}\right) \qquad \text{(low-pass filter)} \qquad (7\text{–}24a)$$

for a receiver that uses an LPF with an equivalent bandwidth of B.

The performance of a matched-filter receiver is obtained by using Eq. (7–20), where the sampling time is $t_0 = T$. The energy in the difference signal is $E_d = A^2T$, so that the BER is

$$P_e = Q\left(\sqrt{\frac{A^2 T}{2N_0}}\right) = Q\left(\sqrt{\frac{E_b}{N_0}}\right) \qquad \text{(matched filter)} \qquad (7\text{–}24b)$$

where the average energy per bit is $E_b = A^2 T/2$ because the energy for a binary 1 is $A^2 T$ and the energy for a binary 0 is 0. For the rectangular pulse shape, the matched filter is an integrator. Consequently, the optimum threshold value is

$$V_T = \frac{s_{01} + s_{02}}{2} = \frac{1}{2}\left(\int_0^T A\,dt + 0\right) = \frac{AT}{2}$$

Often it is desirable to express the BER in terms of E_b/N_0, because it indicates the average energy required to transmit one bit of data over a white (thermal) noise channel. By expressing the BER in terms of E_b/N_0, the performance of different signaling techniques can be easily compared.

A plot of Eq. (7–24b) is shown in Fig. 7–5.

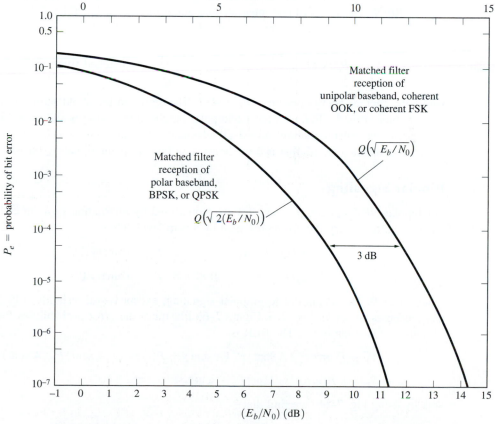

Figure 7–5 P_e for matched-filter reception of several binary signaling schemes.

Polar Signaling

As shown in Fig. 7–4c, the baseband polar signaling waveform is

$$s_1(t) = +A, \quad 0 < t \leq T \quad \text{(binary 1)} \tag{7--25a}$$

$$s_2(t) = -A, \quad 0 < t \leq T \quad \text{(binary 0)} \tag{7--25b}$$

The polar signal is an antipodal signal, since $s_1(t) = -s_2(t)$.

The performance of an LPF receiver system is obtained by using Eq. (7–17). Assuming that the equivalent bandwidth of the LPF is $B \geq 2/T$, we realize that the output signal samples are $s_{01}(t_0) \approx A$ and $s_{02}(t_0) \approx -A$ at the sampling time $t = t_0$. In addition, $\sigma_0^2 = N_0 B$. The optimum threshold setting is now $V_T = 0$. Thus, the BER for polar signaling is

$$P_e = Q\left(\sqrt{\frac{A^2}{N_0 B}}\right) \quad \text{(low-pass filter)} \tag{7--26a}$$

where B is the equivalent bandwidth of the LPF.

The performance of the matched-filter receiver is obtained, once again, by using Eq. (7–20), where $t_0 = T$. (The integrate-and-dump matched filter for polar signaling was given in Fig. 6–17.) Since the energy of the difference signal is $E_d = (2A)^2 T$, the BER is

$$P_e = Q\left(\sqrt{\frac{2A^2 T}{N_0}}\right) = Q\left(\sqrt{2\left(\frac{E_b}{N_0}\right)}\right) \quad \text{(matched filter)} \tag{7--26b}$$

where the average energy per bit is $E_b = A^2 T$. The optimum threshold setting is $V_T = 0$.

A plot of the BER for unipolar and polar baseband signaling is shown in Fig. 7–5. It is apparent that polar signaling has a 3-dB advantage over unipolar signaling, since unipolar signaling requires a E_b/N_0 that is 3 dB larger than that for polar signaling for the same P_e.

Bipolar Signaling

For bipolar NRZ signaling, binary 1's are represented by alternating positive and negative values, and the binary 0's are represented by a zero level. Thus,

$$s_1(t) = \pm A, \quad 0 < t \leq T \quad \text{(binary 1)} \tag{7--27a}$$

$$s_2(t) = 0, \quad 0 < t \leq T \quad \text{(binary 0)} \tag{7--27b}$$

where $A > 0$. This is similar to unipolar signaling, except two thresholds, $+V_T$ and $-V_T$, are needed as shown in Fig. 7–6. Figure 7–6b illustrates the error probabilities for the case of additive Gaussian noise. The BER is

$$P_e = P(\text{error} \mid +A \text{ sent}) P(+A \text{ sent}) + P(\text{error} \mid -A \text{ sent}) P(-A \text{ sent})$$

$$+ P(\text{error} \mid s_2 \text{ sent}) P(s_2 \text{ sent})$$

and, by using Fig. 7–6b, we find that

$$P_e \approx \left[2Q\left(\frac{A - V_T}{\sigma_0}\right)\right]\frac{1}{4} + \left[2Q\left(\frac{V_T}{v_0}\right)\right]\frac{1}{2}$$

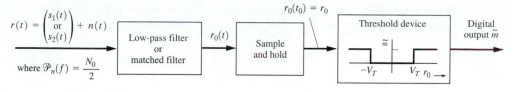

(a) Receiver

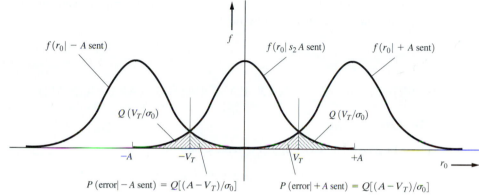

(b) Conditional PDFs

Figure 7–6   Receiver for bipolar signaling.

or

$$P_e \approx Q\left(\frac{V_T}{\sigma_0}\right) + \frac{1}{2}\left(\frac{A - V_T}{\sigma_0}\right)$$

Using calculus, we find that the optimum value of V_T that gives the minimum BER is $V_T = \frac{A}{2} + \frac{\sigma_0^2}{A}\ln2$. For systems with reasonably low (i.e., usable) BERs, $A > \sigma_0$, so that the optimum threshold becomes approximately $V_T = A/2$, the BER is

$$P_e = \frac{3}{2}\,Q\left(\frac{A}{2\sigma_0}\right)$$

For the case of a receiver with a low-pass filter that has a bipolar signal plus white noise at its input, $\sigma_0^2 = N_0 B$. Thus, the BER is

$$P_e = \frac{3}{2}\,Q\left(\sqrt{\frac{A^2}{4N_0 B}}\right) \qquad \text{(low-pass filter)} \qquad (7\text{–}28a)$$

where the PSD of the input noise is $N_0/2$ and the equivalent bandwidth of the filter is B Hz. If a matched filter is used, its output SNR is, using Eq. (6–161),

$$\left(\frac{S}{N}\right)_{out} = \frac{A^2}{\sigma_0^2} = \frac{2E_d}{N_0}$$

For bipolar NRZ signaling, the energy in the different signal is $E_d = A^2T = 2E_b$ where E_b is the average energy per bit. Thus, for a matched-filter receiver, the BER is

$$P_e = \frac{3}{2} \, Q\left(\sqrt{\frac{E_b}{N_0}}\right) \qquad \text{(matched filter)} \qquad (7\text{--}28b)$$

For bipolar RZ signaling, $E_d = A^2T/4 = 2E_b$, so that the resulting BER formula is identical to Eq. (7–28b). These results show that the BER for bipolar signaling is just $\frac{3}{2}$ that for unipolar signaling as described by Eq. (7–24b).

7–3 COHERENT DETECTION OF BANDPASS BINARY SIGNALS

On–Off Keying

From Fig. 5–1c, an OOK signal is represented by

$$s_1(t) = A \cos(\omega_c t + \theta_c), \qquad 0 < t \le T \qquad \text{(binary 1)} \qquad (7\text{--}29a)$$

or

$$s_2(t) = 0, \qquad\qquad\qquad 0 < t \le T \qquad \text{(binary 0)} \qquad (7\text{--}29b)$$

For coherent detection, a product detector is used as illustrated in Fig. 7–7. Actually, in RF applications, a mixer would convert the incoming OOK signal to an IF, so that stable high-gain amplification could be conveniently achieved and then a product detector would translate the signal to baseband. Figure 7–7, equivalently, represents these operations by converting the incoming signal and noise directly to baseband.

Assume that the OOK signal plus white (over the equivalent bandpass) Gaussian noise is present at the receiver input. As developed in Chapter 6, this bandpass noise may be represented by

$$n(t) = x(t) \cos(\omega_c t + \theta_n) - y(t) \sin(\omega_c t + \theta_n)$$

where the PSD of $n(t)$ is $\mathcal{P}_n(f) = N_0/2$ and θ_n is a uniformly distributed random variable that is independent of θ_c.

The receiving filter $H(f)$ of Fig. 7–7 may be some convenient LPF, or it may be a matched filter. Of course, the receiver would be optimized (for the lowest P_e) if a matched filter were used.

First, evaluate the performance of a receiver that uses an LPF where the filter has a dc gain of unity. Assume that the equivalent bandwidth of the filter is $B \ge 2/T$, so that the envelope of the OOK signal is (approximately) preserved at the filter output. The baseband analog output will be

$$r_0(t) = \left.\begin{cases} A, & 0 < t \le T, \quad \text{binary 1} \\ 0, & 0 < t \le T, \quad \text{binary 0} \end{cases}\right\} + x(t) \qquad (7\text{--}30)$$

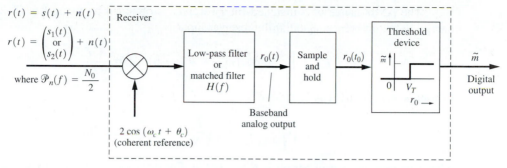

(a) Receiver

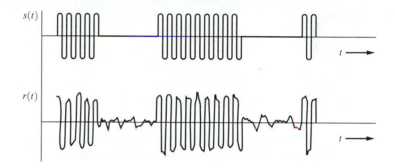

(b) OOK Signaling

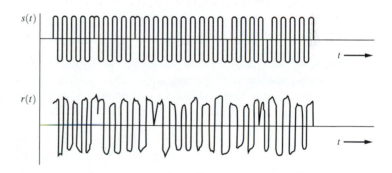

(c) BPSK Signaling

Figure 7–7   Coherent detection of OOK or BPSK signals.

where $x(t)$ is the baseband noise. With the help of Eq. (6–133g), we calculate noise power as $\overline{x^2(t)} = \sigma_0^2 = \overline{n^2(t)} = 2(N_0/2)(2B) = 2N_0B$. Because $s_{01} = A$ and $s_{02} = 0$, the optimum threshold setting is $V_T = A/2$. When we use equation (7–17), the BER is

$$P_e = Q\left(\sqrt{\frac{A^2}{8N_0B}}\right) \qquad \text{(narrowband filter)} \qquad (7\text{–}31)$$

B is the equivalent bandwidth of the LPF. The equivalent bandpass bandwidth of this receiver is $B_p = 2B$.

The performance of a matched-filter receiver is obtained by using Eq. (7–20). The energy in the difference signal at the receiver input is[†]

$$E_d = \int_0^T [A \cos(\omega_c t + \theta_c) - 0]^2 \, dt = \frac{A^2 T}{2} \tag{7–32}$$

Consequently, the BER is

$$P_e = Q\left(\sqrt{\frac{A^2 T}{4N_0}}\right) = Q\left(\sqrt{\frac{E_b}{N_0}}\right) \qquad \text{(matched filter)} \tag{7–33}$$

where the average energy per bit is $E_b = A^2 T/4$. For this case, where $s_1(t)$ has a rectangular (real) envelope, the matched filter is an integrator. Consequently, the optimum threshold value is

$$V_T = \frac{s_{01} + s_{02}}{2} = \frac{1}{2} s_{01} = \frac{1}{2} \left[\int_0^T 2A \cos^2(\omega_c t + \theta_c) \, dt \right]$$

which reduces to $V_T = AT/2$ when $f_c \gg R$. Note that the performance of OOK is exactly the same as that for baseband unipolar signaling, as illustrated in Fig. 7–5.

Binary-Phase-Shift Keying

Referring to Fig. 7–7, we see that the BPSK signal is

$$s_1(t) = A \cos(\omega_c t + \theta_c), \qquad 0 < t \le T \qquad \text{(binary 1)} \tag{7–34a}$$

and

$$s_2(t) = -A \cos(\omega_c t + \theta_c), \qquad 0 < t \le T \qquad \text{(binary 0)} \tag{7–34b}$$

BPSK signaling is also called *phase-reversal keying* (PRK). The BPSK signal is an antipodal signal because $s_1(t) = -s_2(t)$.

Once again, first evaluate the performance of a receiver that uses an LPF having a gain of unity and an equivalent bandwidth of $B \ge 2/T$. The baseband analog output is

$$r_0(t) = \begin{cases} A, & 0 < t \le T \quad \text{binary 1} \\ -A, & 0 < t \le T, \quad \text{binary 0} \end{cases} + x(t) \tag{7–35}$$

where $\overline{x^2(t)} = \sigma_0^2 = \overline{n^2(t)} = 2N_0 B$. Because $s_{01} = A$ and $s_{02} = -A$, the optimum threshold is $V_T = 0$. When we use Eq. (7–17), the BER is

$$P_e = Q\left(\sqrt{\frac{A^2}{2N_0 B}}\right) \qquad \text{(narrowband filter)} \tag{7–36}$$

When BPSK is compared with OOK on a *peak envelope power* (PEP) basis for a given value of N_0, 6 dB less (peak) signal power is required for BPSK signaling to give the same P_e

---

[†] Strictly speaking, f_c needs to be an integer multiple of half the bit rate, $R = 1/T$, to obtain exactly $A^2 T/2$ for E_d. However, because $f_c \gg R$, for all practical purposes $E_d = A^2 T/2$, regardless of whether or not $f_c = nR/2$.

as that for OOK. However, if the two are compared on an average (signal) power basis, the performance of BPSK has a 3-dB advantage over OOK, since the average power of OOK is 3 dB below its PEP (equally likely signaling), but the average power of BPSK is equal to its PEP.

The performance of the matched-filter receiver is obtained by using Eq. (7–20). Recall the matched filter for BPSK signaling that was illustrated in Fig. 6–19, where a correlation processor using an integrate-and-dump filter was shown. The energy in the difference signal at the receiver input is

$$E_d = \int_0^T [2A \cos(\omega_c t + \theta_c)]^2 \, dt = 2A^2T \tag{7-37}$$

Thus, the BER is

$$P_e = Q\left(\sqrt{\frac{A^2T}{N_0}}\right) = Q\left(\sqrt{2\left(\frac{E_b}{N_0}\right)}\right) \qquad \text{(matched filter)} \tag{7-38}$$

where the average energy per bit is $E_b = A^2T/2$ and $V_T = 0$. The performance of BPSK is exactly the same as that for baseband polar signaling; however, it is 3 dB superior to OOK signaling. (See Fig. 7–5.)

Frequency-Shift Keying

The FSK signal can be coherently detected by using two product detectors. This is illustrated in Fig. 7–8, where identical LPFs at the output of the product detectors have been replaced by only one of the filters, since the order of linear operations may be exchanged without affecting the results. The mark (binary 1) and space (binary 0) signals are

$$s_1(t) = A \cos(\omega_1 t + \theta_c), \qquad 0 < t \le T \qquad \text{(binary 1)} \tag{7-39a}$$

$$s_2(t) = A \cos(\omega_2 t + \theta_c), \qquad 0 < t \le T \qquad \text{(binary 0)} \tag{7-39b}$$

where the frequency shift is $2 \Delta F = f_1 - f_2$, assuming that $f_1 > f_2$. This FSK signal plus white Gaussian noise is present at the receiver input. The PSD for $s_1(t)$ and $s_2(t)$ is shown in Fig. 7–8b.

First, evaluate the performance of a receiver that uses an LPF $H(f)$ with a dc gain of 1. Assume that the equivalent bandwidth of the filter is $2/T \le B < \Delta F$. The LPF, when combined with the frequency translation produced by the product detectors, acts as dual bandpass filters—one centered at $f = f_1$ and the other at $f = f_2$, where each has an equivalent bandwidth of $B_p = 2B$. Thus, the input noise that affects the output consists of two narrowband components $n_1(t)$ and $n_2(t)$, where the spectrum of $n_1(t)$ is centered at f_1 and the spectrum of $n_2(t)$ is centered at f_2, as shown in Fig. 7–8. Furthermore, $n(t) = n_1(t) + n_2(t)$, where, using Eq. (6–130),

$$n_1(t) = x_1(t) \cos(\omega_1 t + \theta_c) - y_1(t) \sin(\omega_1 t + \theta_c) \tag{7-40a}$$

and

$$n_2(t) = x_2(t) \cos(\omega_2 t + \theta_c) - y_2(t) \sin(\omega_2 t + \theta_c) \tag{7-40b}$$

The frequency shift is $2\Delta F > 2B$, so that the mark and space signals may be separated by the filtering action. The input signal and noise that pass through the upper channel in Fig. 7–8a as described by

$$r_1(t) = \begin{Bmatrix} s_1(t), & \text{binary 1} \\ 0, & \text{binary 0} \end{Bmatrix} + n_1(t) \tag{7–41}$$

and the signal and noise that pass through the lower channel are described by

$$r_2(t) = \begin{Bmatrix} 0, & \text{binary 1} \\ s_2(t), & \text{binary 0} \end{Bmatrix} + n_2(t) \tag{7–42}$$

where $r(t) = r_1(t) + r_2(t)$. The noise power of $n_1(t)$ and $n_2(t)$ is $\overline{n_1^2(t)} = \overline{n_2^2(t)}$ $= (N_0/2)(4B) = 2N_0B$. Thus, the baseband analog output is

$$r_0(t) = \begin{Bmatrix} +A, & 0 < t \le T, & \text{binary 1} \\ -A, & 0 < t \le T, & \text{binary 0} \end{Bmatrix} + n_0(t) \tag{7–43}$$

where $s_{01} = +A$, $s_{02} = -A$ and $n_0(t) = x_1(t) - x_2(t)$. The optimum threshold setting is $V_T = 0$. Furthermore, the bandpass noise processes $n_1(t)$ and $n_2(t)$ are independent, since they have spectra in nonoverlapping frequency bands (see Fig. 7–8) and they are white. (See Prob. 7–29 for the verification of this statement.) Consequently, the resulting baseband noise processes $x_1(t)$ and $x_2(t)$ are independent, and the output noise power is

$$\overline{n_0^2(t)} = \sigma_0^2 = \overline{x_1^2(t)} + \overline{x_2^2(t)} = \overline{n_1^2(t)} + \overline{n_2^2(t)} = 4N_0B \tag{7–44}$$

Substituting for s_{01}, s_{02}, and σ_0 into Eq. (7–17), we have

$$P_e = Q\left(\sqrt{\frac{A^2}{4N_0B}}\right) \qquad \text{(bandpass filters)} \tag{7–45}$$

Comparing the performance of FSK with that of BPSK and OOK on a PEP basis, we see that FSK requires 3 dB more power than BPSK for the same P_e, but 3 dB less power than OOK. Comparing the performance on an average power basis, we observe that FSK is 3 dB worse than BPSK, but is equivalent to OOK (since the average power of OOK is 3 dB below its PEP).

The performance of FSK signaling with matched-filter reception is obtained from Eq. (7–20). The energy in the difference signal is

$$E_d = \int_0^T [A \cos(\omega_1 t + \theta_c) - A \cos(\omega_2 t + \theta_c)]^2 \, dt$$

$$= \int_0^T [A^2 \cos^2(\omega_1 t + \theta_c) - 2A^2 \cos(\omega_1 t + \theta_c)$$

$$\times \cos(\omega_2 t + \theta_c) + A^2 \cos^2(\omega_2 t + \theta_c)] \, dt$$

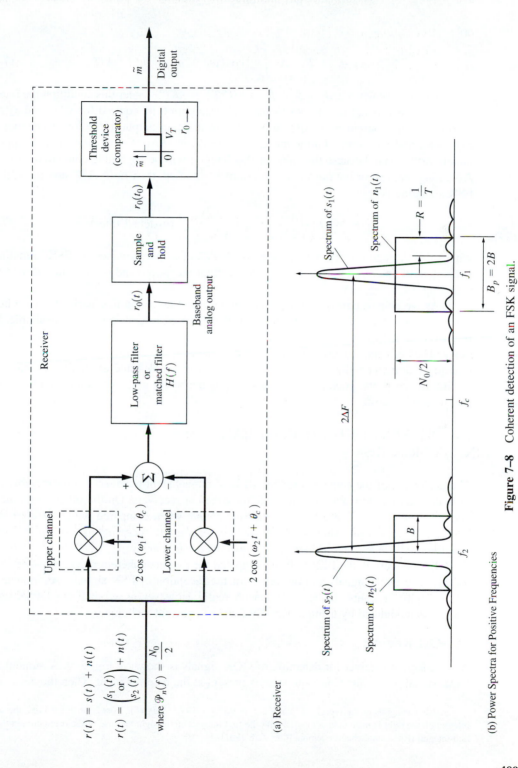

Figure 7–8 Coherent detection of an FSK signal.

or[†]

$$E_d = \tfrac{1}{2}A^2T - A^2 \int_0^T \left[\cos\left((\omega_1 - \omega_2)t\right)\right] dt + \tfrac{1}{2}A^2T \qquad (7\text{--}46)$$

Consider the case when $2\Delta F = f_1 - f_2 = n/(2T) = nR/2$. Under this condition the integral (i.e., the cross-product term) goes to zero. This condition is required for $s_1(t)$ and $s_2(t)$ to be orthogonal. Consequently, $s_1(t)$ will not contribute an output to the lower channel (see Fig. 7–8), and vice versa. Furthermore, if $(f_1 - f_2) \gg R$, $s_1(t)$ and $s_2(t)$ will be approximately orthogonal, because the value of this integral will be negligible compared with A^2T. Assuming that one or both of these conditions is satisfied, then $E_d = A^2T$, and the BER for FSK signaling is

$$P_e = Q\left(\sqrt{\frac{A^2T}{2N_0}}\right) = Q\left(\sqrt{\frac{E_b}{N_0}}\right) \qquad \text{(matched filter)} \qquad (7\text{--}47)$$

where the average energy per bit is $E_b = A^2T/2$. The performance of FSK signaling is equivalent to that of OOK signaling (matched-filter reception) and is 3 dB inferior to BPSK signaling. (See Fig. 7–5.)

As we demonstrate in the next section, coherent detection is superior to noncoherent detection. However, for coherent detection, the coherent reference must be available. This reference is often obtained from the noisy input signal so that the reference itself is also noisy. This, of course, increases P_e over those values given by the preceding formulas. The circuitry that extracts the carrier reference is usually complex and expensive. Often, one is willing to accept the poorer performance of a noncoherent system to simplify the circuitry and reduce the cost.

7–4 NONCOHERENT DETECTION OF BANDPASS BINARY SIGNALS

The *derivation* of the equations for the BER of noncoherent receivers is considerably more difficult than the derivation of the BER for coherent receivers. On the other hand, the *circuitry* for noncoherent receivers is *relatively simple* when compared with that used in coherent receivers. For example, OOK *with noncoherent reception is the most popular signaling technique used in fiber-optic communication systems.*

In this section, the BER will be computed for two noncoherent receivers—one for the reception of OOK signals and the other for the reception of FSK signals. As indicated in Chapter 5, BPSK cannot be detected noncoherently. However, as we shall see, DPSK signals may be demodulated by using a partially (quasi-) coherent technique.

On–Off Keying

A noncoherent receiver for detection of OOK signals is shown in Fig. 7–9. Assume that an OOK signal plus white Gaussian noise is present at the receiver input. Then the noise at the

---

[†] For integrals of the type $\int_0^T A^2 \cos^2(\omega_1 t + \theta_c)\, dt = \tfrac{1}{2}A^2 \left[\int_0^T dt + \int_0^T \cos(2\omega_1 t + 2\theta_c)\, dt\right]$, the second integral on the right is negligible compared with the first integral on the right because of the oscillation in the second integral (Riemann–Lebesque lemma [Olmsted, 1961]).

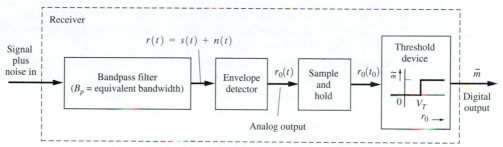

Figure 7–9   Noncoherent detection of OOK.

filter output $n(t)$ will be bandlimited Gaussian noise and the total filter output, consisting of signal plus noise, is

$$r(t) = \begin{cases} r_1(t), & 0 < t \le T, \quad \text{binary 1 sent} \\ r_2(t), & 0 < t \le T, \quad \text{binary 0 sent} \end{cases} \tag{7-48}$$

Let the bandwidth of the filter be B_p, where B_p is at least as large as the transmission bandwidth of the OOK signal, so that the signal waveshape is preserved at the filter output. Then for the case of a binary 1, $s_1(t) = A \cos(\omega_c t + \theta_c)$, so

$$r_1(t) = A \cos(\omega_c t + \theta_c) + n(t), \qquad 0 < t \le T$$

or

$$r_1(t) = [A + x(t)] \cos(\omega_c t + \theta_c) - y(t) \sin(\omega_c t + \theta_c), \qquad 0 < t \le T \tag{7-49}$$

For a binary 0, $s_2(t) = 0$ and

$$r_2(t) = x(t) \cos(\omega_c t + \theta_c) - y(t) \cos(\omega_c t + \theta_c), \qquad 0 < t \le T \tag{7-50}$$

The BER is obtained by using Eq. (7–8), which, for the case of equally likely signaling, is

$$P_e = \tfrac{1}{2} \int_{-\infty}^{V_T} f(r_0|s_1)\, dr_0 + \tfrac{1}{2} \int_{V_T}^{\infty} f(r_0|s_2)\, dr_0 \tag{7-51}$$

We need to evaluate the conditional PDFs for the output of the envelope detector, $f(r_0|s_1)$ and $f(r_0|s_2)$. $f(r_0|s_1)$ is the PDF for $r_0 = r_0(t) = r_{01}$ that occurs when $r_1(t)$ is present at the input of the envelope detector, and $f(r_0|s_2)$ is the PDF for $r_0 = r_0(t_0) = r_{02}$ that occurs when $r_2(t)$ is present at the input of the envelope detector.

We will evaluate $f(r_0|s_2)$ first. When $s_2(t)$ is sent, the input to the envelope detector, $r_2(t)$, consists of bandlimited bandpass Gaussian noise as seen from Eq. (7–50). In Example 6–10, we demonstrated that for this case, the PDF of the envelope is Rayleigh distributed. Of course, the output of the envelope detector is the envelope, so $r_0 = R = r_{02}$. Thus, the PDF for the case of noise alone is

$$f(r_0|s_2) = \begin{cases} \dfrac{r_0}{\sigma^2}\, e^{-r_0^2/(2\sigma^2)}, & r \ge 0 \\ 0, & r_0 \text{ otherwise} \end{cases} \tag{7-52}$$

The parameter σ^2 is the variance of the noise at the input of the envelope detector. Thus, $\sigma^2 = (N_0/2)(2B_p) = N_0 B_p$, where B_p is the effective bandwidth of the bandpass filter and $N_0/2$ is the PSD of the white noise at the receiver input.

For the case of $s_1(t)$ being transmitted, the input to the envelope detector is given by Eq. (7–49). Since $n(t)$ is a Gaussian process (that has no delta functions in its spectrum at $f = \pm f_c$) the in-phase baseband component, $A + x(t)$, is also a Gaussian process with a mean value of A instead of a zero mean, as is the case in Eq. (7–50). The PDF for the envelope $r_0 = R = r_{01}$ is evaluated using the same technique descibed in Example 6–10 and the result cited in Prob. 6–46. Thus, for this case of a sinusoid plus noise at the envelope detector input,

$$f(r_0|s_1) = \begin{cases} \dfrac{r_0}{\sigma^2}\, e^{-(r_0^2 + A^2)/(2\sigma^2)}\, I_0\!\left(\dfrac{r_0 A}{\sigma^2}\right) & r_0 \ge 0 \\[2mm] 0, & r_0 \text{ otherwise} \end{cases} \qquad (7\text{–}53)$$

which is a Rician PDF, where

$$I_0(z) \triangleq \frac{1}{2\pi} \int_0^{2\pi} e^{z\cos\theta}\, d\theta \qquad (7\text{–}54)$$

is the modified Bessel function of the first kind of zero order.

The two conditional PDFs, $f(r_0|s_2)$ and $f(r_0|s_1)$, are shown in Fig. 7–10. Actually, $f(r_0|s_2)$ is a special case of $f(r_0|s_1)$ when $A = 0$, because, for this condition, we have noise only at the detector input and Eq. (7–53) becomes Eq. (7–52). Two other plots of Eq. (7–53) are also given—one for $A = 1$ and another for $A = 4$. It is seen that for $A/\sigma \gg 1$, the mode of the distribution [i.e., where $f(r_0|s_1)$ is a maximum] occurs at the value $r_0 = A$. In addition, note that for $A/\sigma \gg 1$, $f(r_0|s_1)$ takes on a Gaussian shape (as demonstrated subsequently).

The BER for the noncoherent OOK receiver is obtained by substituting Eqs. (7–52) and (7–53) into Eq. (7–51):

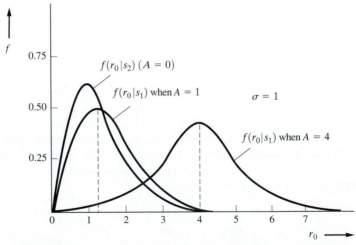

Figure 7–10   Conditional PDFs for noncoherent OOK reception.

$$P_e = \frac{1}{2} \int_0^{V_T} \frac{r_0}{\sigma^2} e^{-(r_0^2 + A^2)/(2\sigma^2)} I_0\left(\frac{r_0 A}{\sigma^2}\right) dr_0 + \frac{1}{2} \int_{V_T}^{\infty} \frac{r_0}{\sigma^2} e^{-r_0^2/(2\sigma^2)} dr_0 \quad (7\text{–}55)$$

The optimum threshold level is the value of V_T for which P_e is a minimum. For $A/\sigma \gg 1$, the optimum threshold is close to $V_T = A/2$, so we will use this value to simplify the mathematics.[†] The integral involving the Bessel function cannot be evaluated in closed form. However, $I_0(z)$ can be approximated by $I_0(z) = e^z/\sqrt{2\pi z}$, which is valid for $z \gg 1$. Then, for $A/\sigma \gg 1$, the left integral in Eq. (7–55) becomes

$$\frac{1}{2} \int_0^{V_T} \frac{r_0}{\sigma^2} e^{-(r_0^2 + A^2)/(2\sigma^2)} I_0\left(\frac{r_0 A}{\sigma^2}\right) dr_0 \approx \frac{1}{2} \int_0^{A/2} \sqrt{\frac{r_0}{2\pi\sigma^2 A}} e^{(r_0 - A)^2/(2\sigma^2)} dr_0$$

Because $A/\sigma \gg 1$, the integrand is negligible except for values of r_0 in the vicinity of A, so the lower limit may be extended to $-\infty$, and $\sqrt{r_0/(2\pi\sigma^2 A)}$ can be replaced by $\sqrt{1/(2\pi\sigma^2)}$. Thus,

$$\frac{1}{2} \int_0^{V_T} \frac{r_0}{\sigma^2} e^{-(r_0^2 + A^2)/(2\sigma^2)} I_0\left(\frac{r_0 A}{\sigma^2}\right) dr_0 \approx \frac{1}{2} \int_{-\infty}^{V_T} \frac{1}{\sqrt{2\pi}\sigma} e^{-(r_0 - A)^2/(2\sigma^2)} dr_0 \qquad (7\text{–}56)$$

When we substitute this equation into Eq. (7–55), the BER becomes

$$P_e = \frac{1}{2} \int_{-\infty}^{A/2} \frac{1}{\sqrt{2\pi}\sigma} e^{-(r_0 - A)^2/(2\sigma^2)} dr_0 + \frac{1}{2} \int_{A/2}^{\infty} \frac{r_0}{\sigma^2} e^{-r_0^2/(2\sigma^2)} dr_0$$

or

$$P_e = \tfrac{1}{2} Q\left(\frac{A}{2\sigma}\right) + \tfrac{1}{2} e^{-A^2/(8\sigma^2)} \qquad (7\text{–}57)$$

Using $Q(z) = e^{-z^2/2}/\sqrt{2\pi z^2}$ for $z \gg 1$, we have

$$P_e = \frac{1}{\sqrt{2\pi}(A/\sigma)} e^{-A^2/(8\sigma^2)} + \tfrac{1}{2} e^{-A^2/(8\sigma^2)}$$

Because $A/\sigma \gg 1$, the second term on the right dominates over the first term. *Finally*, we obtain the approximation for the BER for noncoherent detection of OOK. It is

$$P_e = \tfrac{1}{2} e^{-A^2/(8\sigma^2)}, \qquad \frac{A}{\sigma} \gg 1$$

or

$$P_e = \tfrac{1}{2} e^{-[1/(2TB_p)](E_b/N_0)}, \qquad \frac{E_b}{N_0} \gg \frac{TB_p}{4} \qquad (7\text{–}58)$$

where the average energy per bit is $E_b = A^2 T/4$ and $\sigma^2 = N_0 B_p$. $R = 1/T$ is the bit rate of the OOK signal, and B_p is the equivalent bandwidth of the bandpass filter that precedes the envelope detector.

[†] Most practical systems operate with $A/\sigma \gg 1$.

Equation (7–58) indicates that the BER depends on the bandwidth of the bandpass filter and that P_e becomes smaller as B_p is decreased. Of course, this result is valid only when the ISI is negligible. Referring to Eq. (3–74), we realize that the minimum bandwidth allowed (i.e., for no ISI) is obtained when the rolloff factor is $r = 0$. This implies that the minimum bandpass bandwidth that is allowed is $B_p = 2B = R = 1/T$. A plot of the BER is given in Fig. 7–14 for this minimum-bandwidth case of $B_p = 1/T$.

Frequency-Shift Keying

A noncoherent receiver for detection of frequency-shift-keyed (FSK) signals is shown in Fig. 7–11. The input consists of an FSK signal as described by Eq. (7–39) plus white Gaussian noise with a PSD of $N_0/2$. A sketch of the spectrum of the FSK signal and noise was given in Fig. 7–8b, where the bandpass filter bandwidth is B_p. It is assumed that the frequency shift, $2\Delta F = f_1 - f_2$, is sufficiently large so that the spectra of $s_1(t)$ and $s_2(t)$ have negligible overlap.

The BER for this receiver is obtained by evaluating Eq. (7–8). For signal alone at the receiver input, the output of the summing junction is $r_0(t) = +A$ when a mark (binary 1) is transmitted, and $r_0(t) = -A$ when a space (binary 0) is transmitted. Because of this symmetry and because the noise out of the upper and lower receiver channels is similar, the optimum threshold is $V_T = 0$. Similarly, it is realized that the PDF of $r_0(t)$ conditioned on s_1 and that the PDF of $r_0(t)$ conditioned on s_2 are similar. That is,

$$f(r_0|s_1) = f(-r_0|s_2) \tag{7–59}$$

Substituting Eq. (7–59) into Eq. (7–8), we find the BER to be

$$P_e = \tfrac{1}{2} \int_{-\infty}^{0} f(r_0|s_1)\, dr_0 + \tfrac{1}{2} \int_{0}^{\infty} f(r_0|s_2)\, dr_0$$

or

$$P_e = \int_{0}^{\infty} f(r_0|s_2)\, dr_0 \tag{7–60}$$

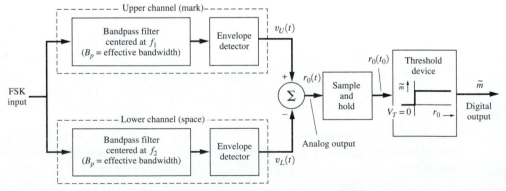

Figure 7–11 Noncoherent detection of FSK.

As shown in Fig. 7–11, $r_0(t)$ is positive when the upper channel output $v_U(t)$ exceeds the lower channel output $v_L(t)$. Thus,

$$P_e = P(v_U > v_L|s_2) \tag{7–61}$$

For the case of a space signal plus noise at the receiver input, we know that the output of the upper bandpass filter is only Gaussian noise (no signal). Thus, the output of the upper envelope detector v_U is noise having a Rayleigh distribution

$$f(v_U|s_2) = \begin{cases} \dfrac{v_U}{\sigma^2} e^{-v_U^2/(2\sigma^2)}, & v_U \geq 0 \\ 0, & v_U < 0 \end{cases} \tag{7–62}$$

where $\sigma^2 = N_0 B_p$. On the other hand, v_L has a Rician distribution, since a sinusoid (the space signal) plus noise appears at the input to the lower envelope detector:

$$f(v_L|s_2) = \begin{cases} \dfrac{v_L}{\sigma^2} e^{-(v_L^2 + A^2)/(2\sigma^2)} I_0\left(\dfrac{v_L A}{\sigma^2}\right), & v_L \geq 0 \\ 0, & v_L < 0 \end{cases} \tag{7–63}$$

where $\sigma^2 = N_0 B_p$ also. Using Eqs. (7–62) and (7–63) in Eq. (7–61), we obtain

$$P_e = \int_0^\infty \frac{v_L}{\sigma^2} e^{-(v_L^2 + A^2)/(2\sigma^2)} I_0\left(\frac{v_L A}{\sigma^2}\right) \left[\int_{v_L}^\infty \frac{v_U}{\sigma^2} e^{-v_U^2/(2\sigma^2)} \, dv_U\right] dv_L$$

When we evaluate the inner integral, the BER becomes

$$P_e = e^{-A^2/(2\sigma^2)} \int_0^\infty \frac{v_L}{\sigma^2} e^{-v_L^2/\sigma^2} I_0\left(\frac{v_L A}{\sigma^2}\right) dv_L \tag{7–64}$$

This integral can be evaluated by using the integral table in Appendix A. Thus, for noncoherent detection of FSK, the BER is

$$P_e = \tfrac{1}{2} e^{-A^2/(4\sigma^2)}$$

or

$$P_e = \tfrac{1}{2} e^{-[1/(2TB_p)](E_b/N_0)} \tag{7–65}$$

where the average energy per bit is $E_b = A^2 T/2$ and $\sigma^2 = N_0 B_p$. $N_0/2$ is the PSD of the input noise, and B_p is the effective bandwidth of each of the bandpass filters. (See Fig. 7–11.) Comparing Eq. (7–65) with Eq. (7–58), we see that OOK and FSK are equivalent on an E_b/N_0 basis. A plot of Eq. (7–65) is given in Fig. 7–14 for the case of the minimum filter bandwidth allowed, $B_p = R = 1/T$, for no ISI. When comparing the error performance of noncoherently detected FSK with that of coherently detected FSK, it is seen that noncoherent FSK requires, at most, only 1 dB more E_b/N_0 than that for coherent FSK if P_e is 10^{-4} or less. The noncoherent FSK receiver is considerably easier to build since the coherent reference signals do not have to be generated. Thus, in practice, almost all of the FSK receivers use noncoherent detection.

Differential Phase-Shift Keying

Phase-shift-keyed signals cannot be detected incoherently. However, a partially coherent technique can be used whereby the phase reference for the present signaling interval is provided by a delayed version of the signal that occurred during the previous signaling interval. This is illustrated by the receivers shown in Fig. 7–12, where differential decoding is provided by the (one-bit) delay and the multiplier. If a BPSK signal (no noise) were applied to the receiver input, the output of the sample-and-hold circuit, $r_0(t_0)$, would be positive (binary 1) if the present data bit and the previous data bit were of the same sense; $r_0(t_0)$ would be negative (binary 0) if the two data bits were different. Consequently, if the data on the BPSK signal are differentially encoded (e.g., see the illustration in Table 3–4), the *decoded* data sequence will be recovered at the output of this receiver. This signaling technique consisting of transmitting a differentially encoded BPSK signal is known as DPSK.

The BER for these DPSK receivers can be derived under the following assumptions:

- The additive input noise is white and Gaussian.
- The phase perturbation of the composite signal plus noise varies slowly so that the phase reference is essentially a constant from the past signaling interval to the present signaling interval.
- The transmitter carrier oscillator is sufficiently stable so that the phase during the present signaling interval is the same as that from the past signaling interval.

The BER for the suboptimum demodulator of Fig. 7–12a has been obtained by J. H. Park for the case of a large input signal-to-noise ratio and for $B_T > 2/T$, but yet not too large. The result is [Park, 1978]

$$P_e = Q\left(\sqrt{\frac{(E_b/N_0)}{1 + [(B_T T/2)/(E_b/N_0)]}}\right) \qquad (7\text{–}66a)$$

For typical values of B_T and E_b/N_0 in the range of $B_T = 3/T$ and $E_b/N_0 = 10$, this BER can be approximated by

$$P_e = Q\left(\sqrt{E_b/N_0}\right) \qquad (7\text{–}66b)$$

Thus the performance of the suboptimum receiver of Fig. 7–12a is similar to that obtained for OOK and FSK as plotted in Fig. 7–14.

Figure 7–12b shows one form of an optimum DPSK receiver that can be obtained [Couch, 1993, Fig. 8–25]. The BER for optimum demodulation of DPSK is

$$P_e = \tfrac{1}{2}e^{-(E_b/N_0)} \qquad (7\text{–}67)$$

Other alternative forms of optimum DPSK receivers are possible [Lindsey and Simon, 1973; Simon, 1978].

A plot of this error characteristic for the case of optimum demodulation of DPSK, Eq. (7–67), is shown in Fig. 7–14. In comparing the error performance of BPSK and DPSK

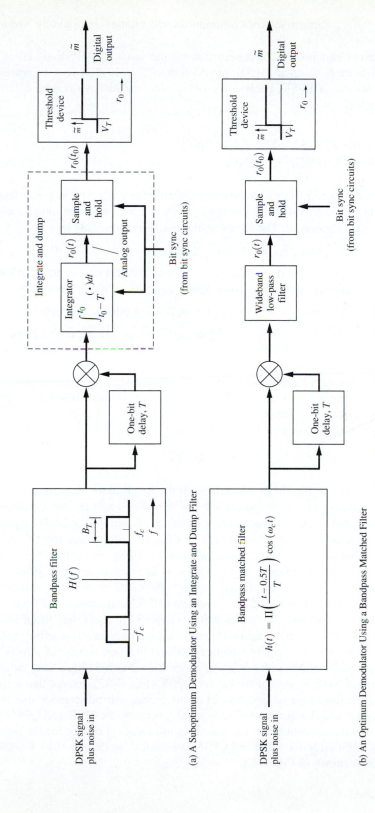

Figure 7–12 Demodulation of DPSK.

(a) A Suboptimum Demodulator Using an Integrate and Dump Filter

(b) An Optimum Demodulator Using a Bandpass Matched Filter

497

with optimum demodulation, it is seen that for the same P_e, DPSK signaling requires, at most, 1 dB more E_b/N_0 than BPSK, provided that $P_e = 10^{-4}$ or less. In practice, DPSK is often used instead of BPSK, because the DPSK receiver does not require a carrier synchronization circuit.

7–5 QUADRATURE PHASE-SHIFT KEYING
AND MINIMUM-SHIFT KEYING

As described in Sec. 5–10, quadrature phase-shift keying (QPSK) is a multilevel signaling technique that uses $L = 4$ levels per symbol. Thus, 2 bits are transmitted during each signaling interval (T seconds). The QPSK signal may be represented by

$$s(t) = (\pm A) \cos(\omega_c t + \theta_c) - (\pm A) \sin(\omega_c t + \theta_c), \qquad 0 < t \le T \qquad (7\text{–}68)$$

where the $(\pm A)$ factor on the cosine carrier is one bit of data and the $(\pm A)$ factor on the sine carrier is another bit of data. The relevant input noise is represented by

$$n(t) = x(t) \cos(\omega_c t + \theta_n) - y(t) \sin(\omega_c t + \theta_n)$$

The QPSK signal is equivalent to two BPSK signals—one using a cosine carrier and the other using a sine carrier. The QPSK signal is detected by using the coherent receiver shown in Fig. 7–13. (This is an application of the IQ detector that was first given in Fig. 4–31.) Because both the upper and lower channels of the receiver are BPSK receivers, the BER is the same as that for a BPSK system. Thus, from Eq. (7–38), the BER for the QPSK receiver is

$$P_e = Q\left(\sqrt{2\left(\frac{E_b}{N_0}\right)}\right) \qquad (7\text{–}69)$$

This is also shown in Fig. 7–14. The BERs for the BPSK and QPSK signaling are identical, but for the same bit rate R, the bandwidths of the two signals are *not* the same. The bandwidth of the QPSK signal is exactly one-half the bandwidth of the BPSK signal for a given bit rate. This result is given by Eq. (5–103) or Eq. (5–106), where the QPSK signal transmits one symbol for every 2 bits of data, whereas the BPSK signal transmits one symbol for each bit of data. The bandwidth of $\pi/4$ QPSK is identical to that for QPSK. For the same BER, differentially detected $\pi/4$ QPSK requires about 3 dB more E_b/N_0 than that for QPSK, but coherently detected $\pi/4$ QPSK has the same BER performance as QPSK.

In Chapter 5, it was shown that MSK is essentially equivalent to QPSK, except that the data on the $x(t)$ and $y(t)$ quadrature modulation components are offset and their equivalent data pulse shape is a positive part of a cosine function instead of a rectangular pulse. (This gives a PSD for MSK that rolls off faster than that for PSK.) The optimum receiver for detection of MSK is similar to that for QPSK (Fig. 7–13), except that a matched filter with a cosine pulse shape is used instead of the rectangular pulse shape that is synthesized by the integrate-and-dump circuit. Consequently, because the MSK and QPSK signal representations and the optimum receiver structures are identical except for the pulse shape, the probability of bit error for MSK and QPSK is identical, as described by Eq. (7–69). A plot of this BER is shown in Fig. 7–14.

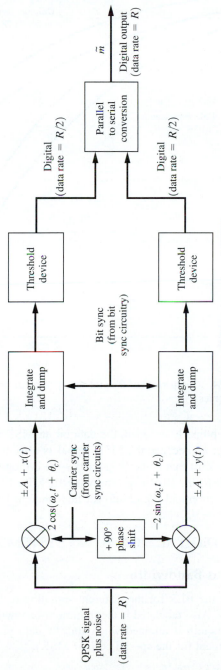

Figure 7–13 Matched-filter detection of QPSK.

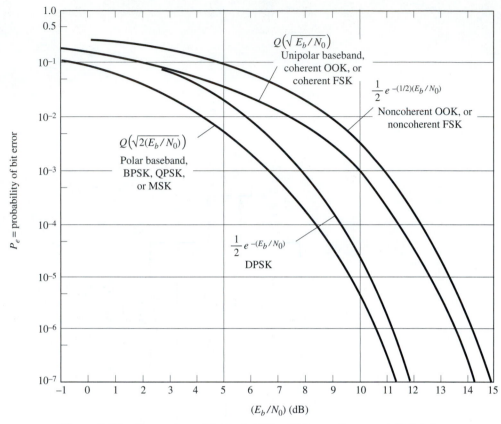

Figure 7–14   Comparison of the probability of bit error for several digital signaling schemes.

If the data are properly encoded, the data on an MSK signal can also be detected by using FM-type detectors because the MSK signal is also an FSK signal with a minimum amount of frequency shift that makes $s_1(t)$ and $s_2(t)$ orthogonal signals. Thus, for suboptimum detection of MSK, the BER is given by the BER for FSK, as described by Eq. (7–47) for coherent FM detection and Eq. (7–65) for noncoherent FM detection.

7–6 COMPARISON OF DIGITAL SIGNALING SYSTEMS

Bit-Error Rate and Bandwidth

Table 7–1 compares the BER for the different signaling techniques that were described in the previous sections. Also tabulated is the minimum bandwidth for these signals. The minimum absolute bandwidth is attained when $\sin x/x$ data pulses are used, as described in Chapters 3 and 5 (except for the special case of MSK).

TABLE 7–1 COMPARISON OF DIGITAL SIGNALING METHODS

Type of Digital Signaling	Minimum Transmission Bandwidth Required[a] (Where R Is the Bit Rate)		Error Performance		
Baseband signaling					
Unipolar	$\frac{1}{2}R$	(5-105)	$Q\left[\sqrt{\left(\frac{E_b}{N_0}\right)}\right]$		(7-24b)
Polar	$\frac{1}{2}R$	(5-105)	$Q\left[\sqrt{2\left(\frac{E_b}{N_0}\right)}\right]$		(7-26b)
Bipolar	$\frac{1}{2}R$	(5-105)	$\frac{3}{2}Q\left[\sqrt{\left(\frac{E_b}{N_0}\right)}\right]$		(7-28b)
Bandpass signaling			*Coherent detection*	*Noncoherent detection*	
OOK	R	(5-106)	$Q\left[\sqrt{\left(\frac{E_b}{N_0}\right)}\right]$ (7-33)	$\frac{1}{2}e^{-(1/2)(E_b/N_0)}$, $\left(\frac{E_b}{N_0}\right) > \frac{1}{4}$	(7-58)
BPSK	R	(5-106)	$Q\left[\sqrt{2\left(\frac{E_b}{N_0}\right)}\right]$ (7-38)	Requires coherent detection	
FSK	$2\Delta F + R$ where $2\Delta F = f_2 - f_1$ is the frequency shift	(5-89)	$Q\left[\sqrt{\left(\frac{E_b}{N_0}\right)}\right]$ (7-47)	$\frac{1}{2}e^{-(1/2)(E_b/N_0)}$	(7-65)
DPSK	R	(5-106)	Not used in practice	$\frac{1}{2}e^{-(E_b/N_0)}$	(7-67)
QPSK	$\frac{1}{2}R$	(5-106)	$Q\left[\sqrt{2\left(\frac{E_b}{N_0}\right)}\right]$ (7-69)	Requires coherent detection	
MSK	$1.5R$ (null bandwidth)	(5-115)	$Q\left[\sqrt{2\left(\frac{E_b}{N_0}\right)}\right]$ (7-69)	$\frac{1}{2}e^{-(1/2)(E_b/N_0)}$	(7-65)

[a] Typical bandwidth specifications by ITU are larger than these minima [Jordan, 1985].

In Fig. 7–14, the BER curves are plotted using the equations presented in Table 7–1. Except for the curves describing the noncoherent detection cases, all of these results assume that the optimum filter—the matched filter—is used in the receiver. In practice, simpler filters work almost as well as the matched filter. For example, in a computer simulation of a BPSK system with a three-pole Butterworth receiving filter having a bandwidth equal to the bit rate, the E_b/N_0 needs to be increased no more than 0.4 dB to obtain the same BER as that obtained when a matched filter is used (for error rates above 10^{-12}).

Comparing the various bandpass signaling techniques, we see that QPSK and MSK give the best overall performance in terms of the minimum bandwidth required for a given signaling rate and one of the smallest P_e for a given E_b/N_0. However, QPSK is relatively expensive to implement, since it requires coherent detection. Figure 7–14 shows that for the same P_e, DPSK (using a noncoherent receiver) requires only about 1 dB more E_b/N_0 than that for QPSK or BPSK, for error rates of 10^{-4} or less. Because DPSK is much easier to receive than BPSK, it is often used in practice in preference to BPSK. Similarly, the BER performance of a noncoherent FSK receiver is very close to that of a coherent FSK receiver. Because noncoherent FSK receivers are simpler than coherent FSK receivers, they are often used in practice. In some applications, there are multiple paths of different lengths between the transmitter and receiver (e.g., caused by reflection). This causes fading and noncoherence in the received signal (phase). In this case, it is very difficult to implement coherent detection, regardless of the complexity of the receiver, and one resorts to the use of noncoherent detection techniques.

Channel coding can be used to reduce the P_e below the values given in Fig. 7–14. This concept was developed in Chapter 1. By using Shannon's channel capacity formula, it was found that, theoretically, the BER would approach zero as long as E_b/N_0 was above -1.59 dB when optimum (unknown) coding was used on an infinite bandwidth channel. With practical coding, it was found that coding gains as large as 9 dB could be achieved on BPSK and QPSK systems. That is, for a given BER the E_b/N_0 requirements in Fig. 7–14 could be reduced by as much as 9 dB when coding was used.

Symbol Error and Bit Error for Multilevel Signaling

For *multilevel* systems (as compared to binary signaling systems), simple closed-form formulas for the probability of *bit error* (also known as the bit error rate, BER) are impossible to obtain. This is why BER results for multilevel systems are not included in Table 7–1 and Fig. 7–14 except for the special case of QPSK (which, as discussed in Sec. 7–5, consists of two orthogonal binary PSK signals). Of course, the BER for multilevel signals can be obtained by simulation or by measurement.

In some cases, simple formulas for the upper bounds on the probability of *symbol error* of multilevel systems can be obtained. For MPSK, a bound on the symbol error is [Wilson, 1996]

$$P(E) \leq Q\left[\sqrt{2\left(\frac{E_b}{N_0}\right)(\log_2 M)\sin^2\left(\frac{\pi}{M}\right)}\right], \text{ for MPSK}$$

for the case of AWGN. The bound becomes tight for reasonable sizes of M and E_b/N_0 such as $M = 8$ and $(E_b/N_0)_{db} = 10$ dB. For QAM signaling (AWGN case), a symbol error bound is [Wilson, 1996]

$$P(E) \leq 4Q \left[\sqrt{2\left(\frac{E_b}{N_0}\right) \eta_M} \right], \text{ for MQAM}$$

where the efficiency factor η_M is -4 dB for 16 QAM, -6 dB for 32 QAM, -8.5 dB for 64 QAM, -10.2 dB for 128 QAM and -13.3 dB for 256 QAM.

The probability of symbol error (also called the word error rate, WER) is not easily related to the BER. However, bounds on the relationship between the BER and the WER are [Couch, 1993]

$$\frac{1}{K} P(E) \leq P_e \leq \frac{(M/2)}{M - 1} P(E)$$

where P_e is the BER, $P(E)$ is the WER and $M = 2^K$. When errors are made under the usual operating conditions of low error rate (say $P_e < 10^{-3}$), the error symbol selected is usually the "nearest neighbor" to the correct symbol on the signal constellation. This results in a BER near the lower bound. For this case, the BER is almost equal to the lowest bound if the bit-to-symbol mapping is a Gray code (see Table 3–1) since there is only a one bit change (error) for the nearest neighbor symbol. For example, if $M = 128$ ($K = 7$), then

$$0.143 \, P(E) \leq P_e \leq 0.504 \, P(E)$$

Under the usual operating conditions of low BER, say $P_e < 10^{-3}$, the BER would be near the lower bound, so $P_e \approx 0.143 \, P(E)$ for $M = 128$.

Synchronization

As we have seen, three levels of synchronization are needed in digital communication systems:

1. Bit synchronization.
2. Frame, or word, synchronization.
3. Carrier synchronization.

Bit and frame synchronizations were discussed in Chapter 3.

Carrier synchronization is required in receivers that use coherent detection. If the spectrum of the digital signal has a discrete line at the carrier frequency, such as in equally likely OOK signaling, a PLL can be used to recover the carrier reference from the received signal. This was described in Chapter 4 and shown in Fig. 4–24. In BPSK signaling, there is no discrete carrier term, but the spectrum is symmetrical about the carrier frequency. Consequently, a Costas loop or a squaring loop may be used for carrier sync recovery. These loops were illustrated in Figs. 5–3 and P5–51. As indicated in Sec. 5–4, these loops may lock up with a 180° phase error, which must be resolved to ensure that the recovered data will not be complemented. For QPSK signals, the carrier reference may be obtained by a more generalized Costas loop or, equivalently, by a fourth-power loop [Spilker, 1977]. These loops have a four-phase ambiguity of 0, ±90, or 180°, which must also be resolved to obtain correctly demodulated data. These facts illustrate once again why noncoherent reception techniques (which can be used for OOK, FSK, and DPSK) are so popular.

Bit synchronizers are needed at the receiver to provide the bit sync signal for clocking the sample-and-hold circuit and, if used, the matched filter circuit. The bit synchronizer was illustrated in Fig. 3–20.

All the BER formulas that we have obtained assume that noise-free bit sync and carrier sync (for coherent detection) are available at the receiver. Of course, if these sync signals are obtained from noisy signals that are present at the receiver input, the reference signals are also noisy. Consequently, the P_e will be larger than that given for the ideal case when noise-free sync is assumed.

The third type of synchronization that is required by most digital systems is frame sync, or word sync. In some systems, this type of sync is used simply to demark the serial data into digital words or *bytes*. In other systems, block coding or convolutional coding is used at the transmitter so that some of the bit errors at the output of the threshold device of the receiver can be detected and corrected by using decoding circuits. In these systems, word sync is needed to clock the receiver decoding circuits. In addition, frame sync is required in time-division multiplex (TDM) systems. This was described in Fig. 3–37. Higher levels of synchronization, such as network synchronization, may be required when data are received from several sources. For example, multiple-access satellite communication systems require network synchronization, as illustrated in Chapter 8.

7–7 OUTPUT SIGNAL-TO-NOISE RATIO FOR PCM SYSTEMS

In the previous sections, we studied how the P_e for various digital systems depends on the energy per bit E_b of the signal at the receiver input and on the level of the input noise spectrum $N_0/2$. Now we look at applications of these signaling techniques where an analog signal is encoded into a PCM signal is composed of the data that are transmitted over the digital system having a BER of P_e. This is illustrated in Fig. 7–15. The digital transmitter and receiver may be any one of those associated with the digital signaling systems studied in the previous sections. For example, $s(t)$ might be an FSK signal, and the receiver would be an FSK receiver. The recovered PCM signal has some bit errors (caused by channel noise). Consequently, the decoded analog waveform at the output of the PCM decoder will have noise because of these bit errors, as well as quantizing noise. The question is: What is the peak signal to average noise ratio $(S/N)_{\text{out}}$ for the analog output? If the input analog signal has a uniform PDF over $-V$ to $+V$ and there are M steps in the uniform quantizer, the answer is

$$\left(\frac{S}{N}\right)_{\text{pk out}} = \frac{3M^2}{1 + 4(M^2 - 1)P_e} \tag{7–70}$$

Applications of this result were first studied in Sec. 3–3.

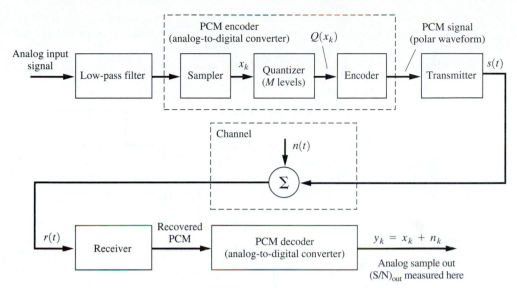

Figure 7–15   PCM communications system.

Equation (7–70) will now be derived. As shown in Fig. 7–15, the analog sample x_k is obtained at the sampling time $t = kT_s$. This sample is quantized to a value $Q(t_k)$, which is one of the M possible levels, as indicated in Fig. 7–16. The quantized sample $Q(x_k)$ is encoded into an n-bit PCM word $(a_{k1}, a_{k2}, \ldots, a_{kn})$, where $M = 2^n$. If polar signaling is used, the a_k's take on values of $+1$ or -1. For simplicity, we assume that the PCM code words are related to the quantized values by[†]

$$Q(x_k) = V \sum_{j=1}^{n} a_{kj} \left(\tfrac{1}{2}\right)^j \tag{7–71}$$

For example, if the PCM word for the kth sample happens to be $(+1, +1, \ldots, +1)$, then the value of the quantized sample will be

$$Q(x_k) = V\left(\frac{1}{2} + \frac{1}{2} + \cdots + \frac{1}{2^n}\right) = \frac{V}{2}\left(1 + \frac{1}{2} + \frac{1}{4} + \cdots + \frac{1}{2^{n-1}}\right)$$

From Appendix A, we find that the sum of this finite series is

$$Q(x_k) = \frac{V}{2}\left[\frac{\left(\tfrac{1}{2}\right)^n - 1}{\tfrac{1}{2} - 1}\right] = V - \frac{V}{2^n}$$

or, because $V = (\delta/2)2^n$,

$$Q(x_k) = V - \frac{\delta}{2}$$

[†] For mathematical simplicity, natural binary coding levels are used in Eq. (7–71), not the Gray coding used in Table 3–1.

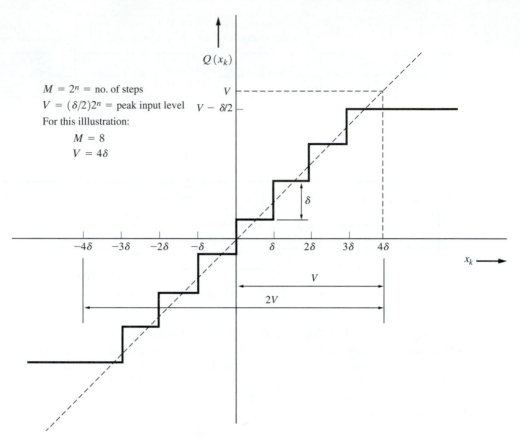

Figure 7–16   Uniform quantizer characteristic for $M = 8$ (with $n = 3$ bits in each PCM word).

where δ is the step size of the quantizer (Fig. 7–16). Thus, the PCM word $(+1, +1, \ldots, +1)$ represents the maximum value of the quantizer, as illustrated in Fig. 7–16. Similarly, the level of the quantizer corresponding to the other code words can be obtained.

Referring to Fig. 7–15 again, we note that the analog sample output of the PCM system for the kth sampling time is

$$y_k = x_k + n_k$$

where x_k is the signal (same as the input sample) and n_k is noise. The output peak signal power to average noise power is then

$$\left(\frac{S}{N}\right)_{\text{pk out}} = \frac{[(x_k)_{\text{max}}]^2}{\overline{n_k^2}} = \frac{V^2}{\overline{n_k^2}} \tag{7-72}$$

where $(x_k)_{\text{max}} = V$, as is readily seen from Fig. 7–16. As indicated in Chapter 3, it is assumed that the noise n_k consists of two uncorrelated effects:

- Quantizing noise that is due to the quantizing error:

$$e_q = Q(x_k) - x_k \qquad (7\text{--}73)$$

- Noise due to bit errors that are caused by the channel noise:

$$e_b = y_k - Q(x_k) \qquad (7\text{--}74)$$

Thus,

$$\overline{n_k^2} = \overline{e_q^2} + \overline{e_b^2} \qquad (7\text{--}75)$$

First, evaluate the quantizing noise power. For a uniformly distributed signal, the quantizing noise is uniformly distributed. Furthermore, as indicated by Fig. 3-8c, the interval of the uniform distribution is $(-\delta/2, \delta/2)$, where δ is the step size ($\delta = 2$ in Fig. 3-8c). Thus, with $M = 2^n = 2V/\delta$ (from Fig. 7-16),

$$\overline{e_q^2} = \int_{-\infty}^{\infty} e_q^2 f(e_q)\, de_q = \int_{-\delta/2}^{\delta/2} e_q^2 \frac{1}{\delta}\, de_q = \frac{\delta^2}{12} = \frac{V^2}{3M^2} \qquad (7\text{--}76)$$

The noise power due to bit errors is evaluated by the use of Eq. (7-74), yielding

$$\overline{e_b^2} = \overline{[y_k - Q(x_k)]^2} \qquad (7\text{--}77)$$

where $Q(x_k)$ is given by Eq. (7-71). The recovered analog sample y_k is reconstructed from the received PCM code word using the same algorithm as that of Eq. (7-71). Assuming that the received PCM word for the kth sample is $(b_{k1}, b_{k2}, \ldots, b_{kn})$, we see that

$$y_k = V \sum_{j=1}^{n} b_{kj} (\tfrac{1}{2})^j \qquad (7\text{--}78)$$

The b's will be different from the a's whenever there is a bit error in the recovered digital (PCM) waveform. By using Eqs. (7-78) and (7-71), Eq. (7-77) becomes

$$\overline{e_b^2} = V^2 \overline{\left[\sum_{j=1}^{n} (b_{kj} - a_{kj})(\tfrac{1}{2})^j \right]^2}$$

$$= V^2 \sum_{j=1}^{n} \sum_{\ell=1}^{n} \left(\overline{b_{kj}b_{k\ell}} - \overline{a_{kj}b_{k\ell}} - \overline{b_{kj}a_{k\ell}} + \overline{a_{kj}a_{k\ell}} \right) 2^{-j-\ell} \qquad (7\text{--}79)$$

where b_{kj} and $b_{k\ell}$ are two bits in the received PCM word that occur at different bit positions when $j \neq \ell$. Similarly, a_{kj} and $b_{k\ell}$ are the transmitted (Tx) and received (Rx) bits in two different bit positions, where $j \neq \ell$ (for the PCM word corresponding to the kth sample). The encoding process produces bits that are independent if $j \neq \ell$. Furthermore, the bits have a zero-mean value. Thus, $\overline{b_{kj}b_{k\ell}} = \overline{b_{kj}b_{k\ell}} = 0$ for $j \neq \ell$. Similar results are obtained for the other averages on the right-hand side of Eq. (7-79). Equation (7-79) becomes

$$\overline{e_b^2} = V^2 \sum_{j=1}^{n} \left(\overline{b_{kj}^2} - 2\overline{a_{kj}b_{kj}} + \overline{a_{kj}^2} \right) 2^{-2j} \qquad (7\text{--}80)$$

Evaluating the averages in this equation, we obtain[†]

$$\overline{b_{kj}^2} = (+1)^2 P(+1\text{Rx}) + (-1)^2 P(-1\text{Rx}) = 1$$

$$\overline{a_{kj}^2} = (+1)^2 P(+1\text{Tx}) + (-1)^2 P(-1\text{Tx}) = 1$$

$$\overline{a_{kj} b_{kj}} = (+1)(+1) P(+1\text{Tx}, +1\text{Rx}) + (-1)(-1) P(-1\text{Tx}, -1\text{Rx})$$

$$+ (-1)(+1) P(-1\text{Tx}, +1\text{Rx}) + (+1)(-1) P(+1\text{Tx}, -1\text{Rx})$$

$$= [P(+1\text{Tx}, +1\text{Rx}) + P(-1\text{Tx}, -1\text{Rx})]$$

$$- [P(-1\text{Tx}, +1\text{Rx}) + P(+1\text{Tx}, -1\text{Rx})]$$

$$= [1 - P_e] - [P_e] = 1 - 2P_e$$

Thus, Eq. (7–80) reduces to

$$\overline{e_b^2} = 4V^2 P_e \sum_{j=1}^{n} \left(\tfrac{1}{4}\right)^j$$

which, from Appendix A, becomes

$$\overline{e_b^2} = V^2 P_e \frac{\left(\tfrac{1}{4}\right)^n - 1}{\tfrac{1}{4} - 1} = \frac{4}{3} V^2 P_e \frac{(2^n)^2 - 1}{(2^n)^2}$$

or

$$\overline{e_b^2} = \tfrac{4}{3} V^2 P_e \frac{M^2 - 1}{M^2} \tag{7–81}$$

Substituting Eqs. (7–76) and (7–81) into Eq. (7–72) with the help of Eq. (7–75), we have

$$\left(\frac{S}{N}\right)_{\text{pk out}} = \frac{V^2}{(V^2/3M^2) + (4V^2/3M^2) P_e (M^2 - 1)}$$

which reduces to Eq. (7–70).

　　　Equation (7–70) is used to obtain the curves shown in Fig. 7–17. For a PCM system with M quantizing steps, $(S/N)_{\text{out}}$ is given as a function of the BER of the digital receiver. For $P_e < 1/(4M^2)$, the analog signal at the output is corrupted primarily by quantizing noise. In fact, for $P_e = 0$, $(S/N)_{\text{out}} = 3M^2$, and all the noise is quantizing noise. Conversely, for $P_e > 1/(4M^2)$, the output is corrupted primarily by channel noise.

　　　It is also stressed that Eq. (7–70) is the *peak* signal to average noise ratio. The *average*-signal-to-average-noise ratio is obtained easily from the results just presented. The average-signal-to-average-noise ratio is

$$\left(\frac{S}{N}\right)_{\text{out}} = \frac{\overline{(x_k)^2}}{n_k^2} = \frac{V^2}{3n_k^2} = \frac{1}{3}\left(\frac{S}{N}\right)_{\text{pk out}}$$

[†] The notation $+1$Tx denotes a binary 1 transmitted, -1Tx denotes a binary 0 transmitted, $+1$Rx denotes a binary 1 received, and -1Rx denotes a binary 0 received.

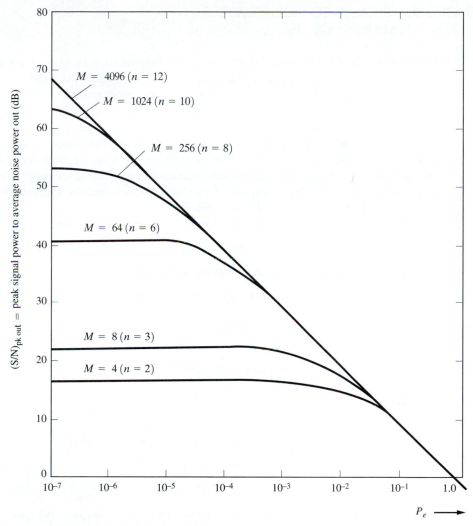

Figure 7–17 $(S/N)_{out}$ of a PCM system as a function of P_e and the number of quantizer steps M.

where $\overline{x_k^2} = V^2/3$ because x_k is uniformly distributed from $-V$ to $+V$. Thus, using Eqs. (7–72) and (7–70), we calculate

$$\left(\frac{S}{N}\right)_{out} = \frac{M^2}{1 + 4(M^2 - 1)P_e} \tag{7–82}$$

when $(S/N)_{out}$ is the average-signal-to-average-noise power ratio at the output of the system.

7–8 OUTPUT SIGNAL-TO-NOISE RATIOS FOR ANALOG SYSTEMS

In Chapters 4 and 5, generalized modulation and demodulation techniques for analog and digital signals were studied. Specific techniques were shown for AM, DSB-SC, SSB, PM, and FM signaling, and the bandwidths of these signals were evaluated. Here we evaluate the output signal-to-noise ratios for these systems as a function of the input signal, noise, and system parameters. Once again, we will find that the mathematical analysis of noncoherent systems is more difficult than that for coherent systems and that approximations are often used to obtain simplified results. However, noncoherent systems are often found to be more prevalent in practice, since the receiver cost is usually lower. This is the case for overall system cost in applications involving one transmitter and thousands or millions of receivers, such as in FM, AM, and analog TV broadcasting.

For systems with additive noise channels the input to the receiver is

$$r(t) = s(t) + n(t)$$

For bandpass communication systems having a transmission bandwidth of B_T,

$$r(t) = \text{Re}\{g_s(t)e^{j(\omega_c t + \theta_c)}\} + \text{Re}\{g_n(t)e^{j(\omega_c t + \theta_c)}\}$$
$$= \text{Re}\{[g_s(t) + g_n(t)]e^{j(\omega_c t + \theta_c)}\}$$

or

$$r(t) = \text{Re}\{g_T(t)e^{j(\omega_c t + \theta_c)}\} \tag{7–83a}$$

where

$$g_T(t) \triangleq g_s(t) + g_n(t)$$
$$= [x_s(t) + x_n(t)] + j[y_s(t) + y_n(t)]$$
$$= x_T(t) + jy_T(t)$$
$$= R_T(t)e^{j\theta_T(t)} \tag{7–83b}$$

$g_T(t)$ denotes the total (i.e., composite) complex envelope at the receiver input; it consists of the complex envelope of the signal plus the complex envelope of the noise. The properties of the total complex envelope, as well as those of Gaussian noise, were given in Chapter 6. The complex envelopes for a number of different types of modulated signals $g_s(t)$ were given in Table 4–1.

Comparison with Baseband Systems

The noise performance of the various types of bandpass systems is examined by evaluating the signal-to-noise power ratio at the receiver output, $(S/N)_{\text{out}}$, when a modulated signal plus noise is present at the receiver input. We would like to see if $(S/N)_{\text{out}}$ is larger for an AM system, a DSB-SC system, or an FM system. To compare these SNRs, the power of the

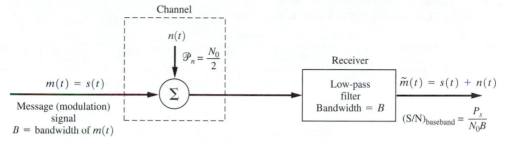

Figure 7–18   Baseband system.

modulated signals at the inputs of these receivers is set to the same value and the PSD of the input noise is $N_0/2$. (That is, the input noise is white with a spectral level set to $N_0/2$.)

To compare the output signal-to-noise ratio $(S/N)_{\text{out}}$ for various bandpass systems, we need a common measurement criterion for the receiver input. For analog systems, the criterion is the received signal power P_s divided by the amount of power in the white noise that is contained in a bandwidth equal to the message (modulation) bandwidth. This is equivalent to the $(S/N)_{\text{out}}$ of a baseband transmission system, as illustrated in Fig. 7–18. That is,

$$\left(\frac{S}{N}\right)_{\text{baseband}} = \frac{P_s}{N_0 B} \qquad (7\text{--}84)$$

We can compare the performance of different modulated systems by evaluating $(S/N)_{\text{out}}$ for each system as a function of $(P_s/N_0 B) = (S/N)_{\text{baseband}}$, where P_s is the power of the AM, DSB-SC, or FM signal at the receiver input. B is chosen to be the bandwidth of the *baseband* (modulating) signal where the same baseband modulating signal is used for all cases so that the same basis of comparison will be realized. [If B were chosen to be the bandwidth of the input-modulated signal, B_T, the comparison would not be on an equal noise PSD basis of $N_0/2$ for a fixed value of $(S/N)_{\text{baseband}}$ because the B_T values for AM and FM signals are different.]

The SNR at the receiver input can also be obtained; it is

$$\left(\frac{S}{N}\right)_{\text{in}} = \frac{P_s}{N_0 B_T} = \left(\frac{S}{N}\right)_{\text{baseband}} \left(\frac{B}{B_T}\right) \qquad (7\text{--}85)$$

where B_T is the bandwidth of the bandpass signal at the receiver input.

$(S/N)_{\text{out}}$ will now be evaluated for several different systems.

AM Systems with Product Detection

Figure 7–19 illustrates the receiver for an AM system with coherent detection. From Eq. (5–3), the complex envelope of the AM signal is

$$g_s(t) = A_c[1 + m(t)]$$

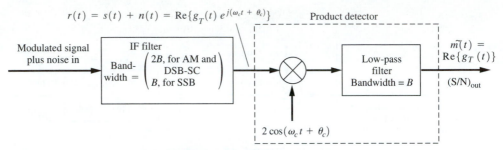

Figure 7–19   Coherent receiver.

The complex envelope of the composite received signal plus noise is

$$g_T(t) = [A_c + A_c m(t) + x_n(t)] + jy_n(t) \qquad (7\text{–}86)$$

For product detection, using Eq. (4–71), we find that the output is

$$\tilde{m}(t) = \text{Re}\{g_T(t)\} = A_c + A_c m(t) + x_n(t)$$

Here A_c is the dc voltage at the detector output that occurs because of the discrete AM carrier,[†] $A_c m(t)$ is the detected modulation, and $x_n(t)$ is the detected noise. The output SNR is

$$\left(\frac{S}{N}\right)_{\text{out}} = \frac{A_c^2 \overline{m^2(t)}}{\overline{x_n^2(t)}} = \frac{A_c^2 \overline{m^2(t)}}{2N_0 B} \qquad (7\text{–}87)$$

where, from Eq. (6–133g), $\overline{x_n^2} = \overline{n^2} = 2(N_0/2)(2B)$. The input signal power is

$$P_s = \frac{A_c^2}{2}\,\overline{[1 + m(t)]^2} = \frac{A_c^2}{2}\,[1 + \overline{m^2}]$$

where it is assumed that $\overline{m(t)} = 0$ (i.e., no dc on the modulating waveform). Thus, the input SNR is

$$\left(\frac{S}{N}\right)_{\text{in}} = \frac{(A_c^2/2)(1 + \overline{m^2})}{2N_0 B} \qquad (7\text{–}88)$$

Combining Eqs. (7–87) and (7–88) yields

$$\frac{(S/N)_{\text{out}}}{(S/N)_{\text{in}}} = \frac{2\overline{m^2}}{1 + \overline{m^2}} \qquad (7\text{–}89)$$

For 100% sine-wave modulation, $\overline{m^2} = \tfrac{1}{2}$ and $(S/N)_{\text{out}}/(S/N)_{\text{in}} = \tfrac{2}{3}$.

    For comparison purposes, Eq. (7–87) can be evaluated in terms of $(S/N)_{\text{baseband}}$ by substituting for P_s:

$$\frac{(S/N)_{\text{out}}}{(S/N)_{\text{baseband}}} = \frac{\overline{m^2}}{1 + \overline{m^2}} \qquad (7\text{–}90)$$

[†] This dc voltage is often used to provide automatic gain control (AGC) of the preceding RF and IF stages.

For 100% sine-wave modulation, $(S/N)_{\text{out}}/(S/N)_{\text{baseband}} = \frac{1}{3}$. This illustrates that this AM system is 4.8 dB worse than a baseband system that uses the same amount of signal power because of the additional power in the discrete AM carrier.[†] This is shown in Fig. 7–27.

AM Systems with Envelope Detection

The envelope detector produces $KR_T(t)$ at its output, where K is a proportionality constant. Then, for additive signal plus noise at the input, the output is

$$KR_T(t) = K|g_s(t) + g_n(t)|$$

Substituting Eq. (7–86) for $g_s(t) + g_n(t)$ gives

$$KR_T(t) = K|[A_c + A_c m(t) + x_n(t)] + j[y_n(t)]| \qquad (7\text{–}91)$$

The power is

$$\overline{[KR_T(t)]^2} = K^2 A_c^2 \left\{ \overline{\left[1 + m(t) + \frac{x_n(t)}{A_c}\right]^2 + \left[\frac{y_n(t)}{A_c}\right]^2} \right\} \qquad (7\text{–}92)$$

For the case of large $(S/N)_{\text{in}}$, $(y_n/A_c)^2 \ll 1$, so we have

$$\overline{[KR_T(t)]^2} = (KA_c)^2 + K^2 A_c^2 \, \overline{m^2} + K^2 \overline{x_n^2} \qquad (7\text{–}93)$$

where $(KA_c)^2$ is the power of the AGC term, $K^2 A_c^2 \, \overline{m^2}$ is the power of the detected modulation (signal), and $K^2 \overline{x_n^2}$ is the power of the detected noise. Thus, for large $(S/N)_{\text{in}}$,

$$\left(\frac{S}{N}\right)_{\text{out}} = \frac{A_c^2 \, \overline{m^2}}{\overline{x_n^2}} = \frac{A_c^2 \, \overline{m^2}}{2N_0 B} \qquad (7\text{–}94)$$

Comparing this with Eq. (7–87), we see that *for large* $(S/N)_{\text{in}}$*, the performance of the envelope detector is identical to that of the product detector.*

For small (S/N)$_{\text{in}}$*, the performance of the envelope detector is much inferior to that of the product detector.* Referring to Eq. (7–91), we note that the detector output is

$$KR_T(t) = K|g_T(t)| = K|A_c[1 + m(t)] + R_n(t)e^{j\theta_n(t)}|$$

For $(S/N)_{\text{in}} < 1$, as seen from Fig. 7–20, the magnitude of $g_T(t)$ can be approximated by

$$KR_T(t) \approx K\{A_c[1 + m(t)] \cos \theta_n(t) + R_n(t)\} \qquad (7\text{–}95)$$

Thus, for the case of a Gaussian noise channel, the output of the envelope detector consists of Rayleigh distributed noise $R_n(t)$, plus a signal term that is *multiplied* by a random noise factor $\cos \theta_n(t)$. This multiplicative effect corrupts the signal to a much larger extent than the additive Rayleigh noise. This produces a *threshold* effect. That is, $(S/N)_{\text{out}}$ becomes *very*

[†] The AM carrier does not contribute to the information ability of the signal but permits AM receivers to use economical envelope detectors.

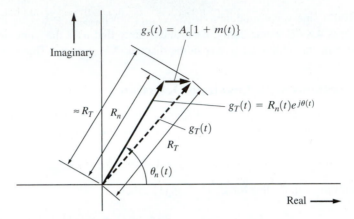

Figure 7–20   Vector diagram for AM, $(S/N)_{in} \ll 1$.

small when $(S/N)_{in} < 1$. In fact, it can be shown that $(S/N)_{out}$ is proportional to the *square* of $(S/N)_{in}$ for the case of $(S/N)_{in} < 1$ [Schwartz, Bennett, and Stein, 1966]. Although the envelope detector is greatly inferior to the product detector for small $(S/N)_{in}$, this deficiency is seldom noticed in practice for AM broadcasting applications. This is because the AM listener is usually interested in listening to stations only if they have reasonably good $(S/N)_{out}$, say 25 dB or more. Under these conditions the envelope detector performance is equivalent to that of the product detector. Moreover, the envelope detector is inexpensive and does not require a coherent reference. For these reasons the envelope detector is used almost exclusively in AM broadcast receivers. For other applications, such as listening to weak AM stations or for AM data transmission systems, product detection may be required to eradicate the multiplicative noise that would occur with envelope detection of weak systems.

DSB-SC Systems

As indicated in Chapter 5, the DSB-SC signal is essentially an AM signal in which the discrete carrier term has been suppressed (i.e., equivalent to infinite percent AM). The modulating waveform $m(t)$ is recovered from the DSB-SC signal by using coherent detection, as shown in Fig. 7–19. For DSB-SC,

$$g_s(t) = A_c m(t) \tag{7-96}$$

Following the development leading to Eq. (7–89), the SNR for DSB-SC is

$$\frac{(S/N)_{out}}{(S/N)_{in}} = 2 \tag{7-97}$$

Using Eq. (7–90), we obtain

$$\frac{(S/N)_{out}}{(S/N)_{baseband}} = 1 \tag{7-98}$$

Thus, the noise performance of a DSB-SC system is the same as that of baseband signaling systems, although the bandwidth requirement is twice as large (i.e., $B_T = 2B$).

SSB Systems

The receiver for an SSB signal is also shown in Fig. 7–19, where the IF bandwidth is now $B_T = B$. The complex envelope for SSB is

$$g_s(t) = A_c[m(t) \pm j\hat{m}(t)] \tag{7–99}$$

where the upper sign is used for USSB and the lower sign is used for LSSB. The complex envelope for the (total) received SSB signal plus noise is

$$g_T(t) = [A_c m(t) + x_n(t)] + j[\pm A_c \hat{m}(t) + y_n(t)] \tag{7–100}$$

The output of the product detector is

$$\tilde{m}(t) = \text{Re}\{g_T(t)\} = A_c m(t) + x_n(t) \tag{7–101}$$

The corresponding SNR is

$$(S/N)_{\text{out}} = \frac{A_c^2 \, \overline{m^2(t)}}{\overline{x_n^2(t)}} = \frac{A_c^2 \, \overline{m^2(t)}}{N_0 B} \tag{7–102}$$

where $\overline{x_n^2} = \overline{n^2} = 2(N_0/2)(B)$. Using Eq. (6–133g), we see that the input signal power is

$$P_s = \tfrac{1}{2}\overline{|g_s(t)|^2} = \frac{A_c^2}{2}[\overline{m^2} + \overline{(\hat{m})^2}] = A_c^2 \overline{m^2} \tag{7–103}$$

and that the input noise power is $P_n = \overline{n^2(t)} = N_0 B$. Thus,

$$\frac{(S/N)_{\text{out}}}{(S/N)_{\text{in}}} = 1 \tag{7–104}$$

Similarly,

$$\frac{(S/N)_{\text{out}}}{(S/N)_{\text{baseband}}} = 1 \tag{7–105}$$

SSB is exactly equivalent to baseband signaling, in terms of both the noise performance and the bandwidth requirements (i.e., $B_T = B$). Furthermore, Eqs. (7–98) and (7–105) show that DSB, SSB, and baseband signaling systems are all equivalent in output SNR.

PM Systems

As shown in Fig. 7–21, the modulation on a PM signal is recovered by a receiver that uses a (coherent) phase detector. (In Chapter 4, it was found that a phase detector could be realized by using a limiter that followed a product detector when β_p is small.) The PM signal has a complex envelope of

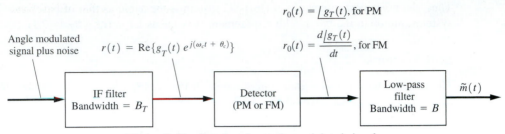

$$r_0(t) = \underline{\big/}\, g_T(t), \text{ for PM}$$

Angle modulated
signal plus noise $r(t) = \operatorname{Re}\{g_T(t)\, e^{j(\omega_c t + \theta_c)}\}$ $r_0(t) = \dfrac{d\underline{\big/}g_T(t)}{dt}, \text{ for FM}$

| IF filter Bandwidth $= B_T$ | Detector (PM or FM) | Low-pass filter Bandwidth $= B$ |

$\tilde{m}(t)$

Figure 7–21 Receiver for angle-modulated signals.

$$g_s(t) = A_c e^{j\theta_s(t)} \tag{7–106a}$$

where

$$\theta_s(t) = D_p m(t) \tag{7–106b}$$

The complex envelope of the composite signal plus noise at the detector input is

$$g_T(t) = |g_T(t)| e^{j\theta_T(t)} = [g_s(t) + g_n(t)] \tag{7–107}$$
$$= A_c e^{j\theta_s(t)} + R_n(t) e^{j\theta_n(t)}$$

When the input is Gaussian noise (only), $R_n(t)$ is Rayleigh distributed and $\theta_n(t)$ is uniformly distributed, as demonstrated in Example 6–10.

The phase detector output is proportional to $\theta_T(t)$:

$$r_0(t) = K \underline{\big/} g_T(t) = K\theta_T(t)$$

Here K is the gain constant of the detector. For large $(S/N)_{\text{in}}$, the phase angle of $g_T(t)$ can be approximated with the help of a vector diagram for $g_T = g_s + g_n$. This is shown in Fig. 7–22. Then, for $A_c \gg R_n(t)$, the composite phase angle is approximated by

$$r_0(t) = K\theta_T(t) \approx K\left\{\theta_s(t) + \frac{R_n(t)}{A_c}\sin[\theta_n(t) - \theta_s(t)]\right\} \tag{7–108}$$

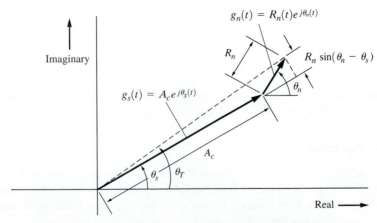

Figure 7–22 Vector diagram for angle modulation, $(S/N)_{\text{in}} \gg 1$.

For the case of no phase modulation (the carrier signal is still present), the equation reduces to

$$r_0(t) \approx \frac{K}{A_c} y_n(t), \qquad \theta_s(t) = 0 \tag{7-109}$$

where, using Eq. (6–127e), we find that $y_n(t) = R_n(t) \sin \theta_n(t)$. This shows that the presence of an unmodulated carrier (at the input to the PM receiver) suppresses the noise on the output. This is called the *quieting effect*, and it occurs when the input signal power is above the *threshold* [i.e., when $(S/N)_{in} \gg 1$]. Furthermore, when phase modulation is present and $(S/N)_{in} \gg 1$, we can replace $R_n(t) \sin[\theta_n(t) - \theta_s(t)]$ by $R_n(t) \sin \theta_n(t)$. This can be done because $\theta_s(t)$ can be considered to be deterministic, and, consequently, $\theta_s(t)$ is a constant for a given value of t. Then, $\theta_n(t) - \theta_s(t)$ will be uniformly distributed over some 2π interval, since, from Eq. (6–153), $\theta_n(t)$ is uniformly distributed over $(0, 2\pi)$. That is, $\cos[\theta_n(t) - \theta_s(t)]$ will have the same PDF as $\cos[\theta_n(t)]$, so that the replacement can be made. Thus, for large $(S/N)_{in}$, the relevant part of the PM detector output is approximated by

$$r_0(t) \approx s_0(t) + n_0(t) \tag{7-110}$$

where

$$s_0(t) = K\theta_s(t) = KD_p m(t) \tag{7-111a}$$

$$n_0(t) = \frac{K}{A_c} y_n(t) \tag{7-111b}$$

The PSD of this output noise, $n_0(t)$, is obtained by the use of Eq. (6–133l), and we get

$$\mathcal{P}_{n_0}(f) = \begin{cases} \dfrac{K^2}{A_c^2} N_0, & |f| \leq B_T/2 \\ 0, & f \text{ otherwise} \end{cases} \tag{7-112}$$

where the PSD of the bandpass input noise is $N_0/2$ over the bandpass of the IF filter and zero outside the IF bandpass. Equation (7–112) is plotted in Fig. 7–23a, and the portion of the spectrum that is passed by the LPF is denoted by the hatched lines.

The receiver output consists of the low-pass filtered version of $r_0(t)$. However, the spectrum of $s_0(t)$ is contained within the bandpass of the filter, so

$$\tilde{m}(t) = s_0(t) + \tilde{n}_0(t) \tag{7-113}$$

where $\tilde{n}_0(t)$ is bandlimited white noise of the hatched portion of Fig. 7–23a. The noise power of the receiver is

$$\overline{[\tilde{n}_0(t)]^2} = \int_{-B}^{B} \mathcal{P}_{n_0}(f) \, df = \frac{2K^2 N_0 B}{A_c^2} \tag{7-114}$$

The output SNR is now easily evaluated with the help of Eqs. (7–111a) and (7–114):

$$\left(\frac{S}{N}\right)_{out} = \frac{\overline{s_0^2}}{\overline{\tilde{n}_0^2}} = \frac{A_c^2 D_p^2 \, \overline{m^2}}{2N_0 B}$$

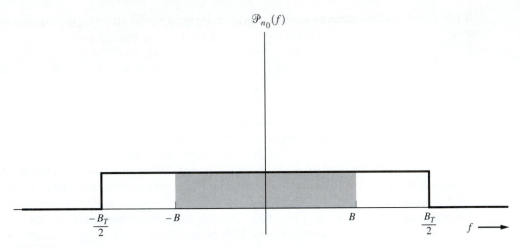

(a) PM Detector

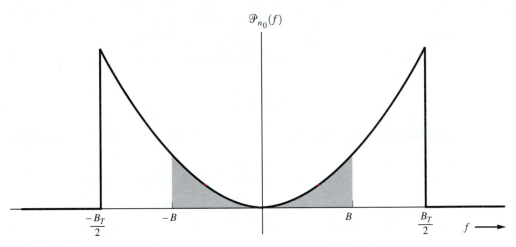

(b) FM Detector

Figure 7–23 PSD for noise out of detectors for receivers of angle-modulated signals.

Using Eqs. (5–46) and (5–47), we can write the sensitivity constant of the PM transmitter as

$$D_p = \frac{\beta_p}{V_p}$$

where β_p is the PM index and V_p is the peak value of $m(t)$. Thus, the output SNR becomes

$$\left(\frac{S}{N}\right)_{\text{out}} = \frac{A_c^2 \beta_p^2 \overline{(m/V_p)^2}}{2N_0 B} \qquad (7\text{–}115)$$

The input SNR is

$$\left(\frac{S}{N}\right)_{\text{in}} = \frac{A_c^2/2}{2(N_0/2)B_T} = \frac{A_c^2}{2N_0 B_T} \tag{7-116}$$

where B_T is the transmission bandwidth of the PM signal (and also the IF filter bandwidth). The transmission bandwidth of the PM signal is given by Carson's rule, Eq. (5-61),

$$B_T = 2(\beta_p + 1)B \tag{7-117}$$

where β_p is the PM index. The PM index is identical to the peak angle deviation of the PM signal, as indicated in Eq. (5-47). Thus,

$$\left(\frac{S}{N}\right)_{\text{in}} = \frac{A_c^2}{4N_0(\beta_p + 1)B} \tag{7-118}$$

Combining this with Eq. (7–115), we obtain the ratio of output to input SNR:

$$\frac{(S/N)_{\text{out}}}{(S/N)_{\text{in}}} = 2\beta_p^2(\beta_p + 1)\overline{\left(\frac{m}{V_p}\right)^2} \tag{7-119}$$

The output SNR can also be expressed in terms of the equivalent baseband system by substituting Eq. (7–84) into Eq. (7–115), where $P_s = A_c^2/2$:

$$\frac{(S/N)_{\text{out}}}{(S/N)_{\text{baseband}}} = \beta_p^2 \overline{\left(\frac{m}{V_p}\right)^2} \tag{7-120}$$

This equation shows that the improvement of a PM system over a baseband signaling system depends on the amount of phase deviation that is used. It seems to indicate that we can make the improvement as large as we wish simply by increasing β_p. This depends on the types of circuits used. If the peak phase deviation exceeds π radians, special "phase un-wrapping" techniques have to be used in some circuits to obtain the true value (as compared with the relative value) of the phase at the output. Thus, the maximum value of $\beta_p m(t)/V_p = D_p m(t)$ might be taken to be π. For sinusoidal modulation, this would provide an improvement of $D_p^2 m^2 = \pi^2/2$, or 6.9 dB, over baseband signaling.

It is emphasized that the results obtained previously for $(S/N)_{\text{out}}$ are valid only when the input signal is above the threshold [i.e., when $(S/N)_{\text{in}} > 1$].

FM Systems

The procedure that we will use to evaluate the output SNR for FM systems is essentially the same as that used for PM systems, except that the output of the FM detector is proportional to $d\theta_T(t)/dt$, whereas the output of the PM detector is proportional to $\theta_T(t)$. The detector in the angle modulated receiver of Fig. 7–21 is now an FM detector. The complex envelope of the FM signal (only) is

$$g_s(t) = A_c e^{j\theta_s(t)} \tag{7-121a}$$

where

$$\theta_s(t) = D_f \int_{-\infty}^{t} m(\lambda)\, d\lambda \tag{7-121b}$$

It is assumed that an FM signal plus white noise is present at the receiver input.

The output of the FM detector is proportional to the derivative of the composite phase at the detector input:

$$r_0(t) = \left(\frac{K}{2\pi}\right)\frac{d\,/\underline{g_T(t)}}{dt} = \left(\frac{K}{2\pi}\right)\frac{d\theta_T(t)}{dt} \tag{7-122}$$

In this equation, K is the FM detector gain. Using the same procedure as that leading to Eqs. (7–108), (7–110), and (7–111), we can approximate the detector output by

$$r_0(t) \approx s_0(t) + n_0(t) \tag{7-123}$$

where, for FM,

$$s_0(t) = \left(\frac{K}{2\pi}\right)\frac{d\theta_s(t)}{dt} = \left(\frac{KD_f}{2\pi}\right)m(t) \tag{7-124a}$$

and

$$n_0(t) = \left(\frac{K}{2\pi A_c}\right)\frac{dy_n(t)}{dt} \tag{7-124b}$$

This result is valid only when the input signal is above the threshold [i.e., when $(S/N)_{in} \gg 1$]. The derivative of the noise, Eq. (7–124b), makes the PSD of the FM output noise different from that for the PM case. For FM, we have

$$\mathcal{P}_{n_0}(f) = \left(\frac{K}{2\pi A_c}\right)^2 |j2\pi f|^2 \mathcal{P}_{y_n}(f)$$

or

$$\mathcal{P}_{n_0}(f) = \begin{cases} \left(\dfrac{K}{A_c}\right)^2 N_0 f^2, & |f| < B_T/2 \\[2mm] 0, & f \ \text{otherwise} \end{cases} \tag{7-125}$$

This shows that the PSD for the noise out of the FM detector has a parabolic shape, as illustrated in Fig. 7–23b.

The receiver output consists of the low-pass filtered version of $r_0(t)$. The noise power for the filtered noise is

$$\overline{[\tilde{n}_0(t)]^2} = \int_{-B}^{B} \mathcal{P}_{n_0}(f)\, df = \frac{2}{3}\left(\frac{K}{A_c}\right)^2 N_0 B^3 \tag{7-126}$$

The output SNR is now easily evaluated using Eqs. (7–124a) and (7–126):

$$\left(\frac{S}{N}\right)_{out} = \frac{\overline{s_0^2}}{\overline{[\tilde{n}_0]^2}} = \frac{3A_c^2[D_f/(2\pi B)]^2 \overline{m^2}}{2N_0 B}$$

From Eqs. (5–44) and (5–48), it is realized that

$$\frac{D_f}{2\pi B} = \frac{\beta_f}{V_p}$$

where V_p is the peak value of $m(t)$. Then, the output SNR becomes

$$\left(\frac{S}{N}\right)_{\text{out}} = \frac{3A_c^2\beta_f^2\overline{(m/V_p)^2}}{2N_0 B} \tag{7–127}$$

The input SNR is

$$\left(\frac{S}{N}\right)_{\text{in}} = \frac{A_c^2}{4N_0(\beta_f + 1)B} \tag{7–128}$$

Combining Eqs. (7–127) and (7–128), we obtain the ratio of output to input SNR

$$\frac{(S/N)_{\text{out}}}{(S/N)_{\text{in}}} = 6\beta_f^2(\beta_f + 1)\overline{\left(\frac{m}{V_p}\right)^2} \tag{7–129}$$

where β_f is the FM index and V_p is the peak value of the modulating signal $m(t)$.

The output SNR can be expressed in terms of the equivalent baseband SNR by substituting Eq. (7–84) into Eq. (7–127), where $P_s = A_c^2/2$,

$$\frac{(S/N)_{\text{out}}}{(S/N)_{\text{baseband}}} = 3\beta_f^2\overline{\left(\frac{m}{V_p}\right)^2} \tag{7–130}$$

For the case of sinusoidal modulation, $\overline{(m/V_p)^2} = \frac{1}{2}$, and Eq. (7–130) becomes

$$\frac{(S/N)_{\text{out}}}{(S/N)_{\text{baseband}}} = \frac{3}{2}\beta_f^2 \quad \text{(sinusoidal modulation)} \tag{7–131}$$

At first glance, these results seem to indicate that the performance of FM systems can be increased without limit simply by increasing the FM index β_f. However, as β_f is increased, the transmission bandwidth increases, and, consequently, $(S/N)_{\text{in}}$ decreases. These equations for $(S/N)_{\text{out}}$ are *valid only* when $(S/N)_{\text{in}} \gg 1$ (i.e., when the input signal power is above the threshold), so $(S/N)_{\text{out}}$ *does not* increase to an excessively large value simply by increasing the FM index β_f. Plots of Eq. (7–131) are given by the dashed lines in Fig. 7–24.

The threshold effect was first analyzed in 1948 [Rice, 1948; Stumpers, 1948]. An excellent tutorial treatment has been given by Taub and Schilling [1986]. They have shown that Eq. (7–131) can be generalized to describe $(S/N)_{\text{out}}$ near the threshold. For the case of sinusoidal modulation, the output SNR for a FM discriminator is shown to be

$$\left(\frac{S}{N}\right)_{\text{out}} = \frac{\frac{3}{2}\beta_f^2\,(S/N)_{\text{baseband}}}{1 + \left(\dfrac{12}{\pi}\,\beta_f\right)\left(\dfrac{S}{N}\right)_{\text{baseband}}\,e\left\{-\left[\dfrac{1}{2(\beta_f+1)}\left(\dfrac{S}{N}\right)_{\text{baseband}}\right]\right\}} \tag{7–132}$$

(No deemphasis is used in obtaining the result.) This output SNR characteristic showing the threshold effect of an FM discriminator is plotted by the solid lines in Fig. 7–24. This figure

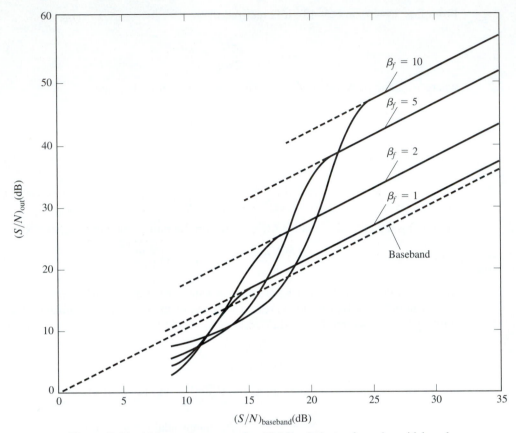

Figure 7–24   Noise performance of an FM discriminator for a sinusoidal modulated FM signal plus Gaussian noise (no deemphasis).

illustrates that the FM noise performance can be substantially better than baseband performance. For example, for $\beta_f = 5$ and $(S/N)_{\text{baseband}} = 25$ dB, the FM performance is 15.7 dB better than the baseband performance. The performance can be improved even further by the use of deemphasis, as we will demonstrate in a later section.

FM Systems with Threshold Extension

Any one of several techniques may be used to lower the threshold below that provided by a receiver that uses only an FM discriminator. For example, a PLL FM detector could be used to extend the threshold below that provided by an FM discriminator. However, when the input SNR is large, all the FM receiving techniques provide the same performance—namely, that predicted by Eq. (7–129) or Eqs. (7–130).

An *FM receiver with feedback* (FMFB) is shown in Fig. 7–25. This is another threshold extension technique. The FMFB receiver provides threshold extension by lowering the modulation index for the FM signal that is applied to the discriminator input. That is, the modulation index of $\tilde{e}(t)$ is smaller than that for $v_{\text{in}}(t)$, as we will show. Thus, the threshold

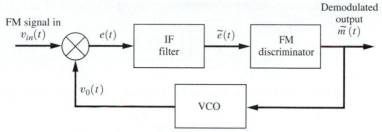

Figure 7–25   FMFB receiver.

will be lower than that illustrated in Fig. 7–24. The calculation of the exact amount of threshold extension that is realized by an FMFB receiver is somewhat involved [Taub and Schilling, 1986]. However, we can easily show that the FMFB technique does indeed reduce the modulation index of the FM signal at the discriminator input. Referring to Fig. 7–25, we find that the FM signal at the receiver input is

$$v_{\text{in}}(t) = A_c \cos[\omega_c t + \theta_i(t)]$$

where

$$\theta_i(t) = D_f \int_{-\infty}^{t} m(\lambda) \, d\lambda$$

The output of the VCO is

$$v_0(t) = A_0 \cos[\omega_0 t + \theta_0(t)]$$

where

$$\theta_0(t) = D_v \int_{-\infty}^{t} \tilde{m}(\lambda) \, d\lambda$$

With these representations for $v_{\text{in}}(t)$ and $v_0(t)$, the output of the multiplier (mixer) is

$$e(t) = A_c A_0 \cos[\omega_c t + \theta_i(t)] \cos[\omega_0 t + \theta_0(t)]$$
$$= \tfrac{1}{2} A_c A_0 \cos[(\omega_c - \omega_0)t + \theta_i(t) - \theta_0(t)]$$
$$+ \tfrac{1}{2} A_c A_0 \cos[(\omega_c + \omega_0)t + \theta_i(t) + \theta_0(t)]$$

If the IF filter is tuned to pass the band of frequencies centered about $f_{\text{if}} \triangleq f_c - f_0$, the IF output is

$$\tilde{e}(t) = \frac{A_c A_0}{2} \cos[\omega_{\text{if}} t + \theta_i(t) - \theta_0(t)] \qquad (7\text{--}133\text{a})$$

or

$$\tilde{e}(t) = \frac{A_c A_0}{2} \cos\left\{\omega_{\text{if}} t + \int_{-\infty}^{t} [D_f m(\lambda) - D_v \tilde{m}(\lambda)] \, d\lambda\right\} \qquad (7\text{--}133\text{b})$$

The FM discriminator output is proportional to the derivative of the phase deviation:

$$\tilde{m}(t) = \frac{K}{2\pi} \frac{d\left\{ \int_{-\infty}^{t} [D_f\, m(\lambda) - D_v \tilde{m}(\lambda)]\, d\lambda \right\}}{dt}$$

Evaluating the derivative and solving the resulting equation for $\tilde{m}(t)$, we obtain

$$\tilde{m}(t) = \left(\frac{KD_f}{2\pi + KD_v} \right) m(t)$$

Substituting this expression for $\tilde{m}(t)$ into Eq. (7–133), we get

$$\tilde{e}(t) = \frac{A_c A_0}{2} \cos\left[\omega_{\text{if}} t + \left(\frac{1}{1 + (K/2\pi)D_v} \right) D_f \int_{-\infty}^{t} m(\lambda)\, d\lambda \right] \qquad (7\text{–}134)$$

This demonstrates that the modulation index of $\tilde{e}(t)$ is exactly $1/[1 + (K/2\pi)D_v]$ of the modulation index of $v_{\text{in}}(t)$. The threshold extension provided by the FMFB receiver is on the order of 5 dB, whereas that of a PLL receiver is on the order of 3 dB (when both are compared with the threshold of an FM discriminator). Although this is not a fantastic improvement, it can be quite significant for systems that operate near the threshold, such as satellite communication systems. A system that uses a threshold extension receiver instead of a conventional receiver may be much less expensive than the system that requires a double-sized antenna to provide the 3-dB signal gain. Other threshold extension techniques have been described and analyzed in the literature [Klapper and Frankle, 1972].

FM Systems with Deemphasis

The noise performance of an FM system can be improved by preemphasizing the higher frequencies of the modulation signal at the transmitter input and deemphasizing the output of the receiver (as first illustrated in Fig. 5–16). This improvement occurs because the PSD of the noise at the output of the FM detector has a parabolic shape, as shown in Fig. 7–23b.

Referring to Fig. 7–21, we incorporate deemphasis into the receiver by including a deemphasis response in the LPF characteristic. Assume that the transfer function of the LPF is

$$H(f) = \frac{1}{1 + j(f/f_1)} \qquad (7\text{–}135)$$

over the message bandwidth of B Hz. For standard FM broadcasting, a 75-μs deemphasis filter is used so that $f_1 = 1/[(2\pi)(75 \times 10^{-6})] = 2.1$ kHz. Using Eq. (7–125), we see that the noise power out of the receiver with deemphasis is

$$\overline{[\tilde{n}_0(t)]^2} = \int_{-B}^{B} |H(f)|^2 \mathcal{P}_{n_0}(f)\, df = \left(\frac{K}{A_c} \right)^2 N_0 \int_{-B}^{B} \left[\frac{1}{1 + (f/f_1)^2} \right] f^2\, df$$

or

$$\overline{[\tilde{n}_0(t)]^2} = 2\left(\frac{K}{A_c} \right)^2 N_0 f_1^3 \left[\frac{B}{f_1} - \tan^{-1}\left(\frac{B}{f_1} \right) \right] \qquad (7\text{–}136)$$

In a typical application, $B/f_1 \gg 1$, and so $\tan^{-1}(B/f_1) \approx \pi/2$, which is negligible when compared with B/f_1. Thus, Eq. (7–136) becomes

$$\overline{[\tilde{n}_0(t)]^2} = 2\left(\frac{K}{A_c}\right)^2 N_0 f_1^2 B \tag{7–137}$$

for $B/f_1 \gg 1$.

The output *signal* power for the preemphasis–deemphasis system is the same as that when preemphasis–deemphasis is not used because the overall frequency response of the system to $m(t)$ is flat (constant) over the bandwidth of B Hz.[†] Thus, from Eqs. (7–137) and (7–124a), the output SNR is

$$\left(\frac{S}{N}\right)_{\text{out}} = \frac{\overline{s_0^2}}{\overline{\tilde{n}_0^2}} = \frac{A_c^2[D_f/(2\pi f_1)]^2\overline{m^2}}{2N_0 B}$$

In addition, $D_f/(2\pi B) = \beta_f/V_p$, so the output SNR reduces to

$$\left(\frac{S}{N}\right)_{\text{out}} = \frac{A_c^2\beta_f^2(B/f_1)^2\,\overline{(m/V_p)^2}}{2N_0 B} \tag{7–138}$$

Using Eq. (7–128), we find that the output to input SNR is

$$\frac{(S/N)_{\text{out}}}{(S/N)_{\text{in}}} = 2\,\beta_f^2\,(\beta_f + 1)\left(\frac{B}{f_1}\right)^2\overline{\left(\frac{m}{V_p}\right)^2} \tag{7–139}$$

where β_f is the FM index, B is the bandwidth of the baseband (modulation) circuits, f_1 is the 3-dB bandwidth of the deemphasis filter, V_p is the peak value of the modulating signal $m(t)$, and $\overline{(m/V_p)^2}$ is the square of the rms value of $m(t)/V_p$.

The output SNR is expressed in terms of the equivalent baseband SNR by substituting Eq. (7–84) into Eq. (7–138):

$$\frac{(S/N)_{\text{out}}}{(S/N)_{\text{baseband}}} = \beta_f^2\left(\frac{B}{f_1}\right)^2\overline{\left(\frac{m}{V_p}\right)^2} \tag{7–140}$$

When a sinusoidal test tone is transmitted over this FM system, $\overline{(m/V_p)^2} = \frac{1}{2}$ and Eq. (7–140) becomes

$$\frac{(S/N)_{\text{out}}}{(S/N)_{\text{baseband}}} = \frac{1}{2}\beta_f^2\left(\frac{B}{f_1}\right)^2 \quad \text{(sinusoidal modulation)} \tag{7–141}$$

Of course, each of these results is valid only when the FM signal at the receiver input is above the threshold.

[†] For a fair comparison of FM systems with and without preemphasis, the peak deviation ΔF needs to be the same for both cases. With typical audio program signals, preemphasis does not increase ΔF appreciably, because the low frequencies dominate in the spectrum of $m(t)$. Thus, this analysis is valid. However, if $m(t)$ is assumed to have a flat spectrum over the audio passband, the gain of the preemphasis filter needs to be reduced so that the peak deviation will be the same with and without preemphasis. In the latter case, there is less improvement in performance when preemphasis is used.

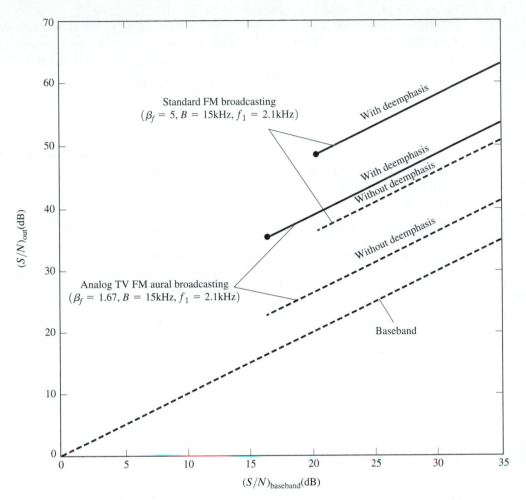

Figure 7–26   Noise performance of standard FM systems for sinusoidal modualtion.

It is interesting to compare the noise performance of commercial FM systems. As shown in Table 5–4, for standard FM broadcasting $\beta_f = 5$, $B = 15$ kHz, and $f_1 = 2.1$ kHz. Using these parameters in Eq. (7–141), we obtain the noise performance of an FM broadcasting system, as shown in Fig. 7–26 by a solid curve. The corresponding performance of the same system, but without preemphasis–deemphasis, is shown by a dashed curve [from Eq. (7–131)]. Similarly, the results are shown for the performance of the conventional analog TV FM aural transmission system where $\beta_f = 1.67$, $B = 15$ kHz, and $f_1 = 2.1$ kHz.

Figure 7–26 also illustrates that the FM noise performance with deemphasis can be substantially better than that of FM without deemphasis. For example, for standard FM broadcasting ($\beta_f = 5$, $B = 15$ kHz, and $f_1 = 2.1$ kHz) with $(S/N)_{baseband} = 25$ dB, the FM performance is superior by 13.3 dB.

7–9 COMPARISON OF ANALOG SIGNALING SYSTEMS

Table 7–2 compares the analog systems that were analyzed in the previous sections. It is seen that the nonlinear modulation systems provide significant improvement in the noise performance, *provided* that the input signal is above the threshold. Of course, the improvement in the noise performance is obtained at the expense of having to use a wider transmission bandwidth. If the input SNR is very low, the linear systems outperform the nonlinear systems. SSB is best in terms of small bandwidth, and it has one of the best noise characteristics at low input SNR.

The selection of a particular system depends on the transmission bandwidth that is allowed and the available receiver input SNR. A comparison of the noise performance of these systems is given in Fig. 7–27, with $V_p = 1$ and $\overline{m^2} = \frac{1}{2}$. For the nonlinear systems a bandwidth spreading ratio of $B_T/B = 12$ is chosen for system comparisons. This corresponds to a $\beta_f = 5$ for the FM systems cited in the figure and corresponds to commercial FM broadcasting.

Note that, except for the "wasted" carrier power in AM, all of the linear modulation methods have the same SNR performance as that for the baseband system. (SSB has the same performance as DSB, because the coherence of the two sidebands in DSB compensates for the half noise power in SSB due to bandwidth reduction.) These comparisons are made on the basis of signals with equal *average* powers. If comparisons are made on equal *peak* powers (i.e., equal peak values for the signals), then SSB has a $(S/N)_{\text{out}}$ that is 3 dB better than DSB and 9 dB better than AM with 100% modulation (as demonstrated by Prob. 7–34). Of course, when operating above the threshold, all of the nonlinear modulation systems have better SNR performance than linear modulation systems, because the nonlinear systems have larger transmission bandwidths.

Ideal System Performance

What is the best noise performance that is theoretically possible? How can wide transmission bandwidth be used to gain improved noise performance? The answer is given by Shannon's channel capacity theorem. The *ideal system* is defined as one that does not lose channel capacity in the detection process. Thus,

$$C_{\text{in}} = C_{\text{out}} \tag{7–142}$$

where C_{in} is the bandpass channel capacity and C_{out} is the channel capacity after detection. Using Eq. (1–10) in Eq. (7–142), we have

$$B_T \log_2[1 + (S/N)_{\text{in}}] = B \log_2[1 + (S/N)_{\text{out}}] \tag{7–143}$$

where B_T is the transmission bandwidth of the bandpass signal at the receiver input and B is the bandwidth of the baseband signal at the receiver output. Solving for $(S/N)_{\text{out}}$, we get

$$\left(\frac{S}{N}\right)_{\text{out}} = \left[1 + \left(\frac{S}{N}\right)_{\text{in}}\right]^{B_T/B} - 1 \tag{7–144}$$

TABLE 7-2 COMPARISON OF ANALOG SIGNALING TECHNIQUES[a]

Type	Linearity	Transmission Bandwidth Required[b]	$\dfrac{(S/N)_{out}}{(S/N)_{baseband}}$		Comments
Baseband	L	B	1	(7-84)	No modulation
AM	L[c]	$2B$	$\dfrac{m^2}{1+m^2}$	(7-90)	Valid for all $(S/N)_{in}$ with coherent detection; valid above the threshold for envelope detection and $|m(t)| \leq 1$
DSB-SC	L	$2B$	1	(7-98)	Coherent detection required
SSB	L	B	1	(7-105)	Coherent detection required; performance identical to baseband system
PM	NL	$2(\beta_p + 1)B$	$\beta_p^2 \left(\dfrac{m}{V_p}\right)^2$	(7-120)	Coherent detection required; valid for $(S/N)_{in}$ above the threshold
FM	NL	$2(\beta_f + 1)B$	$3\beta_f^2 \left(\dfrac{m}{V_p}\right)^2$	(7-130)	Valid for $(S/N)_{in}$ above the threshold
FM with deemphasis	NL	$2(\beta_f + 1)B$	$\beta_f^2 \left(\dfrac{B}{f_1}\right)^2 \left(\dfrac{m}{V_p}\right)^2$	(7-140)	Valid for $(S/N)_{in}$ above the threshold
PCM	NL	d	$M^2/(S/N)_{baseband}$	(7-82)	Valid for $(S/N)_{in}$ above the threshold (i.e., $P_e \to 0$)

[a] B, absolute bandwidth of the modulating signal; f_1, 3-dB bandwidth of the deemphasis signal; L, linear; $m = m(t)$ is the modulating signal; $M = 2^n$, number of quantizing steps where n is the number of bits in the PCM word; NL, nolinear; V_p, peak value of $m(t)$; β_p, PM index; β_f, FM index.

[b] Typical bandwidth specifications by ITU are larger than these minima [Jordan, 1985].

[c] In the strict sense, AM signaling is not linear because of the carrier term. (See Table 4–1.)

[d] The bandwidth depends on the type of digital system used (e.g., OOK and FSK).

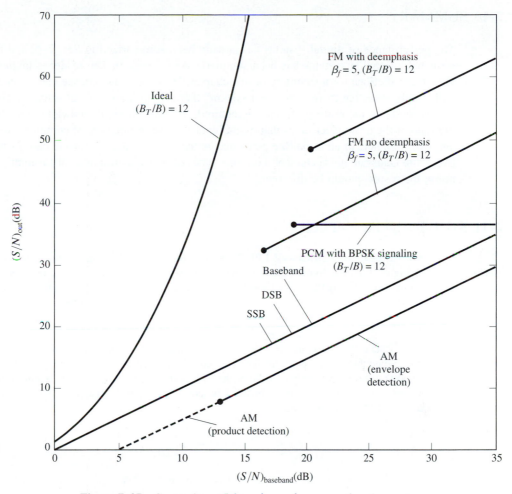

Figure 7–27   Comparison of the noise performance of analog systems.

But

$$\left(\frac{S}{N}\right)_{\text{in}} = \frac{P_s}{N_0 B_T} = \left(\frac{P_s}{N_0 B}\right)\left(\frac{B}{B_T}\right) = \left(\frac{B}{B_T}\right)\left(\frac{S}{N}\right)_{\text{baseband}} \qquad (7\text{--}145)$$

where Eq. (7–84) was used. Thus Eq. (7–144) becomes

$$\left(\frac{S}{N}\right)_{\text{out}} = \left[1 + \left(\frac{B}{B_T}\right)\left(\frac{S}{N}\right)_{\text{baseband}}\right]^{B_T/B} - 1 \qquad (7\text{--}146)$$

Equation (7–146), which describes the ideal system performance, is plotted in Fig. 7–27 for the case of $B_T/B = 12$. As expected, none of the practical signaling systems equals the performance of the ideal system. However, some of the nonlinear systems (near threshold) approach the performance of the ideal system.

7–10 SUMMARY

The performance of digital systems has already been summarized in Sec. 7–6, and the performance of analog systems has been summarized in Sec. 7–9. The reader is invited to review these sections for a summary of this chapter. However, to condense these sections, we will simply state that there is no "best system" that provides a universal solution. The solution depends on the noise performance required, the transmission bandwidth available, and the state of the art in electronics that might favor the use of one type of communication system over another. In addition, the performance has been evaluated for the case of an additive white Gaussian noise channel. For other types of noise distributions or for multiplicative noise, the results would be different.

7–11 STUDY-AID EXAMPLES

SA7–1 BER for a Brickwall LPF Receiver   Referring to Section 7–2 and parts (a) and (b) of Fig. 7–4, let a unipolar signal plus white Gaussian noise be the input to a receiver that uses a low-pass filter (LPF). Using the assumptions leading up to Eq. (7–24a), it is shown that the BER for the data out of a receiver that uses a LPF (not the optimum matched filter) is approximately $P_e = Q\left(\sqrt{\dfrac{A^2}{4N_0 B}}\right)$, where A is the peak value of the input unipolar signal, $N_0/2$ is the PSD of the noise, and B is the equivalent bandwidth of the LPF. In obtaining this result, it is argued that the sample value for a filtered binary-1 signal, $s_{01}(t_0)$, is approximately A, provided that the equivalent bandwidth of the LPF is $B > 2/T$, where T is the width of the transmitted rectangular pulse when a binary-1 is sent. $R = 1/T$ is the data rate. If the receiver uses an ideal LPF (i.e., a brickwall LPF) with

$$H(f) = \Pi\left(\frac{f}{2B}\right) \tag{7–147}$$

show that the approximation $s_{01}(t_0) \approx A$ is valid for $B \geq 2/T$.

Solution   The MATLAB solution is shown by the plots in Fig. 7–28. Fig. 7–28a shows the unfiltered rectangular pulse of amplitude $A = 1$ and width $T = 1$. Figs. 7–28b, 7–28c, and 7–28d show the filtered pulse when the bandwidth of the brickwall LPF is $B = 1/T$, $B = 2/T$, or $B = 3/T$, respectively. For $B \geq 2/T$, it is seen that the sample value is approximately A, where $A = 1$, when the sample is taken near the middle of the bit interval. Also, note that there is negligible ISI provided that $B \geq 2/T$. Furthermore, as shown in Fig. 7–28b for $B = 1/T$, it is seen that $s_{01}(t_0) \approx 1.2 \cdot 1 = 1.2A$. Consequently, it is tempting to use $s_{01}(t_0) \approx 1.2A$ (which will give a lower BER) in the formula for the BER and specify the filter equivalent bandwidth to be exactly $B = 1/T$. However, if this is done, the BER formula will not be correct for the cases when $B > 1/T$. In addition (as shown in SA7–2), if a RC LPF is used (instead of a brickwall LPF, which is impractical to build), we do not get $s_{01}(t_0) = 1.2A$ for $B = 1/T$. That is, if $s_{01}(t_0) = 1.2A$ is used to obtain the BER formula, the formula would not be valid for the RC LPF. Consequently, Eq. (7–24a), which assumes $s_{01}(t_0) \approx A$ for $B \geq 2/T$, is approximately correct for all types of practical LPFs that might be used. Furthermore, if $s_{01}(t_0)$ is not equal to A for a particular filter, then just replace A in (7–24a) by the correct sample value for that filter, and the exact (i.e., not an approximate) result will be obtained.

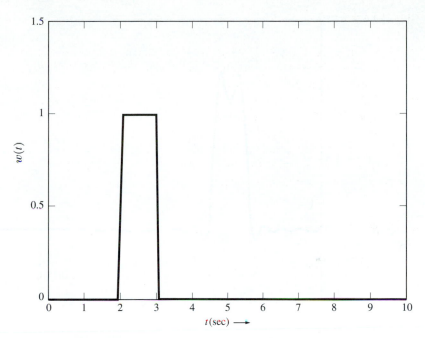

(a) Unfiltered Pulse with Amplitude $A = 1$ and Pulse Width $T = 1$

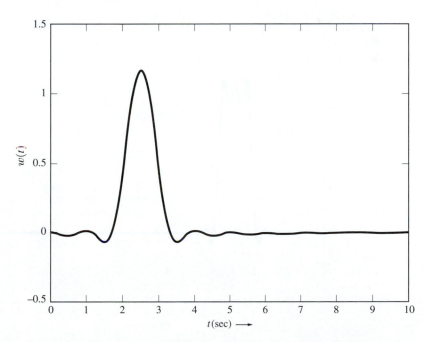

(b) Filtered Pulse with Equivalent Bandwidth $B = 1/T = 1$ Hz

Figure 7–28   Solution for SA7–1 showing the effect of brickwall low-pass filtering.

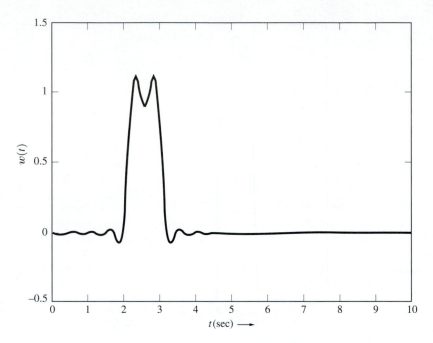

(c) Filtered Pulse with Equivalent Bandwidth $B = 2/T = 2$ Hz

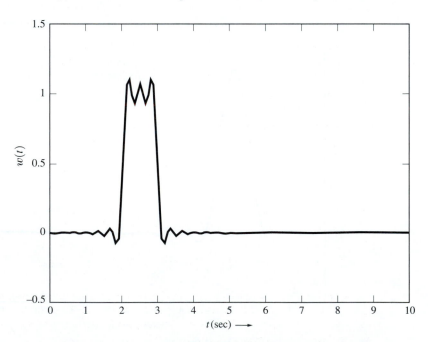

(d) Filtered Pulse with Equivalent Bandwidth $B = 3/T = 3$ Hz

Figure 7–28   (*continued*)

SA7–2 BER for an RC LPF Receiver   Rework SA7–1 for the case when the LPF is a RC filter with

$$H(f) = \frac{1}{1 + j\left(\dfrac{f}{B_{3dB}}\right)} \tag{7-148}$$

where $B_{3dB} = 1/(2\pi RC)$. Note that, using Eq. (6–104), we can see that the equivalent bandwidth of the RC LPF is $B = (\pi/2)B_{3dB}$.

Solution   The MATLAB solution is shown by the plots in Fig. 7–29. Fig. 7–29a shows the unfiltered rectangular pulse of amplitude $A = 1$ and width $T = 1$. Figs. 7–29b, 7–29c, and 7–29d show the filtered pulse when the equivalent bandwidth of the RC LPF is $B = 1/T$, $B = 2/T$, or $B = 3/T$, respectively. For $B = 1/T$, $s_{01}(t_0) = 1 = A$ only if the sampling time, t_0, can be set near the end of the pulse precisely at the point where the filtered pulse is a maximum. This requires the use of an accurate bit synchronizer. For $B \geq 2/T$ it is seen that $s_{01}(t_0) = 1 = A$ and that the sampling time, t_0, is not too critical because the filtered pulse is flat (with a value of A) over a significant portion of the bit (pulse) interval. Also, the ISI is negligible when $B \geq 2/T$; that is, the solution to homework Problem 7–9d demonstrates that the worst case signal to ISI ratio is about 70 dB. Consequently, Eq. (7–24a), which assumes $s_{01}(t_0) \approx A$ is approximately correct for $B \geq 2/T$.

SA7–3 Comparison of BER for RC and MF Receivers   Compare the performance of a digital receiver that uses a RC LPF with the performance of a receiver that uses a matched filter (MF). Referring to Fig. 7–4a, let the input to the receiver be a unipolar signal plus white Gaussian noise. Assume that the unipolar signal has a data rate of $R = 9600$ bits/s and a peak value of $A = 5$. The noise has a PSD of $\mathcal{P}_n(f) = 3 \times 10^{-5}$.

(a) If the receiver filter is a RC LPF with an equivalent bandwidth of $B = 2/T = 2R$, evaluate the SNR at the RC filter input where the bandwidth of the noise is taken to be the equivalent bandwidth of the RC LPF. Also, evaluate the BER for the data at the receiver output.

(b) Repeat (a) for the case of $B = 1/T = R$.

(c) If the receiver filter is a MF, evaluate the SNR at the MF input where the bandwidth of the noise is the equivalent bandwidth of the MF. Also, evaluate the BER for the data at the receiver output.

(d) Compare the BER performance of the receiver that uses the RC LPF with the performance of the MF receiver.

Solution   The average energy per bit for the signal at the receiver input is $E_b = (A^2/2)T = A^2/(2R)$. $\mathcal{P}_n(f) = N_0/2 = 3 \times 10^{-5}$ so $N_0 = 6 \times 10^{-5}$. Thus,

$$\left(\frac{E_b}{N_0}\right)_{dB} = 10 \ \log\left(\frac{A^2}{2N_0R}\right) = 10 \ \log\left(\frac{(5)^2}{2(6 \times 10^{-5}) \ (9600)}\right) = 13.4 \ dB \tag{7-149}$$

(a) For a receiver that uses a RC LPF with $B = 2/T = 2R$, the noise power in the bandwidth B is $P_n = (N_0/2)(2B)$. The rms signal or noise voltage is related to its average power by $V_{rms}^2 = \langle v^2(t) \rangle = P$. Because the average input signal power is $P_s = A^2/2$, the rms signal voltage at the receiver input is

$$V_s = \sqrt{A^2/2} = A/\sqrt{2} = 5/\sqrt{2} = 3.54 \ V \ rms \ signal$$

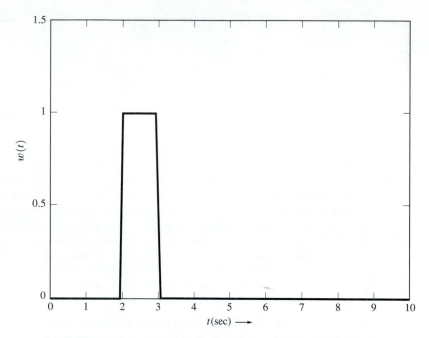

(a) Unfiltered Pulse with Amplitude $A = 1$ and Pulse Width $T = 1$

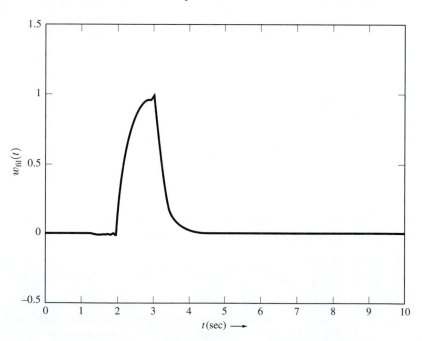

(b) Filtered Pulse with Equivalent Bandwidth $B = 1/T = 1$ Hz

Figure 7–29   Solution for SA7–2 showing the effect of RC low-pass filtering.

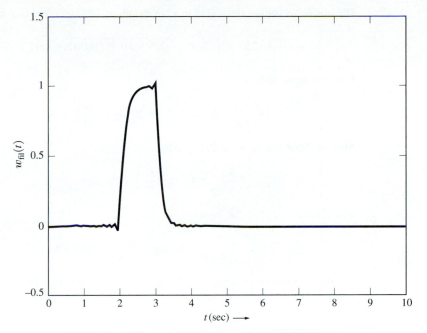

(c) Filtered Pulse with Equivalent Bandwidth $B = 2/T = 2$ Hz

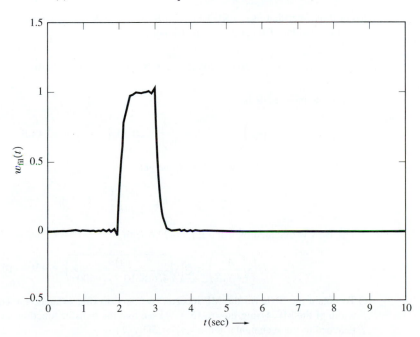

(d) Filtered Pulse with Equivalent Bandwidth $B = 3/T = 3$ Hz

Figure 7–29   (*continued*)

The rms noise voltage (measured in a bandwidth of $2R$) at the receiver input is

$$V_n = \sqrt{\frac{N_0}{2}(2B)} = \sqrt{2N_0R} = \sqrt{2(6 \times 10^{-5})(9600)} = 1.07 \text{ V rms noise}$$

Thus, the input SNR is

$$\left(\frac{S}{N}\right)_{dB} = 20 \log\left(\frac{V_s}{V_n}\right) = 20 \log\left(\frac{3.54}{1.04}\right) = 10.4 \text{ dB}$$

Also, the SNR can be evaluated by using

$$\left(\frac{S}{N}\right)_{dB} = 10 \log\left(\frac{P_s}{P_n}\right) = 10 \log\left(\frac{A^2}{2N_0B}\right) \qquad (7\text{-}150)$$

which, for $B = 2/T = 2R$, becomes

$$\left(\frac{S}{N}\right)_{dB} = 10 \log\left(\frac{A^2}{4N_0R}\right) = 10.4 \text{ dB}$$

The BER is obtained by using Eq. (7–24a) for the case of a RC LPF, where $s_{01}(t_0) \approx A$ and $B = 2/T = 2R$. So, we get

$$P_e = Q\left(\sqrt{\frac{A^2}{8N_0R}}\right) = Q\left(\sqrt{\frac{(5)^2}{8(6 \times 10^{-5})(9600)}}\right) = 9.9 \times 10^{-3}$$

(b) For a receiver that uses a RC LPF with $B = 1/T = R$, the rms noise voltage (measured in a bandwidth of R) at the receiver input is

$$V_n = \sqrt{\frac{N_0}{2}(2B)} = \sqrt{N_0R} = \sqrt{(6 \times 10^{-5})(9600)} = 0.76 \text{ V rms noise}$$

Thus, the input SNR is

$$\left(\frac{S}{N}\right)_{dB} = 20 \log\left(\frac{V_s}{V_n}\right) = 20 \log\left(\frac{3.54}{0.76}\right) = 13.4 \text{ dB}$$

Also, using Eq. (7–150) with $B = R$, we get

$$\left(\frac{S}{N}\right)_{dB} = 10 \log\left(\frac{A^2}{2N_0R}\right) = 13.4 \text{ dB}$$

The BER is obtained by using Eq. (7–24a) with $B = 1/T = R$. So, we get

$$P_e = Q\left(\sqrt{\frac{A^2}{4N_0R}}\right) = Q\left(\sqrt{\frac{(5)^2}{4(6 \times 10^{-5})(9600)}}\right) = 4.9 \times 10^{-4}$$

(c) To obtain the SNR for the MF receiver, we need to first evaluate the equivalent band-width of the MF. Using Eq. (6–155), we see that the transfer function for the MF that is matched to the rectangular pulse $s(t) = 5\Pi(t/T)$ is

$$H(f) = K\frac{S^*(f)}{\mathcal{P}_n(f)} e^{-j\omega t_0} = CT\left(\frac{\sin \pi Tf}{\pi Tf}\right)e^{-j\omega t_0}$$

where $C = 5K/(N_0/2)$. From Eq. (2–192), the equivalent bandwidth of the MF is

$$B = \frac{1}{|H(0)|^2} \int_0^\infty |H(f)|^2 \, df = \int_0^\infty \left(\frac{\sin \pi T f}{\pi T f}\right)^2 df$$

To evaluate this integral, make a change in variable. Let $x = \pi T f$, so that with the aid of Appendix A, we get

$$B = \frac{1}{\pi T} \int_0^\infty \left(\frac{\sin x}{x}\right)^2 dx = \left(\frac{1}{\pi T}\right)\left(\frac{\pi}{2}\right) = \frac{1}{2T} = \frac{1}{2} R \qquad (7\text{–}151)$$

Thus, the rms noise voltage (measured in a bandwidth of $B = R/2$) at the receiver input is

$$V_n = \sqrt{\frac{N_0}{2}} (2B) = \sqrt{N_0 R/2} = \sqrt{(6 \times 10^{-5})(9600)/2} = 0.54 \text{ V rms noise}$$

As previously obtained, the rms signal voltage is $V_s = 3.54$ V. So, the input SNR is

$$\left(\frac{S}{N}\right)_{dB} = 20 \log\left(\frac{V_s}{V_n}\right) = 20 \log\left(\frac{3.54}{0.54}\right) = 16.4 \text{ dB}$$

Also, using Eq. (7–150) with $B = R/2$, we get

$$\left(\frac{S}{N}\right)_{dB} = 10 \log\left(\frac{A^2}{N_0 R}\right) = 16.4 \text{ dB}$$

The BER is obtained by using Eq. (7–24b):

$$P_e = Q\left(\sqrt{\frac{E_b}{N_0}}\right) = Q\left(\sqrt{\frac{A^2 T}{2N_0}}\right) = Q\left(\sqrt{\frac{A^2}{2N_0 R}}\right)$$

$$= Q\left(\sqrt{\frac{(5)^2}{2(6 \times 10^{-5})(9600)}}\right) = 1.6 \times 10^{-6}$$

This result may be verified by using Fig. 7–14 for the case of unipolar signaling with $(E_b/N_0)_{dB} = 13.4$ dB.

(d) As discussed in the solution for SA7–2, if an inexpensive RC LPF is used with $B = 2R$, then an inexpensive bit synchronizer (with imprecise sampling times) is adequate for use in the receiver. As found in (a), this gives a BER of 9.9×10^{-3}.

If better performance (i.e., a lower BER) is desired, the bandwidth of the LPF can be reduced to $B = R$, and a precise (more expensive) bit synchronizer is required. This lowers the BER to 4.9×10^{-4}, as found in part (b).

If even better performance is desired, an MF (more expensive than an LPF) can be used. This is implemented by using an integrate and dump filter as previously shown in Fig. 6-17. This MF requires a bit synchronizer to provide the clocking signals that re-set the integrator and dump the sampler. This MF receiver gives the optimum (lowest BER) performance. In this case, the BER is reduced to 1.6×10^{-6}, as found in part (c). However, it is realized that in some applications, such as some fiber-optic systems, the noise is negligible. In such a case, the simple LPF of part (a) with $B = 2/T = 2R$ is adequate because the BER will be approach zero.

SA7–4 Output S/N Improvement for FM with Preemphasis   In FM signaling systems, preemphasis of the modulation at the transmitter input and deemphasis at the receiver output is often used to improve the output SNR. For 75 μs emphasis, the 3-dB corner frequency of the receiver deemphasis LPF is $f_1 = 1/(2\pi 75\mu s) = 2.12$ kHz. The audio bandwidth is $B = 15$ kHz. Derive a formula for the improvement in dB, I_{dB}, for the SNR of a FM system with preemphasis–deemphasis when compared with the SNR for a FM system without preemphasis–deemphasis. Compute I_{dB} for $f_1 = 2.12$ kHz and $B = 15$ kHz.

Solution   Refer to Eqs. (7–121)–(7–127) for the development of the $(S/N)_{\text{out}}$ equation for FM with no deemphasis and to Eqs. (7–135)–(7–138) for the development of the $(S/N)_{\text{out}}$ equation for FM with deemphasis. Then,

$$I = \frac{(S/N)_{\text{with deemphasis}}}{(S/N)_{\text{no deemphasis}}} = \frac{\left(\dfrac{\overline{s_0^2}}{\overline{n_0^2}}\right)_{\text{with deemphasis}}}{\left(\dfrac{\overline{s_0^2}}{\overline{n_0^2}}\right)_{\text{no deemphasis}}} \qquad (7\text{--}152)$$

From Sec. 5–6 and Fig. 5–15, it is seen that the inclusion of two filters—a deemphasis filter at the receiver, $H_d(f)$, and a preemphasis filter at the transmitter, $H_p(f)$—has no overall effect on the frequency response (and power) of the output audio *signal*, $s_0(t)$, because $H_p(f)H_d(f) = 1$ over the audio bandwidth, B, where $B < f_2$. Thus, $(\overline{s_0^2})_{\text{with deemphasis}} = (\overline{s_0^2})_{\text{no deemphasis}}$, and Eq. (7–152) reduces to

$$I = \frac{(\overline{n_0^2})_{\text{no deemphasis}}}{(\overline{n_0^2})_{\text{with deemphasis}}} \qquad (7\text{--}153)$$

The output noise is different for the receiver with deemphasis (compared with a receiver without deemphasis) because the deemphasis filter attenuates the noise at the higher audio frequencies. (The transmitter preemphasis filter has no effect on the received noise.) Using Eq. (7–126) for $(\overline{n_0^2})_{\text{no deemphasis}}$ and Eq. (7–136) for $(\overline{n_0^2})_{\text{with deemphasis}}$, we see that Eq. (7–153) becomes

$$I = \frac{\dfrac{2}{3}\left(\dfrac{K}{A_c}\right)^2 N_0 B^3}{2\left(\dfrac{K}{A_c}\right)^2 N_0 f_1^3 \left[\dfrac{B}{f_1} - \tan^{-1}\left(\dfrac{B}{f_1}\right)\right]}$$

or

$$I = \frac{B^3}{3 f_1^3 \left[\dfrac{B}{f_1} - \tan^{-1}\left(\dfrac{B}{f_1}\right)\right]} \qquad (7\text{--}154)$$

Because $I_{dB} = 10 \log(I)$, we get

$$I_{dB} = 30 \log\left(\frac{B}{f_1}\right) - 10 \log\left\{3\left[\frac{B}{f_1} - \tan^{-1}\left(\frac{B}{f_1}\right)\right]\right\} \qquad (7\text{--}155)$$

For $f_1 = 2.1$ kHz and $B = 15$ kHz, Eq. (7–155) gives

$$I_{dB} = 13.2 \text{ dB}$$

This checks with the value for I_{dB} that is obtained from Fig. 7–26.

PROBLEMS

7–1 In a binary communication system the receiver test statistic, $r_0(t_0) = r_0$, consists of a polar signal plus noise. The polar signal has values $s_{01} = +A$ and $s_{02} = -A$. Assume that the noise has a *Laplacian* distribution, which is

$$f(n_0) = \frac{1}{\sqrt{2}\,\sigma_0}\, e^{-\sqrt{2}|n_0|/\sigma_0}$$

where σ_0 is the rms value of the noise.

(a) Find the probability of error P_e as a function of A/σ_0 for the case of equally likely signaling and V_T having the optimum value.

(b) Plot P_e as a function of A/σ_0 decibels. Compare this result with that obtained for Gaussian noise as given by Eq. (7–26a).

7–2 Using Eq. (7–8), show that the optimum threshold level for the case of antipodal signaling with additive white Gaussian noise is

$$V_T = \frac{\sigma_0^2}{2s_{01}} \ln \left[\frac{P(s_2 \text{ sent})}{P(s_1 \text{ sent})} \right]$$

Here the receiver filter has an output with a variance of σ_0^2. s_{01} is the value of the sampled binary 1 signal at the filter output. $P(s_1 \text{ sent})$ and $P(s_2 \text{ sent})$ are the probabilities of transmitting a binary 1 and a binary 0, respectively.

7–3 A baseband digital communication system uses polar signaling with matched filter in the receiver. The probability of sending a binary 1 is p, and the probability of sending a binary zero is $1 - p$.

(a) For $E_b/N_0 = 10$ dB, plot P_e as a function of p using a log scale.

(b) Referring to Eq. (1-8), plot the entropy, H, as a function of p. Compare the shapes of these two curves.

7–4 A *whole* binary communication system can be modeled as an *information channel*, as shown in Fig. P7–4. Find equations for the four transition probabilities $P(\tilde{m}|m)$, where both $\tilde{m}$ and m can be binary 1's or binary 0's. Assume that the test statistic is a linear function of the receiver input and that additive white Gaussian noise appears at the receiver input. [*Hint:* Look at Eq. (7–15).]

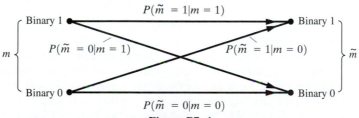

Figure P7–4

7–5 A baseband digital communication system uses unipolar signaling (rectangular pulse shape) with matched filter detection. The data rate is $R = 9,600$ bits/sec.

(a) Find an expression for bit error rate (BER), P_e, as a function of $(S/N)_{\text{in}}$. $(S/N)_{\text{in}}$ is the signal-to-noise power ratio at the receiver input where the noise is measured in a bandwidth corresponding to the equivalent bandwidth of the matched filter. [*Hint:* First find an expression of E_b/N_0 in terms of $(S/N)_{\text{in}}$.]

(b) Plot P_e vs. $(S/N)_{\text{in}}$ in dB units on a log scale over a range of $(S/N)_{\text{in}}$ from 0 to 15 dB.

7–6 Rework Prob. 7–5 for the case of polar signaling.

7–7 Examine how the performance of a baseband digital communication system is affected by the receiver filter. Equation (7–26a) describes the BER when a low-pass filter is used and the bandwidth of the filter is large enough that the signal level at the filter output is $s_{01} = +A$ or $s_{02} = -A$. Instead, suppose that a RC low-pass filter with a restricted bandwidth is used where $T = 1/f_0 = 2\pi RC$. T is the duration (pulse width) of one bit, and f_0 is the 3-dB bandwidth of the RC low-pass filter as described by Eq. (2–147). Assume that the initial conditions of the filter are reset to zero at the beginning of each bit interval.
(a) Derive an expression for P_e as a function of E_b/N_0.
(b) On a log scale, plot the BER obtained in part (a) for E_b/N_0 over a range of 0 to 15 dB.
(c) Compare this result with that for a matched-filter receiver (as shown in Fig. 7–5).

7–8 Consider a baseband digital communication system that uses polar signaling (rectangular pulse shape) where the receiver is shown in Fig. 7–4a. Assume that the receiver uses a second-order Butterworth filter with a 3-dB bandwidth of f_0 . The filter impulse response and transfer function are

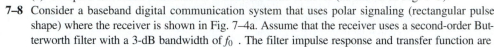

$$h(t) = \left[\sqrt{2}\,\omega_0 e^{-(\omega_0/\sqrt{2})t}\,\sin\!\left(\frac{\omega_0}{\sqrt{2}}\,t\right)\right]u(t)$$

$$H(f) = \frac{1}{(jf/f_0)^2 + \sqrt{2}(jf/f_0) + 1}$$

where $\omega_0 = 2\pi f_0$. Let $f_0 = 1/T$, where T is the bit interval (i.e., pulse width), and assume that the initial conditions of the filter are reset to zero at the beginning of each bit interval.
(a) Derive an expression for P_e as a function of E_b/N_0.
(b) On a log scale, plot the BER obtained in part (a) for E_b/N_0 over a range of 0 to 15 dB.
(c) Compare this result with that for a matched-filter receiver (as shown in Fig. 7–5).

7–9 Consider a baseband unipolar communication system with equally likely signaling. Assume that the receiver uses a simple RC LPF with a time constant of $RC = \tau$ where $\tau = T$ and $1/T$ is the bit rate. (By "simple," it is meant that the initial conditions of the LPF are *not* reset to zero at the beginning of each bit interval.)
(a) For signal alone at the receiver input, evaluate the approximate worst-case signal to ISI ratio (in decibels) out of the LPF at the sampling time $t = t_0 = nT$, where n is an integer.
(b) Evaluate the signal to ISI ratio (in decibels) as a function of the parameter K, where $t = t_0 = (n + K)T$ and $0 < K \le 1$.
(c) What is the optimum sampling time to use to maximize the signal-to-ISI power ratio out of the LPF?
(d) Repeat part (a) for the case when the equivalent bandwidth of the RC LPF is $2/T$.

7–10 Examine a baseband polar communication system with equally likely signaling and no channel noise. Assume that the receiver uses a simple RC LPF with a time constant of $\tau = RC$. (By "simple," it is meant that the initial conditions of the LPF are *not* reset to zero at the beginning of each bit interval.) Evaluate the worst-case approximate signal to ISI ratio (in decibels) out of the LPF at the sampling time $t = t_0 = nT$, where n is an integer. This approximate result will be valid for $T/\tau > \frac{1}{2}$. Plot this result as a function of T/τ for $\frac{1}{2} \le T/\tau \le 5$.

7–11 Consider a baseband unipolar system described in Prob. 7–9d. Assume that white Gaussian noise is present at the receiver input.
(a) Derive an expression for P_e as a function of E_b/N_0 for the case of sampling at the times $t = t_0 = nT$.
(b) Compare the BER obtained in part (a) with the BER characteristic that is obtained when a matched-filter receiver is used. Plot both of these BER characteristics as a function of $(E_b/N_0)_{dB}$ over the range 0 to 15 dB.

7–12 Rework Prob. 7–11 for the case of *polar* baseband signaling.

7–13 For bipolar signaling, the discussion leading up to Eq. (7-28) indicates that the optimum threshold at the receiver is $V_T = \dfrac{A}{2} + \dfrac{\sigma_0^2}{A} \ln 2$.

    (a) Prove that this is the optimum threshold value.

    (b) Show that $A/2$ approximates the optimum threshold if $P_e < 10^{-3}$.

7–14 For unipolar baseband signaling as described by Eq. (7–23),

    (a) Find the matched-filter frequency response and show how the filtering operation can be implemented by using an integrate-and-dump filter.

    (b) Show that the equivalent bandwidth of the matched filter is $B_{eq} = 1/(2T) = R/2$.

7–15 Equally likely polar signaling is used in a baseband communication system. Gaussian noise having a PSD of $N_0/2$ W/Hz plus a polar signal with a peak level of A volts is present at the receiver input. The receiver uses a matched-filter circuit having a voltage gain of 1,000.

    (a) Find the expression for P_e as a function of A, N_0, T, and V_T, where $R = 1/T$ is the bit rate and V_T is the threshold level.

    (b) Plot P_e as a function of V_T for the case of $A = 8 \times 10^{-3}$ V, $N_0/2 = 4 \times 10^{-9}$ W/Hz, and $R = 1200$ bits/sec.

7–16 Consider a baseband polar communication system with matched-filter detection. Assume that the channel noise is white and Gaussian with a PSD of $N_0/2$. The probability of sending a binary 1 is $P(1)$ and the probability of sending a binary 0 is $P(0)$. Find the expression for P_e as a function of the threshold level V_T when the signal level out of the matched filter is A, and the variance of the noise out of the matched filter is $\sigma^2 = N_0/(2T)$, where $R = 1/T$ is the bit rate.

7–17 Design a receiver for detecting the data on a bipolar RZ signal that has a peak value of $A = 5$ volts. In your design assume that an RC low-pass filter will be used and the data rate is 2,400 (bits/sec).

    (a) Draw a block diagram of your design and explain how it works.

    (b) Give the values for the design parameters R, C, and V_T.

    (c) Calculate the PSD level for the noise N_0 that is allowed if P_e is to be less than 10^{-6}.

7–18 A BER of 10^{-5} or less is desired for an OOK communication system where the bit rate is $R = 10$ Mb/s. The input to the receiver consists of the OOK signal plus white Gaussian noise.

    (a) Find the minimum transmission bandwidth required.

    (b) Find the minimum E_b/N_0 required at the receiver input for coherent matched-filter detection.

    (c) Rework part (b) for the case of noncoherent detection.

7–19 Rework Prob. 7–18 for the case when FSK signaling is used. Let $2\Delta F = f_2 - f_1 = 1.5R$.

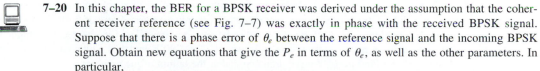

7–20 In this chapter, the BER for a BPSK receiver was derived under the assumption that the coherent receiver reference (see Fig. 7–7) was exactly in phase with the received BPSK signal. Suppose that there is a phase error of θ_e between the reference signal and the incoming BPSK signal. Obtain new equations that give the P_e in terms of θ_e, as well as the other parameters. In particular,

    (a) Obtain a new equation that replaces Eq. (7–36).

    (b) Obtain a new equation that replaces Eq. (7–38).

    (c) Plot results from part (b) where the log plot of P_e is given as a function of θ_e over a range $-\pi < \theta_e < \pi$ for the case when $E_b/N_0 = 10$ dB.

7–21 Digital data are transmitted over a communication system that uses nine repeaters plus a receiver, and BPSK signaling is used. The P_e for each of the regenerative repeaters (see Sec. 3–5) is 5×10^{-8}, assuming additive Gaussian noise.

(a) Find the overall P_e for the system.

(b) If each repeater is replaced by an ideal amplifier (no noise or distortion), what is the P_e of the overall system?

7–22 Digital data are to be transmitted over a toll telephone system using BPSK. Regenerative repeaters are spaced 50 miles apart along the system. The total length of the system is 600 miles. The telephone lines between the repeater sites are equalized over a 300- to 2,700-Hz band and provide an E_b/N_0 (Gaussian noise) of 15 dB to the repeater input.

(a) Find the largest bit rate R that can be accommodated with no ISI.

(b) Find the overall P_e for the system. (Be sure to include the receiver at the end of the system.)

7–23 A BPSK signal is given by

$$s(t) = A \sin[\omega_c t + \theta_c + (\pm 1)\beta_p], \qquad 0 < t \le T$$

The binary data are represented by (± 1), where $(+1)$ is used to transmit a binary 1 and (-1) is used to transmit a binary 0. β_p is the phase modulation index as defined by Eq. (5–47).

(a) For $\beta_p = \pi/2$, show that this BPSK signal becomes the BPSK signal as described by Eq. (7–34).

(b) For $0 < \beta_p < \pi/2$, show that a discrete carrier term is present in addition to the BPSK signal as described by Eq. (7–34).

7–24 Referring to the BPSK signal described in Prob. 7–23, find P_e as a function of the modulation index β_p, where $0 < \beta_p \le \pi/2$.

(a) Find P_e as a function of A, β_p, N_0, and B for the receiver that uses a narrowband filter.

(b) Find P_e as a function of E_b, N_0, and β_p for the receiver that uses a matched filter. (E_b is the average BPSK signal energy that is received during one bit.)

7–25 Referring to the BPSK signal described in Prob. 7–23, let $0 < \beta_p < \pi/2$.

(a) Show a block diagram for the detection of the BPSK signal where a PLL is used to recover the coherent reference signal from the BPSK signal.

(b) Explain why Manchester-encoded data are often used when the receiver uses a PLL for carrier recovery, as in part (a). (*Hint:* Look at the spectrum of the Manchester-encoded PSK signal.)

7–26 In obtaining the P_e for FSK signaling with coherent reception, the energy in the difference signal E_d was needed, as shown in Eq. (7–46). For *orthogonal* FSK signaling the cross-product integral was found to be zero. Suppose that f_1, f_2, and T are chosen so that E_d is maximized.

(a) Find the relationship as a function of f_1, f_2, and T for maximum E_d.

(b) Find the P_e as a function of E_b and N_0 for signaling with this FSK signal.

(c) Sketch the P_e for this type of FSK signal and compare it with a sketch of the P_e for the orthogonal FSK signal that is given by Eq. (7–47).

7–27 An FSK signal with $R = 110$ bits/sec is transmitted over an RF channel that has white Gaussian noise. The receiver uses a noncoherent detector and has a noise figure of 6 dB. The impedance of the antenna input of the receiver is 50 Ω. The signal level at the receiver input is 0.05 μV, and the noise level is $N_0 = kT_0$, where $T_0 = 290$ K and k is Boltzmann's constant. (See Sec. 8–6.) Find the P_e for the digital signal at the output of the receiver.

7–28 Rework Prob. 7–27 for the case of DPSK signaling.

7–29 An analysis of the noise associated with the two channels of an FSK receiver precedes Eq. (7–44). In this analysis, it is stated that $n_1(t)$ and $n_2(t)$ are independent when they arise from a common white Gaussian noise process because $n_1(t)$ and $n_2(t)$ have nonoverlapping spectra. Prove that this statement is correct. [*Hint:* $n_1(t)$ and $n_2(t)$ can be modeled as the outputs of two linear filters that have nonoverlapping transfer functions and the same white noise process, $n(t)$, at their inputs.]

7–30 In most applications, communication systems are designed to have a BER of 10^{-5} or less. Find the minimum E_b/N_0 decibels required to achieve an error rate of 10^{-5} for the following types of signaling.
  (a) Polar baseband.
  (b) OOK.
  (c) BPSK.
  (d) FSK.
  (e) DPSK.

7–31 Digital data are to be transmitted over a telephone line channel. Suppose that the telephone line is equalized over a 300- to 2,700-Hz band and that the signal-to-Gaussian-noise (power) ratio at the output (receive end) is 25 dB.
  (a) Of all the digital signaling techniques studied in this chapter, choose the one that will provide the largest bit rate for a P_e of 10^{-5}. What is the bit rate R for this system?
  (b) Compare this result with the bit rate R that is possible when an ideal digital signaling scheme is used, as given by the Shannon channel capacity stated by Eq. (1–10).

7–32 An analog baseband signal has a uniform PDF and a bandwidth of 3500 Hz. This signal is sampled at an 8 samples/s rate, uniformly quantized, and encoded into a PCM signal having 8-bit words. This PCM signal is transmitted over a DPSK communication system that contains additive white Gaussian channel noise. The signal-to-noise ratio at the receiver input is 8 dB.
  (a) Find the P_e of the recovered PCM signal.
  (b) Find the peak signal/average noise ratio (decibels) out of the PCM system.

7–33 A spread spectrum (SS) signal is often used to combat narrowband interference and for communication security. The SS signal with direct sequence spreading is (see Sec. 5–13)

$$s(t) = A_c c(t) m(t) \cos(\omega_c t + \theta_c)$$

where θ_c is the start-up carrier phase, $m(t)$ is the polar binary data baseband modulation, and $c(t)$ is a polar baseband spreading waveform that usually consists of a pseudonoise (PN) code. The PN code is a binary sequence that is N bits long. The "bits" are called *chips*, since they do not contain data and since many chips are transmitted during the time that it takes to transmit 1 bit of the data [in $m(t)$]. The same N-bit code word is repeated over and over, but N is a large number, so the chip sequence in $c(t)$ looks like digital noise. The PN sequence may be generated by using a clocked r-stage shift register having feedback so that $N = 2^r - 1$. The autocorrelation of a long sequence is approximately

$$R_c(\tau) = \Lambda\!\left(\frac{\tau}{T_c}\right)$$

where T_c is the duration of one chip (the time it takes to send one chip of the PN code). $T_c \ll T_b$, where T_b is the duration of a data bit.
  (a) Find the PSD for the SS signal $s(t)$. [*Hint:* Assume that $m(t)$, $c(t)$, and θ_c are independent. In addition, note that the PSD of $m(t)$ can be approximated by a delta function, since the spectral width of $m(t)$ is very small when compared to that for the spreading waveform $c(t)$.]
  (b) Draw a block diagram for an optimum coherent receiver. Note that $c(t)m(t)$ is first coherently detected and then the data, $m(t)$, are recovered by using a correlation processor.
  (c) Find the expression for P_e.

7–34 Examine the performance of an AM communication system where the receiver uses a product detector. For the case of a sine-wave modulating signal, plot the ratio of $[(S/N)_{\text{out}}/(S/N)_{\text{in}}]$ as a function of the percent modulation.

7–35 An AM transmitter is modulated 40% by a sine-wave audio test tone. This AM signal is transmitted over an additive white Gaussian noise channel. Evaluate the noise performance of this system and determine by how many decibels this system is inferior to a DSB-SC system.

7–36 A phasing-type receiver for SSB signals is shown in Fig. P7–36. (This was the topic of Prob. 5-18.)

(a) Show that this receiver is or is not a linear system.

(b) Derive the equation for SNR out of this receiver when the input is an SSB signal plus white noise with a PSD of $N_0/2$.

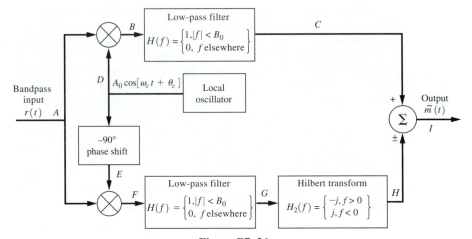

Figure P7–36

7–37 Referring to Fig. P7–36, suppose that the receiver consists only of the upper portion of the figure, so that point C is the output. Let the input be an SSB signal plus white noise with a PSD of $N_0/2$. Find $(S/N)_{out}$.

7–38 Consider the receiver shown in Fig. P7–38. Let the input be a DSB-SC signal plus white noise with a PSD of $N_0/2$. The mean value of the modulation is zero.

(a) For A_0 large, show that this receiver acts like a product detector.

(b) Find the equation for $(S/N)_{out}$ as a function of A_c, $\overline{m^2}$, N_0, A_0, and B_T when A_0 is large.

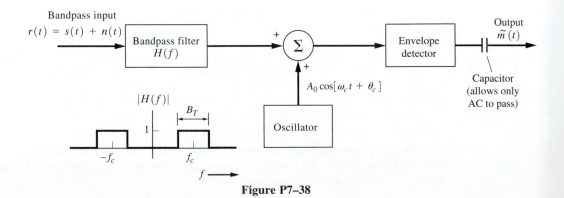

Figure P7–38

7–39 Compare the performance of AM, DSB-SC, and SSB systems when the modulating signal $m(t)$ is a Gaussian random process. Assume that the Gaussian modulation has a zero mean value and a peak value of $V_p = 1$, where $V_p \approx 4\sigma_m$. Compare the noise performance of these three systems by plotting $(S/N)_{\text{out}}/(S/N)_{\text{baseband}}$ for
(a) The AM system.
(b) The DSB-SC system.
(c) The SSB system.

7–40 If linear modulation systems are to be compared on an equal *peak* power basis (i.e., all have equal peak signal values), show that
(a) SSB has a $(S/N)_{\text{out}}$ that is 3 dB better than DSB.
(b) SSB has a $(S/N)_{\text{out}}$ that is 9 dB better than AM.
(*Hint:* See Prob. 5–8.)

7–41 Using Eq. (7–132), plot the output SNR threshold characteristic of a discriminator for the parameters of a conventional TV FM aural system. ($\Delta F = 25$ kHz and $B = 15$ kHz.) Compare plots of the $(S/N)_{\text{out}}$ vs. $(S/N)_{\text{baseband}}$ for this system with those shown in Fig. 7–24.

7–42 An FM receiver has an IF bandwidth of 25 kHz and a baseband bandwidth of 5 kHz. The noise figure of the receiver is 12 dB, and it uses a 75-μsec deemphasis network. An FM signal plus white noise is present at the receiver input, where the PSD of the noise is $N_0/2 = kT/2$. $T = 290$ K. (See Sec. 8–6.) Find the minimum input signal level (in dBm) that will give a SNR of 35 dB at the output when sine-wave test modulation is used.

7–43 Referring to Table 5–4, note that a two-way FM mobile radio system uses the parameters $\beta_f = 1$ and $B = 5$ kHz.
(a) Find $(S/N)_{\text{out}}$ for the case of no deemphasis.
(b) Find $(S/N)_{\text{out}}$ if deemphasis is used with $f_1 = 2.1$ kHz. It is realized that B is not much larger than f_1 in this application.
(c) Plot $(S/N)_{\text{out}}$ vs. $(S/N)_{\text{baseband}}$ for this system, and compare the plot with the result for FM broadcasting as shown in Fig. 7–26.

7–44 Compare the performance of two FM systems that use different deemphasis characteristics. Assume that $\beta_f = 5$, $\overline{(m/V_p)^2} = \frac{1}{2}$, $B = 15$ kHz and that an additive white Gaussian noise channel is used. Find $(S/N)_{\text{out}}/(S/N)_{\text{baseband}}$ for:
(a) 25-μsec deemphasis.
(b) 75-μsec deemphasis.

7–45 A baseband signal $m(t)$ that has a Gaussian (amplitude) distribution frequency modulates a transmitter. Assume that the modulation has a zero-mean value and a peak value of $V_p = 4\sigma_m$. The FM signal is transmitted over an additive white Gaussian noise channel. Let $\beta_f = 3$ and $B = 15$ kHz. Find $(S/N)_{\text{out}}/(S/N)_{\text{baseband}}$ when
(a) No deemphasis is used.
(b) 75-μsec deemphasis is used.

7–46 In FM broadcasting, a preemphasis filter is used at the audio input of the transmitter, and a deemphasis filter is used at the receiver output to improve the output SNR. For 75-μsec emphasis, the 3-dB bandwidth of the receiver deemphasis LPF is $f_1 = 2.1$ kHz. The audio bandwidth is $B = 15$ kHz. Define the improvement factor I as a function of B/f_1 as

$$I = \frac{(S/N)_{\text{out}} \text{ for system with preemphasis-deemphasis}}{(S/N)_{\text{out}} \text{ for system without preemphasis-deemphasis}}$$

For $B = 15$ kHz, plot the decibel improvement that is realized as a function of the design parameter f_1 where 50 Hz $< f_1 <$ 15 kHz.

7–47 In FM broadcasting, preemphasis is used, and yet $\Delta F = 75$ kHz is defined as 100% modulation. Examine the incompatibility of these two standards. For example, assume that the amplitude of a 1-kHz audio test tone is adjusted to produce 100% modulation (i.e., $\Delta F = 75$ kHz).

    (a) If the frequency is changed to 15 kHz, find the ΔF that would be obtained ($f_1 = 2.1$ kHz). What is the percent modulation?

    (b) Explain why this phenomenon does not cause too much difficulty when typical audio program material (modulation) is broadcast.

7–48 Stereo FM transmission was studied in Sec. 5–7. At the transmitter, the left-channel audio, $m_L(t)$, and the right-channel audio, $m_R(t)$, are each preemphasized by an $f_1 = 2.1$-kHz network. These preemphasized audio signals are then converted into the composite baseband modulating signal $m_b(t)$, as shown in Fig. 5–17. At the receiver, the FM detector outputs the composite baseband signal that has been corrupted by noise. (Assume that the noise comes from a white Gaussian noise channel.) This corrupted composite baseband signal is demultiplexed into corrupted left- and right-channel audio signals, $\tilde{m}_L(t)$ and $\tilde{m}_R(t)$, each having been deemphasized by a 2.1-kHz filter. The noise on these outputs arises from the noise at the output of the FM detector that occurs in the 0- to 15-kHz and 23- to 53-kHz bands. The subcarrier frequency is 38 kHz. Assuming that the input SNR of the FM receiver is large, show that the stereo FM system is 22.2 dB more noisy than the corresponding monaural FM system.

7–49 An FDM signal, $m_b(t)$, consists of five 4-kHz-wide channels denoted by C1, C2, ..., C5, as shown in Fig. P7–49. The FDM signal was obtained by modulating five audio signals (each with 4-kHz bandwidth) onto USSB (upper single-sideband) subcarriers. This FDM signal, $m_b(t)$, modulates a DSB-SC transmitter. The DSB-SC signal is transmitted over an additive white Gaussian noise channel. At the receiver the average power of the DSB-SC signal is P_s and the noise has a PSD of $N_0/2$.

    (a) Draw a block diagram for a receiving system with five outputs, one for each audio channel.

    (b) Calculate the output SNR for each of the five audio channels.

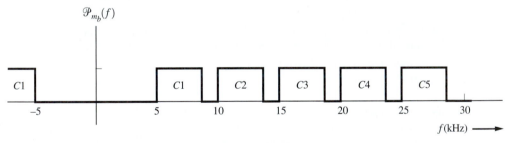

Figure P7–49

7–50 Rework Prob. 7–49 for the case when the FDM signal is frequency modulated onto the main carrier. Assume that the rms carrier deviation is denoted by ΔF_{rms} and that the five audio channels are independent. No deemphasis is used.

7–51 Refer to Fig. 7–27 and

    (a) Verify that the PCM curve is correct.

    (b) Find the equation for the PCM curve for the case of QPSK signaling over the channel.

    (c) Compare the performance of the PCM/QPSK system with that for PCM/BPSK and for the ideal case by sketching the $(S/N)_{out}$ characteristic.

CHAPTER 8

Chapter Objectives

- Telephone systems, digital subscriber lines, and fiber-optic loops
- Satellite communication systems
- Link budget analysis and design
- Cellular and personal communication systems
- Digital and analog television

WIRE AND WIRELESS COMMUNICATION SYSTEMS

8–1 THE EXPLOSIVE GROWTH OF TELECOMMUNICATIONS

The explosive growth of both wire and wireless telecommunications has been sparked by the need for personal communication of voice, video, and data and the availability of low-cost integrated circuits and microprocessors. There is a relentless push for higher speed data transmission over wire and wireless systems. This is caused not only by the growth of traditional voice, video, and data systems, but also by the convenience of personal cellular phones, fax machines, e-mail with file attachments, and the use of the Internet.

The goal of this chapter is to examine the design of some practical communication systems. These systems are based on the communication principles that have been developed in previous chapters. Also, we see how modern designs are influenced by industry standards and how some sophisticated systems are the evolution of historical systems. This should be a "fun" chapter.

Results are presented for telephone systems, satellite systems, fiber-optic systems, cellular telephone systems, personal communication service (PCS), television systems (including digital TV), and link budget analysis for wireless systems. Link budget analysis is concerned with the design of a system to meet a required performance specification as a function of trade-off in the transmitted power, antenna gain, and noise figure of the receiving systems. This performance specification is the maximum allowed probability of error for digital systems and the minimum allowed output SNR for analog systems.

8–2 TELEPHONE SYSTEMS

Modern telephone systems have evolved from the telegraph and telephone systems of the 1800s. Telephone companies that provide services for a large number of users over their *public switched telephone networks* (PSTN) on a for-hire basis are known as *common carriers.* The term is applied to widely diverse businesses, such as mail, airline, trucking, telephone, and data services. The common carriers are usually regulated by the government for the general welfare of the public, and, in some countries, certain common carrier services are provided by the government. The information from multiple users is transmitted over these systems, primarily by using time-division multiplex (TDM) transmission or packet data transmission (such as ATM, which is described in Appendix C).

Historically, telephone systems were designed only to reproduce voice signals that originated from a distant location. Today, modern telephone systems are very sophisticated. They use large digital computers at the *central office* (CO) to switch calls and to monitor the performance of the telephone system. The modern CO routes TDM PCM voice data, video data, and computer data to remote terminals and to other central offices.

Digital service is provided to the customer in any of three ways: (1) a *dedicated leased circuit,* such as a T1 circuit, that is available for use at all times without dialing or switching, (2) a *circuit-switched service* that is available on a dial-up basis, and (3) a *packet-switched service* that is "always on" and is used only when packets are exchanged.

Historical Basis

Modern telephone systems have evolved from the relatively simple analog circuit that was invented by Alexander Graham Bell in 1876. This circuit is shown in Fig. 8–1, where two telephone handsets are connected together by a twisted-pair (i.e., two-wire) telephone line and the telephone handsets are powered by a battery located at the CO. (Historically, the telephone-wire connection between the two parties was made by a telephone switchboard operator.) The battery produces a dc current around the telephone-wire loop. A carbon microphone element is used in each telephone handset. It consists of loosely packed carbon granules in a box that has one flexible side—the diaphragm. As sound pressure waves strike the diaphragm, the carbon granules are compressed and decompressed. This creates a variable resistance that causes the dc loop current to be modulated. Thus an ac audio current signal is produced, as shown in the figure. The handset earphone consists of an electromagnet with a paramagnetic diaphragm placed within the magnetic field. The ac current passing through the electromagnet causes the earphone diaphragm to vibrate and sound is reproduced.

The simple two-wire telephone system shown in Fig. 8–1 has three important advantages: (1) it is inexpensive; (2) the telephone sets are powered from the CO via the telephone

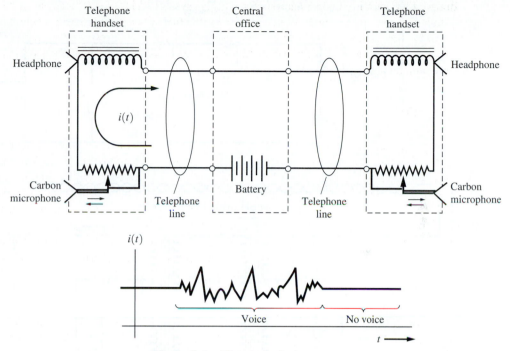

Figure 8–1   Historical telephone system.

line, so no power supply is required at the user's location; and (3) the circuit is full duplex.[†] The two-wire system has one main disadvantage: Amplifiers cannot be used, since they amplify the signal in only one direction. Consequently, for distant telephone connections, a more advanced technique—called a four-wire circuit—is required. In a four-wire circuit, one pair (or one optical fiber) is used for signals sent in the transmit direction, and another is used for arriving signals in the receive direction.

Modern Telephone Systems and Remote Terminals

A simplified diagram for an analog local-loop system that is used today is shown in Fig. 8–2. The local switching office connects the two parties by making a hardwire connection between the two appropriate local loops. This is essentially a series connection, with the earphone and carbon microphone of each telephone handset wired in a series with a battery (located at the telephone plant). Figure 8–2 illustrates this analog local-loop system, which is said to supply *plain old telephone service* (POTS). The wire with positive voltage from the CO is called the *tip* lead, and it is color coded *green*. The wire with negative voltage is called the *ring* lead and is color coded *red*. The terms *tip* and *ring* are derived from the era when a plug switchboard was used at the CO. These leads were connected to a plug with tip and ring contacts. This jack is similar to a stereo headphone jack that has tip, ring, and sleeve contacts. The Earth ground lead was connected to the sleeve contact.

[†] With full-duplex circuits, both parties can talk and listen simultaneously to the other person.

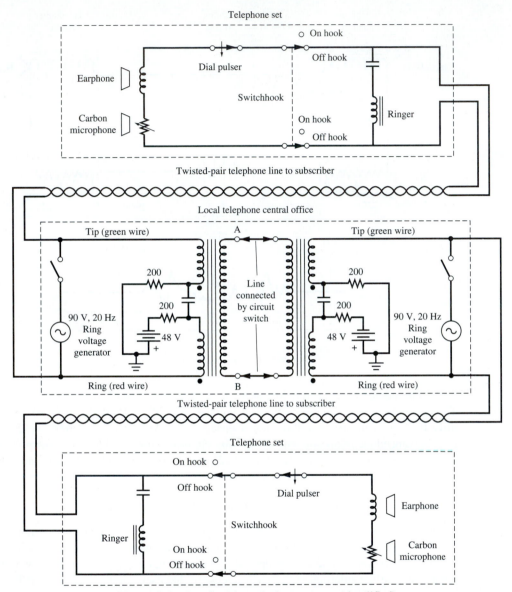

Figure 8–2 Local analog telephone system (simplified).

The sequence of events that occur when placing a local phone call will now be described with the aid of Fig. 8–2 and Table 8–1. The calling party—the upper telephone set in Fig. 8–2—removes the telephone handset; this action closes the switchhook contact (off-hook) so that dc current flows through the caller's telephone line. The current, about 40 mA, is sensed at the CO and causes the CO to place a dial-tone signal (approximately 400 Hz) on the calling party's line. The calling party dials the number by using either pulse or touchtone

TABLE 8–1   TELEPHONE STANDARDS FOR THE SUBSCRIBER LOOP

Item	Standard
On hook (idle status)	Line open circuit, minimum dc resistance 30 kΩ
Off hook (busy status)	Line closed circuit, maximum dc resistance 200 Ω
Battery voltage	48 V
Operating current	20–80 mA, 40 mA typical
Subscriber-loop resistance	0–1300Ω, 3600Ω (max)
Loop loss	8 dB (typical), 17 dB (max)
Ringing voltage	90 V rms, 20 Hz (typical) (usually pulsed on 2 s, off 4 s)
Ringer equivalence number (REN)[a]	0.2 REN (minimum), 5.0 REN (maximum)
Pulse dialing	Momentary open-circuit loop
Pulsing rate	10 pulses/sec ±10%
Duty cycle	58–64% break (open)
Time between digits	600 ms minimum
Pulse code	1 pulse = 1, 2 pulses = 2, …, 10 pulses = 0
Touch-tone[b] dialing	Uses two tones, a low-frequency tone and a high-frequency tone, to specify each digit:

Low Tone	High Tone (Hz)		
	1209	1336	1477
697 Hz	1	2	3
770	4	5	6
852	7	8	9
941	*	0	#

Item	Standard
Level each tone	−6 to −4 dBm
Maximum difference in levels	4 dB
Maximum level (pair)	+2 dBm
Frequency tolerance	±1.5%
Pulse width	50 ms
Time between digits	45 ms minimum
Dial tone	350 plus 440 Hz
Busy signal	480 plus 620 Hz, with 60 interruptions per minute
Ringing signal tone	440 plus 480 Hz, 2 s on, 4 s off
Caller ID[c]	1.2-kbit/s FSK signal between first and second rings (Bell 202 modem standard; see Table C–5)[d]

[a] Indicates the impedance loading caused by the telephone ringer. An REN of 1.0 equals about 8 kΩ, 0.2 REN equals 40 kΩ, and 5.0 REN equals 1.6 kΩ.

[b] Touch tone was a registered trademark of AT&T. It is also known as *dual-tone multiple-frequency* (DTMF) signaling.

[c] Other display services are also proposed [Schwartz, 1993].

[d] [Lancaster, 1991].

dialing. If pulse dialing is used, the dc line current is interrupted for the number of times equal to the digit dialed (at a rate of 10 pulses/s). For example, there are five interruptions of the line current when the number 5 is dialed. Upon reception of the complete number sequence for the called party, the CO places the ringing generator (90 V rms, 20 Hz, on 2 s, off 4 s) on the line corresponding to the number dialed. This rings the phone. When the called party answers, dc current flows in that line to signal the CO to disconnect the ringing generator and to connect the two parties together via the CO circuit switch. Direct current is now flowing in the lines of both the called and the calling party, and there is a connection between the two parties via transformer coupling,[†] as shown in Fig. 8–2. When either person speaks, the sound vibrations cause the resistance of the carbon microphone element to change in synchronization with the vibrations so that the dc line current is modulated (varied). This produces the ac audio signal that is coupled to the headphone of the other party's telephone. Note that both parties may speak and hear each other simultaneously. This is full-duplex operation.

In most modern telephone sets, the carbon microphone element is replaced with an electret or dynamic microphone element and an associated IC amplifier that is powered by the battery voltage from the CO, but the principle of operation is the same as previously described.

The telephone system depicted in Fig. 8–2 is satisfactory as long as the resistance of the twisted-pair loops is 1300 Ω or less. This limits the distance that telephones can be placed from this type of CO to about 15,000 ft for 26-gauge wire, 24,000 ft for 24-gauge, and 38,000 ft or about 7 miles, if 22-gauge twisted-pair wire is used. Historically, 19-gauge wire was used in rural areas so that phones may be located up to 20 miles away from the CO without remote terminals.

Supplying a dedicated wire pair from each subscriber all the way to the CO is expensive. In applications where a large number of subscribers are clustered some distance from the CO, costs can be substantially reduced by using *remote terminals* (RT). (See Fig. 8–3.) Remote terminals also allow telephones to be placed at any distance from the CO, as we will describe subsequently.

The circuit for a typical RT is illustrated in Fig. 8–4. The POTS line card supplies battery voltage and ringing current to the subscriber's telephone. The *two-wire circuit* that carries the duplex VF signals to and from the subscriber is converted into a *four-wire circuit* that carries two one-way (simplex) transmit and receive signals by the use of a *hybrid* circuit. The hybrid circuit is a balanced transformer circuit (or its equivalent electronic circuit) that provides isolation for the transmit and receive signals. As shown in Fig. 8–4b, the hybrid circuit acts as a balanced Wheatstone bridge where $Z_1/Z_3 = Z_2/Z_4$. Thus, the voltage on the receive line (shown at the bottom right of Fig. 8–4b) is balanced out and does not appear on the transmit line (upper right). Consequently, self-oscillation (ringing feedback) is prevented, even though there may be some coupling of the amplified transmitted signal to the receive line at the distant end of the four-wire line or coupling from the transmit line to the receive line along the four-wire path. As shown in Fig. 8–4a, the transmit VF signal is converted to a DS-0 PCM signal, which is time-division-multiplexed with the PCM signals from other subscribers attached to the RT. The TDM signal is sent over a DS-1 trunk

---

[†] The transformers are called *repeat coils* in telephone parlance.

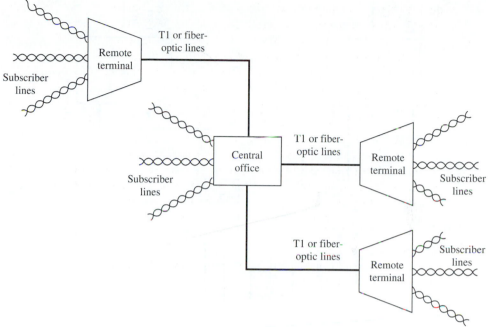

Figure 8–3   Telephone system with remote terminals.

to the CO. Similarly, the received DS-1 signal from the CO is demultiplexed and decoded to obtain the received VF audio for the subscriber. In a popular RT system manufactured by AT&T called SLC-96, 96 VF subscriber lines are digitized and multiplexed onto four T1 lines, and one additional T1 line is on standby in case one of the other T1 lines fails [Chapman, 1984]. From Chapter 3, it is recalled that a T1 line requires two twisted pairs (one for the transmit data and one for the receive data) and that each T1 line (1.544 Mb/s) carries the equivalent of 24 VF signals.

Let us now compare the twisted-pair requirements for systems with and without RTs. If no RT is used, 96 pairs will be required to the CO for 96 subscribers, but if a SLC-96 RT is used, only 10 pairs (5 T1 lines) are needed to the CO. This provides a pair savings (also called pair gain) of 9.6 to 1. Furthermore, the RT may be located at any distance from the CO (there is no 1300-Ω limit) because the pairs are used for DS-1 signaling with repeaters spaced about every mile. Of course, fiber-optic lines can also be used to connect the RT to the CO. For example, if two 560-Mb/s fiber-optic lines are used (one for data sent in each direction), this has a capacity of 8064 VF channels for DS-5 signaling. (See Table 3-8 and Fig. 3-40.) Thus, an RT with a 560-Mb/s fiber-optic link to the CO could serve 8064 subscribers. Furthermore, in Sec. 8–7, we see that no optical repeaters will be required if the RT is located within 35 miles (typically) of the CO.

Telephone companies have replaced their analog switches at the CO with digital switches. Historically, at the CO an analog circuit switch was used to connect the wire pairs of the calling and the called party (as shown in Fig. 8–2). These switches were controlled

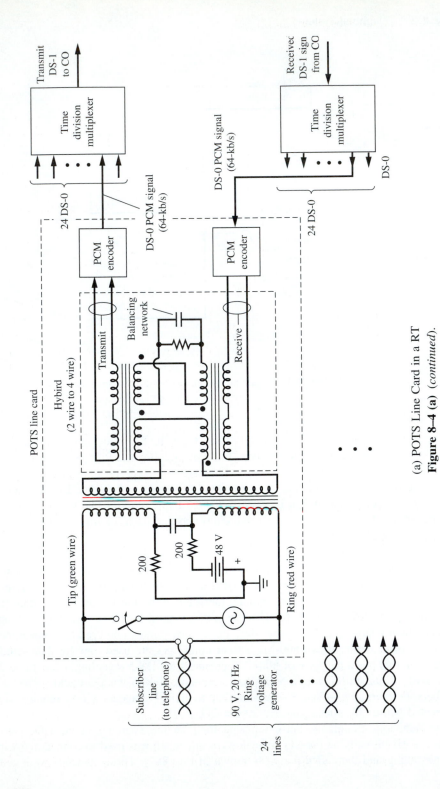

(a) POTS Line Card in a RT

Figure 8–4 (a) *(continued)*.

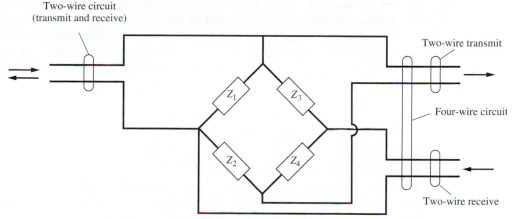

(b) Balanced Wheatstone Bridge Model for a Hybrid Circuit

Figure 8–4 (b) Remote terminal (RT).

by hardwired relay logic. Now, telephone offices use *electronic switching systems* (ESS). With ESS, a digital computer controls the switching operation as directed by software called *stored program control.* Moreover, modern ESS switches use *digital switching* instead of *analog switching.* Examples of these newer switches in use are AT&T's No. 5 ESS, and Northern Telcom's DMS-100 switches. In a digitally switched CO, the customer's VF signal is converted to PCM and is time-division multiplexed with other PCM signals onto a high-speed digital line. (When RTs are used, the conversion of the VF signal to PCM is carried out at the RT.) The digital CO switches a call by placing the PCM data of the calling party into the TDM time slot that has been assigned for the destination (called) party. This is called *time-slot interchange* (TSI). The digital switch is less expensive on a per-customer basis than the analog switch (for a large number of customers), and it allows switching of data and digitized video as well as PCM audio.

For toll calls, the local CO uses trunk lines that are connected to a distant CO. Multiple calls are accommodated using either TDM or packets. A local digital CO acts much like a RT attached to a distant CO. Interestingly, long-distance trunks account for less than 5% of the total cost of the telephone network; most of the cost comes from the switching equipment.

8–3 DIGITAL SUBSCRIBER LINES (DSL)

How can high-speed data and video service be economically provided via wire technology? High-speed data are easily transported from the CO to the RT via fiber-optic cables. The problem is transporting the data to the customer over the "last mile." It is usually not economical to install a dedicated fiber or coaxial line from the RT directly to each customer. However, in most cases, twisted-pair lines are already installed, and short lines of, say, up to 1,500 ft, have useful bandwidths as large as 30 MHz. In conventional applications, the POTS line cards in the RT restrict the bandwidth to the VF range of 3.3 kHz. This limits the data rate to around 50 kb/s for a SNR of 50 dB (V.90 modem). But, even worse, the line may have a lower SNR, so that the VF modem typically falls back to an operating speed of about 24 kb/s in practice (as demonstrated by SA1–3). However, *digital subscriber line* (DSL) cards can be used in the RT to take advantage of the wide bandwidth of short "last mile" twisted pair lines and provide data rates on the order of 6 Mb/s. There are many DSL methods, so we will denote this family of methods by the notation xDSL. All use one or two twisted pairs to serve the "last mile" customer from an RT.

1. *HDSL (high-bit-rate digital subscriber line)* uses two twisted pairs (one transmit and one receive) to support 1.544 Mb/s at full duplex for a distance of up to 12,000 ft from the RT. It uses a 2B1Q line code of ISDN (described below) or a suppressed carrier version of QAM called *carrierless amplitude-phase* (CAP).

2. *SDSL (symmetrical digital subscriber line)* is a one-pair version of HDSL. It provides full duplex to support 768 kb/s in each direction using a hybrid or echo canceller to separate data transmitted from data received.

3. *ADSL (asymmetrical digital subscriber line)* uses one twisted pair to support 6 Mb/s sent downstream to the customer and 640 kb/s sent upstream over a distance of up to 12,000 ft. The ASDL spectrum is above 25 kHz. The band below 4 kHz is used for a VF POTS signal. Two variations of ADSL—called G.DMT and G.Lite—are discussed in the next section.

4. *VDSL (very-high-bit-rate digital subscriber line)* uses one pair of wires to support 25 Mb/s downstream for distances of up to 3,000 ft from the RT or 51 Mb/s downstream for distances of up to 1,000 ft from the RT. Up to 3.2 Mb/s can be supported upstream.

5. *ISDN (integrated service digital network)* uses one twisted-pair line to provide a subscriber data rate of up to 144 kb/s in each direction at distances of up to 18,000 ft from the RT. This technology has been available since 1990 and has been popular in Europe and Japan, but not too popular in the United States. More details about ISDN are discussed in a later section.

G.DMT and G.Lite Digital Subscriber Lines

G.DMT and G.Lite use "always on" (packet) ADSL technology with a spectrum above the 4 kHz voice band. This allows a VF POTS signal to be supported along with the data signal on a single twisted-pair line.

As shown in Fig. 8–5a, G.DMT uses a splitter so that the telephone handsets will not "short out" the data signal and vice versa. G.DMT uses discrete multitone modulation (DMT) with up to 256 carriers and up to 15 bits of data, 32,768 QAM, modulated on each carrier.[†] This provides a maximum data rate of 6.1 Mb/s downstream and 640 kb/s upstream for line length up to 12,000 ft. As shown in Fig. 8–5b, the DMT carrier spacing is 4.3125 kHz. The upstream data are on carriers in the 26–138 kHz band and the downstream carriers are placed in the 138–1,100 kHz band. In an alternative implementation, both the upstream and the downstream carriers are placed in the 26–1,100 kHz band, and a hybrid (or

---

[†] DMT is Orthogonal Frequency Division Multiplexing (OFDM) and is described in Sec. 5–12.

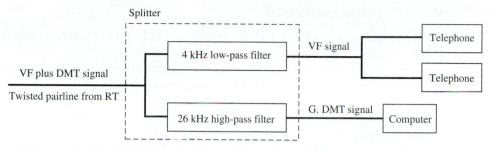

(a) Customer Premises Equipment (CPE)

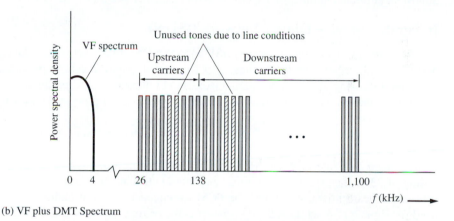

(b) VF plus DMT Spectrum

Figure 8–5   G.DMT digital subscriber line.

echo canceller) is used to separate the upstream and downstream carriers. DMT has the advantage of allowing the modulation to be customized to accommodate the transmission and noise characteristics of a particular line. That is, carriers are deleted at frequencies where there are nulls in the frequency response of the line caused by reflections and at frequencies where the noise is large. This allows a BER of 10^{-7} to be achieved at a data rate that the line can support.

G.Lite uses no splitter. (The telephone and data wires are wired directly together in parallel.) This has the advantage of "no truck roll" (i.e., no installation of a splitter) for the TELCO, but has the disadvantage of old telephone sets loading down (shorting out) the data signal and transmission nulls in the spectrum caused by reflections from telephone line bridge taps (i.e., extension phones). Consequently, G.Lite may not work for some customers unless offending telephones are replaced or a splitter is installed. G.Lite uses fewer carriers than G.DMT to avoid the spectral impairments at the higher frequencies. G.Lite has a data rate of 1.5 Mb/s downstream and 512 kb/s upstream over an 18,000-ft line. It uses 128 DMT carriers with up to 8 bits, 256 QAM, modulated on each carrier.

G.DMT and G.Lite may be the answer to the problem of how to provide economical high-speed internet access. G.DMT is described in ANSI standard T.413 and ITU document G.992.1. G.Lite is described in ITU document G.992.2.

Video On Demand (VOD)

VDSL technology will allow telephone companies (TELCO) to provide a cable-TV type of program service to the consumer using twisted pair lines. An HDTV (high-definition TV) signal requires a data rate of 20 Mb/s for compressed video, and SDTV (standard-definition TV) can be delivered with a 6 Mb/s data rate. (See Sec. 8–9 on digital TV for more details.) With VOD, the subscriber selects the desired program using a set-top box (STB). The TELCO then sends the data for the selected program to the user via VDSL, and the STB converts the data to the video signal that is viewed on the TV set. First-run movies and other pay-per-view programs could also be provided. The VOD subscriber can have access to almost an unlimited number of TV channels (one channel at a time) and other video services.

Integrated Service Digital Network (ISDN)

The *integrated service digital network* (ISDN) uses a DSL to deliver digital data (no VF signal) to the subscriber. The ISDN subscriber can demultiplex the data to provide any one or all of the following applications simultaneously: (1) decode data to produce VF signals for telephone handsets, (2) decode data for a video display, or (3) process data for telemetry or PC applications.

There are two categories of ISDN: (1) *narrowband* or *"basic rate,"* ISDN, denoted by N-ISDN, and (2) *broadband,* or *"primary rate,"* ISDN, denoted by B-ISDN. B-ISDN has an aggregate data rate of 1.536 Mb/s (approximately the same rate as for T1 lines), consisting of 23 B channels (64 kb/s each) and one D channel (64 kb/s). The B channels carry user (or bearer) data, which may be PCM for encoded video or audio. The D channel is used for signaling to set up calls, disconnect calls, or route data for the 23 B channels.

The standard implementation of N-ISDN is shown in Fig. 8–6. The N-ISDN subscriber is connected to the RT of the telephone company by a twisted-pair telephone line. For N-ISDN service, the line must be no longer than 18 kft (3.4 miles) for the 160-kb/s N-ISDN aggregate data rate. (If the subscriber is located within 18 kft of the CO, an RT is not necessary.) The data rate available to the N-ISDN subscriber is 144 kb/s, consisting of data partitioned into two B channels of 64 kb/s each and one D channel of 16 kb/s. In addition to the 2B + D data, the telephone company adds 12 kb/s for framing/timing plus 4 kb/s for overhead to support network operations. This gives an overall data rate of 160 kb/s on the DSL in both the transmit and receive directions (simultaneously) for full-duplex operation. The DSL is terminated on the subscriber's end at the U interface as shown on Fig. 8–6. The NT1 (network termination) incorporates a hybrid circuit to convert the two-wire U interface to a four-wire T interface that has two two-wire buses, one for the transmit and one for the receive data signals. The NT1 is a slave transceiver that derives its clocking signals from the DSL signal that is transmitted from the RT. The NT1 passes the 2B + D data between the U and T interfaces and also processes additional bits that appear on the transmit and receive lines at the T interface. The additional bits are needed for addressing, control, and supervision of terminal equipment. The data rate on each transmit and receive T line is 192 kb/s. The NT2 may or may not be needed. When used, it provides the appropriate higher level protocols for a terminal cluster controller or for a local area network (LAN) access node. If not used, the NT1 four-wire interface becomes the S/T interface. The S and T interfaces are electrically identical and use an RJ45 8–pin modular connector. The S/T interface bus provides

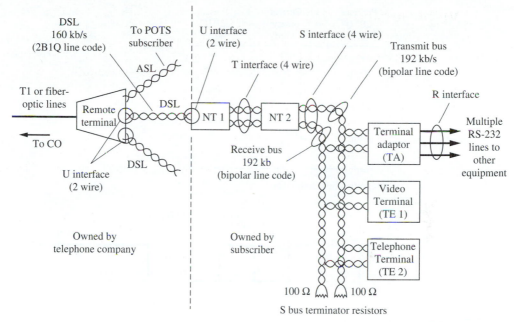

Figure 8–6   N-ISDN system with subscriber equipment attached.

the connection medium for ISDN terminals (video, audio, and data) or a terminal adapter (TA). Up to eight terminals may be bridged onto the S/T bus. The TA allows non-ISDN terminals with RS-232 ports to be used on the ISDN network. For example, a personal computer (PC) can be connected to the N-ISDN network via its serial port and the R interface of the TA. Also, using an alternative approach, a PC adapter card with a built-in NTI and a telephone codec can be used. That is, the DSL U interface modular connector is plugged directly into this adapter card on the PC. The PC then acts as an ISDN video terminal with a built-in telephone. For details on the protocols used at the U, S, and T interfaces, the reader is referred to application notes for ISDN chips such as Motorola's MC 145472 U interface transceiver, MC 145474 S/T interface, and MC 145554 telephone codec.

N-ISDN service to the subscriber via two-wire twisted-pair DSL up to 18 kft from an RT is made possible by the use of multilevel signaling. Referring to Fig. 5–33 with $R = 160$ kb/s, a four-level (i.e., $\ell = 2$ bits) 80-kbaud line code has a null bandwidth of only 80 kHz instead of the 160-kHz null bandwidth for a binary ($\ell = 1$) line code. The 80-kHz bandwidth is supported by 26-gauge twisted-pair cable if it is less than about 18 kft in length. The particular four-level signal used is the 2B1Q line code (for two binary digits encoded into one quadrenary symbol) shown in Fig. 8–7. Note that the 2B1Q line code is a differential symbol code. Thus, if (due to wiring error) the twisted pair is "turned over" so that the tip wire is connected to the ring terminal and the ring to the tip so that the 2BIQ signal polarity is inverted, the decoded binary data will still have the correct polarity (i.e., not complemented).

The primary advantage of ISDN is obvious: It provides TDM data channels directly to the subscriber. It also has disadvantages. For example, since the ISDN subscriber

Previous 2B1Q level	Current binary word	Present 2B1Q level
+1 or +3	00	+1
	01	+3
	10	−1
	11	−3
−1 or −3	00	−1
	01	−3
	10	+1
	11	+3

(a) 2B1Q differential code table

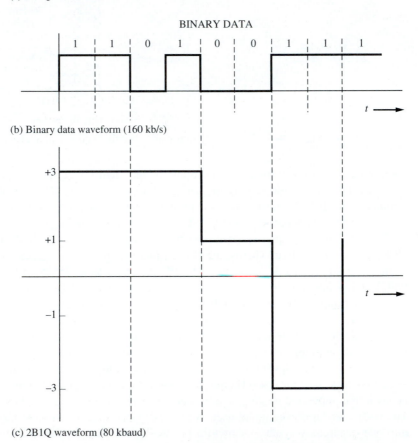

BINARY DATA

(b) Binary data waveform (160 kb/s)

(c) 2B1Q waveform (80 kbaud)

Figure 8–7   2B1Q Line code.

equipment is not powered by the DSL, it will not function when local ac power fails unless a backup power supply is provided by the subscriber. If only POTS is required, the conventional analog telephone is usually much cheaper and more reliable.

8-4 CAPACITIES OF PUBLIC SWITCHED TELEPHONE NETWORKS

The wideband channels used to connect the toll offices consist of one predominant type: fiber-optic cable. Table 8-2 lists some of the wideband systems that are used, or have been used in the past, and indicates the capacity of these systems in terms of the number of VF channels they can handle and their bit rate.

Historically, open-wire pairs, which consist of individual bare wires supported by glass insulators on the cross arms of telephone poles, provided wideband service via FDM/SSB signaling. Occasionally, some open-wire lines can still be seen along railroad tracks.

Fiber-optic cable with OOK signaling is rapidly overtaking twisted-pair cable, coaxial cable, and microwave relay because of its tremendous capacity and relatively low cost. As shown in Table 8-2, data rates on fiber-optic links continue to increase as technology advances. Typically, a 2.5-Gb/s data rate is employed on an OOK optical carrier, as is the case with the FT-2000 system. This is equivalent to 32,000 VF telephone circuits. (See Sec. 8-7 for a description of the FT-2000 system.) Data rates of 10 Gb/s on each carrier are also possible. For even higher capacity, multiple carriers at different wavelengths are used on a single fiber. *Dense wavelength division multiplexing* (DWDM) systems include as many as 40 optical carriers on one fiber. For example, the WaveStar DWDM system achieves a capacity of 400 Gb/s, or 6.5 million VF circuits, by using 40 carriers. However, fiber-optic cable provides service only from one fixed point to another. Conversely, communication satellites provide wideband connections to any point on the globe. Service to isolated locations can be provided almost instantaneously by the use of portable ground stations. This is described in more detail in the next section.

8-5 SATELLITE COMMUNICATION SYSTEMS

The number of satellite communication systems has increased tremendously over the last few years. Satellites have made transoceanic relaying of television signals possible. Satellite communications provide the relaying of data, telephone, and television signals and now provide national direct-into-the-home television transmission via satellite.

Most communication satellites are placed in *geostationary orbit* (GEO). This is a circular orbit in Earth's equatorial plane. The orbit is located 22,300 miles above the equator so that the orbital period is the same as that of the Earth. Consequently, from Earth these satellites appear to be located at a stationary point in the sky, as shown in Fig. 8-8. This enables the Earth station antennas to be simplified, since they are pointed in a fixed direction and do not have to track a moving object. (For communication to the polar regions of the Earth, satellites in polar orbits are used, which require Earth stations with tracking antennas.) To prevent the satellite from tumbling, one of two spin stabilization techniques—spin stabilization or three-axis stabilization—is used. For spin stabilization, the outside cylinder of the satellite is spun to create a gyroscopic effect that stabilizes the satellite. For three-axis stabilization, internal gyroscopes are used to sense satellite movement, and the satellite is stabilized by firing appropriate thruster jets.

TABLE 8-2 CAPACITY OF PUBLIC-SWITCHED TELEPHONE NETWORKS

Transmission Medium	Name	Developer	Year in Service	Number of Voice-Frequency Channels	Bit Rate (Mb/s)	Repeater Spacing (miles)	Operating Frequency (MHz)	Modulation[a] D/A	Method
Open-wire pair	A	Bell	1918	4			0.005–0.025	A	FDM/SSB
	C	Bell	1924	3		150	0.005–0.030	A	FDM/SSB
		CCITT		12		50	0.036–0.140	A	FDM/SSB
	J	Bell	1938	12		50	0.036–0.143	A	FDM/SSB
		CCITT		28			0.003–0.300	A	FDM/SSB
Twisted-pair cable	K	Bell	1938	12		17	0.012–0.060	A	FDM/SSB
		CCITT		12		19	0.012–0.060	A	FDM/SSB
	N1	Bell	1950	12			0.044–0.260	A	FDM/SSB
	N3	Bell	1964	24			0.172–0.268	A	FDM/SSB
	T1[b]	Bell	1962	24	1.544 (DS-1)	1		D	Bipolar
	T1G[b]	AT&T	1985	96	6.312 (DS-2)			D	Four-level
	T2[b]	Bell		96	6.312 (DS-2)	2.3		D	B6ZS
Coaxial cable	L1	Bell	1941	600		8	0.006–2.79	A	FDM/SSB
	L3	Bell	1953	1,860		4	0.312–8.28	A	FDM/SSB
	L4	Bell	1967	3,600		2	0.564–17.55	A	FDM/SSB
	L5	Bell	1974	10,800		1	3.12–60.5	A	FDM/SSB
	T4[b]	Bell		4,032	274.176 (DS-4)	1		D	Polar
	T5[b]	Bell		8,064	560.16 (DS-5)	1		D	Polar
Fiber-optic cable	FT3	Bell	1981	672	44.763 (DS-3)	4.4	0.82 μm	D	TDM/OOK
		Brit. Telcom	1984		140	6	1.3 μm	D	TDM/OOK
		Sask. Telcom	1985		45	6–18	0.84 and 1.3 μm	D	TDM/OOK
	F-400M	Nippon	1985		400	12	1.3 μm	D	TDM/OOK
	FT3C-90	AT&T	1985	1,344	90.254 (DS-3C)	16	1.3 μm	D	TDM/OOK
	FT4E-432	AT&T	1986	6,048	432 (DS-432)	16	1.3 μm	D	TDM/OOK
	LaserNet	Microtel	1985	6,048	417.79 (9DS-3)	25	1.3 μm	D	TDM/OOK
	FTG1.7	AT&T	1987	24,192	1,668 (36DS-3)	29	1.3 μm	D	TDM/OOK
	FT-2000	AT&T	1995	32,256	2,488 (OC-48)	100		D	TDM/OOK
	WaveStar	Lucent	1999	6.25×10^6	400,000	25	1.5 and 1.3 μm	D	DWDM[d]

562

Transoceanic	TAT-1 (SB)	Bell	1956	48		20	0.024–0.168	A	FDM/SSB
	TAT-3 (SD)	Bell	1963	138		11	0.108–1.05	A	FDM/SSB
	TAT-5 (SF)	Bell	1970	845		6	0.564–5.88	A	FDM/SSB
	TAT-6 (SG)	Bell	1976	4,200		3	0.5–30	A	FDM/SSB
	TAT-8 (3 fibers)		1988	8,000	280		1.3 μm	D	TDM/OOK
	TAT-9 (3 fibers)		1991	16,000	565		1.55 μm	D	TDM/OOK
	TAT-10 (6 fibers)		1992	80,000[e]	565		1.55 μm	D	TDM/OOK
	TAT-12 (6 fibers)	Alcatel	1995	200,000	5,000	30	1.48 μm	D	TDM/OOK
	OALW160	Alcatel	2000		160,000	35	1.55 μm	D	DWDM
	TAT-14 (4 fibers)		2001		640,000		1.55 μm	D	DWDM
Microwave relay	TD-2	Bell	1948	600 (1954)	plus 1.544[c]	30	3700–4200	A	FDM/FM
	TH-1	Bell	1961	2,400 (1979)	plus 1.544[c]	30	5925–6245	A	FDM/FM
	TD-3	Bell	1967	1,800 (1979)	plus 1.544[c]	30	3700–4200	A	FDM/FM
	TN-1	Bell	1974	1,800	plus 1.544[c]	30	K band, 18 GHz	A	FDM/FM
	AR6A	Bell	1980	6,000		30	5925–6425	A	FDM/SSB
	18G274	NEC	1974	4,032	274.176 (DS-4)	7	18 GHz	D	4 PSK
	6G90	NEC	1979	1,344	90 (2DS-3)	30	6 GHz	D	16 QAM
	11G135	NEC	1980	2,016	135 (3 DS-3)	30	11 GHz	D	16 QAM
	6G135	NEC	1983	2,016	135 (3 DS-3)	30	6 GHz	D	64 QAM
	MDR-2306	Collins	1983	2,016	135 (3 DS-3)	30	6 GHz	D	64 QAM
	RD-6A	Nortel	1984	2,016	135 (3 DS-3)	30	6 GHz	D	64 QAM
	TN-X/40	Nortel	1996	4,032	310 (2 STS-3)	30	6 GHz	D	512 QAM
Communication satellite	Intelsat IV	COMSAT	1970	8,000		22,300	6 GHz up/4 GHz down	A/D	FDM/FM, QPSK/SCPC
	Intelsat V	COMSAT	1980	25,000		22,300	6/4 and 14/11 GHz	A/D	FDM/FM, QPSK/SCPC
	Intelsat VI	COMSAT	1986	80,000		22,300	6/4 and 14/11 GHz	A/D	FDM/FM, QPSK/SCPC
	Intelsat VIII	COMSAT	1998	112,500		22,300	6/4 and 14/11 GHz	A/D	FDM/FM, QPSK/SCPC

[a] A-analog; D-digital; DWDM-dense wavelength division multiplexing.

[b] See Table 3–9 for more details on the T-carrier system.

[c] Since 1974 data under voice were added to give a 1.544-Mb/s (DS-1) data channel in addition to the stated VF FDM capacity.

[d] 40 wavelength channels at 10 Gb/s each.

[e] VF capacity with statistical multiplexing.

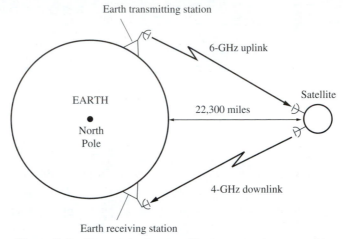

Figure 8–8 Communications satellite in geosynchronous orbit.

The most desired frequency band for satellite communication systems is 6 GHz on the uplink (Earth-to-satellite transmission) and 4 GHz on the downlink (satellite-to-Earth transmission). In this frequency range, the equipment is relatively inexpensive, the cosmic noise is small, and the frequency is low enough that rainfall does not appreciably attenuate the signals. Other losses, such as ionospheric scintillation and atmospheric absorption, are small at these frequencies [Spilker, 1977]. (Absorption occurs in specific frequency bands and is caused by the signal exciting atmospheric gases and water vapor.) However, existing terrestrial microwave radio relay links are already assigned to operate within the 6- and 4-GHz bands (see Table 8–2), so the FCC limits the power density on Earth from the satellite transmitters. One also has to be careful in locating the Earth station satellite-receiving antennas so that they do not receive interfering signals from the terrestrial microwave radio links that are using the same frequency assignment. In the 6/4-GHz band, synchronous satellites are assigned an orbital spacing of 2° (U.S. standard).

Newer satellites operate in higher frequency bands, because there are so few vacant spectral assignments in the 6/4-GHz band (C band). The Ku band satellites use 14 GHz on the uplink and 12 GHz on the downlink, with an orbital spacing of 3°. Some new Ku-band satellites have high-power amplifiers that feed 120 to 240 W into their transmitting antenna, as compared with 20 to 40 W for low- or medium-power satellites. High-power satellites—called direct-broadcast satellites (DBS)—provide TV service directly to the homeowner, who has a small receiving antenna (2 ft or less in diameter).

Each satellite has a number of *transponders* (receiver-to-transmitter) aboard to amplify the received signal from the uplink and to down-convert the signal for transmission on the downlink. (See Fig. 8–9.) Figure 8–9 shows a "bent-pipe transponder" that does not demodulate the received signal and perform signal processing but acts as a high-power-gain down converter. Most transponders are designed for a bandwidth of 36, 54, or 72 MHz, with 36 MHz being the standard used for C-band (6/4-GHz) television relay service. As technology permits, processing transponders will come into use since an improvement in error performance (for digital signaling) can be realized.

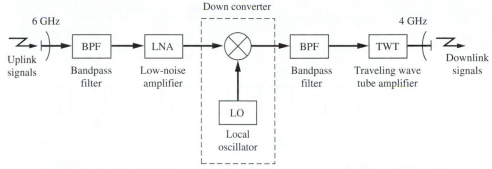

Figure 8–9   Simplified block diagram of a communications satellite transponder.

Each satellite is assigned a synchronous orbit position and a frequency band in which to operate. In the 6/4-GHz band, each satellite is permitted to use a 500-MHz-wide spectral assignment, and a typical satellite has 24 transponders aboard, with each transponder using 36 MHz of the 500-MHz bandwidth assignment. The satellites reuse the same frequency band by having 12 transponders operating with vertically polarized radiated signals and 12 transponders with horizontally polarized signals.[†] A typical 6/4-GHz frequency assignment for satellites is shown in Fig. 8–10. The transponders are denoted by C1 for channel 1, C2 for channel 2, and so on. These satellites are used mainly to relay signals for CATV systems.

Digital and Analog Television Transmission

TV may be relayed via satellite using either digital or analog transmission techniques.

For digital transmission, the baseband video signal is sampled and digitized. The data are usually compressed to conserve the bandwidth of the modulated satellite signal. The data are compressed by removing redundant video samples within each frame of the picture and removing redundant samples that occur frame to frame. For example, the Hughes Digital Satellite System (DSS)[‡] provides more than 200 channels directly to the home subscriber in the United States using two Hughes HS601 satellites. These geostationary satellites are located above the equator at 101° west longitude. Each DSS satellite contains 16 high-power (120 W) transponders operating in the Ku-band (12.2–12.7 GHz). The bandwidth of each transponder is 24 MHz; the effective radiated power of each transponder emitted from the satellite antenna is 48 to 53 dBw directed over the continental United States and southern Canada. Thus, a subscriber can receive the satellite signal using a relatively small receiving antenna consisting of an 18-in parabolic dish. The baseband video for each TV channel is digitized and compressed using the Motion Pictures Experts Group (MPEG) standard [Pancha and Zarki, 1994]. This compression gives an average video data rate of 3 to 6 Mb/s for each channel, depending on the amount of motion in the video. Data for about six TV channels (video plus multichannel audio) are TDM for transmission over each satellite transponder using QPSK modulation [Thomson, 1994]. Furthermore, adaptive data compression is

---

[†] A vertically polarized satellite signal has an E field that is oriented vertically (parallel with the axis of rotation of the Earth). In horizontal polarization, the E field is horizontal.

[‡] DSS is a trademark of DirecTV, Inc., a unit of GM Hughes Electronics, El Segundo, CA.

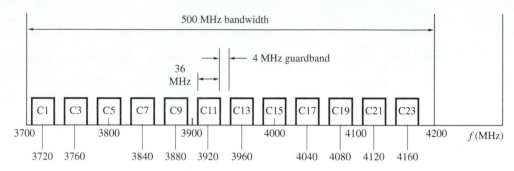

(a) Horizontal Polarization[a]

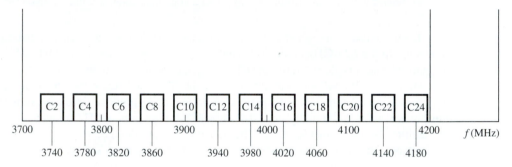

(b) Vertical Polarization[a]

[a]These are the polarizations used for the Galaxy Satellites. Some of the other satellites use opposite polarization assignments.

Figure 8–10   6/4-GHz satellite transponder frequency plan for the downlink channels. (For the uplink frequency plan, add 2225 MHz to the numbers given above.)

used to minimize the data rate of the TDM signal because the data for some of the TV channels in the TDM data stream may be encoded at a lower data rate than others, depending on the amount of motion (and other properties) of each video source. More details about the DSS system are given in Study-Aid Examples SA8–1 and SA8–2. Digital encoding for digital TV (DTV) is discussed in Sec. 8–9.

For analog TV transmission via satellite, the baseband video for a single TV channel is frequency modulated onto a carrier. For example, to relay TV signals to the head end of CATV systems, C-band satellites with 24 transponders are often used (as shown in Fig. 8–10). For each transponder, the 4.5-MHz bandwidth baseband composite video signal of a single TV channel is frequency modulated onto a 6-GHz carrier, as shown in Fig. 8–11. The composite visual signal consists of the black-and-white video signal, the color subcarrier signals, and the synchronizing pulse signal, as discussed in Sec. 8–9. The aural signal is also relayed over the same transponder by frequency modulating it onto a 6.8-MHz subcarrier that is frequency-division-multiplexed with the composite video signal. The resulting wideband signal frequency modulates the transmitter.

The bandwidth of the 6-GHz FM signal may be evaluated by using Carson's rule. The peak deviation of the composite video is 10.5 MHz, and the peak deviation of the subcarrier

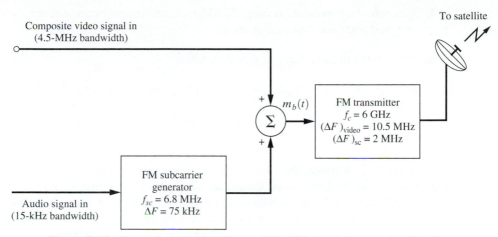

Figure 8-11   Transmission of broadcast-quality TV signals from a ground station.

is 2 MHz, giving an overall peak deviation of $\Delta F = 12.5$ MHz. The baseband bandwidth is approximately 6.8 MHz. The transmission bandwidth is

$$B_T = 2(\Delta F + B) = 2(12.5 + 6.8) = 38.6 \text{ MHz} \tag{8-1}$$

which is accepted by the 36-MHz transponder. In addition, other wideband aural signals (0 to 15 kHz) can also be relayed by using FM subcarriers. Some typical subcarrier frequencies that are used are 5.58, 5.76, 6.2, 6.3, 6.48, 6.8, 7.38, and 7.56 MHz.

Data and Telephone Signal Multiple Access

Satellite relays provide a channel for data and VF (telephone) signaling similar to conventional terrestrial microwave radio links. That is, data may be time-division multiplexed into DS-1, DS-2, and so on, types of signals and (digitally) modulated onto a carrier for transmission via a satellite. The data may also be frequency-division multiplexed. In this case, modems are used to convert the data into signals that are compatible with VF group, supergroup, or mastergroup bandwidths, which are (analog) modulated onto the carrier for transmission to the satellite.

Satellite communication systems do differ from terrestrial microwave links in the techniques used for *multiple access* of a single transponder by multiple uplink and multiple downlink stations. Four main methods are used for multiple access:

1. *Frequency-division multiple access* (FDMA), which is similar to FDM.
2. *Time-division multiple access* (TDMA), which is similar to TDM.[†]
3. *Code-division multiple access* (CDMA) or *spread-spectrum multiple access* (SSMA).

[†] *Satellite-switched time-division multiple access* (SS-TDMA) can also be used. With SS-TDMA satellites, different narrow-beam antennas are switched in at the appropriate time in the TDMA frame period to direct the transmit and receive beams to the desired direction. Thus, with multiple SS-TDMA beams, the same transmit and receive frequencies may be reused simultaneously, so that the capacity of the SS-TDMA satellite is much larger than that of a TDMA satellite. The *Intelsat* VIII is an SS-TDMA satellite.

4. *Space-division multiple access* (SDMA) where narrow-beam antenna patterns are switched from one direction to another.

In addition, either of the following may be used:

1. A *fixed-assigned multiple-access* (FAMA) mode using either FDMA, TDMA, or CDMA techniques.
2. A *demand-assigned multiple-access* (DAMA) mode using either FDMA, TDMA, or CDMA.

In the FAMA mode, the FDMA, TDMA, or CDMA format does not change, even though the traffic load of various Earth stations changes. For example, there is more telephone traffic between Earth stations during daylight hours (local time) than between these stations in the hours after midnight. In the FAMA mode, a large number of satellite channels would be idle during the early morning hours, since they are fixed assigned. In the DAMA mode, the FDMA and TDMA formats are changed as needed, depending on the traffic demand of the Earth stations involved. Consequently, the DAMA mode uses the satellite capacity more efficiently, but it usually costs more to implement and maintain.

In CDMA the different users share the same frequency band simultaneously in time, as opposed to FDMA and TDMA, where users are assigned different frequency slots or time slots. With CDMA, each user is assigned a particular digitally encoded waveform $\varphi_j(t)$ that is (nearly) orthogonal to the waveforms used by the others. (See Secs. 2–4 and 5–13.) Data may be modulated onto this waveform, transmitted over the communication system, and recovered. For example, if one bit of data, m_j, is modulated onto the waveform, the transmitted signal from the jth user might be $m_j \varphi_j(t)$, and the composite CDMA signal from all users would be $w(t) = \sum_j m_j \varphi_j(t)$. The data from the jth user could be recovered from the CDMA waveform by evaluating $\int_0^t w(t)\varphi_j(t)\, dt = m_j$, where T is the length of encoded waveform $\varphi_j(t)$. Gold codes are often used to obtain the encoded waveforms.

Example 8–1 FIXED ASSIGNED MULTIPLE-ACCESS MODE USING AN FDMA FORMAT

A good example of the fixed-assigned multiple-access technique is the FDM/FM format used with the *Intelsat* series satellites. Suppose that ground station A is sharing one 36-MHz transponder in the FDM mode with stations B, C, D, E, F, and G, which are all transmitting simultaneously through the transponder, as shown by their frequency assignments in Fig. 8–12a. Station A has an assigned band of 6237.5 to 6242.5 MHz in which it can transmit 60 VF channels using FDM/FM. This is the capacity of one supergroup. Suppose that station A has traffic as follows: 24 VF channels for station B, 24 VF channels for station D, and 12 VF channels for station E; then the equipment configuration would be as shown in Fig. 8–12b. In a similar way, the remaining spectrum of the 36-MHz transponder is divided among the other Earth stations according to their traffic needs on a fixed assignment basis.

Example 8–2 SPADE SYSTEM

The *Intelsat* series satellites may also be operated in a DAMA mode using an FDMA format consisting of a single QPSK carrier for each telephone (VF) channel. This type of signaling is

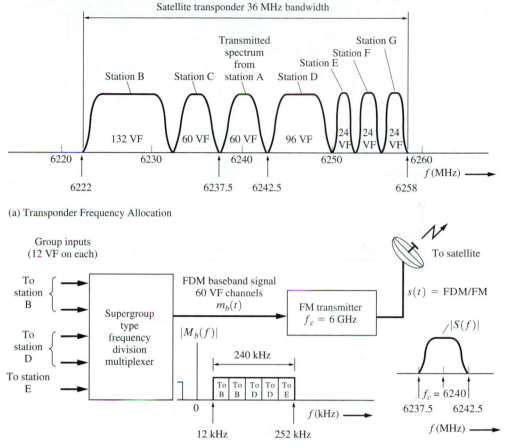

(a) Transponder Frequency Allocation

(b) Station A Ground Transmitting Equipment

Figure 8–12   Fixed-assigned FDMA format for satellite communications.

called *single channel per carrier* (SCPC), in which 800 QPSK signals may be accommodated in the 36-MHz bandwidth of the transponder, as shown in Fig. 8–13a. Thus 800 VF messages may be transmitted simultaneously through one satellite transponder, where each QPSK signal is modulated by a 64-kb/s PCM (digitized) voice signal (studied earlier in Example 3–1 and used as inputs to the North American digital hierarchy shown in Fig. 3–40). This SCPC-DAMA technique, illustrated in Fig. 8–13, is called the *SPADE system*, which is an acronym for single channel per carrier, pulse code modulation, multiple access, demand assignment equipment [Edelson and Werth, 1972]!

The demand assignment of the carrier frequency for the QPSK signal to a particular Earth station uplink is carried out by TDM signaling in the *common signaling channel* (CSC) (see Fig. 8–13a). The CSC consists of a 128-kb/s PSK signal that is time shared among the Earth stations by using a TDMA format, as shown in Fig. 8–13c. PA denotes the synchronizing preamble that occurs at the beginning of each frame, and A, B, C, and so on denote the 1-msec time slots that are reserved for transmission by Earth stations A, B, C, and so on. In this way 49 different Earth stations may be accommodated in the 50-msec frame. For example, if Earth station

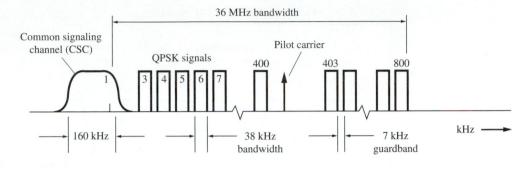

(a) Transponder Frequency Allocation

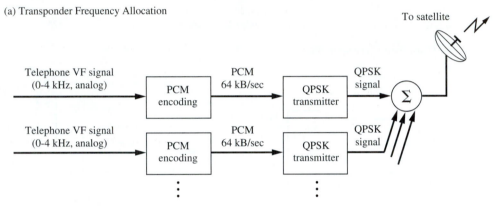

(b) Possible QPSK SCPC Transmitter Configuration

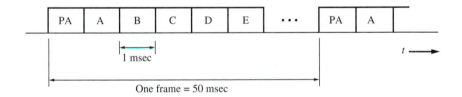

(c) TDMA CSC Signaling Format

Figure 8–13   SPADE satellite communication system for telephone VF message transmission.

B wishes to initiate a call to Earth station D, station B selects a QPSK carrier frequency randomly from among the idle channels that are available and transmits this frequency information along the address of station D (the call destination) in the station B-TDMA time slot. If it is assumed that the frequency has not been selected by another station for another call station D will acknowledge the request in its TDMA time slot. This acknowledgement would be heard by station B about 600 msec after its TDMA signaling request since the round trip time delay to the satellite is 240 ms, plus equipment delays and the delay to the exact time slot assigned to station D with respect to that of station B. If another station, say station C, had selected the same frequency during the request period, a busy signal would be received at station B, and station B would randomly select another available frequency and try again. When the call is over, disconnect signals are transmitted in the TDMA time slot and that carrier frequency is returned for reuse. Because the CSC signaling rate is 128 kb/s and each time slot is 1 ms in duration, 128 bits are available for each accessing station to use for transmitting address information, frequency request information, and disconnect signaling.

In practice, because only 49 time slots are available for the TDMA mode, a number of the SCPC frequencies are assigned on a fixed basis.

In FDMA, such as the SPADE system, when the carriers are turned on and off on demand, there is amplitude modulation on the composite 36-MHz-wide signal that is being amplified by the transponder TWT. Consequently, the drive level of the TWT has to be "backed off" so that the amplifier is not saturated and will be sufficiently linear. Then, the intermodulation products will be sufficiently low. On the other hand, if a single constant-amplitude signal (such as a single-wideband FM signal used in relaying television signals) had been used, IM products would not be a consideration, and the amplifier could be driven harder to provide the fully saturated power output level.

As indicated earlier, TDMA is similar to TDM, where the different Earth stations send up bursts of RF energy that contain packets of information. During the time slot designated for a particular Earth station, that station's signal uses the bandwidth of the whole transponder. (See Fig. 8–14.) Since the Earth stations use a constant envelope modulation technique,

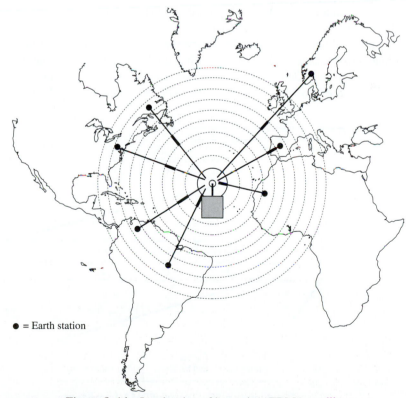

● = Earth station

Figure 8–14   Interleaving of bursts in a TDMA satellite.

such as QPSK, and since only one high-rate modulated signal is being passed through the transponder during any time interval, no interfering IM products are generated (as compared with the FDMA technique described earlier). Thus, the final TWT amplifier in the satellite may be driven into saturation for more power output. This advantage of TDMA over FDMA may be compared with the main disadvantage of the TDMA technique. The disadvantage of TDMA is that strict burst timing is required at the Earth station in order to prevent collision of the bursts at the satellite. In other words, the burst from a particular Earth station has to arrive at the satellite in the exact time slot designated for that station, so that its signal will not interfere with the bursts that are arriving from other Earth stations that are assigned adjacent time slots. Because the ground stations are located at different distances from the satellite and may use different equipment configurations, the time delay from each Earth station to the satellite will be different, and this must be taken into account when the transmission time for each Earth station is computed. In addition, the satellite may be moving with respect to the Earth station, which means that the time delay is actually changing with time. Another disadvantage of TDMA is that the Earth stations are probably transmitting data that have arrived from synchronous terrestrial lines; consequently, the Earth station equipment must contain a large memory in which to buffer the data, which are read out at high speed when the packet of information is sent to the satellite.

A typical TDMA frame format for the data being relayed through a satellite is shown in Fig. 8–15. In this example, station B is sending data to stations A, E, G, and H. One frame consists of data arriving from each Earth station. At any time, only one Earth station provides the time reference signal for the other Earth stations to use for computation of their transmitting time for their data bursts (frame synchronization). The burst lengths from the various stations could be different depending on the traffic volume. The second part of the figure shows an exploded view of a typical burst format that is transmitted from station B. This consists of two main parts, the preamble and data that are being sent to other Earth

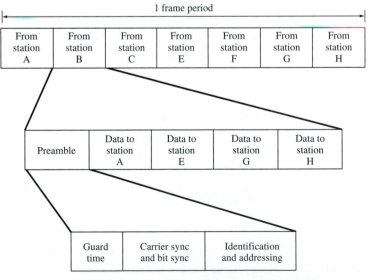

Figure 8–15   Typical TDMA frame format.

stations from station B. The preamble includes a guard time before transmission is begun. Then, a string of synchronization characters is transmitted that give the carrier sync recovery loops and bit timing recovery loops (in the ground station receivers) time to lock onto this burst from station B. The end of the preamble usually contains a unique word that identifies the burst as coming from station B and might indicate the addresses (stations) for which the data are intended.

Another multiple-access method that is similar to TDMA is the *ALOHA* technique [Lam, 1979]. Here the multiple users send bursts of data, called *packets*, whenever they wish. When two or more of the bursts overlap in time, there is a *collision*. When a collision occurs, the users involved retransmit their packets after a random time delay, and, hopefully, a second collision will not occur. If it does occur, the retransmission process is repeated until each party is successful. This technique has the advantage of being relatively inexpensive to implement, but it will not work if there is heavy traffic loading on the satellite, in which case the satellite becomes saturated with colliding packets, and the collisions can be avoided only by stopping all new transmissions and increasing the random delay required before retransmission. A more elaborate technique is called *slotted ALOHA*. With this method, the packets are transmitted at random, but only in certain time slots. This avoids collisions due to partial packet overlap.

Very small aperture terminals (VSATs) have become popular with the availability of Ku-band satellites and with the recent advances that have made low-cost terminals possible. "Very small aperture" implies that these systems use Earth-terminal antennas that are relatively small (about 1 or 2 m in diameter). Solid-state power amplifiers (1 to 2 W), low-cost frequency converters, digital processing, and VLSI circuits have made VSATs feasible. The objective of VSAT systems is to provide low-cost data and voice transmission directly to users such as auto dealerships, banks (automatic teller machines), brokerage firms, pipeline companies (monitor and control), hotels and airlines (reservations), retail stores (data networks), and corporations (point-to-multipoint transmission). Typically, VSATs offer high-quality transmission (BERs of less than 10^{-7} for 99.5% of the time) at data rates from 100 b/s to 64 kb/s [Chakraborty, 1988; Maral, 1995]. Many users share a single satellite transponder via SCPC, TDMA, or CDMA so that the user's cost can be substantially less than that for the same type of service provided by the long-distance public telephone network [Dorf, 1993, p. 2201; Maral, 1995; Rana, McCoskey, and Check, 1990].

Personal Communications via Satellite

Satellite systems are now being developed for communication directly to *personal communication system* (PCS) devices, such as hand-held portable telephones and mobile data terminals. In general, these PCS devices use small nondirectional antennas that have very little gain. Consequently, the signal from the satellite needs to be relatively strong at the user's location. Strong user signals can be achieved if the distance to the satellite is relatively small—around 400 to 1,200 miles. Thus, *low-Earth-orbit* (LEO) satellites, which are not geosyn- chronous, provide a solution to this problem. Other solutions are *medium-Earth-orbit* (MEO) satellites (5,000-mile altitude) and *high-Earth-orbit* (HEO) satellites (10,000 mile altitude) with large high-gain antennas [Balduino, 1995; Wu, Miller, Pritchard, and Pickholtz, 1994]. Table 8–3 shows some of the LEO/MEO/HEO systems that are designed for

TABLE 8–3 PERSONAL COMMUNICATION SATELLITE SYSTEMS

	System		
	Globalstar	**ICO**	**Iridium**[†]
Owner(s)	Loral/Qualcomm	ICO Global Comm. Ltd.	Motorola/ SATcom
System cost (billion $)	2.5	4.6	4.7
Orbit	LEO	MEO	LEO
Orbit altitude (stat. miles)	880	6,450	485
No. of satellites	48	10	66
No. of orbit planes	6	2	6
Inclination of planes	52°	45°	86.4°
Frequencies, up/down (GHz)	1.6/2.4	2.0/2.2	1.6/1.6
No. of VF channels/satellite	2,400	4,500	1,100
Access technique	CDMA/FDM	TDMA/FDM	TDMA/FDM
Modulation	CDMA/QPSK	QPSK	DQPSK
RF power (W), PCS phones	0.5	0.625	0.45
Bandwidth (kHz), PCS Tx	1250	25.2	31.5
Bit rate (kb/s), PCS Tx	2.4	36	50
Cost/min VF channel ($)	0.50	1.25	2.00
Year in Service	2000	2001	1998

use with hand-held PCS devices. These systems provide voice, data, and facsimile (FAX) service. The satellites are distributed around the globe in inclined (nonequatorial) orbits so that worldwide communication will be provided. These systems use cellular telephone technology (see Sec. 8–8) so that service can be provided to a large number of users. For example, the Motorola Iridium satellites employ 48 spot beams so that 1,100 users can be accommodated by each satellite. Eight users share a FDM channel via TDMA with 45 ms transmit and 45 ms receive frames. The bandwidth of the PCS phone transmit signal is 31.5 kHz, and the FDM channel spacing is 41.67 kHz [Evans, 1998]. These PCS systems are designed to seek local TELCO service first (via local cellular phone or local wireless systems) and, if local service is not available, connect to the satellite system for service.

These satellite-based PCS systems may not be successful, due to a lack of customers because of competition from land-based cellular and PCS systems.[†] The land-based systems now provide economical service in most populated areas of the world. The land-based systems are custom designed for use inside airports and in high-rise buildings with miniature PCS phones. However, satellite-based systems are needed to provide service to maritime interests and to disaster areas (where local communication facilities are not functional).

In the next section, a procedure is developed for calculating the signal strength of a received satellite signal and for evaluating the amount of noise that is present in a receiving system. This will be used to evaluate the BER at the output of the digital receivers and the SNR at the output of analog receivers.

[†] The Iridium satellite phone service was terminated in March, 2000 because of bankruptcy due to lack of customers.

8–6 LINK BUDGET ANALYSIS

In this section, formulas will be developed for the signal-to-noise ratio at the detector input as a function of the transmitted *effective isotropic radiated power* (EIRP), the free-space loss, the receiver antenna gain, and the receiver system noise figure. These results will allow us to evaluate the quality of the receiver output (as measured by the probability of bit error for digital systems and output signal-to-noise ratio for analog systems), provided that the noise characteristics of the receiver detector circuit are known.

Signal Power Received

In communication systems, the received signal power (as opposed to the voltage or current level) is the important quantity. The *power* of the received signal is of prime importance when trying to *minimize* the effect of *noise sources* that feed into the system and are amplified. The voltage gain, current gain, and impedance levels have to be such that the required power gain is achieved. In FET circuits, this accomplished by using relatively large voltage levels and small currents (high-impedance circuitry). With bipolar transistors, power gain is achieved with relatively small voltages and large currents (low-impedance circuitry).

A block diagram of a communication system with a free-space transmission channel is shown in Fig. 8–16. The overall power gain (or power transfer function) of the channel is

$$\frac{P_{Rx}}{P_{Tx}} = G_{AT} G_{FS} G_{AR} \tag{8–2}$$

where P_{Tx} is the signal power into the transmitting antenna, G_{AT} is the transmitting antenna power gain, G_{FS} is the free-space power gain [†] (which is orders of magnitude less than one in typical communication systems), G_{AR} is the receiving antenna power gain, and P_{Rx} is the signal power into the receiver.

To use this relationship, these gains should be expressed in terms of useful antenna and free-space parameters [Kraus, 1986]. Here, G_{AT} and G_{AR} are taken to be the power gains with respect to an isotropic antenna. [‡] The EIRP is

$$P_{EIRP} = G_{AT} P_{Tx} \tag{8–3}$$

The antenna power gain is defined as

$$G_A = \frac{\text{radiation power density of the actual antenna in the direction of maximum radiation}}{\text{radiation power density of an isotropic antenna with the same power input}}$$

where the power density (measured in W/m^2) is evaluated at the same distance, d, for both antennas. The gains are for some practical antennas is given in Table 8–4.

[†] A gain transfer function is the output quantity divided by the input quantity, whereas a loss transfer function is the input quantity divided by the output quantity.

[‡] An isotropic antenna is a nonrealizable theoretical antenna that radiates equally well in all directions and is a useful reference for comparing practical antennas.

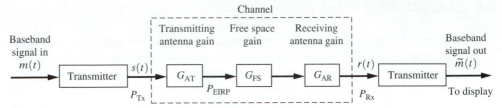

Figure 8–16 Block diagram of a communication system with a free-space transmission channel.

TABLE 8–4　ANTENNA GAINS AND EFFECTIVE AREAS

Type of Antenna	Power Gain, G_A (absolute units)	Effective Area, A_e (m²)
Isotropic	1	$\lambda^2/4\pi$
Infinitesimal dipole or loop	1.5	$1.5\lambda^2/4\pi$
Half-wave dipole	1.64	$1.64\lambda^2/4\pi$
Horn (optimized), mouth area, A	$10A/\lambda^2$	$0.81A$
Parabola or "dish" with face area, A	$7.0A/\lambda^2$	$0.56A$
Turnstile (two crossed dipoles fed 90° out of phase)	1.15	$1.15\lambda^2/4\pi$

The power density (W/m²) of an isotropic antenna at a distance d from the antenna is

$$\text{power density at } d = \frac{\text{transmitted power}}{\text{area of a sphere with radius } d} = \frac{P_{\text{EIRP}}}{4\pi d^2} \qquad (8\text{--}4)$$

The FCC and others often specify the strength of an electromagnetic field by the field intensity, $\mathcal{E}$ (V/m), instead of power density (W/m²). The two are related by

$$\text{power density} = \frac{\mathcal{E}^2}{377} \qquad (8\text{--}5)$$

where the power density and the field strength are evaluated at the same point in space and 377 Ω is the *free-space intrinsic impedance*. If the receiving antenna is placed at d meters from the transmitting antenna, it will act like a catcher's mitt and intercept the power in an effective area of $(A_e)_{\text{Rx}}$ (m²), so that the received power will be

$$P_{\text{Rx}} = G_{\text{AT}} \left(\frac{P_{\text{Tx}}}{4\pi d^2}\right) (A_e)_{\text{Rx}} \qquad (8\text{--}6)$$

where the gain of the transmitting antenna (with respect to an isotropic antenna), G_{AT}, has been included. Table 8–4 also gives the effective area for several types of antennas. The gain and the effective area of an antenna are related by

$$G_A = \frac{4\pi A_e}{\lambda^2} \qquad (8\text{--}7)$$

where $\lambda = c/f$ is the wavelength, c being the speed of light $(3 \times 10^8 \text{ m/s})$ and f the operating frequency in Hz. An antenna is a *reciprocal element*. That is, it has the same gain properties whether it is transmitting or receiving. Substituting Eq. (8–7) into Eq. (8–6), we obtain

$$\frac{P_{\text{Rx}}}{P_{\text{Tx}}} = G_{\text{AT}} \left(\frac{\lambda}{4\pi d}\right)^2 G_{\text{AR}} \tag{8–8}$$

where the free-space gain is

$$G_{\text{FS}} = \left(\frac{\lambda}{4\pi d}\right)^2 = \frac{1}{L_{\text{FS}}} \tag{8–9}$$

and L_{FS} is the free-space path loss (absolute units). The channel gain, expressed in dB, is obtained by taking 10 log [·] of both sides of Eq. (8–2):

$$(G_{\text{channel}})_{\text{dB}} = (G_{\text{AT}})_{\text{dB}} - (L_{\text{FS}})_{\text{dB}} + (G_{\text{AR}})_{\text{dB}} \tag{8–10}$$

In Eq. (8–10), the free-space loss[†] is

$$(L_{\text{FS}})_{\text{dB}} = 20 \log \left(\frac{4\pi d}{\lambda}\right) \text{ dB} \tag{8–11}$$

For example, the free-space loss at 4 GHz for the shortest path to a synchronous satellite from Earth (22,300 miles) is 195.6 dB.

Thermal Noise Sources

The noise power generated by a thermal noise source will be studied because the receiver noise is evaluated in terms of this phenomenon. A conductive element with two terminals may be characterized by its resistance, R ohms. This resistive, or lossy, element contains free electrons that have some random motion if the temperature of the resistor is above absolute zero. This random motion causes a noise voltage to be generated at the terminals of the resistor. Although the noise is small, when the noise is amplified by a high-gain receiver it can become a problem. (If no noise were present, we could communicate to the edge of the universe with infinitely small power, since the signal could always be amplified without having noise introduced.)

This physical lossy element, or physical resistor, can be modeled by an equivalent circuit that consists of a noiseless resistor in series with a noise voltage source. (See Fig. 8–17.) From quantum mechanics, it can be shown that the (normalized) power spectrum corresponding to the voltage source is [Van der Ziel, 1986]

$$\mathcal{P}_v(f) = 2R \left[\frac{h|f|}{2} + \frac{h|f|}{e^{h|f|/(kT)} - 1}\right] \tag{8–12}$$

where

$R = $ value of the physical resistor (ohms),

[†]This free-space path loss expression can be modified to include effects of a multipath channel within an urban building environment. [See Eq. (8–67).]

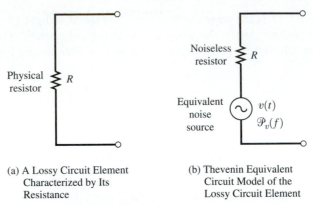

(a) A Lossy Circuit Element
Characterized by Its
Resistance

(b) Thevenin Equivalent
Circuit Model of the
Lossy Circuit Element

Figure 8–17   Thermal noise source.

$h = 6.2 \times 10^{-34}$ J-sec is Planck's constant,

$k = 1.38 \times 10^{-23}$ J/K is Boltzmann's constant, where K is kelvin,

$T = (273 + C)$ is the absolute temperature of the resistor (kelvin).

At room temperature for frequencies below 1,000 GHz, $[h|f|/(kT)] < 1/5$, so that $e^x = 1 + x$ is a good approximation. Then, Eq. (8–12) reduces to

$$\mathscr{P}_v(f) = 2RkT \qquad (8\text{–}13)$$

This equation will be used to develop other formulas in this text since the RF frequencies of interest are usually well below 1,000 GHz and since we are not dealing with temperatures near absolute zero.

If the open-circuit noise voltage that appears across a physical resistor is read by a true rms voltmeter having a bandwidth of B hertz, then, using Eq. (2–67), the reading would be

$$V_{\text{rms}} = \sqrt{\langle v^2 \rangle} = \sqrt{2 \int_0^B \mathscr{P}_v(f)\, df} = \sqrt{4kTBR} \qquad (8\text{–}14)$$

Characterization of Noise Sources

Noise sources may be characterized by the maximum amount of noise power or PSD that can be passed to a load.

> **DEFINITION.**   The *available noise power* is the *maximum*[†] actual (i.e., not normalized) power that can be drawn from a source. The *available PSD* is the *maximum* actual (i.e., not normalized) PSD that can be obtained from a source.

For example, the available PSD for a thermal noise source is easily evaluated using Fig. 8–18 and Eq. (2–142):

[†] The maximum power or maximum PSD is obtained when $Z_L(f) = Z_s^*(f)$, where $Z_L(f)$ is the load impedance and $Z_s^*(f)$ is the conjugate of the source impedance.

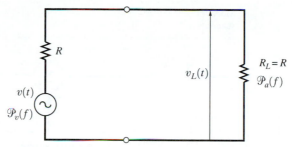

Figure 8-18   Thermal source with a matched load.

$$\mathcal{P}_a(f) = \frac{\mathcal{P}_v(f)|H(f)|^2}{R} = \tfrac{1}{2}kT \quad \text{W/Hz} \tag{8-15}$$

where $H(f) = \tfrac{1}{2}$ for the resistor divider network. The available power from a thermal source in a bandwidth of B hertz is

$$P_a = \int_{-B}^{B} \mathcal{P}_a(f)\, df = \int_{-B}^{B} \frac{1}{2} kT \, df$$

or

$$P_a = kTB \tag{8-16}$$

This equation indicates that the available noise power from a thermal source does *not* depend on the value of R, even though the open-circuit rms voltage does.

The available noise power from a source (not necessarily a thermal source) can be specified by a number called the noise temperature.

DEFINITION.   The *noise temperature* of a source is given by

$$T = \frac{P_a}{kB} \tag{8-17}$$

where P_a is the available power from the source in a bandwidth of B hertz.

In using this definition, it is noted that if the source happens to be of thermal origin, T will be the temperature of the device in kelvin, but if the source is not of thermal origin, the number obtained for T may have nothing to do with the physical temperature of the device.

Noise Characterization of Linear Devices

A linear device with internal noise generators may be modeled as shown in Fig. 8-19. Any device that can be built will have some internal noise sources. As shown in the figure, the device could be modeled as a noise-free device having a power gain $G_a(f)$ and an excess noise source at the output to account for the internal noise of the actual device. Some examples of linear devices that have to be characterized in receiving systems are lossy transmission lines, RF amplifiers, down converters, and IF amplifiers.

The power gain of the devices is the available power gain.

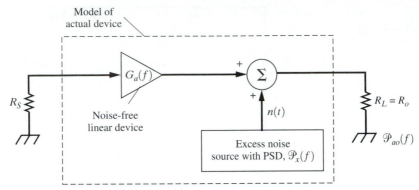

Figure 8–19   Noise model for an actual device.

DEFINITION.   The *available power gain* of a linear device is

$$G_a(f) = \frac{\text{available PSD out of the device}}{\text{available PSD out of the source}} = \frac{\mathscr{P}_{ao}(f)}{\mathscr{P}_{as}(f)} \qquad (8\text{–}18)$$

When $\mathscr{P}_{ao}(f)$ is measured to obtain $G_a(f)$, the source noise power is made large enough that the amplified *source* noise that appears at the output dominates any other noise. In addition, note that $G_a(f)$ is defined in terms of actual (i.e., not normalized) PSD. In general, $G_a(f)$ will depend on the driving source impedance, as well as on elements within the device itself, but it does *not* depend on the load impedance. If the source impedance and the output impedance of the device are equal, $G_a(f) = |H(f)|^2$, where $H(f)$ is the voltage or current transfer function of the linear device.

To characterize the goodness of a device, a figure of merit is needed that compares the actual (noisy) device with an ideal device (i.e., no internal noise sources). Two figures of merit, both of which tell us the same thing—namely, how bad the noise performance of the actual device is—are universally used. They are noise figure and effective input-noise temperature.

DEFINITION.   The *spot noise figure* of a linear device is obtained by terminating the device with a thermal noise source of temperature T_0 on the input and a matched load on the output as indicated in Fig. 8–19. The spot noise figure is

$$F_s(f) = \frac{\text{measured available PSD out of the actual device}}{\text{available PSD out of an ideal device with the same available gain}}$$

or

$$F_s(f) = \frac{\mathscr{P}_{ao}(f)}{(kT_0/2)G_a(f)} = \frac{(kT_0/2)G_a(f) + \mathscr{P}_x(f)}{(kT_0/2)G_a(f)} > 1 \qquad (8\text{–}19)$$

The value of R_s is the same as the source resistance that was used when $G_a(f)$ was evaluated. A standard temperature of $T_0 = 290$ K is used, as this is the value adopted by the IEEE [Haus, 1963].

$F_s(f)$ is called the *spot noise figure* because it refers to the noise characterization at a particular "spot" or frequency in the spectrum. Note that $F_s(f)$ is always greater than unity for an actual device, but it is nearly unity if the device is almost an ideal device. $F_s(f)$ is a function of the source temperature, T_0. Consequently, when the noise figure is evaluated, a standard temperature of $T_0 = 290$ K is used. This corresponds to a temperature of 62.3°F.

Often an average noise figure instead of a spot noise figure is desired. The average is measured over some bandwidth B.

DEFINITION.   The *average noise figure* is

$$F = \frac{P_{ao}}{kT_0 \displaystyle\int_{f_0-B/2}^{f_0+B/2} G_a(f)\,df} \tag{8–20}$$

where

$$P_{ao} = 2\int_{f_0-B/2}^{f_0+B/2} \mathscr{P}_{ao}(f)\,df$$

is the measured available output in a bandwidth B hertz wide centered on a frequency of f_0 and $T_0 = 290$ K.

If the available gain is constant over the band so that $G_a(f) = G_a$ over the frequency interval $(f_0 - B/2) \le f \le (f_0 + B/2)$, the noise figure becomes

$$F = \frac{P_{ao}}{kT_0 B G_a} \tag{8–21a}$$

The noise figure is often measured by using the Y-factor method. This technique is illustrated by Prob. 8–15. The Hewlett-Packard HP8970B noise figure meter uses the Y-factor method for measuring the noise figure of devices.

The noise figure can also be specified in decibel units:[†]

$$F_{dB} = 10\log(F) = 10\log\left(\frac{P_{ao}}{kT_0 B G_a}\right) \tag{8–21b}$$

For example, suppose that the manufacturer of an RF preamplifier specifies that an RF preamp has a 2-dB noise figure. This means that the actual noise power at the output is 1.58 times the power that would occur due to amplification of thermal noise from the input. The other figure of merit for evaluating the noise performance of a linear device is the effective input-noise temperature. This is illustrated in Fig. 8–20.

DEFINITION.   The *spot effective input-noise temperature*, $T_{es}(f)$, of a linear device is the *additional* temperature required for an input source, which is driving an ideal (noise-free) device, to produce the same available PSD at the ideal device output as is obtained from the actual device when it is driven by the input source of temperature T_i kelvin. That is, $T_{es}(f)$ is defined by

[†] Some authors call F the *noise factor* and F_{dB} the *noise figure*.

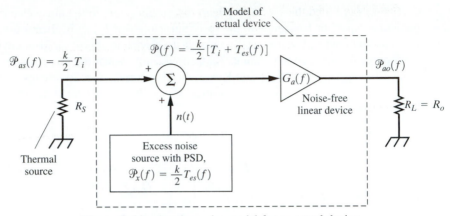

Figure 8–20   Another noise model for an actual device.

$$\mathscr{P}_{ao}(f) = G_a(f)\,\frac{k}{2}\,[T_i + T_{es}(f)] \tag{8–22}$$

where $\mathscr{P}_{ao}(f)$ is the available PSD out of the actual device when driven by an input source of temperature T_i and $T_{es}(f)$ is the spot effective input-noise temperature.

The *average effective input-noise temperature*, T_e, is defined by the equation

$$P_{ao} = k(T_i + T_e)\int_{f_0-B/2}^{f_0+B/2} G_a(f)\,df \tag{8–23}$$

where the measured available noise power out of the device in a band B hertz wide is

$$P_{ao} = 2\int_{f_0-B/2}^{f_0+B/2} \mathscr{P}_{ao}(f)\,df \tag{8–24}$$

Because $G_a(f)$ depends on the source impedance, as well as on the device parameters, $T_e(f)$ will depend on the source impedance used, as well as on the characteristics of the device itself, but it is independent of the value of T_i used. In the definition of T_e, note that the IEEE standards do not specify that $T_i = T_0$, since the value of T_e obtained does not depend on the value of T_i that is used. However, $T_i = T_0 = 290$ may be used for convenience. The effective input-noise temperature can also be evaluated by the Y-factor method, as illustrated by Prob. 8–15.

When the gain is flat (constant) over the frequency band, $G_a(f) = G_a$, the effective input-noise temperature is simply

$$T_e = \frac{P_{ao} - kT_i G_a B}{kG_a B} \tag{8–25}$$

Note that $T_{es}(f)$ and T_e are greater than zero for an actual device, but if the device is nearly ideal (small internal noise sources), they will be very close to zero.

When T_e was evaluated with Eq. (8–23), an input source was used with some convenient value for T_i. However, when the device is used in a system, the available noise power from the source will be different if T_i is different. For example, suppose that the device is an RF preamplifier and the source is an antenna. The available power out of the amplifier when it is connected to the antenna is now[†]

$$P_{ao} = 2 \int_{f_0-B/2}^{f_0+B/2} \mathscr{P}_{as}(f) G_a(f) \, df + kT_e \int_{f_0-B/2}^{f_0+B/2} G_a(f) \, df \qquad (8\text{–}26)$$

where $\mathscr{P}_{as}(f)$ is the available PSD out of the source (antenna). T_e is the average effective input temperature of the amplifier that was evaluated by using the input source T_i. If the gain of the amplifier is approximately constant over the band, this reduces to

$$P_{ao} = G_a P_{as} + kT_e B \, G_a \qquad (8\text{–}27)$$

where the available power from the source (antenna) is

$$P_{as} = 2 \int_{f_0-B/2}^{f_0+B/2} \mathscr{P}_{as}(f) \, df \qquad (8\text{–}28)$$

Furthermore, the available power from the source might be characterized by its noise temperature, T_s, so that, using Eq. (8–17), we get

$$P_{as} = kT_s B \qquad (8\text{–}29)$$

In satellite Earth station receiving applications, the antenna (source) noise temperature might be $T_s = 32$ K at 4 GHz for a parabolic antenna where the noise from the antenna is due to cosmic radiation and to energy received from the ground as the result of the sidelobe beam pattern of the antenna. (The Earth acts as a blackbody noise source with $T = 280$ K.) Note that the $T_s = 32$ K of the antenna is "caused" by *radiation* resistance, which is not the same as a loss resistance (I^2R losses) associated with a thermal source and that T_s has no relation to the physical temperature of the antenna.

In summary, two figures of merit have been defined: noise figure and effective input-noise temperature. By combining Eqs. (8–19) and (8–22), where $T_i = T_0$, a relationship between these two figures of merit for the spot measures is obtained:

$$T_{es}(f) = T_0[F_s(f) - 1] \qquad (8\text{–}30a)$$

Here $T_i = T_0$ is required, because $T_i = T_0$ is used in the definition for the noise figure that precedes Eq. (8–19).

Using Eqs. (8–21a) and (8–25), where $T_i = T_0$, we obtain the same relationship for the average measures:

$$T_e = T_0(F - 1) \qquad (8\text{–}30b)$$

Example 8–3 T_e AND F FOR A TRANSMISSION LINE

The effective input-noise temperature, T_e, and the noise figure, F, for a lossy transmission line (a linear device) will now be evaluated.[‡] This can be accomplished by terminating the transmission line with a source and a load resistance (all having the same physical temperature) that are both

---

[†] The value of P_{ao}, in Eqs. (8–26) and (8–27) is different from that in Eqs. (8–23), (8–24), and (8–25).

[‡] These results also hold for the T_e and F of (impedance) matched attenuators.

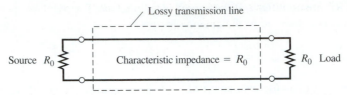

Figure 8–21   Noise figure measurement of a lossy transmission line.

equal to the characteristic impedance of the line, as shown in Fig. 8–21. The gain of the transmission line is $G_a = 1/L$, where L is the transmission line loss (power in divided by power out) in absolute units (i.e., not dB units). Looking into the output port of the transmission line, one sees an equivalent circuit that is resistive (thermal source) with a value of R_0 ohms because the input of the line is terminated by a resistor of R_0 ohms (the characteristic impedance). Assume that the physical temperature of the transmission line is T_L as measured on the Kelvin scale. Since the line acts as a thermal source, the available noise power at its output is $P_{ao} = kT_LB$. Using Eq. (8–25) where the source is at the same physical temperature $T_e = T_L$ we get

$$T_e = \frac{kT_LB - kT_LG_aB}{kG_aB} = T_L\left(\frac{1}{G_a} - 1\right)$$

Thus the effective input-noise temperature for the transmission line is

$$T_e = T_L(L - 1) \qquad\qquad (8\text{–}31\text{a})$$

where T_L is the physical temperature (Kelvin) of the line and L is the line loss.
If the physical temperature of the line happens to be T_0, this becomes

$$T_e = T_0(L - 1) \qquad\qquad (8\text{–}31\text{b})$$

The noise figure for the transmission line is obtained by using Eq. (8–30b) to convert T_e to F. Thus, substituting Eq. (8–31a) into (8–31b), we get

$$T_L(L - 1) = T_0(F - 1)$$

Solving for F, we obtain the noise figure for the transmission line

$$F = 1 + \frac{T_L}{T_0}(L - 1) \qquad\qquad (8\text{–}32\text{a})$$

where T_L is the physical temperature (Kelvin) of the line, $T_0 = 290$, and L is the line loss. If the physical temperature of the line is 290 K (63°F), Eq. (8–32a) reduces to

$$F = \frac{1}{G_a} = L \qquad\qquad (8\text{–}32\text{b})$$

In decibel measure, this is $F_{dB} = L_{dB}$. In other words, if a transmission line has a 3-dB loss, it has a noise figure of 3 dB provided that it has a physical temperature of 63°F. If the temperature is 32°F (273 K), the noise figure, using Eq. (8–32a), will be 2.87 dB. Thus, F_{dB} is approximately L_{dB}, if the transmission line is located in an environment (temperature range) that is inhabitable by humans.

Noise Characterization of Cascaded Linear Devices

In a communication system several linear devices (supplied by different vendors) are often cascaded together to form an overall system, as indicated in Fig. 8–22. In a receiving system, these devices might be an RF preamplifier connected to a transmission line that feeds a down converter and an IF amplifier. (As discussed in Sec. 4–11, the down converter is a linear device and may be characterized by its conversion power gain and its noise figure.) For system performance calculations, we need to evaluate the overall power gain, G_a, and the overall noise characterization (which is given by the overall noise figure or the overall effective input-noise temperature) from the specifications for the individual devices provided by the vendors.

The overall available power gain is

$$G_a(f) = G_{a1}(f)G_{a2}(f)G_{a3}(f)G_{a4}(f)\cdots \tag{8–33}$$

since, for example, for a four-stage system,

$$G_a(f) = \frac{\mathscr{P}_{ao4}}{\mathscr{P}_{as}} = \left(\frac{\mathscr{P}_{ao1}}{\mathscr{P}_{as}}\right)\left(\frac{\mathscr{P}_{ao2}}{\mathscr{P}_{ao1}}\right)\left(\frac{\mathscr{P}_{ao3}}{\mathscr{P}_{ao2}}\right)\left(\frac{\mathscr{P}_{ao4}}{\mathscr{P}_{ao3}}\right)$$

THEOREM.   *The overall noise figure for cascaded linear devices is*[†]

$$F = F_1 + \frac{F_2 - 1}{G_{a1}} + \frac{F_3 - 1}{G_{a1}G_{a2}} + \frac{F_4 - 1}{G_{a1}G_{a2}G_{a3}} + \cdots \tag{8–34}$$

as shown in Fig. 8–22 (for a four-stage system).

Proof. This result may be obtained by using an excess-noise model of Fig. 8–19 for each stage. We will prove the result for a two-stage system as modeled in Fig. 8–23. The overall noise figure is

$$F = \frac{P_{ao2}}{(P_{ao2})_{\text{ideal}}} = \frac{P_{x2} + P_{ao1}G_{a2}}{G_{a1}G_{a2}P_{as}}$$

which becomes

$$F = \frac{P_{x2} + G_{a2}(P_{x1} + G_{a1}P_{as})}{G_{a1}G_{a2}P_{as}} \tag{8–35}$$

Overall system G_a, F, T_e

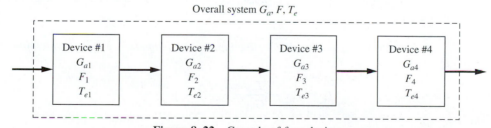

Figure 8–22   Cascade of four devices.

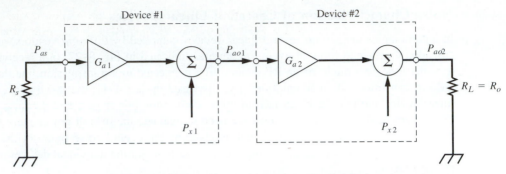

Figure 8–23   Noise model for two cascaded devices.

where $P_{as} = kT_0B$ is the available power from the thermal source. P_{x1} and P_{x2} can be obtained from the noise figures of the individual devices by using Fig. 8–19, so that for the *i*th device,

$$F_i = \frac{P_{aoi}}{G_{ai}P_{as}} = \frac{P_{xi} + G_{ai}P_{as}}{G_{ai}P_{as}}$$

or

$$P_{xi} = G_{ai}P_{as}(F_i - 1) \tag{8-36}$$

Substituting this into Eq. (8–35) for P_{x1} and P_{x2}, we obtain

$$F = F_1 + \frac{F_2 - 1}{G_{a1}}$$

which is identical to Eq. (8–34) for the case of two cascaded stages. In a similar way Eq. (8–34) can be shown to be true for any number of stages.

Looking at Eq. (8–34), we see that if the terms G_{a1}, $G_{a1}G_{a2}$, $G_{a1}G_{a2}G_{a3}$, and so on, are relatively large, F_1 will dominate the overall noise figure. Thus, in receiving system design, it is important that the first stage have a low noise figure and a large available gain, so that the noise figure of the overall system will be as small as possible.

The overall effective input-noise temperature of several cascaded stages can also be evaluated.

THEOREM.   *The overall effective input-noise temperature for cascaded linear devices is*

$$T_e = T_{e1} + \frac{T_{e2}}{G_{a1}} + \frac{T_{e3}}{G_{a1}G_{a2}} + \frac{T_{e4}}{G_{a1}G_{a2}G_{a3}} + \cdots \tag{8-37}$$

as shown in Fig. 8–22.

A proof of this result is left for the reader as an exercise.

For further study concerning the topics of effective input-noise temperature and noise figure, the reader is referred to an authoritative monograph [Mumford and Scheibe, 1968].

Link Budget Evaluation

The performance of a communication system depends on how large the SNR is at the detector input in the receiver. It is engineering custom to call the signal-to-noise ratio before detection the *carrier-to-noise* ratio (CNR). Thus, in this section, we will use CNR to denote the signal-to-noise ratio before detection (bandpass case) and SNR to denote the signal-to-noise ratio after detection (baseband case). Here we are interested in evaluating the detector input CNR as a function of the communication link parameters, such as transmitted EIRP, space loss, receiver antenna gain, and the effective input-noise temperature of the receiving system. The formula that relates these system link parameters to the CNR at the detector input is called the *link budget*.

The communication system may be described by the block diagram shown in Fig. 8–24. In this model, the receiving system from the receiving-antenna output to the detector input is modeled by one linear block that represents the cascaded stages in the receiving system, such as a transmission line, a low-noise amplifier (LNA), a down

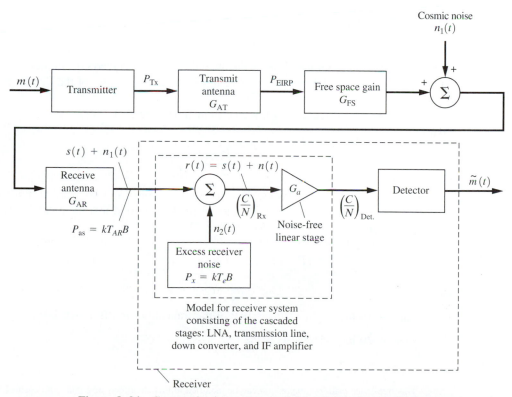

Figure 8–24   Communication system model for link budget evaluation.

converter, and an IF amplifier. This linear block describes the overall available power gain and effective input-noise temperature of these cascaded devices, as described in the preceding section and modeled in Fig. 8–20.

As shown in Fig. 8–24, the CNR at the ideal amplifier input (with gain G_a) is identical to that at the detector input, because the ideal amplifier adds no excess noise and it amplifies the signal and noise equally well in a bandwidth of B hertz (the IF bandwidth). Thus, denoting this simply by CNR, we get

$$\left(\frac{C}{N}\right) \triangleq \left(\frac{C}{N}\right)_{Rx} = \left(\frac{C}{N}\right)_{det} \tag{8–38}$$

where these carrier-to-noise power ratios are indicated on the figure.

Using Eqs. (8–2) and (8–3), we obtain the received signal power

$$C_{Rx} = (P_{EIRP})G_{FS}G_{AR} \tag{8–39}$$

where P_{EIRP} is the EIRP from the transmitter, G_{FS} is the free-space gain, and G_{AR} is the receiving antenna power gain.

When we use Eq. (8–17), the available noise power at the input to the ideal amplifier in the model (Fig. 8–24) is

$$N = kT_{syst}B \tag{8–40}$$

where B is the IF equivalent bandwidth. The receiving system noise temperature is

$$T_{syst} = T_{AR} + T_e \tag{8–41}$$

where T_{AR} is the noise temperature of the antenna (due to received cosmic noise and Earth blackbody radiation) and T_e is the effective input-noise temperature of the overall receiving system.

When Eqs. (8–39) and (8–40) are combined, the carrier-to-noise ratio at the detector input is

$$\frac{C}{N} = \frac{P_{EIRP}G_{FS}G_{AR}}{kT_{syst}B} \tag{8–42}$$

For engineering applications, this formula is converted to decibel units. Using Eq. (8–9) and taking $10 \log [\cdot]$ of both sides of (8–42), we find that the received carrier-to-noise ratio at the detector input in decibels is

$$\left(\frac{C}{N}\right)_{dB} = (P_{EIRP})_{dBw} - (L_{FS})_{dB} + \left(\frac{G_{AR}}{T_{syst}}\right)_{dB} - k_{dB} - B_{dB} \tag{8–43}$$

where

$(P_{EIRP})_{dBw} = 10 \log (P_{EIRP})$ is the EIRP of the transmitter in dB above 1 W,

$(L_{FS})_{dB} = 20 \log [(4\pi d)/\lambda]$ is the path loss,[†]

[†] This free-space path-loss expression can be modified to include effects of a multipath channel within an urban building environment. [See Eq. (8–47) and Eq. (8–67)].

$$k_{dB} = 10 \log(1.38 \times 10^{-23}) = -228.6,$$

$$B_{dB} = 10 \log(B) \quad (B \text{ is the IF bandwidth in hertz}).$$

For analog communication systems, the SNR at the detector output can be related to the CNR at the detector input. The exact relationship depends on the type of detector, as well as the modulation used. These relationships were developed in Chapter 7. They are summarized in Table 7–2 and Fig. 7–27 with the use of Eq. (7–85), where $C/N = (S/N)_{in}$. Example 8–4 uses Eq. (8–43) to evaluate the performance of a CATV satellite receiving system.

For digital communications systems, the BER at the digital output is a measure of performance. The BER is related to E_b/N_0 via the CNR. The E_b/N_0-to-CNR relationship is developed in the next section. Study aid problems SA8–1 and 8–2 evaluate the BER for a DSS television receiving system.

E_b/N_0 Link Budget for Digital Systems

In digital communication systems, the probability of bit error P_e for the digital signal at the detector output describes the quality of the recovered data. P_e, also called the BER, is a function of the ratio of the energy per bit to the noise PSD, (E_b/N_0), as measured at the detector input. The exact relationship between P_e and E_b/N_0 depends on the type of digital signaling used, as shown in Table 7–1 and Fig. 7–14. In this section, we evaluate the E_b/N_0 obtained at the detector input as a function of the communication link parameters.

The energy per bit is given by $E_b = CT_b$, where C is the signal power and T_b is the time required to send one bit. Using Eq. (8–40), we see that the noise PSD (one sided) is $N_0 = kT_{syst}$. Thus,

$$\frac{C}{N} = \frac{E_b/T_b}{N_0 B} = \frac{E_b R}{N_0 B} \qquad (8\text{--}44)$$

where $R = 1/T_b$ is the data rate (b/s). Using Eq. (8–44) in Eq. (8–42), we have

$$\frac{E_b}{N_0} = \frac{P_{EIRP} G_{FS} G_{AR}}{k T_{syst} R} \qquad (8\text{--}45)$$

In decibel units, the E_b/N_0 received at the detector input in a digital communications receiver is related to the link parameters by

$$\left(\frac{E_b}{N_0}\right)_{dB} = (P_{EIRP})_{dBw} - (L_{FS})_{dB} + \left(\frac{G_{AR}}{T_{syst}}\right)_{dB} - k_{dB} - R_{dB} \qquad (8\text{--}46)$$

where $(R)_{dB} = 10 \log(R)$ and R is the data rate (b/s).

For example, suppose that we have BPSK signaling and that an optimum detector is used in the receiver; then $(E_b/N_0)_{dB} = 8.4$ dB is required for $P_e = 10^{-4}$. (See Fig. 7–14.)[†] Using Eq. (8–46), we see that the communication link parameters may be selected to give the required $(E_b/N_0)_{dB}$ of 8.4 dB. Note that as the bit rate is increased, the transmitted power has to be increased or the receiving system performance—denoted by $(G_{AR}/T_{syst})_{dB}$—has

---

[†] If, in addition, coding were used with a coding gain of 3 dB (see Sec. 1–11), an E_b/N_0 of 5.4 dB would be required for $P_e = 10^{-4}$.

to be improved to maintain the required $(E_b/N_0)_{\mathrm{dB}}$ of 8.4 dB. Examples of evaluating the E_b/N_0 link budget to obtain the BER for a DSS television receiving system are shown in SA8–1 and SA8–2.

Path Loss for Urban Wireless Environments

Path loss for a non-free-space wireless environment, such as a path with obstructions consisting of buildings and trees, is difficult to model. This is the case for terrestrial cellular and wireless systems where multipath signals are received that consist of reflected path signals in addition to the direct path signal, as well as additional signal attenuation as the signals pass through foliage and building walls. Often, measurements of received signal strength along the path are made to validate predicted results. For a free-space path, the exponent of the path loss is $n = 2$; that is, the path loss varies as the square $(n = 2)$ of the distance as shown by Eq. (8–9). When obstructions are present, n is larger than 2. Usually it is in the range of 2 to 6 with a typical value being $n = 3$. When the path loss is expressed in dB between the transmit and receive antennas, this gives a log-distance path loss model. Thus, the obstructed path loss is [Rappaport, 1996, p. 126; Sklar, 1997]

$$L_{\mathrm{dB}}(d) = L_{\mathrm{FSdB}}(d_0) + 10n \log\left(\frac{d}{d_0}\right) + X_{\mathrm{dB}} \qquad (8\text{--}47a)$$

where

$$L_{\mathrm{FSdB}}(d_0) = 20 \log\left[(4\pi d_0)/\lambda\right] \qquad (8\text{--}47b)$$

$L_{\mathrm{dB}}(d)$ is the path loss in dB for a distance of d between the antennas, $L_{\mathrm{LFSdB}}(d_0)$ is the free space loss for a distance d_0 that is close to the transmitter, but in the far field, and $d > d_0$. n is the path loss exponent, and X_{dB} is a zero-mean Gaussian random variable representing the variations in the path loss caused by multiple reflections. Typically, d_0 is taken to be 1 km for large urban mobile systems, 100 m for microcell systems, and 1 m for indoor wireless systems. An example of using Eq. (8–47) to evaluate the link budget and the BER for a wireless personal communication device (PCD) is shown in SA8–3.

Example 8–4 Link Budget Evaluation for a Television Receive-Only Terminal for Satellite Signals

The link budget for a television receive-only (TVRO) terminal used for receiving TV signals from a satellite will be evaluated. It is assumed that this receiving terminal is located at Washington. D.C., and is receiving signals from a Hughes *Galaxy* satellite that is in a geostationary orbit at 134° W longitude above the equator. The specifications on the proposed receiving equipment are given in Table 8–5 together with the receiving site coordinates and satellite parameters. This TVRO terminal is typical of those used at the head end of CATV systems for reception of TV signals that are relayed via satellite. As discussed in Sec. 8–5, the composite NTSC baseband video signal is relayed via satellite by frequency-modulating this signal on a 6-GHz carrier that is radiated to the satellite. The satellite down-converts the FM signal to 4 GHz and retransmits this FM signal to the TVRO terminal.

TABLE 8–5   TYPICAL SATELLITE TELEVISION RELAY PARAMETERS

Item	Parameter Value
Hughes *Galaxy I* satellite	
Orbit	Geostationary
Location (above equator)	134° W longitude
Uplink frequency band	6 GHz
Downlink frequency band	4 GHz
$(P_{EIRP})_{dBw}$	36 dBw
TVRO terminal	
Site location	Washington, D.C., 38.8° N latitude, 77° W longitude
Antenna	
Antenna type	10 ft diameter parabola
Noise temperature	32 K (at feed output) for 16.8° elevation
Feedline gain	0.98
Low-noise amplifier	
Noise temperature	40 K
Gain	50 dB
Receiver	
Manufacturer	Microdyne Corp., Model 1100-TVR (×24)
Noise temperature	2610 K
IF bandwidth	30 MHz
FM threshold	8 dB CNR

The transmit antenna EIRP footprint for the *Galaxy I* satellite is shown in Fig. 8–25. From this figure, it is seen that the EIRP directed toward Washington, D.C., is approximately 36 dBw, as obtained for inclusion in Table 8–5.

It can be shown that the *look angles* of the TVRO antenna can be evaluated using the following formulas [Davidoff, 1984, Chap. 9]. The elevation of the TVRO antenna is

$$E = \tan^{-1}\left[\frac{1}{\tan \beta} - \frac{R}{(R + h)\sin \beta}\right], \qquad (8\text{–}48a)$$

where

$$\beta = \cos^{-1}[\cos \varphi \cos \lambda] \qquad (8\text{–}48b)$$

and λ is the difference between the longitude of the TVRO site and the longitude of the *sub-station point* (i.e., the point directly below the geostationary satellite along the equator). φ is the latitude of the TVRO site, $R = 3963$ statute miles is the Earth radius, and $h = 22{,}242$ statute miles is the altitude of the synchronous satellite. The distance between the TVRO site and satellite, called the *slant range*, is given by the cosine law:

$$d = \sqrt{(R + h)^2 + R^2 - 2R(R + h)\cos \beta} \qquad (8\text{–}49)$$

The azimuth of the TVRO antenna is

$$A = \cos^{-1}\left(-\frac{\tan \varphi}{\tan \beta}\right) \qquad (8\text{–}50)$$

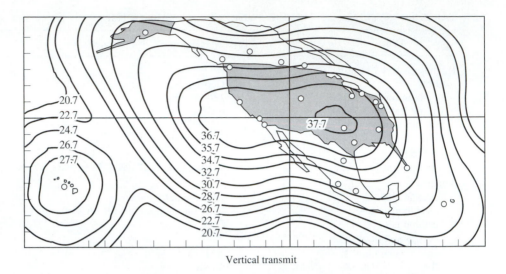

Vertical transmit

Horizontal transmit

Figure 8–25 *Galaxy I* (134° W longitude) antenna pattern EIRP footprint (contours given in dBw units). [Courtesy of Hughes Communications, Inc., a wholly owned subsidiary of Hughes Aircraft Company.]

where the true azimuth, as measured clockwise from the north, is given by either A or $360° - A$, as appropriate, since a calculator gives a value between 0 and 180° for $\cos^{-1}(x)$. Using these formulas for the Washington, D.C., TVRO site,[†] we get $\lambda = 134 - 77 = 57°$ and

$$\beta = \cos^{-1}[\cos(38.8)\cos(57)] = 64.9° \qquad (8\text{–}51)$$

[†] Here for convenience and for ease in interpretation of answers, the calculator is set to evaluate all trigonometric functions in degrees (not radians).

The slant range is

$$d = \sqrt{(26{,}205)^2 + (3963)^2 - 2(3963)(26{,}205)\cos 64.9}$$

$$= 24{,}784 \text{ miles} \tag{8-52}$$

The elevation angle of the TVRO antenna is

$$E = \tan^{-1}\left[\frac{1}{\tan(64.9)} - \frac{3963}{(26{,}205)\sin(64.9)}\right] = 16.8° \tag{8-53}$$

The azimuth angle of the TVRO antenna is

$$A = \cos^{-1}\left[\frac{-\tan(38.8)}{\tan(64.9)}\right] = 247.9° \tag{8-54}$$

Thus, the look angles of the TVRO antenna at Washington, D.C., to the *Galaxy I* satellite are $E = 16.8°$ and $A = 247.9°$.

A block diagram describing the receiving system is shown in Fig. 8–26. Using (8–43), the CNR at the detector input of the receiver is

$$\left(\frac{C}{N}\right)_{\text{dB}} = (P_{\text{EIRP}})_{\text{dBw}} - (L_{\text{FS}})_{\text{dB}} + \left(\frac{G_{\text{AR}}}{T_{\text{syst}}}\right)_{\text{dB}} - k_{\text{dB}} - B_{\text{dB}} \tag{8-55}$$

The corresponding path loss for $d = 24{,}784$ mi at a frequency of 4 GHz is

$$(L_{\text{FS}})_{\text{dB}} = 20\log\left(\frac{4\pi d}{\lambda}\right) = 196.5 \text{ dB}$$

For a 10-ft (3.05-m) parabola, using Table 8–4, we find that the receiving antenna gain is

$$(G_{\text{AR}})_{\text{dB}} = 10\log\left[\frac{7\pi(3.05/2)^2}{\lambda^2}\right] = 10\log(9085) = 39.6 \text{ dB}$$

Using Eqs. (8–41) and (8–37), we obtain the system noise temperature

$$T_{\text{syst}} = T_{\text{AR}} + T_{\text{feed}} + \frac{T_{\text{LNA}}}{G_{\text{feed}}} + \frac{T_{\text{receiver}}}{G_{\text{feed}}G_{\text{LNA}}} \tag{8-56}$$

where the specified antenna noise temperature, including the feed, is $T_A = (T_{\text{AR}} + T_{\text{feed}}) = 32$ K.

The T_{AR} is *not* the ambient temperature of the antenna (which is about 290 K) because the effective resistance of the antenna does not consist of a thermal resistor but of a radiation resistor. That is, if RF power is fed into the antenna, it will not be dissipated in the antenna but instead will be radiated into space. Consequently, T_{AR} is called the *sky noise temperature* and

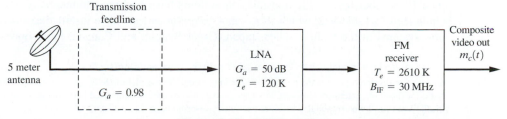

Figure 8–26   TVRO terminal.

is a measure of the amount of cosmic noise power that is received from outer space and noise caused by atmospheric attenuation, where this power is $P_{AR} = kT_{AR}B$. Graphs giving the measured sky noise temperature as a function of the RF frequency and the elevation angle of the antenna are available [Jordan, 1985, Chap. 27; Pratt and Bostian, 1986].

Substituting the receive system parameters into Eq. (8–56), we have

$$T_{\text{syst}} = 32 + \frac{40}{0.98} + \frac{2610}{(0.98)(100,000)} = 72.8 \text{ K}$$

Thus,[†]

$$\left(\frac{G_{AR}}{T_{\text{syst}}}\right)_{\text{dB}} = 10 \log\left(\frac{9,085}{72.8}\right) = 21.0 \text{ dB/K}$$

Also,

$$(B)_{\text{dB}} = 10 \log(30 \times 10^6) = 74.8$$

Substituting these results into Eq. (8–55) yields

$$\left(\frac{C}{N}\right)_{\text{dB}} = 36 - 196.5 + 21.0 - (-228.6) - 74.8$$

Thus, the CNR at the detector input (inside the FM receiver) is

$$\left(\frac{C}{N}\right)_{\text{dB}} = 14.3 \text{ dB} \qquad\qquad (8–57)$$

A CNR of 14.3 dB does not seem to be a very good, does it? Actually, the question is: Does a CNR of 14.3 dB at the detector input give a high-quality video signal at the detector output? This question can be answered easily if we have detector performance curves like those shown in Fig. 8–27. Two curves are plotted in this figure. One is for the output of a receiver that uses an FM discriminator for the FM detector, and the other is for a receiver (Microdyne 1100TVR) that uses a PLL FM detector to provide threshold extension.[‡] For a CNR of 14.3 dB on the input, both receivers will give a 51-dB output SNR, which corresponds to a high-quality picture. However, if the receive signal from the satellite fades because of atmospheric conditions by 6 dB, the output SNR of the receiver with an FM discriminator will be only about 36 dB, whereas the receiver with threshold extension detection will have an output SNR of about 45 dB.

8–7 FIBER-OPTIC SYSTEMS

Since 1980, fiber-optic cable transmission systems have become commercially viable. This is indicated in Table 8–2, which shows that common carriers have installed extensive fiber-optic systems in the United States, Japan, Great Britain, and Canada. Fiber-optic cable is

[†] Although it is a misnomer, engineers specify dB/K as the units for $(G/T)_{\text{dB}}$.

[‡] The threshold point is the point where the knee occurs in the plot of $(S/N)_{\text{out}}$ vs. $(C/N)_{\text{in}}$. In the characteristic for threshold extension receivers, this knee is extended to the left (improved performance) when compared with the knee for the corresponding receiver that uses a discriminator as a detector.

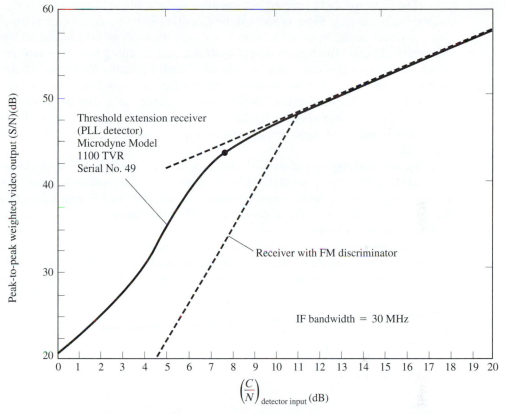

Figure 8–27   Output SNR as a function of FM detector input CNR. [Curve for the Microdyne receiver, Model 1100TVR, courtesy of the Microdyne Corporation, Ocala, FL, 1981.]

the preferred underground transmission medium. Also, as shown in Table 8–2, transoceanic fiber-optic cable systems are now popular. Digital fiber-optic systems use simple OOK modulation of an optical source to produce a modulated light beam. The optical source has a wavelength in the range of 0.85 to 1.6 μm, which is about 190 to 350 THz[†]. (Analog AM fiber-optic systems can also be built.)

The optical sources can be classified into two categories: (1) light-emitting diodes (LED) that produce noncoherent light and (2) solid-state lasers that produce coherent (single-carrier-frequency) light. The LED source is rugged and inexpensive. It has a relatively low power output of about -15 dBm (including coupling loss) and a small modulation bandwidth of about 50 MHz. The laser diode has a relatively high power output of about $+5$ dBm (including coupling loss) and a large modulation bandwidth of about 1 GHz. The laser diode is preferred over the LED, since it produces coherent light and has high output power.

[†] One THz is 10^{12} Hz.

The light source is coupled into the fiber-optic cable, and this cable is the channel transmission medium. Fiber cables can be classified into two categories: (1) multimodal and (2) single mode. Multimodal fiber has a core diameter of 50 μm and a cladding diameter of 125 μm. The light is reflected off the core-cladding boundary as it propagates down the fiber to produce multiple paths of different lengths. This causes pulse dispersion of the OOK signal at the receiving end and thus severely limits the transmission bit rate that can be accommodated. The single-mode fiber has a core diameter of approximately 8 μm and propagates a single wave. Consequently, there is little dispersion of the received pulses of light. Because of its superior performance, single-mode fiber is preferred.

At the receiving end of the fiber, the receiver consists of a PIN diode, an avalanche photodiode (APD), or a GaAsMESFES transistor used as an optical detector. In 0.85-μm systems, any of these devices may be used as a detector. In 1.3- and 1.55-μm systems, the APD is usually used. These detectors act like a simple envelope detector. As shown in Fig. 7–14, better performance (lower probability of bit error) could be obtained by using a coherent detection system that has a product detector. This requires a coherent light source at the receiver that is mixed with the received OOK signal via the nonlinear action of the photodetector. It is possible to have coherent PSK and FSK systems [Basch and Brown, 1985], but they are usually not cost-effective. Dense wavelength division multiplexing (DWDM) systems have a capacity on the order of 400 Gb/s, as shown in Table 8–2.

Example 8–5 LINK BUDGET FOR A FIBER-OPTIC SYSTEM

Figure 8–28 shows the configuration of the AT&T FT-2000 fiber-optic system. It consists of two bidirectional fiber-optic rings with terminals located at service points (such as towns) within a given geographical area. The ring configuration is used to provide redundant data paths; if a cable is cut anywhere around the ring, service is still provided to the terminals by the remaining in-service fiber. The terminals are called add/drop terminals because they add/drop data from the ring for the subscribers that the terminal serves from its location. As shown in Fig. 8–28, two fibers are required in the ring for full duplex service. That is, one fiber provides for the transmitted (Tx) data, and the other fiber provides the simultaneous received (Rx) data.

Specifications for the AT&T FT-2000 system are given in Table 8–6. This system has OC-48 capacity, which is equivalent to 48 DS-3 circuits or 32,256 full-duplex VF circuits as described in Sec. 3–9. (See Table 3–10.)

A link budget analysis for a FT-2000 line span operating at the 1.5 μm wavelength is given in Table 8–7. It shows that a maximum line span of 92 km (57.5 miles) may be used between optical repeaters (or add/drop terminals). Moreover, an optional lightwave booster amplifier may be used at the transmitter end to increase the transmitted optical power to +16 dBm. This allows the maximum line span to be increased to 140 km (87.5 miles) using standard fiber or to 160 km (100 miles) if dispersion shifted fiber is used [AT&T, 1994].

TABLE 8–6   SPECIFICATIONS FOR THE FT-2000 OC-48 LIGHTWAVE SYSTEM

Maximum capacity for two-fiber bidirectional rings	24 DS-3 equivalents (16,128 full-duplex VF circuits per fiber pair) per span 48 DS-3 equivalents (32,256 full-duplex VF circuits) per terminal
Line rate	2.488 Gb/s
Line code	Scrambled unipolar NRZ
Wavelength	1310 nm $\pm$ 20 nm or 1550 nm $\pm$ 15 nm
Optical source	Distributed feedback (DFB) laser
Optical detector	Avalanche photodiode (APD)
Optical fiber	Single mode

Transmitted power
  OC-48 1.31 μm std. performance
  OC-48 1.3 μm high performance
  OC-48 1.55 μm std. performance
	
	Max: +2.5 dBm, Min: −2.0 dBm [†]
	Max: +5.5 dBm, Min: +1.0 dBm[†]
	Max: +4.0 dBm, Min: −2.0 dBm[†]

Receiver sensitivity	−27.0 dBm
Maximum receiver power (no overload)	−10 dBm for 1.31 μm, −9.0 dBm for 1.55 μm
Bit error rate (BER)	$<10^{-10}$ for −27 dBm receiver input $<10^{-9}$ accumulated BER for systems up to 400 km (250 miles)

Maximum repeater spacing
  OC-48 1.31 μm high performance
  OC-48 1.55 μm std. performance
	
	60 km (0.45 dB/km loss fiber)
	92 km (0.25 dB/km loss fiber), dispersion limited

[†] Includes transmitter/receiver connector losses of 0.7 dB each (worst case) and system margins.

Source: AT&T 365-575-100 Manual, *FT-2000 OC-48 Lightwave System*, December 1994.

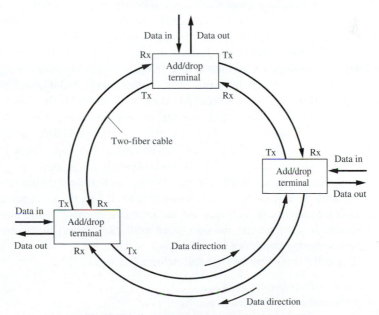

Figure 8–28   FT-2000 optical-fiber ring system.

TABLE 8–7   LINK BUDGET ANALYSIS FOR A 1.55-μm FT-2000 FIBER SPAN

Description	Value
Maximum transmitted power	+4.0 dBm
Receiver sensitivity	−27.0 dBm
Available margin	31.0 dB
Losses in a 57-mile fiber link	
Optical path penalty	2.0 dB
Transmitter/receiver connector loss (0.7 dB each)	1.4 dB
Fiber attenuation (92 km × 0.25 dB/km)[†]	23.0 dB
System margin	4.6 dB
Total loss	31.0 dB

[†]The 0.25-dB/km loss includes splicing losses.

8–8 CELLULAR TELEPHONE SYSTEMS

This section describes cellular telephone and PCS (personal communication service) systems and includes standards for many analog and digital systems in both the 900-MHz and 1,900-MHz bands.

Since the invention of wireless systems, the goal of telephone engineers has been to provide telephone service to individuals by using radio systems to link phone lines with persons in their cars or on foot. There is not enough spectral space to assign a permanent radio channel for each phone subscriber to use over a wide geographical area. However, it is possible to accommodate a large number of subscribers if each radio channel covers only a small geographical area and if the channels are shared (via FDMA, TDMA, or CDMA).

This *cellular radio* concept is illustrated in Fig. 8–29. Each user communicates via radio from a cellular telephone set to the cell-site base station. This base station is connected via telephone lines or a microwave link to the *mobile switching center* (MSC). The MSC connects the user to the called party. If the called party is land based, the connection is via the central office (CO) to the terrestrial telephone network. If the called party is mobile, the connection is made to the cellular site that covers the area in which the called party is located, using an available radio channel in the cell associated with the called party. Theoretically, this cellular concept allows any number of mobile users to be accommodated for a given set of radio channels. That is, if more channels are needed, the existing cell sizes are decreased, and additional small cells are inserted, so that the existing channels can be reused more efficiently. The critical consideration is to design the cells for acceptable levels of cochannel interference [Lee, 1986]. As the mobile user travels from one cell to another, the MSC automatically switches the user to an available channel in the new cell, and the telephone conversation continues uninterrupted.

The cellular concept has the following advantages:

- Large subscriber capacity.
- Efficient use of the radio spectrum.
- Service to hand-held portables, as well as vehicles.
- High-quality telephone and data service to the mobile user at relatively low cost.

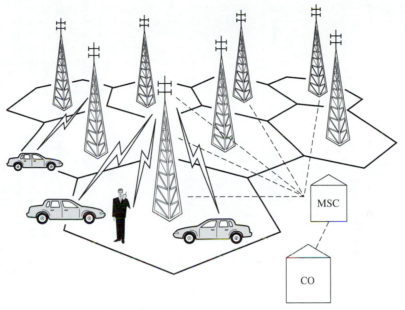

Figure 8–29   Cellular telephone system.

First Generation (1G)—The AMPS Analog System

As shown in Table 8–8, the first generation (1G) cellular telephone system used in the United States was the Advanced Mobile Phone System (AMPS), which was developed by AT&T and Motorola. It is an analog system, since it uses the VF audio signal to frequency modulate (FM) a carrier. To implement this AMPS concept in the United States, the FCC had to find spectral space for assignment. It did so by using the 806- to 890-MHz band that was once assigned to TV channels 70 to 83 (see Table 8–12), but discontinued in 1974. Part of this band was assigned for cellular service, as shown in Table 8–8. Furthermore, the FCC decided to license two competing cell systems in each geographical area; one licensed to a conventional telephone company and another to a nontelephone company common carrier. The nontelephone system is called the *A service,* or *nonwireline service*, and the telephone company system is called the *B service,* or *wireline service.* Thus, subscribers to a 1G cellular system have the option of renting service from the wireline company or the nonwireline company. In addition, as shown in the Table 8–8, the standards provide for full duplex service. That is, one carrier frequency is used for mobile to cell base-station transmission, and another carrier frequency is used for base-station to mobile transmission. Control is provided by FSK signaling. Of the 416 channels that may be used by a cellular service licensee, 21 are used for paging (i.e., control) with FSK signaling. FSK signaling is also used at the beginning and end of each call on a VF channel. When a cellular telephone is turned on, its receiver scans the paging channels looking for the strongest cell-site signal and then locks onto it. The cell site transmits FSK data continuously on a paging channel, and, in particular,

TABLE 8–8 MAJOR ANALOG FDMA CELLULAR TELEPHONE SYSTEMS

	System Name[a] and Where Used				
	AMPS North America			ETACS United Kingdom	JTACS Japan
Item	A Service (Nonwire Line)	B Service (Wire Line)			
Year introduced	1983	1983		1985	1988
Base cell station					
Transmit bands (MHz)	869–880, 890–891.5	880–890, 891.5–894		917–933, 935–960	860–870
Mobile station					
Transmit bands (MHz)	824–835, 845–846.5	835–845, 846.5–849		872–888, 890–915	915–925
Maximum power (watts)	3	3		3	2.8
Cell size, radius (km)	2–20	2–20		2–20	2–20
Number of duplex channels	416[b]	416[b]		1,000	400
Channel bandwidth (kHz)	30	30		25	25
Modulation					
Voice	FM 12-kHz peak deviation	FM 12-kHz peak deviation		FM 9.5-kHz peak deviation	FM 9.5-kHz peak deviation
Control signal	FSK 8-kHz peak deviation	FSK 8-kHz peak deviation		FSK 6.4-kHz peak deviation	FSK 6.4-kHz peak deviation
(Voice and paging channels)	10 kbits/s Manchester line code	10 kbits/s Manchester line code		8 kbits/s Manchester line code	8 kbits/s Manchester line code

[a] AMPS, advanced mobile phone system; ETACS, extended total access communication system; JTACS, Japan total access communication system.
[b] 21 of the 416 channels are used exclusively for paging (i.e., control channels).

it sends control information addressed to a particular cellular phone when a call comes in for it. That is, the FSK control signal tells the phone which channel to use for a call.

Each cellular telephone contains a PROM (programmable read-only memory) or EPROM (erasable programmable read-only memory)—called a *numeric assignment module* (NAM). The NAM is programmed to contain the telephone number—also called the *electronic service number* (ESN)—of the phone and the serial number of the phone, which is assigned by the manufacturer.

When the phone is "on the air" it automatically transmits its serial number to the MSC. The serial number is used by the MSC to lock out phone service to any phone that has been stolen. This feature, of course, discourages theft of the units. The MSC uses the telephone number of the unit to provide billing information. When the phone is used in a remote city, it can be placed in the *roam* mode, so that calls can be initiated or received, but still allow the service to be billed via the caller's "hometown" company.

When a call is placed, the following sequence of events occurs:

1. The cellular subscriber initiates a call by keying in the telephone number of the called party and then presses the *send* key.
2. The MSC verifies that the telephone number is valid and that the user is authorized to place a call.
3. The MSC issues instructions to the user's cellular phone indicating which radio channel to use.
4. The MSC sends out a signal to the called party to ring his or her phone. All of these operations occur within 10s of initiating the call.
5. When the called party answers, the MSC connects the trunk lines for the two parties and initiates billing information.
6. When one party hangs up, the MSC frees the radio channel and completes the billing information.

While a call is in progress, the cellular subscriber may be moving from one cell area to another, so the MSC does the following:

- Monitors the signal strength from the cellular telephone as received at the cell base station. If the signal drops below some designated level, the MSC initiates a "hand off" sequence.
- For "hand off," the MSC inquires about the signal strength received at adjacent cell sites.
- When the signal level becomes sufficiently large at an adjacent cell site, the MSC instructs the cellular radio to switch over to an appropriate channel for communication with that new cell site. This switching process takes less than 250 ms and usually is unnoticed by the subscriber.

There are many different cellular standards. Table 8–8 shows three major analog standards that are used around the world. Moreover, since cellular telephones are so popular, it is clear that these wideband FM analog systems do not have the capacity to accommodate the expected new subscribers. Consequently, narrowband FM analog systems with higher capacity,

TABLE 8–9   NARROWBAND ANALOG FDMA CELLULAR TELEPHONE SYSTEMS

	System Name[a] and Where Used:	
Item	NAMPS North America	NTACS Japan
Year introduced	1992	1993
Base cell station		
Transmit band (MHz)	869–894	843–846, 860–870
Mobile station		
Transmit band (MHz)	824–849	898–901, 915–925
Maximum power (watts)	3	2.8
Number of duplex channels	2496	1040
Channel bandwidth		
Voice (kHz)	10	12.5
Paging (kHz)	30	25
Modulation		
Voice	FM	FM
	5-kHz peak deviation	5-kHz peak deviation
Signaling (voice channel)	FSK	FSK
	0.7-kHz peak deviation	0.7-kHz peak deviation
	100 bits/s, Manchester	100 bits/s, Manchester
Signaling (paging channel)	FSK	FSK
	8-kHz peak deviation	6.4-kHz peak deviation
	10 kbits/s, Manchester	8 kbits/s, Manchester

[a]NAMPS, narrowband advanced mobile phone system; NTACS, narrowband total access communication system.

shown in Table 8–9, are replacing wideband systems on a channel-by-channel basis as needed. For example, one wideband 30-kHz AMPS channel can be replaced by three narrowband 10-kHz NAMPS channels to triple the capacity. For larger capacity, new digital cellular systems, shown in Table 8–10, are used. The GSM digital system is used in Europe to provide one uniform standard that replaces the numerous analog systems that are being phased out.

Second Generation (2G)—The Digital Systems

Table 8–10 shows standards for the popular second-generation (2G) cellular telephone systems. They are all digital. The analog VF signal is converted to a compressed digital bit stream, which is modulated onto a carrier. The bit rate of the compressed data is designed to be relatively small so that a large number of users can be accommodated.

    The digital approach has many advantages. It provides private conversations, resistance to cellular fraud, and allows special features, such as caller ID, that can be easily implemented. The digital approach provides noise-free telephone conversations out to the edge of the cell fringe area because of the sharp threshold effect of digital systems. That is, the data are received either error-free for recovery of noise-free audio or with so many errors that the audio is turned off. Compare this with the noisy audio of analog FM systems when fringe-area signals are received.

TABLE 8-10 MAJOR DIGITAL CELLULAR TELEPHONE SYSTEMS

Item	System Name[a] and Where Used				
	GSM Europe, Asia	IS-136, NADC North America	IS-95 North America, Asia	iDEN North America	
Year introduced	1990	1991	1993	1994	
Base cell station[h]					
Transmit bands (MHz)	935–960	869–894	869–894	851–866,	
Mobile station[h]					
Transmit bands (MHz)	890–915	824–849	824–849	806–821	
Maximum power (watts)	20	3	0.2	3	
Number of duplex channels	125	832	20	600	
Channel bandwidth (kHz)	200	30	1250	25	
Channel access method	TDMA	TDMA	CDMA[f]	TDMA	
Users per channel	8	3	35[g]	3	
Modulation[b]					
Type	GMSK	$\pi/4$ DQPSK	QPSK	quad-16QAM	
Data rate	270.833 kb/s	48.6 kb/s	9.6 kb/s[f]	64 kb/s	
Filter	0.3R Gaussian[c]	$r = 0.35$ raised cosine[d]		$r = 0.2$ raised cosine[d]	
Speech coding[e]	RPE-LTP	VSELP	QCELP	VSELP	
	13 kb/s	8 kb/s	8 kb/s (variable)	8 kb/s	

[a]GSM, group special mobile; NADC, North American digital cellular; iDEN, Integrated digital enhanced network.

[b]GMSK, Gaussian MSK (i.e., $h = 0.5$ FSK, also called FFSK, with Gaussian premodulation filter; see Sec. 5-11 and footnote c; $\pi/4$ DQPSK, differential QPSK with $\pi/4$ rotation of reference phase for each symbol transmitted; quad-16QAM is four 16QAM carriers for each signal.

[c]Baseband Gaussian filter with 3 dB bandwidth equal to 0.3 of the bit rate.

[d]Square root of raised cosine characteristic at transmitter and receiver.

[e]RPE-LTP, regular pulse excitation—long-term prediction; VSELP, vector sum excited linear prediction filter; QCELP, Qualcomm's codebook excited linear prediction.

[f]The spread spectrum chip rate is 1.2288 Mchips/s, for a spreading factor of 128 or 21-dB processing gain.

[g]For CDMA, channel frequencies may be reused in adjacent cells; other methods (TDMA and FDMA) require a seven-cell reuse pattern.

[h]For 1,900-MHz-band PCS system, the base station transmit band is 1,930–1,990 MHz and the mobile station transmit band is 1,850–1,910 MHz.

603

The *Group Special Mobile (GSM)* system (see Table 8–10) is the 2G system that replaces many different 1G analog systems in Europe. Now, one can travel around Europe and remain in communication via the use of a GSM phone, instead of needing up to 15 different types of 1G cell phones. GSM uses TDMA that provides time slots to accommodate up to eight users on each 200-kHz-wide channel. The GSM phone is programmed for a particular user's account by an attached *smart card*, which contains the user's phone number and other account information. The smart card allows the subscriber to clone any GSM phone, whether it is a handset or a fixed mobile phone, to his/her account by attaching the card. If a user travels to a country where a different frequency band is used, a rented GSM phone can be instantly programmed to the user's account by attaching the card. GSM has been adopted by many countries around the world and is used by about 50% of the worldwide cellular subscribers. Details of GSM system design are published in a book by Garg and Wilkes [Garg and Wilks, 1999].

The *IS-54, IS-136 North American Digital Cellular (NADC)* system was the first 2G standard adopted for use in the United States. It uses TDMA to accommodate up to three users on each 30-kHz-wide channel. It is designed to replace AMPS analog signaling on a channel-for-channel basis. Thus, an incumbent AMPS service provider can gradually convert some channels to IS-54 digital service as needed, depending on customer demand.

The *IS-95* standard was developed by Qualcomm and adopted for use in the United States 900-MHz cellular band in 1993. It uses CDMA to accommodate up to 35 users in a 1.2-MHz-wide channel. One IS-95 channel can be used to replace 41 30-kHz-wide AMPS channels. The advantage of CDMA is that the same channel frequencies may be reused in adjacent cell sites, because the users are distinguished by their different spreading codes. (A different set of spreading codes is used in adjacent cell sites.) The CDMA approach has many advantages. For example, the number of allowed users for each channel bandwidth is not a hard number (as it is in TDMA, where there are a fixed number of available time slots), since additional users with new spreading codes can be added. The additional users cause a graceful increase in noise level. Users can be added until the errors become intolerable (due to the increased noise level). It is expected that the CDMA approach will become more and more popular and become the dominant type of system. Details of CDMA are given in books by Rhee [Rhee, 1998] and Garg [Garg, 2000].

The *Integrated Digital Enhanced Network (iDEN)* was developed by Motorola. It has been adopted by Nextel in their system, which uses specialized mobile radio (SMR) frequencies in the 800 MHz band. The iDEN system uses TDMA to accommodate up to 3 users in each 25 kHz-wide channel. Each channel uses four 16QAM carriers so that $4 \times 4 = 16$ data bits are sent in parallel during each symbol time-interval. This is similar to OFDM signaling, which was discussed in Sec. 5–12. Thus, the time duration of a symbol is relatively long, so that the transmitted signal is resistant to multipath fading. In addition to the usual cellular phone features, the iDEN protocol allows for a *virtual private network* (VPN) to be established for business subscribers. Thus, co-workers of the business can direct connect a private call or a group call to each other over the VPN.

As indicated in Table 8–10, all of these 2G systems use some sort of low-bit-rate speech coding of the VF signal so that a larger number of users can be accommodated in the limited RF bandwidth. See Sec. 3–8 for a discussion of speech coding. For example, the vector sum excited linear prediction filter (VSELP) encoder used for the NADC and iDEN systems is a modified version of the codebook excited linear prediction (CELP) technique developed by B. Atal at the AT&T Bell Laboratories [Gerson and Jasiuk, 1990]. This technique mimics the two main parts of the human speech system: the vocal cords and the vocal tract. The vocal tract is simulated by a time-varying linear prediction filter, and the sounds of the vocal cord used for the filter excitation are simulated with a database (i.e., a codebook) of possible excitations. The VSELP encoder (at the transmitter) partitions the talker's VF signal into 20-ms segments. The encoder sequences through the possible codebook excitation patterns and the possible values for the filter parameters to find the synthesized speech segment that gives the best match to the VF segment. The encoder parameters that specify this best match are then transmitted via digital data to the receiver, where the received data establish the parameters for the receiver speech synthesizer so that the VF signal may be reproduced for the listener.

The 1,900-MHZ Band PCS Systems

In order to provide more cellular competition in the United States, the FCC has reallocated frequencies in the 1.9 GHz band from point-to-point microwave service to cellular PCS. The following describes the PCS band:

> 1,850–1,910 MHz   PCS Mobile Transmit
>
> 1,930–1,990 Mhz   Base Station Transmit to PCS Mobile

Furthermore, this band is divided into six duplex blocks for assignment to six PCS providers in each service area. This is in addition to the two 900-MHz cellular providers for each service area. The choice of which cellular standard to use for service is decided by each licensed service provider. Since 1997, the licensed companies for these new PCS blocks have been coming "on the air." They provide service with either the GSM (TDMA), the IS-54 (TDMA), or the IS-95 (CDMA) standard, which are described in Table 8–10 (using the 1.9-GHz PCS frequencies).

There are now multiple PCS companies providing wireless telephone service in cities and along major highways that connect these cities. This competition has greatly reduced the rates paid by subscribers for wireless telephone service. Each PCS phone is customized for the PCS standard that has been selected for use by his/her service provider. In other words, there is no common digital standard in use by all companies. However, there is universal 900-MHz AMPS cellular service almost everywhere in the United States, including low-population rural areas. If a digital phone user roams to another service area, his or her phone will not work unless it is compatible with the digital standard used in that roaming area. Consequently, most companies offer dual-mode or triple-mode phones that have fall-back coverage to a 900-MHz AMPS service provider.

Third Generation (3G) Systems

From the discussion in the previous section, it is clear that a worldwide compatible cellular/PCS standard is needed for the third-generation (3G) system. It would allow wireless users to roam worldwide with one phone. This has been under discussion for some time, but countries and equipment manufacturers have noncompatible political, technical, and patent issues. The European Standards Institute (ETSI) has proposed a Universal Mobile Telecommunications System (UMTS), and the International Telecommunications Union (ITU) is developing the International Mobile Telecommunications 2000 (IMT2000) system, which includes UMTS as a subset. In addition to voice, the 3G system should support data transmission with rates selectable from 9.6 kb/s to 2 Mb/s. Many believe that both GSM and CDMA should be allowed and that the 3G system should be backward compatible with existing systems. It should also provide a seamless extension of the Internet. The 3G system needs to be cost-effective, and frequencies must be found for worldwide compatibility. Time will tell if such a 3G system will become a reality.

8–9 TELEVISION

Television (TV) is a method of reproducing fixed or moving visual images by the use of electronic signals. There are numerous kinds of TV systems, and different standards have been adopted around the world. Moreover, as discussed subsequently, TV is in a state of flux since there is a worldwide push for *digital television* (DTV). First, we will concentrate on black-and-white broadcast analog TV, as it has developed in the United States since the 1930s, and then color television and DTV will be examined.

Black-and-White Television

The black-and-white image of a scene is essentially the intensity of the light as a function of the x- and y-coordinates of the scene and of time. However, an electrical waveform is only a function of time, so that some means of encoding the x- and y-coordinates associated with the intensity of light from a single point in the scene must be used. In broadcast TV, this is accomplished by using *raster scanning*, which is shown in Fig. 8–30. It is assumed that the whole scene is scanned before any object in the scene has moved appreciably. Thus, moving images can be transmitted by sending a series of still images, just as in motion pictures. As the scene is scanned, its intensity is signified by the amplitude of the video signal. A synchronizing pulse is inserted at the end of each scan line by the electronic switch, as shown in Fig. 8–30, to tell the TV receiver to start another scan line. For illustrative purposes, a black triangle is located in the upper left portion of the scene. In Figs. 8–30b and 8–30c, numbers are used to indicate the corresponding times on the video waveform, with the location of the point being scanned in the scene at that instant. The composite video signal is actually a hybrid signal that consists of a digital waveform during the synchronizing interval and an analog waveform during the video interval. Operating personnel monitor the quality of the video by looking at a *waveform monitor*, which displays the composite video waveform, as shown in Fig. 8–30c.

　　The picture (as viewed) is generated by scanning left to right to create a line, and the lines move from the top to the bottom of the screen. During retrace, indicated by the dashed lines from right to left, the CRT is turned off by the video level that is "blacker than black" during the sync pulse interval.[†] The scene is actually covered by scanning alternate lines

---

[†] In addition, most TV sets switch off the electron beam during retrace by using the sync pulse obtained from the horizontal output (flyback) transformer.

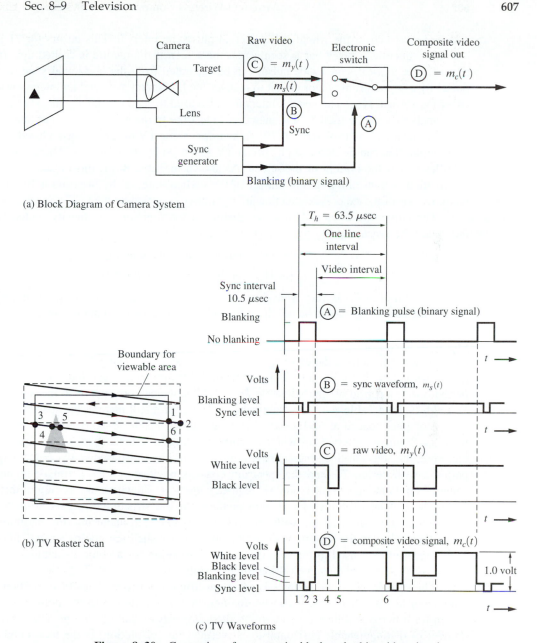

(a) Block Diagram of Camera System

(b) TV Raster Scan

(c) TV Waveforms

Figure 8–30   Generation of a composite black-and-white video signal.

from the top to the bottom (one field) and returning to the top to scan the missed lines (the next field). This technique is called *interlacing* and reduces the flicker effect on moving pictures, giving two fields (passes down the screen) for every frame. In the United States there are approximately 30 frames/s transmitted, which is equivalent to 60 fields/s for the 2:1 interlacing that is used. When the scan reaches the bottom of a field, the sync pulse is widened out to cover the whole video interval plus sync interval. This occurs for 21 lines, and, consequently, a black bar is seen across the screen if the picture is "rolled up." In the receiver, the vertical sync (wide pulse) can be separated from the horizontal sync (narrow pulse) by using a differentiator circuit to recover the horizontal sync and an integrator to recover the vertical sync. The vertical interval is also used to send special signals, such as testing signals that can be used by TV engineers to test the frequency response and linearity of the equipment on-line [Solomon, 1979]. In addition, some TV programs have closed-caption information that can be decoded on special TV sets used by deaf viewers. These sets insert subtitles across the bottom of the screen. As shown in Table 8–11, this closed-captioning information is transmitted during line 21 of the vertical interval by inserting a 16-bit code word. Details of the exact sync structure are rigorously specified by the FCC. (See Fig. 8–37.)

The composite baseband signal consisting of the *luminance* (intensity) video signal $m_y(t)$ and the *synchronizing* signal $m_s(t)$ is described by

$$m_c(t) = \begin{cases} m_s(t), & \text{during the sync interval} \\ m_y(t), & \text{during the video interval} \end{cases} \tag{8–58}$$

The spectrum of $m_c(t)$ is very wide since $m_c(t)$ contains rectangular synchronizing pulses. In fact, the bandwidth would be infinite if the pulses had perfect rectangular shapes. In order to reduce the bandwidth, the pulses are rounded as specified by FCC standards. (See Fig. 8–37.) This allows $m_c(t)$ to be filtered to a bandwidth of $B = 4.2$ MHz (U.S. standard). For a still picture, the exact spectrum can be calculated using Fourier series analysis (as shown in Sec. 2–5). For a typical picture all fields are similar, and thus $m_c(t)$ would be periodic with a period of approximately $T_0 = 1/60$ s, which corresponds to the field rate of 60 fields/s. (See Table 8–11 for exact values.) Thus, for a still picture, the spectrum would consist of lines spaced at 60-Hz intervals. However, from Fig. 8–30, we can observe that there are dominant intervals of width T_h corresponding to scanned lines of the frame. Furthermore, the adjacent lines usually have similar waveshapes. Consequently, $m_c(t)$ is quasi-periodic with period T_h, and the spectrum consists of clusters of spectral lines that are centered on harmonics of the scanning frequency $nf_h = n/T_h$, where the spacing of the lines within these clumps is 60 Hz. For moving pictures, the line structure of the spectrum "fuzzes out" to a continuous spectrum with spectral clusters centered about nf_h. Furthermore, between these clusters the spectrum is nearly empty. As we shall see, these "vacant" intervals in the spectrum are used to transmit the color information for a color TV signal. This, of course, is a form of frequency-division multiplexing.

The resolution of TV pictures is often specified in terms of lines of resolution. The number of horizontal lines that can be distinguished from the top to the bottom of a TV screen for a horizontal line test pattern is called the *vertical-line resolution*. The maximum number of distinguishable horizontal lines (vertical line resolution) is the total number of scanning lines in the raster less those not used for the image. That is, the vertical resolution is

$$n_v = (N_f - N_v) \quad \text{lines} \tag{8–59a}$$

where N_f is the total number of scanning lines per frame and N_v is the number of lines in the vertical interval (not image lines) for each frame. For U.S. standards (Table 8–11), the maximum vertical resolution is

TABLE 8–11   U.S. ANALOG TELEVISION BROADCASTING STANDARDS

Item	FCC Standard
Channel bandwidth	6 MHz
Visual carrier frequency	1.25 MHz $\pm$ 1000 Hz above the lower boundary of the channel
Aural carrier frequency	4.5 MHz $\pm$ 1000 Hz above the visual carrier frequency
Chrominance subcarrier frequency	3.579545 MHz $\pm$ 10 Hz
Aspect (width-to-height) ratio	Four units horizontally for every three units vertically
Modulation-type visual carrier	AM with negative polarity (i.e., a decrease in light level causes an increased real envelope level)
Aural carrier	FM with 100% modulation being $\Delta F = 25$ kHz with a frequency response of 50 to 15,000 Hz using 75-μs preemphasis
Visual modulation levels	
Blanking level	75% $\pm$ 2.5% of peak real envelope level (sync tip level)
Reference black level	7.5% $\pm$ 2.5% (of the video range between blanking and reference white level) below blanking level; this is called the *setup* level by TV engineers[a]
Reference white level	12.5% $\pm$ 2.5% of sync tip level[a]
Scanning	
Number of lines	525 lines/frame, interlaced 2:1
Scanning sequence	Horizontally: left to right; vertically: top to bottom
Horizontal scanning frequency, f_h	15,734.264 $\pm$ 0.044 Hz (2/455 of chrominance frequency); 15,750 Hz may be used during monochrome transmissions
Vertical scanning frequency, f_v	59.94 Hz (2/525 of horizontal scanning frequency); 60 Hz may be used during monochrome transmissions; 21 equivalent horizontal lines occur during the vertical sync interval of each field
Vertical interval signaling	Lines 13, 14, 15, 16. Teletext
	Lines 17, 18. Vertical interval test signals (VITS)
	Line 19. Vertical interval reference (VIR)
	Line 20, Field 1. Station identification
	Line 21, Field 1. Captioning data
	Line 21, Field 2. Captioning framing code ($\frac{1}{2}$ line)

[a] See Fig. 8–31b.

$$n_v = 525 - 42 = 483 \quad \text{lines} \tag{8–59b}$$

The resolution in the horizontal direction is limited by the frequency response allowed for $m_c(t)$. For example, if a sine-wave test signal is used during the video interval, the highest sine-wave frequency that can be transmitted through the system will be $B = 4.2$ MHz (U.S. standard), where B is the system video bandwidth. For each peak of the sine wave, a dot will appear along the horizontal direction as the CRT beam sweeps from left to right. Thus, the horizontal resolution for 2:1 interlacing is

$$n_h = 2B(T_h - T_b) \quad \text{pixels} \tag{8–60a}$$

where B is the video bandwidth, T_h is the horizontal interval, and T_b is the blanking interval. (See Fig. 8–30.) For U.S. standards, $B = 4.2$ MHz, $T_h = 63.5$ μs, and $T_b = 10.5$ μs, so that the horizontal resolution is

$$n_h = 445 \quad \text{pixels} \tag{8–60b}$$

Furthermore, because of poor video bandwidth and the poor interlacing characteristics of consumer TV sets, this horizontal resolution of 445 is usually not obtained in practice.[†] At best, the U.S. standard of 445 $\times$ 483 does not provide very good resolution, and this poor resolution is especially noticeable on large-screen TV sets. [For comparison, the super VGA (SVGA) standard for computer monitors provides a resolution as large as 1024 $\times$ 768.]

For broadcast TV transmission (U.S. standard), the composite video signal of Eq. (8–58) is inverted and amplitude modulated onto an RF carrier so that the AM signal is

$$s_v(t) = A_c[1 - 0.875m_c(t)] \cos \omega_c t \tag{8–61}$$

This is shown in Fig. 8–31. The lower sideband of this signal is attenuated so that the spectrum will fit into the 6-MHz TV channel bandwidth. This attenuation can be achieved by using a vestigial sideband filter, which is a bandpass filter that attenuates most of the lower sideband. The resulting filtered signal is called a *vestigial sideband* (VSB) signal, $s_{v_0}(t)$, as described in Sec. 5–5. The discrete carrier term in the VSB signal changes when the picture is switched from one scene to another, because the composite baseband signal has a nonzero dc level that depends on the particular scene being televised. The audio signal for the TV program is transmitted via a separate FM carrier located exactly 4.5 MHz above the visual carrier frequency.

The rated power of a TV visual signal is the effective isotropic radiated peak envelope power (EIRP, i.e., effective sync tip average power), and it is usually called simply the *effective radiated power* (ERP). The EIRP is the power that would be required to be fed into an isotropic antenna to get the same field strength as that obtained from an antenna that is actually used as measured in the direction of its maximum radiation. The antenna pattern of an omnidirectional TV station is shaped like a doughnut with the tower passing through the hole. (The antenna structure is located at the top of the tower for maximum line-of-sight coverage.) The ERP of the TV station (visual or aural) signal is

$$P_{\text{ERP}} = P_{\text{PEP}} G_A G_L \tag{8–62}$$

where

P_{PEP} = peak envelope power out of the transmitter,

G_A = power gain of the antenna (absolute units as opposed to decibel units) with respect to an isotropic antenna,

G_L = total gain of the transmission line system from the transmitter output to the antenna (including the duplexer gain).

For a VHF TV station operating on channel 5, some typical values are $G_A = 5.7$, $G_L = 0.873$ (850 ft of transmission line plus duplexer), and a visual PEP of 20.1 kW. This

---

[†] Typically, a horizontal resolution of about 300 lines is obtained.

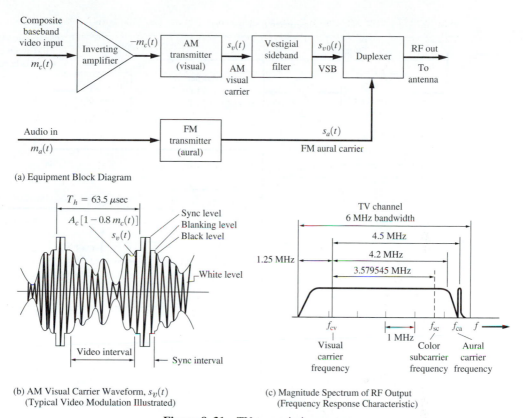

(a) Equipment Block Diagram

(b) AM Visual Carrier Waveform, $s_v(t)$
(Typical Video Modulation Illustrated)

(c) Magnitude Spectrum of RF Output
(Frequency Response Characteristic)

Figure 8-31   TV transmission system.

gives an overall ERP of 100 kW, which is the maximum power licensed by the FCC for channels 2 to 6 (low-VHF band). For a UHF station operating on channel 20, some typical values are $P_{\text{PEP}} = 21.7$ kW, $G_A = 27$, $G_L = 0.854$ (850 ft of transmission line), which gives an ERP of 500 kW.

A simplified block diagram of a black-and-white TV receiver is shown in Fig. 8-32. The composite video, $m_c(t)$, as described earlier, plus a 4.5-MHz FM carrier containing the audio modulation, appears at the output of the envelope detector. The 4.5-MHz FM carrier is present because of the nonlinearity of the envelope detector and because the detector input (IF signal) contains, among other terms, the FM aural carrier plus a discrete (i.e., sinusoidal) term, which is the video carrier, located 4.5 MHz away from the FM carrier signal. The intermodulation product of these two signals produces the 4.5-MHz FM signal, which is called the *intercarrier signal*, and contains the aural modulation. Of course, if either of these two input signal components disappears, the 4.5-MHz output signal will also disappear. This will occur if the white level of the picture is allowed to go below, say, 10% of the peak envelope level (sync tip level) of the AM visual signal. This is why the FCC specifies that the white level cannot be lower than $(12.5 \pm 2.5)\%$ of the peak envelope level (see Table 8-11 and Fig. 8-31b). When this occurs, a buzz will be heard in the sound signal since

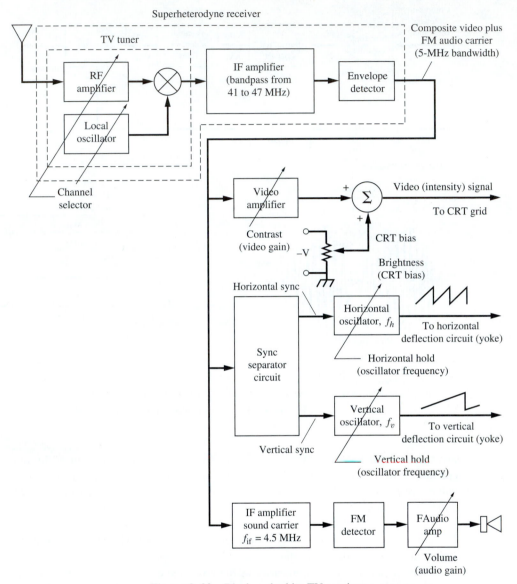

Figure 8–32   Black-and-white TV receiver.

the 4.5-MHz FM carrier is disappearing at a 60-Hz (field) rate during the white portion of the TV scene. The sync separator circuit consists of a lossy integrator to recover the (wide) vertical sync pulses and a lossy differentiator to recover the (narrow) horizontal pulses. The remainder of the block diagram is self-explanatory. The user-available adjustments for a black-and-white TV are also shown.

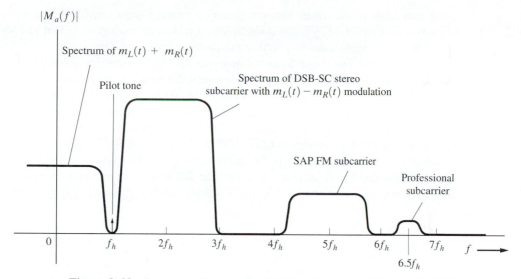

Figure 8–33 Spectrum of composite FDM baseband modulation on the FM aural carrier of U.S. TV stations.

MTS Stereo Sound

Stereo audio, also known as *multichannel television sound* (MTS), is also available and is accomplished using the FDM technique shown in Fig. 8–33. This FDM technique is similar to that used in standard FM broadcasting, studied in Sec. 5–7 and illustrated by Fig. 5–18. For compatibility with monaural audio television receivers, the left plus right audio, $m_L(t) + m_R(t)$, is frequency modulated directly onto the aural carrier of the television station. Stereo capability is obtained by using a DSB-SC subcarrier with $m_L(t) - m_R(t)$ modulation. The subcarrier frequency is $2f_h$, where $f_h = 15.734$ kHz is the horizontal sync frequency for the video. (See Table 8–11.) Capability is also provided for a *second audio program* (SAP), such as audio in a second language. This is accomplished by using an FM subcarrier of frequency $5f_h$. Furthermore, a *professional channel* subcarrier at $6.5f_h$ is allowed for the transmission of voice or data on a subscription service basis to paying customers. For further details on the FDM aural standards that have been adopted by television broadcasters and for other possible FDM formats, see a paper by Eilers [1985].

Color Television

Color television pictures are synthesized by combining red, green, and blue light in the proper proportion to produce the desired color. This is done by using a color CRT that has three types of phosphors: one type that emits red light, one that emits green light, and a third type that emits blue light when they are struck by electrons. Therefore, three electronic video signals—$m_R(t)$, $m_G(t)$, and $m_B(t)$—are needed to drive the circuits that create the red, green, and blue light. Furthermore, a compatible color TV transmission system is needed such that black-and-white TV sets will receive the luminance video signal, $m_y(t)$, which corresponds to the gray scale of the scene. This compatibility between color and black-and-white transmission is accomplished by using a multiplexing technique analogous to that of FM stereo, except that three signals are involved in the problem instead of two.

The National Television System Committee (NTSC) compatible color TV system that was developed in the United States and adopted for U.S. transmissions in 1954 will be discussed. This technique is illustrated in Fig. 8–34. To accomplish compatible transmission, the waveforms that correspond to the intensity of the red, green, and blue components of the scene—$m_R(t)$, $m_G(t)$, and $m_B(t)$, respectively—are linearly combined to synthesize the equivalent black-and-white signal, $m_y(t)$, and two other independent video signals, called the *in-phase* and *quadrature* components, $m_i(t)$ and $m_q(t)$. (The names of these components arise from the quadrature modulation technique that is used to modulate them onto a subcarrier.) The exact equations used for these signals can be expressed in matrix form:

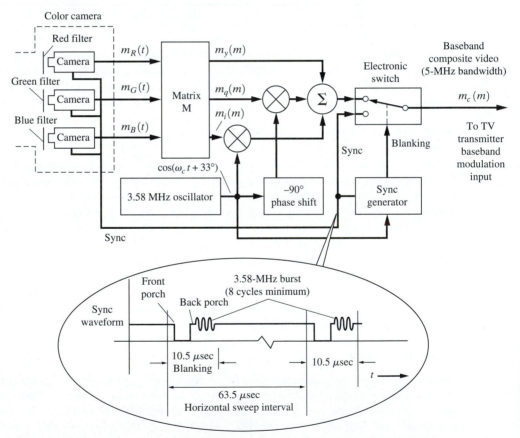

Figure 8–34 Generation of the composite NTSC baseband video signal. [Technically, $m_q(t)$ is low-pass filtered to an 0.5-MHz bandwidth and $m_i(t)$ is low-pass filtered to a 1.5-MHz bandwidth.]

$$\begin{bmatrix} m_y(t) \\ m_i(t) \\ m_q(t) \end{bmatrix} = \underbrace{\begin{bmatrix} 0.3 & 0.59 & 0.11 \\ 0.6 & -0.28 & -0.32 \\ 0.21 & -0.52 & 0.31 \end{bmatrix}}_{M} \begin{bmatrix} m_R(t) \\ m_G(t) \\ m_B(t) \end{bmatrix} \tag{8–63}$$

where M is a 3×3 matrix that translates the red, green, and blue signals to the brightness, in-phase, and quadrature phase video (baseband) signals. For example, from Eq. (8–63), the equation for the *luminance* (black-and-white) signal is

$$m_y(t) = 0.3 m_R(t) + 0.59 m_G(t) + 0.11 m_B(t) \tag{8–64}$$

Similarly, the equations for $m_i(t)$ and $m_q(t)$, which are called the *chrominance* signals, are easy to obtain. In addition, note that *if* $m_R(t) = m_G(t) = m_B(t) = 1$, which corresponds to maximum red, green, and blue, then $m_y(t) = 1$, which corresponds to the white level in the black-and-white picture.

The chrominance components are two other linearly independent components, and if they are transmitted to the color TV receiver together with the luminance signal, these three signals can be used to recover the red, green, and blue signals using an inverse matrix operation. The bandwidth of the luminance signal needs to be maintained at 4.2 MHz to preserve sharp (high-frequency) transition in the intensity of the light that occurs along edges of objects in the scene. However, the bandwidth of the chrominance signals does not need to be this large, since the eye is not as sensitive to color transitions in a scene as it is to black-and-white transitions. According to NTSC standards, the bandwidth of the $m_i(t)$ signal is 1.5 MHz, and the bandwidth of $m_q(t)$ is 0.5 MHz. When they are quadrature-modulated onto a subcarrier, the resulting composite baseband signal will then have a bandwidth of 4.2 MHz. (The upper sideband of the in-phase subcarrier signal is attenuated to keep a 4.2-MHz bandwidth.)

The composite NTSC baseband video signal is

$$m_c(t) = \begin{cases} m_s(t), & \text{during the sync interval} \\ m_y(t), + \mathrm{Re}\{g_{sc}(t)e^{j\omega_{sc}t}\}, & \text{during the video interval} \end{cases} \tag{8–65}$$

where

$$g_{sc}(t) = [m_i(t) - jm_q(t)]e^{j33°} \tag{8–66}$$

Equation (8–66) indicates that the chrominance information is quadrature modulated onto a subcarrier (as described in Table 4–1). $m_s(t)$ is the sync signal, $m_y(t)$ is the luminance signal, and $g_{sc}(t)$ is the complex envelope of the subcarrier signal. The subcarrier frequency is $f_{sc} = 3.579545$ MHz $\pm$ 10 Hz, which, as we will see later, is chosen so that the subcarrier signal will not interfere with the luminance signal even though both signals fall within the 4.2-MHz passband.

In analog broadcast TV, this composite NTSC baseband color signal is amplitude modulated onto the visual carrier, as described by Eq. (8–61) and shown in Fig. 8–31. In microwave relay applications and satellite relay applications, this NTSC baseband signal is frequency modulated onto a carrier, as described in Sec. 8–5. These color baseband signals can be transmitted over communication systems that are used for black-and-white TV, although the frequency response and linearity tolerances that are specified are tighter for color-TV transmission.

As indicated before, the complex envelope for the subcarrier, as described by Eq. (8–66), contains the color information. The magnitude of the complex envelope, $|g_{sc}(t)|$ is the *saturation,* or amount of color, in the scene as it is scanned with time. The angle of the complex envelope, $/g_{sc}(t)$, indicates the hue, or tint, as a function of time. Thus, $g_{sc}(t)$ is a vector that moves about in the complex plane as a function of time as the lines of the picture are scanned. This vector may be viewed on an instrument called a *vectorscope,* which has a CRT that displays the vector $g_{sc}(t)$, as shown in Fig. 8–35. The vectorscope is a common piece of test equipment that is used to calibrate color-TV equipment and to provide an on-line measure of the quality of the color signal being transmitted. The usual vectorscope presentation has the x-axis located in the vertical direction, as indicated on the figure. Using Eq. (8–66), we see that the positive axis directions for the $m_i(t)$ and $m_q(t)$ signals are also indicated on the vectorscope presentation, together with the vectors for the saturated red, green, and blue colors. For example, if saturated red is present, then $m_R = 1$ and $m_G = m_B = 0$. Using Eq. (8–63), we get $m_i = 0.6$ and $m_q = 0.21$. Thus, from Eq. (8–66), the saturated red vector is

$$(g_{sc})_{red} = (0.60 - j0.21)e^{j33°} = 0.64 \ /13.7°$$

Similarly, the saturated green vector, is $(g_{sc})_{green} = 0.59 \ /151°$, and the saturated blue vector is $(g_{sc})_{blue} = 0.45 \ /257°$. These vectors are shown in Fig. 8–35. It is interesting to note that the complementary colors have opposite polarities from each other on the vectorscope, where cyan is the complement of red, magenta is the complement of green, and yellow is the complement of blue.

For color subcarrier demodulation at the receiver, a phase reference is needed. This is provided by keying in an eight-cycle (or more) burst of the 3.58-MHz subcarrier sinusoid on the "back porch" of the horizontal sync pulse, as shown in Fig. 8–34. The phase of the reference sinusoid is $+90°$ with respect to the x-axis, as indicated on the vectorscope display (Fig. 8–35).

A color receiver is similar to a black-and-white TV receiver except that it has color demodulator circuits and a color CRT. This difference is shown in Fig. 8–36, where the color demodulator circuitry is indicated. The in-phase and quadrature signals, $m_i(t)$ and $m_q(t)$, are recovered (demodulated) by using product detectors where the reference signal is obtained from a PLL circuit that is locked onto the color burst (which was transmitted on the back porch of the sync pulse). M^{-1} is the inverse matrix of that given in Eq. (8–63) so that the three video waveforms corresponding to the red, green, and blue intensity signals are recovered. Note that the hue control on the TV set adjusts the phase of the reference that sets the tint of the picture on the TV set. The color control sets the gain of the chrominance subcarrier signal. If the gain is reduced to zero, no subcarrier will be present at the input to the product detectors. Consequently, a black-and-white picture will be reproduced on the color-TV CRT.

Of course, it is also possible to use different reference phases on the balanced detector inputs, *provided* that the decoding matrix is changed to produce m_R, m_G, and m_B at its output. For example, one popular technique uses the R-Y and B-Y phases on the input to the balanced modulators [Grob, 1975]. Here the reference phase out of the hue control circuit

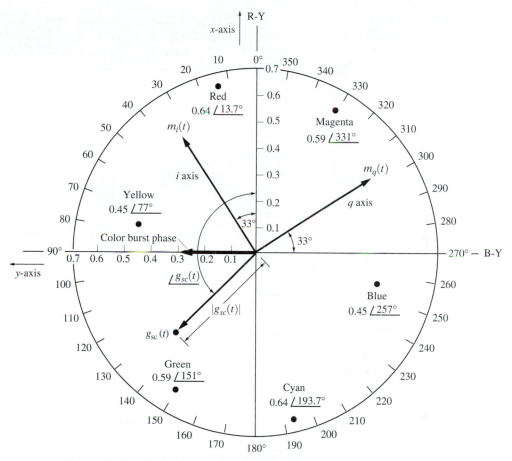

Figure 8–35 Vectorscope presentation of the complex envelope for the 3.58-MHz chrominance subcarrier.

is 0° instead of 33°, as shown in Fig. 8–36. This corresponds to detecting the subcarrier along the R-Y and B-Y axes, as indicated on the vectorscope presentation (Fig. 8–35). In this case, $m_R(t) - m_Y(t)$ is obtained at the output of the upper product detector, and if a gain of two is used on the lower product detector, $m_B(t) - m_Y(t)$ is obtained at its output. These two outputs may be added in the proper proportions to obtain $m_G(t) - m_Y(t)$. These three signals, $m_R(t) - m_Y(t)$, $m_G(t) - m_Y(t)$, and $m_B(t) - m_Y(t)$, may be applied to the three control grids of a three-gun color CRT (one gun activating the red phosphor, one gun for the green, and one for the red). The $-m_Y(t)$ signal is applied to the common cathode of the three guns so that the effective signals on the guns are $m_R(t)$, $m_G(t)$, and $m_B(t)$, respectively, which are the desired signals. Furthermore, for black-and-white operation, it is only necessary to remove the three signals from the control grid since the luminance drive signal is applied at the cathode.

As stated earlier, both the luminance and chrominance subcarrier signals are contained in the 4.2-MHz composite NTSC baseband signal. This gives rise to interfering signals in

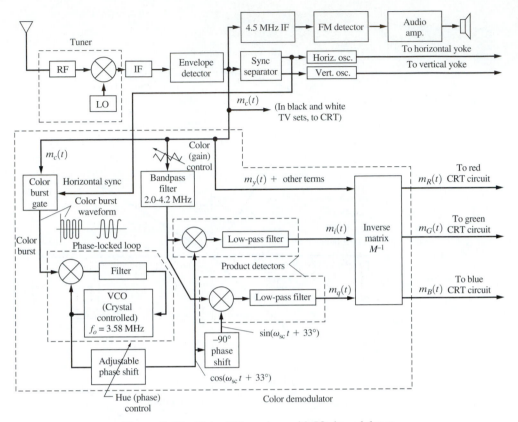

Figure 8–36   Color TV receiver with IQ demodulators.

the receiver video circuits. For example, in Fig. 8–36 there is an interfering signal (i.e., other terms) on the input to the inverse matrix circuit. The interference is averaged out on a line-by-line scanning basis by the viewer's vision if the chrominance subcarrier frequency is an odd multiple of half of the horizontal scanning frequency. For example, from Table 8–11, $f_{sc} = 3,579,545$ Hz and $f_h = 15,734$ Hz. Consequently, there are 227.5 cycles of the 3.58–MHz chrominance interference across one scanned line of the picture. Because there is a half cycle left over, the interference on the *next* scanned line will be *180° out of phase* with that of the present line, and the interference will be *canceled out* by the eye when one views the picture.

As we have seen, the NTSC color system is an ingenious application of basic engineering principles, and this ingenuity is certainly not appreciated by the average color-TV viewer.

Standards for TV and CATV Systems

A summary of some of the U.S. analog TV transmission standards is shown in Table 8–11, and details of the synchronizing waveform are given in Fig. 8–37. A listing of the frequencies

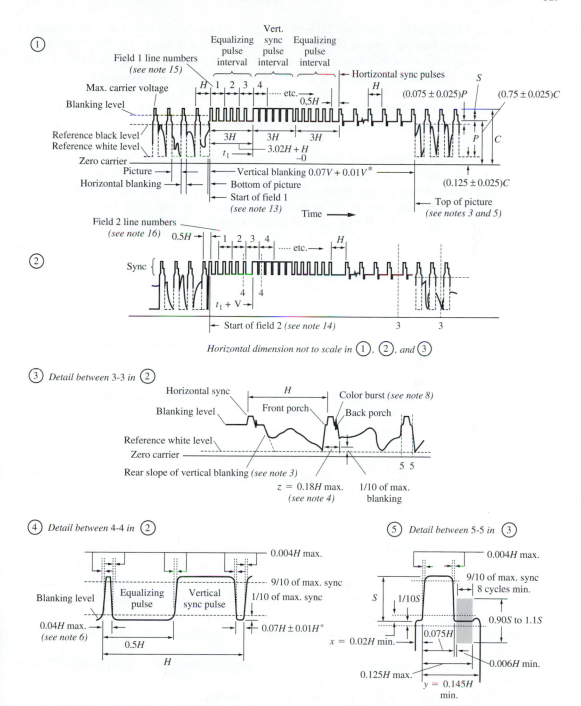

Figure 8–37   U.S. television synchronizing waveform standards for color transmission. (From Part 73 of FCC Rules and Regulations.)

for the TV channels is given in Table 8–12. In addition, a comparison of some broadcast TV standards that are used by various countries is shown in Table 8–13. The CCIR system B standard, which is a 625-lines/frame, 25-frames/sec, negative modulation system, is the dominant European TV standard. In addition to these standards for conventional TV shown in this table, there are U.S. standards for digital TV, and these as will be discussed in the next section. When TV signals are relayed from one country to another via satellite, the standards of the originating country are used, and, if needed, the signal is converted to the standards of the receiving country at the receiving ground station.

U.S. CATV channel standards are presented in Table 8–14. CATV systems use coaxial cable with amplifiers to distribute TV and other signals from the "head end" to the subscriber. Since the cable signals are not radiated, the frequencies that are norrnally assigned to two-way radio and other on-the-air services can be used for additional TV channels on the cable. Consequently, the CATV channel assignments have been standardized as shown in

---

NOTES TO FIGURE 8–37

1. H = time from start of one line to start of next line = $1/f_h$. (See Table 8–11.)
2. V = time from start of one field to start of next field = $1/f_v$. (See Table 8–11.)
3. Leading and trailing edges of vertical blanking should be complete in less than 0.1 H.
4. Leading and trailing slopes of horizontal blanking must be steep enough to preserve minimum and maximum values of $(x + y)$ and (z) under all conditions of picture content.
5. Dimensions marked with an asterisk indicate that tolerances given are permitted only for long-time variations and not for successive cycles.
6. Equalizing pulse area shall be between 0.45 and 0.5 of area of a horizontal sync pulse.
7. Color burst follows each horizontal pulse, but is omitted following the equalizing pulses and during the broad vertical pulses.
8. Color bursts to be omitted during monochrome transmission.
9. The burst frequency shall be 3.579545 MHz. The tolerance on the frequency shall be ±10 Hz, with a maximum rate of change of frequency not to exceed $1/10$ Hz/s.
10. The horizontal scanning frequency shall be 2/455 times the burst frequency.
11. The dimensions specified for the burst determine the times of starting and stopping the burst but not its phase. The color burst consists of amplitude modulation of a continuous sine wave.
12. Dimension P represents the peak excursion of the luminance signal from blanking level, but does not include the chrominance signal. Dimension S is the sync amplitude above blanking level. Dimension C is the peak carrier amplitude.
13. Start of field 1 is defined by a whole line between first equalizing pulse and preceding H sync pulses.
14. Start of field 2 is defined by a half line between first equalizing pulse and preceding H sync pulses.
15. Field 1 line numbers start with first equalizing pulse in field 1.
16. Field 2 line numbers start with second equalizing pulse in field 2.
17. Refer to Table 8–11 for further explanations and tolerances.

TABLE 8–12   U.S. TELEVISION CHANNEL FREQUENCY ASSIGNMENTS

	Channel Number	Band (MHz)	UHF Channel Number	Band	UHF Channel Number	Band
Radio astronomy	2	54–60	29	560–566	57	728–734
Aeronautical	3	60–66	30	566–572	58	734–740
Two-way radio	4	66–72	31	572–578	59	740–746
	5	76–82	32	578–584	60[a]	746–752
FM broadcasting	6	82–88	33	584–590	61[a]	752–758
Aeronautical	7	174–180	34	590–596	62[a]	758–764
Two-way radio	8	180–186	35	596–602	63[a]	764–770
	9	186–192	36	602–608	64[a]	770–776
	10	192–198	37	608–614	65[a]	776–782
	11	198–204	38	614–620	66[a]	782–788
VHF	12	204–210	39	620–626	67[a]	788–794
channels	13	210–216	40	626–632	68[a]	794–800
			41	632–638	69[a]	800–806
UHF channels	14	470–476	42	638–644		
	15	476–482	43	644–650	70[b]	806–812
	16	482–488	44	650–656	71[b]	812–818
	17	488–494	45	656–662	72[b]	818–824
	18	494–500	46	662–668	73[b]	824–830
	19	500–506	47	668–674	74[b]	830–836
	20	506–512	48	674–680	75[b]	836–842
	21	512–518	49	680–686	76[b]	842–848
	22	518–524	50	686–692	77[b]	848–854
	23	524–530	51	692–698	78[b]	854–860
	24	530–536	52	698–704	79[b]	860–866
	25	536–542	53	704–710	80[b]	866–872
	26	542–548	54	710–716	81[b]	872–878
	27	548–554	55	716–722	82[b]	878–884
	28	554–560	56	722–728	83[b]	884–890

[a]These frequencies will be reassigned to non-TV services as existing analog TV stations cease transmitting after the conversion to digital TV broadcasting on Channels 2 through 59.

[b]These frequencies are now allocated to two-way radio and cellular telephone service, but were allocated to UHF TV broadcasting before 1974.

this table. To receive these added channels, the subscriber has to have a TV set with CATV channel capability or use a down converter to convert the CATV channel to a standard channel such as channel 3 or channel 4. CATV systems can also be *bidirectional*, allowing some subscribers to transmit signals to the head end. In low-band split systems, frequencies below 50 MHz (T-band channels) are used for transmitting to the head end. In mid-band, split system frequencies below 150 MHz are used for transmitting to the head end.

Premium TV programs are often available to subscribers for an additional cost. The premium channels are usually scrambled by modifying the video sync or by using digital encryption techniques. Subscribers to the premium services are provided with appropriate decoders that, in some systems, are addressable and remotely programmable by the CATV company.

TABLE 8–13   COMPARISON OF SOME BROADCAST TELEVISION STANDARDS

	Where Used			
Item	North America, South America, Japan	Spain, Italy, England, Germany, CCIR System B[a]	France	USSR
Lines/frame	525	625	625	625
Lines/second	15,750	15,625	15,625	15,625
Frames/second	30	25	25	25
Baseband video bandwidth (MHz)	4.2	5	6	6
Channel bandwidth (MHz)	6	7	8	8
Polarity of AM video modulation	Negative	Negative	Positive	Negative
Type of aural carrier	FM	FM	AM	FM
Color system	NTSC[b]	PAL[c]	SECAM[d]	SECAM[d]
Color subcarrier frequency (MHz)	3.58	4.43	4.43	4.43

[a]The 625-line/frame, 25-frame/s system is the CCIR (International Radio Consultative Committee) system B standard and is the dominant TV standard used on the European continent.

[b]NTSC, National Television System Committee (United States).

[c]PAL, Phase Alternation Line (Europe).

[d]SECAM, Sequential Couleur à Mémoire (French).

Digital TV (DTV)

Since 1987, the FCC has been encouraging the development of a *high-definition television* (HDTV) system to replace the NTSC system. The HDTV system has high resolution that approaches the quality of 35-mm film and a widescreen aspect ratio (width to height) of 16:9 instead of the narrowscreen 4:3 aspect ratio of NTSC. More than 20 HDTV methods were proposed [Jurgen, 1988; Jurgen, 1991]. By 1993, the number of competing proposals were reduced to four digital systems [Challapali et al., 1995; Harris, 1993; Zou, 1991]. In May of 1993, the proponents of these four competing systems joined forces to develop a single digital television (DTV) system that used the best ideas of the four competing systems [Challapali et al., 1995; Hopkins, 1994; and Petajan, 1995]. This joint group is called the Grand Alliance (GA), and its members are: AT&T, General Instrument Corporation, Massachusetts Institute of Technology (MIT), Philips Electronics North America Corporation, David Sarnoff Research Center, Thomson Consumer Electronics, and Zenith Electronics Corporation. This resulted in the 1996 FCC DTV standard, and (DTV) stations are now on the air in major U.S. cities.

TABLE 8–14   CATV CHANNEL FREQUENCIES

CATV Channel Number	Band (MHz)	Letter Channel Designator	CATV Channel Number	UHF Band (MHz)
Sub-VHF	5.75–11.75	T–7	37	300–306
	11.75–17.75	T–8	38	306–312
	17.75–23.75	T–9	39	312–318
	23.75–29.75	T–10	40	318–324
	29.75–35.75	T–11	41	324–330
	35.75–41.75	T–12	42	330–336
	41.75–47.55	T–13	43	336–342
			44	342–348
Low VHF			45	348–354
	2	54–60	46	354–360
	3	60–66	47	360–366
	4	66–72	48	366–372
	5	76–82	49	372–378
	6	82–88	50	378–384
FM Broadcasting	88–108		51	384–390
			52	390–396
High VHF	7	174–180	53	396–402
	8	180–186	54	402–408
	9	186–192	55	408–414
	10	192–198	56	414–420
	11	198–204	57	420–426
	12	204–210	58	426–432
	13	210–216	59	432–438
			60	438–444
Midband	14	120–126	A	61	444–450
	15	126–132	B	⋮	
	16	132–138	C		
	17	138–144	D	89	612–618
	18	144–150	E	90	618–624
	19	150–156	F	91	624–630
	20	156–162	G	92	630–636
	21	162–168	H	93	636–642
	22	168–174	I	94	642–648
				95	90–96
Superband	23	216–222	J	96	96–102
	24	222–228	K	97	102–108
	25	228–234	L	98	108–114
	26	234–240	M	99	114–120
	27	240–246	N		
	28	246–252	O		
	29	252–258	P		
	30	258–264	Q		
	31	264–270	R		
	32	270–276	S		
	33	276–282	T		
	34	282–288	U		
	35	288–294	V		
	36	294–300	W		

Table 8–15 summarizes the U.S. DTV standard. The standard is actually several standards combined, allowing for different resolutions and aspect ratios. There are five major formats. Two are *high definition* (HD), and three are *standard definition* (SD). Widescreen pictures (16:9 aspect ratio) are specified in both the HD formats and in one SD format. The other two SD formats use conventional-width pictures (4:3). In addition, the frame rates may be 24, 30, or 60 frames/s with interlace scan or progressive scan. The 24-frames/s mode is included to more easily accommodate motion picture film (which is 24 frames/s). These variations give a total of 18 possible picture formats. All DTV receivers are designed to decode and display pictures for all 18 formats (although the resolution may not be preserved). The broadcaster may switch formats from program to program (or from program to commercial). However, it is not cost effective for broadcasters to be able to produce all formats. For HDTV, the NBC and CBS television networks use 1080i (i.e., 1080 × 1980 pixels with interlace screening). ABC uses 720p for HDTV, since they believe that progressively scanning 720 lines is visually superior to 1080i. For SDTV, NBC and ABC use 480p, and CBS uses 480i. Note that the vertical resolution for SDTV, which is 480 lines, is equivalent to the vertical resolution for U.S. analog TV, which is 483 lines, as given by Eq. (8–59b). The horizontal resolution for 4:3 SDTV, 704 or 640 pixels, is larger than that of 445 pixels, as given by Eq. (8–60b) for analog TV.

The bit rate for uncompressed video data is tremendous. For example, referring to Table 8–15, for 1080 active lines, 1920 samples (pixels) per line, 8 bits per sample, 30 frames (pictures) per second, and 3 primary colors (RGB), the bit rate is $1{,}080 \times 1{,}920 \times 8 \times 30 \times 3 = 1{,}500$ Mb/s. However, a TV channel with a 6–MHz bandwidth can only support a data rate of about 20 Mb/s if 8-level (3 bits/symbol) multilevel signaling is used. (From Fig. 5–33, $B_T \approx D = R/\ell = 20/3 = 6.67$ MHz.) Consequently a data-compression factor of about 75 (1,500/20) is needed. As shown in Table 8–15, this compression factor is achieved by using a *Motion Pictures Experts Group* (MPEG) encoding technique. This technique consists of taking the *discrete cosine transform* (DCT) of 8 × 8 blocks of pixels within each frame and digitizing (using) only the significant DFT coefficients of each block for transmission. Furthermore, for each new frame, data are sent only when there is a change (motion) within the 8 × 8 pixel block frame to frame.

Frame-to-frame redundancy is removed by motion estimation and motion compensation. There are three types of encoded frames. The *interframe* (*I*-frame) is coded in isolation without prediction and is used as a reference frame. The *predictive frame* (*P*-frame) is forward predicted from the last *I*-frame. The *bidirectional frame* (*B*-frame) is predicted for either past or future *I* and *P* frames. A sequence of one *I*-frame and one or more *P* and *B* frames form a MPEG *group of pictures* (GOP) [Gibson, 1997].

As indicated in Table 8–15, this produces a payload (compressed) data rate of 19.39 Mb/s. This payload data plus parity bits for the FEC codes are fed into a 3-bit digital-to-analog converter (DAC) to produce an 8-level baseband line code that has a symbol (baud) rate of $D = 10.76$ Msymbols/sec. As shown in Fig. 8–38, the 8-level data are partitioned into segments consisting of 832 symbols with a 4-symbol synchronization (sync) pulse included at the beginning of each segment to provide a reference for the transmitted data. Additional training data is inserted over a whole segment after 312 segments. The additional data can be used to train the receiver to adjust for channel fading and cancel multipath interference. The composite 8-level baseband signal is amplitude modulated onto the carrier. In the mod-

TABLE 8–15 U.S. DTV SYSTEM

Video specifications

Format type	HD	HD	SD	SD	SD
Aspect ratio	16:9	16:9	4:3	16:9	4:3
Active scan lines/frame	1080	720	480	480	480
Pixels/line	1920	1280	704	704	640
Frame rates (Hz)[a]	24p, 30p, 30i	24p, 30p, 60p	24p, 30p, 30i, 60p	24p, 30p, 30i, 60p	24p, 30p, 30i, 60p

Compression standard[b]	MPEG-2
Compression technique[c]	DCT, 8×8 block

Audio specifications

Method	Dolbey AC-3
Audio bandwidth	20 kHz
Sampling frequency	48 kHz
Dynamic range	100 dB
Number of surround channels[d]	5.1
Compressed data rate	384 kb/s
Multiple languages	Via multiple AC-3 data streams

Data transport system

Type	Packet
TDM technique	MPEG-2, system layer
Packet size	188 bytes
Packet header size	4 bytes
Encryption	Provision descrambling by authorized decoders
Special features support	Closed captioning and private data

Transmission specifications for terrestrial mode broadcasting[e]

Modulation[f]	8VSB
Bits/symbol	3
Channel bandwidth	6 MHz
Channel filter	Raised cosine-rolloff, $r = 0.115/2 = 0.0575$
Symbol (baud) rate (with sync)	10.76 Msymbols/s
Payload data rate (with sync)	19.39 Mb/s
Coding (FEC)	Rate 2/3 TCM and (207,187) Reed-Solomon
CNR threshold	14.9 dB for a BER of 3×10^{-6}
Pilot (for carrier sync)	310 kHz above lower band edge of RF channel

[a]p = progressive scan, i = Interlace scan.

[b]MPEG = Moving Picture Experts Group of the International Standards Organization (ISO).

[c]DCT = Two-dimensional (horizontal and vertical) Discrete Cosine Transform taken over an 8×8 block of pixels. The insignificant DCT values are ingored to provide data compression.

[d]Left, center, right, left surround, right surround, and subwoofer. The sixth (subwoofer) channel contains only low audio frequencies so it is considered to be a 0.1 channel.

[e]Terrestrial mode is for off-the-air broadcasting. For transmission via CATV, 16 VSB may be used. This allows a higher payload data rate of 38.6 Mbits/sec over a 6-MHz bandwidth channel. However, a minimum (CNR) of 28.3 dB is required.

[f]8VSB = 8 level Vestigial Sideband Modulation. That is, an 8 level baseband line code is amplitude modulated onto a suppressed carrier (DSB-SC) signal, and the lower sideband is filtered off to produce the VSB signal.

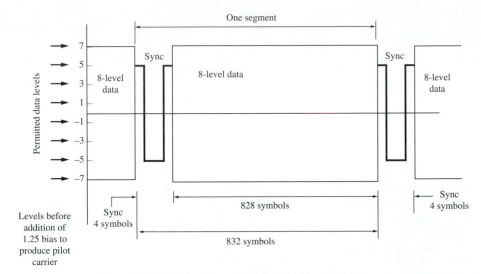

Figure 8–38   8-level baseband signal with segment sync.

ulation process, a dc bias voltage is added to the baseband signal so that there will be a discrete carrier term in the spectrum. This pilot tone at the carrier frequency, f_c, provides synchronization for the carrier oscillator circuits in the DTV receiver. A vestigial sideband filter (described in Sec. 5–5 and Fig. 5–6) is used to remove the lower sideband of the AM signal and produce the vestigial sideband signal. The VSB filter is designed such that the band edges roll off with a square root raised-cosine rolloff characteristic (at the transmitter and the receiver). As discussed in Sec. 3–6, this rolloff is used to filter off the sidelobes of the $\sin(x)/x$ spectrum (Fig. 5–33) that occur for rectangular-shaped data symbols and yet not introduce ISI on the filtered symbols. Fig. 8–39 shows the spectrum of the resulting 8VSB signal where the carrier frequency, f_c, is 309.44 kHz above the lower band edge.

The complete US DTV standards are available on the *Advanced Television System Committee* (ATSC) Web site at www.atsc.org. In Europe, the *Digital Video Broadcasting* (DVB) standard has been adopted for digital TV. DVB uses *Coded Orthogonal Frequency-*

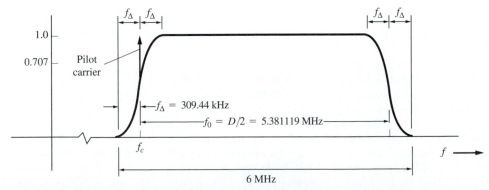

Figure 8–39   Spectrum of 8-VSB DTV signal with square root raised cosine-rolloff.

Division Multiplexing (COFDM) for modulation instead of 8-VSB. See www.dvb.org for details.

Because the number of pixels in a SD frame ($480 \times 704 = 337{,}920$) are about $1/6$ the number of pixels in a HD frame ($1080 \times 1920 = 2{,}073{,}600$), the DTV broadcaster has the very interesting option of time-division multiplexing data for multiple SD programs, since this can be achieved without exceeding the 19.39 Mb/s payload data rate. That is, over a 6-MHz TV channel, four to six SDTV programs may be transmitted simultaneously in place of one 1080i HD program. This multiple SDTV option may be so profitable for broadcasters that they may offer very few HDTV programs, especially since HDTV programs are much more costly to produce.

Digital TV has the following advantages:

- Use of digital signal processing (DSP) circuits.
- Error-free regeneration of relayed and recorded TV pictures, since binary data represent the picture.
- Multiple sound channels (four to six) of CD quality for stereo multilingual capability.
- Data may be multiplexed for text captioning screen graphics and for control of TV recording and other equipment.
- Multipath (ghost) images and ignition noise can be canceled using DSP circuits.
- Lower TV transmitter power required because digital modulation is used.
- Co-channel digital HDTV signals interfere less with each other than co-channel NTSC analog signals.
- Several SDTV programs may be transmitted simultaneously over each channel.

The DTV system is implemented by assigning a new DTV channel for each existing NTSC TV station. During the transition period, each station will simultaneously broadcast a NTSC signal on its existing channel and a DTV signal on its new channel. Channels 2 through 59 will be used for the DTV stations. By the year 2011, the FCC expects everyone to be using DTV, and the NTSC stations will cease transmission. (Also, Channels 60 through 69 will be reassigned for mobile radio and PCS use). DTV sets cost much more than analog TV sets, especially if the DTV sets have large high-resolution displays. Obviously, the economics of the marketplace will dictate how fast (or if) this transition to DTV will occur.

8–10 SUMMARY

A wide range of wire and wireless systems were studied. Modem telephone systems with digital central offices and remote terminals were described. Standards for data transmission via DSL, as well as standards used for POTS, were given. Specifications of PSTN systems were listed for transmission via fiber-optic cable and via satellite. Noise figure and effective input-noise temperature for wireless receivers were defined. Case studies of link budgets were presented. AMPS, GSK TDMA, iDEN, and CDMA standards for cellular phone systems were examined. Cellular 900-MHz and PCS 1,900-MHz bands were described. Finally, analog and digital TV systems were studied. Standards were given for analog NTSC TV, for digital SDTV and digital HDTV.

8–11 STUDY-AID EXAMPLES

SA8–1 Link Budget for a DSS Receiver   Compute the link budget for a digital TV satellite system that is similar to the GM Hughes DSS system that was described in Sec. 8–5. The DSS satellite is located in a geostationary orbit at 101° W longitude above the equator. Assume that the downlink receiving site is Gainesville, FL, located at 82.43° W longitude and 29.71° N latitude. The DSS satellite transmits downlink signals in the Ku band (12.2–12.7 GHz) using 16 transponders. Each transponder has a bandwidth of 24 MHz and an EIRP of 52 dBw in the direction of the United States and, in particular, to Gainesville, FL. Each transponder radiates a QPSK signal with a data rate of 40 Mb/s. The receiving system consists of (1) an 18-inch-diameter parabolic antenna with an attached (2) low-noise block downconverter (LNB) that converts the Ku band input down to 950 to 1450 MHz, (3) a transmission line that connects the LNB to (4) a receiver that is located on the top of the subscriber's TV set [Thomson, 1994]. The LNB has a gain of 40 dB and a noise figure (NF) of 0.6 dB. The RG6/U coaxial transmission line is 110 feet in length and has a loss of 8 dB/100 ft in the 950–1450-MHz band. The receiver detects the data packets, decodes them, and converts the data to analog video and audio signals using the built-in digital-to-analog converters. The receiver has a NF of 10 dB and an IF bandwidth of 24 MHz. Assume that the antenna source temperature is 20 K.

Compute the $(C/N)_{dB}$, $(E_b/N_0)_{dB}$, and the BER for this receiving system.

Solution.   The receiving antenna pointing parameters (azimuth and elevation) from Gainesville, FL, to the satellite are evaluated using Eqs. (8–47) to (8–54).

$$\beta = \cos^{-1}\left[\cos(29.71)\cos(101 - 82.43)\right] = 34.58°$$

The azimuth is

$$A = 360 - \cos^{-1}\left(-\frac{\tan(29.71)}{\tan(34.58)}\right) = 214.13°$$

The elevation is

$$E = \tan^{-1}\left[\frac{1}{\tan(34.58)} - \frac{3963}{(26{,}205)\sin(34.58)}\right] = 49.82°$$

Using Eq. (8–49), we find that the slant range is

$$d = \sqrt{(26{,}205)^2 + (3963)^2 - 2(3693)(26{,}205)\cos(34.58)}$$

Thus,

$$d = 23{,}052 \text{ miles}, \quad \text{or} \quad 3.709 \times 10^7 \text{ m}$$

The overall NF for the receiving system is evaluated using Eq. (8–34), where $F_1 = 0.6$ dB $= 1.15$, $G_1 = 40$ dB $= 10^4$, $F_2 = 110$ ft $\times$ 8 dB/100 ft $= 8.8$ dB $= 7.59$, $G_2 = -8.8$ dB $= 0.13$, and $F_3 = 10$ dB $= 10$. Then,

$$F = F_1 + \frac{F_2 - 1}{G_{a1}} + \frac{F_3 - 1}{G_{a1}G_{a2}} = 1.15 + \frac{7.59 - 1}{10^4} + \frac{10 - 1}{(10^4)(0.13)}$$

or $F = 1.15 + 6.59 \times 10^{-4} + 6.83 \times 10^{-3} = 1.15 = 0.6$ dB. Thus, $T_e = (F - 1)T_0 = (1.15 - 1)(290) = 43.18K$.

Note: The gain of the LNB is designed to be sufficiently large so that the NF contributions of the transmission line and the receiver are negligible.

The receiving antenna is computed using Table 8–4 for an 18–in = 0.46-m-diameter parabola, where $\lambda = c/f = 3 \times 10^8/12.45 \times 10^9 = 0.0241$ m.

$$(G_{AR})_{dB} = 10 \log\left[\frac{7\pi(0.46/2)^2}{(0.0241)^2}\right] = 32.96 \text{ dB}$$

$$(T_{syst})_{dB} = 10 \log(T_{AR} + T_e) = 10 \log(20 + 43.18) = 18.01 \text{ dBK}$$

Thus,

$$\left(\frac{G_{AR}}{T_{syst}}\right)_{dB} = 32.96 - 18.01 = 14.96 \text{ dB/K}$$

The $(C/N)_{dB}$ is evaluated using Eq. (8–55), where

$$(L_{FS})_{dB} = 20 \log\left(\frac{4\pi d}{\lambda}\right) = 20 \log\left(\frac{4\pi(3.709 \times 10^7)}{(0.0241)}\right) = 205.73 \text{ dB}$$

and

$$(B)_{dB} = 10 \log(B) = 10 \log(24 \times 10^6) = 73.8 \text{ dB}$$

Thus,

$$\left(\frac{C}{N}\right)_{dB} = (P_{EIRP})_{dBw} - (L_{FS})_{dB} + \left(\frac{G_{AR}}{T_{syst}}\right)_{dB} - k_{dB} - B_{dB}$$

$$= 52 - 205.73 + 14.96 - (-228.6) - 73.8$$

or

$$(C/N)_{dB} = 16.03 \text{ dB}$$

$(E_b/N_0)_{dB}$ may be evaluated using Eq. (8–44), where $B = 24$ MHz, and $R = 40$ Mb/s. Then,

$$\left(\frac{E_b}{N_0}\right)_{dB} = \left(\frac{C}{N}\right)_{dB} + \left(\frac{B}{R}\right)_{dB} = 16.03 - 2.22 = 13.81 \text{ dB}$$

An $(E_b/N_0)_{dB}$ of 13.81 dB = 24.05 gives negligible errors for QPSK signaling. That is, if no coding is used, the QPSK BER is given by Eq. (7–69) as

$$P_e = Q\left(\sqrt{2\left(\frac{E_b}{N_0}\right)}\right) = Q\left(\sqrt{2(24.05)}\right) = 2.0 \times 10^{-12}$$

or one error every 3.4 hours. However, if there is signal fading (because of rain or other atmospheric conditions), significant errors may occur. This is examined in Example SA8–2.

SA8–2 DSS Link Budget with Fading   Repeat Example SA8–1, and assume that there is a 4-dB signal fade of the Ku band signal because of rain. Compute $(C/N)_{dB}$, $(E_b/N_0)_{dB}$ and the BER with and without coding. For the coding case, assume a 3-dB coding gain.

Solution

$$\left(\frac{C}{N}\right)_{\text{fade dB}} = \left(\frac{C}{N}\right)_{dB} - (L_{\text{fade}})_{dB} = 16.03 - 4.0 = 12.03 \text{ dB}$$

and

$$\left(\frac{E_b}{N_0}\right)_{\text{fade dB}} = 9.81 \text{ dB} = 9.57$$

Then, the BER with fading and no coding is

$$P_e = Q\left(\sqrt{2(9.57)}\right) = 6.04 \times 10^{-6}$$

or one error in 4.1 ms. This performance is not acceptable; consequently FEC coding is needed. Using coding with a coding gain of 3dB, compute the effective (E_b/N_0) by referring to Fig. 1–8:

$$\left(\frac{E_b}{N_0}\right)_{dB} = 9.81 + 3.0 = 12.81 \text{ dB} = 19.10$$

Then, using Eq. (7–69), we see that the BER with fading and coding becomes

$$P_e = Q\left(\sqrt{2(19.10)}\right) = 3.2 \times 10^{-10}$$

or one error every 78 s.

SA8–3 BER for a Wireless Device Link   Evaluate the BER of a wireless personal communication device (PCD). Assume that a portable computer/telephone/video terminal is connected to the outside world via a wireless link within a building. The wireless link operates between the PCD and a base station unit located within the building. The wireless link uses OOK signaling on a carrier frequency of 2.4 GHz, and the data rate is 2 Mb/s. The PCD transmit power on the uplink is 0.5 mW. The base station receiver has a noise figure of 8 dB and an IF bandwidth of 4 MHz. It incorporates an envelope detector to detect the data. External noise at the receiver input is negligible when compared with the internal noise of the receiver. Assume that the transmit and receive antennas are simple dipoles; each has a gain of 2.15 dB. The path loss between the transmit and receive antenna within the building environment is modeled by Eq. (8–47) as

$$L_{dB}(d) = L_{\text{FSdB}}(d_0) + 10n \log\left(\frac{d}{d_0}\right) + X_{dB} \qquad (8\text{–}67)$$

where $L_{dB}(d)$ is the path loss in dB for a distance of d between the antennas, $L_{\text{FSdB}}(d_0)$ is the free-space loss for a distance d_0 that is close to the transmitter, but in the far field, and $d > d_0$. n is the path loss exponent, and X_{dB} represents the loss margin due to variations in the path loss caused by multiple reflections. For this example, choose $d_0 = 50$ feet, $n = 3$, and $X_{dB} = 7$ dB. (The exponent, n, would be 2 for the free-space case and 4 for a 2-ray ground-reflection case.)

For a distance of 200 feet between the PCD and the base station, calculate the CNR at the detector input of the base station receiver and the BER for the detected data at the receiver output.

Solution.   The CNR is obtained by using Eq. (8–67) to replace $(L_{FS})_{dB}$ in Eq. (8–43). Using the values just given, we get $(P_{EIRP})_{dBw} = -30.86$ dBw. $T_{system} = 1{,}540$ K, and $L_{dB}(200) = 88.18$ dB. From Eq. (8–43) with the PCD located 200 feet from the base station, the CNR is

$$\left(\frac{C}{N}\right)_{dB} = 13.22 \text{ dB}, \qquad 200 \text{ feet spacing}$$

Also, using Eq. (8–44), we get $(E_b/N_0)_{dB} = 16.23$ dB. For this case of OOK with (noncoherent) envelope detection, the BER is obtained by using Eq. (7–58). The bit rate is $R = 1/T = 2$ Mb/s, and the IF bandwidth is $B_P = 4$ MHz. For a spacing of 200 feet between the PCD and the base station, the BER of the base station output data stream is

$$P_e = 1.36 \times 10^{-5}, \qquad 200 \text{ feet spacing}$$

Using other values for d in these equations, MATLAB can compute the CNR and BER over a whole range of spacing from 50 feet to 500 feet. The MATLAB plots are shown in Figures 8–40 and 8–41.

The range of this PCD wireless link could be increased by increasing the transmit power, reducing the receiver noise figure, or implementing a spread spectrum (SS) system. If direct sequence SS is used with a $r = 4$ stage shift register, as shown in Fig. 5–37, then the PN code length is $N = 15$. Assuming that code length spans one bit of data, then the chip rate is

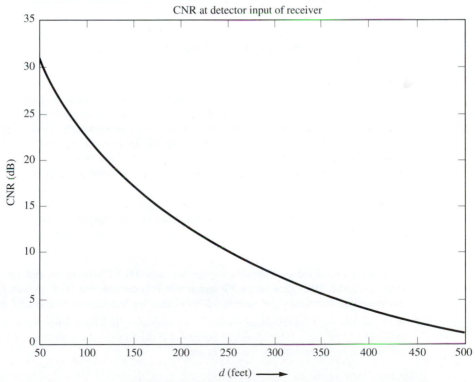

Figure 8–40   CNR for a PCD wireless link.

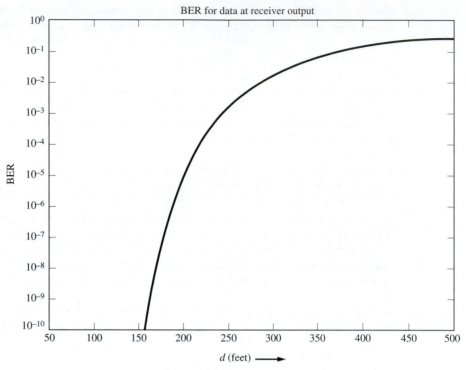

Figure 8–41   BER for a PCD wireless link.

$R_c = NR$. From Eq. (5–131), this SS system would provide a processing gain of $G_p = R_c/R = N = 15$, or $G_{p\,dB} = 11.76$ dB. If despreading at the receiver is implemented after the IF stage (i.e., after the internal noise source of the receiver), this processing gain would increase the CNR at the detector input by 11.76 dB. This would increase the useful range to around 500 feet. That is, referring to Fig. 8–40 at $d = 500$ feet, we know that a processing gain of 11.76 dB would result in a CNR of 13.0 dB, which corresponds to a BER of 2.1×10^{-5}.

PROBLEMS

8–1 A remote terminal for a telephone company services 300 VF subscribers and 150 G.Lite DSL subscribers (1.5 Mb/s data plus a VF signal that is converted to a DS-0 signal). Compute the minimum data rate needed for the receive fiber-optic line that terminates at the RT from the CO.

8–2 A 50-pair line provides telephone service to 50 subscribers in a rural subdivision via local loops to the CO. How many subscribers can be served if the 50 pairs are converted to T1 lines and a remote terminal is installed in the subdivision?

8–3 Assume that a telephone company has remote terminals connected to its CO via T1 lines. Draw a block diagram that illustrates how the T1 lines are interfaced with the CO switch if the CO uses:

(a) An analog switch.

(b) An integrated digital switch.

8–4 Indicate whether a conference-call connection is better or may be worse than a single-party call if a digital switch is used (at the CO) instead of an analog switch. Explain your answer.

8–5 Full-duplex data of 24 kb/s in each direction from a personal computer is sent via a twisted-pair telephone line having a bandpass from 300–2700 Hz. Explain why modems are needed at each end of the line.

8–6 A satellite with twelve 36-MHz bandwidth transponders operates in the 6/4-GHz bands with 500 MHz bandwidth and 4-MHz guardbands on the 4-GHz downlink, as shown in Fig. 8–10. Calculate the percentage bandwidth that is used for the guardbands.

8–7 An Earth station uses a 3-m-diameter parabolic antenna to receive a 4-GHz signal from a geosynchronous satellite. If the satellite transmitter delivers 10 W into a 3-m-diameter transmitting antenna and the satellite is located 36,000 km from the receiver, what is the power received?

8–8 Figure 8–12b shows an FDM/FM ground station for a satellite communication system. Find the peak frequency deviation needed to achieve the allocated spectral bandwidth for the 6240-MHz signal.

8–9 A microwave transmitter has an output of 0.1 W at 2 GHz. Assume that this transmitter is used in a microwave communication system where the transmitting and receiving antennas are parabolas, each 4 ft in diameter.

(a) Evaluate the gain of the antennas.

(b) Calculate the EIRP of the transmitted signal.

(c) If the receiving antenna is located 15 miles from the transmitting antenna over a free-space path, find the available signal power out of the receiving antenna in dBm units.

8–10 Using MATLAB or MathCAD, plot the PSD for a thermal noise source with a resistance of 10 kΩ over a frequency range of 10 through 100,000 GHz where $T = 300$ K.

8–11 Given the RC circuit shown in Fig. P8–11, where R is a physical resistor at temperature T, find the rms value of the noise voltage that appears at the output in terms of k, T, R, and C.

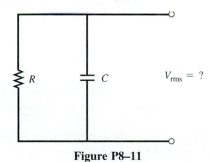

R C $V_{rms} = ?$

Figure P8–11

8–12 A receiver is connected to an antenna system that has a noise temperature of 100 K. Find the noise power that is available from the source over a 20-MHz band.

8–13 A bipolar transistor amplifier is modeled as shown in Fig. P8–13. Find a formula for the available power gain in terms of the appropriate parameters.

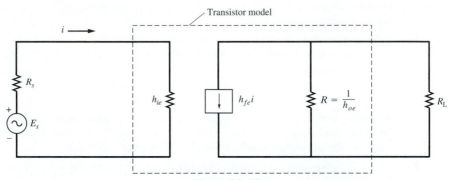

Figure P8–13

8–14 Using the definition for the available power gain, $G_a(f)$, as given by (8–18), show that $G_a(f)$ depends on the driving source impedance as well as the elements of the device and that $G_a(f)$ does not depend on the load impedance. [*Hint*: Calculate $G_a(f)$ for a simple resistive network.]

8–15 Show that the effective input-noise temperature and the noise figure can be evaluated from measurements that use the Y-factor method. With this method the device under test (DUT) is first connected to a noise source that has a relatively large output denoted by its source temperature, T_h, where the subscript h denotes "hot," and then the available noise power at the output of the DUT, P_{aoh} is measured with a power meter. Next, the DUT is connected to a source that has relatively low source temperature, T_c, where the subscript c denotes "cold," and noise power at the output of the DUT is measured, P_{aoc}. Show that
(a) The effective input noise temperature of the DUT is

$$T_e = \frac{T_h - YT_c}{Y - 1}$$

where $Y = P_{aoh}/P_{aoc}$ is obtained from the measurements.
(b) The noise figure of the DUT is

$$F = \frac{[(T_h/T_0) - 1] - Y[(T_c/T_0) - 1]}{Y - 1}$$

where $T_0 = 290$ K.

8–16 If a signal plus noise is fed into a linear device, show that the noise figure of that device is given by $F = (S/N)_{in}/(S/N)_{out}$ (*Hint*: Start with the basic definition of noise figure that is given in this chapter.)

8–17 An antenna is pointed in a direction such that it has a noise temperature of 30 K. It is connected to a preamplifier that has a noise figure of 1.6 dB and an available gain of 30 dB over an effective bandwidth of 10 MHz.
(a) Find the effective input noise temperature for the preamplifier.
(b) Find the available noise power out of the preamplifier.

8–18 A 10-MHz SSB-AM signal, which is modulated by an audio signal that is bandlimited to 5 kHz, is being detected by a receiver that has a noise figure of 10 dB. The signal power at the receiver input is 10^{-10} mW, and the PSD of the input noise, $\mathscr{P}(f) = kT/2$, is 2×10^{-21}. Evaluate

(a) The IF bandwidth needed.

(b) The SNR at the receiver input.

(c) The SNR at the receiver output, assuming that a product detector is used.

8–19 An FSK signal with $R = 110$ b/s is transmitted over an RF channel that has white Gaussian noise. The receiver uses a noncoherent detector and has a noise figure of 6 dB. The impedance of the antenna input of the receiver is 50 Ω. The signal level at the receiver input is 0.05 μV, and the noise level is $N_0 = kT_0$, where $T_0 = 290$ K and k is Boltzmann's constant. Find the P_e for the digital signal at the output of the receiver.

8–20 Work Prob. 8–19 for the case of DPSK signaling.

8–21 Prove that the overall effective input-noise temperature for cascaded linear devices is given by Eq. (8–37).

8–22 A TV set is connected to an antenna system as shown in Fig. P8–22. Evaluate

(a) The overall noise figure.

(b) The overall noise figure if a 20-dB RF preamp with a 3-dB noise figure is inserted at point B.

(c) The overall noise figure if the preamp is inserted at point A.

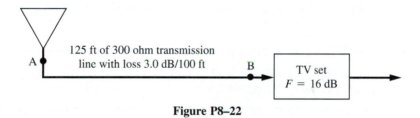

Figure P8–22

8–23 An Earth station receiving system consists of a 20-dB gain antenna with $T_{AR} = 80$ K, an RF amplifier with $G_a = 40$ dB and $T_e = 30$ K, and a down converter with $T_e = 15,000$ K. What is the overall effective input-noise temperature of this receiving system?

8–24 A low-noise amplifier (LNA), a down converter, and an IF amplifier are connected in cascade. The LNA has a gain of 40 dB and a T_e of 25 K. The down converter has a noise figure of 8 dB and a conversion gain of 6 dB. The IF amplifier has a gain of 60 dB and a noise figure of 14 dB. Evaluate the overall effective input-noise temperature for this system.

8–25 A geosynchronous satellite transmits 13.5 dBW EIRP on a 4-GHz downlink to an Earth station. The receiving system has a gain of 60 dB, an effective input-noise temperature of 30 K, an antenna source noise temperature of 60 K, and an IF bandwidth of 36 MHz. If the satellite is located 24,500 miles from the receiver, what is the $(C/N)_{dB}$ at the input of the receiver detector circuit?

8–26 An antenna with $T_{AR} = 160$ K is connected to a receiver by means of a waveguide that has a physical temperature of 290 K and a loss of 2 dB. The receiver has a noise bandwidth of 1 MHz, an effective input-noise temperature of 800 K, and a gain of 120 dB from its antenna input to its IF output. Using MATLAB or MathCAD, find

(a) The system noise temperature at the input of the receiver.

(b) The overall noise figure.

(c) The available noise power at the IF output.

8–27 The *Intelsat IV* satellite uses 36-MHz transponders with downlinks operating in the 4-GHz band. Each satellite transponder has a power output of 3.5 W and may be used with a 17° global coverage antenna that has a gain of 20 dB.

(a) For users located at the subsatellite point (i.e., directly below the satellite), show that $(C/N)_{dB} = (G_{AR}/T_{syst})_{dB} - 17.1$, where $(G_{AR}/T_{syst})_{dB}$ is the Earth receiving station figure of merit.

(b) Design a ground receiving station (block diagram) showing reasonable specifications for each block so that the IF output CNR will be 12 dB. Discuss the trade-offs that are involved.

8–28 The efficiency of a parabolic antenna is determined by the accuracy of the parabolic reflecting surface and other factors. The gain is $G_A = 4\pi\eta A/\lambda^2$, where η is the antenna efficiency. In Table 8–4 an efficiency of 56% was used to obtain the formula $G_A = 7A/\lambda^2$. In a ground receiving system for the *Intelsat IV* satellite, assume that a (G_{AR}/T_{syst}) of 40 dB is needed. Using a 30-m antenna, give the required antenna efficiency if the system noise temperature is 85 K. How does the required antenna efficiency change if a 25-m antenna is used?

8–29 Evaluate the performance of a TVRO system. Assume that the TVRO terminal is located in Miami, Florida (26.8° N latitude, 80.2° W longitude). A 10-ft-diameter parabolic receiving antenna is used and is pointed toward the *Galaxy I* satellite. Other TVRO parameters are given in Table 8–5, Fig. 8–25, and Fig. 8–27.

(a) Find the TVRO antenna look angles to the satellite and find the slant range.

(b) Find the overall receiver system temperature.

(c) Find the $(C/N)_{dB}$ into the receiver detector.

(d) Find the $(S/N)_{dB}$ out of the receiver.

8–30 Repeat Prob. 8–29 for the case when the TVRO terminal is located at Anchorage, Alaska (61.2° N latitude, 149.8° W longitude), and a 8-m-diameter parabolic antenna is used.

8–31 A TVRO receive system consists of an 8-ft-diameter parabolic antenna that feeds a 50-dB, 25K LNA. The sky noise temperature (with feed) is 32 K. The system is designed to receive signals from the *Galaxy I* satellite. The LNA has a post-mixer circuit that down-converts the satellite signal to 70 MHz. The 70-MHz signal is fed to the TVRO receiver via a 120-ft length of 75-Ω coaxial cable. The cable has a loss of 3 dB/100 ft. The receiver has a bandwidth of 36 MHz and a noise temperature of 3800 K. Assume that the TVRO site is located in Los Angeles, California (34° N latitude, 118.3° W longitude), and vertical polarization is of interest.

(a) Find the TVRO antenna look angles to the satellite and find the slant range.

(b) Find the overall system temperature.

(c) Find the $(C/N)_{dB}$ into the receiver detector.

8–32 An Earth station receiving system operates at 11.95 GHz and consists of a 20-m antenna with a gain of 65.53 dB and $T_{AR} = 60$ K, a waveguide with a loss of 1.28 dB and a physical temperature of 290 K, a LNA with $T_e = 50$ K and 60 dB gain, and a down converter with $T_e = 11,000$ K. Using MATLAB or MathCAD, find $(G/T)_{dB}$ for the receiving system evaluated at

(a) The input to the LNA.

(b) The input to the waveguide.

(c) The input to the down converter.

8–33 Rework Ex. 8–4 for the case of reception of TV signals from a direct broadcast satellite (DBS). Assume that the system parameters are similar to those given in Table 8–5, except that the satellite power is 316 kW EIRP radiated in the 12-GHz band. Furthermore, assume that a 0.5-m parabolic receiving antenna is used and that the LNA has a 50-K noise temperature.

8–34 The most distant planet from our sun is Pluto, which is located at a (greatest) distance from the Earth of 7.5×10^9 km. Assume that an unmanned spacecraft with a 2-GHz, 10-W transponder is in the vicinity of Pluto. A receiving Earth station with a 64-m antenna is available that has a system noise temperature of 16 K at 2 GHz. Calculate the size of a spacecraft antenna that is required for a 300-b/s BPSK data link to Earth that has a 10^{-3} BER (corresponding to E_b/N_0 of 9.88 dB). Allow 2 dB for incidental losses.

8–35 Using MATLAB, MathCAD, or some spreadsheet program that will run on a personal computer (PC), design a spreadsheet that will solve Prob. 8–31. Run the spreadsheet on your PC, and verify that it gives the correct answers. Print out your results. Also, try other parameters, such as those appropriate for your location, and print out the results.

8–36 Assume that you want to analyze the overall performance of a satellite relay system that uses a "bent pipe" transponder. Let $(C/N)_{up}$ denote the carrier-to-noise ratio for the transponder as evaluated in the IF band of the satellite transponder. Let $(C/N)_{dn}$ denote the IF CNR ratio at the downlink receiving ground station when the satellite is sending a perfect (noise-free) signal. Show that the overall operating CNR ratio at the IF of the receiving ground station, $(C/N)_{ov}$, is given by

$$\frac{1}{(C/N)_{ov}} = \frac{1}{(C/N)_{up}} + \frac{1}{(C/N)_{dn}}$$

8–37 A PCS cellular site base station operates with 10W into an antenna with 18 dB gain at 1800 MHz. The path loss reference distance is $d_0 = 0.25$ miles, $X_{dB} = 0$, and the PCS phone antenna has 0 dB gain. Find the received power in dBm at the antenna output of the PCS phone when the PCS phone is located at distances of 1 mile, 2 miles, 5 miles, and 10 miles from the BS if the path loss exponent is
(a) $n = 2$ (free space condition).
(b) $n = 3$.
(c) $n = 4$.

8–38 The transmitter of a PCS CDMA cellular phone feeds 200 mw into its 0 dB gain antenna. The PCS signal is spread over a bandwidth of 1.25 MHz in the 1900 MHz band. Calculate the signal power received in dBm at the output of the base station antenna where the PCS phone is located 2 km from the BS. Assume a path loss model given by Eq. (8–47), where $d_0 = 100$ m, $n = 3$ and $X_{dB} = 0$. The BS antenna has a gain of 16 dB.

8–39 Given a CDMA cellular phone link as described in Prob. 8–38 and by the IS-95 standard shown in Table 8–10. The base station receiving system has a noise figure of 8 dB and multiple access interference (MAI) from other phones adds 20 dB to the noise level at the front end of the BS receiver. Compute the CNR, $(C/N)_{dB}$, after despreading (i.e., at the data detector input) in the BS receiver. Also, find $(E_b/N_0)_{dB}$ after despreading. Assume that the bandwidth after despreading is $B = 19.2$ kHz, since the data rate with coding is 19.2 kb/s (9.6 kb/s payload data rate with rate $\frac{1}{2}$ coding) on each I and Q BPSK carrier of the QPSK signal.

8–40 A low-power TV station is licensed to operate con TV Channel 35 with an effective radiated visual power of 1000 W. The tower height is 400 ft, and the transmission line is 450 ft long. Assume that a $1\frac{5}{8}$-in.-diameter semirigid 50-Ω coaxial transmission line will be used that has a loss of 0.6 dB/100 ft (at the operating frequency). The antenna has a gain of 5.6 dB, and the duplexer loss is negligible. Find the PEP required at the visual transmitter output.

8–41 An analog TV visual transmitter is tested by modulating it with a periodic test signal. The envelope of the transmitter output is viewed on an oscilloscope as shown in Fig. P8–41, where K is an unknown constant. The output of the transmitter is connected to a 50-Ω dummy load that has a calibrated average reading wattmeter. The wattmeter reads 6.9 kW. Find the PEP of the transmitter output.

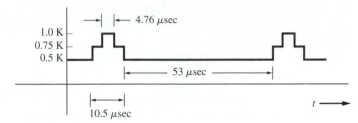

Figure P8–41

8–42 Specifications for the MTS stereo audio system for analog TV are given in Fig. 8–33. Using this figure,

  (a) Design a block diagram for TV receiver circuitry that will detect the stereo audio signals and the second audio program (SAP) signal.

  (b) Describe how the circuit works. That is, give mathematical expressions for the signal at each point on your block diagram and explain in words. Be sure to specify all filter transfer functions, VCO center frequencies, and so on.

8–43 An analog TV broadcaster wants to send digital data to subscribers concerning the stock market using the professional aural subcarrier. Propose two digital signaling techniques that will satisfy the spectral requirements of Fig. 8–33 where the available bandwidth is $0.5f_h$. Specifically, design and draw block diagrams of transmit and receive modems and give the specifications for transmitting data at a rate of **(a)** 1200 b/s and **(b)** 9600 b/s. Discuss why you think you have a good design.

8–44 In the R–Y, B–Y color-TV subcarrier demodulation system, the G–Y signal is obtained from the R–Y and B–Y signals. That is,

$$[m_G(t) - m_Y(t)] = K_1[m_R(t) - m_Y(t)] + K_2[m_B(t) - m_Y(t)]$$

  (a) Find the values for K_1 and K_2 that are required.

  (b) Draw a possible block diagram for a R–Y, B–Y system, and explain how it works.

8–45 Referring to the digital DTV standards in Table 8–15, the payload data rate with sync is 19.39 Mb/s. Show that with the addition of coding bits and training bits, the 8VSB signal has a symbol rate of 10.76 Mbaud.

8–46 8VSB signaling is used in the US for DTV transmission. To prevent ISI, the spectrum of the 8VSB signal is designed to roll off with a square root raised cosine-rolloff filtering characteristic, as shown in Fig. 8–39.

  (a) Show that the absolute bandwidth of a VSB signal with raised cosine-rolloff filtering is

$$B_{\text{VSB}} = \frac{1}{2}(1 + 2r)D$$

  where D is the symbol rate and r is the rolloff factor. [*Hint*: The derivation of this formula is similar to that for Eq. (3–74).]

  (b) Using the parameters for DTV from Table 8–15 in the formula developed in part **(a)**, show that the absolute bandwidth for DTV is 6 MHz.

MATHEMATICAL TECHNIQUES, IDENTITIES, AND TABLES

A–1 TRIGONOMETRY AND COMPLEX NUMBERS

Definitions

$$\sin x = \frac{e^{jx} - e^{-jx}}{2j}$$

$$\cos x = \frac{e^{jx} + e^{-jx}}{2}$$

Euler's formula

$$\tan x = \frac{\sin x}{\cos x} = \frac{e^{jx} - e^{-jx}}{j(e^{jx} + e^{-jx})}$$

Trigonometric Identities and Complex Numbers

$e^{\pm jx} = \cos x \pm j \sin x$ (Euler's theorem)

$$e^{\pm j\pi/2} = \pm j \qquad\qquad e^{\pm jn\pi} = \begin{cases} 1, & n \text{ even} \\ -1, & n \text{ odd} \end{cases}$$

$$x + jy = Re^{j\theta}, \quad \text{where} \quad R = \sqrt{x^2 + y^2}, \quad \theta = \tan^{-1}(y/x)$$

$$(Re^{j\theta})^y = R^y e^{jy\theta} \quad (R_1 e^{j\theta_1})(R_2 e^{jy\theta_2}) = R_1 R_2 e^{j(\theta_1 + \theta_2)}$$

$$\cos(x \pm y) = \cos x \cos y \mp \sin x \sin y \qquad 2\cos x \cos y = \cos(x - y) + \cos(x + y)$$

$$\sin(x \pm y) = \sin x \cos y \pm \cos x \sin y \qquad 2\sin x \sin y = \cos(x - y) - \cos(x + y)$$

$$\qquad\qquad\qquad\qquad\qquad\qquad\qquad 2\sin x \cos y = \sin(x - y) + \sin(x + y)$$

$$\cos\left(x \pm \frac{\pi}{2}\right) = \mp\sin x \qquad\qquad 2\cos^2 x = 1 + \cos 2x$$

$$\qquad\qquad\qquad\qquad\qquad 2\sin^2 x = 1 - \cos 2x$$

$$\sin\left(x \pm \frac{\pi}{2}\right) = \pm\cos x \qquad\qquad 4\cos^3 x = 3\cos x + \cos 3x$$

$$\qquad\qquad\qquad\qquad\qquad 4\sin^3 x = 3\sin x - \sin 3x$$

$$\cos 2x = \cos^2 x - \sin^2 x \qquad\qquad 8\cos^4 x = 3 + 4\cos 2x + \cos 4x$$

$$\sin 2x = 2\sin x \cos x \qquad\qquad 8\sin^4 x = 3 - 4\cos 2x + \cos 4x$$

$$R\cos(x + \theta) = A\cos x - B\sin x$$

$$\text{where} \quad R = \sqrt{A^2 + B^2}, \quad \theta = \tan^{-1}(B/A), \quad A = R\cos\theta, \quad B = R\sin\theta$$

A–2 DIFFERENTIAL CALCULUS

Definition

$$\frac{df(x)}{dx} = \lim_{\Delta x \to 0} \frac{f(x + (\Delta x/2)) - f(x - (\Delta x/2))}{\Delta x}$$

Differentiation Rules

$$\frac{du(x)\,v(x)}{dx} = u(x)\frac{dv(x)}{dx} + v(x)\frac{du(x)}{dx} \quad \text{(products)} \qquad \frac{du[v(x)]}{dx} = \frac{du}{dv}\frac{dv}{dx} \quad \text{(chain rule)}$$

$$\frac{d\left(\dfrac{u(x)}{v(x)}\right)}{dx} = \frac{v(x)\dfrac{du(x)}{dx} - u(x)\dfrac{dv(x)}{dx}}{v^2(x)} \quad \text{(quotient)}$$

Derivative Table

$$\frac{d[x^n]}{dx} = nx^{n-1} \qquad\qquad\qquad \frac{d\tan^{-1}ax}{dx} = \frac{a}{1 + (ax)^2}$$

$$\frac{d\sin ax}{dx} = a\cos ax$$

$$\frac{d[e^{ax}]}{dx} = ae^{ax}$$

$$\frac{d\cos ax}{dx} = -a\sin ax$$

$$\frac{d[a^x]}{dx} = a^x \ln a$$

$$\frac{d\tan ax}{dx} = \frac{a}{\cos^2 ax}$$

$$\frac{d(\ln x)}{dx} = \frac{1}{x}$$

$$\frac{d\sin^{-1}ax}{dx} = \frac{a}{\sqrt{1-(ax)^2}}$$

$$\frac{d(\log_a x)}{dx} = \frac{1}{x}\log_a e$$

$$\frac{d\cos^{-1}ax}{dx} = -\frac{a}{\sqrt{1-(ax)^2}}$$

$$\frac{d\left[\int_{a(x)}^{b(x)} f(\lambda, x)\, d\lambda\right]}{dx} = f(b(x), x)\frac{db(x)}{dx} - f(a(x), x)\frac{da(x)}{dx}$$

$$+ \int_{a(x)}^{b(x)} \frac{\partial f(\lambda, x)}{\partial x}\, d\lambda \qquad \text{(Leibniz's rule)}$$

A–3 INDETERMINATE FORMS

If $\lim_{x\to a} f(x)$ is of the form

$$\frac{0}{0}, \ \frac{\infty}{\infty}, \ 0\cdot\infty, \ \infty - \infty, \ 0^\circ, \ \infty^\circ, \ 1^\infty$$

then

$$\lim_{x\to a} f(x) = \lim_{x\to a}\left[\frac{N(x)}{D(x)}\right] = \lim_{x\to a}\left[\frac{(dN(x)/dx)}{(dD(x)/dx)}\right] \qquad \text{(L'Hôpital's rule)}$$

where $N(x)$ is the numerator of $f(x)$, $D(x)$ is the denominator of $f(x)$, $N(a) = 0$, and $D(a) = 0$.

A–4 INTEGRAL CALCULUS

Definition

$$\int f(x)\, dx = \lim_{\Delta x \to 0}\left\{\sum_n [f(n\,\Delta x)]\,\Delta x\right\}$$

Integration Techniques

1. Change in variable. Let $v = u(x)$:

$$\int_a^b f(x)\, dx = \int_{u(a)}^{u(b)} \left(\frac{f(x)}{dv/dx} \bigg|_{x = u^{-1}(v)} \right) dv$$

3. Integral tables.

4. Complex variable techniques.

5. Numerical methods.

2. Integration by parts

$$\int u\, dv = uv - \int v\, du$$

A–5 INTEGRAL TABLES

Indefinite Integrals

$$\int (a + bx)^n\, dx = \frac{(a + bx)^{n+1}}{b(n + 1)}, \qquad 0 < n$$

$$\int \frac{dx}{a + bx} = \frac{1}{b} \ln|a + bx|$$

$$\int \frac{dx}{(a + bx)^n} = \frac{-1}{(n - 1)b(a + bx)^{n-1}}, \qquad 1 < n$$

$$\int \frac{dx}{c + bx + ax^2} = \begin{cases} \dfrac{2}{\sqrt{4ac - b^2}} \tan^{-1}\left(\dfrac{2ax + b}{\sqrt{4ac - b^2}}\right), & b^2 < 4ac \\[3mm] \dfrac{1}{\sqrt{b^2 - 4ac}} \ln\left|\dfrac{2ax + b - \sqrt{b^2 - 4ac}}{2ax + b + \sqrt{b^2 - 4ac}}\right|, & b^2 > 4ac \\[3mm] \dfrac{-2}{\sqrt{2ax + b}}, & b^2 = 4ac \end{cases}$$

$$\int \frac{x\, dx}{c + bx + ax^2} = \frac{1}{2a} \ln|ax^2 + bx + c| - \frac{b}{2a} \int \frac{dx}{c + bx + ax^2}$$

$$\int \frac{dx}{a^2 + b^2x^2} = \frac{1}{ab} \tan^{-1}\left(\frac{bx}{a}\right)$$

$$\int \frac{x\, dx}{a^2 + x^2} = \frac{1}{2} \ln(a^2 + x^2)$$

$$\int \cos x\, dx = \sin x \qquad\qquad\qquad \int \sin x\, dx = -\cos x$$

$$\int x \cos x\, dx = \cos x + x \sin x \qquad\qquad \int x \sin x\, dx = \sin x - x \cos x$$

$$\int x^2 \cos x \, dx = 2x \cos x + (x^2 - 2) \sin x$$

$$\int x^2 \sin x \, dx = 2x \sin x - (x^2 - 2) \cos x$$

$$\int e^{ax} \, dx = \frac{e^{ax}}{a}$$

$$\int x^3 e^{ax} \, dx = e^{ax} \left(\frac{x^3}{a} - \frac{3x^2}{a^2} + \frac{6x}{a^3} - \frac{6}{a^4} \right)$$

$$\int x e^{ax} \, dx = e^{ax} \left(\frac{x}{a} - \frac{1}{a^2} \right)$$

$$\int e^{ax} \sin x \, dx = \frac{e^{ax}}{a^2 + 1} (a \sin x - \cos x)$$

$$\int x^2 e^{ax} \, dx = e^{ax} \left(\frac{x^2}{a} - \frac{2x}{a^2} + \frac{2}{a^3} \right)$$

$$\int e^{ax} \cos x \, dx = \frac{e^{ax}}{a^2 + 1} (a \cos x - \sin x)$$

Definite Integrals

$$\int_0^\infty \frac{x^{m-1}}{1 + x^n} \, dx = \frac{\pi/n}{\sin(m\pi/n)}, \qquad n > m > 0$$

$$\int_0^\infty x^{\alpha-1} e^{-x} \, dx = \Gamma(\alpha), \quad \alpha > 0, \quad \text{where} \qquad
\begin{aligned}
& \Gamma(\alpha + 1) = \alpha\Gamma(\alpha) \\
& \Gamma(1) = 1; \Gamma(\tfrac{1}{2}) = \sqrt{\pi} \\
& \Gamma(n) = (n - 1)! \quad \text{if } n \text{ is a positive integer}
\end{aligned}$$

$$\int_0^\infty x^{2n} e^{-ax^2} \, dx = \frac{1 \cdot 3 \cdot 5 \cdots (2n - 1)}{2^{n+1} a^n} \sqrt{\frac{\pi}{a}}$$

$$\int_{-\infty}^\infty e^{-a^2 x^2 + bx} \, dx = \frac{\sqrt{\pi}}{a} e^{b^2/(4a^2)}, \qquad a > 0$$

$$\int_0^\infty e^{-ax} \cos bx \, dx = \frac{a}{a^2 + b^2}, \qquad a > 0$$

$$\int_0^\infty e^{-ax} \sin bx \, dx = \frac{b}{a^2 + b^2}, \qquad a > 0$$

$$\int_0^\infty e^{-a^2 x^2} \cos bx \, dx = \frac{\sqrt{\pi} \, e^{-b^2/4a^2}}{2a}, \qquad a > 0$$

$$\int_0^\infty x^{\alpha-1} \cos bx \, dx = \frac{\Gamma(\alpha)}{b^\alpha} \cos \tfrac{1}{2}\pi\alpha, \qquad 0 < \alpha < 1, \, b > 0$$

$$\int_0^\infty x^{\alpha-1} \sin bx \, dx = \frac{\Gamma(\alpha)}{b^\alpha} \sin \tfrac{1}{2}\pi\alpha, \qquad 0 < |\alpha| < 1, \, b > 0$$

$$\int_0^\infty x e^{-ax^2} I_k(bx)\, dx = \frac{1}{2a}\, e^{b^2/4a}, \qquad \text{where} \quad I_k(bx) = \frac{1}{\pi}\int_0^\pi e^{bx\cos\theta}\cos k\theta\, d\theta$$

$$\int_0^\infty \frac{\sin x}{x}\, dx = \int_0^\infty \mathrm{Sa}(x)\, dx = \frac{\pi}{2} \qquad \int_0^\infty \frac{\cos ax}{b^2 + x^2}\, dx = \frac{\pi}{2b}\, e^{-ab}, a > 0,\ b > 0$$

$$\int_0^\infty \left(\frac{\sin x}{x}\right)^2 dx = \int_0^\infty \mathrm{Sa}^2(x)\, dx = \frac{\pi}{2} \qquad \int_0^\infty \frac{x\sin ax}{b^2 + x^2}\, dx = \frac{\pi}{2}\, e^{-ab}, a > 0,\ b > 0$$

$$\int_{-\infty}^\infty e^{\pm j2\pi yx}\, dx = \delta(y)$$

A–6 SERIES EXPANSIONS

Finite Series

$$\sum_{n=1}^N n = \frac{N(N+1)}{2} \qquad\qquad \sum_{n=0}^N a^n = \frac{a^{N+1} - 1}{a - 1}$$

$$\sum_{n=1}^N n^2 = \frac{N(N+1)(2N+1)}{6} \qquad \sum_{n=0}^N \frac{N!}{n!\,(N-n)!}\, x^n y^{N-n} = (x + y)^N$$

$$\sum_{n=1}^N n^3 = \frac{N^2(N+1)^2}{4} \qquad \sum_{n=0}^N e^{j(\theta + n\phi)} = \frac{\sin\left[(N+1)\phi/2\right]}{\sin(\phi/2)}\, e^{j[\theta + (N\phi/2)]}$$

$$\sum_{k=0}^N \binom{N}{k} a^{N-k} b^k = (a + b)^N, \qquad \text{where} \quad \binom{N}{k} = \frac{N!}{(N-k)!\,k!}$$

Infinite Series

$$f(x) = \sum_{n=0}^\infty \left(\frac{f^{(n)}(a)}{n!}\right)(x - a)^n \qquad \text{(Taylor's series)}$$

$$f(x) = \sum_{n=-\infty}^\infty c_n e^{jn\omega_0 x}, \qquad \text{for } a \le x \le a + T \qquad \text{(Fourier series)}$$

$$\text{where} \quad c_n = \frac{1}{T}\int_a^{a+T} f(x)e^{-jn\omega_0 x}\, dx \qquad \text{and} \quad \omega_0 = \frac{2\pi}{T}$$

$$e^x = \sum_{n=0}^\infty \frac{x^n}{n!}$$

$$\sin x = \sum_{n=0}^{\infty} \frac{(-1)^n x^{2n+1}}{(2n+1)!} \qquad\qquad \cos x = \sum_{n=0}^{\infty} \frac{(-1)^n x^{2n}}{(2n)!}$$

A–7   HILBERT TRANSFORM PAIRS[†]

Definition of Hilbert Transform: $\hat{x}(t) \triangleq x(t) * \dfrac{1}{\pi t} = \dfrac{1}{\pi} \displaystyle\int_{-\infty}^{\infty} \dfrac{x(\lambda)}{t - \lambda} d\lambda$

Function	Hilbert Transform
1. $x(at + b)$	$\hat{x}(at + b)$
2. $x(t) + y(t)$	$\hat{x}(t) + \hat{y}(t)$
3. $\dfrac{d^n x(t)}{dt^n}$	$\dfrac{d^n}{dt^n} \hat{x}(t)$
4. A constant	0
5. $\dfrac{1}{t}$	$-\pi \delta(t)$
6. $\sin(\omega_0 t + \theta)$	$-\cos(\omega_0 t + \theta)$
7. $\dfrac{\sin at}{at} = \text{Sa}(at)$	$-\dfrac{1}{2\pi} at\, \text{Sa}^2(at)$
8. $e^{\pm j\omega_0 t}$	$\mp j e^{\pm j\omega_0 t}$
9. $\delta(t)$	$\dfrac{1}{\pi t}$
10. $\dfrac{a}{\pi(t^2 + a^2)}$	$\dfrac{t}{\pi(t^2 + a^2)}$
11. $\Pi\left(\dfrac{t}{T}\right) \triangleq \begin{cases} 1, & \lvert t \rvert \le T/2 \\ 0, & t \text{ elsewhere} \end{cases}$	$\dfrac{1}{\pi} \ln \left\lvert \dfrac{2t + T}{2t - T} \right\rvert$

A–8   THE DIRAC DELTA FUNCTION

DEFINITION.   The Dirac *delta function* $\delta(x)$, also called the *unit impulse function*, satisfies *both* of the following conditions:

$$\int_{-\infty}^{\infty} \delta(x)\, dx = 1, \qquad \text{and} \qquad \delta(x) = \begin{cases} \infty, & x = 0 \\ 0, & x \ne 0 \end{cases}$$

Consequently, $\delta(x)$ is a "singular" function.[‡]

[†] Fourier transform theorems are given in Table 2-1, and Fourier transform pairs are given in Table 2-2.

[‡] The Dirac delta function is not an ordinary function since it is really undefined at $x = 0$. However, it is described by the mathematical theory of distributions [Bremermann, 1965].

Properties of Dirac Delta Functions

1. $\delta(x)$ can be expressed in terms of the limit of some ordinary functions such that (in the limit of some parameter) the ordinary function satisfies the definition for $\delta(x)$. For example,

$$\delta(x) = \lim_{\sigma \to 0} \left(\frac{1}{\sqrt{2\pi}\,\sigma} e^{-x^2/(2\sigma^2)} \right) \quad \text{or} \quad \delta(x) = \lim_{a \to \infty} \left[\frac{a}{\pi} \left(\frac{\sin ax}{ax} \right) \right]$$

For these two examples, $\delta(-x) = \delta(x)$, so for *these* cases $\delta(x)$ is said to be an *even-sided delta function*. The even-sided delta function is used throughout this text, except when specifying the PDF of a discrete random variable.

$$\delta(x) = \begin{cases} \lim_{a \to \infty} (ae^{ax}), & x \leq 0 \\ 0, & x > 0 \end{cases}$$

This is an example of a single-sided delta function; in particular, this is a *left-sided delta function*. This type of delta function is used to specify the PDF of a discrete point of a random variable. (See Appendix B.)

2. Sifting property:

$$\int_{-\infty}^{\infty} w(x)\, \delta(x - x_0)\, dx = w(x_0)$$

3. For *even-sided* delta functions

$$\int_{a}^{b} w(x)\, \delta(x - x_0)\, dx = \begin{cases} 0, & x_0 < a \\ \frac{1}{2} w(a), & x_0 = a \\ w(x_0), & a < x_0 < b \\ \frac{1}{2} w(b), & x_0 = b \\ 0, & x_0 > b \end{cases}$$

where $b > a$.

4. For *left-sided* delta functions,

$$\int_{a}^{b} w(x)\delta(x - x_0)\, dx = \begin{cases} 0, & x_0 \leq a \\ w(x_0), & a < x_0 \leq b \\ 0, & x_0 > b \end{cases}$$

where $b > a$.

5. $\displaystyle \int_{-\infty}^{\infty} w(x)\, \delta^{(n)}(x - x_0)\, dx = (-1)^n\, w^{(n)}(x_0)$

where the superscript (n) denotes the nth derivative with respect to x.

6. The Fourier transform of $\delta(x)$ is unity. That is,

$$\mathcal{F}[\delta(x)] = 1$$

Conversely,

$$\delta(x) = \mathcal{F}^{-1}[1]$$

7. The scaling property is

$$\delta(ax) = \frac{1}{|a|}\,\delta(x)$$

8. For *even-sided* delta functions,

$$\delta(x) = \int_{-\infty}^{\infty} e^{\pm j2\pi xy}\,dy$$

A–9 TABULATION OF Sa$(x) = (\sin x)/x$

x	Sa(x)	Sa$^2(x)$	x	Sa(x)	Sa$^2(x)$
0.0	1.0000	1.0000	5.2	−0.1699	0.0289
0.2	0.9933	0.9867	5.4	−0.1431	0.0205
0.4	0.9735	0.9478	5.6	−0.1127	0.0127
0.6	0.9411	0.8856	5.8	−0.0801	0.0064
0.8	0.8967	0.8041	6.0	−0.0466	0.0022
1.0	0.8415	0.7081	6.2	−0.0134	0.0002
1.2	0.7767	0.6033	2π	0.0000	0.0000
1.4	0.7039	0.4955	6.4	0.0182	0.0003
1.6	0.6247	0.3903	6.6	0.0472	0.0022
1.8	0.5410	0.2927	6.8	0.0727	0.0053
2.0	0.4546	0.2067	7.0	0.0939	0.0088
2.2	0.3675	0.1351	7.2	0.1102	0.0122
2.4	0.2814	0.0792	7.4	0.1214	0.0147
2.6	0.1983	0.0393	7.6	0.1274	0.0162
2.8	0.1196	0.0143	7.8	0.1280	0.0164
3.0	0.0470	0.0022	8.0	0.1237	0.0153
π	0.0000	0.0000	8.2	0.1147	0.0132
3.2	−0.0182	0.0003	8.4	0.1017	0.0104
3.4	−0.0752	0.0056	8.6	0.0854	0.0073
3.6	−0.1229	0.0151	8.8	0.0665	0.0044
3.8	−0.1610	0.0259	9.0	0.0458	0.0021
4.0	−0.1892	0.0358	9.2	0.0242	0.0006
4.2	−0.2075	0.0431	9.4	0.0026	0.0000
4.4	−0.2163	0.0468	3π	0.0000	0.0000
4.6	−0.2160	0.0467	9.6	−0.0182	0.0003
4.8	−0.2075	0.0431	9.8	−0.0374	0.0014
5.0	−0.1918	0.0368	10.0	−0.0544	0.0030

A–10 TABULATION OF $Q(z)$

$$Q(z) \triangleq \frac{1}{\sqrt{2\pi}} \int_z^{\infty} e^{-\lambda^2/2} \, d\lambda$$

For $z \geq 3$, $Q(z) \approx \dfrac{1}{\sqrt{2\pi} \, z} e^{-z^2/2}$ (See Fig. B–7.)

Also,

$$Q(-z) = 1 - Q(z)$$

$$Q(z) = \tfrac{1}{2}\text{erfc}\left(\frac{z}{\sqrt{2}}\right) = \frac{1}{2}\left[1 - \text{erf}\left(\frac{z}{\sqrt{2}}\right)\right]$$

where $\text{erfc}(x) \triangleq \dfrac{2}{\sqrt{\pi}} \int_x^{\infty} e^{-\lambda^2} \, d\lambda$ and $\text{erf}(x) = \dfrac{2}{\sqrt{\pi}} \int_0^{x} e^{-\lambda^2} \, d\lambda$

For $z \geq 0$, a rational function approximation is [Abramowitz and Stegun, 1964; Ziemer and Tranter, 1995]

$$Q(z) = \frac{e^{-z^2/2}}{\sqrt{2\pi}} \left(b_1 t + b_2 t^2 + b_3 t^3 + b_4 t^4 + b_5 t^5\right)$$

where $t = 1/(1 + pz)$, with $p = 0.2316419$,

$b_1 = 0.31981530$ $b_2 = -0.356563782$ $b_3 = 1.781477937$

$b_4 = -1.821255978$ $b_5 = 1.330274429$

Another approximation for $Q(z)$ for $z \geq 0$ is [Börjesson and Sunberg, 1979; Peebles, 1993]

$$Q(z) = \left[\frac{1}{(1 - 0.339)z + 0.339\sqrt{z^2 + 5.510}}\right] \frac{e^{-z^2/2}}{\sqrt{2\pi}}$$

This approximation has a maximum absolute error of 0.27% for $z \geq 0$.

z	$Q(z)$	z	$Q(z)$
0.0	0.50000	2.0	0.02275
0.1	0.46017	2.1	0.01786
0.2	0.42074	2.2	0.01390
0.3	0.38209	2.3	0.01072
0.4	0.34458	2.4	0.00820
0.5	0.30854	2.5	0.00621
0.6	0.27425	2.6	0.00466
0.7	0.24196	2.7	0.00347
0.8	0.21186	2.8	0.00256
0.9	0.18406	2.9	0.00187
1.0	0.15866	3.0	0.00135
1.1	0.13567	3.1	0.00097
1.2	0.11507	3.2	0.00069
1.3	0.09680	3.3	0.00048
1.4	0.08076	3.4	0.00034
1.5	0.06681	3.5	0.00023
1.6	0.05480	3.6	0.00016
1.7	0.04457	3.7	0.00011
1.8	0.03593	3.8	0.00007
1.9	0.02872	3.9	0.00005
		4.0	0.00003

Also see Fig. B–7 for a plot of $Q(z)$.

PROBABILITY AND RANDOM VARIABLES

B–1 INTRODUCTION

The need for probability theory arises in every scientific discipline, since it is impossible to be exactly sure of values that are obtained by measurement. For example, we might say that we are 90% sure that a voltage is within ± 0.1 V of a 5-V level. This is a statistical description of the voltage parameter as opposed to a deterministic description, whereby we might define the voltage as being exactly 5 V.

This appendix is intended to be a short course in probability and random variables. Numerous excellent books cover this topic in more detail [Childers, 1997; Papoulis, 1991; Peebles, 1993; Shanmugan and Breipohl, 1988]. This appendix will provide a good introduction to the topic for the new student or a quick review for the student who is already knowledgeable in this area.

If probability and random variables are new topics for you, you will soon realize that to understand them, you will need to master many new definitions that seem to be introduced all at once. It is important to memorize these definitions in order to have a vocabulary that can be used when conversing with others about statistical results. In addition, you must develop a feeling for the engineering application of these definitions and theorems. We will accomplish this at the beginning, and you will grasp more complicated statistical ideas easily.

B–2 SETS

DEFINITION.   A *set is* a collection (or class) of objects.

The largest set or the all-embracing set of objects under consideration in an experiment is called the *universal set.* All other sets under consideration in the experiment are subsets, or *events,* of the universal set. This is illustrated in Fig. B–1a, where a *Venn diagram* is

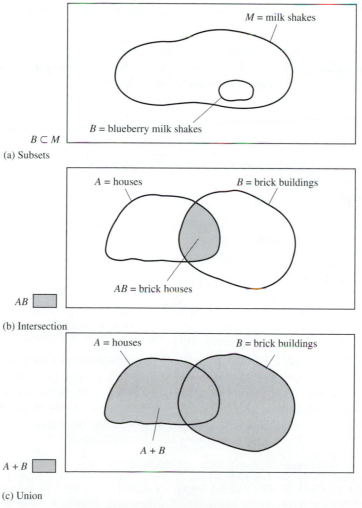

Figure B–1   Venn diagram.

given. For example, M might denote the set containing all types of milk shakes, and B might denote the subset of blueberry milk shakes. Thus, B *is* contained in M, which is denoted by $B \subset M$. In parts b and c of this figure, two sets—A and B—are shown. There are two basic ways of describing the combination of set A and set B. These are called *intersection* and *union*.

> **DEFINITION.** The *intersection* of set A and set B, denoted by AB, is the set of elements that is common to A and B. (Mathematicians use the notation $A \cap B$.)

The intersection of A and B is analogous to the AND operation used in digital logic. For example, if A denotes houses and B denotes brick buildings, then the event $C = AB$ would denote only brick houses. This is illustrated by Fig. B–1b.

> **DEFINITION.** The *union* of set A and set B, denoted by $A + B$, is the set that contains all the elements of A or all the elements of B or both. (Mathematicians use the notation $A \cup B$.)

The union of A and B is analogous to the OR operation used in digital logic. Continuing with the example just given, $D = A + B$ would be the set that contains all brick buildings, all houses, and all brick houses. This is illustrated in Fig. B–1c.

The events A and B are called *simple events,* and the events $C = AB$ and $D = A + B$ are called *compound events,* since they are logical functions of simple events.

B–3 PROBABILITY AND RELATIVE FREQUENCY

Simple Probability

The probability of an event A, denoted by $P(A)$, may be defined in terms of the relative frequency of A occurring in n trials.

> **DEFINITION.** [†]

$$P(A) = \lim_{n \to \infty} \left(\frac{n_A}{n} \right) \tag{B–1}$$

where n_A is the number of times that A occurs in n trials.

In practice, n is taken to be some reasonable number such that a larger value for n would give approximately the same value for $P(A)$. For example, suppose that a coin is tossed 40 times and that the heads event, denoted by A, occurs 19 times. Then, the probability of a head would be evaluated as approximately $P(A) = \frac{19}{40}$, where the true value of $P(A) = 0.5$ would have been obtained if $n = \infty$ had been used.

[†] Here an engineering approach is used to define probability. Strictly speaking, statisticians have developed probability theory based on three axioms: (1) $P(A) > 0$, (2) $P(S) = 1$, where S is the sure event, and (3) $P(A + B) = P(A) + P(B)$, provided that AB is a null event (i.e., $AB = \emptyset$). Statisticians define $P(A)$ as being any function of A that satisfies these axioms. The engineering definition is consistent with this approach since it satisfies these axioms.

From the definition of probability, as given by Eq. (B–1), it is seen that all probability functions have the property

$$0 \leq P(A) \leq 1 \qquad \text{(B–2)}$$

where $P(A) = 0$ if the event A is a *null event* (never occurs) and $P(A) = 1$ if the event A is a *sure event* (always occurs).

Joint Probability

DEFINITION. The probability of a *joint event*, AB, is

$$P(AB) = \lim_{n \to \infty} \left(\frac{n_{AB}}{n} \right) \qquad \text{(B–3)}$$

where n_{AB} is the number of times that the event AB occurs in n trials.

In addition, two events, A and B, are said to be *mutually exclusive* if AB is a null event, which implies that $P(AB) \equiv 0$.

Example B–1 EVALUATION OF PROBABILITIES

Let event A denote an auto accident blocking a certain street intersection during a 1-min interval. Let event B denote that it is raining at the street intersection during a 1-min period. Then, event $E = AB$ would be a blocked intersection while it is raining, as evaluated in 1-min increments.

Suppose that experimental measurements are tabulated continuously for 1 week, and it is found that $n_A = 25$, $n_B = 300$, $n_{AB} = 20$, and there are $n = 10{,}080$ 1-min intervals in the week of measurements. ($n_A = 25$ does not mean that there were 25 accidents in a 1-week period, but that the intersection was blocked for 25 1-min periods because of car accidents; this is similarly true for n_B and n_{AB}.) These results indicate that the probability of having a blocked intersection is $P(A) = 0.0025$, the probability of rain is $P(B) = 0.03$, and the probability of having a blocked intersection and it is raining is $P(AB) = 0.002$.

The probability of the union of two events may be evaluated by measuring the compound event directly, or it may be evaluated from the probabilities of simple events as given by the following theorem:

THEOREM. *Let $E = A + B$; then,*

$$P(E) = P(A + B) = P(A) + P(B) - P(AB) \qquad \text{(B–4)}$$

Proof. Let the event A only occur n_1 times out of n trials, the event B only occur n_2 times out of n trials, and the event AB occur n_{AB} times. Thus,

$$P(A + B) = \lim_{n \to \infty} \left(\frac{n_1 + n_2 + n_{AB}}{n} \right)$$

$$= \lim_{n \to \infty} \left(\frac{n_1 + n_{AB}}{n} \right) + \lim_{n \to \infty} \left(\frac{n_2 + n_{AB}}{n} \right) - \lim_{n \to \infty} \left(\frac{n_{AB}}{n} \right)$$

which is identical to Eq. (B–4), since

$$P(A) = \lim_{n\to\infty}\left(\frac{n_1 + n_{AB}}{n}\right),$$

$$P(B) = \lim_{n\to\infty}\left(\frac{n_2 + n_{AB}}{n}\right), \text{ and } P(AB) = \lim_{n\to\infty}\left(\frac{n_{AB}}{n}\right)$$

Example B–1 (Continued)

The probability of having a blocked intersection or rain occurring or both is then

$$P(A + B) = 0.0025 + 0.03 - 0.002 \approx 0.03 \qquad \text{(B–5)}$$

Conditional Probabilities

DEFINITION. The probability that an event A occurs, given that an event B has also occurred, is denoted by $P(A|B)$, which is defined as

$$P(A|B) = \lim_{n_B\to\infty}\left(\frac{n_{AB}}{n_B}\right) \qquad \text{(B–6)}$$

Example B–1 (Continued)

The probability of a blocked intersection when it is raining is approximately

$$P(A|B) = \tfrac{20}{300} = 0.066 \qquad \text{(B–7)}$$

THEOREM. *Let* $E = AB$; then

$$P(AB) = P(A)P(B|A) = P(B)P(A|B) \qquad \text{(B–8)}$$

This is known as Bayes' theorem.

Proof.

$$P(AB) = \lim_{n\to\infty}\left(\frac{n_{AB}}{n}\right) = \lim_{\substack{n\to\infty \\ n_A\to\text{large}}}\left(\frac{n_{AB}}{n_A}\frac{n_A}{n}\right) = P(B|A)P(A) \qquad \text{(B–9)}$$

It is noted that the values obtained for $P(AB)$, $P(B)$, and $P(A|B)$ in Example B–1 can be verified by using Eq. (B–8).

DEFINITION. Two events, A and B, are said to be *independent* if either

$$P(A|B) = P(A) \qquad \text{(B–10)}$$

or

$$P(B|A) = P(B) \qquad \text{(B–11)}$$

Using this definition, we can easily demonstrate that events A and B of Example B–1 are not independent. Conversely, if event A had been defined as getting heads on a coin toss, while B was the event that it was raining at the intersection, then A and B would be independent. Why?

Using Eqs. (B–8) and (B–10), we can show that if a set of events $A_1, A_2, \ldots, A_n$, are independent, then a necessary condition is[†]

$$P(A_1 A_2 \cdots A_n) = P(A_1) P(A_2) \cdots P(A_n) \qquad \text{(B–12)}$$

B–4 RANDOM VARIABLES

DEFINITION.   A real-valued *random variable* is a real-valued function defined on the events (elements) of the probability system.

An understanding of why this definition is needed is fundamental to the topic of probability theory. So far, we have defined probabilities in terms of events A, B, C, and so on. This method is awkward to use when the sets are objects (apples, oranges, etc.) instead of numbers. It is more convenient to describe sets by numerical values, so that equations can be obtained as a function of numerical values instead of functions of alphanumeric parameters. This method is accomplished by using the random variable.

Example B–2 RANDOM VARIABLE

Referring to Fig. B–2, we can show the mutually exclusive events A, B, C, D, E, F, G, and H by a Venn diagram. These are all the possible outcomes of an experiment, so the sure event is $S = A + B + C + D + E + F + G + H$. Each of these events is denoted by some convenient value of the random variable x, as shown in the table in this figure. The assigned values for x may be positive, negative, fractions, or integers as long as they are real numbers. Since all the events are mutually exclusive, using (B–4) yields

$$P(S) = 1 = P(A) + P(B) + P(C) + P(D) + P(E) + P(F) + P(G) + P(H) \qquad \text{(B–13)}$$

That is, the probabilities have to sum to unity (the probability of a sure event), as shown in the table, and the individual probabilities have been given or measured as shown. For example, $P(C) = P(-1.5) = 0.2$. These values for the probabilities may be plotted as a function of the random variable x, as shown in the graph of $P(x)$. This is a *discrete* (or *point*) distribution since the random variable takes on only discrete (as opposed to continuous) values.

B–5 CUMULATIVE DISTRIBUTION FUNCTIONS AND PROBABILITY DENSITY FUNCTIONS

DEFINITION.   The *cumulative distribution function* (CDF) of the random variable x is given by $F(a)$, where

$$F(a) \triangleq P(x \le a) \equiv \lim_{n \to \infty} \left(\frac{n_{x \le a}}{n} \right) \qquad \text{(B–14)}$$

where $F(a)$ *is a unitless function.*

[†] Equation (B–12) is not a sufficient condition for $A_1, A_2, \ldots, A_n$ to be independent [Papoulis, 1984, p. 34].

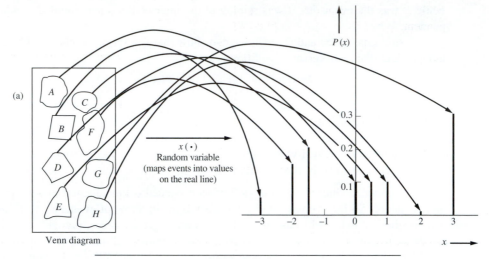

Event [·]	Value of Random Variable $x[·]$	Probability of Event $P(x)$
A	0.0	0.10
B	−3.0	0.05
C	−1.5	0.20
D	−2.0	0.15
E	+0.5	0.10
F	+1.0	0.10
G	+2.0	0.00
H	+3.0	0.30
		Total = 1.00

Figure B–2   Random variable and probability functions for Example B–2.

DEFINITION.   The *probability density function* (PDF) of the random variable x is given by $f(x)$, where

$$f(x) = \frac{dF(a)}{da}\bigg|_{a=x} = \frac{dP(x \le a)}{da}\bigg|_{a=x} = \lim_{\substack{n\to\infty \\ \Delta x \to 0}} \left[\frac{1}{\Delta x}\left(\frac{n_{\Delta x}}{n}\right) \right] \qquad \text{(B–15)}$$

where $f(x)$ has units of $1/x$.

Example B–2 (Continued)

The CDF for this example that was illustrated in Fig. B–2 is easily obtained using Eq. (B–14). The resulting CDF is shown in Fig. B–3. Note that the CDF starts at a zero value on the left ($a = -\infty$), and that the probability is accumulated until the CDF is unity on the right ($a = +\infty$).

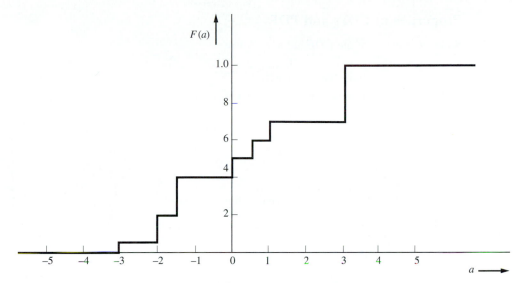

Figure B–3   CDF for Example B–2.

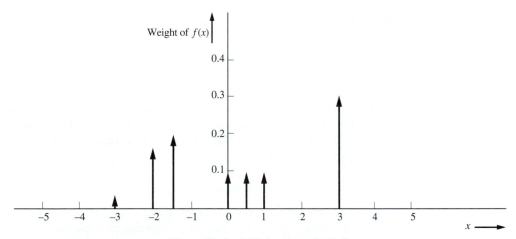

Figure B–4   PDF for Example B–2.

Using Eq. (B–15), we obtain the PDF by taking the derivative of the CDF. The result is shown in Fig. B–4. The PDF consists of Dirac delta functions located at the assigned (discrete) values of the random variable and having weights equal to the probabilities of the associated event.[†]

[†] Left-sided delta functions are used here so that if $x = a$ happens to be a discrete point, $F(a) = P(x \leq a)$ includes all the probability from $x = -\infty$ up to and *including* the point $x = a$. See Sec. A–8 (Appendix A) for properties of Dirac delta functions.

Properties of CDFs and PDFs

Some *properties of the CDF* are as follows:

1. $F(a)$ is a nondecreasing function.
2. $F(a)$ is *right-hand* continuous. That is,

$$\lim_{\substack{\varepsilon \to 0 \\ \varepsilon > 0}} F(a + \varepsilon) = F(a)$$

3.

$$F(a) = \lim_{\substack{\varepsilon \to 0 \\ \varepsilon > 0}} \int_{-\infty}^{a+\varepsilon} f(x)\, dx \qquad\qquad \text{(B–16)}$$

4. $0 \le F(a) \le 1$.
5. $F(-\infty) = 0$.
6. $F(+\infty) = 1$.

Note that the ε is needed to account for a discrete point that might occur at $x = a$. If there is no discrete point at $x = a$, the limit is not necessary.

Some *properties of the PDF* are as follows:

1. $f(x) \ge 0$. That is, $f(x)$ *is* a nonnegative function.
2. $\int_{-\infty}^{\infty} f(x)\, dx = F(+\infty) = 1$. $\qquad\qquad$ (B–17)

As we will see later, $f(x)$ may have values larger than unity; however, the area under $f(x)$ is equal to unity. These properties of the CDF and PDF are very useful in checking results of problems. That is, if a CDF or a PDF violates any of these properties, you *know* that an error has been made in the calculations.

Discrete and Continuous Distributions

Example B–2 is an example of a *discrete*, or *point*, *distribution*. That is, the random variable has M discrete values $x_1, x_2, x_3, \ldots, x_M$. ($M = 7$ in this example.) Consequently, the CDF increased only in jumps [i.e., $F(a)$ was discontinuous] as a increased, and the PDF consisted of delta functions located at the discrete values of the random variable. In contrast to this example of a discrete distribution, there are continuous distributions, one of which is illustrated in the next example. If a random variable is allowed to take on any value in some interval, it is a *continuously distributed* random variable in that interval.

Example B–3 A CONTINUOUS DISTRIBUTION

Let the random variable denote the voltages that are associated with a collection of a large number of flashlight batteries (1.5-V cells). If the number of batteries in the collection were infinite, the number of different voltage values (events) that we could obtain would be infinite, so that the distributions (PDF and CDF) would be continuous functions. Suppose that, by measurement, the CDF is evaluated first by using $F(a) = P(x \le a) = \lim_{n \to \infty} (n_{x \le a}/n)$, where n is the number of batteries in the whole collection and $n_{x \le a}$ is the number of batteries in the collection with voltages less than or equal to a V, where a is a parameter. The CDF that might be obtained is illustrated in Fig. B–5a. The associated PDF is obtained by taking the derivative of the CDF, as shown in Fig. B–5b. Note that $f(x)$ exceeds unity for some values of x, but that the area under $f(x)$ is unity (a PDF property). (You might check to see that the other CDF and PDF properties are also satisfied.)

THEOREM.

$$F(b) - F(a) = P(x \le b) - P(x \le a) = P(a < x \le b)$$

$$= \lim_{\substack{\varepsilon \to 0 \\ \varepsilon > 0}} \left[\int_{a+\varepsilon}^{b+\varepsilon} f(x)\, dx \right] \qquad \text{(B–18)}$$

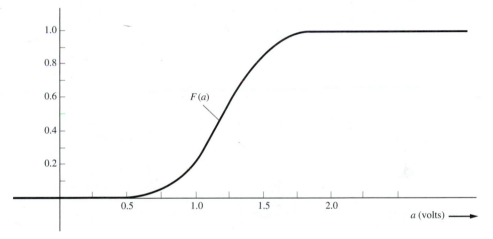

(a) Cumulative Distribution Function

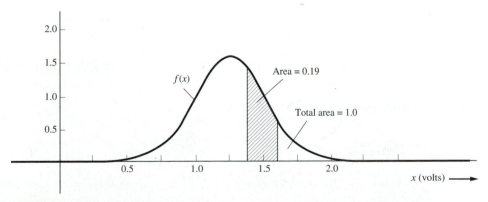

(b) Probability Density Function

Figure B–5   CDF and PDF for a continuous distribution (Example B–3).

Proof.

$$F(b) - F(a) = \lim_{\substack{\varepsilon \to 0 \\ \varepsilon > 0}} \left[\int_{-\infty}^{b+\varepsilon} f(x)\, dx - \int_{-\infty}^{a+\varepsilon} f(x)\, dx \right]$$

$$= \lim_{\substack{\varepsilon \to 0 \\ \varepsilon > 0}} \left[\int_{-\infty}^{a+\varepsilon} f(x)\, dx + \int_{a+\varepsilon}^{b+\varepsilon} f(x)\, dx - \int_{-\infty}^{a+\varepsilon} f(x)\, dx \right]$$

$$= \lim_{\substack{\varepsilon \to 0 \\ \varepsilon > 0}} \left[\int_{a+\varepsilon}^{b+\varepsilon} f(x)\, dx \right]$$

Example B–3 (Continued)

Suppose that we wanted to calculate the probability of obtaining a battery having a voltage between 1.4 and 1.6 V. Using this theorem and Fig. B–5, we compute

$$P(1.4 < x \le 1.6) = \int_{1.4}^{1.6} f(x)\, dx = F(1.6) - F(1.4) = 0.19$$

We also realize that the probability of obtaining a 1.5-V battery is zero. Why? However, the probability of obtaining a 1.5 V $\pm$ 0.1 V battery is 0.19.

In communication systems, there are digital signals that may have discrete distributions (one discrete value for each permitted level in a multilevel signal), and there are analog signals and noise that have continuous distributions. There may also be mixed distributions, which contain discrete, as well as continuous, values. For example, these occur when a continuous signal and noise are clipped by an amplifier that is driven into saturation.

THEOREM. *If x is discretely distributed, then*

$$f(x) = \sum_{i=1}^{M} P(x_i)\, \delta(x - x_i) \tag{B–19}$$

where M is the number of discrete events and $P(x_i)$ is the probability of obtaining the discrete event x_i.

This theorem was illustrated by Example B–2, where the PDF of this discrete distribution is plotted in Fig. B–4.

THEOREM. *If x is discretely distributed, then†*

$$F(a) = \sum_{i=1}^{L} P(x_i) \tag{B–20}$$

---

† Equation (B–20) is identical to $F(a) = \sum_{i=1}^{M} P(x_i) u(a - x_i)$, where $u(y)$ is a unit step function [defined in Eq. (2–49)].

L is the largest integer such that $x_L \leq a$, $L \leq M$, and M is the number of points in the discrete distribution. Here it is assumed that the discrete points x_i are indexed so that they occur in ascending order of the index. That is, $x_1 < x_2 < x_3 < \cdots < x_M$.

    This theorem was illustrated in Fig. B–3, which is a plot of Eq. (B–20) for Ex. B–2.

    In electrical engineering problems, the CDFs or PDFs of waveforms are relatively easy to obtain by using the relative frequency approach, as described by Eqs. (B–14) and (B–15). For example, the PDFs for triangular and square waves are given in Fig. B–6. These are obtained by sweeping a narrow horizontal window, Δx volts wide, vertically across the waveforms and measuring the relative frequency of occurrence of voltages in the Δx window. The time axis is divided into n intervals, and the waveform appears $n_{\Delta x}$ times in these intervals in the Δx window. A rough idea of the PDF of a waveform can also be obtained by looking at the waveform on an analog oscilloscope. Using *no* horizontal sweep, the intensity of the presentation as a function of the voltage (y-axis) gives the PDF. (This assumes that the intensity of the image is proportional to the time that the waveform dwells in the window Δy units wide.) The PDF is proportional to the intensity as a function of y (the random variable).

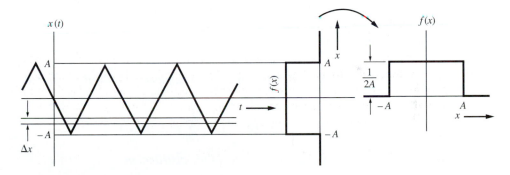

(a) Triangular Waveform and Its Associated PDF

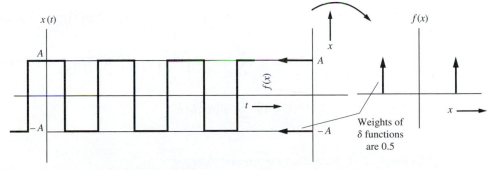

(b) Square Waveform and Its Associated PDF

Figure B–6   PDFs for triangular waves and square waves.

B–6 ENSEMBLE AVERAGE AND MOMENTS

Ensemble Average

One of the primary uses of probability theory is to evaluate the average value of a random variable (which represents some physical phenomenon) or to evaluate the average value of some function of the random variable. In general, let the function of the random variable be denoted by $y = h(x)$.

> **DEFINITION.** The *expected value*, which is also called the *ensemble average*, of $y = h(x)$ *is* given by

$$\bar{y} = \overline{[h(x)]} \triangleq \int_{-\infty}^{\infty} [h(x)] f(x)\, dx \qquad (B–21)$$

This definition may be used for discrete, as well as continuous, random variables. Note that the operator is linear. The operator is

$$\overline{[\cdot]} = \int_{-\infty}^{\infty} [\cdot] f(x)\, dx \qquad (B–22)$$

Note that other authors may denote the ensemble average of y by $E[y]$ or $\langle y \rangle$. We will use the $\bar{y}$ notation, since it is easier to write and more convenient to use when long equations are being evaluated with a large number of averaging operators.

> **THEOREM.** *If x is a **discretely** distributed random variable, the expected value can be evaluated by using*

$$\bar{y} = \overline{[h(x)]} = \sum_{i=1}^{M} h(x_i) P(x_i) \qquad (B–23)$$

where M is the number of discrete points in the distribution.

> **Proof.** Using Eqs. (B–19) in (B–21), we get
>
> $$\overline{[h(x)]} = \int_{-\infty}^{\infty} h(x) \left[\sum_{i=1}^{M} P(x_i)\, \delta(x - x_i) \right] dx$$
>
> $$= \sum_{i=1}^{M} P(x_i) \int_{-\infty}^{\infty} h(x)\, \delta(x - x_i)\, dx$$
>
> $$= \sum_{i=1}^{M} P(x_i) h(x_i)$$

Example B–4 EVALUATION OF AN AVERAGE

We will now show that Eq. (B–23) and, consequently, the definition of expected value as given by Eq. (B–21) are consistent with the way in which we usually evaluate averages. Suppose that we have a class of $n = 40$ students who take a test. The resulting test scores are 1 paper with a score of 100, 2 with scores of 95, 4 with 90, 6 with 85, 10 with 80, 10 with 75, 5 with 70, 1 with 65, and 1 paper with a score of 60. Then, the class average is

$$\bar{x} = \frac{100(1) + 95(2) + 90(4) + 85(6) + 80(10) + 75(10) + 70(5) + 65(1) + 60(1)}{40}$$

$$= 100\left(\frac{1}{40}\right) + 95\left(\frac{2}{40}\right) + 90\left(\frac{4}{40}\right) + 85\left(\frac{6}{40}\right) + 80\left(\frac{10}{40}\right) + 75\left(\frac{10}{40}\right)$$

$$+ 70\left(\frac{5}{40}\right) + 65\left(\frac{1}{40}\right) + 60\left(\frac{1}{40}\right)$$

$$= \sum_{i=1}^{9} x_i P(x_i) = 79.6 \tag{B–24}$$

Moments

Moments are defined as ensemble averages of some specific functions used for $h(x)$. For example, for the rth moment (defined subsequently), let $y = h(x) = (x - x_0)^r$.

DEFINITION. The rth *moment* of the random variable x taken about the point $x = x_0$ is given by

$$\overline{(x - x_0)^r} = \int_{-\infty}^{\infty} (x - x_0)^r f(x)\, dx \tag{B–25}$$

DEFINITION. The *mean m* is the first moment taken about the origin (i.e., $x_0 = 0$). Thus,

$$m \triangleq \bar{x} = \int_{-\infty}^{\infty} x f(x)\, dx \tag{B–26}$$

DEFINITION. The *variance* σ^2 is the second moment taken about the mean. Thus,

$$\sigma^2 = \overline{(x - \bar{x})^2} = \int_{-\infty}^{\infty} (x - \bar{x})^2 f(x)\, dx \tag{B–27}$$

DEFINITION. The *standard deviation* σ is the square root of the variance. Thus,

$$\sigma = \sqrt{\sigma^2} = \sqrt{\int_{-\infty}^{\infty} (x - \bar{x})^2 f(x)\, dx} \tag{B–28}$$

As an engineer, you may recognize the integrals of Eqs. (B–26) and (B–27) as related to applications in mechanical problems. The mean is equivalent to the center of gravity of a mass that is distributed along a single dimension, where $f(x)$ denotes the mass density as a function of the x-axis. The variance is equivalent to the moment of inertia about the center of gravity. However, you might ask, "What is the significance of the mean, variance, and other moments in electrical engineering problems?" In Chapter 6, it is shown that if x

represents a voltage or current waveform, the mean gives the dc value of the waveform. The second moment $(r = 2)$ taken about the origin $(x_0 = 0)$, which is $\overline{x^2}$, gives the normalized power. $\overline{\sigma^2}$ gives the normalized power in the corresponding ac coupled waveform. Consequently, $\sqrt{\overline{x^2}}$ is the rms value of the waveform, and σ is the rms value of the corresponding ac coupled waveform.

In statistical terms, m gives the *center of gravity* of the PDF, and σ gives us the *spread* of the PDF about this center of gravity. For example, in Fig. B–5 the voltage distribution for a collection of flashlight batteries is given. The mean is $\overline{x} = 1.25$ V, and the standard deviation is $\sigma = 0.25$ V. This figure illustrates a Gaussian distribution that will be studied in detail in Sec. B–7. For this Gaussian distribution, the area under $f(x)$ from $x = 1.0$ to 1.5 V, which corresponds to the interval $\overline{x} \pm \sigma$, is 0.68. Thus, we conclude that 68% of the batteries have voltages within one standard deviation of the mean value (Gaussian distribution).

There are several ways to specify a number that is used to describe the typical, or most common, value of x. The *mean m* is one such measure that gives the center of gravity. Another measure is the *median*, which corresponds to the value $x = a$, where $F(a) = \frac{1}{2}$. A third measure is called the *mode*, which corresponds to the value of x where $f(x)$ is a maximum, assuming that the PDF has only one maximum. For the Gaussian distribution, all these measures give the same number, namely, $x = m$. For other types of distributions, the values obtained for the mean, median, and mode will usually be nearly the same number. The variance is also related to the second moment about the origin and the mean, as described by the following theorem:

THEOREM.

$$\sigma^2 = \overline{x^2} - (\overline{x})^2 \tag{B–29}$$

A proof of this theorem illustrates how the ensemble average operator notation is used.

Proof.

$$\sigma^2 = \overline{(x - \overline{x})^2}$$

$$= \overline{[x^2]} - \overline{[2x\overline{x}]} + \overline{[(\overline{x})^2]} \tag{B–30}$$

Because $\overline{[\cdot]}$ is a linear operator, $\overline{[2x\overline{x}]} = 2\overline{x}\,\overline{x} = 2(\overline{x})^2$. Moreover, $(\overline{x})^2$ is a constant, and the average value of a constant is the constant itself. That is, for the constant c,

$$\overline{c} = \int_{-\infty}^{\infty} c f(x)\, dx = c \int_{-\infty}^{\infty} f(x)\, dx = c \tag{B–31}$$

So, using Eq. (B–31) in Eq. (B–30), we obtain

$$\sigma^2 = \overline{x^2} - 2(\overline{x})^2 + (\overline{x})^2$$

which is equivalent to Eq. (B–29).

Thus, there are two ways to evaluate the variance: (1) by use of the definition as given by Eq. (B–27) or (2) by use of the theorem that is Eq. (B–29).

B–7 EXAMPLES OF IMPORTANT DISTRIBUTIONS

There are numerous types of distributions. Some of the more important ones used in communication and statistical problems are summarized in Table B–1. Here, the equations for their PDF and CDF, a sketch of the PDF, and the formula for the mean and variance are given. These distributions will be studied in more detail in the paragraphs that follow.

Binomial Distribution

The binomial distribution is useful for describing digital, as well as other statistical problems. Its application is best illustrated by an example.

Assume that we have a binary word n bits long and that the probability of sending a binary 1 is p. Consequently, the probability of sending a binary 0 is $1 - p$. We want to evaluate the probability of obtaining n-bit words that contain k binary 1s. One such word is k binary 1s followed by $n - k$ binary 0s. The probability of obtaining this word is $p^k(1 - p)^{n-k}$. There are also other n-bit words that contain k binary 1s. In fact, the number of different n-bit words containing k binary 1s is

$$\binom{n}{k} = \frac{n!}{(n - k)!k!} \tag{B–32}$$

(This can be demonstrated by taking a numerical example, such as $n = 8$ and $k = 3$.) The symbol $\binom{n}{k}$ is used in algebra to denote the operation described by Eq. (B–32) and is read "the combination of n things taken k at a time." Thus, the probability of obtaining an n-bit word containing k binary 1s is

$$P(k) = \binom{n}{k} p^k(1 - p)^{n-k} \tag{B–33}$$

If we let the random variable x denote these discrete values, then $x = k$, where k can take on the values $0, 1, 2, \ldots, n$, and we obtain the *binomial PDF*

$$f(x) = \sum_{k=0}^{n} P(k)\delta(x - k) \tag{B–34}$$

where $P(k)$ is given by Eq. (B–33).

The name *binomial* comes from the fact that the $P(k)$ are the individual terms in a binomial expansion. That is, letting $q = 1 - p$, we get

$$(p + q)^n = \sum_{k=0}^{n} \binom{n}{k} p^k q^{n-k} = \sum_{k=0}^{n} P(k) \tag{B–35}$$

TABLE B-1 SOME DISTRIBUTIONS AND THEIR PROPERTIES

Name of Distribution	Type	Sketch of PDF	Equation for: Cumulative Distribution Function (CDF)	Probability Density Function (PDF)	Mean	Variance
Binomial	Discrete	$n=3$, $p=0.6$	$F(a) = \sum_{\substack{k=0 \\ m \le a}}^{m} P(k)$ where $P(k) = \binom{n}{k} p^k (1-p)^{n-k}$	$f(x) = \sum_{k=0}^{n} P(k)\,\delta(x-k)$ where $P(k) = \binom{n}{k} p^k (1-p)^{n-k}$	np	$np(1-p)$
Poisson	Discrete	$\lambda = 2$	$F(a) = \sum_{\substack{k=0 \\ m \le a}}^{m} P(k)$ where $P(k) = \dfrac{\lambda^k}{k!} e^{-\lambda}$	$f(x) = \sum_{k=0}^{\infty} P(k)\,\delta(x-k)$ where $P(k) = \dfrac{\lambda^k}{k!} e^{-\lambda}$	λ	λ
Uniform	Continuous		$F(a) = \begin{cases} 0, & a < \left(\dfrac{2m-A}{2}\right) \\[2mm] \dfrac{1}{A}\left[a - \left(\dfrac{2m-A}{2}\right)\right], & |a-m| \le \dfrac{A}{2} \\[2mm] 1, & a \ge \left(\dfrac{2m-A}{2}\right) \end{cases}$	$f(x) = \begin{cases} 0, & x < \left(\dfrac{2m-A}{2}\right) \\[2mm] \dfrac{1}{A}, & |x-m| \le \dfrac{A}{2} \\[2mm] 0, & x > \left(\dfrac{2m+A}{2}\right) \end{cases}$	m	$\dfrac{A^2}{12}$
Gaussian	Continuous		$F(a) = Q\!\left(\dfrac{m-a}{\sigma}\right)$ where $Q(\sigma) \triangleq \dfrac{1}{\sqrt{2\pi}} \displaystyle\int_{a}^{\infty} e^{-x^2/2}\,dx$	$f(x) = \dfrac{1}{\sqrt{2\pi}\,\sigma} \exp\!\left[-(x-m)^2/2\sigma^2\right]$	m	σ^2
Sinusoidal	Continuous		$F(a) = \begin{cases} 0, & a \le -A \\[2mm] \dfrac{1}{\pi}\left[\dfrac{\pi}{2} + \sin^{-1}\!\left(\dfrac{a}{A}\right)\right], & |a| \le A \\[2mm] 1, & a \ge A \end{cases}$	$f(x) = \begin{cases} 0, & x < -A \\[2mm] \dfrac{1}{\pi\sqrt{A^2 - x^2}}, & |x| \le A \\[2mm] 0, & x > A \end{cases}$	0	$\dfrac{A^2}{2}$

The combinations $\binom{n}{k}$, which are also the binomial coefficients, can be evaluated by using Pascal's triangle:

$$
\begin{array}{cccccccccccc}
n = 0 & & & & & & 1 & & & & & \\
n = 1 & & & & & 1 & & 1 & & & & \\
n = 2 & & & & 1 & & 2 & & 1 & & & \\
n = 3 & & & 1 & & 3 & & 3 & & 1 & & \\
n = 4 & & 1 & & 4 & & 6 & & 4 & & 1 & \\
n = 5 & 1 & & 5 & & 10 & & 10 & & 5 & & 1 \\
\vdots & & & & & & \vdots & & & & &
\end{array}
$$

For a particular value of n, the combinations $\binom{n}{k}$, for $k = 0, 1, \ldots, n$, are the elements in the nth row. For example, for $n = 3$,

$$
\binom{3}{0} = 1, \quad \binom{3}{1} = 3, \quad \binom{3}{2} = 3, \quad \text{and} \quad \binom{3}{3} = 1
$$

The mean value of the binomial distribution is evaluated by the use of Eq. (B–23):

$$
m = \bar{x} = \sum_{k=0}^{n} x_k P(x_k) = \sum_{k=0}^{n} k P(k)
$$

or

$$
m = \sum_{k=1}^{n} k \binom{n}{k} p^k q^{n-k} \tag{B–36}
$$

Using the identity

$$
k \binom{n}{k} = \frac{kn!}{(n-k)!k!} = \frac{n(n-1)!}{(n-k)!(k-1)!}
$$

$$
= n \left[\frac{(n-1)!}{((n-1)-(k-1))!(k-1)!} \right]
$$

or

$$
k \binom{n}{k} = n \binom{n-1}{k-1} \tag{B–37}
$$

we have

$$
m = \sum_{k=1}^{n} n \binom{n-1}{k-1} p^k q^{n-k} \tag{B–38}
$$

Making a change in the index, let $j = k - 1$. Thus,

$$
m = \sum_{j=0}^{n-1} n \binom{n-1}{j} p^{j+1} q^{n-(j+1)}
$$

$$
= np \left[\sum_{j=0}^{n-1} \binom{n-1}{j} p^j q^{(n-1)-j} \right] = np \left[(p+q)^{n-1} \right] \tag{B–39}
$$

Recalling that p and q are probabilities and $p + q = 1$, we see that $(p + q)^{n-1} = 1$. Thus, Eq. (B–39) reduces to

$$m = np \tag{B–40}$$

Similarly, it can be shown that the variance is $np(1 - p)$ by using $\sigma^2 = \overline{x^2} - (\overline{x})^2$.

Poisson Distribution

The Poisson distribution (Table B–1) is obtained as a limiting approximation of a binomial distribution when n is very large and p is very small, but the product $np = \lambda$ is some reasonable size [Thomas, 1969].

Uniform Distribution

The uniform distribution is

$$f(x) = \begin{cases} 0, & x < \left(\dfrac{2m - A}{2}\right) \\[2ex] \dfrac{1}{A}, & |x - m| \leq \dfrac{A}{2} \\[2ex] 0, & x > \left(\dfrac{2m + A}{2}\right) \end{cases} \tag{B–41}$$

where A is the peak-to-peak value of the random variable. This is illustrated by the sketch shown in Table B–1. The mean of this distribution is

$$\int_{-\infty}^{\infty} x\, f(x)\, dx = \int_{m-(A/2)}^{m+(A/2)} x\, \frac{1}{A}\, dx = m \tag{B–42}$$

and the variance is

$$\sigma^2 = \int_{m-(A/2)}^{m+(A/2)} (x - m)^2\, \frac{1}{A}\, dx \tag{B–43}$$

Making a change in variable, let $y = x - m$:

$$\sigma^2 = \frac{1}{A} \int_{-A/2}^{A/2} y^2\, dy = \frac{A^2}{12} \tag{B–44}$$

The uniform distribution is useful in describing quantizing noise that is created when an analog signal is converted into a PCM signal, as discussed in Chapter 3. In Chapter 6, it is shown that it also describes the noise out of a phase detector when the input is Gaussian noise (as described subsequently).

Gaussian Distribution

The Gaussian distribution, which is also called the *normal distribution*, is one of the most important distributions, if not *the* most important. As discussed in Chapter 5, thermal noise has a Gaussian distribution. Numerous other phenomena can also be described by Gaussian statistics, and many theorems have been developed by statisticians that are based on Gauss-ian assumptions. It cannot be overemphasized that the Gaussian distribution is very impor-tant in analyzing both communication problems and problems in statistics. It can also be shown that the Gaussian distribution can be obtained as the limiting form of the binomial distribution when n becomes large, while holding the mean $m = np$ finite, and letting the variance $\sigma^2 = np(1 - p)$ be much larger than unity [Feller, 1957; Papoulis, 1984].

DEFINITION.   The Gaussian distribution is

$$f(x) = \frac{1}{\sqrt{2\pi}\,\sigma}\, e^{-(x-m)^2/(2\sigma^2)} \tag{B–45}$$

where m is the mean and σ^2 is the variance.

A sketch of Eq. (B–45) is given in Fig. B–5, together with the CDF for the Gaussian random variable where $m = 1.25$ and $\sigma = 0.25$. The Gaussian PDF is symmetrical about $x = m$, with the area under the PDF being $\frac{1}{2}$ for $(-\infty \le x \le m)$ and $\frac{1}{2}$ for $(m \le x \le \infty)$. The peak value of the PDF is $1/(\sqrt{2\pi}\,\sigma)$, so that as $\sigma \to 0$, the Gaussian PDF goes into a δ function located at $x = m$ (since the area under the PDF is always unity).

We will now show that Eq. (B–45) is properly normalized, (i.e., that the area under $f(x)$ is unity). This can be accomplished by letting I represent the integral of the PDF:

$$I \triangleq \int_{-\infty}^{\infty} f(x)\,dx = \int_{-\infty}^{\infty} \frac{1}{\sqrt{2\pi}\,\sigma}\, e^{-(x-m)^2/(2\sigma^2)}\,dx \tag{B–46}$$

Making a change in variable, we let $y = (x - m)/\sigma$. Then,

$$I = \frac{1}{\sqrt{2\pi}\,\sigma} \int_{-\infty}^{\infty} e^{-y^2/2}\,(\sigma dy) = \frac{1}{\sqrt{2\pi}} \int_{-\infty}^{\infty} e^{-y^2/2}\,dy \tag{B–47}$$

The integral I can be shown to be unity by showing that I^2 is unity:

$$I^2 = \left[\frac{1}{\sqrt{2\pi}} \int_{-\infty}^{\infty} e^{-x^2/2}\,dx \right]\left[\frac{1}{\sqrt{2\pi}} \int_{-\infty}^{\infty} e^{-y^2/2}\,dy \right]$$

$$= \frac{1}{2\pi} \int_{-\infty}^{\infty} \int_{-\infty}^{\infty} e^{-(x^2+y^2)/2}\,dx\,dy \tag{B–48}$$

Make a change in the variables to polar coordinates from Cartesian coordinates. Let $r^2 = x^2 + y^2$, and let $\theta = \tan^{-1}(y/x)$; then,

$$I^2 = \frac{1}{2\pi} \int_0^{2\pi} \left[\int_0^{\infty} e^{-r^2/2}\, r\, dr \right] d\theta = \frac{1}{2\pi} \int_0^{2\pi} d\theta = 1 \tag{B–49}$$

Thus, $I^2 = 1$, and, consequently, $I = 1$.

Up to now, we have *assumed* that the parameters m and σ^2 of Eq. (B–45) were the mean and the variance of the distribution. We need to *show* that indeed they are! This may be accomplished by writing the Gaussian form in terms of some arbitrary parameters α and β:

$$f(x) = \frac{1}{\sqrt{2\pi}\,\beta}\, e^{-(x-\alpha)^2/(2\beta^2)} \tag{B–50}$$

This function is still properly normalized, as demonstrated by Eq. (B–49). First, we need to show that the parameter α is the mean:

$$m = \int_{-\infty}^{\infty} x f(x) \, dx = \frac{1}{\sqrt{2\pi}\,\beta} \int_{-\infty}^{\infty} x e^{-(x-\alpha)^2/(2\beta^2)} \, dx \tag{B–51}$$

Making a change in variable, we let $y = (x - \alpha)/\beta$; then,

$$m = \frac{1}{\sqrt{2\pi}} \int_{-\infty}^{\infty} (\beta y + \alpha) e^{-y^2/2} \, dy$$

or

$$m = \frac{\beta}{\sqrt{2\pi}} \int_{-\infty}^{\infty} (y e^{-y^2/2}) \, dy + \alpha \left(\int_{-\infty}^{\infty} \frac{1}{\sqrt{2\pi}} e^{-y^2/2} \, dy \right) \tag{B–52}$$

The first integral on the right of Eq. (B–52) is zero, because the integrand is an odd function and the integral is evaluated over symmetrical limits. The second integral on the right is the integral of a properly normalized Gaussian PDF, so the integral has a value of unity. Thus, Eq. (B–52) becomes

$$m = \alpha \tag{B–53}$$

and we have shown that the parameter α is the mean value.

The variance is

$$\sigma^2 = \int_{-\infty}^{\infty} (x - m)^2 f(x) \, dx$$

$$= \frac{1}{\sqrt{2\pi}\,\beta} \int_{-\infty}^{\infty} (x - m)^2 e^{-(x-m)^2/(2\beta^2)} \, dx \tag{B–54}$$

Similarly, we need to show that $\sigma^2 = \beta^2$. This will be left as a homework exercise.

The next question to answer is, "What is the CDF for the Gaussian distribution?"

THEOREM. *The cumulative distribution function (CDF) for the Gaussian distribution is*

$$F(a) = Q\left(\frac{m - a}{\sigma}\right) = \tfrac{1}{2} \operatorname{erfc}\left(\frac{m - a}{\sqrt{2}\,\sigma}\right) \tag{B–55}$$

where the Q function is defined by

$$Q(z) \triangleq \frac{1}{\sqrt{2\pi}} \int_{z}^{\infty} e^{-\lambda^2/2} \, d\lambda \tag{B–56}$$

and the complementary error function (erfc) is defined as

$$\operatorname{erfc}(z) \triangleq \frac{2}{\sqrt{\pi}} \int_{z}^{\infty} e^{-\lambda^2} \, d\lambda \tag{B–57}$$

It can also be shown that

$$\text{erfc}(z) = 1 - \text{erf}(z) \tag{B–58}$$

where the error function is defined as

$$\text{erf}(z) \triangleq \frac{2}{\sqrt{\pi}} \int_0^z e^{-\lambda^2} \, d\lambda \tag{B–59}$$

The Q function and the complementary error function, as used in Eq. (B–55), give the same curve for $F(a)$. Because neither of the corresponding integrals, given by Eqs. (B–56) and (B–57), can be evaluated in closed form, math tables (see Sec. A–10, Appendix A), numerical integration techniques, or closed-form approximations must be used to evaluate them. The equivalence between the two functions is

$$Q(z) = \tfrac{1}{2}\,\text{erfc}\left(\frac{z}{\sqrt{2}}\right) \tag{B–60}$$

Communication engineers often prefer to use the Q function instead of the erfc function, since solutions to problems written in terms of the Q function do not require the writing of the $\tfrac{1}{2}$ and $1/\sqrt{2}$ factors. Conversely, the advantage of using the $\text{erf}(z)$ or $\text{erfc}(z)$ functions is that they are one of the standard functions in MATLAB and MathCAD, and these functions are also available on some hand calculators. However, textbooks in probability and statistics usually give a tabulation of the normalized CDF. This is a tabulation of $F(a)$ for the case of $m = 0$ and $\sigma = 1$, and it is equivalent to $Q(-a)$ and $\tfrac{1}{2}\,\text{erfc}\,(-a/\sqrt{2})$. Since $Q(z)$ and $\tfrac{1}{2}\,\text{erfc}(z/\sqrt{2})$ are equivalent, which one is used is a matter of personal preference. We use the Q-function notation in this book.

Proof. Proof of a Theorem for the Gaussian CDF

$$F(a) = \int_{-\infty}^a f(x) \, dx = \frac{1}{\sqrt{2\pi}\,\sigma} \int_{-\infty}^a e^{-(x-m)^2/(2\sigma^2)} \, dx \tag{B–61}$$

Making a change in variable, let $y = (m - x)/\sigma$:

$$F(a) = \frac{1}{\sqrt{2\pi}\,\sigma} \int_{\infty}^{(m-a)/\sigma} e^{-y^2/2}\,(-\sigma dy) \tag{B–62}$$

or

$$F(a) = \frac{1}{\sqrt{2\pi}} \int_{(m-a)/\sigma}^{\infty} e^{-y^2/2}\, dy = Q\left(\frac{m - a}{\sigma}\right) \tag{B–63}$$

Similarly, $F(a)$ may be expressed in terms of the complementary error function.

As mentioned earlier, it is unfortunate that the integrals for $Q(z)$ or $\text{erfc}(z)$ cannot be evaluated in closed form. However, for large values of z, very good closed-form approximations can be obtained, and for small values of z numerical integration techniques can be applied easily. A plot of $Q(z)$ is shown in Fig. B–7 for $z \geq 0$, and a tabulation of $Q(z)$ is given in Sec. A–10.

A relatively simple closed-form upper bound for $Q(z)$, $z > 0$, is

$$Q(z) < \frac{1}{\sqrt{2\pi}\,z}\, e^{-z^2/2}, \qquad z > 0 \tag{B–64}$$

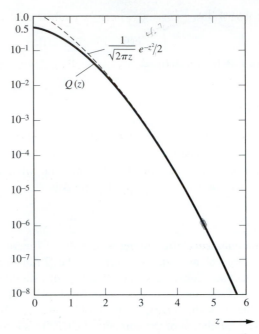

Figure B–7   The function $Q(z)$ and an overbound, $\dfrac{1}{\sqrt{2\pi}\,z}\,e^{-z^2/2}$.

This is also shown in Fig. B–7. It is obtained by evaluating the $Q(z)$ integral by parts:

$$Q(z) = \int_z^\infty \frac{1}{\sqrt{2\pi}}\, e^{-\lambda^2/2}\,d\lambda = \int_z^\infty u\,dv = uv \Big|_z^\infty - \int_z^\infty v\,du$$

where $u = 1/(\sqrt{2\pi}\,\lambda)$ and $dv = \lambda e^{-\lambda^2/2}\,d\lambda$. Thus,

$$Q(z) = \left(\frac{1}{\sqrt{2\pi}\,\lambda}\right)\left(- e^{-\lambda^2/2}\right)\Big|_z^\infty - \int_z^\infty \left(- e^{-\lambda^2/2}\right)\left(- \frac{1}{\sqrt{2\pi}\,\lambda^2}\,d\lambda\right)$$

Dropping the integral, which is a positive quantity, we obtain the upper bound on $Q(z)$, as given by Eq. (B–64). If needed, a lower bound may also be obtained [Wozencraft and Jacobs, 1965]. A rational function approximation for $Q(z)$ is given in Sec. (A–10), and a closed-form approximation has an error of less than 0.27%.

For values of $z \geq 3$, this upper bound has an error of less than 10% when compared with the actual value for $Q(z)$. That is, $Q(3) = 1.35 \times 10^{-3}$ and the upper bound has a value of 1.48×10^{-3} for $z = 3$. This is an error of 9.4%. For $z = 4$, the error is 5.6%, and for $z = 5$, the error is 3.6%. In evaluating the probability of error for digital systems, as discussed in Chapters 6, 7, and 8, the result is often found to be a Q function. Since most useful digital systems have a probability of error of 10^{-3} or less, this upper bound becomes very useful for evaluating $Q(z)$. At any rate, if the upper bound approximation is used, we know that the value obtained indicates slightly poorer performance than is theoretically possible. In this sense, this approximation will give a worst-case result.

For the case of z negative, $Q(z)$ can be evaluated by using the identity

$$Q(-z) = 1 - Q(z) \tag{B–65}$$

where the Q value for positive z is used (as obtained from Fig. B–7) to compute the Q of the negative value of z.

Sinusoidal Distribution

THEOREM.   *If $x = A \sin \psi$, where ψ has the uniform distribution,*

$$f_\psi(\psi) = \begin{cases} \dfrac{1}{2\pi}, & |\psi| \le \pi \\[2ex] 0, & \text{elsewhere} \end{cases} \tag{B–66}$$

then the PDF for the sinusoid is given by

$$f_x(x) = \begin{cases} 0, & x < -A \\[2ex] \dfrac{1}{\pi\sqrt{A^2 - x^2}}, & |x| \le A \\[2ex] 0, & x > A \end{cases} \tag{B–67}$$

A proof of this theorem will be given in Sec. B–8.

A sketch of the PDF for a sinusoid is given in Table B–1, along with equations for the CDF and the variance. Note that the standard deviation, which is equivalent to the rms value as discussed in Chapter 6, is $\sigma = A/\sqrt{2}$. This should *not* be a surprising result.

The sinusoidal distribution can be used to model observed phenomena. For example, x might represent an oscillator voltage where $\psi = \omega_0 t + \theta_0$. Here the frequency of oscillation is f_0, and ω_0 and t are assumed to be deterministic values. θ_0 represents the random startup phase of the oscillator. (When the power of an unsynchronized oscillator is turned on, the oscillation builds up from a noise voltage that is present in the circuit.) In another oscillator model, time might be considered to be a uniformly distributed random variable, where $\psi = \omega_0 t$. Here, once again, we would have a sinusoidal distribution for x.

B–8 FUNCTIONAL TRANSFORMATIONS OF RANDOM VARIABLES

As illustrated by the preceding sinusoidal distribution, we often need to evaluate the PDF for a random variable that is a function of another random variable for which the distribution is known. This is illustrated pictorially in Fig. B–8. Here the input random variable is denoted by x, and the output random variable is denoted by y. Because several PDFs are involved, *subscripts* (such as x in f_x) will be used to indicate with which random variable the PDF is associated. The arguments of the PDFs may change, depending on what substitutions are made, as equations are reduced.

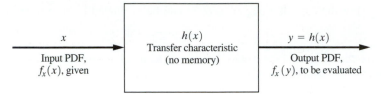

Figure B–8   Functional transformation of random variables.

THEOREM.   *If* $y = h(x)$, *where* $h(\cdot)$ *is the output-to-input (transfer) characteristic of a device **without memory**,[†] then the PDF of the output is*

$$f_y(y) = \sum_{i=1}^{M} \frac{f_x(x)}{|dy/dx|}\bigg|_{x=x_i=h_i^{-1}(y)} \tag{B–68}$$

where $f_x(x)$ *is the PDF of the input,* x. M *is the number of real roots of* $y = h(x)$. *That is, the inverse of* $y = h(x)$ *gives* $x_1, x_2, \ldots, x_M$ *for a single value of* y. $|\cdot|$ *denotes the absolute value and the single vertical line denotes the evaluation of the quantity at* $x = x_i = h_i^{-1}(y)$.

Two examples will now be worked out to demonstrate the application of this theorem, and then the proof will be given.

Example B–5 SINUSOIDAL DISTRIBUTION

Let

$$y = h(x) = A \sin x \tag{B–69}$$

where x is uniformly distributed over $-\pi$ to $+\pi$, as given by Eq. (B–66). This is illustrated in Fig. B–9. For a given value of y, say, $-A < y_0 < A$, there are two possible inverse values for x, namely, x_1 and x_2, as shown in the figure. Thus $M = 2$, provided that $|y| < A$. Otherwise, $M = 0$. Evaluating the derivative of Eq. (B–69), we obtain

$$\frac{dy}{dx} = A \cos x$$

and for $0 \leq y \leq A$, we get

$$x_1 = \text{Sin}^{-1}\left(\frac{y}{A}\right)$$

and

$$x_2 = \pi - x_1$$

where the uppercase S in $\text{Sin}^{-1}(\cdot)$ denotes the principal angle. A similar result is obtained for $-A \leq y \leq 0$. Using Eq. (B–68), we find that the PDF for y is

---

[†] The output-to-input characteristic $h(x)$ should not be confused with the impulse response of a linear network, which was denoted by $h(t)$.

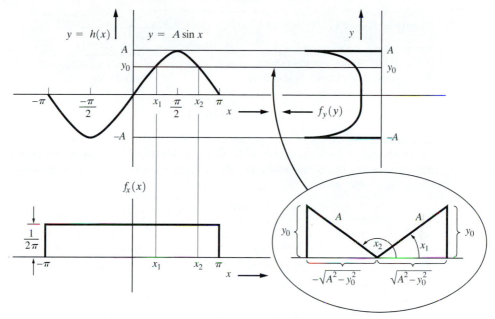

Figure B–9   Evaluation of the PDF of a sinusoid (Ex. B–5).

$$f_y(y) = \begin{cases} \dfrac{f_x(x_1)}{|A\cos x_1|} + \dfrac{f_x(x_2)}{|-A\cos x_2|}, & |y| \le A \\[4mm] 0, & y \text{ elsewhere} \end{cases} \tag{B–70}$$

The denominators of these two terms are evaluated with the aid of the triangles shown in the insert of Fig. B–9. By using this result and substituting the uniform PDF for $f_x(x)$, (B–70) becomes

$$f_y(y) = \begin{cases} \dfrac{1/2\pi}{|\sqrt{A^2 - y^2}|} + \dfrac{1/2\pi}{|-\sqrt{A^2 - y^2}|}, & |y| \le A \\[4mm] 0, & y \text{ elsewhere} \end{cases}$$

or

$$f_y(y) = \begin{cases} 0, & y < -A \\[3mm] \dfrac{1}{\pi\sqrt{A^2 - y^2}}, & |y| \le A \\[3mm] 0, & y > A \end{cases} \tag{B–71}$$

which is the PDF for a sinusoid as given first by Eq. (B–67). This result is intuitively obvious, since we realize that a sinusoidal waveform spends most of its time near its peak values and passes through zero relatively rapidly. Thus, the PDF should peak up at $+A$ and $-A$ V.

Example B–6 PDF FOR THE OUTPUT OF A DIODE CHARACTERISTIC

Assume that a diode current-voltage characteristic is modeled by the ideal characteristic shown in Fig. B–10, where y is the current through the diode and x is the voltage across the diode. This type of characteristic is also called *half-wave linear rectification.*

$$y = \begin{cases} Bx, & x > 0 \\ 0, & x \leq 0 \end{cases} \tag{B-72}$$

where $B > 0$. For $y > 0$, $M = 1$; and for $y < 0$, $M = 0$. However, if $y = 0$, there are an infinite number of roots for x (i.e., all $x \leq 0$). Consequently, there will be a discrete point at $y = 0$ if the area under $f_x(x)$ is nonzero for $x \leq 0$ (i.e., for the values of x that are mapped into $y = 0$). Using Eq. (B–68), we get

$$f_y(y) = \begin{cases} \dfrac{f_x(y/B)}{B}, & y > 0 \\ 0, & y < 0 \end{cases} + P(y = 0)\,\delta(y) \tag{B-73}$$

where

$$P(y = 0) = P(x \leq 0) = \int_{-\infty}^{0} f_x(x)\,dx = F_x(0) \tag{B-74}$$

Suppose that x has a Gaussian distribution with zero mean; then these equations reduce to

$$f_y(y) = \begin{cases} \dfrac{1}{\sqrt{2\pi}B\sigma}\,e^{-y^2/(2B^2\sigma^2)}, & y > 0 \\ 0, & y < 0 \end{cases} + \tfrac{1}{2}\,\delta(y) \tag{B-75}$$

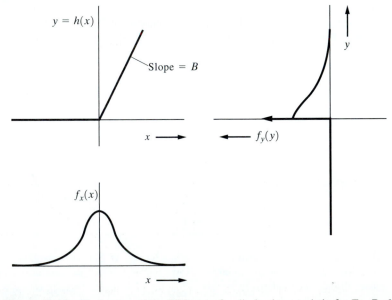

Figure B–10   Evaluation of the PDF out of a diode characteristic for Ex. B–6.

A sketch of this result is shown in Fig. B–10. For $B = 1$, note that the output is the same as the input for x positive (i.e., $y = x > 0$), so that the PDF of the output is the same as the PDF of the input for $y > 0$. For $x < 0$, the values of x are mapped into the point $y = 0$, so that the PDF of y contains a δ function of weight $\frac{1}{2}$ at the point $y = 0$.

Proof. We will demonstrate that Eq. (B–68) is valid by partitioning the x-axis into intervals over which $h(x)$ is monotonically increasing, monotonically decreasing, or a constant. As we have seen in Ex. B–6, when $h(x)$ is a constant over some interval of x, a discrete point at y equal to that constant is possible. In addition, discrete points in the distribution of x will be mapped into discrete points in y, even in regions where $h(x)$ is not a constant.

Now we will demonstrate that the theorem, as described by Eq. (B–68), is correct by taking the case, for example, where $y = h(x)$ is monotonically decreasing for $x < x_0$ and monotonically increasing for $x > x_0$. This is illustrated in Fig. B–11. The CDF for y is then

$$F_y(y_0) = P(y \leq y_0) = P(x_1 \leq x \leq x_2)$$
$$= P[(x = x_1) + (x_1 < x \leq x_2)] \tag{B–76}$$

where the $+$ sign denotes the union operation. Then, using Eq. (B–4), we get

$$F_y(y_0) = P(x_1) + P(x_1 < x \leq x_2)$$

or

$$F_y(y_0) = P(x_1) + F_x(x_2) - F_x(x_1) \tag{B–77}$$

The PDF of y is obtained by taking the derivative of both sides of this equation:

$$\frac{dF_y(y_0)}{dy_0} = \frac{dF_x(x_2)}{dx_2}\frac{dx_2}{dy_0} - \frac{dF_x(x_1)}{dx_1}\frac{dx_1}{dy_0} \tag{B–78}$$

where $dP(x_1)/dy_0 = 0$ since $P(x_1)$ is a constant. Because

$$dF_x(x_2)/dx_2 = f_x(x_2) \qquad \text{and} \qquad dF_x(x_1)/dx_1 = f_x(x_1)$$

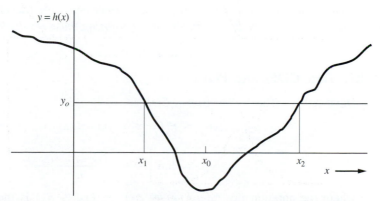

Figure B–11   Example of a function $h(x)$ that monotonically decreases for $x < x_0$ and monotonically increases for $x > x_0$.

Eq. (B–78) becomes

$$f_y(y_0) = \frac{f_x(x_2)}{dy_0/dx_2} + \frac{f_x(x_1)}{-\,dy_0/dx_1} \tag{B–79}$$

At the point $x = x_1$, the slope of y is negative, because the function is monotonically decreasing for $x < x_0$; thus, $dy_0/dx_1 < 0$, and Eq. (B–79) becomes

$$f_y(y_0) = \sum_{i=1}^{M=2} \frac{f_x(x)}{|dy_0/dx|}\Bigg|_{x=x_i=h_i^{-1}(y_0)} \tag{B–80}$$

When there are more than two intervals during which $h(x)$ is monotonically increasing or decreasing, this procedure may be extended so that Eq. (B–68) is obtained.

In concluding this discussion on the functional transformation of a random variable, it should be emphasized that the description of the mapping function $y = h(x)$ assumes that the output y, at any instant, depends on the value of the input x only at that same instant and not on previous (or future) values of x. Thus, this technique is applicable to devices that contain no memory (i.e., no inductance or capacitance) elements; however, the device may be nonlinear, as we have seen in the preceding examples.

B–9 MULTIVARIATE STATISTICS

In Sec. B–3, the probabilities of simple, events, joint probabilities, and conditional probabilities were defined. In Secs. B–4 and B–5, using the probability of simple events, we developed the concepts of PDFs and CDFs. The PDFs and CDFs involved only one random variable, so these are one-dimensional problems. Similarly, the moments involved only one-dimensional integration.

In this section, multiple-dimensional problems, also called *multivariate statistics*, will be developed. These involve PDFs and CDFs that are associated with probabilities of intersecting events and with conditional probabilities. In addition, multiple-dimensional moments will be obtained as an extension of the one-dimensional moments that were studied in Sec. B–6. *If the reader clearly understands the one-dimensional case* (developed in the preceding sections), *there will be little difficulty in generalizing those results to the N-dimensional case.*

Multivariate CDFs and PDFs

DEFINITION. The *N-dimensional CDF* is

$$F(a_1, a_2, \ldots, a_N) = P[(x_1 \le a_1)(x_2 \le a_2) \cdots (x_N \le a_N)]$$

$$= \lim_{n\to\infty} \left[\frac{n_{(x_1 \le a_1)\,(x_2 \le a_2)\cdots(x_N \le a_N)}}{n} \right] \tag{B–81}$$

where the notation $(x_1 \le a_1)(x_2 \le a_2) \cdots (x_N \le a_N)$ is the intersection event consisting of the intersection of the events associated with $x_1 \le a_1$, $x_2 \le a_2$, etc.

DEFINITION. The *N-dimensional PDF* is

$$f(x_1, x_2, \ldots, x_N) = \frac{\partial^N F(a_1, a_2, \ldots, a_N)}{\partial a_1 \partial a_2 \cdots \partial a_N}\bigg|_{\mathbf{a} = \mathbf{x}} \tag{B–82}$$

where **a** and **x** are the row vectors, $\mathbf{a} = (a_1, a_2, \ldots, a_N)$, and $\mathbf{x} = f(x_1, x_2, \ldots, x_N)$.

DEFINITION. The expected value of $y = h(\mathbf{x})$ is

$$\overline{[y]} = \overline{h(x_1, x_2, \ldots, x_N)}$$

$$= \int_{-\infty}^{\infty} \int_{-\infty}^{\infty} \cdots \int_{-\infty}^{\infty} h(x_1, x_2, \ldots, x_N)$$

$$\times f(x_1, x_2, \ldots, x_N)\, dx_1\, dx_2 \cdots dx_N \tag{B–83}$$

Some *properties* of N-dimensional random variables are

1. $f(x_1, x_2, \ldots, x_N) \geq 0$ $\hspace{5cm}$ (B–84a)

2. $\displaystyle \int_{-\infty}^{\infty} \int_{-\infty}^{\infty} \cdots \int_{-\infty}^{\infty} f(x_1, x_2, \ldots, x_N)\, dx_1\, dx_2 \cdots dx_N = 1$ $\hspace{1cm}$ (B–84b)

3. $F(a_1, a_2, \ldots, a_N)$

$$= \lim_{\substack{\varepsilon \to 0 \\ \varepsilon > 0}} \int_{-\infty}^{a_1 + \varepsilon} \int_{-\infty}^{a_2 + \varepsilon} \cdots \int_{-\infty}^{a_N + \varepsilon} f(x_1, x_2, \ldots, x_N)\, dx_1\, dx_2 \cdots dx_N \tag{B–84c}$$

4. $F(a_1, a_2, \ldots, a_N) \equiv 0$ if *any* $a_i = -\infty$, $\quad i = 1, 2, \ldots, N$ $\hspace{1cm}$ (B–84d)

5. $F(a_1, a_2, \ldots, a_N) = 1$ when *all* $a_i = +\infty$, $\quad i = 1, 2, \ldots, N$ $\hspace{1cm}$ (B–84e)

6. $P[(a_1 < x_1 \leq b_1)(a_2 < x_2 \leq b_2) \cdots (a_N < x_N \leq b_N)]$

$$= \lim_{\substack{\varepsilon \to 0 \\ \varepsilon > 0}} \int_{a_1 + \varepsilon}^{b_1 + \varepsilon} \int_{a_2 + \varepsilon}^{b_2 + \varepsilon} \cdots \int_{a_N + \varepsilon}^{b_N + \varepsilon} f(x_1, x_2, \ldots, x_N)\, dx_1\, dx_2 \cdots dx_N \tag{B–84f}$$

The definitions and properties for multivariate PDFs and CDFs are based on the concepts of joint probabilities, as discussed in Sec. B–3. In a similar way, conditional PDFs and CDFs can be obtained [Papoulis, 1984]. Using the property $P(AB) = P(A)P(B|A)$ from Eq. (B–8), we find that the joint PDF of x_1 and x_2 is

$$f(x_1, x_2) = f(x_1)\, f(x_2|x_1) \tag{B–85}$$

where $f(x_2|x_1)$ is the conditional PDF of x_2 given x_1. Generalizing further, we obtain

$$f(x_1, x_2|x_3) = f(x_1|x_3)\, f(x_2|x_1, x_3) \tag{B–86}$$

Many other expressions for relationships between multiple-dimensional PDFs should also be apparent. When x_1 and x_2 are *independent*, $f(x_2|x_1) = f_{x_2}(x_2)$ and

$$f_{\mathbf{x}}(x_1, x_2) = f_{x_1}(x_1) f_{x_2}(x_2) \tag{B–87}$$

where the subscript x_1 denotes the PDF associated with x_1, and the subscript x_2 denotes the PDF associated with x_2. For N independent random variables, this becomes

$$f_\mathbf{x}(x_1, x_2, \ldots, x_N) = f_{x_1}(x_1) f_{x_2}(x_2) \cdots f_{x_N}(x_N) \qquad (B\text{–}88)$$

THEOREM.   *If the Nth-dimensional PDF of x is known, then the Lth-dimensional PDF of x can be obtained when $L < N$ by*

$$f(x_1, x_2, \ldots, x_L)$$

$$= \underbrace{\int_{-\infty}^{\infty} \int_{-\infty}^{\infty} \cdots \int_{-\infty}^{\infty}}_{N\,-\,L\ \text{integrals}} f(x_1, x_2, \ldots, x_N) \, dx_{L+1} \, dx_{L+2} \cdots dx_N \qquad (B\text{–}89)$$

This Lth-dimensional PDF, where $L < N$, is sometimes called the **marginal PDF,** *since it is obtained from a higher dimensional (Nth) PDF.*

Proof.   First, show that this result is correct if $N = 2$ and $L = 1$:

$$\int_{-\infty}^{\infty} f(x_1, x_2) \, dx_2 = \int_{-\infty}^{\infty} f(x_1) f(x_2|x_1) \, dx_2$$

$$= f(x_1) \int_{-\infty}^{\infty} f(x_2|x_1) \, dx_2 = f(x_1) \qquad (B\text{–}90)$$

since the area under $f(x_2|x_1)$ is unity. This procedure is readily extended to prove the Lth-dimensional case of Eq. (B–89).

Bivariate Statistics

Bivariate (or joint) distributions are the $N = 2$-dimensional case. In this section, the definitions from the previous section will be used to evaluate two-dimensional moments. As shown in Chapter 6, bivariate statistics have some very important applications to electrical engineering problems, and some additional definitions need to be studied.

DEFINITION.   The *correlation* (or *joint mean*) of x_1 and x_2 is

$$m_{12} = \overline{x_1 x_2} = \int_{-\infty}^{\infty} \int_{-\infty}^{\infty} x_1 x_2 \, f(x_1, x_2) \, dx_1 \, dx_2 \qquad (B\text{–}91)$$

DEFINITION.   Two random variables x_1 and x_2 are said to be *uncorrelated* if

$$m_{12} = \overline{x_1 x_2} = \overline{x_1}\,\overline{x_2} = m_1 m_2 \qquad (B\text{–}92)$$

If x_1 and x_2 are independent, it follows that they are also uncorrelated, but the converse is not generally true. However, as we will see, the converse is true for bivariate Gaussian random variables.

DEFINITION.   Two random variables are said to be *orthogonal* if

$$m_{12} = \overline{x_1 x_2} \equiv 0 \tag{B–93}$$

Note the similarity of the definition of orthogonal random variables to that of orthogonal functions given by Eq. (2-73).

DEFINITION.   The *covariance* is

$$u_{11} = \overline{(x_1 - m_1)(x_2 - m_2)}$$

$$= \int_{-\infty}^{\infty} \int_{-\infty}^{\infty} (x_1 - m_1)(x_2 - m_2) f(x_1, x_2)\, dx_1\, dx_2 \tag{B–94}$$

It should be clear that if x_1 and x_2 are independent, the covariance is zero (and x_1 and x_2 are uncorrelated). The converse is not generally true, but it is true for the case of bivariate Gaussian random variables.

DEFINITION.   The *correlation coefficient* is

$$\rho = \frac{u_{11}}{\sigma_1 \sigma_2} = \frac{\overline{(x_1 - m_1)(x_2 - m_2)}}{\sqrt{\overline{(x_1 - m_1)^2}}\,\sqrt{\overline{(x_2 - m_2)^2}}} \tag{B–95}$$

This is also called the *normalized covariance*. The correlation coefficient is always within the range

$$-1 \leq \rho \leq +1 \tag{B–96}$$

For example, suppose that $x_1 = x_2$; then $\rho = +1$. If $x_1 = -x_2$, then $\rho = -1$; and if x_1 and x_2 are independent, $\rho = 0$. Thus the correlation coefficient tells us, on the average, how likely a value of x_1 is to being proportional to the value for x_2. This subject is discussed in more detail in Chapter 6, where these results are extended to include random processes (time functions). There the dependence of the value of a waveform at one time is compared with the value of the waveform that occurs at another time. This will bring the concept of frequency response into the problem.

Gaussian Bivariate Distribution

A good example of a joint ($N = 2$) distribution that is of great importance is the bivariate Gaussian distribution. The bivariate Gaussian PDF is

$$f(x_1, x_2) = \frac{1}{2\pi\sigma_1\sigma_2\sqrt{1-\rho^2}}\, e^{-\frac{1}{2(1-\rho^2)}\left[\frac{(x_1 - m_1)^2}{\sigma_1^2} - 2\rho\frac{(x_1 - m_1)(x_2 - m_2)}{\sigma_1\sigma_2} + \frac{(x_2 - m_2)^2}{\sigma_2^2}\right]} \tag{B–97}$$

where σ_1^2 is the variance of x_1, σ_2^2 is the variance of x_2, m_1 is the mean of x_1, and m_2 is the mean of x_2. Examining Eq. (B–97), we see that if $\rho = 0$, $f(x_1, x_2) = f(x_1)f(x_2)$, where $f(x_1)$ and $f(x_2)$ are the one-dimensional PDFs of x_1 and x_2. Thus, if bivariate *Gaussian* random variables are uncorrelated (which implies that $\rho = 0$), they are independent.

A sketch of the bivariate (two-dimensional) Gaussian PDF is shown in Fig. B–12.

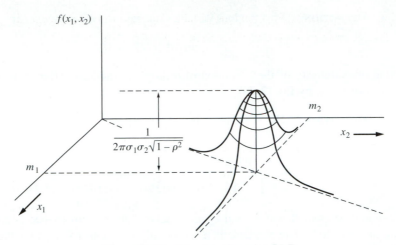

Figure B–12   Bivariate Gaussian PDF.

Multivariate Functional Transformation

Section B–8 will now be generalized for the multivariate case. Referring to Fig. B–13, we will obtain the PDF for $\mathbf{y}$, denoted by $f_y(\mathbf{y})$ in terms of the PDF for $\mathbf{x}$, denoted by $f_x(\mathbf{x})$.

THEOREM.   *Let* $\mathbf{y} = \mathbf{h}(\mathbf{x})$ *denote the transfer characteristic of a device (no memory) that has N inputs, denoted by* $\mathbf{x} = (x_1, x_2, \ldots, x_N)$; *N outputs, denoted by* $\mathbf{y} = (y_1, y_2, \ldots, y_N)$; *and* $y_i = h_i(\mathbf{x})$. *That is,*

$$y_1 = h_1(x_1, x_2, \ldots, x_N)$$

$$y_2 = h_2(x_1, x_2, \ldots, x_N)$$

$$\vdots \qquad\qquad\qquad (\text{B–98})$$

$$y_N = h_N(x_1, x_2, \ldots, x_N)$$

Furthermore, let $\mathbf{x}_i$, $i = 1, 2, \ldots, M$ *denote the real roots (vectors) of the equation* $\mathbf{y} = \mathbf{h}(\mathbf{x})$. *The PDF of the output is then*

$$f_\mathbf{y}(\mathbf{y}) = \sum_{i=1}^{M} \frac{f_\mathbf{x}(\mathbf{x})}{|J(\mathbf{y}/\mathbf{x})|}\bigg|_{\mathbf{x} = \mathbf{x}_i = \mathbf{h}_i^{-1}(\mathbf{y})} \qquad (\text{B–99})$$

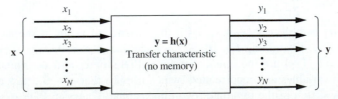

Figure B–13   Multivariate functional transformation of random variables.

where $|\cdot|$ denotes the absolute value operation and $J(\mathbf{y}/\mathbf{x})$ is the Jacobian of the coordinate transformation to $\mathbf{y}$ from $\mathbf{x}$. The Jacobian is defined as

$$J\left(\frac{\mathbf{y}}{\mathbf{x}}\right) = \mathrm{Det} \begin{bmatrix} \dfrac{\partial h_1(\mathbf{x})}{\partial x_1} & \dfrac{\partial h_1(\mathbf{x})}{\partial x_2} & & \dfrac{\partial h_1(\mathbf{x})}{\partial x_N} \\[2ex] \dfrac{\partial h_2(\mathbf{x})}{\partial x_1} & \dfrac{\partial h_2(\mathbf{x})}{\partial x_2} & \cdots & \dfrac{\partial h_2(\mathbf{x})}{\partial x_N} \\[1ex] & & \cdots & \\[1ex] \vdots & \vdots & & \vdots \\[2ex] \dfrac{\partial h_N(\mathbf{x})}{\partial x_1} & \dfrac{\partial h_N(\mathbf{x})}{\partial x_2} & \cdots & \dfrac{\partial h_N(\mathbf{x})}{\partial x_N} \end{bmatrix} \quad \text{(B–100)}$$

where $\mathrm{Det}[\cdot]$ denotes the determinant of the matrix $[\cdot]$.

A proof of this theorem will not be given, but it should be clear that it is a generalization of the theorem for the one-dimensional case that was studied in Sec. B–8. The coordinate transformation relates differentials in one coordinate system to those in another [Thomas, 1969]:

$$dy_1 \, dy_2 \, \cdots \, dy_N = J\left(\frac{\mathbf{y}}{\mathbf{x}}\right) dx_1 \, dx_2 \, \cdots \, dx_N \quad \text{(B–101)}$$

Example B–7 PDF FOR THE SUM OF TWO RANDOM VARIABLES

Suppose that we have a circuit configuration (such as an operational amplifier) that sums two inputs—x_1 and x_2—to produce the output

$$y = A(x_1 + x_2) \quad \text{(B–102)}$$

where A is the gain of the circuit. Assume that $f(x_1, x_2)$ is known and that we wish to obtain a formula for the PDF of the output in terms of the joint PDF for the inputs.

We can use the theorem described by Eq. (B–99) to solve this problem. However, for two inputs, we need two outputs in order to satisfy the assumptions of the theorem. This is achieved by defining an auxiliary variable for the output, (say, y_2). Thus,

$$y_1 = h_1(\mathbf{x}) = A(x_1 + x_2) \quad \text{(B–103)}$$

$$y_2 = h_2(\mathbf{x}) = Ax_1 \quad \text{(B–104)}$$

The choice of the equation to use for the auxiliary variable, Eq. (B–104), is immaterial, provided that it is an independent equation so that the determinant, $J(\mathbf{y}/\mathbf{x})$, is not zero. However, the equation is usually selected to simplify the ensuing mathematics. Using Eqs. (B–103) and (B–104), we get

$$J = \mathrm{Det} \begin{bmatrix} A & A \\ A & 0 \end{bmatrix} = -A^2 \quad \text{(B–105)}$$

Substituting this into Eq. (B–99) yields

$$f_{\mathbf{y}}(y_1, y_2) = \left. \frac{f_x(x_1, x_2)}{|-A^2|} \right|_{x = h^{-1}(\mathbf{y})}$$

or

$$f_{\mathbf{y}}(y_1, y_2) = \frac{1}{A^2} f_x\left(\frac{y_2}{A}, \frac{1}{A}(y_1 - y_2)\right) \tag{B-106}$$

We want to find a formula for $f_{y_1}(y_1)$, since $y_1 = A(x_1 + x_2)$. This is obtained by evaluating the marginal PDF from Eq. (B-106).

$$f_{y_1}(y_1) = \int_{-\infty}^{\infty} f_{\mathbf{y}}(y_1, y_2)\, dy_2$$

or

$$f_{y_1}(y_1) = \frac{1}{A^2} \int_{-\infty}^{\infty} f_x\left(\frac{y_2}{A}, \frac{1}{A}(y_1 - y_2)\right) dy_2 \tag{B-107}$$

This general result relates the PDF of $y = y_1$ to the joint PDF of $\mathbf{x}$ where $y = A(x_1 + x_2)$. *If x_1 and x_2 are independent and $A = 1$*, Eq. (B-107) becomes

$$f(y) = \int_{-\infty}^{\infty} f_{x_1}(\lambda) f_{x_2}(y - \lambda)\, d\lambda$$

or

$$f(y) = f_{x_1}(y) * f_{x_2}(y) \tag{B-108}$$

where $*$ denotes the convolution operation. Similarly, if we sum N independent random variables, the PDF for the sum is the $(N - 1)$-fold convolution of the one-dimensional PDFs for the N random variables.

Central Limit Theorem

If we have the sum of a number of independent random variables with arbitrary one-dimensional PDFs, the *central limit theorem* states that the PDF for the sum of these independent random variables approaches a Gaussian (normal) distribution under very general conditions. Strictly speaking, the central limit theorem does not hold for the PDF if the independent random variables are discretely distributed. In this case, the PDF for the sum will consist of delta functions (not Gaussian, which is continuous); however, if the delta functions are "smeared out" (e.g., if the delta functions are replaced with rectangles that have corresponding areas), the resulting PDF will be approximately Gaussian. Regardless, the cumulative distribution function (CDF) for the sum will approach that of a Gaussian CDF.

The central limit theorem is illustrated in the following example.

Example B–8 PDF FOR THE SUM OF THREE INDEPENDENT, UNIFORMLY DISTRIBUTED RANDOM VARIABLES

The central limit theorem will be illustrated by evaluating the exact PDF for the sum of three independent uniformly distributed random variables. This exact result will be compared with the Gaussian PDF, as predicted by the central limit theorem.

Let each of the independent random variables x_i have a uniform distribution, as shown in Fig. B–14a. The PDF for $y_1 = x_1 + x_2$, denoted by $f(y_1)$, is obtained by the convolution operation described by Eq. (B–108) and Fig. 2–7. This result is shown in Fig. B–14b. It is seen that after only one convolution operation the PDF for the sum (which is a triangle) is going toward the Gaussian shape. The PDF for $y_2 = x_1 + x_2 + x_3$, $f(y_2)$, is obtained by convolving the triangular PDF with another uniform PDF. The result is

$$f(y_2) = \begin{cases} 0, & y_2 \leq -\dfrac{3}{2}A \\[2ex] \dfrac{1}{2A^3}\left(\dfrac{3}{2}A + y_2\right)^2, & -\dfrac{3}{2}A \leq y_2 \leq -\dfrac{1}{2}A \\[2ex] \dfrac{1}{2A^3}\left(\dfrac{3}{2}A^2 - 2y_2^2\right), & |y_2| \leq \dfrac{1}{2}A \\[2ex] \dfrac{1}{2A^3}\left(\dfrac{3}{2}A - y_2\right)^2, & \dfrac{1}{2}A \leq y_2 \leq \dfrac{3}{2}A \\[2ex] 0, & y_2 \geq \dfrac{3}{2}A \end{cases} \tag{B–109}$$

This curve is plotted by the solid line in Fig. B–14c. For comparison purposes, a matching Gaussian curve, with $1/(\sqrt{2\pi}\,\sigma) = 3/(4A)$, is also shown by the dashed line. It is seen that $f(y_2)$ is very close to the Gaussian curve for $|y_2| < \frac{3}{2}A$, as predicted by the central limit theorem. Of course, $f(y_2)$ is certainly not Gaussian for $|y_2| > \frac{3}{2}A$, because $f(y_2) \equiv 0$ in this region, whereas the Gaussian curve is not zero except at $y = \pm\infty$. Thus, we observe that the Gaussian approximation (as predicted by the central limit theorem) is not very good on the *tails* of the distribution. In Chapter 7, it is shown that the probability of bit error for digital systems is obtained by evaluating the area under the tails of a distribution. If the distribution is not known to be Gaussian and the central limit theorem is used to approximate the distribution by a Gaussian PDF, the results are often not very accurate, as shown by this example. However, if the area under the distribution is needed near the mean value, the Gaussian approximation may be very good.

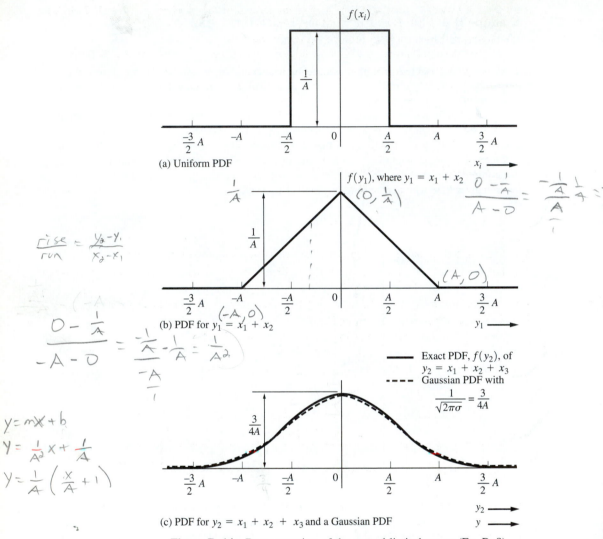

(a) Uniform PDF

(b) PDF for $y_1 = x_1 + x_2$

(c) PDF for $y_2 = x_1 + x_2 + x_3$ and a Gaussian PDF

Figure B–14   Demonstration of the central limit theorem (Ex. B–8).

PROBLEMS

B–1 A long binary message contains 1,428 binary 1s and 2,668 binary 0s. What is the probability of obtaining a binary 1 in any received bit?

B–2 (a) Find the probability of getting an 8 in the toss of two dice.
(b) Find the probability of getting either a 5, 7, or 8 in the toss of two dice.

B–3 Show that

$$P(A + B + C) = P(A) + P(B) + P(C)$$
$$- P(AB) - P(AC) - P(BC) + P(ABC)$$

B–4 A die is tossed. The probability of getting any face is $P(x) = \frac{1}{6}$, where $x = k = 1, 2, 3, 4, 5,$ or 6.
   (a) Find the probability of getting an odd-numbered face.
   (b) Find the probability of getting a 4 when an even-numbered face is obtained on a toss.

B–5 Which of the following functions satisfy the properties for a PDF. Why?

   (a) $f(x) = \dfrac{1}{\pi}\left(\dfrac{1}{1 + x^2}\right)$

   (b) $f(x) = \begin{cases} |x|, & |x| < 1 \\ 0, & x \text{ otherwise} \end{cases}$

   (c) $f(x) = \begin{cases} \frac{1}{6}(8 - x), & 4 \le x \le 10 \\ 0, & x \text{ otherwise} \end{cases}$

   (d) $f(x) = \displaystyle\sum_{k=0}^{\infty} \frac{3}{4}\left(\frac{1}{4}\right)^k \delta(x - k)$

B–6 Show that all cumulative distribution functions must satisfy the properties given in Sec. B–5.

B–7 Let $f(x) = Ke^{-b|x|}$, where K and b are positive constants. Find the mathematical expression for the CDF and sketch the results.

B–8 Find the probability that $-\frac{1}{4}A \le y_1 \le \frac{1}{4}A$ for the triangular distribution shown in Fig. B–14b.

B–9 A triangular PDF is shown in Fig. B–14b.
   (a) Find a mathematical expression that describes the CDF.
   (b) Sketch the CDF.

B–10 Evaluate the PDFs for the two waveforms shown in Fig. PB–10.

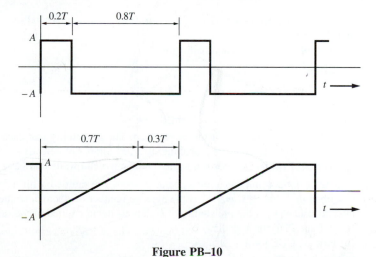

Figure PB–10

B–11 Let a PDF be given by $f(x) = Ke^{-bx}$, for $x \ge 0$, and $f(x) = 0$, for $x < 0$, where K and b are positive constants.
   (a) Find the value required for K in terms of b.
   (b) Find m in terms of b.
   (c) Find σ^2 in terms of b.

B–12 A random variable x has a PDF

$$f(x) = \begin{cases} \frac{3}{32}(-x^2 + 8x - 12), & 2 < x < 6 \\ 0, & x \text{ elsewhere} \end{cases}$$

(a) Demonstrate that $f(x)$ is a valid PDF.
(b) Find the mean.
(c) Find the second moment.
(d) Find the variance.

B–13 Determine the standard deviation for the triangular distribution shown in Fig. B–14b.

B–14 (a) Find the terms in a binomial distribution for $n = 7$, where $p = 0.5$.
(b) Sketch the PDF for this binomial distribution.
(c) Find and sketch the CDF for this distribution.
(d) Find $\overline{x^3}$ for this distribution.

B–15 For a binomial distribution, show that $\sigma^2 = np(1 - p)$.

B–16 A binomial random variable x_k has values of k, where

$$k = 0, 1, \ldots, n; \qquad P(k) = \binom{n}{k} p^k q^{n-k}; \qquad q = 1 - p$$

Assume that $n = 160$ and $p = 0.1$.
(a) Plot $P(k)$.
(b) Compare the plot of part (a) with a plot $P(k)$ using the Gaussian approximation

$$\binom{n}{k} p^k q^{n-k} \approx \frac{1}{\sqrt{2\pi}\,\sigma} e^{-(k-m)^2/2\sigma^2}$$

which is valid when $npq \gg 1$ and $|k - np|$ is in the neighborhood of $\sqrt{npq}$, where $\sigma = \sqrt{npq}$ and $m = np$.
(c) Also plot the Poisson approximation

$$\binom{n}{k} p^k q^{n-k} \approx \frac{\lambda^k}{k!} e^{-\lambda}$$

where $\lambda = np$, n is large, and p is small.

B–17 An order of $n = 3{,}000$ transistors is received. The probability that each transistor is defective is $p = 0.001$. What is the probability that the number of defective transistors in the batch is 6 or less? (*Note*: The Poisson approximation is valid when n *is* large and p is small.)

B–18 In a fiber-optic communication system, photons are emitted with a Poisson distribution as described in Table B–1. $m = \lambda$ is the average number of photons emitted in an arbitrary time interval, and $P(k)$ *is* the probability of k photons being emitted in the same interval.
(a) Plot the PDF for $\lambda = 0.5$.
(b) Plot the CDF for $\lambda = 0.5$.
(c) Show that $m = \lambda$.
(d) Show that $\sigma = \sqrt{\lambda}$.

B–19 Let x be a random variable that has a Laplacian distribution. The Laplacian PDF is $f(x) = (1/2b)e^{-|x-m|/b}$, where b and m are real constants and $b > 0$.
(a) Find the mean of x in terms of b and m.
(b) Find the variance of x in terms of b and m.

B–20 Given the Gaussian PDF

$$f(x) = \frac{1}{\sqrt{2\pi}\beta} e^{-(x-m)^2/(2\beta^2)}$$

show that the variance of this distribution is β^2.

B–21 In a manufacturing process for resistors, the values obtained for the resistors have a Gaussian distribution where the desired value is the mean value. If we want 95% of the manufactured 1-kΩ resistors to have a tolerance of ±10%, what is the required value for σ?

B–22 Assume that x has a Gaussian distribution. Find the probability that
(a) $|x - m| < \sigma$.
(b) $|x - m| < 2\sigma$.
(c) $|x - m| < 3\sigma$.
Obtain numerical results by using MATLAB, MathCAD, or tables if necessary.

B–23 Show that

(a) $Q(z) = \frac{1}{2} \text{erfc}\left(\frac{z}{\sqrt{2}}\right)$.

(b) $Q(-z) = 1 - Q(z)$.

(c) $Q(z) = \frac{1}{2}\left[1 - \text{erf}\left(\frac{z}{\sqrt{2}}\right)\right]$.

B–24 For a Gaussian distribution, show that

(a) $F(a) = \frac{1}{2} \text{erfc}\left(\frac{m-a}{\sqrt{2}\,\sigma}\right)$.   (b) $F(a) = \frac{1}{2}\left[1 + \text{erfc}\left(\frac{a-m}{\sqrt{2}\,\sigma}\right)\right]$.

B–25 A noise voltage has a Gaussian distribution. The rms value is 5 V, and the dc value is 1.0 V. Find the probability of the voltage having values between −5 and +5 V.

B–26 Suppose that x is a Gaussian random variable with $m = 5$ and $\sigma = 0.6$.
(a) Find the probability that $x \leq 1$.
(b) Find the probability that $x \leq 6$.

B–27 The Gaussian random variable x has a zero mean and a variance of 2. Let A be the event such that $|x| < 3$.
(a) Find an expression for the conditional PDF $f(x|A)$.
(b) Plot $f(x|A)$ over the range $|x| < 5$.
(c) Plot $f(x)$ over the range $|x| < 5$ and compare these two plots.

B–28 Let x have a sinusoidal distribution with a PDF as given by Eq. (B–67). Show that the CDF is

$$F(a) = \begin{cases} 0, & a \leq -A \\[2mm] \dfrac{1}{\pi}\left[\dfrac{\pi}{2} + \sin^{-1}\left(\dfrac{a}{A}\right)\right], & |a| \leq A \\[2mm] 1, & a \geq A \end{cases}$$

B–29 (a) If x has a sinusoidal distribution with the peak value of x being A, show that the rms value is $\sigma = A/\sqrt{2}$. [*Hint*: Use Eq. (B–67).]
(b) If $x = A \cos \psi$, where ψ is uniformly distributed between $-\pi$ and $+\pi$, show that the rms value of x is $\sigma = A/\sqrt{2}$.

B–30 Given that $y = x^2$ and x is a Gaussian random variable with mean value m and variance σ^2, find a formula for the PDF of y in terms of m and σ^2.

B–31 x is a uniformly distributed random variable over the range $-1 \le x \le 1$ plus a discrete point at $x = \frac{1}{2}$ with $P(x = \frac{1}{2}) = \frac{1}{4}$.

(a) Find a mathematical expression for the PDF for x, and plot your result.

(b) Find the PDF for y, where

$$y = \begin{cases} x^2, & x \ge 0 \\ 0, & x < 0 \end{cases}$$

Sketch your result.

B–32 A saturating amplifier is modeled by

$$y = \begin{cases} Ax_0, & x > x_0 \\ Ax, & |x| \le x_0 \\ -Ax_0, & x < -x_0 \end{cases}$$

Assume that x is a Gaussian random variable with mean value m and variance σ^2. Find a formula for the PDF of y in terms of A, x_0, m, and σ^2.

B–33 A sinusoid with a peak value of 8 V is applied to the input of a quantizer. The quantizer characteristic is shown in Fig. 3–8a. Calculate and plot the PDF for the output.

B–34 A voltage waveform that has a Gaussian distribution is applied to the input of a full-wave rectifier circuit. The full-wave rectifier is described by $y(t) = |x(t)|$, where $x(t)$ is the input and $y(t)$ is the output. The input waveform has a dc value of 1 V and an rms value of 2 V.

(a) Plot the PDF for the input waveform.

(b) Plot the PDF for the output waveform.

B–35 Refer to Example B–6 and Eq. (B–75), which describe the PDF for the output of an ideal diode (half-wave rectifier) characteristic. Find the mean (dc) value of the output.

B–36 Given the joint density function,

$$f(x_1, x_2) = \begin{cases} e^{-(1/2)(4x_1 + x_2)}, & x_1 \ge 0,\ x_2 \ge 0 \\ 0, & \text{otherwise} \end{cases}$$

(a) Verify that $f(x_1, x_2)$ is a density function.

(b) Show that x_1 and x_2 are either independent or dependent.

(c) Evaluate $P(1 \le x_1 \le 2, x_2 \le 4)$.

(d) Find ρ.

B–37 A joint density function is

$$f(x_1, x_2) = \begin{cases} K(x_1 + x_1 x_2), & 0 \le x_1 \le 1,\ 0 \le x_2 \le 4 \\ 0, & \text{elsewhere} \end{cases}$$

(a) Find K.   (b) Determine if x_1 and x_2 are independent.

(c) Find $F_{x_1 x_2}(0.5, 2)$.   (d) Find $F_{x_2|x_1}(x_2|x_1)$.

B–38 Let $y = x_1 + x_2$, where x_1 and x_2 are uncorrelated random variables. Show that

(a) $\bar{y} = m_1 + m_2$, where $m_1 = \overline{x_1}$ and $m_2 = \overline{x_2}$.

(b) $\sigma_y^2 = \sigma_1^2 + \sigma_2^2$, where $\sigma_1^2 = \overline{(x_1 - m_1)^2}$ and $\sigma_2^2 = \overline{(x_2 - m_2)^2}$.

[Hint: Use the ensemble operator notation similar to that used in the proof for Eq. (B–29).]

B–39 Let $x_1 = \cos\theta$ and $x_2 = \sin\theta$, where θ is uniformly distributed over $(0, 2\pi)$. Show that

(a) x_1 and x_2 are uncorrelated.

(b) x_1 and x_2 are not independent.

B–40 Two random variables x_1 and x_2 are jointly Gaussian. The joint PDF is described by Eq. (B–97), where $m_1 = m_2 = 0$, $\sigma_{x_1} = \sigma_{x_2} = 1$ and $\rho = 0.5$. Plot $f(x_1, x_2)$ for x_1 over the range $|x_1| < 5$ and $x_2 = 0$. Also give plots for $f(x_1, x_2)$ for $|x_1| < 5$ and $x_2 = 0.4, 0.8, 1.2$, and 1.6.

B–41 Show that the marginal PDF of a bivariate Gaussian PDF is a one-dimensional Gaussian PDF. That is, evaluate

$$f(x_1) = \int_{-\infty}^{\infty} f(x_1, x_2)\, dx_2$$

where $f(x_1, x_2)$ is given by Eq. (B–97). [*Hint*: Factor some terms outside the integral containing x_1 (but not x_2). Complete the square on the exponent of the remaining integrand so that a Gaussian PDF form is obtained. Use the property that the integral of a properly normalized Gaussian PDF is unity.]

B–42 (a) $y = A_1 x_1 + A_2 x_2$, where A_1 and A_2 are constants and the joint PDF of x_1 and x_2 is $f_x(x_1, x_2)$. Find a formula for the PDF of y in terms of the (joint) PDF of $\mathbf{x}$.
(b) If x_1 and x_2 are independent, how can this formula be simplified?

B–43 Two independent random variables—x and y—have the PDFs $f(x) = 5e^{-5x}\, u(x)$ and $f(y) = 2e^{-2y}\, u(y)$. Plot the PDF for w where $w = x + y$.

B–44 Two Gaussian random variables x_1 and x_2 have a mean vector $\mathbf{m}_x$ and a covariance matrix $\mathbf{C}_x$ as shown. Two new random variables y_1 and y_2 are formed by the linear transformation $\mathbf{y} = \mathbf{Tx}$.

$$\mathbf{m}_x = \begin{bmatrix} 2 \\ -1 \end{bmatrix} \qquad \mathbf{C}_x = \begin{bmatrix} 5 & -2/\sqrt{5} \\ -2/\sqrt{5} & 4 \end{bmatrix} \qquad \mathbf{T} = \begin{bmatrix} 1 & 1/2 \\ 1/2 & 1 \end{bmatrix}$$

(a) Find the mean vector for $\mathbf{y}$, which is denoted by $\mathbf{m}_y$.
(b) Find the covariance matrix for $\mathbf{y}$, which is denoted by $\mathbf{C}_y$.
(c) Find the correlation coefficient for y_1 and y_2. (*Hint*: See Sec. 6–6.)

B–45 Three Gaussian random variables x_1, x_2, and x_3 have zero mean values. Three new random variables y_1, y_2, and y_3 are formed by the linear transformation $\mathbf{y} = \mathbf{Tx}$, where

$$\mathbf{C}_x = \begin{bmatrix} 6.0 & 2.3 & 1.5 \\ 2.3 & 6.0 & 2.3 \\ 1.5 & 2.3 & 6.0 \end{bmatrix} \qquad \mathbf{T} = \begin{bmatrix} 5 & 2 & -1 \\ -1 & 3 & 1 \\ 2 & -1 & 2 \end{bmatrix}$$

(a) Find the covariance matrix for $\mathbf{y}$, which is denoted by $\mathbf{C}_y$.
(b) Write an expression for the PDF $f(y_1, y_2, y_3)$. (*Hint*: See Sec. 6–6.)

B–46 (a) Find a formula for the PDF of $y = Ax_1 x_2$, where x_1 and x_2 are random variables having the joint PDF $f_x(x_1, x_2)$.
(b) If x_1, and x_2 are independent, reduce the formula obtained in part (a) to a simpler result.

B–47 $y_2 = x_1 + x_2 + x_3$, where x_1, x_2, and x_3 are independent random variables. Each of the x_i has a one-dimensional PDF that is uniformly distributed over $-(A/2) \le x_i \le (A/2)$. Show that the PDF of y_2 is given by Eq. (B–109).

B–48 Use the built-in random number generator of MATLAB or MathCAD to demonstrate the central limit theorem. That is,
(a) Compute samples of the random variable y, where $y = \Sigma x_i$ and the x_i values are obtained from the random number generator.
(b) Plot the PDF for y by using the histogram function of MATLAB or MathCAD.

١ ٣ ٥

STANDARDS AND TERMINOLOGY FOR COMPUTER COMMUNICATIONS

C–1 CODES

Baudot

In 1875, Emile Baudot developed a "multiplex telegraph" system. The Baudot code has evolved into what is now known as the *International Telegraph Alphabet (ITA) Number 2.* This Baudot code is given in Table C–1.

The Baudot code has two serious disadvantages: (1) It has no provisions for parity bits, and (2) it is a *sequential* code. That is, the Letters (down shift) control character is sent to place the printer in the letter-printing mode, and the Figure (up shift) control character is used to shift the printer into the figure-printing mode. If an error causes a control character to be received that shifts the printer into a false mode, the printer will print an incorrect string of characters until a proper control-shift character is received.

ASCII

The American National Standard Code for Information Interchange (ASCII) was first adopted in 1963 and updated in 1967. It is now the most popular code used by computer terminals. The code is shown in Table C–2. It has a total of eight bits per character. Seven bits

TABLE C–1   BAUDOT CODE[a]

Character Case		Bit Pattern					Character Case		Bit Pattern				
Lower	Upper	5	4	3	2	1	Lower	Upper	5	4	3	2	1
A	–	0	0	0	1	1	Q	1	1	0	1	1	1
B	?	1	1	0	0	1	R	4	0	1	0	1	0
C	:	0	1	1	1	0	S	'	0	0	1	0	1
D	$	0	1	0	0	1	T	5	1	0	0	0	0
E	3	0	0	0	0	1	U	7	0	0	1	1	1
F	!	0	1	1	0	1	V	;	1	1	1	1	0
G	&	1	1	0	1	0	W	2	1	0	0	1	1
H	#	1	0	1	0	0	X	/	1	1	1	0	1
I	8	0	0	1	1	0	Y	6	1	0	1	0	1
J	Bell	0	1	0	1	1	Z	"	1	0	0	0	1
K	(	0	1	1	1	1	Letters (shift) ↓		1	1	1	1	1
L	)	1	0	0	1	0	Figures (shift) ↑		1	1	0	1	1
M	.	1	1	1	0	0	Space (SP)		0	0	1	0	0
N	,	0	1	1	0	0	Carriage return		0	1	0	0	0
O	9	1	1	0	0	0	Line feed		0	0	0	1	0
P	0	0	1	0	1	0	Blank		0	0	0	0	0

[a] 1 = mark = punch hole, and 0 = space = no punch hole for paper tape.

are completely specified as shown. The eighth bit is a parity bit that may be even or odd parity, depending on the choice selected for use in a particular installation.

C–2 DTE/DCE AND ETHERNET INTERFACE STANDARDS

A general block diagram for a computer communication system is shown in Fig. C–1. To facilitate the use of equipment supplied by different vendors, interface standards have been adopted for the connection of data terminal equipment (DTE) and data communications equipment (DCE). The DTE may be a digital computer, a printer, a keyboard, etc., and the DCE is a modem. In addition, one type of DTE, say a printer, may be connected to another type of DTE, say a computer, by the same type of interface.

Numerous organizations are concerned with advancement of practical standards. Internationally, the International Telegraph and Telephone Consultative Committee (CCITT) and the International Organization for Standardization (ISO) are the best known.

In the United States, as well as internationally, four standards groups are the American National Standards Institute (ANSI), the Electronics Industries Association (EIA), the Institute of Electrical and Electronics Engineers (IEEE), and the National Communications System (NCS) of the federal government.

TABLE C–2   ASCII CODE[a]

Bit Position				7 0 6 0 5 0	0 0 1	0 1 0	0 1 1	1 0 0	1 0 1	1 1 0	1 1 1
4	3	2	1								
0	0	0	0	NUL	DLE	SP	0	@	P	\	p
0	0	0	1	SOH	DC1	!	1	A	Q	a	q
0	0	1	0	STX	DC2	"	2	B	R	b	r
0	0	1	1	ETX	DC3	#	3	C	S	c	s
0	1	0	0	EOT	DC4	$	4	D	T	d	t
0	1	0	1	ENQ	NAK	%	5	E	U	e	u
0	1	1	0	ACK	SYN	&	6	F	V	f	v
0	1	1	1	BEL	ETB	'	7	G	W	g	w
1	0	0	0	BS	CAN	(	8	H	X	h	x
1	0	0	1	HT	EM	)	9	I	Y	i	y
1	0	1	0	LF	SUB	*	:	J	Z	j	z
1	0	1	1	VT	ESC	+	;	K	[	k	{
1	1	0	0	FF	FS	'	<	L	\	l	:
1	1	0	1	CR	GS	–	=	M	]	m	}
1	1	1	0	SO	RS	.	>	N	^	n	~
1	1	1	1	SI	US	/	?	O	—	o	DEL

[a]
ACK	Acknowledge	ENQ	Enquiry	RS	Record separator
BEL	Bell, or alarm	EOT	End of transmission	SI	Shift in
BS	Backspace	ESC	Escape	SO	Shift out
CAN	Cancel	ETB	End of transmission block	SOH	Start of heading
CR	Carriage return	ETX	End of text	SP	Space
DC1	Device control 1	FF	Form feed	STX	Start of text
DC2	Device control 2	FS	File separator	SUB	Substitute
DC3	Device control 3	GS	Group separator	SYN	Synchronous idle
DC4	Device control 4	HT	Horizontal tab	US	Unit separator
DEL	Delete	LF	Line feed	VT	Vertical tab
DLE	Data link escape	NAK	Negative acknowledge		
EM	End of medium	NUL	Null, or all zeros		

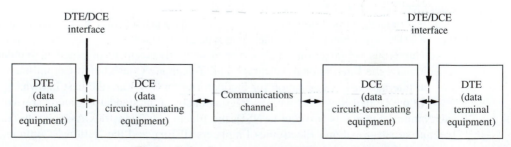

Figure C–1   Computer communication system.

DTE/DCE standards are developed in terms of ① the physical and electrical interface, and ② the link control (software) interface that provides framing and synchronization and error-detection and correction functions.

Universal Serial Bus (USB)

The Universal Serial Bus (USB) was developed by the PC industry to connect peripherals such as CD ROM drives, modems, tape drives, scanners, and printers to a PC. These devices may be "hot swapped" by disconnecting one and connecting another without shutting down the PC. The peripheral will be detected automatically when it is connected.

There are four wires in the USB cable. Two wires provide a differential data signal. This is equivalent to balanced positive and negative signals with respect to the ground wire. Another wire provides +5V dc (with respect to ground) from the PC power supply to supply limited power to the attached devices. Up to 127 devices may be connected in daisy-chain fashion (each peripheral connected to another peripheral, with the first connected to a PC or hub) to hub devices that are connected to the USB port on the PC.

The USB supports a maximum data rate of 12 Mb/s via packet transmission. The PC controls the data transmission by using a "token packet" to poll the connected devices and to specify the direction of data transmission packets from specified devices.

RS-232D, RS-422A, RS-449, and RS-530 Serial Interfaces

The RS-232D serial interface (also known as the *EIA standard interface*) was developed in 1969 by the EIA in cooperation with the Bell System and independent computer and modem manufacturers. The CCITT has adopted an interface standard, called V.24, that is very similar to the RS-232D standard. The military MIL-188c is also similar. The RS232D pin connections are given in Fig. C–2. The standard provides for serial data transmission by

DATA SET READY   6 —
REQUEST TO SEND   7 —
CLEAR TO SEND   8 —
RING INDICATOR   9 —
— 1   DATA CARRIER DETECT
— 2   RECEIVE DATA
— 3   TRANSMIT DATA
— 4   DATA TERMINAL READY
— 5   SIGNAL GROUND

(a) DB9 (9-pin connector)

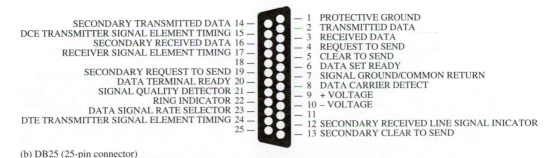

SECONDARY TRANSMITTED DATA   14 —
DCE TRANSMITTER SIGNAL ELEMENT TIMING   15 —
SECONDARY RECEIVED DATA   16 —
RECEIVER SIGNAL ELEMENT TIMING   17 —
18 —
SECONDARY REQUEST TO SEND   19 —
DATA TERMINAL READY   20 —
SIGNAL QUALITY DETECTOR   21 —
RING INDICATOR   22 —
DATA SIGNAL RATE SELECTOR   23 —
DTE TRANSMITTER SIGNAL ELEMENT TIMING   24 —
25 —
— 1   PROTECTIVE GROUND
— 2   TRANSMITTED DATA
— 3   RECEIVED DATA
— 4   REQUEST TO SEND
— 5   CLEAR TO SEND
— 6   DATA SET READY
— 7   SIGNAL GROUND/COMMON RETURN
— 8   DATA CARRIER DETECT
— 9   + VOLTAGE
— 10   – VOLTAGE
— 11
— 12   SECONDARY RECEIVED LINE SIGNAL INICATOR
— 13   SECONDARY CLEAR TO SEND

(b) DB25 (25-pin connector)

Figure C–2 RS-232D male connectors.

using a voltage level of $-V_0$ for a binary 1 (mark) and a level of $+V_0$ for a binary 0 (space), where $3 \leq V_0 \leq 25$ V. A typical value for V_0 is 6 V. The voltage appears on the appropriate lead measured with respect to the common (signal) ground lead. This is called *unbalanced signaling* and is also known as single-ended signaling. One lead is used for transmitting data (with respect to ground), and another lead is used for simultaneously receiving data.

The RS-232D standard defines the electrical characteristics, the functional description of interchange circuits, and a list of standard applications. The type of physical connector is not specified, but a DB25 (25 pins) or DB9 (9 pins) connector is generally used in practice, as shown in Fig. C–2.

The RS-232D interface is intended for data rates up to 20 kb/s and cable lengths up to 50 ft. Longer lengths are possible if twisted-pair cable is used and the loading capacitance is kept below 2500 pF. This interface is one of the most popular serial interfaces used in computer communication equipment today.

For higher data rates and longer cable lengths, the RS-422A interface was developed. This interface uses *balanced signaling* (also known as differential signaling), and, consequently, two wires are required for each circuit. This method gives greater noise and crosstalk immunity, because a common signal return path is not used. The balanced-signaling technique allows the transition region between the mark and space voltage levels to be reduced so that the signal levels are $\pm V_0$, where $0.2 \leq V_0 \leq 25$ V is used for balanced signaling. The RS-422A interface uses DB37 and DB9 connectors (37 and 9 pins, respectively), as specified in the ISO 4902 standard. The pin connections are given in Fig. C–3, where the two-wire connections for each balanced circuit are specified. The RS-422A standard allows for data rates up to 100 kb/s at 4000 ft or 40 Mb/s at 40 ft.

The RS-449 signal interface was intended to be the successor to the RS-232 interface. The RS-449 interface connections are also given in Fig. C–3. A 37-pin D connector is used for the primary channel, and an auxiliary 9-pin D connector is used (when needed) for a secondary channel. The 37-pin connector allows for unbalanced or (optionally) balanced signaling where the signaling levels are those for RS-232D or RS-222A, respectively. Balanced signaling allows the RS-449 interface to support data rates up to 2 Mb/s and cable lengths up to 667 ft.

The RS-530 interface supersedes the RS-449. It is a balanced-signaling complement to the RS-232C (which uses unbalanced signaling). The RS-530 interface uses a DB-25 (25-pin) connector, as described by Fig. C–3. It accommodates data rates up to 2 Mb/s.

Centronics Parallel Interface

Printers are often connected to computing equipment by a Centronics parallel interface. Figure C–4 hows the 36-pin connector for the Centronics parallel interface. This interface has eight data lines that provide eight bits of data in parallel. The data are controlled by the STROB pulse. This interface uses unbalanced signaling with transistor–transistor-logic (TTL) levels.

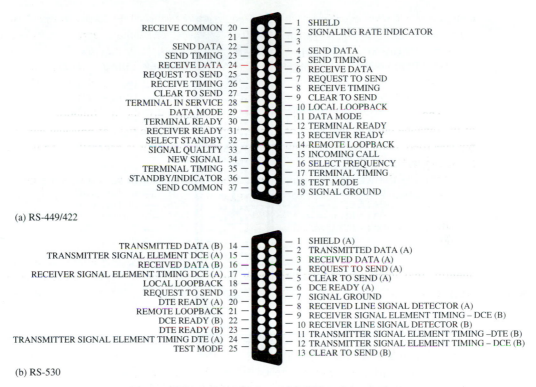

(a) RS-449/422

(b) RS-530

Figure C–3   RS-449/422 and RS-530 male connectors.

IEEE-488 Parallel Interface

The IEEE-488 parallel interface is designed to connect computer equipment with programmable instruments, such as programmable voltmeters, signal sources, and power supplies. It allows programmable instruments to be easily connected together to create an *automatic test equipment* (ATE) system. This interface was first developed in 1965 by Hewlett Packard and was known as the HP Interface Bus (HP-IB), or the General Purpose Interface Bus (GPIB). In 1975, the IEEE adopted an evolved standard, which was named the IEEE-488 interface. Table C–3 presents some details on this parallel interface. The lines are held at +5 V for a binary 0 (false) and pulled to ground for a binary 1 (true). This standard allows the bus to extend over a distance of 67 ft. Every IEEE-488 device is a listener, a talker, or a controller. A *listener* is capable of receiving data from the bus, such as a printer or a programmable power supply. A *talker* is capable of transmitting data over the bus, such as a frequency counter or a voltmeter. There can be only one active talker on the bus at a time. A *controller* is a computer-type device that controls the network (including itself) and specifies which devices are active talkers or listeners. A 24-pin IEEE-488 connector is illustrated in Fig. C–5.

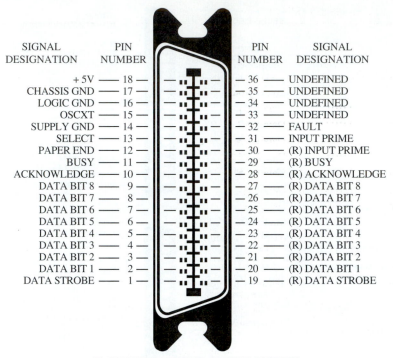

SIGNAL DESIGNATION	PIN NUMBER		PIN NUMBER	SIGNAL DESIGNATION
+ 5V	18		36	UNDEFINED
CHASSIS GND	17		35	UNDEFINED
LOGIC GND	16		34	UNDEFINED
OSCXT	15		33	UNDEFINED
SUPPLY GND	14		32	FAULT
SELECT	13		31	INPUT PRIME
PAPER END	12		30	(R) INPUT PRIME
BUSY	11		29	(R) BUSY
ACKNOWLEDGE	10		28	(R) ACKNOWLEDGE
DATA BIT 8	9		27	(R) DATA BIT 8
DATA BIT 7	8		26	(R) DATA BIT 7
DATA BIT 6	7		25	(R) DATA BIT 6
DATA BIT 5	6		24	(R) DATA BIT 5
DATA BIT 4	5		23	(R) DATA BIT 4
DATA BIT 3	4		22	(R) DATA BIT 3
DATA BIT 2	3		21	(R) DATA BIT 2
DATA BIT 1	2		20	(R) DATA BIT 1
DATA STROBE	1		19	(R) DATA STROBE

(R) INDICATES SIGNAL GROUND RETURN

Figure C–4　Centronics connector.

Ethernet (IEEE 802.3) Interface

Ethernet is a method of connecting multiple computers, servers, and peripherals together to form a *local area network* (LAN). It was first developed by the Xerox Corporation in the 1970s and supported by Digital Equipment Corporation (DEC) and Intel in 1980. In 1988, the Ethernet concept was refined by the IEEE, and the IEEE 802.3 standard was published. As shown in Table C–6, this standard provides a 10-Mb/s serial data bus for a LAN using either coaxial cable or twisted-pair wire.

　　In the 10base5 (thick Ethernet) system, a 50-Ω coaxial backbone cable (0.4 in. diameter) up to 500 m in length is used. The 10 in 10base5 indicates that a 10-Mb/s data rate is supported; the base indicates that baseband signaling (Manchester line code) is used, and 5 indicates that the maximum backbone length is 500 m unless repeaters are used. A DTE device with a Ethernet card is connected to the backbone cable by using a transceiver—also called a *medium attachment unit* (MAU)—in series with the backbone cable at the tap point. A cable drop then connects the transceiver to the DTE Ethernet card.

　　In the 10base2 (thin net, or cheap net) system, a 50-Ω thin (0.25 in. diameter) coaxial cable up to 200 m in length is used. The DTE device is connected to this coaxial bus by inserting a BNC T-connector in series with the cable at the tap point. In this thin-net system the T-connector must be connected directly to the Ethernet card on the DTE. (No drop cable is allowed.)

TABLE C–3   IEEE-488 INTERFACE[a]

24-Pin						
A	B	Circuit	Description	Data Line	Control Line	Handshake Line
	24	GND	Logic ground	×		
1	24	D101	Data bit 1	×		
2	24	D102	Data bit 2	×		
3	24	D103	Data bit 3	×		
4	24	D104	Data bit 4	×		
13	24	D105	Data bit 5	×		
14	24	D106	Data bit 6	×		
15	24	D107	Data bit 7	×		
16	24	D108	Data bit 8	×		
9	21	IFC	Interface Clear: used by controller to place devices into quiescent state		×	
10	22	SRQ	Service Request: used by device to request service		×	
17	24	REN	Remote Enable: used by controller to override device front panel controls		×	
5	24	EOI	End of Identity: (a) used by talker to end message, or (b) used by controller with ATN for parallel pole		×	
11	23	ATN	Attention: used by controller to end its data and listen for a command		×	
6	18	DAV	Data Valid: used by controller to indicate a control byte on data bus		×	
7	19	NRFD	Not Ready For Data: used by listener to indicate it is not ready			×
8	20	NDAC	No Data Accepted: used by listener to indicate it has not yet accepted last byte			×
	12		Shield			

[a] Column B pins are return leads to be grounded at the system ground point.

The 10base-T system uses twisted-pair wire up to 100 m in length to connect each DTE device with a 10base-T Ethernet card directly to a central hub. (There are no taps on the twisted-pair wire.)

Fast ethernet provides a ten-fold improvement in data rate when compared with regular ethernet. As shown in Table C-4, separate transmit and receive lines are required.

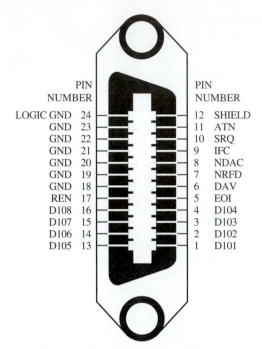

Figure C–5   IEEE-488 connector.

Multiple ethernet systems (segments) may be connected together using a multiport repeater. A gateway may also connect the Ethernet LAN to a wide area network (WAN), which uses either coaxial cable or fiber-optic cable to link to other networks.

C–3 THE ISO OSI NETWORK MODEL

When users are connected together, many things must be considered for orderly, efficient data transmission. This complicated problem can be understood better and the network can be maintained more easily if the different tasks are divided into layered modules. These modules are implemented in both hardware and software. Most computer network designers and manufacturers follow the layered model recommended by the ISO in its Open System Interconnection (OSI) reference model. This model is shown in Fig. C–6. This figure presents an example of two end users (host A and host B) that are connected to a data communications network via the physical channel. The physical channel may be connections via telephone company (TELCO) lines, fiber-optic link, microwave radio, or satellite links.

The layered approach has many attractive features. By adhering to the appropriate hardware or software specifications that define the layer boundaries, many different vendors can supply hardware/software that will work efficiently in the overall network. Furthermore, various parts of the system can be updated and maintained by replacing or modifying the hardware or software at each layer level instead of replacing the whole system.

TABLE C–4 ETHERNET (IEEE 802.3) STANDARDS

Item	Thick (Standard) Ethernet	Thin Ethernet	Twisted-pair Ethernet[a]	Twisted-pair Fast Ethernet[a]	Fiber Fast Ethernet[a]
IEEE designation	10base5	10base2	10base-T	100base-TX	100base-FX
Transmission medium	50-Ω coax	50-Ω coax	Twisted-pair wire	Two pairs (Tx, Rx)	Two fibers (Tx, Rx)
Cable diameter or wire size	0.4 in. diameter	0.25 in. diameter	22 to 26 AWG	Category 5	Single-mode fiber
Data rate	10 Mb/s	10 Mb/s	10 Mb/s	100 Mb/s	100 Mb/s
Maximum segment length	500 m[b]	200 m[b]	100 m	100 m	10,000 m
Maximum length with repeaters	4000 m	4000 m	NA	NA	NA
Minimum distance between taps	2.5 m	0.5 m	NA	NA	NA
Tab configuration	Transceiver	T-connector[c]	NA	NA	NA
Maximum number of taps	100	30	NA	NA	NA
Type of connectors used	N	BNC	NA	NA	NA

[a] NA, not applicable. Each device is connected to a central hub via an unshielded twisted-wire pair.

[b] Each end of the segment is terminated with a 50-Ω resistive load.

[c] T-connector must be connected directly to the Ethernet card on the DTE device.

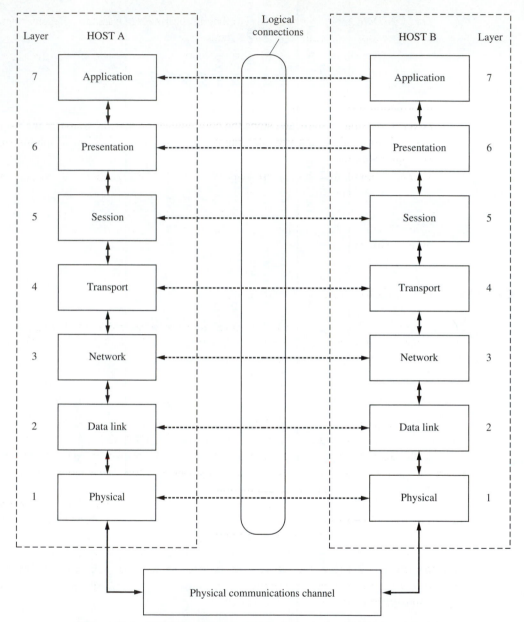

Figure C–6   ISO OSI reference model of physical and logical connections.

The functions of the various layers may be described as follows:

Layer I: Physical.   Concerned with bit transmission. Standards specify signal levels, cable connectors, and cable. For multiple-access packet transmission systems, the methods of carrier sense and collision detection are also specified.

Layer 2: Data Link.   Concerned with beginning message transmission, error detection and correction, and ending message transmission.

Layer 3: Network.   Concerned with choosing the route taken by the message, flow control, and message priorities.

Layer 4: Transport.   Concerned with end-to-end reliability (error detection and information recovery) and maps logical addresses to end-user devices.

Layer 5: Session.   Starts and stops the communication session, transfers the user from one task to another, and provides recovery from communication problems (restart) without losing data.

Layer 6: Presentation.   Converts data to the appropriate syntax for the display devices (character sets and graphics) and encodes or decodes compressed data.

Layer 7: Application.   Provides a log-in procedure, checks passwords, allows file upload and downloads and remote job entry, and tabulates system resources used for billing purposes.

C–4 DATA LINK CONTROL PROTOCOLS

Data link control (DLC) protocols, also called *line protocols,* are a set of rules for operating a computer network or a public switched telephone network (PSTN). They standardize the framing, addressing, and error-control techniques used in the system.

TCP/IP

The Transmission Control Protocol/Internet Protocol (TCP/IP) is used on the Internet. It operates within the transport and application layers of the OSI model.

In the transport layer, it breaks the message or file into packets and attaches a destination IP address. The packet address is read by gateway computers along the network to route the packet to the appropriate destination.

In the applications layer, familiar operations such as the World Wide Web Hypertext Transfer Protocol (HTTT), the File Transfer Protocol (FTP), Telnet (TELNET), and e-mail via the Simple Mail Transfer Protocol (SMTP) are popular applications.

PC users who are connected to the Internet via a dial-up telephone line to an Internet Service Provider (ISP) use the Serial Line Internet Protocol (SLIP) or the Point-to-Point Protocol (PPP). These protocols are used to encapsulate the IP packets so that they will be passed undisturbed over the dial-up connection.

SDLC

Synchronous Data Link Control (SDLC) is a *bit-oriented protocol.* That is, instead of using control characters, it uses a unique flag at the beginning and end of each frame. It was developed by IBM in 1974 to provide a more versatile protocol than BISYNC. It is a full-duplex protocol, which means that data can be transmitted in both directions simultaneously. Furthermore, SDLC permits the transmission to one remote location while receiving from

BEGINNING FLAG 01111110	ADDRESS 8 bits	CONTROL 8 bits	INFORMATION Any number of bits	FRAME CHECK 16 bits	ENDING FLAG 01111110

Figure C–7   SDLC and HDLC formats.

a different remote location on a multidrop line. SDLC also uses ARQ error control; in ARQ, a station has to acknowledge the correct receipt of at least the eighth preceding frame if communication is to continue. The SDLC frame is shown in Fig. C-7.

HDLC

High-Level Data Link Control (HDLC) was approved by ISO in 1975 (ISO 3309) and has been updated since then by other ISO standards. It is a *packet protocol* that has been adopted worldwide by many vendors. IBM's SDLC is a subset of HDLC, and ANSI's Advanced Data Communications Control Procedure (ADCCP) is equivalent to HDLC. As discussed in the next section, X.25 uses the HDLC protocol.

The general structure of the HDLC frame (packet) is shown in Fig. C–7. The address field may be extended recursively to allow more addresses. Two basic types of frames are allowed: control frames and information frames. In control frames, there is no information field, and the control frame is used to set up, take down, or check a virtual link between users. In the information frame, the information field may be of any size. The information frames are numbered in sequence (by bits in the control field), so that the sending station does not have to wait for an acknowledgment of the frame just transmitted before sending another frame. The acknowledgment can be received at a later time (for error detection and recovery), since the frames are numbered. In addition, the packets may arrive out of sequence at the receiver, and the numbering allows the receiver to place the information in the correct order.

HDLC was originally developed for unbalanced operation (ISO 4335), in which one primary station controls secondary stations by sending command frames. The secondary stations respond only when polled in the normal response mode. In the asynchronous mode (not to be confused with asynchronous bit transmission), the secondary stations are allowed to send unsolicited response frames. Newer modes (ISO 6256) allow balanced operation, in which any station may initiate the setting up of a virtual call.

CCITT X.25 Protocol

The CCITT X.25 standard is entitled "Interface Between a DTE and a DCE for Terminals Operating in the Packet Mode on Public Data Networks." It involves the use of data packets as the basis for creating a virtual circuit. With a virtual circuit, two users appear to have a private data connection between themselves, although they are actually sharing the same physical channel with other users. This is accomplished by breaking the serial data stream of each user into "packets" of data of a certain length. Each packet has a header attached, with the destination address for the packet and other control information. The packets from each user are interspersed with those of other users on the X.25 line.

The virtual circuits may be set up temporarily or permanently. A temporary virtual circuit consists of three operations: (1) setting up the call, (2) transferring the data, and (3) taking down (disconnecting) the virtual connection. In setting up the call, a logical channel number is used. With a permanent virtual connection, the logical number is assigned permanently to the customer when the X.25 line is leased, and the virtual connection is always available so that data can be transferred at any instant. Thus, the permanent virtual connection appears to be a private leased line to the end users.

The X.25 standard does not specify how the virtual circuits are implemented. However, X.25 is actually concerned with three layers of protocols. (See the ISO OSI protocol layers as described in Sec. C–3.) The first layer is the DCE/DTE physical layer interface, which uses the X.21 standard for synchronous data transmission, the second is the data link level, which gives the method of attaching address information to the packets according to the HDLC protocol, and the third is the network layer, which sets up the virtual circuits by the use of two different types of packets. The control packets contain control information, such as CALL REQUEST (sent by the caller), CALL ACCEPT (sent by the called party), CALL CONNECT, and CLEAR. The data packets contain the data that are exchanged between the users on a virtual circuit.

Asynchronous Transfer Mode (ATM)

The *asynchronous transfer mode* (ATM) is a packet protocol that is used in some public switched telephone networks (PSTN). ATM allows each user to have a virtual connection with an effective data rate that matched the user's instantaneous data-rate need, whether it be large or small or whether the data-rate requirement changes as time advances.

The ATM packet contains 424 bits (53 bytes) of which 40 bits (5 bytes) are the header (indicating data type and routing) and 384 bits (48 bytes) are the user's data payload. This is similar to the HDLC protocol, except that HDLC was designed for computer communication and ATM is designed to transport a wide variation of data rates and data types, such as digital video, digital audio, and computer data, that occur in the PSTN. In ATM the cell header has a cell loss-priority bit that distinguishes cells that might be discarded by the PSTN (without significantly disturbing that particular user) when there is some network data-overload condition. For example, a user who might tolerate data loss may be sending data on a virtual channel corresponding to the reading of the water height of some river. Some of the height readings could be discarded occasionally without compromising the effectiveness of the table of readings. This data loss for some ATM users allows the ATM system to be more robust for other users that cannot tolerate data loss (and who pay premium rates for the higher quality service).

ATM is an alternative to the *synchronous transfer mode* (STM) technology, such as T1 systems, where the user is locked in to a subscribed data rate whether or not he or she needs it. That is, with ATM, the user can send packets as often as needed, so that the user's effective data rate is matched to the user's needs and not to some fixed prescribed data rate, as is the case for STM technology.

C–5 Modem Standards

Why are modems (modulators–demodulators) needed? The modulator takes the data from a computer system and constructs a transmit signal with the appropriate spectral characteristic that matches the channel. The demodulator decodes the received signal to provide data to the computer. Most modems also provide full-duplex data transfer (i.e., simultaneous data transmission and reception).

Table C-5 summarizes the standards for modems used to transmit and receive data over twisted-pair telephone lines. The modems are listed in order of ascending data rate, and, in general, this is the order in which the standards were developed. To provide full-duplex transmission over dial-up lines, the lower speed modems have transmit and receive signals with spectra located in adjacent frequency bands (i.e., frequency–division multiplexing). Higher speed modems have transmit and receive signals located in the same frequency band, and a hybrid circuit (see Fig. 8–5) is included in the modem. The hybrid separates the transmit and receive signals, which are encoded or decoded to provide transmit and receive data streams from or to the computer.

The data capacity of the telephone line depends on the S/N and the bandwidth as shown by Eq. (1–10). For modern telephone systems with PCM line cards (see Sec. 8–2), the bandwidth is limited by the antialiasing filter on the line card to about 3,600 Hz, since the PCM sampling rate is 8 ksamples/s and a rolloff transmission band is needed for the filter. Also, the frequencies below 300 Hz are not used, because of 60-Hz power line noise and noise harmonics. This gives a usable bandwidth of about 3,300 Hz. Consequently, the capacity of this telephone line system is as low as 27.4 kb/s for a S/N of 25 dB (see SA1-2) and as high as 57 kb/s for a S/N of 52 dB (see Sec. 3–3).

Referring to Table C–6, the popular modems of the early 1980's were the Bell 103 and the Bell 212A, which use FSK modulation to transmit rates of 300 b/s and 1,200 b/s, respectively. Amateur radio operators use the Bell 103 standard for audio tones into and out of SSB transceivers for packet radio transmission on the HF amateur radio bands, and the Bell 202 standard is used for audio into and out of FM transceivers for packet transmission on the VHF and UHF bands.

Later, for higher capacity, QAM signaling (Table C–7) was adopted. Table C–8 shows the 128-point signal constellation for the once-popular V.32bis 14.4 kb/s modem. As described in Table C–5, the popular V.34 + QAM modem uses a 1664 point signal constellation to achieve a 33.6 kb/s data rate [Forney, Brown, et al., 1996].

The latest modem standard is the V.90. It uses PCM signaling technology, which is described in Sec. 3–3, to obtain a data rate of 56 kb/s.

In all cases, the modem drops to a lower speed standard if the telephone line does not have sufficient S/N to support the higher data rate. Measurements have shown that practically all lines in the United States can support a data rate of at least 24 kb/s [Forney, Brown, et al., 1996]. Higher speed modems incorporate *training* when they first connect. With this

TABLE C–5 MODEM STANDARDS

Type	Data Rate (bits/s) Normal	Fallback	Mode[a]	Type of Line	Synch/ Asynch	Modulation[b]	Transmitting Frequencies (Hz)[c] Originate	Answer
Bell 103/113	300		FDX	2W Dial-up	Async	FSK	1,070S / 1,270M	2,025S / 2,225M
V.21	300		FDX	2W Dial-up	Async	FSK	980M / 1,080S	1,650M / 1,850S
Bell 202S	1,200		HDX	2W Dial-up	Async	FSK	1,200M / 2,200S	
Bell 202T	1,200		FDX	4W Lease	Async	FSK	1,200M / 2,200S	
V.23	1,200	600	HDX	2W Dial-up	Async	FSK	1,300M / 2,100S; 1,300M / 1,700S	
Bell 212A	1,200	300	FDX	2W Dial-up	Either	4DPSK; FSK	1,200	2,400
V.22	1,200	600	FDX	2W Dial-up	Either	FSK	Same as Bell 103; Same as Bell 212	
Bell 201C	2,400		FDX	2W Dial-up	Either	2DPSK	1,200	2,400
V.22bis	2,400	1200	FDX	2W Dial-up	Either	4DPSK	Same as V.22	2,400
V.26	2,400		FDX	4W Lease	Sync	4DPSK	1,800 (same as Bell 201C)	
V.26bis	2,400	1200	FDX	4W Lease	Sync	4DPSK	1,800	
Bell 208A	4,800		FDX	4W Lease	Sync	8DPSK	1,800	
Bell 208B	4,800		HDX	2W Dial-up	Sync	8DPSK	1,800	
V.27bis	4,800	2400	FDX	4W Lease	Sync	8DPSK	1,800	
Bell 209	9,600		FDX	4W Lease	Sync	4DPSK	1,800	
V.29	9,600	7200/4800	FDX	4W Lease D	Sync	16QAM	1,700	
V.32	9,600	4800/2400	FDX	4W Lease D	Sync	16QAM	1,700	
V.32bis	14,400	12,000/9600	FDX	2W Dial-up	Either	32QAM[d]	1,800	1,800
V.33	14,400	9600/2400	FDX	2W Dial-up	Either	128QAM[d]	1,800	1,800
V.34	28,800	26,400/24,000	FDX	4W Lease C2	Sync	128QAM[d]	1,800	1,800
V.34+	33,600	V.34/V.32bis	FDX	2W Dial-up	Either	960QAM[d]	1,800	1,800
V.90	56,000	V.34+/V.34	FDX	2W Dial-up	Either	1664QAM[d]	1,800	1,800
				2W Dial-up	Either	PCM[e]	NA	

[a] FDX, full duplex; HDX, half duplex. [b] FSK, frequency-shift keying; BPSK, binary-phase-shift keying; MDPSK, M-ary differential phase-shift keying; VSB, vestigial sideband; MQAM, M-ary quadratic amplitude modulation. [c] M, mark; S, space. [d] Trellis-coded modulation is used. [e] See Sec. 3–3.

TABLE C–6 300, 1200, AND 2400 B/S PERSONAL MODEMS[a]

	Bell System Type			CCITT
	103/113	**202**	**212A**	**V.22 bis**
Data	Serial binary Asynchronous Full duplex (two-wire line)	Dial-up or leased line Serial binary Asynchronous Half-duplex (two-wire line)	Serial binary Asynchronous or synchronous Full-duplex, dial-up lines (also contains a 103 type modem)	Serial binary Asynchronous or synchronous Full-duplex, dial-up (also usually contains a 212A-type modem)
Data rate	0 to 300 b/s	0 to 1,200 b/s (dial-up) 0 to 1,800 b/s (leased C2)	1,200 b/s 600 baud (symbols/s)	2,400 b/s 600 baud (symbols/s)
Modulation	FSK	FSK	QPSK ($M = 4$ phases)	QAM (16-point rectangular-type signal constellation)
Frequency/ phase assignment	*Originate end* / *Answer end* Tx: Space 1,070 Hz / 2,025 Hz Mark 1,270 Hz / 2,225 Hz Rx: Space 2,025 Hz / 1,070 Hz Mark 2,225 Hz / 1,270 Hz	Mark: 1,200 Hz Space: 2,200 Hz	Carrier frequencies Originate end / Answer end Tx: $f_c = 1,200$ Hz / $f_c = 2,400$ Hz Rx: $f_c = 2,400$ / $f_c = 1,200$ Hz *Message (2 bits)* / *Phase Angle* 00 → 90° 01 → 0° 10 → 180° 11 → 270°	Originate end / Answer end Tx: $f_c = 1,200$ Hz / $f_c = 2,400$ Hz Rx: $f_c = 2,400$ Hz / $f_c = 1,200$ Hz
Transmit level	0 to −12 dBm	0 to −12 dBm		
Receive level	0 to −50 dBm	0 to −50 dBm (dial-up line) 0 to −40 dBm (leased line)		

[a] Tx, Transmit; Rx, receive.

TABLE C-7 V.32 MODEM STANDARD

Item		Signal Constellation
Data	Serial binary, Asynchronous or synchronous Full duplex over two-wire line[a]	
Carrier freq.	Transmit[a]: 1,800 Hz Receive[a]: 1,800 Hz	
Option 1 DATA rate	9,600 b/s for high SNR 4,800 b/s for low SNR	*Option 2: 32 QAM or QPSK*
Modulation	32 QAM, 2400 baud, for high SNR using trellis-coded modulation (see Fig. 1–9, where $n = 3$ and $m - k = 2$) with 4 data bits plus 1 coding bit per symbol QPSK, 2,400 baud (states A, B, C, D) for low SNR	
Option 2 DATA rate	9,600 b/s for high SNR 4,800 b/s for low SNR	*Option 2: 16 QAM or QPSK*
Modulation	16 QAM, 2,400 baud, for high SNR QPSK, 2,400 baud (states A, B, C, D) for low SNR	

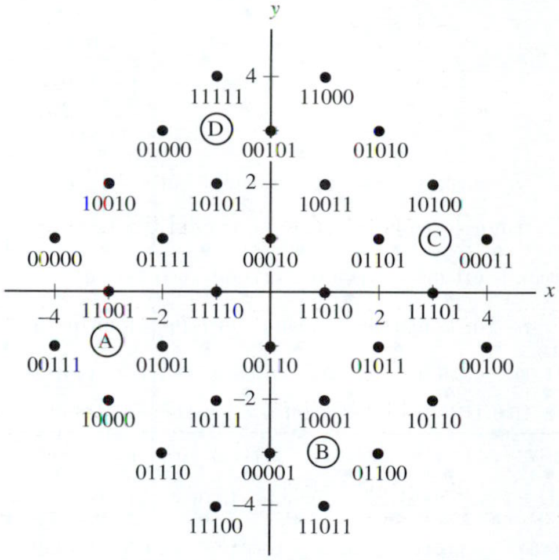

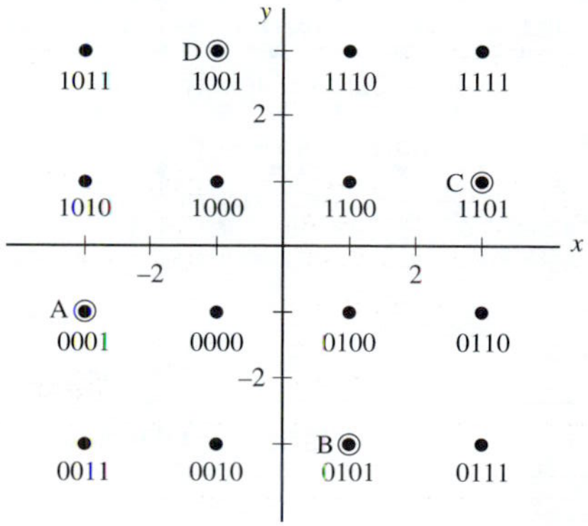

[a] A two-wire to four-wire hybrid is used in this modem to obtain the transmit and receive lines.

TABLE C–8 V.32BIS AND V.33 MODEM STANDARDS

Item	128 QAM Signal Constellation
Data	Serial binary
V.32bis	Synchronous/asynchronous, full duplex over two-wire dial up line[a]
V.33	Synchronous, full duplex over four-wire leased line[b]
Carrier freq.	Transmit: 1,800Hz Receive: 1,800 Hz
Data rate	14,400 b/s
Modulation	128 QAM, 2,400 baud, using trellis-coded modulation (see Fig. 1–9, where $n = 3$ and $m - k = 4$) with 6 data bits plus 1 coding bit per symbol
Fallback mode	12,000 bits/s using 64 QAM, 2,400 baud, (signal constellation not shown) and trellis-coded modulation with 5 data bits plus 1 coding bit per symbol

[a] The V.32bis modem uses a two-wire line and an internal two-wire to four-wire hybrid to obtain the transmit and receive lines.

[b] The V.33 modem uses two wires for transmit and two wires for receive.

procedure, the modems first "talk" to each other to negotiate the highest data rate that is possible for the given noise condition on the line and for the particular modems used. In addition, a procedure called "fastrain" is used to monitor for changes in the line and noise characteristics. When needed, fastrain stops the data transmission and renegotiates for the highest speed possible for reliable data transmission. The highest speed is lower if the noise increases and higher if the noise decreases.

Standards for higher speed services continue to be developed. For data rates in the range of 144 kb/s to 1.5 Mb/s, TELCOs provide ISDN and DSL service. (See Sec. 8–2.) Cable-TV and digital-satellite operators are also offering high-speed data services.

USING MATLAB

MATLAB is a very useful tool for analyzing and designing communication systems using the personal computer (PC). It has become widely used in engineering courses and is often used in industry for system design and simulation.

MATLAB is an acronym for Matrix Laboratory. It treats all constants and variables as matrices, row vectors, or column vectors. Consequently, the default operations in MATLAB are matrix operations. For example, $a * b$ is the matrix multiplication of a and b. This means that MATLAB program code can be very concise. Complicated operations can be expressed using very few lines of code with no "do" loops. The results can be computed very efficiently with the PC. However, this can make the code hard to understand, although it is compact and computationally efficient. The use of matrices and vectors also allows for powerful graphical capabilities.

In our MATLAB programming, we will adopt the concept of "keeping the code simple." This means that loops may appear in our code in order to make it easy to understand, although a more compact code could be used that would run faster. Our goal is to use MATLAB as a tool to make communication systems easy to understand.

To solve a problem using MATLAB, the MATLAB program is first run on the PC. MATLAB is an interpretive program. That is, results are computed after each line of code is entered. One has the option of keying in MATLAB statements one line at a time for immediate execution; or, alternately, a script file containing MATLAB code statements may be called up and run by MATLAB. This script or text file is also called an M-file, because the filename has the form xxxx.M. For programs with more than a couple of lines of code, the M-file method is usually used. The computed results may be shown in tabulated or graphical form. The M-files may be created by the MATLAB text editor or by another text editor, such as Notepad (running under Windows on the PC).

M-files are provided for solving selected equations and study-aid problems in this book. The selected equations are marked with a PC (🖳) symbol. The M-files can be downloaded from the World Wide Web Internet site

http://www.couch.ece.ufl.edu or http://www.prenhall.com/couch

or by anonymous file transfer protocol (FTP), downloaded from

ftp.ece.ufl.edu

under the subdirectory /pub/COUCH/6ed/MATLAB. The M-files may also be obtained from your instructor. For example, the file *e1_006.m* computes the results for Eq. (1–6). *Table2_3.m* is the M-file shown in Table 2–3 and produces the MATLAB plots shown in Fig. 2–21.

D–1 Quick Start for Running M-Files

To quickly get a MATLAB Program running that plots some results, the following steps are suggested:

1. Install MATLAB on your PC, as per the instructions from Math Works, Inc. (A Student Edition of MATLAB is available at a reasonable price.)
2. Download the M-files for *Digital and Analog Communications Systems*, 6th ed.
3. Install the downloaded M-files on your hard disk on a convenient subdirectory such as C:\COUCH
4. Bring up (i.e., run) MATLAB.
5. Sign on to the subdirectory where the M-files are located by entering the change directory command *cd C:\COUCH* (or wherever you saved the M-files) at the MATLAB prompt.
6. Check to see that you are in the subdirectory by entering the directory listing command *dir* at the MATLAB prompt. You should see the directory listing of the M-files.
7. Run one of the M-files. For example, to plot the results shown in Fig. 2–21, enter *fig2_21* at the prompt. (Alternatively, from the File pull-down menu, select Run Script, and enter *fig2_21.*) You should get a plot of the results. If the MATLAB Figure Window does not appear, select Window from the pull-down menu. Use the spacebar on your keyboard to toggle through several plots. You can print the plotted results by selecting Print on the pull-down File menu shown on the Figure window.

8. You can view and edit an M-file by selecting Open under the File pull-down menu on the MATLAB Command Window. For example, to see the program listing that produced the plot of Fig. 2–21, open *fig2_21.m*. Alternatively, you can list the file by entering *type fig2_21* at the MATLAB prompt, or you can edit the file by typing *edit fig2_21* at the prompt.

D–2 Programming in MATLAB

MATLAB provides an excellent on-line help system. At the MATLAB prompt, enter *help* for a list of help topics. For help on a specific topic, such as the use of the colon, enter *help colon*. If the result is so long that the description scrolls off the screen, use the mouse pointer on the scroll bar to move back to the beginning of the description. Math Works also provides a printed manual with the MATLAB program, and there are independent reference books on MATLAB. [See Hanselman and Littlefield, 1998.]

As indicated previously, the basic type of variable used in MATLAB is a matrix. This sets the stage, so to speak, for the notation used in MATLAB programs. To become acquainted with basic MATLAB programming concepts, bring up MATLAB now on your PC. Then, at the MATLAB prompt, key in the one-line examples given in the following list. The up and down cursor keys can be used to scroll through previous lines that you have typed. The list summarizes the basic notation used in MATLAB programs.

1. A row vector is created by a comma-delimited or a space-delimited list. For example, enter *M1 = [1,2,3,4,5,6]* or *M1 = [1 2 3 4 5 6]*. This is a six-element row vector, or a 1×6 matrix (1 row $\times$ 6 columns).

2. A column vector is created by a semicolon delimited list. For example, *M2 = [1;2;3; 4;5;6]* is a six-element column vector, or a 6×1 matrix.

3. Enter *M3 = [1 2 3; 4 5 6]*. This creates a 2×3 matrix. To view the element in the 2nd row and 3rd column, enter *M3(2,3)*. It has a value of 6.

4. A scalar is a 1×1 matrix. For example, enter *M4 = 2.5*.

5. Enter *M5 = 0.5 + 2j*. This specifies a complex number that is a 1×1 matrix.

6. The colon operator is used to index arrays and to create elements of vectors. The notation used is start-value:skip-increment:end-value. For example, enter *t = 1:2:6*. This creates the row vector **t** = [1 3 5]. If the skip increment is deleted, the default increment is 1. For example, *u = 1:6* is the row vector **u** = [1 2 3 4 5 6]. The colon can also serve as a wild card. For example, in Item 3, *M3 (1,3)* denotes the element in the 1st row and 3rd column, but *M3 (1,:)* would denote the row vector coresponding to the 1st row of the matrix M3.

7. Once a variable (i.e., a matrix) is defined, it remains in memory until it is cleared. That is, its value can be obtained by typing in its symbol. For example, enter *M3* at the MATLAB prompt.

8. Enter *whos*. Whos is the MATLAB command that lists all variables (i.e., matrices) that are stored in memory. It also shows the size of each matrix.

9. MATLAB is case sensitive. That is, M3 is a different variable from the variable m3, which has not been defined.

10. The transpose operator is '. For example, enter *M1'*, which is the transpose of M1.

11. Insert a semicolon at the end of a MATLAB statement to suppress the display of the computed result for that statement. For example, enter *y = 6*3;*. The computed result is not displayed on the screen. However, you can display the computed result by entering *y*.

12. If a MATLAB statement is too long to fit on one line, it may be continued on the next line by inserting ... at the end of the original line.

13. Multiple statements may be entered on the same line if the statements are separated by commas or semicolons. For example, enter *I = 1, i = 5.*

14. If MATLAB gets caught in a loop (due to a programming error), the computation can be terminated by pressing control-C. (The control key is abbreviated with ^. Thus, control-c is written here as ^C. For example, enter *for (i = 1:inf), i, end.* Terminate the computation by using *^C*.

15. The usual elementary functions (i.e., trigonometric and logarithmic) are built into MATLAB. For a listing, enter *help elfun*. For a list of specialized functions, such as Bessel functions, enter *help specfun*. User defined functions can also be created via M-files; enter *help function* for details.

16. The plotting of the results shown in Fig. 2–21 were obtained by running the MATLAB M-file program listed in Table 2–3. The function *plot (t, w)* gives a plot of the wave-form, vector **w**, as a function of time vector, **t**, as shown at the top of Fig. 2–21. Multiple plots in a single window can be obtained by using the subplot function. For example, multiple plots arranged in a three-row $\times$ one-column array can be obtained by using *subplot (3, 1, x)*, where *x* is the plot number within the three possible plots. Referring to Fig. 2–21, note that the three plots in the bottom $\frac{3}{4}$ of the figure are created by using *subplot (3, 1, 1)*, *subplot (3, 1, 2)* and *subplot (3, 1, 3)*; or, as shown in Table 2–3, by using *subplot (3 1 1)*, *subplot (3 1 2)* and *subplot (3 1 3)*. In another example, multiple plots arranged in a three-row $\times$ two-column array can be obtained by using *subplot (3, 2, x)* where *x* is the plot number within the six possible plots. That is, *subplot (3, 2, 3)* would be the third of six plots, where the third plot appears in the first column of the second row.

17. The % sign is used to define a comment statement. All text on a line that follows a % sign is interpreted as a comment statement. For example, see the first line of *table2_3.m*.

REFERENCES

ABRAMOWITZ, M., and I. A. STEGUN (Editors), *Handbook of Mathematical Functions*, National Bureau of Standards, Superintendent of Documents, U.S. Government Printing Office, Washington, DC, 1964. Also available in paperback from Dover Publications, New York, 1965.

AMOROSO, F., "The Bandwidth of Digital Data Signals," *IEEE Communications Magazine,* vol. 18, no. 6, November 1980, pp. 13–24.

ANDERSON, R. R., and J. SALZ, "Spectra of Digital FM," *Bell System Technical Journal,* vol. 44, July–August 1965, pp. 1165–1189.

ANVARI, K., and D. Woo, "Susceptibility of $\pi/4$ DQPSK TDMA Channel to Receiver Impairments," *RF Design* February 1991, pp. 49–55.

AOYAMA, T., W. R. DAUMER, and G. MODENA (Editors), Special Issue on Voice Coding for Communications, *IEEE Journal on Selected Areas of Communications*, vol. 6, February 1988.

ARRL, *The 1998 ARRL Handbook for Radio Amateurs,* 75th ed., 1997; *The 1997 ARRL Handbook for Radio Amateurs*, 1996; *The 1992 ARRL Handbook for Radio Amateurs*, 1991, American Radio Relay League, Newington, CT.

AT&T, *FT-2000 OC-48 Lightwave System*, Publication No. AT&T 365-575-100, AT&T Regional Technical Assistance Center (800–432–6600), Holmdel, NJ, December 1994.

716

BAINES, R., "The DSP Bottleneck," *IEEE Communications Magazine*, vol. 33, May 1995, pp. 46–54.

BALDUINO, P. R. H., "Latin America Goes Wireless via Satellite," *IEEE Communications Magazine*, vol. 33, no. 9, September 1995, pp. 114–122.

BASCH, E. E., and T. G. BROWN, "Introduction to Coherent Optical Fiber Transmission," *IEEE Communications Magazine*, vol. 23, May 1985, pp. 23–30.

BEDROSIAN, E. B., and S. O. RICE, "Distortion and Crosstalk of Linearly Filtered Angle-Modulated Signals," *Proceedings of the IEEE*, vol. 56, January 1968, pp. 2–13.

BELL TELEPHONE LABORATORIES, *Transmission Systems for Communications*, 4th ed., Western Electric Company, Winston-Salem, NC, 1970.

BENDAT, J. S., and A. G. PIERSOL, *Random Data: Analysis and Measurement Procedures*, Wiley-Interescience, New York, 1971.

BENEDETTO, S., E. BIGLIERI, and V. CASTELLANI, *Digital Transmission Theory,* Prentice Hall, Upper Saddle River, NJ, 1987.

BENEDETTO, S., M. MONDIN, and G. MONTORSI, "Performance Evaluation of Trellis-Coded Modulation Schemes," *Proceedings of the IEEE*, vol. 82, no. 6, June 1994, pp. 833–855.

BENEDETTO, S., and G. MONTORSI, "Unveiling Turbo Codes: Some Results on Parallel Concatenated Coding Schemes," *IEEE Transactions on Information Theory*, vol. IT-42, March 1996, pp. 409–428.

BENNETT, W. R., and J. R. DAVEY, *Data Transmission,* McGraw–Hill Book Company, New York, 1965.

BENNETT, W. R., and S. O. RICE, "Spectral Density and Autocorrelation Functions Associated with Binary Frequency Shift Keying," *Bell System Technical Journal,* vol. 42, September 1963, pp. 2355–2385.

BENSON, K. B., and J. C. WHITAKER, *Television Engineering Handbook*, Rev. ed., McGraw–Hill Book Company, New York, 1992.

BERGLAND, G. D., "A Guided Tour of the Fast Fourier Transform," *IEEE Spectrum*, vol. 6, July 1969, pp. 41–52.

BEST, R. E., *Phase-Locked Loops*, 4th ed., McGraw–Hill, Inc., New York, 1999.

BHARGAVA, V. K., "Forward Error Correction Schemes for Digital Communications," *IEEE Communications Magazine*, vol. 21, January 1983, pp. 11–19.

BHARGAVA, V. K., D. HACCOUN, R. MATYAS, and P. P. NUSPL, *Digital Communications by Satellite*, Wiley-Interscience, New York, 1981.

BIC, J. C., D. DUPONTEIL, and J. C. IMBEAUX, *Elements of Digital Communication*, John Wiley & Sons, New York, 1991.

BIGLIERI, E., D. DIVSALAR, P. J. MCLANE, and M. K. SIMON, *Introduction to Trellis-Coded Modulation with Applications*, Macmillan Publishing Company, New York, 1991.

BLACKMAN, R. B., and J. W. TUKEY, *The Measurement of Power Spectra*, Dover, New York, 1958.

BLAHUT, R. E., *Theory and Practice of Error Control Codes*, Addison-Wesley Publishing Company, Reading, MA, 1983.

BLANCHARD, A., *Phase-Locked Loops,* Wiley-Interscience, New York, 1976.

BOASHASH, B., "Estimating and Interpreting the Instantaneous Frequency of a Signal—Part 1: Fundamentals" and "Part 2: Algorithms and Applications, *Proceedings of the IEEE*, vol. 80, no. 4, April 1992, pp. 520–538, 540–568.

BÖRJESSON P. O., and C. E. W. SUNDBERG, "Simple Approximations for the Error Function Q(x) for Communication Applications," *IEEE Transactions on Communications*, vol. COM-27, March 1979, pp. 639–643.

BOWRON, P., and F. W. STEPHENSON, *Active Filters for Communications and Instrumentation*, McGraw–Hill Book Company, New York, 1979.

BREMERMANN, H., *Distributions, Complex Variables and Fourier Transforms*, Addison-Wesley Publishing Company, Reading, MA, 1965.

BRILEY, B. E., *Introduction to Telephone Switching*, Addison-Wesley Publishing Company, Reading, MA, 1983.

BROADCASTING, *Broadcasting and Cable Yearbook 1999*, R. R. Bowker, New Providence, NJ, 1999.

BUDAGAVI, M., and J. D. GIBSON, "Speech Coding in Mobile Radio Communications," *Proceeding of the IEEE*, vol. 86, no. 7, July 1998, pp. 1402–1412.

BYLANSKI, P., and D. G. W. INGRAM, *Digital Transmission Systems*, Peter Peregrinus Ltd., Herts, England, 1976.

CAMPANELLA, M., U. LoFaso, and G. MAMOLA, Optimum Pulse Shape for Minimum Spectral Occupancy in FSK Signals," *IEEE Transactions on Vehicular Technology*, vol. VT-33, May 1984, pp. 67–75.

CARLSON, A. B., *Communication Systems*, 3d ed., McGraw–Hill Book Company, New York, 1986.

CATTERMOLE, K. W., *Principles of Pulse-Code Modulation*, American Elsevier, New York, 1969.

CCITT STUDY GROUP XVII, "Recommendation V.32 for a Family of 2-Wire, Duplex Modems Operating on the General Switched Telephone Network and on Leased Telephone-Type Circuits," *Document AP VIII-43E*, May 1984.

CHAKRABORTY, D., "VSAT Communication Networks—An Overview," *IEEE Communications Magazine*, vol. 26, no. 5, May 1988, pp. 10–24.

CHALLAPALI, K., X. LEBEQUE, J. S. LIM, W. H. PAIK, R. SAINT GIRONS, E. PETAJAN, V. SATHE, P. A. SNOPKO, and J. ZDEPSKI, "The Grand Alliance System for US HDTV," *Proceedings of the IEEE*, vol. 83, no. 2, February 1995, pp. 158–173.

CHAPMAN, R. C. (Editor), "The SLC96 Subscriber Loop Carrier System," *AT&T Bell Laboratories Technical Journal*, vol. 63, no. 10, Part 2, December 1984, pp. 2273–2437.

CHESTER, D. B., "Digital IF Filter Technology for 3G Systems: An Introduction," *IEEE Communications Magazine*, vol. 37, February 1999, pp. 102–107.

CHILDERS, D. G., *Probability and Random Processes Using MATLAB*, Irwin, Chicago, 1997.

CHILDERS, D., and A. DURLING, *Digital Filtering and Signal Processing*, West Publishing Company, New York, 1975.

CHORAFAS, D. N., *Telephony, Today and Tomorrow*, Prentice-Hall, Englewood Cliffs, NJ, 1984.

CLARK, G. C., and J. B. CAIN, *Error-Correction Coding for Digital Communications*, Plenum Publishing Corporation, New York, 1981.

CONKLING, C. "Fractional -N Synthesizers Trim Current, Phase Noise," *Microwaves and RF*, February 1998, pp. 126–134.

COOPER, G. R., and C. D. MCGILLEM, *Modern Communications and Spread Spectrum*, McGraw–Hill Book Company, New York, 1986.

COUCH, L. W., "A Study of a Driven Oscillator with FM Feedback by Use of a Phase-Lock-Loop Model," *IEEE Transactions on Microwave Theory and Techniques*, vol. MTT-19, no. 4, April 1971, pp. 357–366.

COUCH, L. W., *Digital and Analog Communication Systems*, 4th ed., Macmillan Publishing Company, New York, 1993.

COUCH, L. W., *Modern Communications Systems*, Prentice Hall, Upper Saddle River, NJ, 1995.

COURANT, R., and D. HILBERT, *Methods of Mathematical Physics*, Vol. 1, Wiley-Interscience, New York, 1953.

DAMMANN, C. L., L. D. McDANIEL, and C. L. MADDOX, "D2 Channel Bank—Multiplexing and Coding," *Bell System Technical Journal*, vol. 51, October 1972, pp. 1675–1700.

DAVENPORT, W. B., JR., and W. L. ROOT, *An Introduction to the Theory of Random Signals and Noise*, McGraw–Hill Book Company, New York, 1958.

DAVIDOFF, M. R., *The Satellite Experimenter's Handbook*, American Radio Relay League, Newington, CT, 1984.

DAVIS, D. W., and D. L. A. BARBER, *Communication Networks for Computers*, John Wiley & Sons, New York, 1973.

DEANGELO, J., "New Transmitter Design for the 80's," *BM/E (Broadcast Management/Engineering)*, vol. 18, March 1982, pp. 215–226.

DECINA, M., and G. MODENA, "CCITT Standards on Digital Speech Processing," *IEEE Journal on Selected Areas of Communications*, vol. 6, February 1988, pp. 227–234.

DEFFEBACH, H. L., and W. O. FROST, "A Survey of Digital Baseband Signaling Techniques," *NASA Technical Memorandum NASATM X-64615*, June 30, 1971.

DEJAGER, F., "Delta Modulation, A Method of PCM Transmission Using a 1-Unit Code," *Phillips Research Report*, no. 7, December 1952, pp. 442–466.

DEJAGER, F., and C. B. DEKKER, "Tamed Frequency Modulation: A Novel Method to Achieve Spectrum Economy in Digital Transmission," *IEEE Transactions on Communications*, vol. COM-26, May 1978, pp. 534–542.

DESHPANDE, G. S, and P. H. WITTKE, "Correlative Encoding of Digital FM," *IEEE Transactions on Communications*, vol. COM-29, February 1981, pp. 156–162.

DHAKE, A. M., *Television Engineering*, McGraw–Hill Book Company, New Delhi, India, 1980.

DIXON, R. C., *Spread Spectrum Systems*, 3d ed., John Wiley & Sons, New York, 1994.

DORF, R. C. (Editor-in-Chief), *The Electrical Engineering Handbook*, CRC Press, Boca Raton, FL, 1993.

DORF, R. C. (Editor), *The Engineering Handbook*, CRC Press, Boca Raton, FL, 1996.

EDELSON, B. I., and A. M. WERTH, "SPADE System Progress and Application," *COM-SAT Technical Review*, Spring 1972, pp. 221–242.

EILERS, C. G., "TV Multichannel Sound—The BTSC System," *IEEE Transactions on Consumer Electronics*, vol. CE-31, February 1985, pp. 1–7.

EVANS, J. V., "Satellite Systems for Personal Communications," *Proceedings of the IEEE*, vol. 86, no. 7, July 1998, pp. 1325–1341.

FASHANO, M., and A. L. STRODTBECK, "Communication System Simulation and Analysis with SYS-TID," *IEEE Journal on Selected Areas of Communications*, vol. SAC-2, January 1984, pp. 8–29.

FEHER, K., *Digital Communications—Satellite/Earth Station Engineering*, Prentice Hall, Englewood Cliffs, NJ, 1981.

FELLER, W., *An Introduction to Probability Theory and Its Applications*, John Wiley & Sons, New York, 1957, p. 168.

FIKE, J. L., and G. E. FRIEND, *Understanding Telephone Electronics*, 2d ed., Texas Instruments, Dallas, 1984.

FINK, D. G., and H. W. BEATY (Editors), *Standard Handbook for Electrical Engineers*, McGraw–Hill Book Company, New York, 1978.

FLANAGAN, J. L., M. R. SCHROEDER, B. S. ATAL, R. E. CROCHIERE, N. S. JAYANT, and J. M. TRIBOLET, "Speech Coding," *IEEE Transactions on Communications*, vol. COM-27, April 1979, pp. 710–737.

FOLEY, J., "Iridium: Key to World Wide Cellular Communications," *Telecommunications*, vol. 25, October 1991, pp. 23–28.

FORNEY, G. D., "The Viterbi Algorithm," *Proceedings of the IEEE*, vol. 61, March 1973, pp. 268–273.

FORNEY, G. D., L. Brown, M. V. Eyuboglu, and J. L. Moran, "The V.34 High-Speed Modem Standard," *IEEE Communications Magazine*, vol. 34, December 1996, pp. 28–33.

FROHNE, R., "A High-Performance, Single-Signal, Direct-Conversion Receiver with DSP Filtering," *QST*, vol. 82, April 1998, pp. 40–46.

GALLAGHER, R. G., *Information Theory and Reliable Communications*, John Wiley & Sons, New York, 1968.

GARDNER, F. M., *Phaselock Techniques*, 2d ed., John Wiley & Sons, New York, 1979.

GARDNER, F. M., and W. C. LINDSEY (Guest Editors), Special Issue on Synchronization, *IEEE Transactions on Communications*, vol. COM-28, no. 8, August 1980.

GARG, V. K., *IS-95 CDMA and cdma 2000*, Prentice Hall PTR, Upper Saddle River, NJ, 2000.

GARG, V. K., and J. E. WILKES, *Principles and Applications of GSM*, Prentice Hall PTR, Upper Saddle River, NJ, 1999.

GERSHO, A., "Advances in Speed and Audio Compression," *Proceedings of the IEEE*, vol. 82, no. 6, June 1994, pp. 900–912.

GERSHO, A., "Charge-Coupled Devices: The Analog Shift Register Comes of Age," *IEEE Communications Society Magazine*, vol. 13, November 1975, pp. 27–32.

GERSON, I. A., and M. A. JASIUK, "Vector Sum Excited Linear Prediction (VSELP) Speech Coding at 8kb/s," *International Conf. on Acoustics, Speech, and Signal Processing*, Albuquerque, NM, April 1990, pp. 461–464.

GIBSON, J. D., Editor, *The Communications Handbook*, CRC Press, Inc., Boca Raton, FL, 1997.

GLISSON, T. L., *Introduction to System Analysis*, McGraw–Hill Book Company, New York, 1986.

GOLDBERG, R. R., *Fourier Transforms*, Cambridge University Press, New York, 1961.

GOLOMB, S. W. (Editor), *Digital Communications with Space Applications*, Prentice Hall, Englewood Cliffs, NJ, 1964.

GOODMAN, D. J., "Trends in Cellular and Cordless Communication," *IEEE Communications Magazine*, vol. 29, June 1991, pp. 31–40.

GORDON, G. D., and W. L. MORGAN, *Principles of Communication Satellites*, Wiley-Interscience, New York, 1993.

GRIFFITHS, J., *Radio Wave Propagation and Antennas*, Prentice Hall, Englewood Cliffs, NJ, 1987.

GROB, B., and C. A. HERNDON, *Basic Television*, 6th ed., McGraw–Hill Book Company, New York, 1998.

GRUBB, J. L., "The Traveller's Dream Come True," *IEEE Communications Magazine*, vol. 29, November 1991, pp. 48–51.

GULSTAD, K., "Vibrating Cable Relay," *Electrical Review* (London), vol. 42, 1898; vol. 51, 1902.

GUPTA, S. C., "Phase Locked Loops," *Proceedings of the IEEE*, vol. 63, February 1975, pp. 291 –306.

HA, T. T., *Digital Satellite Communications*, Macmillan Publishing Company, New York, 1986.

HA, T. T., *Digital Satellite Communications*, 2d ed., John Wiley & Sons, New York, 1989.

HAMMING, R. W., "Error Detecting and Error Correcting Codes," *Bell System Technical Journal*, vol. 29, April 1950, pp. 147–160.

HÄNDEL, R., and M. N. HUBER, *Integrated Broadband Networks*, Addison-Wesley Publishing Company, Reading, MA, 1991.

HANSELMAN, D., and B. LITTLEFIELD, *Mastering MATLAB 5*, Prentice Hall, Upper Saddle River, NJ, 1998.

HARRIS, A., "The New World of HDTV," *Electronics Now*, vol. 64, May 1993, pp. 33–40 and p. 72.

HARRIS, F. J., "On the Use of Windows for Harmonic Analysis with the Discrete Fourier Transform," *Proceedings of the IEEE*, vol. 66, January 1978, pp. 51–83.

HARRIS, F. J., "The Discrete Fourier Transform Applied to Time Domain Signal Processing," *IEEE Communications Magazine*, vol. 20, May 1982, pp. 13–22.

HARTLEY, R. V., "Transmission of Information," *Bell System Technical Journal*, vol. 27, July 1948, pp. 535–563.

HAUS, H. A., Chairman IRE Subcommittee 7.9 on Noise, "Description of Noise Performance of Amplifiers and Receiving Systems," *Proceedings of the IEEE*, vol. 51, no. 3, March 1963, pp. 436–442.

HAYKIN, S., *Communication Systems*, 2d ed., John Wiley & Sons, New York, 1983.

HOLMES, J. K., *Coherent Spread Spectrum Systems*, Wiley-Interscience, New York, 1982.

HOPKINS, R., "Digital Terrestrial HDTV for North America: The Grand Alliance HDTV System," *IEEE Transactions on Consumer Electronics*, vol. 40, no. 3, August 1994, pp. 185–198.

HSING, T. R., C. CHEN, and J. A. BELLISIO, "Video Communications and Services in the Copper Loop," *IEEE Communications Magazine*, vol. 31, January 1993, pp. 62–68.

HSU, H. P., "Chapter 2–Sampling," *The Mobile Communications Handbook*, 2d ed., (J. D. Gibson, Editor), CRC Press, Boca Raton, FL, 1999, pp. 2.1–2.10.

HUANG, D. T., and C. F. VALENTI, "Digital Subscriber Lines: Network Considerations for ISDN Basic Access Standard," *Proceedings of the IEEE*, vol. 79, February 1991, pp. 125–144.

IRMER, T., "An Overview of Digital Hierarchies in the World Today," *IEEE International Conference on Communications*, San Francisco, June 1975, pp. 16–1 to 16–4.

IRWIN, J. D., *Basic Engineering Circuit Analysis*, Macmillan Publishing Company, New York, 5th ed. 1995.

JACOBS, I., "Design Considerations for Long-Haul Lightwave Systems," *IEEE Journal on Selected Areas of Communications*, vol. 4, December 1986, pp. 1389–1395.

JAMES, R. T., and P. E. MUENCH, "A.T.&T. Facilities and Services," *Proceedings of the IEEE*, vol. 60, November 1972, pp. 1342–1349.

JAYANT, N. S., "Digital Encoding of Speech Waveforms," *Proceedings of the IEEE*, May 1974, pp. 611–632.

JAYANT, N. S., "Coding Speech at Low Bit Rates," *IEEE Spectrum*, vol. 23, August 1986, pp. 58–63.

JAYANT, N. S., and P. NOLL, *Digital Coding of Waveforms*, Prentice Hall, Englewood Cliffs, NJ, 1984.

JENKINS, M. G., and D. G. WATTS, *Spectral Analysis and Its Applications*, Holden-Day, San Francisco, CA, 1968.

JERRI, A. J., "The Shannon Sampling Theorem—Its Various Extensions and Applications: A Tutorial Review," *Proceedings of the IEEE*, vol. 65, no. 11, November 1977, pp. 1565–1596.

JORDAN, E. C. (Editor), *Reference Data for Engineers: Radio, Electronics, Computer, and Communications*, 7th ed., Howard W. Sams & Company, Indianapolis, IN, 1985.

JORDAN, E. C., and K. G. BALMAIN, *Electromagnetic Waves and Radiating Systems*, 2d ed., Prentice Hall, Englewood Cliffs, NJ, 1968.

JURGEN, R. K., "High-Definition Television Update," *IEEE Spectrum*, vol. 25, April 1988, pp. 56–62.

JURGEN, R. K., "The Challenges of Digital HDTV," *IEEE Spectrum*, vol. 28, April 1991, pp. 28–30 and 71–73.

KAUFMAN, M., and A. H. SEIDMAN, *Handbook of Electronics Calculations*, McGraw–Hill Book Company, New York, 1979.

KAY, S. M., *Modern Spectral Estimation—Theory and Applications*, Prentice Hall, Englewood Cliffs, NJ, 1986.

KAY, S. M., and S. L. MARPLE, "Spectrum-Analysis—A Modern Perspective," *Proceedings of the IEEE*, vol. 69, November 1981, pp. 1380–1419.

KAZAKOS, D., and P. PAPANTONI-KAZAKOS, *Detection and Estimation*, Computer Science Press, New York, 1990.

KLAPPER, J., and J. T. FRANKLE, *Phase-Locked and Frequency–Feedback Systems*, Academic Press, New York, 1972.

KRAUS, J. D., *Radio Astronomy*, 2d ed., Cygnus-Quasar Books, Powell, OH, 1986.

KRAUSS, H. L., C. W. BOSTIAN, and F. H. RAAB, *Solid State Radio Engineering*, John Wiley & Sons, New York, 1980.

KRETZMER, E. R., "Generalization of a Technique for Binary Data Communications," *IEEE Transactions on Communication Technology*, vol. COM-14, February 1966, pp. 67–68.

LAM, S. S., "Satellite Packet Communications, Multiple Access Protocols and Performance," *IEEE Transactions on Communications*, vol. COM-27, October 1979, pp. 1456–1466.

LANCASTER, D., "Hardware Hacker—Caller Number Delivery," *Radio-Electronics*, vol. 62, August 1991, pp. 69–72.

LATHI, B. P., *Modern Digital and Analog Communication Systems*, 3rd ed., Oxford University Press, Inc., London, 1998.

LEE, W. C. Y., "Elements of Cellular Mobile Radio Systems," *IEEE Transactions on Vehicular Technology*, vol. VT-35, May 1986, pp. 48–56.

LEE, W. C. Y., *Mobile Cellular Telecommunications Systems*, McGraw–Hill Book Company, New York, 1989.

LEIB, H., and S. PASUPATHY, "Error-Control Properties of Minimum Shift Keying," *IEEE Communications Magazine*, vol. 31, January 1993, pp. 52–61.

LENDER, A., "The Duobinary Technique for High Speed Data Transmission," *IEEE Transactions on Communication Electronics*, vol. 82, May 1963, pp. 214–218.

LENDER, A., "Correlative Level Coding for Binary-Data Transmission," *IEEE Spectrum*, vol. 3, February 1966, pp. 104–115.

LI, V. O. K., and X. QUI, "Personal Communication Systems (PCS)," *Proceedings of the IEEE*, vol. 83, no. 9, September 1995, pp. 1210–1243.

LIN, D. W., C. CHEN, and T. R. HSING, "Video on Phone Lines: Technology and Applications," *Proceedings of the IEEE*, vol. 83, no. 2, February 1995, pp. 175–192.

LIN, S., and D. J. COSTELLO, JR., *Error Control Coding*, Prentice Hall, Englewood Cliffs, NJ, 1983.

LINDSEY, W. C., and C. M. CHIE, "A Survey of Digital Phase-Locked Loops," *Proceedings of the IEEE*, vol. 69, April 1981, pp. 410–431.

LINDSEY, W. C., and M. K. SIMON, *Telecommunication System Engineering*, Prentice Hall, Englewood Cliffs, NJ, 1973.

LUCKY, R., "Through a Glass Darkly–Viewing Communication in 2012 from 1961," *Proceedings of the IEEE*, vol. 87, July 1999, pp. 1296–1300.

LUCKY, R. W., and H. R. RUDIN, "Generalized Automatic Equalization for Communication Channels," *IEEE International Communication Conference*, vol. 22, 1966.

LUCKY, R. W., J. SALZ, and E. J. WELDON, *Principles of Data Communication,* McGraw–Hill Book Company, New York, 1968.

MANASSEWITSCH, V., *Frequency Synthesizers*, 3d ed., Wiley-Interscience, New York, 1987.

MARAL, G., *VSAT Networks*, John Wiley & Sons, New York, 1995.

MARKLE, R. E., "Single Sideband Triples Microwave Radio Route Capacity," *Bell Systems Laboratories Record*, vol. 56, no. 4, April 1978, pp. 105–110.

MARKLE, R. E., "Prologue, The AR6A Single-Sideband Microwave Radio System," *Bell System Technical Journal*, vol. 62, December 1983, pp. 3249–3253. (The entire December 1983 issue deals with the AR6A system.)

MARPLE, S. L., *Digital Spectral Analysis*, Prentice Hall, Englewood Cliffs, NJ, 1986.

MCELIECE, R. J., The Theory of Information and Coding *(Encyclopedia of Mathematics and Its Applications*, vol. 3), Addison-Wesley Publishing Company, Reading, MA, 1977.

MCGILL, D. T., F. D. NATALI, and G. P. EDWARDS, "Spread-spectrum Technology for Commercial Applications," *Proceeding of the IEEE*, vol. 82, no. 4, April 1994, 572–584.

MELSA, J. L., and D. L. COHN, *Decision and Estimation Theory*, McGraw–Hill Book Company, New York, 1978.

MENNIE, D., "AM Stereo: Five Competing Options," *IEEE Spectrum*, vol. 15, June 1978, pp. 24–31.

MIYAOKA, S., "Digital Audio Is Compact and Rugged," *IEEE Spectrum*, vol. 21, March 1984, pp. 35–39.

MOTOROLA, "MC145500 Series PCM Codec-Filter Mono-Circuit," *Motorola Semiconductor Technical Data*, available from *www.motorola.com*, 1995.

MOTOROLA, *Telecommunications Device Data*, Motorola Semiconductor Products, Austin, TX, 1985.

MUMFORD, W. W., and E. H. SCHEIBE, *Noise Performance Factors in Communication Systems*, Horizon House-Microwave, Dedham, MA, 1968.

MUROTO, K., "GMSK Modulation for Digital Mobile Radio Telephony," *IEEE Transactions on Communications*, vol. COM-29, July 1981, pp. 1044–1050.

MURPHY, E., "Whatever Happened to AM Stereo?" *IEEE Spectrum*, vol. 25, March 1988, p. 17.

NILSSON, J. W., *Electric Circuits*, 3d ed., Addison-Wesley Publishing Company, Reading, MA, 1990.

NORTH, D. O., "An Analysis of the Factors which Determine Signal/Noise Discrimination in Pulsed-Carrier Systems," *RCA Technical Report*, PTR-6-C, June 1943; reprinted in *Proceedings of the IEEE*, vol. 51, July 1963, pp. 1016–1027.

NYQUIST, H., "Certain Topics in Telegraph Transmission Theory," *Transactions of the AIEE*, vol. 47, February 1928, pp. 617–644.

OKUMURA, Y., and M. SHINJI (Guest Editors) Special Issue on Mobile Radio Communications, *IEEE Communications Magazine*, vol. 24, February 1986.

OLMSTED, J. M. H., *Advanced Calculus*, Appleton-Century-Crofts, New York, 1961.

O'NEAL, J. B., "Delta Modulation and Quantizing Noise, Analytical and Computer Simulation Results for Gaussian and Television Input Signals," *Bell System Technical Journal*, vol. 45, January 1966 (a), pp. 117–141.

O'NEAL, J. B., "Predictive Quantization (DPCM) for the Transmission of Television Signals," *Bell System Technical Journal*, vol. 45, May–June 1966(b), pp. 689–721.

OPPENHEIM, A. V., and R. W. SCHAFER, *Digital Signal Processing*, Prentice Hall, Englewood Cliffs, NJ, 1975.

OPPENHEIM, A. V., and R. W. SCHAFER, *Discrete-Time Signal Processing,* Prentice Hall, Englewood Cliffs, NJ, 1989.

PADGETT, J. E., C. G. GUNTHER, and T. HATTORI, "Overview of Wireless Personal Communications," *IEEE Communications Magazine*, vol. 33, no. 1, January 1995, pp. 28–41.

PANCHA, P., and M. EL ZARKI, "MPEG Coding of Variable Bit Rate Video Transmission," *IEEE Communications Magazine*, vol. 32, no. 5, May 1994, pp. 54–66.

PANDHI, S. N., "The Universal Data Connection," *IEEE Spectrum*, vol. 24, July 1987, pp. 31–37.

PANTER, P. F., *Modulation, Noise and Spectral Analysis*, McGraw–Hill Book Company, New York, 1965.

PAPOULIS, A., *Probability, Random Variables and Stochastic Processes*, McGraw–Hill Book Company, New York, 2d ed., 1984, 3d ed., 1991.

PARK, J. H., JR., "On Binary DPSK Detection," *IEEE Transactions on Communications*, vol. COM-26, April 1978, pp. 484–486.

PASUPATHY, S., "Correlative Coding," *IEEE Communications Society Magazine*, vol. 15, July 1977, pp. 4–11.

PEEBLES, P. Z., *Communication System Principles*, Addison-Wesley Publishing Company, Reading, MA, 1976.

PEEBLES, P. Z., *Digital Communication Systems*, Prentice Hall, Englewood Cliffs, NJ, 1987.

PEEBLES, P. Z., *Probability, Random Variables, and Random Signal Principles*, 3d ed., McGraw–Hill Book Company, New York, 1993.

PEEK, J. B. H., "Communications Aspects of the Compact Disk Digital Audio System," *IEEE Communications Magazine*, vol. 23, February 1985, pp. 7–15.

PERSONICK, S. D., "Digital Transmission Building Blocks," *IEEE Communications Magazine*, vol. 18, January 1980, pp. 27–36.

PETAJAN, E., "The HDTV Grand Alliance System," *Proceedings of the IEEE*, vol. 83, no. 7, July 1995, pp. 1094–1105.

PETERSON, R. L., R. E. ZIEMER, and D. E. BORTH, *Introduction to Spread-Spectrum Communications*, Prentice Hall, Upper Saddle River, NJ, 1995.

PETERSON, W. W., and E. J. WELDON, *Error-Correcting Codes*, MIT Press, Cambridge, MA, 1972.

PETTIT, R. H., *ECM and ECCM Techniques for Digital Communication Systems*, Lifetime Learning Publications (a division of Wadsworth, Inc.), Belmont, CA, 1982.

PRATT, T., and C. W. BOSTIAN, *Satellite Communications*, John Wiley & Sons, New York, 1986.

PRITCHARD, W. L., and C. A. KASE, "Getting Set for Direct-Broadcast Satellites," *IEEE Spectrum*, vol. 18, August 1981, pp. 22–28.

PROAKIS, J. G., *Digital Communications*, 3d ed., McGraw–Hill Book Company, New York, 1995.

QURESHI, S., "Adaptive Equalization," *IEEE Communications Magazine*, vol. 20, March 1982, pp. 9–16.

RAMO, S., J. R. WHINNERY, and T. VANDUZER, *Fields and Waves In Communication Electronics*, 3d ed., John Wiley & Sons, Inc., New York, 1994.

RANA, A. H., J. MCCOSKEY, and W. CHECK, "VSAT Technology, Trends and Applications," *Proceedings of the IEEE*, vol. 78, July 1990, pp. 1087–1095.

RAPPAPORT, T. S., "Characteristics of UHF Multipath Radio Channels in Factory Buildings," *IEEE Transactions on Antennas and Propagation*, vol. 37, August 1989, pp. 1058–1069.

RAPPAPORT, T. S., *Wireless Communications, Principles and Practice*, Prentice Hall PTR, Upper Saddle River, NJ, 1996.

REEVE, W. H., *Subscriber Loop Signaling and Transmission Handbook*, IEEE Press, New York, 1995.

RHEE, M. Y., *CDMA Cellular Mobile Communications and Network Security*, Prentice Hall PTR, Upper Saddle River, NJ, 1998.

RHEE, S. B., and W. C. Y. LEE (Editors), Special Issue on Digital Cellular Technologies, *IEEE Transactions on Vehicular Technology*, vol. 40, May 1991.

RICE, S. O., "Mathematical Analysis of Random Noise," *Bell System Technical Journal*, vol. 23, July 1944, pp. 282–333 and vol. 24, January 1945, pp. 46–156; reprinted in *Selected Papers on Noise and Stochastic Processes*, N. Wax (Editor), Dover Publications, New York, 1954.

RICE, S. O., "Statistical Properties of a Sine-Wave Plus Random Noise," *Bell System Technical Journal*, vol. 27, January 1948, pp. 109–157.

ROCKWELL DEFENSE ELECTRONICS, *Collins Prop Man High Frequency Propagation Software*, Rockwell, Collins Avionics and Communications Division, Cedar Rapids, IA, 1995.

ROOT, W. L., "Remarks, Most Historical, on Signal Detection and Signal Parameter Estimation," *Proceedings of the IEEE*, vol. 75, November 1987, pp. 1446–1457.

ROSE, R. B., "MIMIMUF: A Simplified MUF-Prediction Program for Microcomputers," *QST*, vol. 66, December 1982, pp. 36–38.

ROSE, R. B., "Technical Correspondence—MIMIMUF Revisited," *QST*, vol. 68, March 1984, p. 46.

ROWE, H. E., *Signals and Noise in Communication Systems*, D. Van Nostrand Company, Princeton, NJ, 1965.

RYAN, J. S. (Editor), Special Issue on Telecommunications Standards, *IEEE Communications Magazine*, vol. 23, January 1985.

SABIN, W. E., and E. O. SCHOENIKE, *Single-Sideband Systems and Circuits*, McGraw–Hill Book Company, New York, 1987.

SALTZBERG, B. R., T. R. HSING, J. M. CIOFFI, and D. W. LIN (Editors), Special Issue on High-Speed Digital Subscriber Lines, *IEEE Journal on Selected Areas in Communications*, vol. 9, August 1991.

SCHARF, L. L., *Statistical Signal Processing: Detection, Estimation, and Time Series Analysis*, Addison-Wesley Publishing Company, Reading, MA, 1991.

SCHAUMANN, R., M. S. GHAUSI, and K. R. LAKER, *Design of Analog Filters*, Prentice Hall, Englewood Cliffs, NJ, 1990.

SCHILLING, D. L., R. L. PICKHOLTZ, and L. B. MILSTEIN, "Spread Spectrum Goes Commercial," *IEEE Spectrum*, vol. 27, August 1990, pp. 40–45.

SCHILLING, D. L., L. B. MILSTEIN, R. L. PICKHOLTZ, M. KULLBACK, and F. MILLER, "Spread Spectrum for Commercial Communications," *IEEE Communications Magazine*, vol. 29, April 1991, pp. 66–67.

SCHWARTZ, B. K., "The Analog Display Services Interface," *IEEE Communications Magazine*, vol. 31, April 1993, pp. 70–75.

SCHWARTZ, M., W. R. BENNETT, and S. STEIN, *Communication Systems and Techniques*, McGraw–Hill Book Company, New York, 1966.

SHANMUGAN, K. S., *Digital and Analog Communication Systems*, John Wiley & Sons, New York, 1979.

SHANMUGAN, K. S., and A. M. BREIPOHL, *Random Signals: Detection, Estimation and Data Analysis*, John Wiley & Sons, New York, 1988.

SHANNON, C. E., "A Mathematical Theory and Communication," *Bell System Technical Journal*, vol. 27, July 1948, pp. 379–423, and October 1948, pp. 623–656.

SHANNON, C. E., "Communication in the Presence of Noise," *Proceedings of the IRE*, vol. 37, January 1949, pp. 10–21. (Reprinted in *Proceedings of the IEEE*, vol. 86, no. 2, February 1998, pp. 447–457).

SIMON, M. K., "Comments on 'On Binary DPSK Detection'," *IEEE Transactions on Communications*, vol. COM-26, October 1978, pp. 1477–1478.

SINNEMA, W., and R. MCPHERSON, *Electronic Communications*, Prentice Hall Canada, Inc., Scarborough, Ontario, Canada, 1991.

SIPERKO, C. M., "LaserNet—A Fiber Optic Intrastate Network (Planning and Engineering Considerations)," *IEEE Communications Magazine*, vol. 23, May 1985, pp. 31–45.

SKLAR, B., *Digital Communications*, Prentice Hall, Englewood Cliffs, NJ, 1988.

SKLAR, B., "Rayleigh Fading Channels in Mobile Digital Communications Systems—Part 1: Characterization," *IEEE Communications Magazine*, vol. 35, September 1997, pp. 136–146.

SKLAR, B., "A Primer on Turbo Code Concepts," *IEEE Communications Magazine*, vol. 35, December 1997, pp. 94–101.

SLEPIAN, D., "On Bandwidth," *Proceedings of the IEEE*, vol. 64, no. 3, March 1976, pp. 292–300.

SMITH, B., "Instantaneous Companding of Quantized Signals," *Bell System Technical Journal*, vol. 36, May 1957, pp. 653–709.

SMITH, J. R., *Modern Communication Circuits*, McGraw–Hill Book Company, New York, 2d ed., 1998.

SOLOMON, L., "The Upcoming New World of TV Reception," *Popular Electronics*, vol. 15, no. 5, May 1979, pp. 49–62.

SPANIAS, A. S., "Speech Coding: A Tutorial Review," *Proceedings of the IEEE*, vol. 82, no. 10, October 1994, pp. 1541–1582.

SPILKER, J. J., *Digital Communications by Satellite*, Prentice Hall, Englewood Cliffs, NJ, 1977.

STALLINGS, W., *Data and Computer Communications*, 5th ed., Prentice Hall, Upper Saddle River, NJ, 1996.

STALLINGS, W., *ISDN and Broadband ISDN*, Macmillan Publishing Company, New York, 1992.

STARK, H., F. B. TUTEUR, and J. B. ANDERSON, *Modern Electrical Communications*, Prentice Hall, Englewood Cliffs, NJ, 1988.

STIFFLER, J. J., *Theory of Synchronous Communication*, Prentice Hall, Englewood Cliffs, NJ, 1971.

STUMPERS, F. L., "Theory of Frequency—Modulation Noise," *Proceedings of the IRE*, vol. 36, September 1948, pp. 1081–1092.

SUNDE, E. D., *Communications Engineering Technology*, John Wiley & Sons, New York, 1969.

SWEENEY, P., *Error Control Coding*, Prentice Hall, Englewood Cliffs, NJ, 1991.

TAUB, H., and D. L. SCHILLING, *Principles of Communication Systems*, 2d ed., McGraw–Hill Book Company, New York, 1986.

THOMAS, J. B., *An Introduction to Statistical Communication Theory*, John Wiley & Sons, New York, 1969.

THOMSON CONSUMER ELECTRONICS, Inc., *RCA Self-Installer Manual (for DSS Receiver System)*, Thomson Consumer Electronics, Inc., Indianapolis, IN, 1994.

TURIN, G., "An Introduction to Digital Matched Filters," *Proceedings of the IEEE*, vol. 64, no. 7, July 1976, pp. 1092–1112.

UNGERBOECK, G., "Channel Coding with Multilevel/Phase Signals," *IEEE Transactions on Information Theory*, vol. IT-28, January 1982, pp. 55–67.

UNGERBOECK, G., "Trellis-Coded Modulation with Redundant Signal Sets," Parts 1 and 2, *IEEE Communications Magazine*, vol. 25, no. 2, February 1987, pp. 5–21.

VAN DER ZIEL, A., *Noise in Solid State Devices and Circuits*, Wiley-Interscience, New York, 1986.

VITERBI, A. J., "When Not to Spread Spectrum—A Sequel," *IEEE Communications Magazine*, vol. 23, April 1985, pp. 12–17.

VITERBI, A. J., and J. K. OMURA, *Principles of Digital Communication and Coding*, McGraw–Hill Book Company, New York, 1979.

WALDEN, R. H., "Performance Trends for Analog-to-Digital Converters," *IEEE Communications Magazine*, vol. 37, February 1999, pp. 96–101.

WEAVER, D. K., "A Third Method of Generating and Detection of Single Sideband Signals," *Proceedings of the IRE*, vol. 44, December 1956, pp. 1703–1705.

WEI, L., "Rotationally Invariant Convolutional Channel Coding with Expanded Signal Space—Part II: Nonlinear Codes," *IEEE Journal on Selected Areas in Communications*, vol. SAC-2, no. 2, September 1984, pp. 672–686.

WHALEN, A. D., *Detection of Signals in Noise*, Academic Press, New York, 1971.

WHITAKER, J. C. (Editor), *The Electronics Handbook*, CRC Press, Boca Raton, FL, 1996.

WHITTAKER, E. T., "On Functions which Are Represented by the Expansions of the Interpolation Theory," *Proceedings of the Royal Society* (Edinburgh), vol. 35, 1915, pp. 181–194.

WIENER, N., *Extrapolation, Interpolation, and Smoothing of Stationary Time Series with Engineering Applications*, MIT Press, Cambridge, MA, 1949.

WILSON, S. G., *Digital Modulation and Coding*, Prentice Hall, Upper Saddle River, NJ, 1996.

WOZENCRAFT, J. M., and I. M. JACOBS, *Principles of Communication Engineering*, John Wiley & Sons, New York, 1965.

WU, W. W., E. F. MILLER, W. L. PRITCHARD, and R. L. PICKHOLTZ, "Mobile Satellite Communication," *Proceedings of the IEEE*, vol. 82, no. 9, September 1994, pp. 1431–1448.

WYLIE, C. R., JR., *Advanced Engineering Analysis*, John Wiley & Sons, New York, 1960.

WYNER, A. D., and S. SHAMAI (SHITZ), "Introduction to "Communication in the Presence of Noise" by C. E. Shannon," *Proceedings of the IEEE*, vol. 86, no. 2, February 1998, pp. 442–446.

YOUNG, P. H., *Electronic Communication Techniques*, Merrill Publishing Co., Columbus OH, 1990.

XIONG, F., "Modern Techniques in Satellite Communications," *IEEE Communications Magazine*, vol. 32, no. 8, August 1994, pp. 84–98.

ZIEMER, R. E., and R. L. PETERSON, *Digital Communications and Spread Spectrum Systems*, Macmillan Publishing Company, New York, 1985.

ZIEMER, R. E., and W. H. TRANTER, *Principles of Communications*, 4th ed., John Wiley & Sons, Inc., New York, 1995.

ZIEMER, R. E., W. H. TRANTER, and D. R. FANNIN, *Signals and Systems*, Prentice Hall, Upper Saddle River, NJ, 4th ed., 1998.

ZOU, W. Y., "Comparison of Proposed HDTV Terrestrial Broadcasting Systems," *IEEE Transactions on Broadcasting*, vol. 37, December 1991, pp. 145–147.

ANSWERS TO SELECTED PROBLEMS

Chapter 1

1–5 1.97 bits

1–9 $H = 3.084$ bits

1–11 $H = 3.32$ bits, $T = 1.66$ s

Chapter 2

2–9 36 dB

2–10 **(a)** 4.08×10^{-14} W   **(b)** -103.9 dBm   **(c)** 1.75 μV

2–16 $S(f) = -\dfrac{A}{\omega^2} + Ae^{-j\omega T_0}\left(\dfrac{1}{\omega^2} + j\dfrac{T_0}{\omega}\right),$ where $\omega = 2\pi f$

Note that $S(0) = A\,T_0^2/2$.

2–40 **(a)** $(\sin 4t)/(4t)$ **(b)** 7

2-45 **(a)** $-0.4545A$ **(b)** $0.9129A$

2-47 **(a)** $(A_1 + A_2)/\sqrt{2}$ **(b)** $(\sqrt{A_1^2 + A_2^2})/\sqrt{2}$ **(c)** $|A_2 - A_1|/\sqrt{2}$

       (d) $(\sqrt{A_1^2 + A_2^2})/\sqrt{2}$ **(e)** $(\sqrt{A_1^2 + A_2^2})/\sqrt{2}$

2-56 $c_n = \begin{cases} \dfrac{AT^2}{2T_0}, & n = 0 \\[3ex] \dfrac{A\{e^{-j2\pi n T/T_0}[1 + j2\pi n\,(T/T_0)] - 1\}}{(2\pi n)^2/T_0}, & n \neq 0 \end{cases}$

2-62 $c_n = \begin{cases} 0, & n = \text{even} \\[2ex] \dfrac{4}{n^2\pi^2}, & n = \text{odd} \end{cases}$ **(b)** $\frac{1}{3}$ W

2-82 **(a)** 6.28 ms   **(b)** 160

Chapter 3

3-3 **(a)** $W_s(f) = d \displaystyle\sum_{n=-\infty}^{\infty} \dfrac{\sin \pi n d}{\pi n d} \begin{cases} 1, & |f - nf_s| \leq 2500 \\[1.5ex] \frac{-1}{1500}\,(|f - nf_s| - 4000), & 2500 \leq |f - nf_s| \leq 4000 \\[1.5ex] 0, & f \text{ elsewhere} \end{cases}$

where $d = 0.5$ and $f_s = 10{,}000$

(b) $W_s(f) = 0.5 \left(\dfrac{\sin \pi \tau f}{\pi \tau f}\right) \displaystyle\sum_{k=-\infty}^{\infty} \begin{cases} 1, & |f - kf_s| \leq 2500 \\[1.5ex] \frac{-1}{1500}\,(|f - kf_s| - 4000), & 2500 \leq |f - kf_s| \leq 4000 \\[1.5ex] 0, & f \text{ elsewhere} \end{cases}$

where $\tau = 50 \times 10^{-6}$ and $f_s = 10{,}000$

3-6 $W_s(f) = -j\left(\dfrac{1 - \cos(\pi f/f_s)}{\pi f/f_s}\right)$

$\times \displaystyle\sum_{k=-\infty}^{\infty} \begin{cases} 1, & |f - kf_s| \leq 2500 \\[1.5ex] \frac{-1}{1500}\,(|f - kf_s| - 4000), & 2500 \leq |f - kf_s| \leq 4000 \\[1.5ex] 0, & f \text{ elsewhere} \end{cases}$

3-8 **(a)** 200 samples/s   **(b)** 9 bits/word   **(c)** 1.8 kb/s   **(d)** 900 Hz

3-14 **(a)** 5 bits   **(b)** 27 kHz

3–24 (a) $\mathcal{P}(f) = \dfrac{A^2 T_b}{4} \left[\dfrac{\sin\left(\frac{1}{2}\pi f\, T_b\right)}{\frac{1}{2}\pi f\, T_b} \right]^2$

(b) $\mathcal{P}(f) = \dfrac{A^2 T_b}{4} \left[\dfrac{\sin\left(\frac{1}{4}\pi f\, T_b\right)}{\frac{1}{4}\pi f\, T_b} \right]^2 \left[\sin\left(\frac{1}{4}\pi f\, T_b\right) \right]^2$

3–34 (a) 32.4 kb/s   **(b)** 10.8 ksymbols/s   **(c)** 5.4 kHz

3–35 (a) 3.2 ksymbols/s   **(b)** 0.5

3–45 (a) 5.33 kb/s   **(b)** 667 Hz

3–46 (a) 10.7 kb/s   **(b)** 1.33 kHz

3–53 (a) 0.00534   **(b)** 0.427

3–64 (a) 40 Hz   **(b)** 1280 Hz

Chapter 4

4–9 (a) $g(t) = 500 + 200 \sin \omega_a t$, AM, $m(t) = 0.4 \sin \omega_a t$
(b) $x(t) = 500 + 200 \sin \omega_a t$, $y(t) = 0$
(c) $R(t) = 500 + 200 \sin \omega_a t$, $\theta(t) = 0$
(d) 2.7 kW

4–20 48.3%

4–23 $K\sqrt{m^2(t) + [\hat{m}(t)]^2}$, yes

4–28 (a) $(j2\pi f)/[j2\pi f + K_d K_v F_1(f)]$ **(b)** $(j2\pi f)(f_1 + jf)/[j2\pi f(f_1 + jf) + K_d K_v f_1]$

4–35 (a) 107.6 MHz   **(b)** RF: At least 96.81–96.99 MHz; IF: 10.61–10.79 MHz
(c) 118.3 MHz

Chapter 5

5–1 (a) 6.99 dBk   **(b)** 707 V   **(d)** 7025 W   **(e)** 18,050 W

5–6 $S(f) = \dfrac{1}{2}\left[\displaystyle\sum_{n=-\infty}^{\infty} c_n \delta(f - f_c - n f_m) + \sum_{n=-\infty}^{\infty} c_n^* \delta(f + f_c + n f_m) \right]$

where $c_n = \dfrac{A_c}{2\pi}\left[2\theta_1\left(\dfrac{\sin n\theta_1}{n\theta_1}\right) + 2A_m\,\dfrac{\cos n\theta_1 \sin\theta_1 - n\sin n\theta_1 \cos\theta_1}{1 - n^2} \right]$,

$A_m = 1.2$, and $\theta_1 = 146.4°$

5–7 (a) $s(t) = (\cos \omega_1 t + 2\cos 2\omega_1 t)\cos \omega_c t$,   where $\omega_1 = 1000\pi$
(b) $S(f) = \frac{1}{4}[\delta(f - f_c + f_1) + \delta(f + f_c - f_1) + \delta(f - f_c - f_1)$
$+ \delta(f + f_c + f_1)] + \frac{1}{2}[\delta(f - f_c + 2f_1) + \delta(f + f_c - 2f_1)$
$+ \delta(f - f_c - 2f_1) + \delta(f + f_c + 2f_1)]$
(c) 1.25 W   **(d)** 4.5 W

5–14 (b) $s(t) = \Pi(t) \cos \omega_c t - \dfrac{1}{\pi} \ln\left(\dfrac{\left| t + \frac{1}{2} \right|}{\left| t - \frac{1}{2} \right|} \right) \sin \omega_c t$

(c) $\max[s(t)] = \infty$

5–24 (a) $f_{\mathrm{BPF}} = 12.96$ MHz, $B_{\mathrm{BPF}} = 48.75$ kHz   **(b)** 7.96 or 17.96 MHz
(c) 9.38 kHz

5–26 (a) $m(t) = 0.2 \cos(2000\,\pi t)$, $M_p = 0.2$ V, $f_m = 1$ kHz
(b) $m(t) = -0.13 \sin(2000\,\pi t)$, $M_p = 0.13$ V, $f_m = 1$ kHz
(c) $P_{AV} = 2500$ W, $P_{PEP} = 2500$ W

5–31 $S(f) = \frac{1}{2}[G(f - f_c) + G^*(-f - f_c)]$, where
$G(f) = \sum_{n=-\infty}^{\infty} J_n(0.7854)\,\delta(f - nf_m)$, $f_c = 146.52$ MHz,
and $f_m = 1$ kHz. $B = 3.57$ kHz

5–35 $S(f) = \frac{1}{2}[\sum_{n=-\infty}^{\infty} c_n \delta(f - f_c - nf_m) + \sum_{n=-\infty}^{\infty} c_n^* \delta(f + f_c + nf_m)]$, where

$$c_n = 5(e^{j\beta} - 1)\left(\frac{\sin(n\pi/2)}{n\pi/2}\right), f_m = 1 \text{ kHz}, \beta = 0.7854, \text{ and } f_c = 60 \text{ MHz}$$

5–39 (a) 5.5 Hz

(b) $S(f) = \dfrac{A_c}{2}\left\{\delta(f - f_c) + \delta(f + f_c) + \dfrac{D_f}{2\pi}\sum_{\substack{n=-\infty \\ n=0}}^{\infty}\left(\dfrac{c_n}{nf_m}\right)\right.$

$$\left. \times [\delta(f - f_c - nf_m) - \delta(f + f_c - nf_m)]\right\},$$

where $c_n = 5e^{-jn\pi/2}\left(\dfrac{\sin(n\pi/2)}{n\pi/2}\right)(n = 0)$,

$D_f = 6.98$ rad/V-s, $f_m = 100$ Hz, and $f_c = 30$ MHz

5–42 $\mathcal{P}(f) \approx 0.893[2\delta(f - f_c + 5f_0) + \delta(f - f_c + 3f_0)$
$+ \delta(f - f_c - f_0) + 2\delta(f - f_c + 3f_0) + \delta(f - f_c - 5f_0)$
$+ 2\delta(f + f_c - 5f_0) + \delta(f + f_c - 3f_0) + \delta(f + f_c + f_0)$
$+ 2\delta(f + f_c + 3f_0) + \delta(f + f_c + 5f_0)]$

where $f_0 \triangleq \dfrac{D_f}{2\pi} = 15.9$ kHz and $f_c = 2$ GHz

5–45 (a) 20 kHz

(b) $S(f) = \frac{1}{2}[M_1(f - f_c) + M_1(f - f_c)] + \frac{j}{2}[M_2(f + f_c) - M_2(f - f_c)]$

5–46 (a) $S(f) = \frac{1}{2}[X(f - f_c) + X^*(-f - f_c)]$,

where $X(f) = \dfrac{A_c}{2}\sum_{n=-\infty}^{\infty}\left(\dfrac{\sin(n\pi/2)}{n\pi/2}\right)\delta(f - \frac{1}{2}nR)$, $R = 24$ kb/s

(b) 48 kHz

5–53 (a) 96 kHz   **(b)** 101 kHz
5–67 (a) $-45°$, $+45°$, $+135°$, $-135°$, $-45°$, $-45°$, $-45°$ **(b)** 1.13 Mb/s

Chapter 6

6–2 (a) $\bar{x} = (2\sqrt{2}\,A/\pi)\cos(\omega_0 t + \pi/4)$ **(b)** $x(t)$ is not stationary

6–4 $2/\sqrt{\pi} = 1.128$

6–8 **(a)** 15 W **(b)** 11 W **(c)** 19 W

6–10 **(a)** No **(b)** Yes **(c)** Yes **(d)** No

6–17 **(a)** $\sqrt{B}$ **(b)** $R_x(\tau) = B[(\sin \pi B\tau)/(\pi B\tau)]^2$

6–22 **(a)** $N_0 K^2/(8\pi^2 f^2)$ **(b)** ∞

6–25 $1/(2\pi f_0)$

6–30 **(a)** $H(f) = 10\left|\left[\left(1 + j\dfrac{f}{f_0}\right)^2\right]\right|$

(b) $0.690 f_0$, where $f_0 = 1/(2\pi RC)$

6–35 **(a)** Not WSS **(b)** $\mathcal{P}_x(f) = \dfrac{A_0^2}{4}[\delta(f - f_0) + \delta(f + f_0)]$ **(c)** Yes

6–42 $x(t) = m_1(t) + m_2(t), y(t) = \hat{m}_1(t) - \hat{m}_2(t)$

$\mathcal{P}_{m_1}(f) = A\Pi\left(\dfrac{f}{2(f_3 - f_c)}\right), \mathcal{P}_{m_2}(f) = A\Lambda\left(\dfrac{f}{f_c - f_1}\right)$

6–44 $\mathcal{P}_v(f) = \frac{1}{4}[\mathcal{P}_x(f - f_c) + \mathcal{P}_x(-f - f_c)]$,

where $\mathcal{P}_x(f) = T_b\left(\dfrac{1 - \cos \pi f T_b}{\pi f T_b}\right)^2$

6–50 **(a)** $y(t)=\begin{cases}\frac{1}{2}, & t = 0 \\ 1, & 0 < t < T \\ \frac{1}{2}, & t = T \\ 0, & t \text{ elsewhere}\end{cases}$ **(b)** Same pulse shape as for part (a)

Chapter 7

7–1 $P_e = \frac{1}{2}e^{-\sqrt{2}A/\sigma_0}$

7–11 $P_e = Q\left(\sqrt{0.25(E_b/N_0)}\right)$

7–14 **(a)** $H(f) = Te^{-j\pi fT}\left(\dfrac{\sin \pi f T}{\pi f T}\right)$ **(b)** $B_{eq} = 1/(2T)$

7–16 $P_e = P(1)Q\left(\sqrt{\dfrac{2(-V_T + A)^2 T}{N_0}}\right) + P(0)Q\left(\sqrt{\dfrac{2(V_T + A)^2 T}{N_0}}\right)$

7–21 **(a)** 5×10^{-7} **(b)** 4.42×10^{-2}

7–22 **(a)** 2400 b/s **(b)** $(P_e)_{overall} = 1.11 \times 10^{-14}$

7–27 3.2×10^{-7}

7–31 **(a)** 4800 b/s **(b)** 19,943 b/s

7–32 **(a)** 9.09×10^{-4} **(b)** 29.1 dB

7–35 11.3 dB inferior

7–42 -100.9 dBm

7–47 **(a)** 651% **(b)** Low-frequency components predominate

7–49 $(S/N)_{out} = (5 \times 10^{-5})(P_s/N_0)$ for all five channels

Chapter 8

8–9 (a) 25.6 dB   (b) 36.3 W   (c) -54.9 dBm

8–13 $G_a = \dfrac{h_{fe}^2 R_s}{h_{oe}(R_s + h_{ie})^2}$

8–17 (a) 129 K   (b) -76.6 dBm

8–19 3.2×10^{-7}

8–22 (a) 19.8 dB   (b) 7.52 dB   (c) 4.67 dB

8–34 12.84-m diameter

8–41 20.6 kW

Appendix B

B–1 0.3486

B–2 (a) 5/36   (b) 15/36

B–4 (a) 1/2   (b) 1/3

B–8 0.4375

B–11 (a) b (b) $1/b$ (c) $1/b^2$

B–21 51.0

B–26 (a) 1.337×10^{-11} (b) 0.9520

B–30 $f(y) = \begin{cases} \dfrac{1}{2\sigma\sqrt{2\pi y}}\, [e^{-(\sqrt{y}+m)^2/(2\sigma^2)} + e^{-(\sqrt{y}-m)^2/(2\sigma^2)}], & y \geq 0, \\[2mm] 0, & y < 0 \end{cases}$

B–35 $B\sigma/\sqrt{2\pi}$

B–37 (a) 1/6   (b) Yes   (c) 1/12

(d) $f(x_2|x_1) = \begin{cases} \frac{1}{12}(1 + x_2), & 0 \leq x_1 \leq 1, \quad 0 \leq x_2 \leq 4 \\ 0, & \text{otherwise} \end{cases}$

B–46 (a) $f_y(y) = \displaystyle\int_{-\infty}^{\infty} \frac{1}{|Ax|} f_x\left(\frac{y}{Ax}, x\right) dx$ (b) $f_y(y) = \displaystyle\int_{-\infty}^{\infty} \frac{1}{|Ax|} f_{x_1}\left(\frac{y}{Ax}\right) f_{x_2}(x)\, dx$

INDEX